FLIES AND DISEASE, Volume I

Ecology, Classification, and Biotic Associations

FLIES AND DISEASE

Volume I ECOLOGY, CLASSIFICATION AND BIOTIC ASSOCIATIONS

BY BERNARD GREENBERG

Princeton University Press, Princeton, New Jersey · 1971

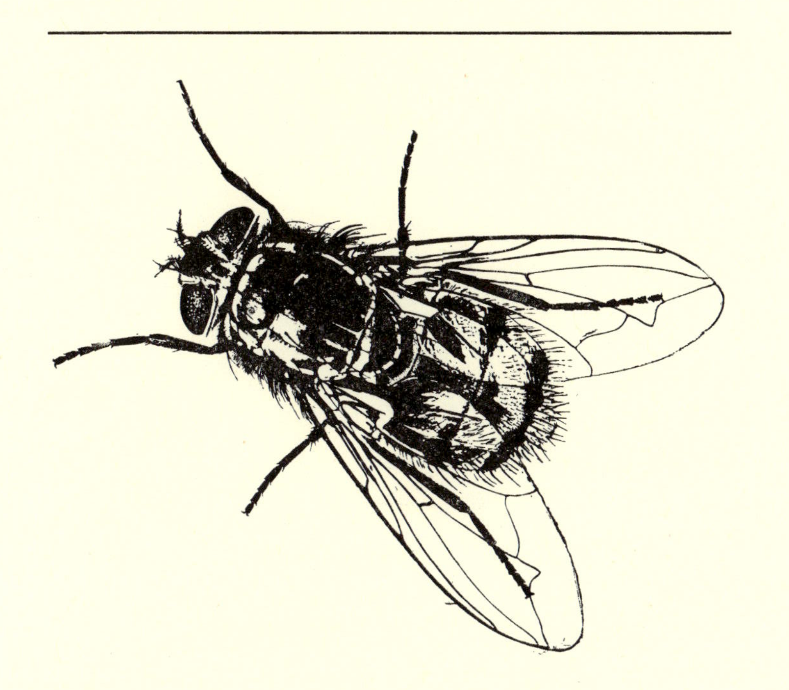

 L.C. Card: 68-56310. ISBN: 0-691-08071-2. Printed in the United States of America by Princeton University Press. This book has been composed in Linotype Times Roman.

Publication of this book has been aided by grants DA-MD-59-193-67-G-9225, DADA-17-69-G-9225, and DADA-17-70-G-9313 from the U.S. Army Medical Research and Development Command.

To my wife, Barbie

Preface

WE HAVE RECENTLY passed the centennial of Davaine's pioneer experiments using stable flies and blowflies to transmit anthrax to healthy animals from sick ones. These studies were performed at a time when many physicians and scientists still doubted the microbial origin of such diseases. The ensuing century has seen such a flowering of bio-medical thought and practice that it has brought the control of many infectious diseases within our reach. Often the control of a disease has rested on the empirical use of this drug or that antibiotic with the consequence that many of the underlying biological interactions between host and pathogen have remained unexplored. This is not only esthetically and epistemologically unsatisfying, but scientifically unsound. The ricocheting of resistant populations of bacteria, protozoa, and insects emphasizes the need for ongoing basic research. Unfortunately, the development of resistance and behavioral changes in vector species have not been the only complications. Who could have predicted thirty years ago the meteoric rise of DDT, its vast successes, and now its possible demise along with other slowly degradable pesticides? Because of these things and also because much of the world's population continues to live in a sanitational limbo with massive microbial challenge, the framework of this book remains conservatively biological.

Flies carry a centuries-old indictment as disseminators of disease. Their case has often come up and been thrown out for lack of evidence. Just as often they have been convicted on inadequate evidence. It is my hope that a rational evaluation of their role can be based on such knowledge of their ecology, phenology, geographic distribution, and biotic associations as is presented in this volume. It is by no means the full picture and another volume will develop more particular aspects of fly life cycles, microbial interactions and epidemiology as they relate to specific diseases.

More than eight years have passed since I began compiling the literature on the biotic associations of synanthropic flies. At the outset, this material was intended as a chapter in a work on flies and disease. But like Aladdin's genie uncorked, the modest idea has expanded far beyond anticipation. The extent of the biological associations recorded in this book speaks eloquently of the "perpetuum mobile" of flies. Few animal groups are endowed with such mobility and often demoniac energy.

Likewise, few groups have elicited such sustained investigative interest from scientists throughout the world. In the face of literature inaccessibility and language barriers, I have endeavored to capture the cosmopolitanism of fly research, because a lesser effort would be unrepresentative of this international interest. This is not to say that my efforts have been uniformly successful. A dead file, several inches thick, remains as a reminder of publications not seen. We must console ourselves with the thought that every fire leaves a residue.

I have also tried to add a time dimension to this work by including fly-mite and fly-fungus observations going back to the eighteenth century. This was done for the historical interest of these observations, not for their impeccable accuracy.

The fly associations are intra-specific and inter-specific, and include cannibalism, hybridization, competition, commensalism, predation, parasitoidism, parasitism, hyperparasitism, and phoresy. Some reports, especially those dealing with micro-organisms, simply record the occurrence of an organism. Others provide details, in varying degree, of the interaction between the organism and the fly. More than seven hundred species belonging to the lower plant and the major animal phyla are involved in these associations. They range from viruses and bacteria to birds and mammals. I hope this book will serve microbiologist, mammalogist, and the many "ologists" between, in addition to those interested in the biology of species interactions and biological control. I am especially hopeful that this book will find its way into the armamentaria of epidemiologists and public health workers concerned with fly-borne disease. The interested reader will find a ready entrée into the literature through either the fly or the organism associated with it. The sections on the ecology and distribution of flies, and the keys to adults and larvae provide the non-Dipterist with the means for studying some of the more important synanthropic flies of his region.

The many specialists who have helped me will be acknowledged in the section which follows. I would like to thank Mrs. Lynn Rubin, Grace Steward, Karin Mede, Joseph Chiu, and Ellyn Millman, who have assisted me at various times during the gestation period of eight years. Their ideas and devotion to the task have contributed materially to the final result. I am also grateful to Edward Shafer for his skillful translations of Russian, Scandinavian, Dutch, and other articles; and to Miss Aletha Kowitz for her cooperation in procuring interlibrary loans—in one year alone, she processed eight hundred of our requests.

I am especially indebted to Lt. Colonels Bruce F. Eldridge and Wallace P. Murdoch (Ret.) of the U.S. Army Medical Research and Development Command.

Bernard Greenberg, *Chicago, Illinois*

Contributors

Dr. Jindra Dušek
University of Agriculture and Forestry
Brno, Czechoslovakia

Dr. František Gregor
Czechoslovak Academy of Science
Brno, Czechoslovakia

Dr. Hugh E. Paterson
University of Western Australia
Nedlands, Australia

Dr. Adrian Pont
British Museum of Natural History
London, England

Professor Dalibor Povolný
University of Agriculture and Forestry
Brno, Czechoslovakia

Dr. Marshall R. Wheeler
Genetics Foundation
The University of Texas
Austin, Texas

Contents

FLIES AND DISEASE, Volume I

Ecology, Classification, and Biotic Associations

1
Introduction

THE BIOTIC ASSOCIATIONS of synanthropic flies are recorded in two sections, Chapter VI and Chapter VII. Chapter VI systematically lists synanthropic flies and the organisms associated with them. Chapter VII systematically lists various organisms and the synanthropic flies which are associated with them. The flies, in both sections, are arranged phylogenetically according to suborder, division, section, superfamily, family, subfamily, and genus, with species and subspecies arranged alphabetically.

Viruses are systematically arranged by groups and subgroups. Divisions within subgroups are organized first according to type number, and second, when an additional name is used, such as Lansing Strain, alphabetically. Schizophyta are systematically arranged by class, order, family, and genus. Species of bacteria are listed alphabetically. The remainder of the phyla included in the associations—Fungi, Protozoa, Platyhelminthes, Aschelminthes, Mollusca, Annelida, Arthropoda (with the exception of the order Diptera), and Chordata—are arranged phylogenetically by class, order, and family, with genera, species, and subspecies arranged alphabetically.

Although an attempt was made to follow one authority in the arrangement of each of the above phyla, in some cases it was necessary to use several authorities.

Within each entry recording an association, the species name and author is followed by the investigator and the year of publication. When an association between two species is recorded in more than one publication, the name of the species and author are given only in the first entry. The other entries of the same association are indented and chronologically listed below. Question marks preceding any of the species names or the information about the association indicate questionable identification or information. In the section listing various organisms and flies associated with them, the nature of the relationship is often included in the main species heading when it is consistent for all the associations listed under it.

In all cases an attempt was made to use the currently accepted species name (a list of the authorities consulted is included at the end of this section). Consequently, the synonymy listed includes only the currently unacceptable names used by those investigators cited. When there is a doubt about the appropriate synonymy of a particular name, it is indicated by placing a question mark before the synonym. Synonyms are placed in brackets in individual entries when an investigator used two different species names in the same article for what is now considered to be one species.

Many of the early papers recording fly-mite associations presented problems in validation. In assembling and analyzing these associations Oudemans' interpretations in "Kritisch Historisch Overzicht Der Acarologie" were leaned upon heavily.

The following examples are included to clarify the description given above:

CHAPTER VI

Musca larvipara Porchinskiĭ	EUBACTERIALES: ENTERO-BACTERIACEAE
	Escherichia coli var. *acidilactici* (Topley and Wilson) Yale = [*Escherichia paraacidilactici* A]
	Zmeev, 1943
	Escherichia coli var. *acidilactici* (Topley and Wilson) Yale = [*Escherichia paraacidilactici* B]
	Zmeev, 1943
	?*Shigella* Castellani and Chalmers
	Sychevskaia͡ et al., 1959a
	ASCHELMINTHES: NEMATODA
	SPIRURIDEA: THELAZIIDAE
	Thelazia gulosa Raillet and Henry
	Klesov, 1951
	Világiová, 1962

CHAPTER VII

Escherichia coli var. *acidilactici*.... (Topley and Wilson) Yale (Synonyms: *Escherichia paraacidilactici* A, *Escherichia paraacidilactici* B)	MUSCIDAE: MUSCINAE
	Musca domestica Linnaeus
	Scott, 1917a
	Musca larvipara Porchinskiĭ
	Zmeev, 1943

The following references were used in the arrangement of phyla, classes and orders:

VIRUSES
Burnet, F. M., and Stanley, W. M., editors
1959. The viruses. Academic Press, New York. Vol. 3, 428pp.

PHYLUM: SCHIZOPHYTA
Breed, R. S., Murray, E.G.D., and Smith, N. R.
1957. Bergey's manual of determinative bacteriology. The Williams and Wilkins Co., Baltimore. 1094pp.

PHYLUM: THALLOPHYTA
Bessey, E. A.
1950. Morphology and taxonomy of fungi. The Blakiston Co., Philadelphia. 791pp.
Clements, F. E., and Shear, C. L.
1954. The genera of fungi. Hafner Publishing Co., New York. 496pp.
Gilman, J. C.
1957. A manual of soil fungi. Second edition. The Iowa State College Press, Ames, Iowa. 450pp.
Lodder, J., and Kreger-Van Rij, N.J.W.
1952. The yeasts, a taxonomic study. Interscience Publishers, Inc., New York, North Holland Publishing Co., Amsterdam. 713pp.

PHYLUM: PROTOZOA
Kudo, R. R.
1966. Protozoology. Charles C. Thomas, Publisher, Springfield, Illinois. 1174pp.

PHYLUM: PLATYHELMINTHES
Yamaguti, S.
1959. Systema helminthum. The cestodes of vertebrates. Interscience Publishers, Inc., New York and London. Vol. 2, 860pp.

PHYLUM: ASCHELMINTHES
Yamaguti, S.
1961. Systema helminthum. The nematodes of vertebrates. Interscience Publishers, Inc., New York and London. Vol. 3, Parts 1 and 2, 1261pp.

ORDER: ACARINA
Baker, W. E., and Wharton, G. W.
1952. An introduction to acarology. The Macmillan Co., New York. 465pp.

CLASS: INSECTA (except Diptera)

Essig, E. O.
1942. College entomology. The Macmillan Co., New York. 900pp.

ORDER: DIPTERA

Stone, A., Sabrosky, C. W., Wirth, W. W., Foote, R. H., and Coulson, J. R.
1965. A catalog of the Diptera of America north of Mexico. Agricultural Research Service, United States Department of Agriculture, United States Government Printing Office, Washington, D.C. 1696pp.

CLASS: AVES

Austin, O. L., Jr.
1961. Birds of the world. Herbert S. Zim, editor, Golden Press, New York. 316pp.

GENERAL

Storer, T. I., and Usinger, R. L.
1965. General zoology. Fourth edition. McGraw-Hill Book Co., New York. 741pp.

The following references were used in the validation of species:

PHYLUM: SCHIZOPHYTA

Breed, R. S., Murray, E.G.D., and Smith, N. R.
1957. Bergey's manual of determinative bacteriology. Seventh edition. The Williams and Wilkins Co., Baltimore. 1094pp.

Buchanan, R. E., Holt, J. G., and Lessel, E. F., Jr.
1966. Index bergeyana. The Williams and Wilkins Co., Baltimore. 1472pp.

Steinhaus, E. A.
1947. Insect microbiology. Comstock Publishing Co., Inc., Ithaca, New York. 763pp.

PHYLUM: THALLOPHYTA

Ainsworth, G. C.
1961. Dictionary of the fungi. Commonwealth Mycological Institute, Kew, Surrey. 547pp.

Bessey, E. A.
1950. Morphology and taxonomy of fungi. The Blakiston Co., Philadelphia. 791pp.

Clements, F. E., and Shear, C. L.
1954. The genera of fungi. Hafner Publishing Co., New York. 496pp.

Gilman, G. C.
1957. A manual of soil fungi. Second edition. The Iowa State College Press, Ames, Iowa. 450pp.

Lodder, J., and Kreger-Van Rij, N.J.W.
1952. The yeasts, a taxonomic study. Interscience Publishers, Inc., New York, North Holland Publishing Co., Amsterdam. 713pp.

Raper, B., and Thom, C.
1949. A manual of the penicillia. The Williams and Wilkins Co., Baltimore, 875pp.

PHYLUM: PROTOZOA

Kudo, R. R.
1966. Protozoology. Charles C. Thomas, Publisher, Springfield, Illinois. 1174pp.

Wallace, F. G.
1966. The trypanosomatid parasites of insects and arachnids. Experimental Parasitology, *18*:124-193.

PHYLUM: PLATYHELMINTHES

Yamaguti, S.
1959. Systema helminthum. The cestodes of vertebrates. Interscience Publishers, Inc., New York and London. 860pp.

PHYLUM: ASCHELMINTHES

Yamaguti, S.
1961. Systema helminthum. The nematodes of vertebrates. Interscience Publishers, Inc., New York and London. 1261pp.

PHYLUM: MOLLUSCA

Connolly, M.
1938. Monographic survey of South African non-marine Mollusca. Annals of the South African Museum. *33* (Part 1):1-659.

Thiele, J.
1929. Handbuch der systematischen Weichtierkunde. Jena Verlag von Gustav Fischer, Berlin. 778pp.

PHYLUM: ANNELIDA

Gates, G. E.
1955. Notes on several species of the earthworm, genus: *Diplocardia* Garmen, 1888. Bulletin of the Museum of Comparative Zoology, at Harvard College, Cambridge, Massachusetts. *113* (No. 3):227-259.

Gates, G. E.
1959. On a taxonomic puzzle in the classification of the earthworms. Bulletin of the Museum of Comparative Zoology, at Harvard College, Cambridge, Massachusetts. *121* (No. 6):227-261.

ORDER: PSEUDOSCORPIONIDEA

Beier, M.
1963. Ordnung Pseudoscorpionidea. Akademie-Verlag, Berlin. 313pp.

ORDER: ARANEAE

Bonnet, P.
1959. Bibliographa araneorum. Tome II. Les Artisans de L'Imprimerie Douladoure. 5058pp.

ORDER: ACARINA

Baker, E. W., and Wharton, G. W.
1952. An introduction to acarology. The Macmillan Co., New York. 465pp.

Oudemans, A. C.
1926. Kritisch historisch overzicht der acarologie. Eerste gedeelte, 850 V.C. tot 1758. Tijdschr. Entom., *69*:Suppl. viii+500pp.

Oudemans, A. C.
1929. Kritisch historisch overzicht der acarologie. Tweede gedeelte, 1759-1804. Tijdschr. Entom., *72*:Suppl. xvii+1097pp.

ORDER: ORTHOPTERA

Bei-Bienko, G. Ya., and Mishchenko, L. L.
1963. Locusts and grasshoppers of the U.S.S.R. and adjacent countries. Keys to the fauna of the U.S.S.R. The Zoological Institute of the U.S.S.R. Academy of Sciences, No. 38. Izdatel'stvo Akademii Nauk S.S.R. Moskva-Leningrad, 1951. Translated from Russian, Israel Program for Scientific Translations, Jerusalem. Part I, 400pp.; Part II, 291pp.

Blatchley, W. S.
1920. Orthoptera of Northeastern America. The Nature Publishing Co., Indianapolis. 784pp.

Johnston, H. B.
1956. Annotated catalogue of African grasshoppers. Published for the Anti-Locust Research Centre at the University Press, Cambridge. 833pp.

Uvarov, B.
1966. Grasshoppers and locusts, a handbook of general acridology. Published for the Anti-Locust Research Centre at the University Press, Cambridge. Vol. I, 481pp.

ORDER: ANOPLURA

Ferris, G. F., and Stojanovich, C. J.
1951. The sucking lice. Pacific Coast Entomological Society Memoirs, San Francisco. Vol. 1, 320pp.

ORDER: ODONATA

Longfield, C.
1937. The dragonflies of the British Isles. Frederick Warne and Co., Ltd., London and New York. 256pp.

Needham, J. G., and Heywood, H. B.
1929. A handbook of the dragonflies of North America. Charles C. Thomas, Publisher, Springfield, Illinois. 378pp.

Needham, J. G., and Westfall, M. J., Jr.
1955. A manual of the dragonflies of North America. University of California Press, Berkeley and Los Angeles. 615pp.

ORDER: HEMIPTERA

Miller, N.C.E.
1956. The Biology of the Heteroptera. Leonard Hill [Books] Ltd., London. 162pp.

Van Duzee, E. P.
1917. Catalogue of the Hemiptera of America north of Mexico. University of California Press, Berkeley. 902pp.

Van Duzee, E. P.
1916. Checklist of the Hemiptera of America north of Mexico. New York Entomological Society. 111pp.

ORDER: LEPIDOPTERA

Forster, W., and Wohlfahrt, T. A.
1960. Die Schmetterlinge Mitteleuropas. Band III. Spinner und Schwärmer. Frankh'sche Verlagsbuchhandlung, W. Keller and Co., Stuttgart, Germany. 239pp.

McDunnough, J.
1938. Checklist of the Lepidoptera of Canada and the United States of America. Part I, Macrolepidoptera. Memoirs of the Southern California Academy of Sciences, Los Angeles, California. 275pp.

McDunnough, J.
1939. Checklist of the Lepidoptera of Canada and the United States of America. Part II, Macrolepidoptera. Memoirs of the Southern California Academy of Sciences, Los Angeles, California. 171pp.

Staudinger, O., and Rebel, H.
1901. Catalog der Lepidopteren des Palaearctischen Faunengebietes. R. Friedländer and Sohn, Berlin. 411pp.

Wagner, H.
1914. Lepidopterorum catalogus. W. Junk, Berlin. Part 19, 63pp. Also, 1934, Part 66, 141pp.; 1936, Part 73, 484pp.

ORDER: COLEOPTERA

Blackwelder, R. E.

1939. Leng catalogue of the Coleoptera of America, north of Mexico. Fourth Supplement, 1933 to 1938 (inclusive). John D. Sherman, Jr., Mount Vernon, New York. 146pp.

Blackwelder, R. E. and R. M.

1948. Leng catalogue of the Coleoptera of America, north of Mexico. Fifth Supplement, 1939 to 1947 (inclusive). John D. Sherman, Jr., Mount Vernon, New York. 87pp.

Leng, C. W.

1920. Catalogue of the Coleoptera of America, north of Mexico. John D. Sherman, Jr., Mount Vernon, New York. 470pp.

Leng, C. W.

1927. Leng catalogue of the Coleoptera of America, north of Mexico. Supplement, 1919 to 1924 (inclusive). John D. Sherman, Jr., Mount Vernon, New York. 78pp.

Leng, C. W.

1933. Leng catalogue of the Coleoptera of America, north of Mexico. Second and Third Supplements, 1925 to 1932 (inclusive). John D. Sherman, Jr., Mount Vernon, New York. 112pp.

Schenkling, S., editor

1910-40. Coleopterorum catalogus. Auspicus et auxilio W. Junk, Uitgevery W. Junk, Verlag für Naturwissenschaften, 's Gravenhage. Vols. 1-31.

Sheerpeltz, O.

1934. Zwei neue Arten der Gattung *Aleochara* Gravh. (Coleopt., Staphylinidae) die aus den Puppen von *Lyperosia* (Dipt.), als Parasiten gezogen wurden. Revue Suisse de Zoologie, Annales de la Société d'Histoire Naturelle de Genève. *41* (No. 6):131-147.

Winkler, A., editor

1924-32. Catalogus coleopterorum regionis palaearcticae. XVIII, Dittesgasse 11. Vol. I: Wien, 1924-27; Vol. II: Wien, 1927-32. 1698pp.

ORDER: HYMENOPTERA

Bouček, Z.

1963. A taxonomic study in *Spalangia* Latr. (Hymenoptera, Chalcidoidea). Acta Entom., Musei Nationalis, Prag. *35*:492-512.

Muesebeck, C.F.W., Krombein, K. V., and Townes, H. K.
1951. Hymenoptera of America north of Mexico—Synoptic catalogue. United States Government Printing Office, Washington, D. C. 1420pp.

Nikol'skaya, M. N.
1963. The chalcid fauna of the U.S.S.R., Izdatel'stvo Akademii Nauk S.S.R., Moskva-Leningrad, 1952. Published for the National Science Foundation, Washington, D. C., by the Israel Program for Scientific Translations, Jerusalem. 593pp.

Tillyard, R. J.
1926. The insects of Australia and New Zealand. Angus and Robertson, Ltd., Sydney, Australia. 559pp.

Zimmerman, E. C.
1948. Insects of Hawaii. University of Hawaii Press, Honolulu. Vol. 2, 475pp.

ORDER: DIPTERA

Aldrich, J. M.
1916. *Sarcophaga* and allies in North America. Murphey-Bivins Co. Press, Lafayette, Indiana. Vol. 1, 301pp.

Bohart, G. E., and Gressitt, J. L.
1951. Filth-inhabiting flies of Guam. Bernice P. Bishop Museum, Honolulu, Hawaii. 152pp.

Curran, C. H.
1965. The families and genera of North American Diptera. Second edition. Henry Tripp, Publisher, Woodhaven, New York. 515pp.

Hall, D. G.
1948. The blowflies of North America. The Thomas Say Foundation, Monumental Printing Co., Baltimore, Maryland. 477pp.

James, M. T.
1947. The flies that cause myiasis in man. United States Government Printing Office, Washington, D.C. 175pp.

Lindner, E.
1925-66. Die Fliegen der palaearktischen Region. Stuttgart, Schweizerbart'sche Verlagsbuchhandlung. (Various fascicles were used.)

Stone, A., Sabrosky, C. W., Wirth, W. W., Foote, R. H., and Coulson, J. R.
1965. A catalog of the Diptera of America north of Mexico. Agricultural Research Service, United States Department of Agriculture, United States Government Printing Office, Washington, D. C. 1696pp.

West, L. S.
1951. The housefly: Its natural history, medical importance and control. Comstock Publishing Co., Inc., associated with Cornell University Press, Ithaca, New York. 584pp.

Wytsman, P. de
1937. Genera insectorum. Family Muscidae, Fascicle 205, 604pp., par E. Séguy, Quatre-Bras, Tervueren, Belgium.

Zumpt, F.
1965. Myiasis in man and animals in the old world. Butterworths & Co., Ltd., London. 267pp.

CLASS: AVES

Austin, O. L., Jr.
1961. Birds of the world. Herbert S. Zim, editor, Golden Press, New York. 316pp.

Mathews, G. M.
1921-22. The birds of Australia, H. F. & G. Witherby Ltd., London. Vol. 9, 518pp. Also, Vol. 11, 1923-24, 593pp.; Vol. 12, 1926-27, 454pp.

Mayr, E., and Greenway, J. C., Jr., editors
1960. A checklist of birds of the world. Museum of Comparative Zoology, Cambridge, Massachusetts. Vol. IX, 506pp.

Ripley, S. D., II
1961. A synopsis of the birds of India and Pakistan. Bombay Natural Historical Society, India. 702pp.

Vaurie, C.
1959. The birds of the Palearctic fauna. H. F. & G. Witherby Ltd., London. 762pp.

Wetmore, A.
1960. A classification for the birds of the world. Smithsonian Institution, Washington, D. C., Smithsonian Miscellaneous Collections. Vol. 139 (No. 11), 37pp.

Witherby, H. F., Jourdain, Rev. F.C.R., Ticehurst, N. F., and Tucker, B. W.
1938. The handbook of British birds. H. F. & G. Witherby, Ltd., London. Vol. I, 326pp.

Wolstenholme, H.
1926. Official checklist of the birds of Australia. Melbourne. 212pp.

CLASSES: AMPHIBIA AND REPTILIA

Boulenger, G. A.
1882. Catalogue of the *Batrachia Salientia s. Ecaudata* in the Collection of the British Museum, London. 2nd edn., 503pp.

Lovebridge, A.
1957. Checklist of the reptiles and amphibians of East Africa. Bulletin of the Museum of Comparative Zoology, at Harvard College, Cambridge, Massachusetts. *117* (No. 2):153-362.

Mertens, R., and Wermuth, H.
1960. Die Amphibien und Reptilien Europas. Verlag Waldemar Kramer, Frankfurt am Main. 264pp.

Pasteur, G., and Bons, J.
1960. Catalogue des reptiles actuels du Maroc. Révision de formes d'Afrique, d'Europe et d'Asie. Travaux de l'Institut Scientifique, Rabat. Chérifien Série Zoologie, No. 21. 132pp.

CLASS: MAMMALIA

Ellerman, J. R., and Morrison-Scott, T.C.S.
1966. Checklist of Palearctic and Indian mammals. British Museum, London. 810pp.

Simpson, G. G.
1945. The principles of classification and a classification of mammals. Bulletin of the American Museum of Natural History, New York. *85*:350pp.

GENERAL

Grassé, P.-P., editor
1949-65. Traité de zoologie. Anatomie systématique, biologie. Libraires de L'Académie de Médecine, Paris. (Various fascicles were used.)

Kloet, G. S., and Hincks, W. D.
1945. A checklist of British insects. Kloet and Hincks. 483 pp.

Grateful appreciation is extended to the following individuals who provided help in validating species and systematically arranging obscure groups.

Viruses: Dr. Joseph L. Melnick, Baylor University.

Thallophyta: Dr. Patricio Poncedeleon, Field Museum of Natural History, Chicago.

Protozoa: Dr. John O. Corliss, University of Maryland; Dr. Norman D. Levine, University of Illinois, Urbana; and Dr. F. G. Wallace, University of Minnesota.

Aschelminthes: Dr. W. R. Nickle, U. S. Department of Agriculture, Crops Protection Research Branch.

Mollusca: Dr. Alan Solem, Field Museum of Natural History, Chicago.

Annelida: Dr. Meredith Jones, Smithsonian Institution, U. S. National Museum.

Acarina: Dr. Richard Axtell, North Carolina State University; Dr. Edward W. Baker, U. S. Department of Agriculture, Introduction Research Branch; and Dr. P. H. Vercammen-Grandjean, University of California.

Lepidoptera: Dr. Edward L. Todd, Smithsonian Institution, U. S. National Museum; and Dr. August Ziemer, Field Museum of Natural History, Chicago.

Coleoptera: The following individuals are associated with the Field Museum of Natural History, Chicago: Mr. Michael Prokop, and Dr. Rupert L. Wenzel.

Hymenoptera: The following individuals are associated with the Smithsonian Institution, U. S. National Museum: Dr. B. D. Burks, Dr. Richard C. Froeschner, Dr. Paul M. Marsh, Mr. David R. Smith, and Luella Walkley.

Diptera: The following individuals are associated with the U. S. Department of Agriculture, Entomology Research Division: Dr. Richard H. Foote, Dr. Raymond J. Gagné, Dr. Curtis W. Sabrosky, Dr. George Steyskal, Dr. Alan Stone, and Dr. Willis W. Wirth. Dr. Harold Dodge, formerly at Washington State University; Dr. H. C. Huckett, Riverhead, New York; Dr. Maurice T. James, Washington State University; and Dr. Selwyn S. Roback, Academy of Natural Sciences, Philadelphia.

Aves: Dr. Emmet R. Blake, Field Museum of Natural History, Chicago.

Reptilia and Amphibia: Dr. Hyman Marx, Field Museum of Natural History, Chicago.

Mammalia: Dr. Joseph C. Moore, Field Museum of Natural History, Chicago.

2

Synanthropy

Dalibor Povolný

DEFINITION, EVOLUTION AND CLASSIFICATION

THE TERM "SYNANTHROPIC" is applied particularly to flies and certain rodents coexisting with man over an extended period. A search for the origin of this term certainly provides differing opinions as to its meaning, but for the present let it suffice that the term "synanthropy" has had long usage in the European literature.

A definition of synanthropy as an ecological phenomenon is based on the general acceptance of a zoocoenosis (or biocoenosis) either as a concrete and specific animal community or an abstraction of one. Such communities generally exist, of course, only on a temporary basis. This is based on the view that biocoenoses also evolve, either naturally or secondarily through man's activity. From this point of view, the existing biocoenoses can be divided into two groups: primary, or natural, biocoenoses, and secondary, or cultural, biocoenoses. The second group comprises all types of biocoenoses whose composition has in a way been influenced by the activity of man—from biocoenoses influenced by occasional utilization to those of a cultivated steppe, with their highly developed monocultures and with all that results from their being rationally cultivated by man. Thus, secondary biocoenoses have developed with the intervention of man in nature, starting with primitive adaptation of vegetation (and the development of agrobiocoenoses), and culminating in the cultivation of highly improved plants (monocultural coenoses) (Fig. 1). Naturally, the animal communities of secondary biocoenoses were selectively evolved from communities of the original eubiocoenoses; hence, they consist of those species able to adapt to the newly developed conditions, in some cases becoming even better adapted to the new environment, compared with the original one. A parallel process developed when the animals domesticated by man became the subject of his intensive economic interest. Man thus commenced an artificial elimination of animals from their eubiocoenosis, while at the same time he confined these animals more or less permanently to his residence, in order to utilize them as sources of food, work, etc. Thus, man has created a special type of secondary biocoenosis, the biocoenosis of human residence, or anthropobiocoenosis, originally being a common residence of man and domesticated animals (Fig. 2). In addition, a third group of animals became members of the

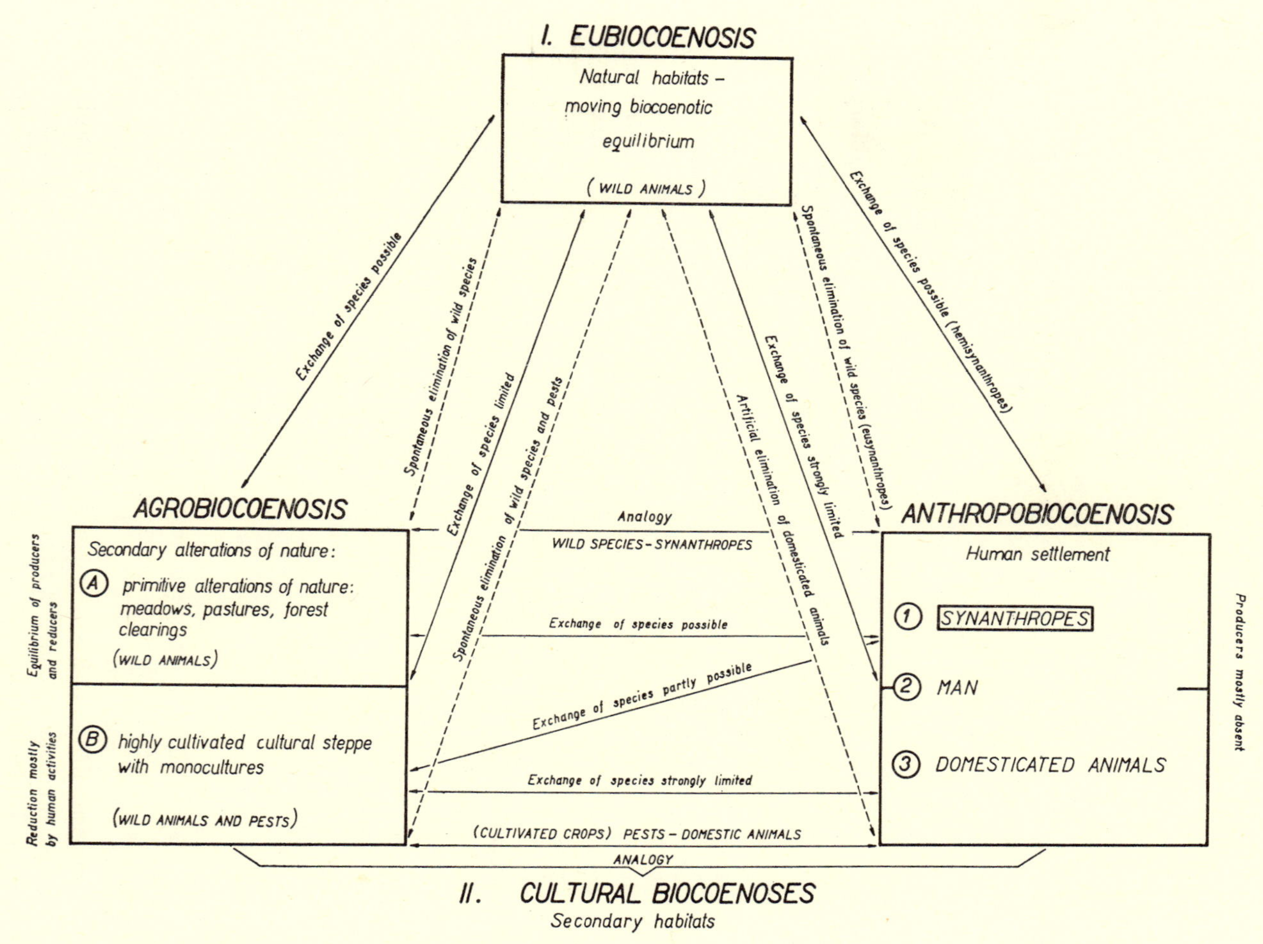

FIG. 1. Dynamic exchange of synanthropic and other organisms between components of the biotic environment

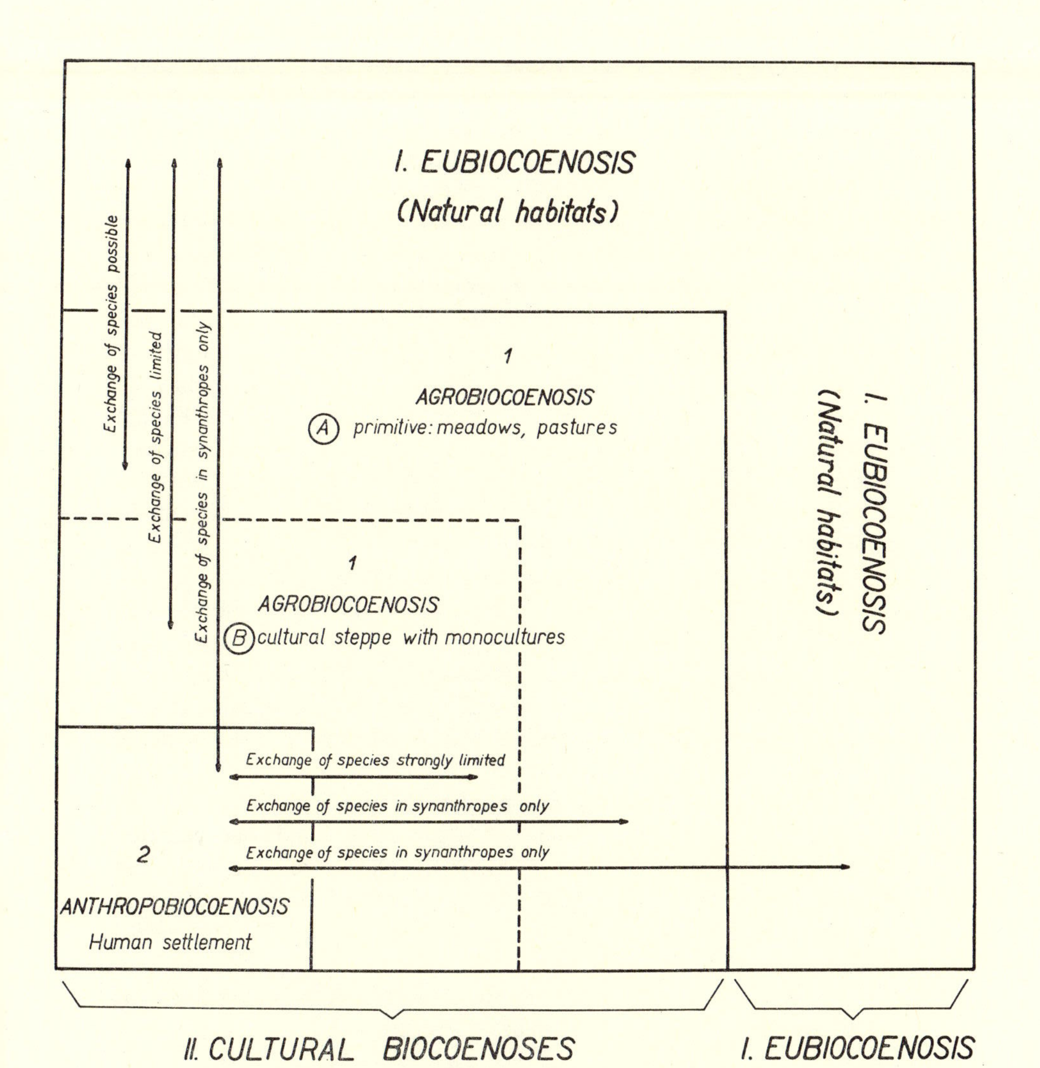

FIG. 2. The anthropobiocoenosis as a component of cultural biocoenoses, and possibilities for exchange of species with the eubiocoenosis

anthropobiocoenosis, without man's sanction. These are the rodents, flies, and other arthropods. The biocoenosis of human residence, consists of the following three components:

(a) man, its creator;
(b) domestic animals, its products; and
(c) synanthropic animals, its spontaneous members.

Thus synanthropy is a spontaneous membership in the anthropobiocoenosis, generally contrary to man's wish. Naturally, there are various stimuli leading synanthropic animals to their spontaneous membership in the anthropobiocoenosis, as far as their relationship to human residences is concerned. This is reflected in the unusually wide range of synanthropic animals. Substantially, we may discern three components again. The first one consists of animals connected with anthropobiocoenoses directly through man (e.g., commensals of man). A second group consists of those animals whose membership in the anthropobiocoenosis is realized through domestic animals (e.g., numerous coprophagous and ectoparasitic flies). Finally, a third group may be formed by those animals which find suitable shelter within the human residence itself. Naturally, the three above-mentioned types of relationship do not quite cover all existing eventualities. However, it must be understood that the stimuli leading the various animals to their membership in the anthropobiocoenosis may be manifold. This must be borne in mind as the synanthropy of animals may appear in various degrees and, also, in variously close or qualitatively different relationships. Thence, the *classification* of synanthropy is a much more complicated task than a general *definition* of synanthropy. Notwithstanding, it is important that an unambiguous definition of this ecological category be based on a consistent conception of synanthropy as a biocoenotic feature.

Certainly, flies are a group whose synanthropy is often particularly marked and, therefore, well defined. On the other hand, one can find various degrees of synanthropy in flies, from total association with man (so that without an anthropocoenosis many species could not exist at all outside their autochthonous area) to species in which this connection is quite loose and facultative.

A classification of synanthropic flies is essentially determined by the following requirements of species:

(1) trophic requirements of adults and larvae for substrate;

(2) ecologic requirements of various species of flies for abiotic factors (temperature and relative humidity) chiefly in the microclimate; and biotic factors (population density, reproductive potential, competition).

Adults of synanthropic flies feed chiefly on excreta of man and domestic animals (coprophagy), decaying organic compounds (saprophagy and necrophagy), and fresh as well as decomposing proteins of animals (carnivorous habit); many also visit flowers. Larvae of synanthropic flies feed on much the same substrates as adults. Moreover, larvae of some synanthropic flies are often predators, feeding facultatively on larvae of other synanthropic species. There is also the natural tendency of certain species of flies to frequent the anthropobiocoenosis as larvae and/or adults because some of its abiotic components (farm buildings, farmyards, cellars, etc.) better satisfy their requirements than similar components of the eubiocoenoses (cavities in trees, rocks and soil, etc.).

The combination of the individual requirements forms the basis for the potential hygienic and epidemiologic significance of flies. These relations are best illustrated by two examples of classical representatives of synanthropic flies.

The primary environment for the larval development of *Phaenicia sericata* is the decaying flesh of carcasses or necrotic tissues of living vertebrates; secondarily, the larvae can develop in various kitchen offal, organic remains and, potentially, even in feces. The food of adults is the same; during the summer it also includes fruit and other sources of saccharides. Thence, it follows that under the conditions of an anthropobiocoenosis, *P. sericata* fulfills all the conditions of a closed circle between microbially contaminated sources of food and man (Fig. 3).

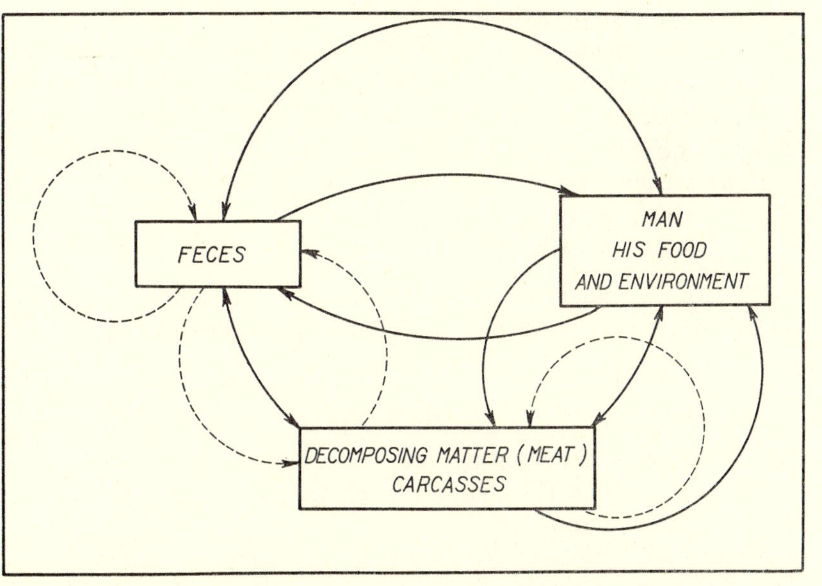

FIG. 3. Schematic representation of communicative and non-communicative species, i.e., of closed (——) and open (– – –) circle in relation to man

On the other hand, *Protophormia terraenovae* is a similar commensal of man with high population densities in mountains and in northern latitudes. Unlike *P. sericata,* this species does not invade human dwellings. Also, its diet is more limited since adults avoid feces, and all its developmental stages are confined to flesh and kitchen offal. These differences indicate that irrespective of population density in the anthropobiocoenosis, *P. terraenovae* does not directly close the circle between microbially contaminated sources of food and man.

In comparison with these two species, the classical synanthropy of *Musca domestica* is many sided and very pronounced. This species seems to have retained its original thermophilous habits. Around human habitations, *M. domestica* is trophically linked to all types of human waste, whereas country populations of this species are symbovine, that is, connected chiefly with excreta of farm animals, and above all, pigs (Fig. 4).

Moreover, its behavior qualifies *M. domestica* as a vector of various microbes. Its vagrant habits enable it to circulate constantly among potential sources of microbes and various human foods. In warmer zones, this liaison is directly effected between human feces and food.

The following classification of synanthropic flies is based on a synthesis of their bionomics and behavior:

(1) Eusynanthropes
  (a) exophilous forms
  (b) endophilous forms
(2) Hemisynanthropes
(3) Asynanthropes
(4) Symbovines
  (a) pasture types
  (b) stable types
(5) Causers of myiasis

The term "communicative" denotes the flies' oscillations between the contaminated environment and man's immediate surroundings.

Eusynanthropes

These flies are closely associated with the anthropobiocoenosis in which their entire development takes place. Many of these species have spread extensively in the human environment, showing a tendency toward cosmopolitanism.

These species primarily frequent households, food processing plants, slaughterhouses, garbage depots, rendering plants, etc. Within this group, one should distinguish between endophilous and exophilous

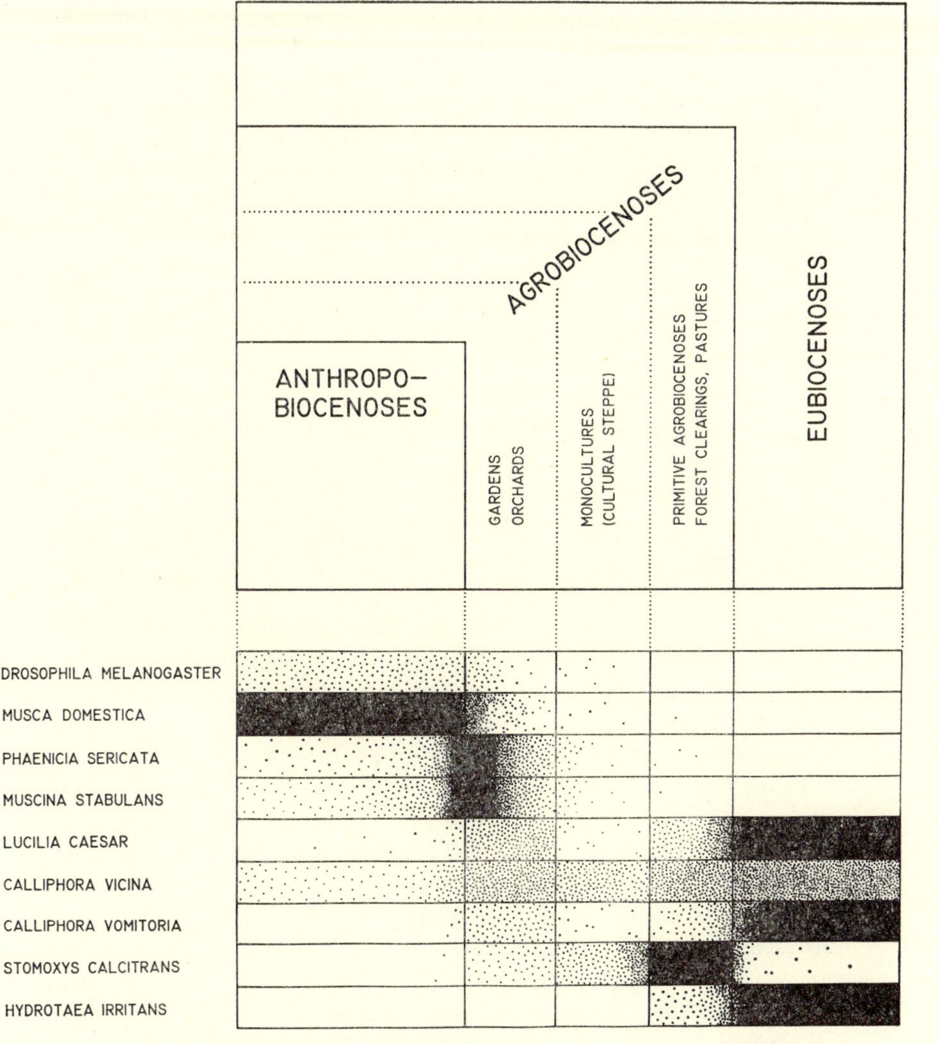

FIG. 4. Differences in synanthropy of some common species of flies expressed by their population density in various biocoenoses

forms. Endophilous species (or endophilous populations of certain species) are tied to the anthropobiocoenoses both trophically and microclimatically. Outside this environment, they are unable to produce populations of high density except during certain seasons of the year (viz. hot summers). Examples of such types are provided by *M. domestica, Drosophila fasciata* and other species of this genus, as well as some *Psychoda* spp. This endophilous habit is particularly conspicuous in *M. domestica.* This species also enables us to recognize synanthropy as an evolutionary phenomenon. *M. domestica* is a species of the subtropics and tropics of the Old World, characterized by exceptional trophic plasticity. The original (autochthonous) populations of this species are primarily coprophagous, feeding on feces of large ruminants. Thus, synanthropy of this species must have started to evolve simultaneously with the domestication of large ruminants during the so-called Agricultural Revolution in the first prehistoric agricultural towns of the Near and Middle East (in the region of the "Fertile Crescent"). From these primarily symbovine pasture forms evolved first the symbovine stable forms and, at last, synanthropic forms, through enlargement of their food range due to the "necessity" of passing the winter directly in human habitations. The populations of *M. domestica,* selected both ecologically and ethologically, spread secondarily over anthropobiocoenoses throughout the world, resulting in the present state.

Endophilous habits are also observed in certain mosquitoes and in some root gnats of the family Sciaridae which develop in the detritus of flower pots, etc. They are not considered synanthropic as a rule, since, their adults are not necessarily connected with the range of human food. However, they are true eusynanthropes, since they are connected with the anthropobiocoenosis.

Exophilous eusynanthropes are associated with the anthropobiocoenosis but do not necessarily require human habitations. They are also less closely connected with man trophically (thence, they are less specialized), and they are less particular as to microclimate. Thus, one does not as a rule encounter them directly in human habitations, but one can observe in these forms a tendency toward abandoning their natural areas, which is characteristic of eusynanthropes. This group includes certain species of the genus *Phaenicia,* particularly *P. sericata.* These substantially southern and thermophilous species are known to occur exclusively in human settlements (that is, in anthropobiocoenoses) in northern regions (frequently, beyond the Arctic Circle); the same is observed in other species of Calliphoridae, and in Piophilidae. In this connection, particular note is given *Fannia canicularis.* Although occurring outdoors away from human settlements, this species usually attains higher population densities under anthropobiocoenotic conditions. Moreover, this species shows seasonal endophilous habits, particularly during summer. *F. canicularis* seems to produce two independent population circles, one living independently of man and the other synanthropic (see also the bionomics of this species). Thus, synanthropy may appear in a segment of a population where it evolved through selective action of the anthropobiocoenosis. This is the case of *P. sericata* in the southern part of Central Europe. Included among exophilous eusynanthropes are *Calliphora vicina, Muscina stabulans,* and related species (Fig. 5).

Hemisynanthropes

This group contains species whose existence is independent of anthropobiocoenoses. Originally, human interference with nature was directed against the natural environments of these species. With the evolution of the anthropobiocoenosis, there appeared in these species synanthropic tendencies resulting in markedly increasing population density in the environs of man. However, unlike eusynanthropes, extinction of anthropobiocoenoses does not necessarily result in extinction of hemisynanthropes, as they do not abandon their primeval area and, thus, remain independent of man. Signs of hemisynanthropy appear during occasional or temporary interference of man with nature, such as latrines in camps, temporary field work, and war campaigns. Among the hemisynanthropes are such species as *Lucilia caesar, L. illustris, B. silvarum, Calliphora vomitoria, Pyrellia cyanicolor, Ophyra leucostoma, Piophila vulgaris, Hylemya strigosa, Hylemya* spp., *Muscina pabulorum,* and *Mydaea urbana.* There is a particularly interesting ecological alternation between the eusynanthrope *P. sericata* and the hemisynanthropes *L. caesar* and *L. illustris,* one group replacing the other under certain circumstances in various biocoenoses. In this respect, the Central European *C. vomitoria* is also interesting (see Fig. 5). Compared to *C. vicina,* it is a much more shade-loving form, living mainly in larger forests and, also, in higher elevations. In the vicinity of mountain chalets, hotels, and sanatoria, *C. vomitoria* attains high population densities and is even capable of actively invading dwellings. There are also those species which come into direct contact with man. A typical communicative one is *Hydrotaea irritans* which gregariously attacks man in forests and feeds upon his conjunctivae. Predatory larvae of this fly develop gregariously in carrion and feces.

It is evident from what has been described above that there is no definite limit between hemisynanthropes and exophilous eusynanthropes. The ecology and behavior of a population may depend on such factors as latitude, climate, character of the biocoenosis, etc. A species which acts as a hemisynanthrope in Central Europe may act as an exophilous eusynanthrope in Northern Europe, occurring there only in the vicinity of anthropobiocoenoses (e.g., *C. vicina* in Scandinavia). For this reason, synanthropy is best considered as a dynamic process in which hemisynanthropy may be a phase of exophilous eusynanthropy. It is thus

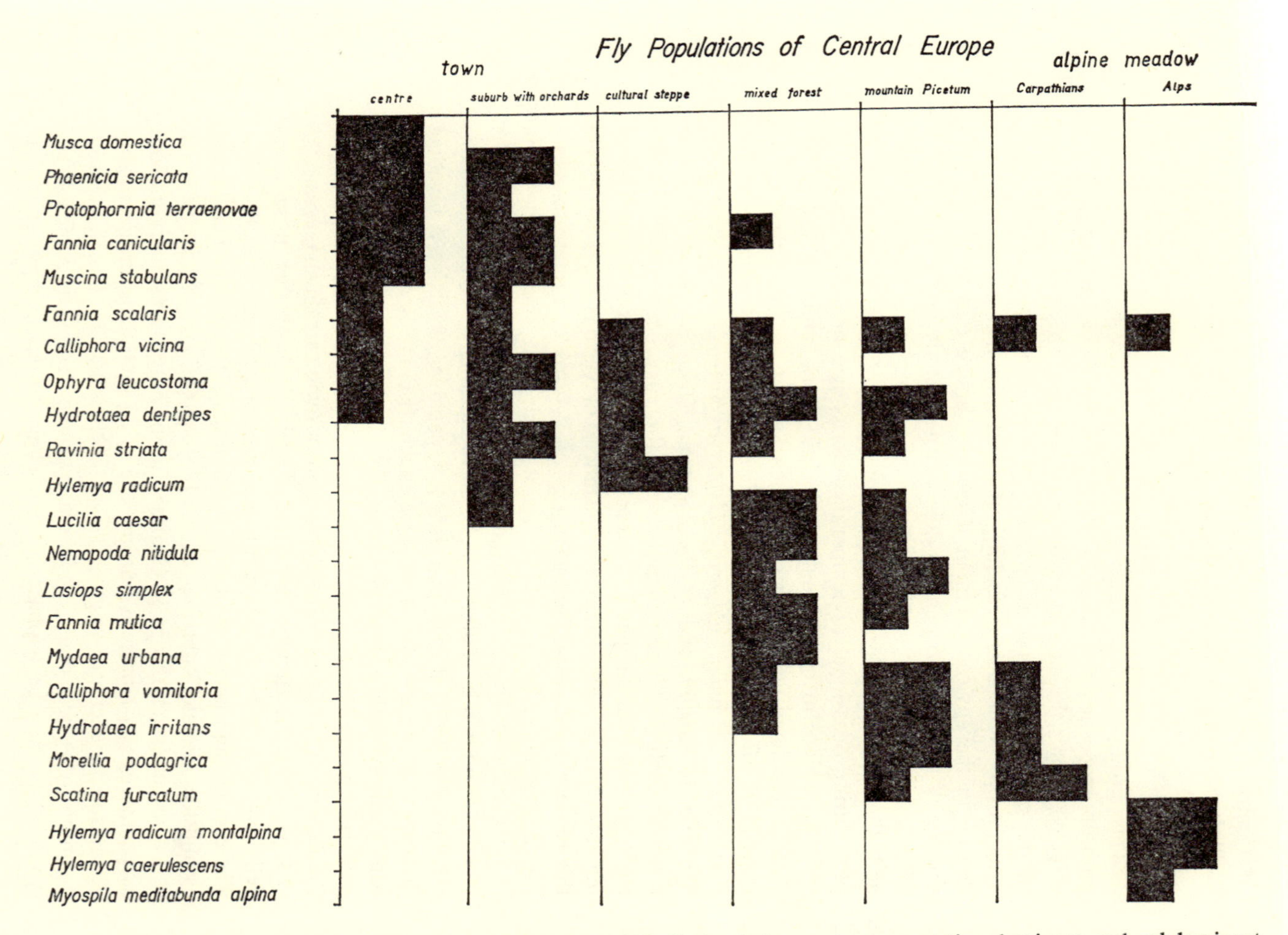

FIG. 5. Schematic representation (based on more than 100,000 flies) of the most representative dominant and subdominant species of synanthropic flies visiting feces in the more characteristic biotypes of Central Europe. (*Scatina furcatum* = *Scatophaga furcata*)

difficult and of limited value to maintain sharply defined and rigid categories.

Asynanthropes

This group includes those groups of flies which do not fulfill the conditions for synanthropic flies.

Symbovines

Symbovines are linked with the anthropobiocoenosis through excreta of domestic ruminants.

Two subgroups are distinguished within this group:

(a) Pasture forms, attaining high population densities in cattle pastures but otherwise resembling hemisynanthropes as they are not immediately dependent on anthropobiocoenoses. This subgroup includes species of the genera *Mesembrina, Orthellia, Morellia, Musca, Hydrotaea,* etc. Particularly important are species whose adults are bloodsuckers, such as species of *Haematobia.*

(b) Stable forms, adapted to excreta of stabled animals and attaining increased population densities directly in and around animal enclosures. This subgroup includes *Stomoxys calcitrans, H. irritans,* and stable populations of *M. domestica.* The latter may differ considerably from typical house populations of this species by their dynamics as well as by a number of additional ecological and populational features.

Causers of myiases

The medical-veterinary importance of this group has led to their separate treatment. However, it is noteworthy that even some causers of myiases (e.g., *P. sericata, Chrysomya* spp., and certain Sarcophagidae) can be true synanthropes.

SOCIETIES OF SYNANTHROPIC FLIES IN THE PALEARCTIC REGION

The aim of this section is to provide a synthesis of information on the composition of synanthropic fly populations in various biotypes and their subdivisions throughout the Palearctic region. This synthesis is derived from the work of a number of investigators who have studied the problem for many years. The diverse methods which they have used have sometimes made it necessary to adapt their data (when comparable) in order to produce the uniform graphic treatment presented here.

Central European cultivated steppe, with remains of original steppes and forest-steppes, in the warmest parts of Central Europe

This biotype is characterized by such typical dominant species as *Agria latifrons, Ravinia striata, Sarcophaga melanura, Calythea albicincta*, and *Hylemya cinerella*, and, in its southernmost part, by *Phaenicia sericata.* As a rule, these leading species are represented in a majority of trappings and their predominance is due to their higher population densities. They are supplemented by certain other species, whose presence is noteworthy though their populations are smaller and more scattered. This group includes certain species of the family Sarcophagidae, such as *Sarcophaga haemorrhoidalis, Parasarcophaga aegyptica, P. jacobsoni, P. portschinskyi*, and *P. aratrix*, as well as *Fannia leucosticta.* The occurrence of certain species may be the result of emigration from temporarily arid regions. For example, variation in the population density of *Chrysomya albiceps* in the subtropics east of the Mediterranean has probably resulted in the invasion of this fly into Central Europe. A similar effect may also be operating to produce the scattered occurrence of *Bellieria maculata.*

A complex of species common to the cultivated steppe and the zone of deciduous forests includes *Fannia canicularis, F. scalaris, Muscina stabulans, Pollenia rudis, Calliphora vicina*, and *Sarcophaga carnaria,* all of which show frequent tendencies toward eusynanthropy. *Muscina assimilis, Hydrotaea dentipes*, and *Phormia regina* also occur but in smaller numbers. At the transition to deciduous forests there is a quantitative increase of *Lucilia caesar*, whereas in the moist meadows of the Pannonian region the numbers of this fly are reduced and brought into balance with populations of *Phaenicia sericata. Sepsis punctum* is commonly found on feces and *Ophyra leucostoma* on meat. Bottomland forests in the alluvial zone around larger rivers are characterized by the frequent occurrence of *Pegomya socia* and the dominant *Lucilia caesar* and *Ophyra leucostoma*, as well as by the more hygrophilous species of the genera *Phaonia, Mydaea*, and *Pyrellia. Platystoma lugubre* is a rare but exclusive inhabitant of bottomland forests and similar habitats.

Zone of deciduous forests, chiefly Querceto-Carpinetum, at warm medium altitudes

This zone is characterized by the dominance of *Lucilia caesar, Pyrellia cyanicolor*, and *Nemopoda nitidula*, in addition to the typical *Hylemya strigosa, Calliphora vomitoria*, and *Hydrotaea dentipes* (Figs. 6, 7, 8). These are joined by the less abundant species *Lasiops*

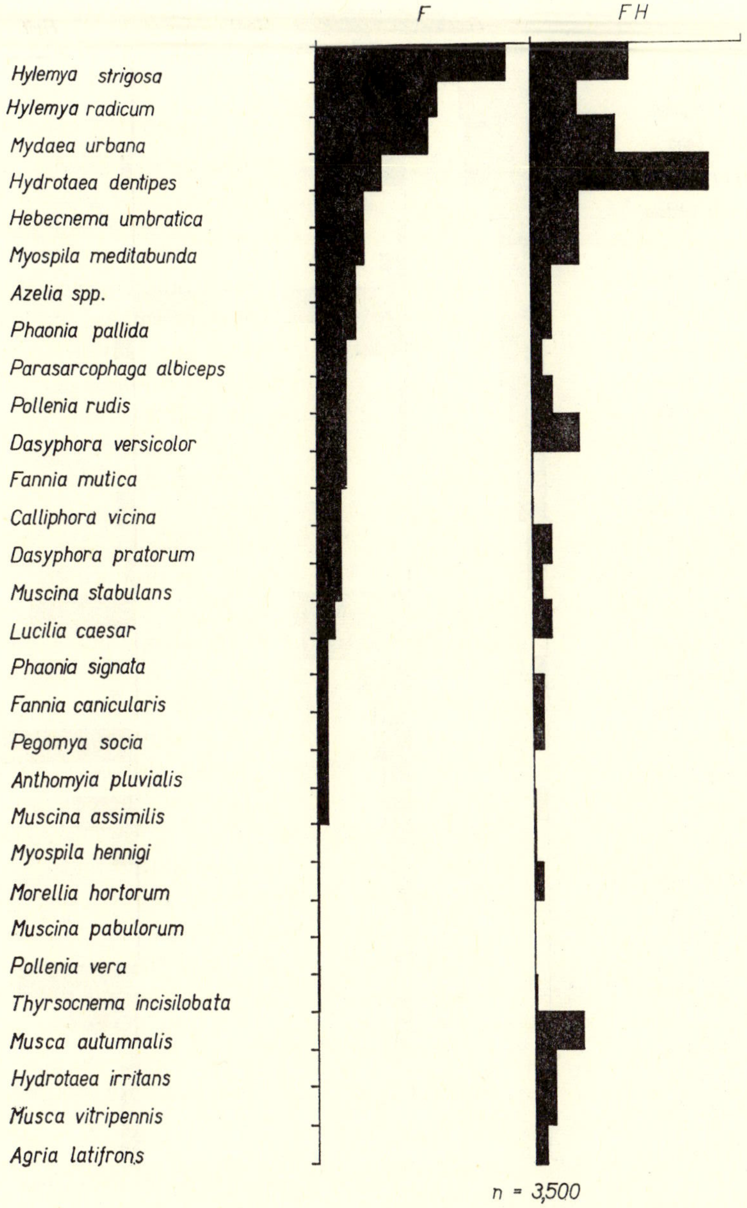

F FH

Hylemya strigosa
Hylemya radicum
Mydaea urbana
Hydrotaea dentipes
Hebecnema umbratica
Myospila meditabunda
Azelia spp.
Phaonia pallida
Parasarcophaga albiceps
Pollenia rudis
Dasyphora versicolor
Fannia mutica
Calliphora vicina
Dasyphora pratorum
Muscina stabulans
Lucilia caesar
Phaonia signata
Fannia canicularis
Pegomya socia
Anthomyia pluvialis
Muscina assimilis
Myospila hennigi
Morellia hortorum
Muscina pabulorum
Pollenia vera
Thyrsocnema incisilobata
Musca autumnalis
Hydrotaea irritans
Musca vitripennis
Agria latifrons

n = 3,500

FIG. 6. Comparison (based on 3,500 flies) of two samples of hemisynanthropic coprophagous flies visiting human feces (F) and feces of horses (FH) taken during the first week of July on the margin of a Querceto-Carpinetum forest in Central Europe. These censuses may be considered as typical for habitat and season in the Western Palearctic

FIG. 7. Comparison (based on 3,500 flies) of three samples of hemisynanthropic flies with nearly the same species composition as in the preceding graph. These samples were collected during the first week of September on the margin of a Querceto-Carpinetum forest in Central Europe. Comparison with the previous scheme shows the following: (a) changes of population density in the single species observed; (b) seasonal influence on the qualitative and quantitative relations of certain species; (c) the communicative nature of some species is manifested by their visit to all three baits—feces (F), meat (M), and fruit (Fr)

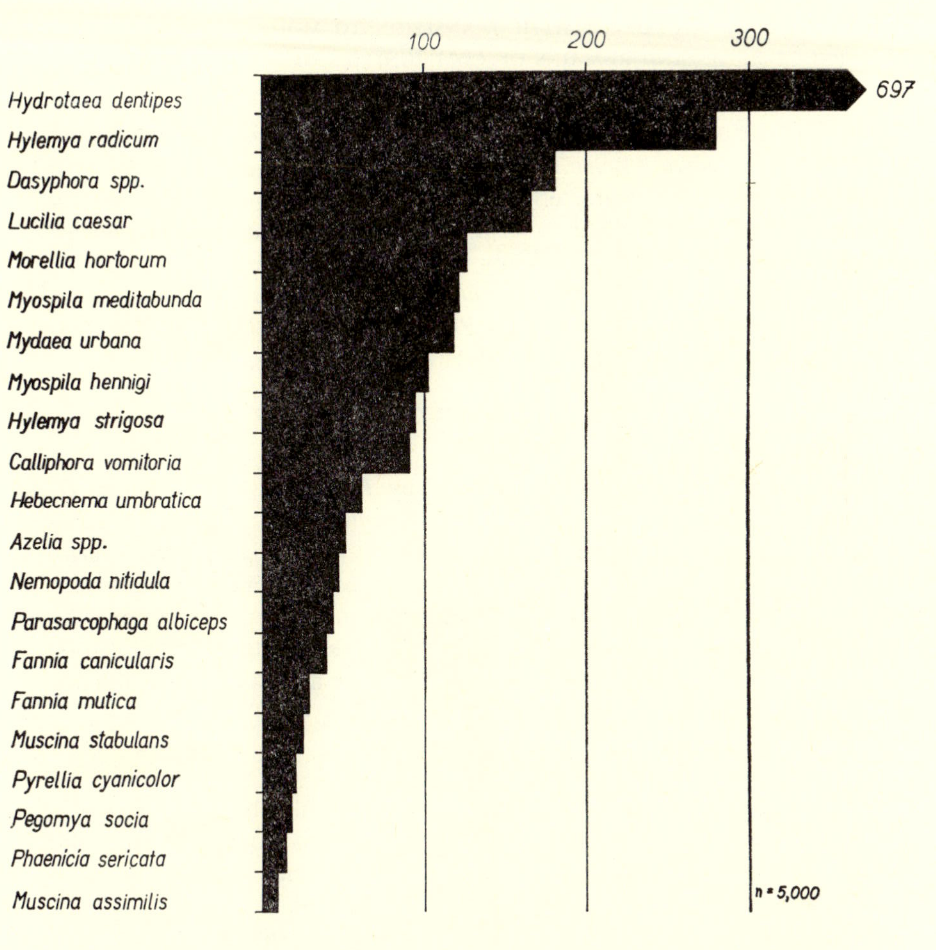

FIG. 8. Schematic representation of an association of dominant species of synanthropic flies sampled on feces once a week during May, June, and July on the margin of a Central European Querceto-Carpinetum forest. Species, e.g., *Sarcophaga melanura, Muscina pabulorum, Phormia regina, Pollenia rudis, Pollenia vera, Scatophaga stercoraria,* and *Cypsela nigra* were uncommon. This sample represents a good average aspect of this type of habitat in the western part of the Palearctic. Seasonal changes are reflected in changes in population densities of individual species

simplex, Fannia mutica, Mesembrina meridiana, Mydaea urbana, Muscina assimilis, Myospila meditabunda and, quite commonly during late summer, by *Polietes lardaria.* The family Sarcophagidae is represented mainly by *Parasarcophaga albiceps, P. similis, P. tuberosa, P. harpax, P. aratrix,* and *P. pseudoscoparia.*

Zone of mixed forest with transition to mountain spruce belts and remains of communities of the *Fagetum abietosum* and *Acereto-Fraxinetum* type

This zone harbors typical populations with *Lasiops simplex, Fannia mutica, Hylemya lasciva, Pegomya socia*, and *Mydaea urbana* predominating. Also characteristic are *Calliphora vomitoria, Hydrotaea dentipes, Phaonia pallida, Myospila hennigi, Hylemya strigosa, Polietes lardaria, Fannia parva* (typically associated with the Fagetums of the Carpathians), and *Neuroctena anilis.* In the mountain forest belts one finds *Mesembrina mystacea, Phaonia variegata, Calliphora uralensis, C. loewi, Hydrotaea irritans*, and *Morellia podagrica.*

Mountain forest belt with Picetum and some autochthonous larch

This zone is characterized mainly by *H. irritans, H. strigosa, H. radicum, M. deserta, M. scutellaris, C. vomitoria, M. podagrica* (missing in the Hercynian part of Central Europe), *Scatophaga stercoraria,* and *Scatophaga furcata* (Fig. 9). Here one also finds most of the species typical of the *Fagetum* and *Acereto-Fraxinetum* forest type. Among the non-dominant species are *Calliphora subalpina, C. alpina, C. uralensis,* and *C. loewi.* This biotype is noteworthy for the almost complete absence of members of the family Sarcophagidae.

Zone of Alpine meadows

This biotype cannot be characterized uniformly, because there is a distinct subnival belt in the Alps which is absent from the Carpathian Mountains. Nevertheless, we may note the common occurrence of *H. radicum* (in the Alps, of the endemic subspecies *H. radicum montalpina*), *H. strigosa, C. mortuorum, S. furcata, S. stercoraria, C. vomitoria, C. alpina, C. subalpina, M. deserta, M. podagrica, Myospila hennigi, Fannia mutica,* and *F. monilis* (Fig. 10). In the Carpathians, *P. lardaria* and *M. scutellaris* extend into this zone from the forest belt. The subnival zone of the Alps is characterized by a number of quite specific forms such as *Steringomyia stylifera, Hylemya caerulescens, H. grisella* (the latter two species often abundant), *Myospila meditabunda alpina, Phaonia tenuiseta, Mydaea rufinervis,* and *Pegohylemya albifacies.* The southern slopes of the Alps and their valleys are exposed to the penetration of thermophilous fauna from deciduous and mixed forests. As a result, these valleys share many species with their neighboring biotypes, which include *L. caesar, M. podagrica, N. nitidula, H. radicum, P. cyanicolor, S. carnaria,* and *Thyrsocnema laciniosa.*

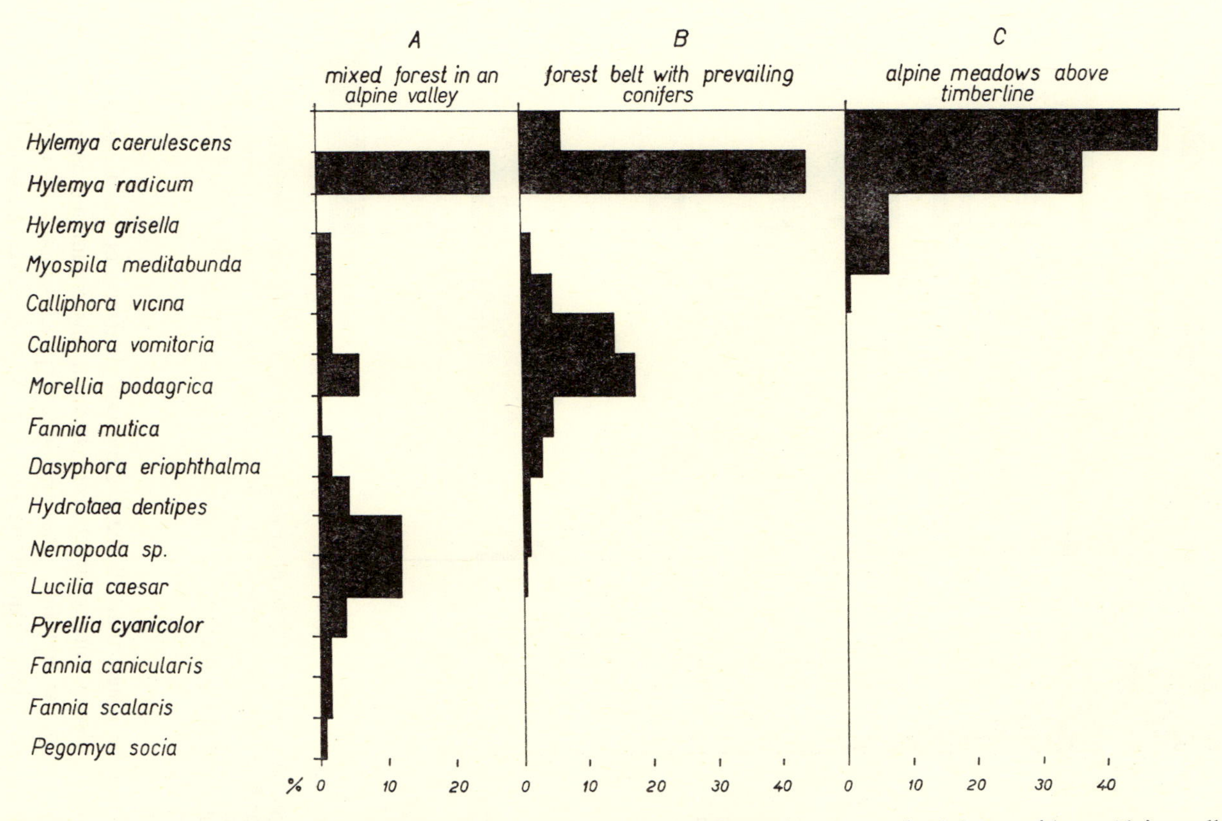

FIG. 9. Representation of sixteen dominant hemisynanthropic flies in the Tyrolean Alps. A = mixed forest with an Alpine valley at about 800 m.; B = forest belt with prevailing conifers between 1700 and 1900 m.; and C = Alpine meadows above timberline (above 1900 m.). Among the non-dominant species of flies the following should be mentioned as characteristic. For zone A: *Lucilia* sp. div., *Parasarcophaga* sp. div., *Bellieria niverca*, *Mydaea urbana*, and *Lasiops simplex*. For zone B: *Mydaea deserta*, *Hydrotaea pandellei*, and *Scatophaga furcata*. For zone C: *Steringomyia stylifera*, *Phaonia tenuiseta*, *Mydaea rufinervis*, and *Pegohylemya albifacies*

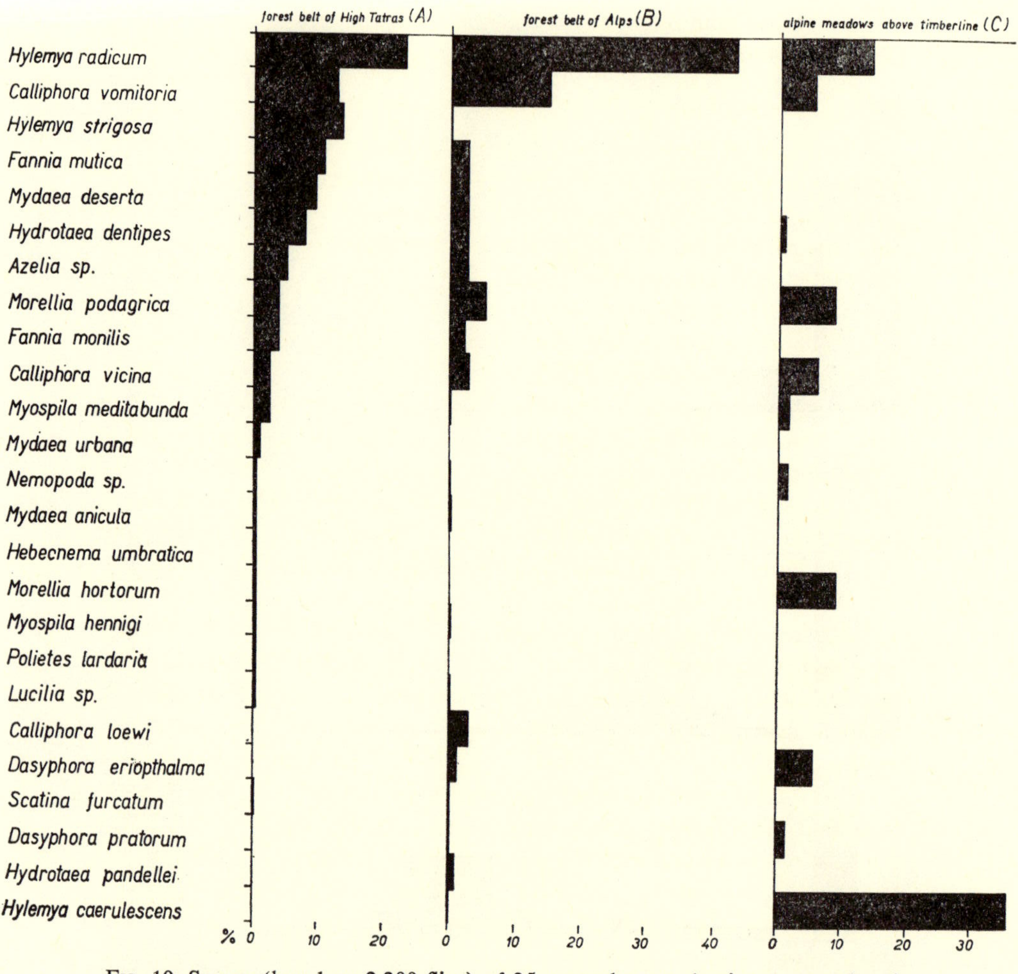

FIG. 10. Survey (based on 2,200 flies) of 25 coprophagous dominant synanthropic flies in the forest belt of the High Tatras at about 1100 m. (A); in the forest belt of the Alps at about 1900 m. (B); in Alpine meadows above timberline of the Alps at about 1800 m. and higher (C). (*Scatina furcatum* = *Scatophaga furcata*)

It should be noted that all these dipterous denizens of the Central European landscape are characterized here in their summer aspect. In early spring or autumn, one encounters psychrophilic forms at much lower elevations than during summer (*S. stercoraria, C. mortuorum, P. rudis,* and *C. uralensis*).

Modifications of the biotypes described above often reflect important differences in the species composition of synanthropic fly populations. Some of these man-made environments foster fly species which directly affect man and animals, and therefore justify additional detail.

ANTHROPOBIOCOENOSES OF LARGELY URBAN TYPE

Alterations in the biotic and abiotic factors in this environment have resulted in a spontaneous elimination of the chiefly hemisynanthropic forms. As a consequence, secondary zoocoenoses of synanthropic flies have arisen, with the predominating eusynanthropic forms showing both exophilous and endophilous habits. The most typical eusynanthropic species tend to occur outside their autochthonous areas as a result of secondary cosmopolitanism. They include *P. sericata, C. vicina, M. stabulans, P. terraenovae,* and *P. casei.* The two most typical endophilous eusynanthropes are *M. domestica* and *F. canicularis.* Anthropobiocoenoses and their immediate vicinity tend to support increased populations of eusynanthropic species. Certain less common but characteristic species also occur in the anthropobiocoenoses, such as *P. demandata, S. haemorrhoidalis,* and *F. leucosticta.* It is noteworthy that the latter species occurs in eusynanthropic populations in Europe but not in Asia. Of equal interest is the discovery that the northern limit of free-living natural populations of *P. demandata* and *P. sericata* are roughly at a level with the northern border of the Great Hungarian Lowland. Only by means of their association with anthropobiocoenoses, are both species able to penetrate far to the north of this border into southern Finland.

A very representative sample taken on the outskirts of Prague indicates that the following communicative and exophilous species are characteristic of anthropobiocoenoses of Central Europe: *P. sericata, H. radicum, O. leucostoma, L. caesar, H. dentipes, P. varipes, P. terraenovae, M. stabulans,* and *H. cinerella* (Fig. 11). *P. sericata* is foremost among the communicative species. These surveys point up the necessity for effective control of synanthropic flies in the environs of food processing plants, rendering plants, and rubbish heaps.

Equally interesting results were obtained from trapping flies in a fruit and vegetable market in Ujpest, in central Hungary (Fig. 12). There was a marked predominance of communicative eusynanthropic species, such as *P. sericata* and *M. stabulans*, and a lower density of *M. domestica* in the open market place. There was also an abundance of *S. melanura, M. assimilis, S. haemorrhoidalis, H. radicum, L. caesar,* and *P. aegyptica.* In gardens in the center of Budapest, the following species predominated: *P. sericata, H. radicum, L. caesar, R. striata, S.*

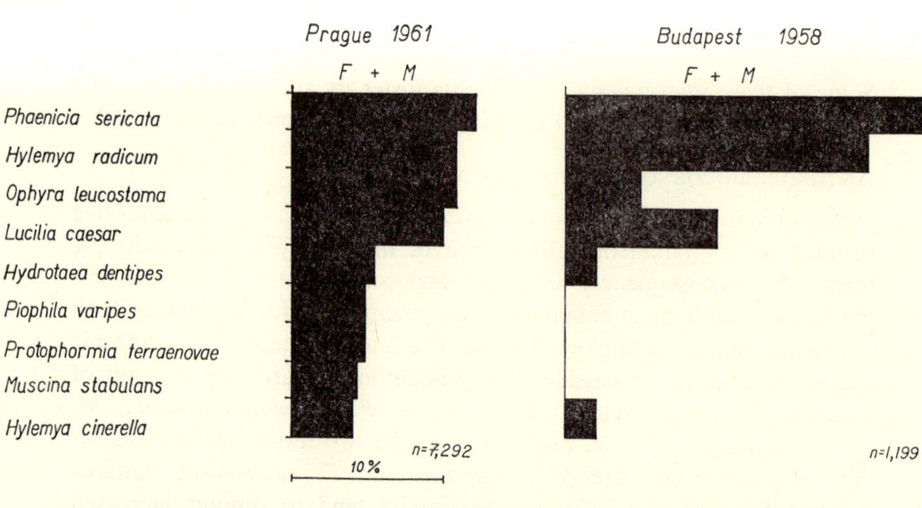

FIG. 11. Schematic representation of hemisynanthropic flies sampled on feces (F) and meat (M) under similar conditions in Prague and in Budapest. The typical increase of *P. sericata* is noteworthy as is the decrease of *P. terraenovae* toward the south. A marked narrowing of the spectrum of species takes place with the predominance of eusynanthropic forms

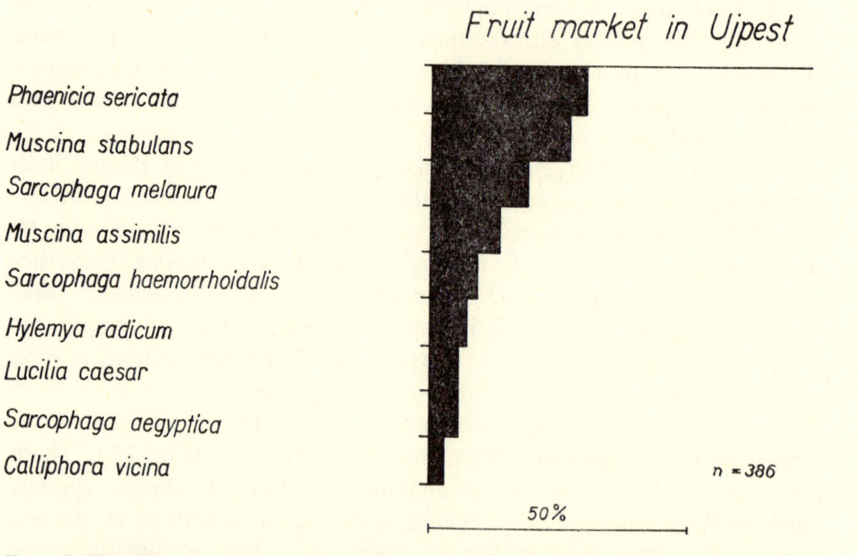

FIG. 12. Sample of hemisynanthropic flies taken in a fruit market in central Hungary. The communicative *P. sericata* and *M. stabulans* prevail. Note the appearance of south European *P. aegyptica* among the dominant species. The strong constriction of the spectrum of flies is due to the anthropobiocenosis, as well as to the density of the eusynanthropic *P. sericata* and *M. stabulans*

melanura, O. leucostoma, H. dentipes, F. canicularis, S. haemorrhoidalis, S. punctum, P. demandata, P. nigrimana, and *P. latipes. M. domestica* was also present and, at times, *C. albiceps* invaded from the south. It is important to note that with the exception of the carnivorous *C. albiceps* and *Piophila* spp., the above species were trapped on both feces and meat, which indicates a significant communicative behavior. In a rendering plant yard in Budapest, a mass occurrence of *M. domestica, P. sericata,* and *Piophila* spp. and a scattered occurrence of *P. regina* and *C. albiceps* were observed.

This situation, typical of the environs of most of the larger towns in the warmer parts of Central Europe, warrants seasonal microbiological examination of food, fruit, and meat for potential contamination, and comparison with the microflora of flies, even during non-epidemic periods.

AGROBIOCOENOSES OF PASTURES

Here we have a typical fauna of symbovine and coprophagous species adapted to excreta of ruminants and horses. At lower elevations in steppes, forest-steppes, and warm deciduous forests *Musca autumnalis, M. meditabunda, D. pratorum, D. versicolor, O. caesarion, O. cornicina,* and numerous *Sepsis* species are common. Less common but still typical are *M. hortorum, H. umbratica* and, in the vicinity of forests, *M. meridiana.* Thermophilous species such as *M. autumnalis* retreat from submontane pastures in mixed forest belts whereas other species appear there, having been eliminated from their original eubiocoenoses surrounding these pastures. Examples of these are *H. strigosa, M. meditabunda, P. lardaria, D. pratorum, D. versicolor, M. podagrica,* and members of the Sphaeroceridae. The species which are common in lowland pastures are scarcer here. *P. lardaria, S. furcata, S. stercoraria,* and *M. scutellaris* are typical of the high mountain pastures and alpine meadows in the Carpathians. *H. radicum montalpina, H. caerulescens, H. grisella* and *M. meditabunda alpina* are typical of the subnival zone in the Alps (Fig. 13).

Censuses of synanthropic flies may also be compiled for smaller areas of the Palearctic region, based on latitude, altitude, proximity of the sea and local character of vegetation.

1. Northern Europe and Asia

Thorough analyses of populations of species of synanthropic flies in northern Europe and Asia are still lacking, but the Calliphoridae have been the focus of considerable study in Scandinavia. In general, it appears true that there are notable densities of certain *Calliphora* in

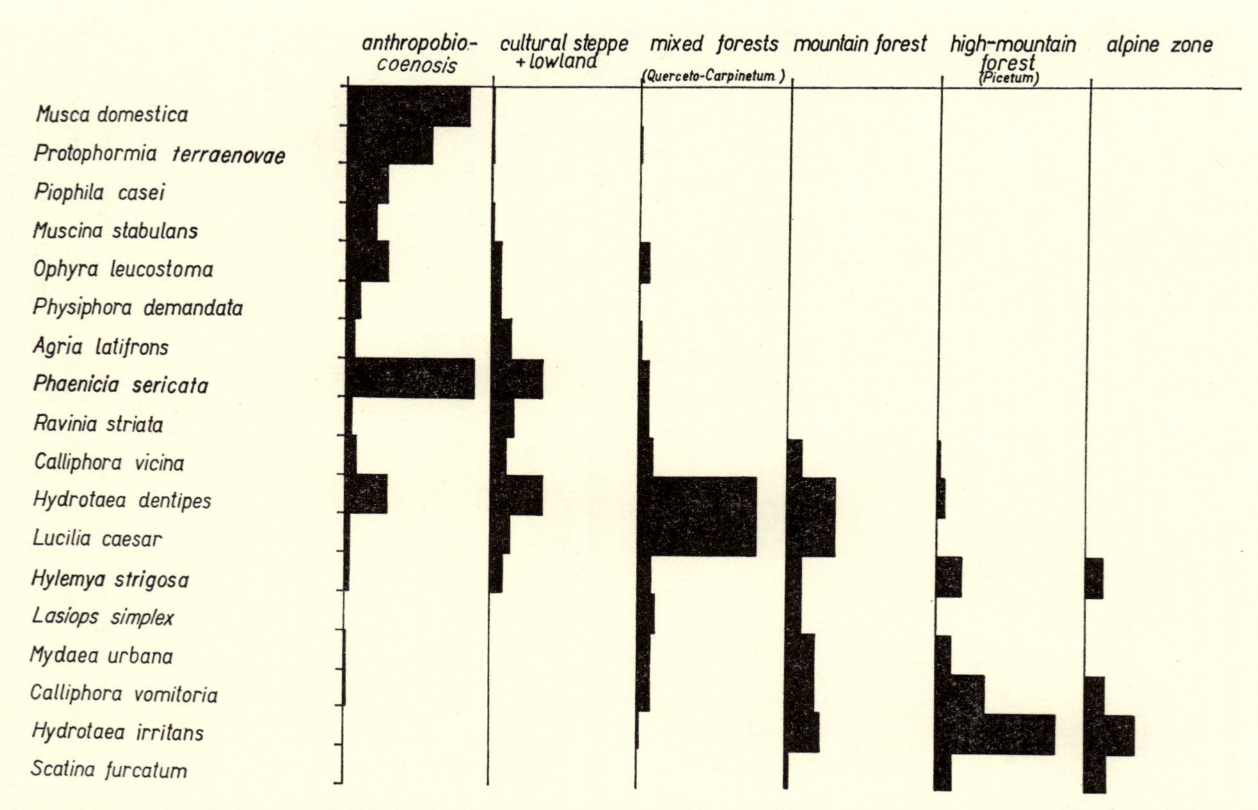

FIG. 13. Representation (based on 55,000 flies) of the population densities of eighteen species of synanthropic flies typical of the mild zone of the western part of the Palearctic. For each biotype only the most representative species, not confined to narrow habitats, have been selected. (*Scatina furcatum* = *Scatophaga furcata*)

northern Europe, whereas in the more southerly part of northern Europe, i.e., in the oak and secondary deciduous forests, certain species of *Lucilia* are more characteristic. The Scandinavian blow flies are further distinguished by the gradual northward decline of *P. sericata* which reaches southern Finland and has a scattered distribution in the north correlated with its eusynanthropic behavior (Fig. 14). *L. caesar* and *B. silvarum* reach central Finland and exhibit similar eusynanthropic tendencies. The highest relative densities in the genus *Lucilia* are attained by *L. illustris* which is the most common species throughout south and central Finland. The central part of northern Scandinavia, with its predominant coniferous forests, is characterized by the almost complete absence of sarcophagids and a marked predominance of *Calliphora*, with *vicina, subalpina* and *loewi* in order of frequency. *C. vicina* is most common in the forest belt where it is eusynanthropic. *C. vomitoria* shows much the same trend but at lower densities. *P. terraenovae* and *C. uralensis* attain their greatest populations as eusynanthropes in northern Finland. The inverse is observed in the decidedly hemisynanthropic *C. lowei* and *C. subalpina* which decrease toward anthropobiocoenoses, and also in *P. sericata* (scarce in human settlements), *L. caesar* (eusynanthropic here), and *B. silvarum* (hemisynanthropic). Farther south, in northern Poland and Germany, *C. subalpina* is confined to the Carpathians and the Alps.

Trappings near the border of the European subarctic have yielded *C. uralensis* (eusynanthropic), *C. vomitoria, C. loewi,* and *C. alpina* (hemisynanthropic), but not *C. vicina.* The Finnish subarctic is characterized by *H. binotata* and *C. mortuorum* among the hemisynanthropes, and *P. terraenovae* among the exophilous eusynanthropes. The more extensive zone of the Eurasian tundra contains a complex of synanthropic flies which includes *C. mortuorum, C. vomitoria, L. illustris, B. atriceps, H. dentipes, H. cinerella,* and a number of *Scatophaga* spp., e.g., *suillum, varipes, perfectum, arcticum,* and *cordylurinum.* Here we also find the non-dominant *Triceratopyga calliphoroides, Acrophaga alpina, Lucilia fuscipalpis, Abonesia genarum, Piophila vulgaris, P. arctica, Themira arctica, Scatophaga furcata, Coelopa frigida,* and *Eristalis tenax.* The chief species of the entire Eurasian tundra are *P. terraenovae* and *C. uralensis.* Thus far, there are no detailed data on population size of the various species, but they seem to be capable of attaining high local densities under conditions of synanthropy.

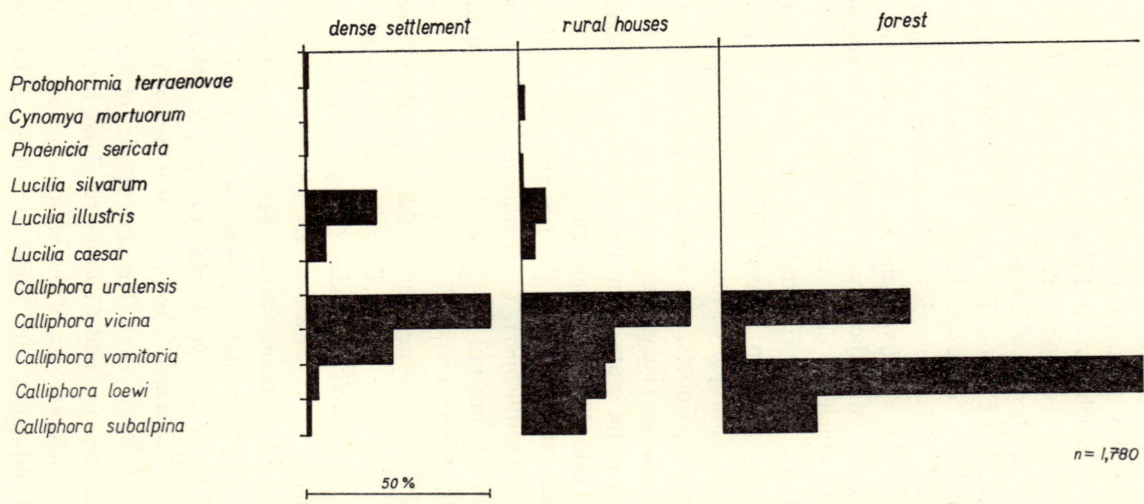

FIG. 14. Representation of the synanthropic blow flies of southern Finland indicating the degree of eusynanthropy and hemisynanthropy

2. *The British Isles*

According to present knowledge, the specific composition of synanthropic flies in central and southern Britain strongly resembles that of the mixed forest belts at medium elevations of the European continent. There is an absence of thermophilous elements, particularly species of Sarcophagidae. Very likely, this is due to the mild and semihumid character of the Atlantic climate. Under natural conditions, at the edge of forests in central Britain, *H. radicum, L. illustris, M. hortorum, B. ater, F. canicularis,* and *M. meditabunda* predominate in summer. *S. stercoraria, H. meteorica, H. armipes, P. lardaria,* and *C. vicina* are also common. Natural habitats along the southern coast of England are characterized in summer by the occurrence of *H. radicum, C. vicina, H. dentipes, M. meditabunda, S. stercoraria,* certain *Helina* spp. and *Hebecnema* spp., scattered *P. sericata,* as well as by numerous species of Sepsidae and Sphaeroceridae. Continuous forests and shaded parks (even those in the center and outskirts of large towns, such as Hyde Park and Kenwood in London itself) are characterized by the predominance of species represented in Figure 15. Due to their park character, British towns frequently offer better conditions for both variety and density of synanthropic flies than nature does. This is demonstrated by a comparison of results of trapping synanthropic flies in Hyde Park in the center of London, with those obtained in the natural reserve of Woodwalton Fen near Ramsey, or in southern England (Fig. 15). In addition to *M. domestica* and *P. sericata,* which are communicative but of low density, *H. radicum* is an important and common synanthropic fly. The dipterous fauna of northern England and Scotland have not been sufficiently investigated, but a general similarity with that of Scandinavia may be expected.

3. *The Mediterranean Region*

Effects opposite to those described above are characteristic of the southern and especially the southeastern part of Europe, e.g., the Balkan Peninsula. Certain species of flies, showing eusynanthropic tendencies in Central and northern Europe, produce hemisynanthropic populations in southern Europe. This is true of *P. sericata, P. demandata,* and *M. domestica* (Fig. 16). Such populations of *P. sericata* and *M. domestica* nevertheless retain a high degree of eusynanthropy. On the other hand, hemisynanthropic forms which have greater latitude with respect to temperature and humidity, e.g., *L. caesar,* become eusynanthropic in the woodless areas of the cultivated steppe and in limestone situations.

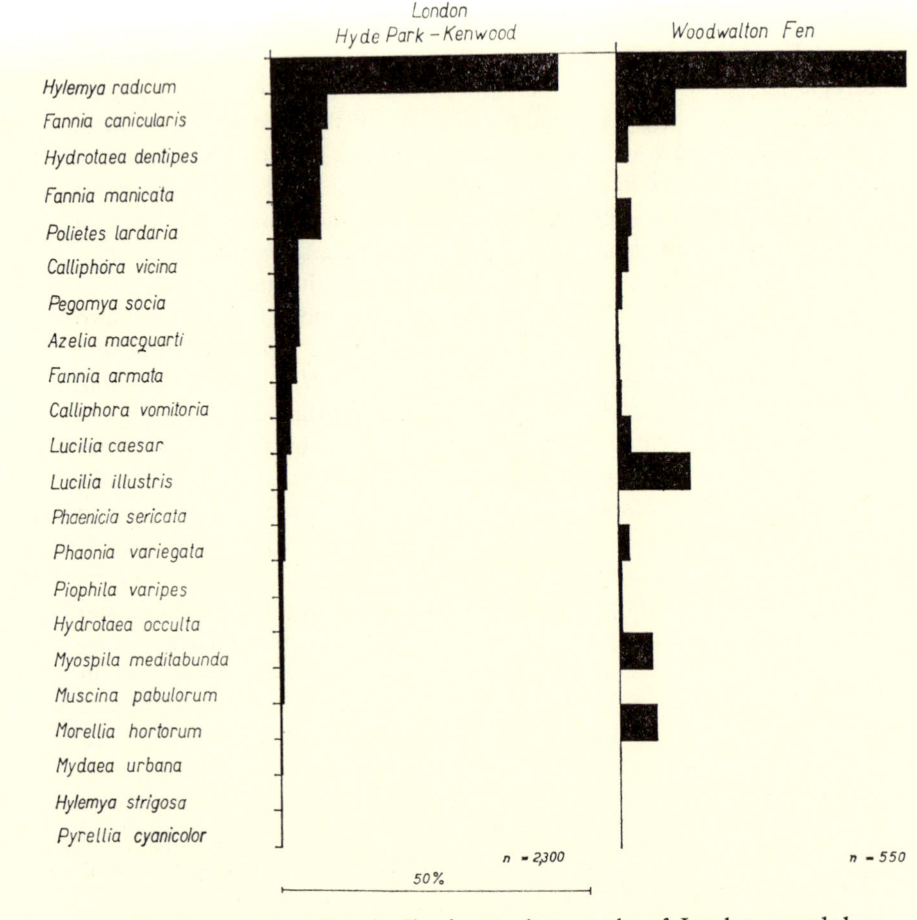

FIG. 15. The hemisynanthropic flies in two large parks of London, sampled on feces. Species composition is strongly influenced by the mezoclimate of a mixed forest. Nevertheless, note the several representatives of eusynanthropic flies, e.g., *P. sericata, C. vicina,* and *L. illustris*

In Central Europe, one finds xerothermophilous species which are less temperature dependent, e.g., *D. cyanella.* These zoocoenological changes are best demonstrated by the increase in population density and number of thermophilous forms of Sarcophagidae: *A. latifrons, R. striata, S. haemorrhoidalis, S. melanura, P. aegyptica, P. jacobsoni, P. portschinskyi, P. exuberans,* and *Wohlfartia magnifica.* As we have already noted, subtropic elements—e.g., *C. albiceps*—may also penetrate into this area from the south. High population densities are

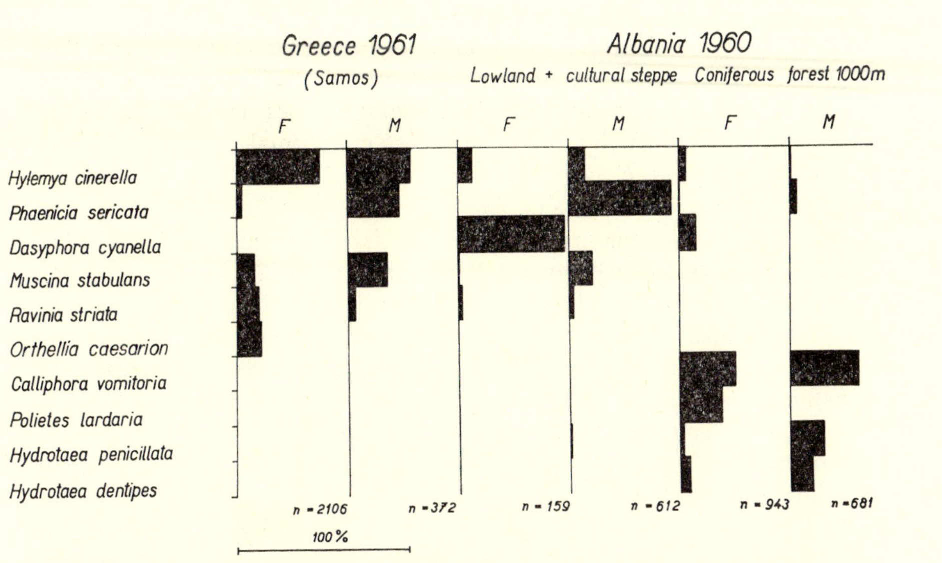

Fig. 16. Typical hemisynanthropic flies of the European Mediterranean, taken on feces (F) and meat (M). The occurrence of *Dasyphora cyanella* and *Hydrotaea penicillata* are characteristic, as well as the considerable densities of *P. sericata* in the wild. Of the non-dominant species, the presence of *Dasyphora albofasciata* is notable

produced by *P. sericata, H. cinerella, M. stabulans, Ulidia erythrophthalma, R. striata,* and *P. regina.*

In the mountain forests of southern Europe and in their remains a reduction takes place in the rich spectrum of species typical of the Central European mixed and coniferous forests. Commonly found are *C. vomitoria, H. radicum,* and *P. lardaria,* joined by certain Mediterranean or rather thermophilous species, such as *H. penicillata* (replacing the retreating *H. irritans*), *H. cyrtoneurina,* and *Dasyphora albofasciata.*

4. Near and Middle East

The synanthropic fly fauna of the Near and Middle East is significant for an understanding of the fly fauna of the western Palearctic. This region served as a refuge for the thermophilous Palearctic fauna during the glacial periods, and it was from this center that populations spread to the north and northwest during the interglacial periods and the Holocene. This fauna is therefore an organic part of the western Palearctic and is distinguished mainly by increases in the density of thermophilous Palearctic, and even Holarctic, species. This is due not only to climatic factors but also to readier availability of food and breeding materials resulting from lower hygienic standards. In this group, we have the following types: species with a summer pupal diapause, *S. stercoraria* and *P. lardaria*; species retaining at least periodic winter activity in the warmest continental areas of Central Asia, *H. cinerella, F. canicularis, F. scalaris, M. stabulans, P. rudis, C. vicina*, and *D. asiatica*; and finally, species showing early spring occurrence, i.e., hibernating as third instar larvae or pupae, *M. assimilis, D. querceti, O. leucostoma,* and *P. regina.* We note the absence or rarity of hygrophilic and psychrophilic elements in this region.

The extensive deforestation of this Middle Eastern region has substantially influenced the ecology of the entire area and emphasized its natural desert character. As a consequence, the xerothermic steppe, semidesert, and desert species come to the fore during the summer months. The key group in these regions is the non-dominant sarcophagids: *P. aegyptica* in western and middle Asia, North Africa, and the Balkan Peninsula; and *Bellieria maculata, Parasarcophaga semenovi, P. hirtipes, P. barbata, P. chivensis, P. jacobsoni, P. securifera* and numerous species of *Wohlfartia—W. balassogloi, W. nuba, W. indigena, W. fedtschenkoi* and *W. trina*—in middle Asia. The dominant sarcophagids, *R. striata, S. haemorrhoidalis, S. melanura* and *S. latifrons*, are here joined by *P. demandata, P. chalybaea, M. domestica vicina, M. sorbens, M. osiris, F. leucosticta, C. albiceps, P. sericata,* and *Sepsis thoracica*, all of which are frequently dominant. Absent from these regions are *L. caesar, C. vomitoria,* and *H. dentipes,* as well as a number of other species present in much of Europe. A shift is observed in favor of the sarcophagids, whereas numerous muscid and calliphorid species of the Temperate Zone are eliminated and replaced by subtropical species.

The public health danger of flies discussed in connection with the Prague and Ujpest surveys is even more pronounced in the Middle East. This is readily observed in the results of fly trapping in the fruit bazaar in Samarkand, where many of the predominating species are highly communicative: *M. domestica vicina, F. scalaris, F. canicularis, D. querceti, M. stabulans, C. vicina, P. sericata, P. regina, M. sorbens,* and *D. asiatica.* There was also the typical occurrence of *P. demandata, C. albiceps,* and *F. leucosticta.* During spring and autumn, these flies are joined by a group of species with wider ecological tolerance, such as *P. rudis, M. assimilis, M. stabulans, O. leucostoma,* and *H. cinerella.* Similar relations were found at Bukhara.

The northern part of middle Asia, i.e., northern Kirghiz and Armenia, corresponds in latitude to southern and Central Europe. Its fly fauna is essentially similar, with some distinctly Continental additions (Fig.

17). Here, *domestica* is present in its nominate form, rather than as the subspecies *vicina.* The Continental character of the climate is evidenced by the occurrence of the purely Continental species *F. leucosticta* and by high summer densities of *P. sericata, R. striata,* and *S. melanura.* At the same time, the rather hygrophilous species, *P. cadaverina, M. stabulans, M. meditabunda,* and *C. vicina,* are less frequent and occur only in moister seasons. *P. terraenovae* is also present.

The high mountains of middle Asia, the Pamirs, have a more southerly aspect than their European counterparts, as shown by the common oc-

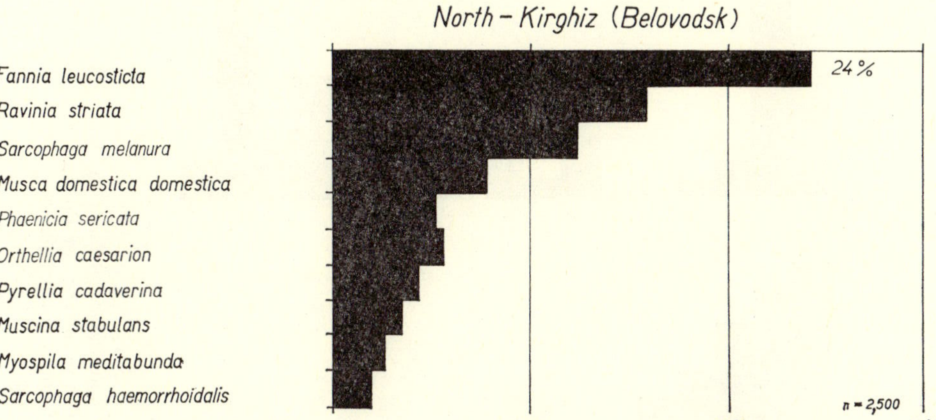

FIG. 17. The hemisynanthropic fly fauna of northern Kirghiz, sampled on feces near a human settlement. Of the non-dominant but quite characteristic species, *Physiphora demandata, Dasyphora asiatica, Musca sorbens, M. osiris, M. tempestiva, M. larvipara, M. autumnalis,* and *Sepsis thoracica* were present. Except for *Fannia leucosticta* (an eastern Continental species), *Musca domestica* (with its relatively high density), and the endemic species of Central Asia, this sample greatly resembles the fauna of the European Mediterranean

currence of thermophilous forms at considerable elevations (Fig. 18). There is also a blending of purely endemic elements with species typical of the hygrophilous fauna of more northern latitudes. For instance, *R. striata* is found at elevations over 3,500 meters and is a predominant species up to 3,000 meters. *H. cinerella* is common up to the subnival zone at over 4,000 meters. *M. domestica domestica* is the most frequent synanthrope up to elevations of about 2,800 meters, followed by *F. canicularis, D. gussakovskii, S. haemorrhoidalis, O. caesarion, M. osiris, P. sericata, M. stabulans, C. vicina, S. melanura, F. leucosticta,* and *C. uralensis.* Between 3,600 and 4,000 meters, *P. ter-*

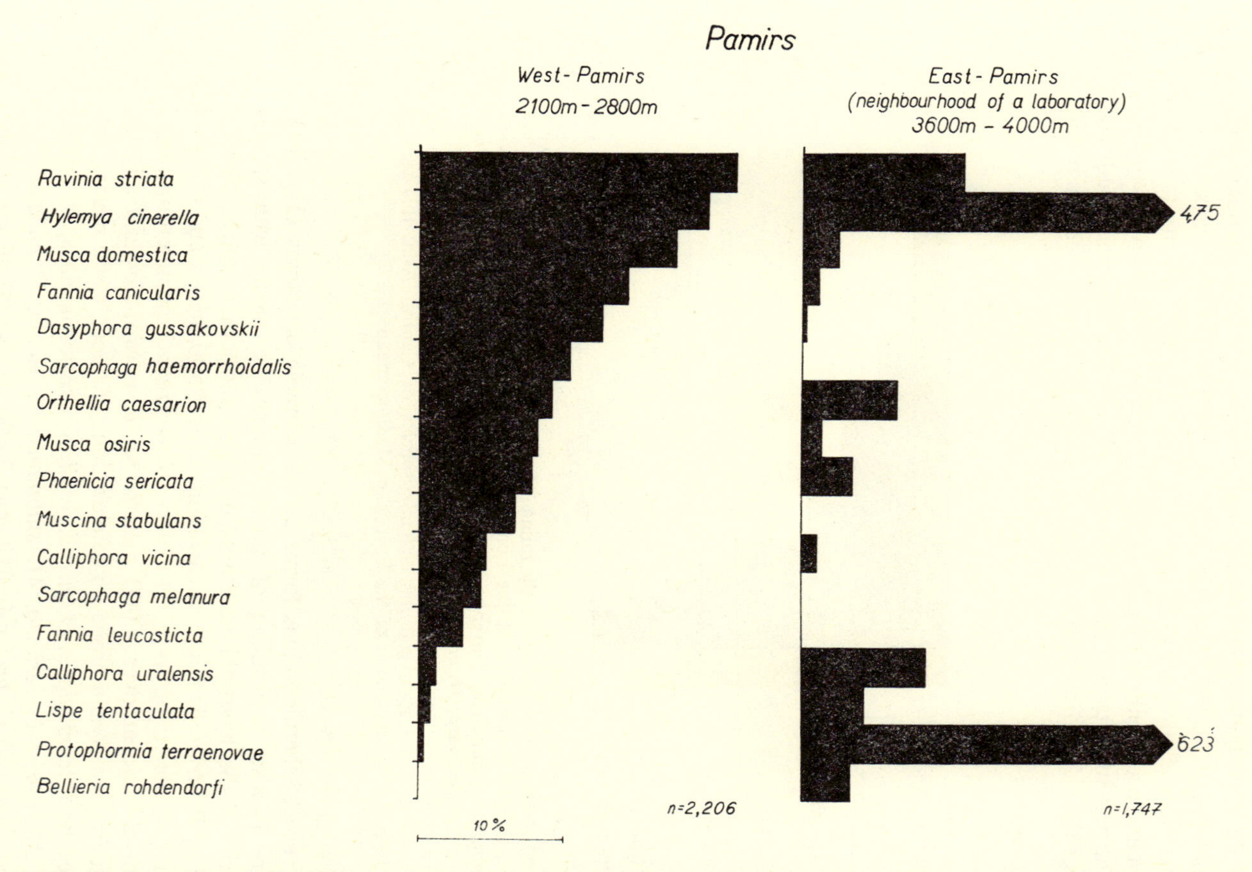

FIG. 18. The hemisynanthropic fly fauna of the mountains of Central Asia attracted to feces. The strong influence of thermophilous elements, reaching elevations over 2000 m., as well as the sudden occurrence of *Protophormia terraenovae* above 3000 m., is characteristic. *Hylemya cinerella* is revealed as a species typical of non-forested habitats

raenovae is the predominant fly. In descending order, we find *H. cinerella striata, O. caesarion, C. uralensis,* and *L. tentaculata.* Not a single thermophilous Asiatic species attains any great density here. There is a probable occurrence of the specific mountain coprophagous and carnivorous species but thus far no evidence is available.

In Afghanistan which is the southeastern Palearctic corner of middle Asia, there is an interesting penetration of the Oriental fauna headed by *Chrysomya megacephala* (Fig. 19). In spring (March and April), a number of species were found in an open-air bazaar at Jalal-Abad,

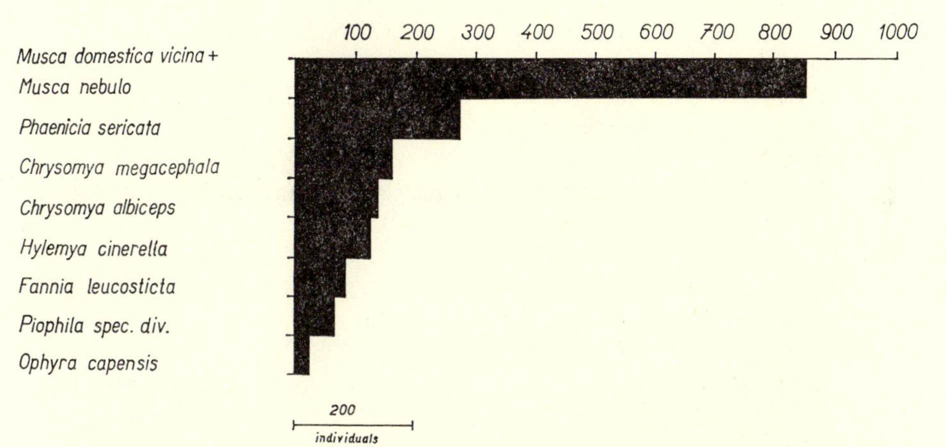

FIG. 19. Population of mostly eusynanthropic flies sampled in a slaughter house on the margin of Jalal-Abad (southwest Afghanistan) at the end of March 1966. The dominance of the *Musca* populations as well as the high density of *P. sericata* and *C. megacephala* are characteristic of this area

of which *M. domestica vicina* and *M. d. nebulo* made up over 80 percent of the population. Due to the early season, the two subspecies could not be reliably separated, as the width of the frons overlapped considerably in both forms. Zoogeographically, however, this area is characteristic of *M. d. nebulo.* Other species included *P. sericata* (over 12 percent) and *C. megacephala* (about 5 percent). Less frequent or scattered were *C. albiceps, H. cinerella, P. casei, O. capensis, M. stabulans, C. vicina, C. vomitoria, F. leucosticta, M. meditabunda, Phorbia platura, Musca lucidula, M. sorbens,* and *R. striata.* In May, *C. megacephala* and *Piophila* spp. were almost as abundant as *M. domestica vicina* and *M. d. nebulo.*

Naturally, this general outline of the synanthropic flies as components of the Palearctic zoocoenoses is not exhaustive. Its aim has been to delineate characteristic fly populations of typical environments. The reader interested in obtaining more detailed information on the bionomics and distribution of this group of flies is referred to the bibliography which follows.

BIBLIOGRAPHY

Cernov, J. I.
1959. Sinantropnye dvukrylye Jugorskogo poluostrova i ostrova Vajgacha. Entom. Obozr., *38*:579-582.

Cernov, J. I.
1961. Komplex sinantropnych dvukrylych v arkticheskich tundrach Jakutti. Nauchn. Dokl. Vys. Shk. Biol. Nauki, *3*:35-38.

Cernov, J. I.
1965. Komplex sinantropnych dvukrylych v tundrovoj zone SSSR. Entom. Obozr., *44*:72-83.

Derbeneva-Ukhova, V. P.
1952. Muchi i ich epidemiologicheskoje znachenije. Gosud. Izdat. Med. Lit., Moskva., Medgiz, 271pp.

Derbeneva-Ukhova, V. P.
1961. K sravnitelnoj ekologii sinantropnych vidov semějstva Muscidae i Calliphoridae. Med. Parasit., *32*:27-37.

Gadzhei, E. F.
1963. Migracija much v uslovijach goroda. Med. Parasit., *32*:465-468.

Gregor, F., and Povolný, D.
1958. Versuch einer Klassifikation der synanthropen Fliegen. Jour. Hyg. Epid. Microb. Immun., *2*:205-216.

Gregor, F., and Povolný, D.
1959. Beitrag zur Kenntnis synanthroper Fliegen Bulgariens. Práce Brněnské Zákl. ČSAV, *31*:377-384.

Gregor, F., and Povolný, D.
1959. Eine Ausbeute von synanthropen Fliegen aus Slowenien. Českoslov. Parasit., *6*:97-112.

Gregor, F., and Povolný, D.
1960. Beitrag zur Kenntnis synanthroper Fliegen Ungarns. Acta Soc. Entom. Čechoslov., *57*:158-177.

Gregor, F., and Povolný, D.
1960. Beitrag zur Kenntnis synanthroper Fliegen Albaniens. Českoslov. Parasit., *7*:115-131.

Nuorteva, P., and Skaren, U.
1960. Studies on the significance of flies in the transmission of poliomyelitis. Ann. Entom. Fenn., *26*:221-226.

Nuorteva, P., Kotimaa, T., Pohjolainen, L., and Räsänen, T.
1964. Blowflies on the refuse depot of the city of Kuopio in Central Finland. Ann. Entom. Fenn., *30*:94-104.

Nuorteva, P., and Laurikainen, E.
1964. Synanthropy of blowflies on the island of Gotland, Sweden. Acta Entom. Fenn., *30*:187-190.

Peters, H.
1959. *Stomoxys,* ein interessantes Spurenelement innerhalb der synanthropen Dipterenfauna. Zeitsch. Angew. Zool., *46*: 356-358.

Petrova, E. F.
1942. Synanthrope Fliegen des Gebirges Alma-Ata. Med. Parasit., *11*:(1-2):86-89.

Petrova, E. F.
1948. Einige Angaben über Mücken und Fliegen des zentralen Teiles der Wüste Betpak-Dala. Med. Parasit., *18*:527-530.

Povolný, D.
1959. Gesichtspunkte der Klassifikation von synanthropen Fliegen. Zeitsch. Angew. Zool., *46*:324-328.

Rohdendorf, B. B.
1959. Die Arten der Sarcophaginae in den Faunenkomplexen synanthroper Zweiflügler der verschiedenen Landschaftszonen der USSR. Zeitsch. Angew. Zool., *46*:348-356.

Shtakelberg, A. A.
1956. Sinantropnye dvukrylye fauny SSSR. Izdat. Akad. Nauk SSSR, Moskva-Leningrad, *60*:164pp.

Steyskal, G. C.
1957. The relative abundance of flies, collected at human faeces. Zeitsch. Angew. Zool., *44*:79-83.

Shura-Bura, B. L., Shaikov, A. D., Ivanova, E. V., Glazunova, A. Ia., Mitriukova, M. S., and Fedorova, K. G.
1958. O charaktere rasseivania nekotorych vidov sinantropnych much mesta vypuska. [The character of dispersion from the point of release in certain species of synanthropic flies.] Entom. Obozr., *37*:336-346.

Shura-Bura, B. L., Sukhomlinova, O. I, and Isarova, B. I.
1962. Primenenije metoda radiomarkirovki k izucheniju sposobnosti sinantropnych much pereletat vodnyje pregrady. [Radioactive tracers as an aid to studying the ability of synanthropic flies

Gregor, F., and Povolný, D.
1960. On a case of mass occurrence of flies *Hydrotaea irritans* and *Morellia podagrica* in High Tatras in 1959. Folia Zool., *23*:323-326.

Gregor, F., and Povolný, D.
1960. Zur Chorologie und hygienisch-epidemiologischen Rolle synanthroper Fliegen in Mitteleuropa. XI. Intl. Kongr. Entom., Wien.

Gregor, F., and Povolný, D.
1961. Resultate stationärer Untersuchungen von synanthropen Fliegen in der Umgebung einer Ortschaft in der Ostslowakei. Folia Zool., *24*:17-44.

Gregor, F., and Povolný, D.
1961. Synanthrope und andere gesundheitlich wichtige Fliegen des Bezirkes von Prešov in der Ostslowakei. Sbornik Kraj. Patol. Vých. Slov., *1*:53-72.

Gregor, F., and Povolný, D.
1964. Eine Ausbeute von synanthropen Fliegen aus Tirol. Folia Zool., *13*:229-248.

Havlík, B., and Baťová, B.
1961. A study of the most abundant synanthropic flies occurring in Prague. Acta Soc. Entom. Čechoslov., *58*:1-11.

Kirchberg, E.
1951. Untersuchungen über die Fliegenfauna an menschlichen Fäkalien. Zeitsch. Hyg. Zool., *39*:129-139.

Kirchberg, E.
1953. Untersuchungen über die Fliegenfauna menschlicher Fäkalien. VI. Congr. Intl. Microb. Rome, *5*:485-487.

Kirchberg, E.
1954. Zur Kenntnis einiger Schmeissfliegen von hygienischer Bedeutung. Verh. Deutsch. Ges. Angew. Entom., *12*:99-103.

Kirchberg, E.
1958. Über einige Musciden von hygienischer Bedeutung. Verh. Deutsch. Ges. Angew. Entom., *14*:36-42.

Kirchberg, E.
1959. Vergleichende Untersuchungen über den Dipterenbesuch an verschiedenen keimhaltigen Medien. Zeitsch. Angew. Zool., *46*:363-368.

Kirchberg, E.
1961. Zur Kenntnis fäkalgebundener Fliegen auf Samos (Griechenland). Zentralb. Bakt. Parasit. Infekt. Hyg., *182*:267-275.

Kirchberg, E., and Bakri, G.
1960. Unterschiedlicher Befall von eiweisshaltigen Lebensmitteln durch Schmeissfliegen. Zeitsch. Angew. Entom., *47*:60-68.

Kühnelt, W.
1956. Gesichtspunkte zur Beurteilung der Grosstadtfauna. Österr. Zool. Ztschr. Wien., *6*:30-54.

Kvasnikova, P. A.
1931. Muchi zhilych i chozjajstvennych postojek cheloveka goroda Tomska. Izv. Tomsk Gos. Univ., *83*:9-47.

Lineva, V. A.
1964. Ekologija sinantropnych much v naselennych punktach na juge Azerbeidzhana. Probl. Med. Parasit., 572-601.

Lobanov, A. M.
1959. Ekologo-fanisticheskij obzor sinantropnych vidov Calliphoridae goroda Ivanova i ego okresnostej. Sbor. Nauchn. Trud. IGMI, *22*:263-268.

Lörincz, F., Szappanos, G., and Makara, G.
1936. Recherches entreprises en Hongrie sur les mouches entrant en contact avec les excréments humains. Bull. Trimestr. Organisation Hyg. Soc. Nations, 5, *2*:251-261.

Lörincz, F., Szappanos, G., and Makara, G.
1936. On flies visiting human feces in Hungary. Quart. Bull. Health Org. League of Nations, Geneva, *2*:228-236.

Lörincz, F., and Mihályi, F.
1938. Vizsgálatok a légykérdés egészségügyi vonatkozásairól (Untersuchungen über die hygienische Bedeutung der Fliegenfrage in Ungarn). Állatt. Közlem., *35*:1-13.

MacLeod, J., and Donnelly, J.
1956. The geographical distribution of blowflies in Great Britain. Bull. Entom. Res., *47*:597-619.

MacLeod, J., and Donnelly, J.
1957. Some ecological relationships of natural populations of calliphorine blowflies. Jour. Anim. Ecol., *26*:135-170.

MacLeod, J., and Donnelly, J.
1958. Local distribution and dispersal paths of blowflies in hill country. Jour. Anim. Ecol., *27*:349-374.

Mihályi, F.
1965. Rearing flies from feces and meat, infected under natural conditions. Acta Zool. Hung., *11*:153-164.

Mihályi, F.
1966. Flies visiting fruit and meat in an open-air market in Budapest. Acta Zool. Acad. Scientiarum Hungaricae, *12*:331-337.

Mourier, H.
1965. The behaviour of house flies (*Musca domestica*) towards "new objects." Vidensk. Medd. Dansk. Naturh. Foren., *128*: 221-231.

Nuorteva, P.
1958. Some peculiarities of the seasonal occurrence of poliomyelitis in Finland. Ann. Med. Exp. Fenn., *36*:335-342.

Nuorteva, P.
1959. Studies on the significance of flies in the transmission of poliomyelitis. Ann. Entom. Fenn., *25*:1-24.

Nuorteva, P.
1959. Studies on the significance of flies in the transmission of poliomyelitis. Ann. Entom. Fenn., *25*:121-136.

Nuorteva, P.
1959. Studies on the significance of flies in the transmission of poliomyelitis. Ann. Entom. Fenn., *25*:137-162.

Nuorteva, P.
1963. Synanthropy of blowflies in Finland. Ann. Entom. Fenn., *29*:1-49.

Nuorteva, P.
1964. Differences in the ecology of *Lucilia caesar* and *L. illustris* in Finland. Wiadom. Parazyt., *10*:583-587.

Nuorteva, P.
1964. The zonal distribution of blowflies on the arctic hill Ailigas in Finland. Ann. Entom. Fenn., *30*:218-226.

Nuorteva, P.
1965. The flying activity of blowflies in subarctic conditions. Ann. Entom. Fenn., *31*:242-245.

Nuorteva, P.
1965. Synanthropy of blowflies. Proc. 12th Intl. Congr. Entom. London, 1964, p. 786.

Nuorteva, P.
1966. The flying activity of *Phormia terrae-novae* R.-D. in subarctic conditions. Ann. Zool. Fenn., *3*:73-81.

Nuorteva, P.
1966. Local distribution of blowflies in relation to human settlements in an area around the town of Forssa in South Finland. Ann. Entom. Fenn., *32*:128-137.

Nuorteva, P.
1966. The occurrence of *Phormia terrae-novae* R.-D. and other blowflies in the archipelago of the subarctic Lake Inarinjärvi. Ann. Entom. Fenn., *32*:240-241.

to fly over water barriers.] Entom. Obozr., *41*:99-108.

Sukhova, M. N.
1957 Synanthrope Fliegen einzelner Landschaftszonen der USSR. Dissertation, 1-26, Moskva.

Sychevskaia͡, V. I.
1956. Synanthrope Fliegen des Kara-Kalpakien. Entom. Obozr., *35*:347-358.

Sychevskaia͡, V. I.
1957. Synanthrope Fliegen der Umgebung von Bjelovodsk (Nord-kirgisien). Entom. Obozr., *36*:108-115.

Sychevskaia͡, V. I.
1957. Über die saisonmässigen Landschaftszonen von Uzbekistan. Zool. Zhurnal, *35*:719-757.

Sychevskaia͡, V. I.
1958. Über die Artenzusammensetzung der synanthropen Fliegen von Uzbekistan. Arb. Inst. Malaria Med. Parasit., *3*:257-261.

Sychevskaia͡, V. I.
1960. On the phenology of synanthropic flies of Uzbekistan. Med. Parasit., *29*:66-72.

Sychevskaia͡, V. I.
1961. Vidovoj sostav sinantropnych much Ferganskoj doliny. [Species of synanthropic flies in the Fergana Valley.] Izv. Otd. Selskochoz. Biol. Nauk AN Tadzh. SSR, *5*:85-89.

Sychevskaia͡, V. I.
1962. Ob izmeneniach sutochnoj dinamiki vidovogo sostava sinantropnych much v techenie sezona. Entom. Obozr., *41*: 545-553.

Sychevskaia͡, V. I.
1963. Sinantropnye muchi iz semejstva Helomyzidae v srednej Asii. [Species of synanthropic flies of the family Helomyzidae in Central Asia.] Zool. Zhurnal, *42*:1659-1665.

Teschner, D.
1959. Hausfliegen als Fäkalienbesucher im Stadtgebiet. Zeitsch. Angew. Zool., *46*:358-363.

Teschner, D.
1961. Beiträge zur Kenntnis der Fauna eines Müllplatzes in Hamburg. Entom. Mitt. Zool. Staatsinst. Zool. Mus. Hamburg, *35*:1-16.

Tischler, W.
1950. Biozoonotische Untersuchungen bei Hausfliegen. Zeitsch. Angew. Entom., *32*:195-207.

Trofimov, G. K., and Engelgart, L. S.
1948. Über die Rolle der synanthropen Fliegen bei Darmwurmepidemien in Baku. Med. Parasit., *17*:247-252.

Zaidenov, A. M.
1961. Opyt izuchenia epidemiologicheskogo znachenija sinantropnych much v uslovijach goroda. [The experience of the study of the epidemiological role of synanthropic flies under the conditions of a town.] Entom. Obozr., *40*:554-567.

Zimin, L. S.
1944. Die synanthropen Fliegen des hissarschen Gebietes-Muscidae, Calliphoridae, Sarcophagidae, Sammlung über Probleme der Darminfektionen, 133-166, Stalinabad.

Zimin, L. S.
1944. Jahreszeitliche und vierundzwanzigsstündliche Schwankungen der Fliegenhäufigkeit in den Wohngebieten, verbunden mit Temperatur und Feuchtigkeitsschwankungen, Sammlung über Probleme der Darminfektionen, *167*:176.

Zimin, L. S.
1944. Synanthrope Fliegen von Südtadschikistan an und ihre medizinische-hygienische Bedeutung, Sammlung über Probleme der Darminfektionen, 133-666, Stalinabad.

3

Bionomics of Flies

Bernard Greenberg and
Dalibor Povolný

THE PURPOSE of this chapter is to provide for the more important synanthropic flies summaries of existing information on geographic and phenologic distribution, and patterns of adult behavior and larval breeding as they relate to man and other organisms. In presenting the synanthropic biology of flies, only chance mention is therefore made of the effects of biotic and abiotic factors on the various stages in development from the egg to the adult. These will be dealt with more thoroughly in the next volume. In such instances, our purpose is to point out the important ecological and ethological moments in the life of various species which merit attention.

Anastellorhina augur

This fly is limited in its distribution to the Australian region where it is an important sheep myiasis fly. It may occur in great numbers during early summer when it is one of the dominant blowflies in the bush and corrals. It also invades the home. Larvae hatch quickly following oviposition and have been recorded as breeding in a great variety of decomposing substrates, including dead insects, snails, sour milk, cheese, fermenting grain, and bone manure. It has not been recorded as breeding naturally in cow or horse dung. As many as 50 flies have been bred from a bird cadaver smaller than a sparrow.

Anthomyia pluvialis

This fly has a Holarctic range, occurring from Alaska to Georgia and throughout Europe and Asia to China and Japan. It disports itself near water and with the approach of rain, males execute an aerial dance. Larvae live in decomposing plant material and occur in the nests of birds.

Atherigona orientalis

This is a cosmopolitan, hemisynanthropic fly in temperate and subtropical regions. The adult is extremely catholic in its tastes and may be taken on carrion, decaying plant material, including fruits and vegetables, and human excrement. It is essentially an exophilic fly, not inclined to enter dark places, including privy pits. Larvae breed in substrates equal in variety to those which attract adults. The larva preys upon other maggots. On Guam, this is the most abundant muscoid fly during most seasons, and it is common around villages and camp kitchens. In a Samoan study of yaws transmission, a collection of wound-feeding flies yielded 0.6 percent of *A. orientalis.*

Blaesoxipha plinthopyga

This is a hemisynanthropic fly which ranges in the Nearctic from southern Canada to California and in the Neotropics from Mexico and the West Indies to Brazil. The larvae are commonly found on carcasses or as internal parasites of insects. The isolation of vaccine strains of polio virus from these flies soon after administration to a human population indicates its attraction to feces as well.

Calliphora terraenovae

This is a hemisynanthropic species. Like other members of the genus, these maggots are saprophagous, breeding in decomposing organic matter. Adults are attracted to carrion. The species is distributed in the Nearctic region from Alaska to Greenland, south to California and northern Florida. There is a close affinity between this species, the Holarctic *vomitoria* and the Palearctic *uralensis.*

Calliphora uralensis

This species is abundant in the forest zone of Eurasia. It is important in eastern Europe, especially in the forest zone of U.S.S.R. where, in the opinion of Soviet authors it is distinctly eusynanthropic and communicative. In Central Europe, it is hemisynanthropic and less abundant. Larvae show a wide food range but are most numerous in liquidy excreta, particularly in latrines and cesspools. Larvae do poorly in meat and compared with those of *C. vicina* are found less frequently. Adults alight on sunlit places and oviposit on walls of privies. The species penetrates far to the north in Europe. The related species *C. loewi* and *C. subalpina* occur mostly in forests, frequently showing disjunct distribution (especially, *C. subalpina*). They are sometimes hemisynanthropic, particularly in northern Europe.

Calliphora vicina

This is a eusynanthropic communicative species, frequently showing endophilous tendencies. It is distributed in the Holarctic region, secondarily following man into neighboring regions (Australia, New Zealand). Adults frequent fruit, decaying meat, and feces; larvae are chiefly carnivorous, developing in decomposing meat. In the subtropics, the species occurs in winter; in the Temperate Zone, chiefly in spring and autumn; in the subpolar zone, in summer. Consequently, it occurs in mountains in the Oriental region. In northern Europe, the species shows particularly strong synanthropy and is often quite numerous.

Calliphora vomitoria

This is a hemisynanthropic to exophilous eusynanthropic species, which is sometimes communicative. Like *C. vicina,* it is markedly saprophagous. In Europe, it is chiefly a forest species ascending the mountains where it sometimes occurs in large numbers, and around mountain chalets and hotels in various refuse. The phenology of this species is characterized by spring and autumn peaks. It is also frequent in forests during summer. Its distribution is Holarctic, but in North America it is more common in the wooded parts of Canada. In northern Europe, it often occurs together with *C. vicina* but is less abundant.

Chrysomya albiceps

This is a Paleotropic to Paleosubtropic species which spreads during late summer of hot, moist years up to Central Europe from the Mediterranean (e.g., during 1958 to 1959). The species prefers high temperatures and humidity. Adults are attracted in large numbers by meat bait. Larvae are predatory, attacking those of other synanthropic flies present in carrion. It is a hemisynanthropic species whose epidemiologic role has not been studied in detail. It does not appear to compete with *P. sericata,* because of its more restrictive temperature requirements and besides, the two species do not coincide phenologically.

Chrysomya chloropyga

This hemisynanthropic species is a common blowfly in the southern parts of Africa, from Tanganyika and the Congo south to the tip of the continent. It is an agent of myiasis in sheep, second in importance to *Phaenicia cuprina.* The flies love sunlight, seldom enter buildings, and go into the shade only in response to a powerful olfactory stimulus or to shelter from bad weather. Adults feed on the liquefied meat produced by the maggots. Eggs are deposited in crevices of carrion and, very rarely, on feces.

Chrysomya marginalis

This species occurs in the southern Palearctic region in Syria, Egypt, and Arabia; and in the Oriental region, in India west of the Indus. It is a very common carrion breeder in the Ethiopian region south of the Sahara to the Cape of Good Hope. The adults rarely enter houses but are common in butcher shops where they are attracted to the meat. They swarm on cow dung and feces during the wet season and feed on exudates from sores and wounds of cattle. It is a hemisynanthropic to symbovine species.

Chrysomya megacephala

It is a hemisynanthropic to eusynanthropic exophilous species, markedly communicative which penetrates the Palearctic (Afghanistan) from the Oriental region. According to the observations of one of us (D. P., Jalal-Abad, 1965, 1966), the species occurs gregariously on carrion where it oviposits intensively. The species also occurs in latrines, on various decaying matter, and on fruit. It deserves the attention of hygienists and epidemiologists.

Chrysomya putoria

This species has the same distribution in subtropical and tropical Africa as *C. marginalis.* Its larvae breed in dung and in outdoor latrines. The occurrence of poliovirus, *Salmonella,* and other enteric pathogens in these flies in Madagascar and other parts of Africa underscores their potential public health significance.

Chrysomya rufifacies

This fly has an Oriental and Australasian distribution. *C. rufifacies* replaces *C. albiceps* in these regions. *C. rufifacies* is more adapted to tropical conditions, shown by the fact that it occurs year-round in the warmer part of its range (Moree, in northern New South Wales) and is absent during winter in more southerly Australian latitudes. This fly, like *Cochliomyia macellaria* in North America, appears to have no cold-resistant stage. Though implicated in myiasis, its nutritional role is more likely that of a scavenger, feeding on dead tissue and even preying upon other maggots. It has been used successfully in osteomyelitis therapy. Oviposition usually occurs in the folds or fur of carrion only after a putrefactive odor has developed, and the female apparently prefers to oviposit where other maggots are present. Adults also find human excrement fairly attractive, ripe fruit only slightly attractive. They feed readily on the nectar of certain flowering trees. On Guam, they are seen occasionally in privies, almost never in houses. Next to *C. megacephala* they are the most common blowfly on this island.

Chrysomya varipes

This fly is known only from the Australian continent. Its biology is not yet fully known. It is a carrion breeder, and its biotic associations thus far recorded concern its biological control by means of parasites and predators.

Cochliomyia macellaria

This species is distributed in the Nearctic region from southern Canada to southern United States, and in the Neotropical region from Mexico to Patagonia. It is hemisynanthropic and extremely communicative. Adults prefer warm, humid weather and the greatest numbers are recorded during rainy periods. Adults are attracted to carrion and garbage. They are also found feeding on the nectar of wild parsnip and are attracted to plants which emit an odor of carrion, e.g., *Aristolochia* sp. The fly is extremely abundant in the environs of slaughter houses where offal is discarded and is common in outdoor markets in the tropics.

It occurs in southern Florida and southern Texas throughout the year. In the northern part of its range, it is killed out by frost in the fall and the region is repopulated by spring emigrants. The importance of this species in myiasis is well documented.

Cynomya mortuorum

This fly prefers the northern Holarctic region of Europe and North America. It is often collected on raw fish, and field studies in Finland showed it is also attracted to human feces but not as strongly. The synanthropy of this species remains uncertain, but there is evidence that the slight synanthropy of this fly in northern Finland diminishes toward the south.

Cynomyopsis cadaverina

This fly is found in the Nearctic region from Ungava Bay in northern Labrador to the border between Texas and Mexico. It is most abundant along the Canadian-United States border. It is a psychrophylic species with seasonal peaks in early spring and late fall. Adults may occasionally appear during warm spells in mid-winter. Flies enter houses readily. About 50 percent of all flies trapped in spring belong to this species. Adults are attracted to human excrement and to carrion in an advanced state of putrefaction. Overwintering is believed to be primarily in the pupal stage but one of us (B. G.) has evidence which indicates that the adult is the overwintering stage, at least in the Chicago region. *C. cadaverina* is characterized as hemisynanthropic and facultatively communicative.

Dasyphora pratorum

Of the numerous coprophagous species of this genus (*D. cyanella, D. albofasciata, D. cyanicolor, D. eriophthalma*) occurring in the Western Palearctic, *D. pratorum* is relatively the most frequent one. It is one of the typical pasture flies and is primarily symbovine. Adults visit feces, dung, fruit, as well as flesh. Females are viviparous and larvae develop chiefly in cattle feces.

Dendrophaonia querceti

This species occurs throughout Europe and Soviet Central Asia; in North America from south central to eastern Canada south to Colorado, Illinois, and Georgia. It has been reared from the nests of birds, e.g., titmice in Vienna, but commonly breeds in manure and human excrement. In France, it has been found breeding in huge numbers in the undisturbed excrement of rabbits. In Germany, a survey of flies attracted to human feces yielded a small percentage of this species, primarily females. This fly is hemisynanthropic and occasionally communicative.

Drosophila funebris

This is a cosmopolitan, eusynanthropic, endophilous species common through much of the world. It breeds in various fruits and especially in animal matter, human excrement, and cesspools. In Soviet Central Asia, it frequents toilets. It has recently been recorded for the first time from Hawaii.

Drosophila melanogaster

This fly is cosmopolitan throughout the warmer and temperate parts of the world. Larvae breed in overripe and decaying fruit and less often in human excrement. In the United States, it has often been found in houses, attracted or carried there, especially in fall, in connection with fruit storage and canning. There is some doubt whether this species can hibernate in the cooler parts of its range, except indoors. Documentation of this intensively studied fly is vast and readily available and need not concern us here, except to note that it is eusynanthropic and communicative.

Drosophila willistoni

This is a Neotropical species taken on St. Vincent Island in the West Indies and in Texas and Florida. The biotic associations recorded for this species are laboratory ones.

Dryomyza anilis

Among the distinctly forest coprophages (*Mydaea, Polietes lardaria,* etc.), this Holarctic hemisynanthropic species has a narrow temperature preferendum. It is found on feces even in dense shrubbery and in shaded and moist places.

Eristalis tenax

This is a non-communicative exophilous synanthrope with secondary cosmopolitan tendency, distributed over a vast area of the Holarctic region. The flies are floricolous, feeding on pollen, but are also attracted by odors of decay. The larvae of this species develop in thin or liquid effluents from dung pits; their presence in water or mud invariably indicates a high degree of pollution. Pupation takes place in dry environments. Larvae are frequently attacked by those of *Ophyra, Muscina, Phaonia,* etc. *E. tenax* is associated with primitive rustic conditions, and need not concern us here, except to note that it is eusynanthropic and communicative.

Fannia canicularis

This is an endophilous, eusynanthropic, cosmopolitan species. With *Musca domestica,* it is among the most abundant flies of human habitations, especially during summer months. Adult males show typical behavior, alighting preferably on wall projections, suspended objects, chandeliers, etc. These flies are frequent in toilets, privies, stables, pigsties, etc. They visit excreta of all kinds as well as food, fruits, and beverages. In more northerly latitudes, the lesser house fly lives in hemisynanthropic populations (especially in birds' nests). However, it is possible that this is a case of a very closely related species of the *canicularis* group (particularly, *F. subpubescens*) which was not separated from *F. canicularis* until recently. Adults are less vagrant and prefer the kind of substrate in which they developed in the larval stage. Larvae develop in excreta of cattle and man, saturated with urine, as well as in poultry manure in henneries. Eggs and larvae show excellent aquatic adaptation for they are able to develop in organic substrates which are quite liquid. There are cases in which larvae occurred in wet diapers and in the urogenital tract of man. The hygienic and epidemiologic importance of *F. canicularis* is still uncertain. Phenologically, *F. canicularis* attains peak populations in the temperate zone during July, whereas *M. domestica* prevails in late summer.

Fannia femoralis

This fly has a temperate distribution in the United States, from West Virginia to Montana and California, and Georgia to Louisiana. Its endemic area is considered by some to be northern South America. In California, it is sometimes reported as the predominant fly species found breeding in chicken droppings at study sites (95 percent of all flies). In this region, its seasonal peak occurs during the coldest part of each year, whereas its parasitization by hymenopterous parasites reaches a peak during the warm months.

Fannia incisurata

This fly has a Holarctic distribution and is found from Quebec to British Columbia, south to Colorado, Mississippi, and Virginia. It occurs from Iceland to China and Japan, and in North Africa. It is also recorded from Mexico and Argentina in the Neotropical region. It has breeding habits similar to *F. scalaris* and the larva appears better equipped for flotation in a semiliquid medium. It has been recorded as the most abundant species breeding in cesspools in the vicinity of Paris.

It also breeds in excrement, cadavers, and in birds' nests. *F. incisurata* is hemisynanthropic, occasionally tending to eusynanthropy, and exophilous.

Fannia leucosticta

Unlike the other species of this genus, *F. leucosticta* prefers a hot, dry climate. For this reason, it shows eusynanthropy mainly in human settlements in arid areas of the entire Holarctic region. The larvae attain unusual population densities particularly in thick and drying masses of excreta (of man and poultry). Adults are common in market places and bazaars.

Fannia manicata

It is hemisynanthropic. This species is widely distributed in the Holarctic region from Alaska to Greenland, south to Colorado and Georgia; and from Scotland to Kamchatka and Egypt. It has been recorded from nests of Hymenoptera, decaying fungi, and various other decomposing organic matter.

Fannia pusio

This is a hemisynanthropic species tending toward eusynanthropy. This fly has a Nearctic and Neotropical distribution from New York to the West Indies, South America, and Hawaii. It was observed breeding in semiliquid hen manure in Hawaii. On Guam, it was abundant in traps baited with carrion and human excrement. It was seldom collected indoors but could be seen with greater frequency soaring in the shade of dwellings and privies. Breeding normally occurs in material in an advanced state of decomposition. The prepupae prefer dry soil in which to pupate.

Fannia scalaris

This is a Holarctic, eusynanthropic fly which is markedly exophilous. It shows endophilous tendencies only under primitive conditions in settlements insufficiently isolated from privies and dung pits. This is the most substantial difference between this species and *F. canicularis*. For mass development of this species, semiliquid masses of feces (human as well as animal, particularly of pigs) are required; in scattered feces, the larvae are scarce and are subject to competition from other coprophagous species. The hygienic and epidemiologic role of this fly is still uncertain. This species is closely related to the Holarctic species *F. incisurata* which has similar bionomics.

Graphomya maculata

This is a Holarctic species distributed in North America from Alaska to Greenland, south to California and Georgia. In the Palearctic, it is found from its northern ocean boundaries to its southern boundaries. In Siberia, its distribution is not clear. It is also known from Africa and South America. Larvae breed in animal excrement and prey upon other maggots.

Haematobia exigua

Distributed in Australasia from Dutch East Indies south to Australia. Its predilection for carabao or buffalo has earned the name "buffalo fly." Occasionally, it bites man. Eggs are laid on grass and other objects next to cow droppings, the maggots developing rapidly and pupating therein. This fly prefers to bite near the corner of the eye and near wounds where *H. irritans* never bites. Apparently, *H. exigua* is attracted by moisture where tears, serum, or blood have formed wet spots on the skin. It also has a decided preference for bulls and for older animals in a herd, as well as for the shorter-haired breeds. This fly may be characterized as symbovine, exophilous, and occasionally hemisynanthropic.

Haematobia irritans

This is a Holarctic species, which is spreading to subtropical and tropical regions. It is a symbovine type, mainly a pasture type which actively sucks the blood of its host. On cool days in Central Europe, it invades stables situated close to pastures where it feeds on cattle even at night. It occurs in large numbers only on very hot days, but unlike the larger members of related species this fly consumes relatively little blood. Eggs are laid on fresh feces. The importance of this species stems from its troubling cattle and thus decreasing milk production. Adults frequently feed on only one animal, so that the epizootiologic importance of this species appears to be less than that of other hematophages.

Haematobia titillans

This species is known as the southern cow fly in Russia. It is well known from the European part of the Soviet Union from Latvia to the southern Ukraine and the Crimea, and also beyond the Caucasus into Central Asia and Mongolia. It is also found in southern Europe. It is most abundant in steppe and semidesert regions and is found mainly in pastures. The larva develops in cow dung. The biology of this fly is not well known, but it is under suspicion as a carrier of pathogens from one animal to another. The adult bites domestic animals as well as people. It attacks on hot, sunny days. Under these conditions flight activity is diminished. It sometimes bites people during the evening in well-lit areas, but it is primarily a symbovine, exophilous type which is only occasionally communicative.

Hermetia illucens

This is a hemisynanthropic species which is rather widely distributed throughout the Western hemisphere, the Australian region from Samoa to Hawaii, and in some areas of the Palearctic region. The larvae breed in a variety of decomposing plant and animal material and even in the waste materials, wax and honey, of beehives. In the southern United States, they are common in the semiliquid contents of outdoor privies and in the Solomons they have been found in a stagnant ditch fed by waste washing water. Having a prolonged larval life, they will develop in manure heaps but not in isolated droppings. The adult enters houses freely. Larvae of this species may under some conditions exert significant natural control over house fly breeding, not by predation but by competition and overgrowth.

Hippelates flavipes

This fly is entirely neotropical, extending from the Bahamas and central Mexico southward through the West Indies and Central America to Paraguay and southern Brazil. It was the dominant chloropid captured in Jamaica, using live or decomposing animals as bait. This and some other *Hippelates* species are attracted to ulcers on people or animals, but not to decaying oranges, tomatoes, or human feces. *H. flavipes* occurs in greatest numbers at elevations varying between 100 and 1500 feet above sea level, the fly being rare at sea level and relatively infrequent above 2000 feet. The fly has a decided preference for feeding on the lower extremities, and the majority of primary yaws lesions in Jamaica occur on the legs. *Hippelates* eye gnats breed in loose soils which contain adequate moisture and quantities of decaying plant material. There is evidence that any disturbance of the soil such as disking, which turns under plant material, increases oviposition activity of the fly and food for the larvae, with consequent increases in adult density. Before Sabrosky, this species was incorrectly referred to as "*pallipes.*"

Hippelates pallipes

According to Sabrosky, this is a Temperate Zone species which is distributed in the United States from Quebec and Maine south to Florida, Texas, and California. It has a south temperate form in Argentina, Chile, and Peru.

Hippelates pusio

On the basis of recent data on bionomics and interbreeding, this may be a complex of species or subspecies. Distribution includes the West Indies and Mexico through southern United States, north from Washington to Pennsylvania. It is a common hemisynanthrope in Jamaica with almost the reverse altitude distribution compared with *H. flavipes*. In Jamaica, it was found near the seashore, but practically disappeared above 1000 feet. There seems to be little evidence that *H. pusio* transmits conjunctivitis in Jamaica, and it probably plays no part in the transmission of yaws either, because it is most plentiful in areas where yaws does not exist. It is a pestiferous insect in the southeastern United States. In California, this fly is present in cooler climates and almost absent from hot dry regions. In the latter, it is replaced by *H. collusor* as the most annoying eye gnat. *H. pusio*, which may be synonymous with *collusor* but differs in behavior and color, occurs throughout the year in the Coachella Valley in California, with spring and fall peaks. Larvae feed on a great variety of decomposing organic matter, including excrement.

Hydrotaea dentipes

A very common hemisynanthropic species of Holarctic distribution, this fly is associated with man under primitive rustic conditions, but it does not trouble man. Adults frequent feces and decaying fruit, and are common in the vicinity of latrines and offal pits. Larvae develop most frequently in accumulated feces, especially of man, chiefly in sandy soil. They prefer compact and drying feces and gregariously abandon thin ones. Third instar larvae change to predatory habits to the detriment of larvae of other coprophagous species (e.g., *Musca domestica, Stomoxys calcitrans*, and *Musca sorbens*) and are capable of decimating their numbers.

Hydrotaea irritans

This is one of the most important forest hemisynanthropes and is particularly communicative. Adults are confined to larger forests, both mixed and coniferous, chiefly in hilly areas but have a vertical distribution on mountains, following zones of spruce. They occur less frequently outside forests, as the adults are hygrophilous and shade-loving. They belong to the most aggressive and troublesome flies. With usually high population densities, this species, together with *Morellia* spp. (esp. *M. podagrica*), can render work or a sojourn in woods insufferable, particularly during warm summer months. The flies irresistibly attack conjunctivae of eyes and the nose and mouth and even enter the auditory canal. They become facultatively hematophagous. Females concentrate in large numbers on feces and carrion. Also, adults can be found on inflorescences of daucaceous plants infested with aphids. Final instar larvae are predatory, preying upon larvae of other coprophagous flies. The species deserves the combined attention of entomologists and microbiologists.

Hylemya cinerella

This symbovid fly is widely distributed in the Holarctic region. In Denmark, adults feed mainly on the liquid of cow droppings. They also visit flowers of *Gagea, Draba, Pastinaca,* and *Achillea.* The larvae develop in cow droppings. The few available observations indicate that adults visit cow droppings which are 1/2- to 1-hour old. In Central Asia, the flies are commonly found on feces, fruits, and other decomposing matter. They are common in nature and also in human settlements where they congregate especially in market places and gardens. They are rarely found inside the home. In this region, larvae are especially numerous in the dung of pigs, horses, and cows, and also in garbage. In the middle zone, flies are most common from May to August. In Central Asia, they have two populational peaks: May to June and September to November. During the hottest part of the summer (July to August) the activity of this species declines and its larvae cannot be found in feces.

Hylemya radicum

Together with the related species *H. cinerella, H. radicum* is among the most frequent visitors of excreta, decaying vegetables, fruit, and meat as well as other decomposing organic matter. These are hemisynanthropic Holarctic flies, common in nature and frequent visitors to open primitive market places. They are communicative hemisynanthropes which never enter human habitations. Larvae are frequently found in feces, especially human, as well as in decaying refuse of all kinds. Species of this genus are abundant in medium to high mountain elevations, disappearing during hot months and in warm areas.

Hylemya strigosa

This is a psychrophilous, and rather hygrophilous (thence, mostly forest species) hemisynanthropic fly. It can be found in pastures during early spring. Later, it migrates to forests where it can attain rather high population densities through utilization of human feces, decaying fruit and meat. Females are viviparous, the larvae develop chiefly in dung and feces.

Lucilia caesar

This species is widely and exclusively distributed in the Palearctic region. It is hemisynanthropic, showing a tendency toward exophilous eusynanthropy only in the subpolar regions. It occurs in eastern Asia as well as in Europe, though it is more common in the latter. It has been mistaken for *L. illustris* in North America. Ecologically, it frequently replaces *P. sericata,* but it is more shade-loving (and therefore occurs in forests) and more psychrophilic (therefore, the natural populations of this species penetrate farther to the north). Larvae are saprophagous even under natural conditions, thus showing a slightly wider range of food in comparison with the originally purely carnivorous natural populations of *P. sericata.* With a few exceptions, *L. caesar* shows no tendency toward eusynanthropy, or these tendencies (and also tendencies toward myiases) are either local (in northern Europe) or quite accidental.

Lucilia illustris

This fly has a Holarctic distribution from North America through Europe to the Far East. In North America, it is found from southern Canada to northern Mexico and is common in the Midwest. It also occurs in the Australian region, but has not been taken in the Mediterranean region. It appears most frequently in the middle zone of the European part of the Soviet Union. It is an open woodland and meadow species, appearing on the first warm days of spring. Adults may be collected in large numbers upon flowers of wild parsnip in the Midwest in July. They are primarily attracted to fresh carrion and are found less often on human feces. Larvae, too, occur most frequently in carrion. Russian writers report the breeding of this fly in cities in garbage, as well. The ecology of this hemisynanthropic species needs further study.

Lyperosiops stimulans

This is an exophilous symbovine species, rarely invading stables. As a rule, it alights on fences, vegetation, and ground in pastures. Adults feed on cattle, preferring dark-colored individuals. Females must consume blood prior to each oviposition and feed on liquidy excreta shortly before ovipositing. Larvae develop in fresh feces of cattle. Like all species of this group, the present species may be of epizootiologic importance.

Morellia hortorum

A common communicative hemisynanthropic species, this fly rarely occurs in large numbers. Adults are frequent on flowering daucaceous plants as well as on other herbs, particularly if infested with aphids. They also appear on feces, carrion, fruit, and decaying offal. Females are facultatively hematophagous, sucking blood of cattle attacked by Stomoxydini and tabanids. They usually attain higher population densities in pastures. Larvae develop in cattle excreta and dung which is old and somewhat desiccated.

Musca autumnalis

This is one of the most frequent symbovines of the pasture, attaining high populations in late summer (August and early September). It was introduced and has been spreading throughout temperate North America since World War II. Adults frequent cattle droppings and attack both cattle and man. While the males are often floricolous, the females are facultatively hematophagous. Due to their habit of actively attacking conjunctivae, they are markedly communicative. *M. autumnalis* is the intermediate host for larvae of *Thelazia rhodesii* which are sucked up from infested eyes of cattle by adult flies. In the fly they continue to develop into the invasive larval stage. Thelasiosis appears as chronic kerato-conjunctivitis sometimes resulting in blindness of the infested individual. Breeding takes place chiefly in cattle feces and rarely in dung or kitchen offal. In farm dwellings in the vicinity of pastures adults may congregate in attics and crevices to hibernate.

Musca crassirostris

The species is distributed throughout Africa and the Oriental region. In Egypt and in East and Central Africa, it has long been known as a voracious blood-sucker. The proboscis of this fly is worthy of note for its feeding habits and role in disease transmission are unquestionally influenced by it. The prestomal teeth are not only reduced in number, but are enormously increased in size, and together with the highly modified interdental armature form a most efficient scratching apparatus. Though superficially suggestive of a piercing organ, such as that of *Stomoxys,* the proboscis of this fly, and of *M. inferior* and *M. lusoria* as well, does not penetrate the skin. On this point, however, there is a difference of opinion, since some have stated that *M. crassirostris* "readily pierces the skin." Larvae breed in cow and horse dung. This was originally a symbovine species which is now moving toward hemisynanthropy.

Musca domestica

Literature on this species is vast. It is a classical eusynanthropic, endophilous, and markedly communicative species, fulfilling all the conditions required of a disease vector. Originally, *M. domestica* was probably coprophagous, adapted to excreta of ungulates. This is evidenced by the fact that the artificial media serving for laboratory breeding of *M. domestica* resemble the excreta of ungulates both chemically and physically. Secondarily, the food range of both adults and larvae increased as a result of synanthropy. Thus, synanthropic populations of this species have become trophically adapted to food of man and to the wastes of his household. The original trophic adaptation of *M. domes-*

tica to feces is preserved only in subtropical asynanthropic populations and geographic forms of this species, e.g., *M. domestica vicina.* Unlike other synanthropic flies (in the broad sense) and despite its thermophilous habits, *M. domestica* is capable of overwintering in all its developmental stages and of developing in the anthropobiocoenoses at temperatures around 18° C. throughout the year. These characters enabled the species to become a permanent member of the anthropobiocoenosis and to attain cosmopolitan distribution. In the Temperate Zone, populations of *M. domestica* are decimated in winter; the highest population densities are attained in late spring and again in August and early September.

In the tropics and subtropics of the Old World, populations of *M. domestica* are not confined to human habitations and even the synanthropic populations are exophilous, especially during hot summers. These populations concentrate in the vicinity of human habitations mainly for trophic reasons. The importance of the house fly results from its (1) eusynanthropy, i.e., close coexistence with man, sharing his artificial biocoenosis (anthropobiocoenosis); (2) consumption of both contaminated and non-contaminated food; (3) great flight activity and dispersal; (4) constant alternation between feces and food.

Musca inferior

This is a hematophagous species distributed in the Oriental region. It is widely distributed in India, but it is never seen in large numbers and is easily missed. It can always be recognized by noting that as soon as it settles on the skin of an animal it begins to suck blood immediately, whereas *Musca bezzii,* which closely resembles it, flits about from one spot to another. Adults visit human feces and the larvae are fecal breeders. The species is symbovine and imagines are facultatively hemisynanthropic and communicative.

Musca larvipara

This is another typical Palearctic symbovine of the pasture. Females gregariously attack grazing cattle and actively follow Stomoxydini and tabanids to suck blood from the minute lesions caused by these flies. Males are floricolous but do not withdraw far from the pasture. Like *M. autumnalis, M. larvipara* is a thermophilous species, showing the highest activity on sunny, hot summer days. The species is distributed in southern and eastern Europe. Females give birth to second instar larvae which they deposit on the surface of excreta of cattle and rarely on that of other animals. Remains of the chorion of the eggs are found on the surface as a rule. Adults, particularly females, actively suck the conjunctivae of cattle and man and also transmit larvae of *Thelazia*; therefore, their hygienic importance is similar to that of *M. autumnalis*.

Musca lusoria

This is a large, hematophagous species widely distributed in the Ethiopian region (central and northern Africa) and in Ceylon. The species is viviparous, with larvae developing in cattle dung. Adults attack eyes and wounds of domestic animals.

Musca domestica nebulo

Recent laboratory studies demonstrating considerable genetic continuity between this type, *M. d. domestica*, and *M. d. vicina* support the subspecific designation though morphological differences exist between populations from various parts of the world. It is a common bazaar and house fly throughout the Ethiopian region and is the common house fly of South India. It breeds in night soil trenches, but can be easily reared in horse dung and in the contents of goat intestines, which are to be found collected outside slaughterhouses in India. Its synanthropy places this fly among the important public health species.

Musca tempestiva

This is a symbovine fly of the pasture which is communicative since it alights on feces as well as on grazing cattle and man. The flies are particularly troublesome on sultry summer days, attacking, together with other species (*Morellia simplex, M. podagrica, Hydrotaea irritans, Musca autumnalis,* and *M. larvipara*), grazing cattle and man. They suck secretions from the conjunctivae and sweat glands and, being a facultatively hematophagous species, also blood from minute lesions of the skin. Larvae occur gregariously in excreta of cattle, horses, and pigs. The ecology and behavior of this species deserve further study by medical entomologists.

Musca terraereginae

This fly is closely related to *M. d. nebulo,* but it is limited to the Australian region where it is a rather uncommon fly. It has been bred from horse dung and may be categorized as hemisynanthropic to symbovine.

Musca sorbens

This fly is one of the most widely distributed species of the genus in the Old World. It is common in the Mediterranean along the North African coast and was observed to board all steamers going east through the Suez Canal. It occurs in the Australasian region and is usually one of the commonest flies on Guam, especially toward the southern end of the island where native habitations and livestock are most abundant. It is common around garbage dumps, pig pens, privies, carrion, and various types of excrement. In the Ethiopian, as in the Oriental regions, it abounds in bazaars, swarming on foodstuffs of all kinds. It commonly feeds on exudations from cuts and sores which it can apparently sense from some distance. The fly rasps a sore until it is perforated and serum collects. It is also quite common around the eyes, a habit which is crucial for the transmission of trachoma, conjunctivitis, and other ophthalmias. Larvae breed in excrement of various domesticated animals and man. The adult does not enter houses as freely as *M. domestica.* In Egypt *M. sorbens*, given the choice of eight kinds of animal dung, preferred human feces for oviposition; aside from pig dung, flies did not oviposit on other animal feces except in very small numbers. Human stools collected in Egyptian villages produced mainly *M. sorbens*. These flies are found in Central Asia, congregating in large numbers in bazaars and are called market flies. They attack people, especially children, and are particularly active during the warm parts of the day. They prefer bright sunlight, only small numbers enter dwellings. In this region they are active from May through November; *M. sorbens* hibernates as larvae and pupae.

Muscina stabulans

This is a Holarctic eusynanthropic species, communicative under rustic conditions and also endophilous under primitive rustic conditions. Adults are invariably found in the immediate vicinity of human settlements and are important members of the anthropobiocoenosis, particularly in the country; they are scarce in towns. They alight on walls of privies and are also frequent on fruit in orchards and vineyards. In nature, they are frequently found on sap flowing from trees as well as on tissues infested with aphids or psyllids. Phenologically, the species is later than *Protophormia terraenovae* and certain *Calliphora* spp., which also suck sap of trees. In northern latitudes they frequent sunlit places, whereas to the south they select habitats with dense growths of shrubs and trees. The food range of larvae is diverse. Preferably, they develop in human feces both scattered and accumulated, as in latrines, etc. In other feces and in decaying fruits they attain lower population densities. They are frequent in both decaying and growing cruciferous plants; in the latter, their occurrence is conditioned by primary infestation of the plants by other flies (esp. *Hylemya brassicae*, etc.). They are rather rare in carrion. Also, they have been found in lepidopterous larvae and dead bodies of other insects, as well as in nests of synanthropic birds (swallows, house sparrows). First instar larvae appear to be mostly saprophagous, those of older instars (particular the third) are predatory, attacking larvae of other flies. The related species *Muscina assimilis* and *M. pabulorum* are markedly hemisynanthropic with similar trophic habits. *M. assimilis* is a particularly communicative hemisynanthrope.

Mydaea urbana

This is a coprophagous, Holarctic, hemisynanthropic species, associated with forest environments as are the related species, *M. modesta, M. ancilla,* and *M. scutellaris* R.-D. Larvae are typically coprophagous at the beginning of their development, becoming predatory as they mature.

Myospila meditabunda

This is a Holarctic, hemisynanthropic and symbovine species. It is a typical coprophagous species which prefers pastures. Larvae are predatory at least in their final instars and attack larvae especially of the genera *Orthellia, Dasyphora,* and *Siphona.* They have also been reared from dead lepidopterous larvae. Adults feed on the conjunctivae of eyes, in nostrils, etc., of cattle and, also, on certain flowering herbs. In the Alps, this species is exceeded by *Myospila alpina.*

Nemopoda nitidula

This frequent Holarctic representative of the family Sepsidae occurs, together with *Sepsis* spp., on feces (chiefly of man), but first instar larvae are broadly saprophagous. In contrast to *Sepsis* spp., adults of *N. nitidula* are abundant under natural conditions where they frequently attain high densities.

Ophyra aenescens

This fly occurs in the United States from Oregon to Arizona and from Illinois to the East Coast and Florida. It is also Neotropical and Oceanic. In Central America it is abundant, entering houses and alighting on food, much like *Musca domestica.* Adults are attracted to carrion and animal manure as well as to human excrement, as indicated by the isolation of vaccine strains of poliovirus from natural fly populations. Larvae breed in all the above materials.

Ophyra anthrax

The biology of this species is similar to that of *O. leucostoma.* In contrast to the latter, *O. anthrax* is a thermophilous species which does not extend far north. It is especially widespread in the Caucasus and in middle Asia. In this latter region adults were recorded in toilets, garbage deposits, on animal carcasses, and also within dwellings. This fly is hemisynanthropic with a eusynanthropic tendency.

Ophyra leucostoma

This species is a common Holarctic hemisynanthrope to eusynanthrope. Adults alight with predilection on vegetation, particularly the lower surface of leaves, on sultry days. They are also common around privies, slaughterhouses, etc. In its northern range this species is markedly heliophilous. Larvae develop in feces of man and domestic animals as well as in kitchen offal and animal carcasses. Second and third instar larvae convert to predatory life habits, frequently attacking larvae of *Musca domestica*, and are capable of considerably decreasing the latter's population density. The related species *Ophyra capensis* is a more southern species in the Palearctic. Its bionomics are similar to those of the preceding species.

Ophyra nigra

This fly has an Australasian range and appears to be closely associated with man in the tropics. It is abundant in Queensland, principally as a carrion feeder. Unlike *M. domestica* and *S. calcitrans*, the larvae of this species were found to pupate right in the excessively moist manure, from which they were able to emerge successfully. They were probably introduced into Hawaii from the Orient by shipping. This species is best characterized as hemisynanthropic to eusynanthropic, and facultatively communicative.

Orthellia caesarion

This fly is found in the Holarctic region, including Asia and Europe, and between both coasts of North America from southern Canada to Georgia and California. It also occurs in the subalpine region of Lapland. This is a symbovine species generally abundant in open fields. Adults feed primarily on the dung liquid of cattle. In Denmark, they are also commonly seen on flowering *Erodium, Bellis*, and *Chrysanthemum* in spring and fall, and on *Heracleum* and *Cotoneaster* in summer. The dark fecal spots and greenish black cow dung in their intestines leave little doubt, however, as to their main diet. Adults are attracted to droppings for feeding and oviposition which are 1/2- to 1-hour old. Larvae develop within the droppings but pupation occurs in a drier situation. It has been reared as a representative species from cow pats in Indiana.

Parasarcophaga albiceps

Adults frequently visit human feces, decaying refuse, and fruit. Larvae are coprophagous.

Phaenicia cuprina

The life history of this fly is similar to that of *P. sericata* with which it was formerly confused. However, it inhabits the drier and warmer parts of Africa and Asia. Some consider its introduction into Australia and the New World probably a recent occurrence. Approximately 90 percent of all sheep "strikes" in Australia are caused by this species. Writers disagree as to whether this species has similar habits in Africa. In South Africa, it has been reported as common in butcher shops and about abattoirs; its nutrition is that of a scavenger. Besides feeding on fallen fruits, the nectar of flowering plants, and on the honeydew of aphids, it is known to feed on feces. On Guam, *P. cuprina* prefers large masses of liquefying garbage to carrion. Adults were also collected from human excrement and privies, and rather commonly around kitchen doorways where the overflow from slop pails had moistened the soil.

Phaenicia pallescens

This fly is closely related to *sericata* and *cuprina.* It is essentially a southeastern species in North America, ranging from Washington, D.C. to Florida and Mexico. It is common around rotting fallen fruits and is abundant in meat-baited traps. Adults are numerous in market areas of cities where decaying produce attracts them. The species is hemisynanthropic to eusynanthropic, communicative and exophilous.

Phaenicia sericata

This is a classical exophilous and communicative eusynanthrope. Very likely, it was originally a Holarctic species distributed over the warmer regions of the Temperate Zone. Due to its unusual adaptability and ability to compete with other coprophagous species, the species has gradually spread outside its original area. It has received considerable attention. The food range of the larva is wide, enabling it to develop on various substrates, although originally a carnivorous form which still prefers slowly decomposing (not fresh) meat. In the latter, it develops most rapidly. The species has become adapted to various kitchen offal from which it can even utilize chiefly carbohydrate substrates. It has spread far to the north where it lives in synanthropic populations only. The species causes primary myiases of sheep (Great Britain, South Africa, Australia).

Phaonia pallidosa

Among the numerous species of the genus *Phaonia, P. pallidosa* is the most frequent visitor of feces and decomposing flesh. The species never attains high population densities. It is a shade-loving fly occurring mainly in forested regions and, also, in the north of Europe. Larvae are widely saprophagous, and their second instars are predacious. The species is mostly a forest hemisynanthrope.

Phormia regina

This is a hemisynanthropic to eusynanthropic exophilous species which sometimes occurs in large numbers. It is substantially Holarctic in distribution. Larvae develop chiefly in decaying carrion but can be facultative producers of myiases of mammals and birds. Adults are frequent on inflorescences of daucaceous plants and also alight on fruit. They are spring-fall flies.

Physiphora demandata

This is a nearly cosmopolitan, hemisynanthropic to exophilous eusynanthropic species. Adults visit feces, decomposing flesh, decaying fruit, as well as plants infested with aphids. Like *Phaenicia sericata,* this species penetrates farther to the north in purely eusynanthropic populations. In Central Europe, as with *P. sericata,* the northern limit of this species passes Hungary and the lowlands of Czechoslovakia. Both these species are good indicators of an average annual isotherm of about 10° C. Larvae are saprophagous, developing in kitchen offal, human feces, excreta of domestic animals, and other decomposing matter.

Piophila casei

This was originally a Palearctic species which secondarily spread through anthropobiocoenoses all over the world—evidence of the eusynanthropy of this species. It attains the highest densities in offal and heaps of bones in rendering plants, slaughterhouses, fish processing plants, carrion pits, etc. Larvae can also be found in lard, cheese, roe, and other protein food. Pupation occurs in the larval medium or close by. Aside from the typical maggot-like movements, larvae are capable of jumping relatively long distances by twisting their bodies. The species causes considerable losses in the meat industry. Little is known of the hygienic importance of this species.

The related hemisynanthropic species of the genus *Piophila,* e.g. *P. vulgaris, P. varipes, P. foveolata, P. latipes,* etc., also attain high population densities, some despite the severe conditions of the north (Greenland, Iceland). To date, little attention has been given to the taxonomic problems of this group, and only one species, *P. casei,* is referred to especially in the applied literature.

Polietes lardaria

The species belongs among typical hemisynanthropes, preferring mountain and submontane pastures, shadowy growths and forests (similar to the more thermophilous *P. albolineata*). Adults appear, especially in autumn, on feces of all kinds and are active even in rather cold weather. In addition to animal and human feces, they also occur on certain flowers (*Solidago virgaurea*), particularly those infested with aphids. Larvae are predatory, attacking those of other coprophagous flies.

Pollenia rudis

This species occurs in the Palearctic region from Ireland, throughout Europe and North Africa, to Siberia and China; in India; and in the Nearctic from Nova Scotia and British Columbia south to Florida and California. The larvae develop as internal parasites of various earthworms. Eggs are laid in the soil and the young larvae seek out earthworms which they enter through the spermiducal and other pores. First instar larvae, about 1 mm. in length, live in the genital segments or inside the seminal vesicles throughout the winter without undergoing any change. At the end of April, parasitic maggots, which have resisted encystment and digestion by the host's amebocytes, begin to develop. About 45 days later, the adult appears. Larvae have not been reared successfully on any normal blowfly medium, nor on manure, dead earthworms, or decomposing plant material. There are four generations of *P. rudis* per year in the United States. In the fall great numbers of the last generation enter homes where the adults hibernate, though more natural sites are animal burrows, hollows in logs, etc. They prefer rooms in upper stories, especially those which are cool and undisturbed. In the Palearctic, this species and its congeners—*P. atramentaria, P. dasypoda, P. intermedia, P. varia, P. vera,* and *P. vespillo*—are frequent visitors to feces, particularly in spring and fall. In summer, adults are found in forests.

Protophormia terraenovae

This fly is very common in the cooler regions of the Holarctic, replacing *Chrysomya* species of the Old World tropics and subtropics. In North America, it is the northern counterpart of *Phormia regina* which is not abundant in Canada. *P. terraenovae* is saprophagous and appears in the early spring. In more northern localities and at higher elevations it may be taken during July and August (viz. about 7,000 feet in Colorado). During July, it is very abundant in the subarctic regions. It has been taken within 550 miles of the North Pole. The considerable list of isolations of fecal organisms from this fly indicates its association with feces. It is generally of secondary importance in myiasis, except in Scotland where it may be the primary fly in sheep myiasis. Laboratory investigations indicate that maggots of this fly successfully overgrow *Musca domestica.* How effective this may be under natural conditions is not known.

This is a hemisynanthropic to exophilous eusynanthropic species which is non-communicative. The activity of adults is influenced by their absorption of the sun's rays. Their body temperature may be as much as 22° C higher than the ambient temperature. Flies become aerial at temperatures below 10° C, when the relative humidity falls below 70 percent. Above 28° C, there is a bimodal diurnal activity curve; below this temperature there is a single peak. Larvae develop chiefly in carrion. Adults are common in slaughter house yards around heaps of organic refuse, and frequently also visit flowers of *Heracleum*. In early spring, they suck sap flowing from wounded trees. Their hygienic and epidemiologic importance is probably slight.

Pyrellia cadaverina

A Palearctic, hemisynanthropic species, this fly frequently visits feces as well as decaying flesh. Larvae are coprophagous, preferring horse feces. Populations of this species attain high densities only sporadically. The species is frequently accompanied by the related *P. ignita*.

Ravinia striata

This species, together with *S. haemorrhoidalis* and *S. melanura*, is a frequent visitor to human feces on the periphery of human settlements. They are scarcer on other decomposing matter (offal, decaying fruit). Larvae develop preferentially in feces, chiefly of man and pigs.

Sarcophaga argyrostoma

This is a Holarctic, hemisynanthropic species which has spread to parts of South America, India, the Marshall Islands, and Hawaii. The larvae are primarily carrion feeders but also occur in human feces. This species, and those sarcophagids which follow, all produce myiasis.

Sarcophaga bullata

This species occurs only in the Nearctic region where it is found from Quebec and British Columbia to Florida and California. It is a hemisynanthropic fly which is commonly used in laboratory studies.

Sarcophaga carnaria

It is widespread throughout most of the Palearctic region from Denmark and Ireland south to Sicily and east through Europe (as far north in European Russia as the Timansk Tundra) to Siberia and Mongolia. The biology of the immature stages in nature has not been sufficiently studied. Some writers erroneously consider them to be exclusive parasites of earthworms. However, we (D. P.) succeeded in rearing a number of larvae in decaying meat, and they are also known to occur in privies and other unsanitary facilities. This species is widespread in the Soviet Union, and, according to Russian writers, the flies are found especially along roads, on vegetation and on flowers (particularly dandelions during the first half of summer). Adults are also found on putrefying fruits and meats and on corpses of animals. The unsanitary aspect of this fly is evidenced by isolations of bacteria and Protozoa of enteric origin from wild flies.

Sarcophaga haemorrhoidalis

This fly has an almost world-wide distribution, occurring on all continents but absent from the Oriental and much of the Australian regions. It is absent or rare in the cooler areas of Europe such as England and Denmark, and not found in northeastern Europe or Siberia, nor in the east coast province of Asiatic Russia. It is very common in the Ethiopian region and for this reason it is considered indigenous there. In the New World, it is distributed from southern Canada to Argentina and east to Hawaii. This is a hemisynanthropic fly which tends to be eusynanthropic especially in the warmer regions where it is common in and around human dwellings. It is a thermophilous species which avoids large forests. Larvae are deposited on carrion, excrement and exposed meats; some consider that they are mainly fecal breeders. In Africa, adults are commonly found indoors and are especially attracted to freshly deposited stools. Aside from the importance of this fly as a cause of myiasis, its habits make it a public health suspect.

Sarcophaga hirtipes

This fly is widely distributed in the Palearctic region from Germany through the Middle East to Szechwan in China. It also occurs in Baluchistan. In the Ethiopian region, it ranges south of the Sahara to the Cape of Good Hope. This species has habits similar to those of *S. haemorrhoidalis.* Besides the occurrence of larvae in wounds of sheep and cattle, they are frequently found in animal and human excrement and in decomposing animal and plant material, including carcasses and melons. It is a hemisynanthropic species.

Sarcophaga melanura

This hemisynanthropic species has a Holarctic distribution, having been reported from northeastern United States and Quebec. Russian writers report these flies are frequently found on fruits and sweets in markets and in places which have uncovered produce. They are also found in human dwellings and frequent human, horse, and cattle excrement and garbage. The larvae develop in feces of pigs and dung of other animals.

Sarcophaga misera

The taxonomic status of this species is confused and therefore, distributional and bionomic data lack finality. This species is distributed in Daghestan and China in the Palearctic, India through the East Indies in the Oriental region, and in the Australian region. The fly is normally a carrion feeder but has been found breeding naturally in horse dung in Australia.

Sarcophaga peregrina

It is recorded from the eastern areas of the Palearctic (China and Japan), from the Oriental region (India to Java and the Philippines), and the Australian region (South Australia and Oceania, including Fiji, Samoa, and Hawaiian Islands). It breeds in all kinds of decomposing matter, and especially in human feces. It is considered hemisynanthropic.

Scatophaga stercoraria

This is a species of Holarctic distribution which is facultatively eusynanthropic under rustic conditions but otherwise hemisynanthropic. It is conspicuous but has little practical importance. Adults and larvae are predators. They are present from early spring (March) till late autumn (November). During the summer months, flies concentrate in the shadow of forests or at higher elevations. Larvae are active at temperatures slightly above 0° C. Adults are found in large numbers in dunghills of cattle, horses, and pigs, as well as human excreta. They are attracted to this substrate to feed, and also to prey upon other coprophagous species of flies, particularly the smaller ones (Sepsidae, Cypselidae). Larvae of these latter families are preyed upon in feces by larvae of *S. stercoraria.* At times, adults occur on flowers of certain plants (especially daucaceous ones) where they prey upon small flies, e.g., Chironomidae.

Sepsis punctum

Adults of this, as well as the related hemisynanthropic species (above all, *S. violacea, S. fulgens, S. thoracica, S. cynipsea,* etc.), are markedly herbicolous. They attain considerably high population densities in dung pits and farmyards. They are facultative exophilous synanthropes, feebly communicative. Larvae are saprophagous and have been found in feces of cattle and man and in a variety of decomposing vegetable and animal matter.

Siphunculina funicola

This is the "eye-fly" of India, Ceylon, and Java. Adults alight on eyes, lips, nose, and ears of men and animals in the tropics, eager to suck up liquid secretions from these organs, and from perspiring skin. This supplements their diet of honey-dew and moisture from plants. The fly occurs along the seacoast and in fairly high elevations (Bangalore). It is absent only during the cold months and has population peaks during the short intervals of warm weather after the southwest monsoon sets in. Adults have the peculiar habit of resting on suspended ropes and strings, preferably old ones, hanging under thatched roofs. The clusters are always found away and protected from the winds. The breeding habits of this hemisynanthropic fly are poorly understood.

Stomoxys calcitrans

This is a secondarily cosmopolitan endophilous symbovine of considerable practical importance. It is closely associated with stables and rustic human dwellings and is most frequent in the former when cattle are present. It occurs outside these buildings only during the day, visiting dry matter of plant origin (especially wooden doors, fences, etc.) and basking in the sunshine. Toward autumn, the flies migrate gregariously to stables and fodder preparation rooms, congregating on warm walls (e.g., chimneys). On hot days, the species also occurs in pastures. The species can be described as thermophilous and heliophilous. In late summer, the flies invade homes and can bite man. They prefer cattle, being less frequent on horses and pigs; under laboratory conditions, the flies consume blood of laboratory mammals and birds. Both sexes suck blood in illuminated stables even late at night. Feeding is repeated frequently. Unlike house fly maggots, larvae of the biting stable fly are fairly restricted to substrates which contain rotting plant material, commonly straw. Thus, they are often found breeding in large numbers in lake weeds piled up on shore. Toward middle and late summer in the United States, these flies constitute a major annoyance to bathers. It is readily seen from the list of its biotic associations that *S. calcitrans* has deservedly received attention as a potential vector of many microbial diseases of man and domestic animals. Also of considerable importance is the hematophagous habit of *S. calcitrans* which results in decreased milk production of cows, as well as weight losses and death of calves.

Stomoxys nigra

Our biotic associations concerning this species are derived from the African continent and the adjoining islands of Madagascar and Mauritius. It occurs as well in the Orient. The larvae of this species and those of *S. calcitrans* breed under identical conditions but the former's development is more rapid. In Madagascar the adult bites with the same tenacity as its European counterpart. *S. nigra* is a symbovine, exophilous species.

(*continued on page 83*)

Chrysomya megacephala (Fabricius), 1794, ♀. Actual size 10.0 mm.
COURTESY F. GREGOR.

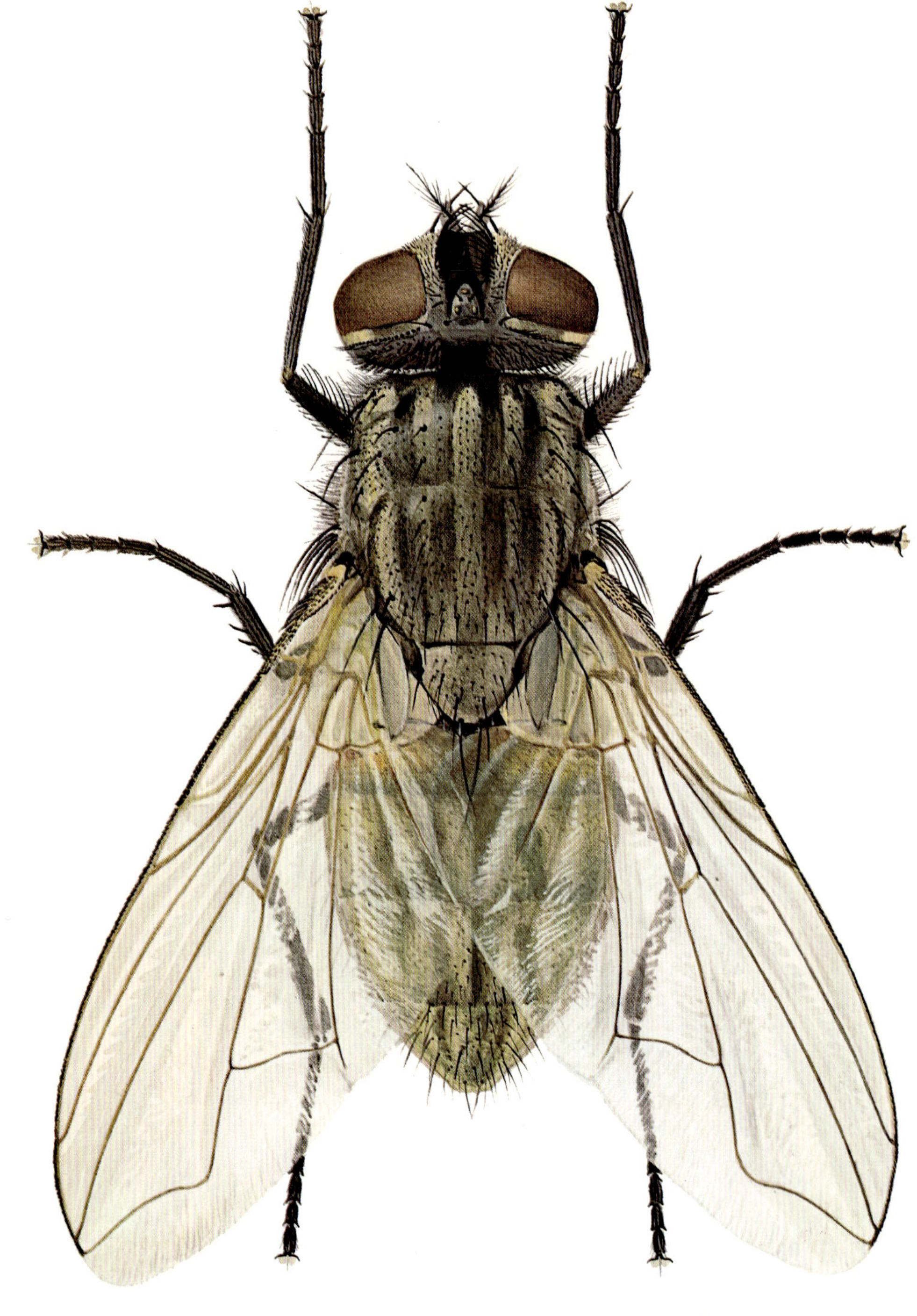

Musca domestica domestica (Linnaeus), 1758, ♀. Actual size 7.0 mm.
COURTESY F. GREGOR.

Hippelates pusio (Loew), 1872, ♀. Actual size 1. 6 mm.
COURTESY F. GREGOR.

Phaenicia sericata (Meigen), 1826, ♀. Actual size 8.0 mm.
COURTESY F. GREGOR.

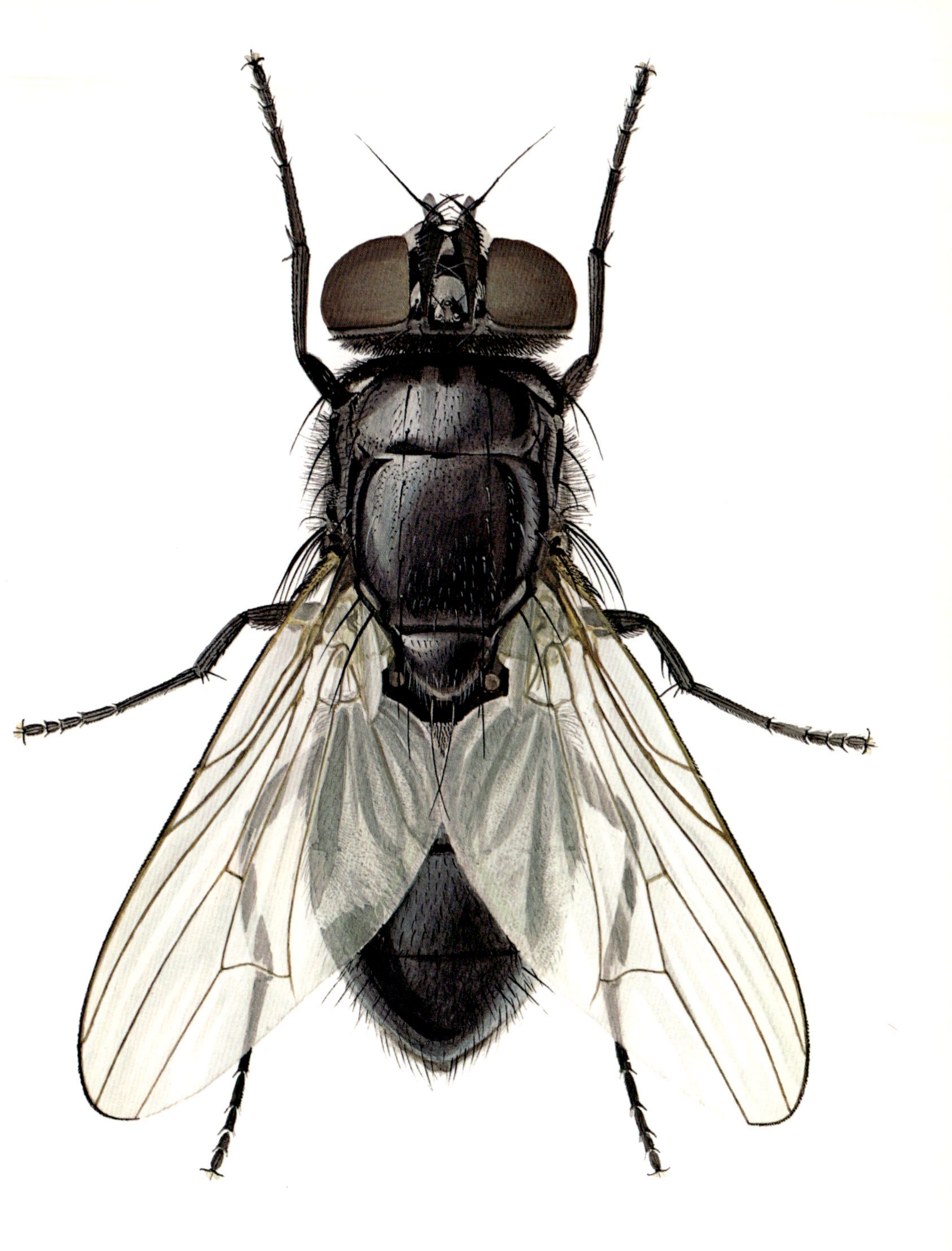

Ophyra leucostoma (Wiedemann), 1817, ♀. Actual size 6.5 mm.
COURTESY F. GREGOR.

Protophormia terraenovae (Robineau-Desvoidy), 1830, ♀.
Actual size 10.0 mm. COURTESY F. GREGOR.

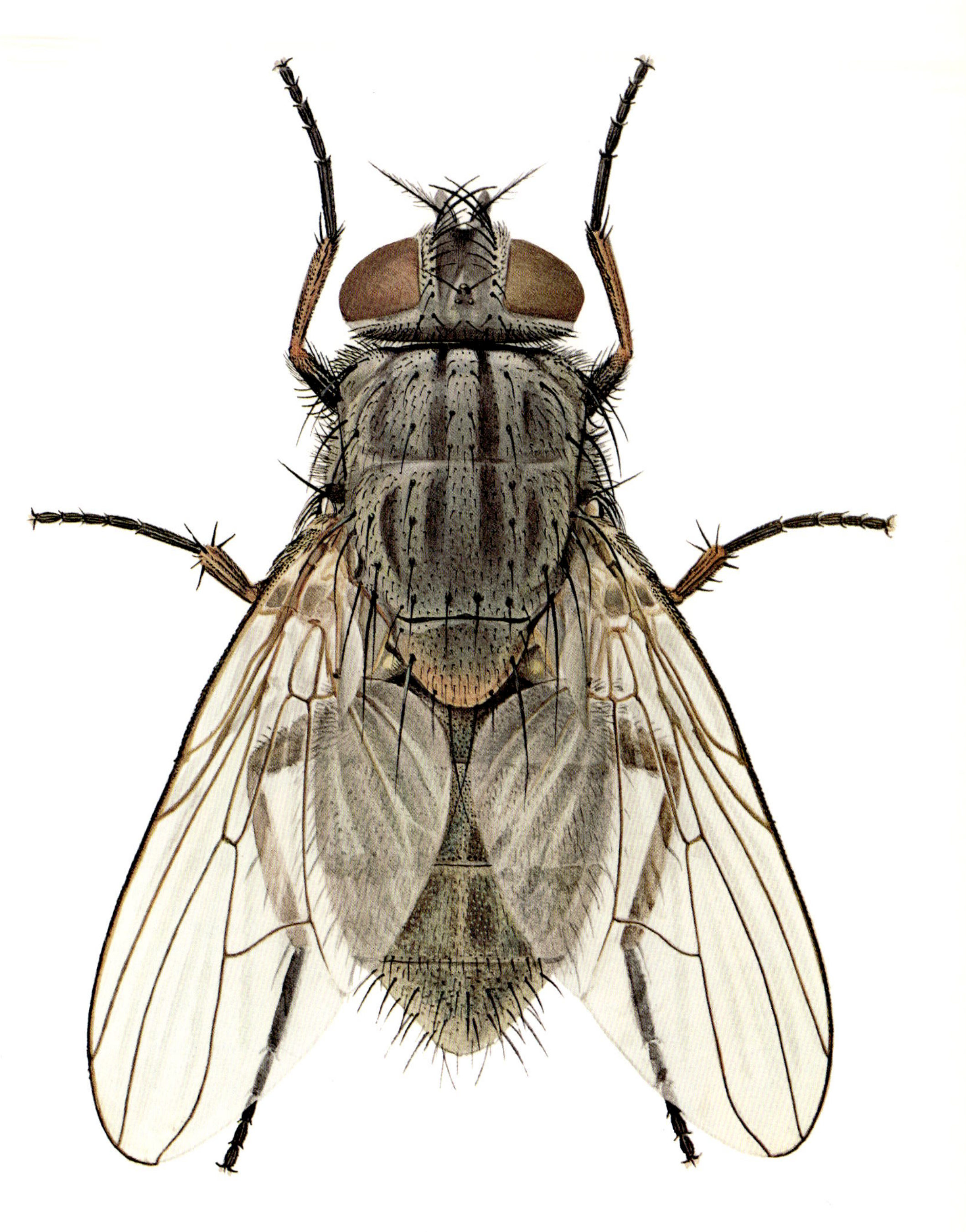

Muscina stabulans (Fallén), 1816, ♀. Actual size 8.5 mm.
COURTESY F. GREGOR.

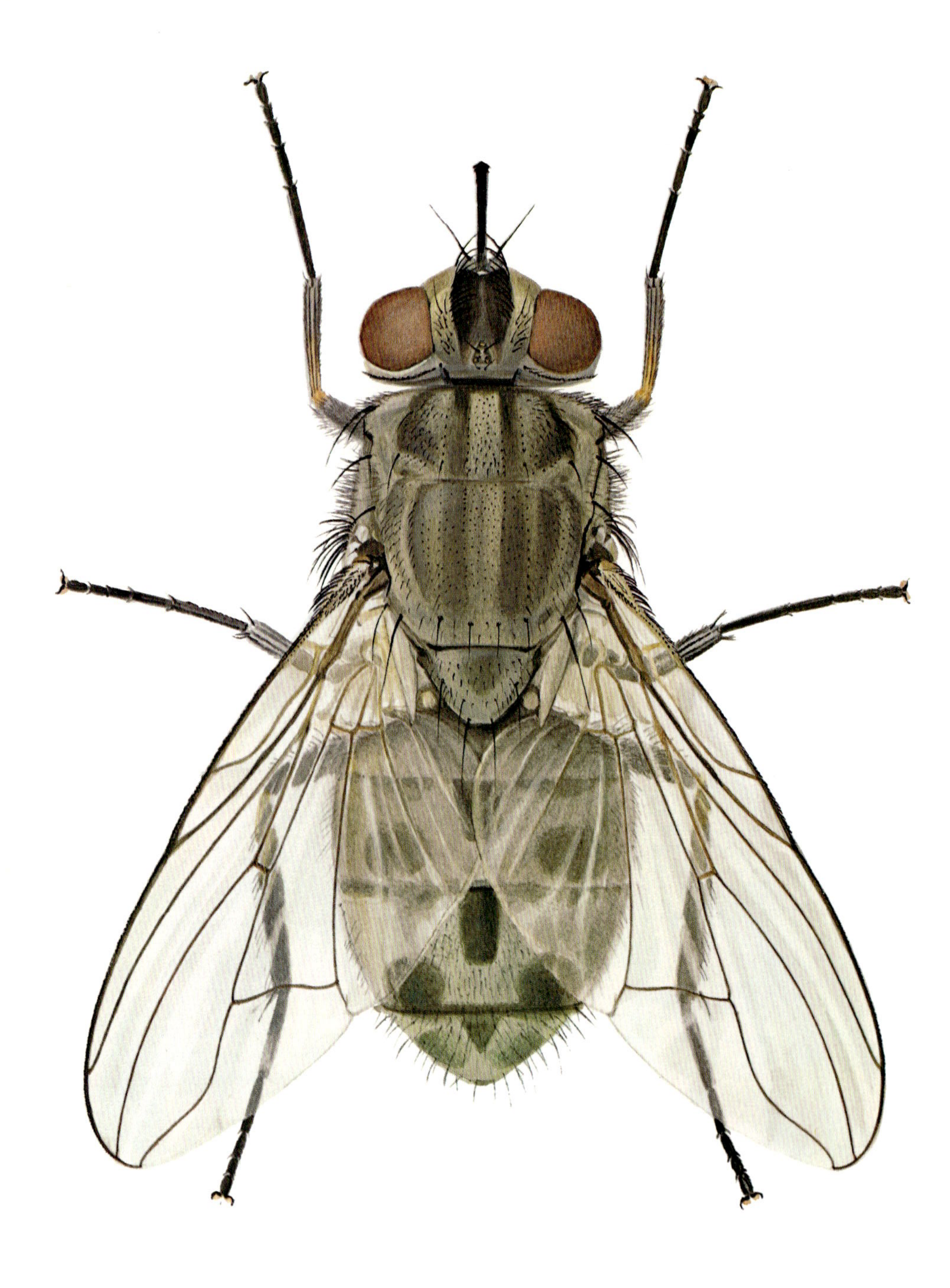

Stomoxys calcitrans (Linnaeus), 1758 ♀. Actual size 7.0 mm.
COURTESY F. GREGOR.

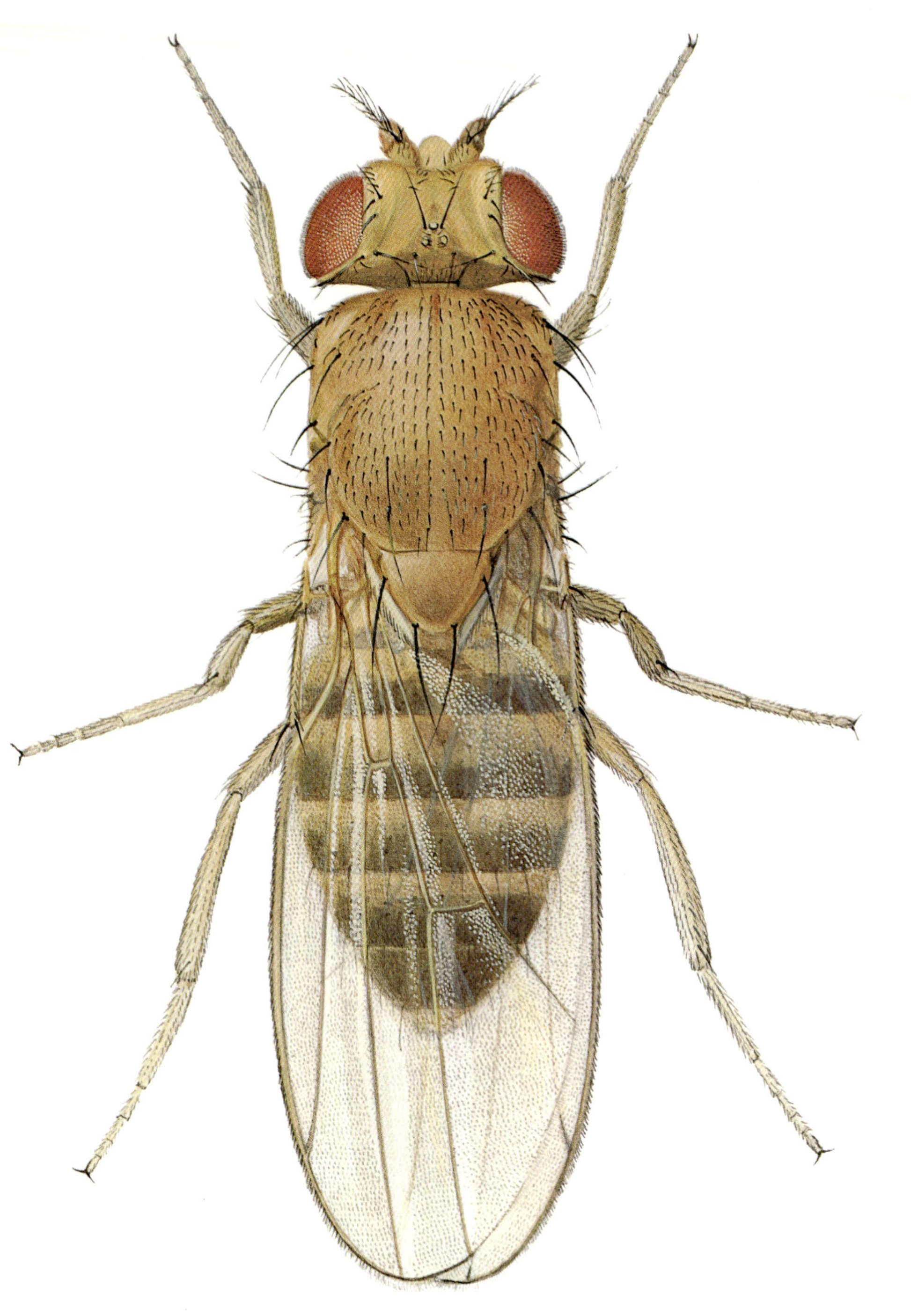

Drosophila melanogaster (Meigen), 1830, ♀. Actual size 2.0 mm.
COURTESY F. GREGOR.

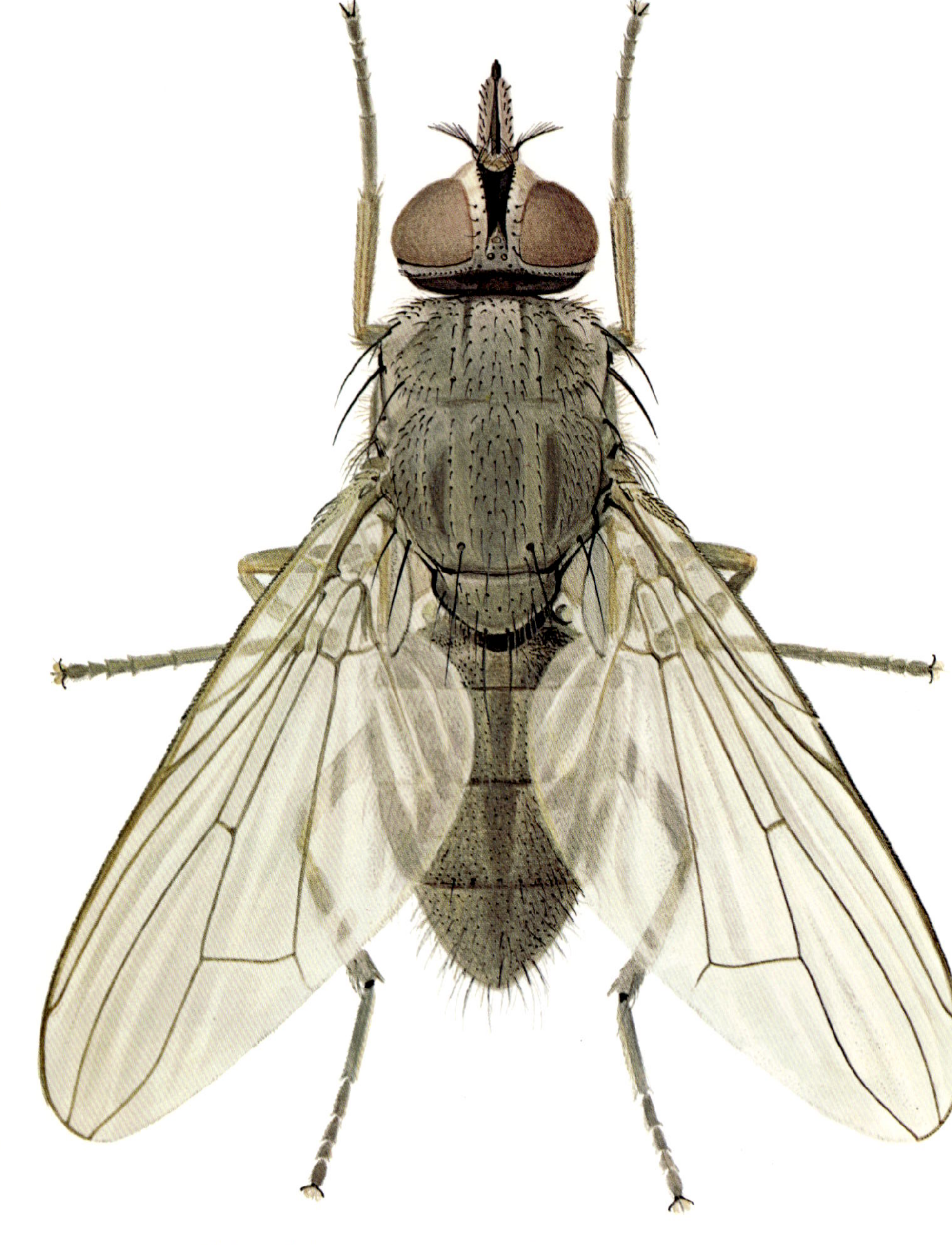

Siphona irritans (Linnaeus), 1758, ♀. Actual size 4.0 mm.
COURTESY F. GREGOR.

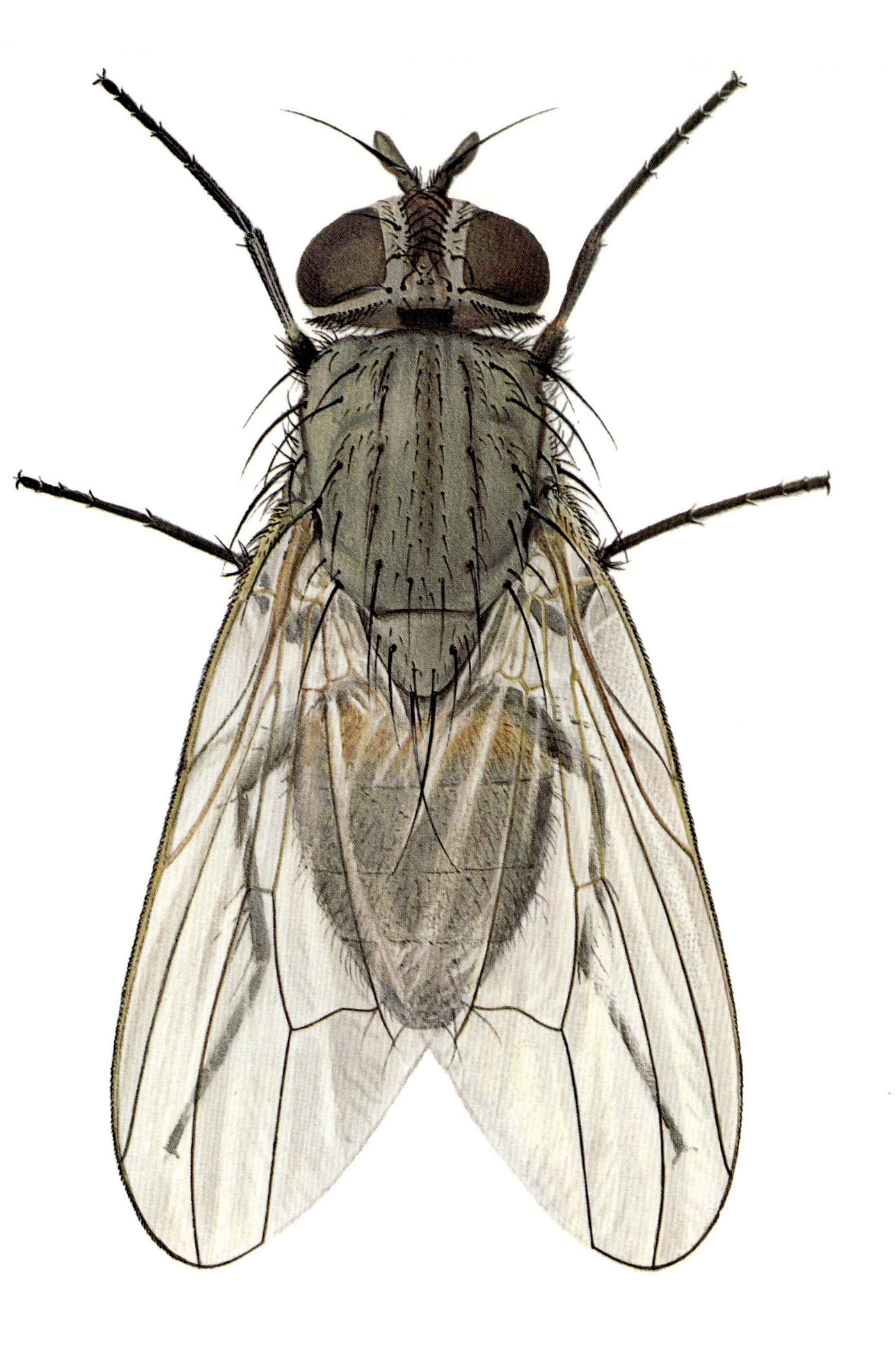

Fannia canicularis (Linnaeus), 1761, ♀. Actual size 5.0 mm.
COURTESY F. GREGOR.

Calliphora vicina (Robineau-Desvoidy), 1830, ♀.
Actual size 11.0 mm. COURTESY F. GREGOR.

Phormia regina (Meigen), 1826, ♀. Actual size 9.0 mm.
COURTESY F. GREGOR.

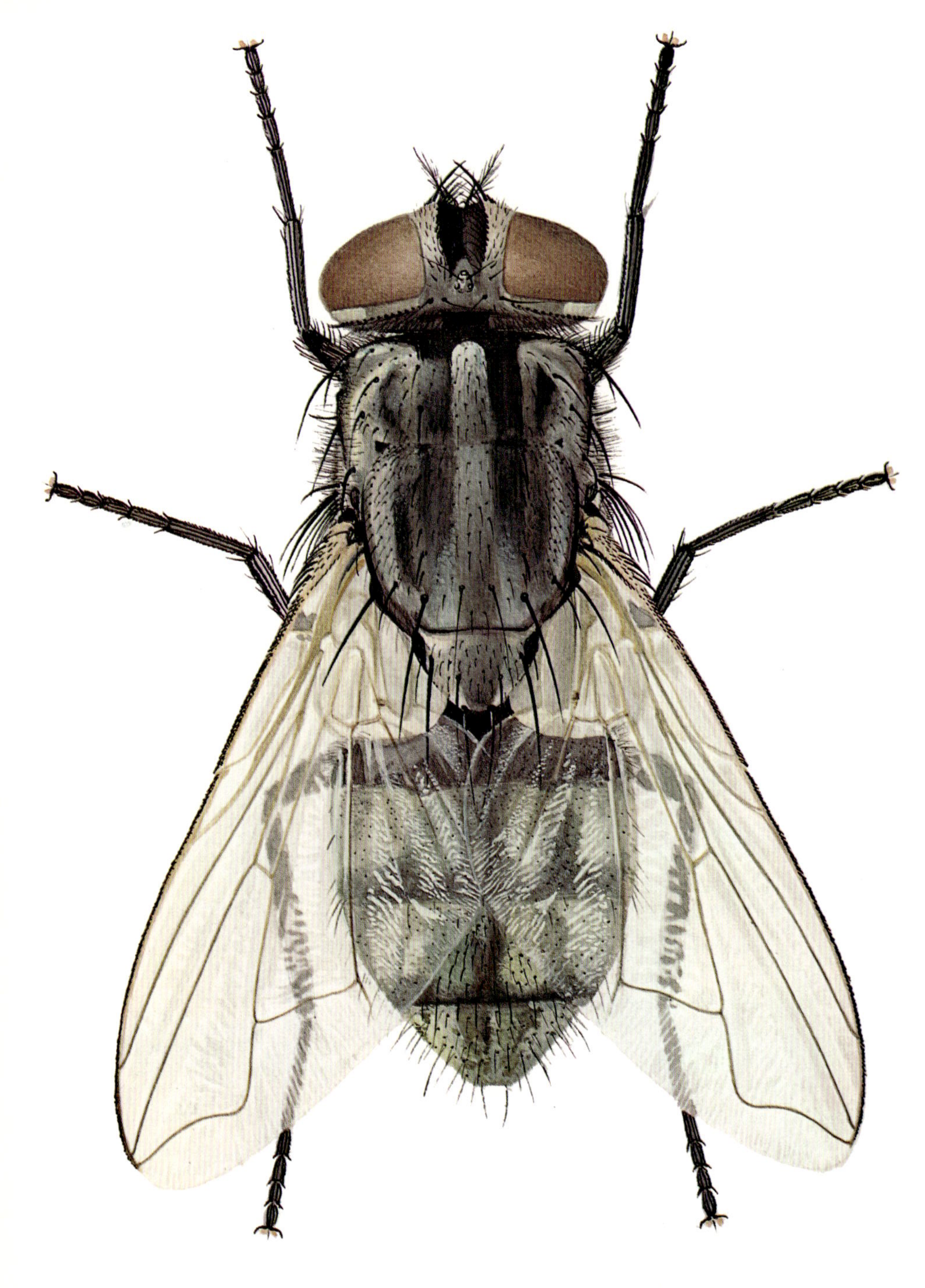

Musca sorbens (Wiedemann), 1830, ♀. Actual size 5.5 mm.
COURTESY F. GREGOR.

Sarcophaga haemorrhoidalis (Fallén), 1817, ♀. Actual size 11.5 mm.
COURTESY F. GREGOR.

Xylota pipiens

This is a largely hemisynanthropic floricolous syrphid, common though not as abundant as *Eristalis tenax*. Larvae develop singly in sewage effluents and are versatile saprophages.

Volucella obesa

Larvae of this genus are scavengers in the nests of bumblebees and wasps (*Vespa*). *V. obesa* adults in Ecuador yielded a large number of *Salmonella* and some *Shigella* types and must be considered among filth flies of public health importance.

4

Keys to Adult Flies

František Gregor, Adrian Pont,
Hugh E. Paterson, and
Marshall R. Wheeler

INTRODUCTION

THE MATERIAL IN this chapter is arranged in three sections. The first section contains keys to the common synanthropic flies of the Palearctic region. Though exclusively New World in their distribution *Cochliomyia hominivorax* and *C. macellaria* are included here because of their importance. Accompanying these keys are generalized drawings of the head, thorax, wing, and leg which are designed to familiarize the non-specialist with the particular vocabulary associated with the highly evolved morphology of flies. Drawings of diagnostic structures and other features will facilitate identification of species. These drawings, 153 in all, and keys have been prepared by Dr. František Gregor. The second section contains a synonymic list of *Musca* species, additional generalized drawings, and keys to the 22 more-or-less common species which are known to occur in the various zoogeographic regions. This section has been prepared by Drs. Adrian Pont and Hugh E. Paterson; Figs. 149 and 150 were executed by Dr. Pont. In the third section, Dr. Marshall Wheeler's brief discussion of synanthropic *Drosophila* is accompanied by a key to the most common cosmopolitan species.

Inevitably, there are some species not dealt with here which are treated elsewhere in the book, and vice versa. The majority of common synanthropes do receive the triple treatment which this volume intended—ecology, classification, and biotic associations.

1. Maxillary palpi long, consisting of 4-6 segments, antenna consisting of more than 3 segments 2

——Maxillary palpi consisting of one segment, antenna apparently 3-segmented, the 3rd segment bearing a dorsally placed arista as a rule 4

2. Small, mothlike flies, with wings broad, pointed, long-haired, and folded rooflike over the body when at rest (Fig. 1): Psychodidae.

——Flies other than above 3

3. Wing venation differentiated as several thick veins near anterior margin of wing and several thin veins inside the wing area (Fig. 7). Antennae subclavate:

  Scatopsidae, e.g., *Scatopse notata* (Linn.) and other species.

——Wing venation as in Fig. 2, antennae not subclavate: Anisopodidae, e.g., *Sylvicola fenestralis* (Scop.).

4. Wings with anterior veins strong and the others oblique and weak, thorax strongly convex (Figs. 3, 4). Minute flies capable of rapid running:
   Phoridae, e.g., *Megaselia* Rond. and other genera.

——Wing venation not differentiated in this way; as a rule one or more transverse veins near the base or in the area of the wing (Figs. 11, 29, 129) 5

5. Anal cell long, frequently reaching the margin of the wing (Fig. 10) 6

——Anal cell about equal in length to the hind-basal cell or slightly longer, rarely absent 7

6. Wing with a spurious vein present as a veinlike fold in the membrane between the radius and medius, traversing cross vein r-m. Large or medium-sized flies, most brightly colored:
   Syrphidae.

——Wing venation of different character (Fig. 5). Antennal segments fused to form a short, bare projection, body bare, abdomen wrinkled:
   Scenopinidae, e.g., *Scenopinus fenestralis* (Linn.).

7. 2nd antennal segment with a longitudinal suture extending along its upper outer edge (Figs. 98, 115). Squamae usually large, postalar callus distinct (Figs. 45, 53). Sexual dimorphism of head present as a rule: eyes of male contiguous on the vertex 17

——2nd antennal segment without a longitudinal suture, squamae usually small, postalar callus not differentiated. No conspicuous sexual dimorphism 8

8. Costal vein interrupted, broken, or thinned before bifurcation into sc and r_1 (Fig. 33) 9

   Costal vein continuous, not interrupted or thinned (Fig. 15) 14

9. 1st tarsal segment of hind-tarsi shorter than 2nd segment (Fig. 35): Sphaeroceridae.

——1st tarsal segment of hind-tarsi longer than 2nd segment 10

10. Costa interrupted in only one place, namely, near where sc arises or near r_1 if sc does not reach the costa (Fig. 41) 11

——Costa interrupted in two places: near origin of humeral transverse vein and near the origin of r_1 or sc (Fig. 39) 13

11. Anal cell and hind-basal cell absent: Chloropidae, e.g., genus *Hippelates*: costal vein prolonged to the fork of m; t_3 with a long curved spine (Fig. 40).

——Anal cell or hind-basal cell present 12

12. Costa with a series of spines all along its length as a rule (Fig. 43): Heleomyzidae.

——Costa bare, anal vein shorter, curved parallel with posterior margin of wing. Body black, glossy:
   Piophilidae.

13. Arista long plumose (in synanthropic species) (Fig. 42). Small yellowish flies with a prominent facial carina and with dark abdominal markings:
   Drosophilidae.

——Arista bare or pubescent. Minute, mostly black flies:
Milichiidae.

14. Palpi vestigial, front smooth and bare in anterior part, abdomen more or less constricted anteriorly. Flies with antlike appearance:
Sepsidae.

——Palpi well developed, abdomen never constricted 15

15. Tibiae with dorsal preapical spines. Prelabrum strongly prominent:
Dryomyzidae.

——Tibiae without dorsal preapical spines 16

16. Veins r_{4+5} and m_{1+2} strongly convergent toward apex of wing; anal cell pointlike, prolonged along the anal vein (Fig. 11): Otitidae.

——No such combination of characters. Vein r_5 setulose, 3 supraalar bristles present, propleura bare, wings frequently spotted:
Platystomatidae.

17. Hypopleural bristles absent 18

——Hypopleura in front of spiracles with a curved, concavity forward, row of setae (Fig. 14) 20

18. Preabdomen consisting of 6 segments (1st and 2nd ones fused); eyes of both sexes broadly separated; front invariably without cruciate bristles; vein m_1 always straight; lower calyptra strongly reduced to form a membraneous fold:
Scatophagidae.

——Preabdomen consisting of 5 segments; eyes of male contiguous as a rule; frons frequently with cruciate bristles, m_1 frequently curved forward, lower calyptra well developed 19

19. Anal vein cu_{1b} + 1a continuous to wing margin even as a fold:
Anthomyiidae.

——Anal vein cu_{1b} + 1a not continuous to wing margin even as a fold:
Muscidae.

20. Postscutellum well developed (Fig. 14): Tachinidae or Larvaevoridae (generally asynanthropic flies).

——Postscutellum absent or weakly developed 21

21. Thorax with pale and dark vittae, abdomen with pearly maculae or dark round maculae. Body densely pollinose, never metallic blue or green. As a rule 3 or more notopleural bristles present. Vestigial m_2 vein or veinlike fold in place of bifurcation of m_1 and m_2 present (Fig. 148): Sarcophagidae.

——Body metallic blue, green or purple: if not so, then thorax with wavy golden hair. Two notopleural bristles, vein m_2 entirely absent and not suggested by a fold of wing membrane (Fig. 129): Calliphoridae.

Scatopsidae

(Only genus *Scatopse* with several synanthropic spp.)

1. Vein cu_2 strongly curved backward before middle, m_1 simple; 1.5-2 mm: *Scatopse fuscipes* Meig.

——Vein cu_2 undulated in middle and then resuming its original direction, m_1 with a vestige of an accessory vein basally and broken in that place (Fig. 7); 2.5-3 mm: *Scatopse notata* (Linn.).

PHORIDAE

1. Abdomen of male hirsute with flattened bristles; body black, legs pale brown; 2-3 mm: *Megaselia rufipes* (Meig.).
——Abdomen of male with short clinging pubescence; body black, legs fulvous; 1-2 mm: *Megaselia pulicaria* (Fall.).

SYRPHIDAE

1. Vein r_{4+5} deeply curved inwardly (Fig. 10) 2
——Vein r_{4+5} nearly straight (Fig. 8). Hind-femora conspicuously incrassate, spiny on lower side, hind-tibiae strongly curved (Fig. 9); 7-9 mm: *Xylota pipiens* (Linn.).
2. R_1 cell open (*Helophilus* Meig.) 3
——R_1 cell closed 5
3. Facial vitta yellow or reddish; 15-18 mm: *Helophilus trivittatus* (Fabr.).
——Facial vitta black 4
4. Yellow maculae of 2nd abdominal tergite reaching its posterior margin; 13-16 mm: *Helophilus hybridus* Loew.
——Yellow maculae of 2nd abdominal tergite not reaching its posterior margin; 11-17 mm: *Helophilus pendulus* (Linn.).
5. Eyes without maculation (*Eristalis* Latr.) 6
——Eyes with distinct maculation 9
6. Arista practically bare; 15-19 mm: *Eristalis tenax* (Linn.).
——Arista pubescent 7
7. Abdomen with short but dense pubescence; 9-12 mm: *Eristalis intricarius* (Linn.).
——Abdomen with weak pubescence, nearly bare 8
8. Face without dark medial vitta; 10-12 mm: *Eristalis arbustorum* (Linn.).
——Face with a dark medial vitta; 11-13 mm: *Eristalis nemorum* (Linn.).
9. Maculae irregular, partly fused; 6-8 mm: *Eristalinus sepulchralis* (Linn.).
——Maculae isolated, small, regular; 10-12 mm: *Eristalis aeneus* (Scop.).

OTITIDAE

1. Antennae situated in sharply defined sockets. Head distinctly broader than thorax. Frons with distinct sculpturing; 3-4 mm: *Ulidia erythrophthalma* Meig.
——Antennal sockets shallow or absent 2
2. Veins r_{4+5} and m_{1+2} meet in a point in apex of wing. Body with green and blue metallic luster, thorax dull, abdomen glossy; 4 mm: *Physiphora demandata* (Fabr.).
——Veins r_{4+5} and m_{1+2} not converging to a point, wings maculate; 2.5-4 mm: *Euxesta* Loew.

Platystomatidae

1. Thorax: 2 sa and 2 pa present. Upper part of scutellum haired at least on sides. Wing as in Fig. 13; 6-10 mm: *Platystoma lugubre* R.-D.

——Thorax: 1 sa, 2 pa present. Scutellum bare above, all tarsi mostly black, 5th tergite of male longer than 3 preceding tergites, 3rd tergite of female as long or longer than 4th tergite; 4.5-7 mm: *Platystoma seminationis* Fabr.

Dryomyzidae

1. Vein r_1 setulose on obverse side. Body testaceous, wings infuscated on transverse veins and at the ends of longitudinal veins at apex (Fig. 15); 6-10 mm: *Dryomyza anilis* Fall.

——Vein r_1 bare. Wings translucent, 3rd antennal segment brown, tarsi black apically; 7-9 mm: *Dryomyza flaveola* (Fabr.).

Sepsidae

1. Apex of wing near origin of r_{2+3} with dark blotch (Fig. 19). (*Sepsis*) 5

——Apex of wing without a dark blotch 2

2. Mesopleura with a distinct seta (Fig. 17) 3

——Mesopleura and vte without a distinct seta, pvt present. (*Themira*) 15

3. ors distinctly developed (Fig. 17), fore-legs as in Fig. 17. Body black, legs testaceous; 3-4 mm: *Meroplius stercorarius* (R.-D.).

——ors reduced, body brown with dull bronze luster, partly fulvous or rufous (*Nemopoda*) 4

4. Male: 3rd trochanter with 2 black spines directed anteriorly (Fig. 18); female: 5th abdominal sternite less than twice as wide as 3rd sternite. Wing apex mostly slightly tinged; 3-4.5 mm: *Nemopoda nitidula* (Fall.).

——Male: 3rd trochanter without black spines; female: 5th abdominal sternite twice as wide as 3rd sternite. Wing apex mostly clear; 3.5-4.5 mm: *Nemopoda pectinulata* Loew.

5. Sternopleura shiningly black, pollinose only on upper margin (Fig. 19); 2.5-4 mm: *Sepsis thoracica* (R.-D.).

——Sternopleura dull, pollinose all over 6

6. Males 7

——Females 12

7. f_1 with a conspicuous tubercle with spines (Fig. 20) situated apically from the middle tubercle or group of spines 8

——f_1 with only 1 larger tubercle or 1 medial group of spines (Fig. 23) 10

8. f, t_2 and t_3 entirely black, 2nd abdominal tergite without longer hairs on sides. Body entirely black; 2.5-3.5 mm: ♂ *Sepsis fulgens* Hffmsgg.

——Legs mostly testaceous, frequently base of abdomen also testaceous, 2nd abdominal tergite with a group of longer hairs on sides 9

9. Prescutellar dc strong, another pair of dc in front of them absent or reduced. Fore-leg of male as in Fig. 21; tibia 3 of female with 1 ad below middle; 3.5-5 mm: ♂ *Sepsis punctum* (Fabr.).

——2 pairs of equally strong dc present, fore-leg of male as in Fig. 22, tibia 3 of female without ad; 3-4.5 mm: ♂ *Sepsis violacea* Meig.

10. t_1 with a deep excision in basal third (Fig. 23); 3-4 mm: ♂ *Sepsis cynipsea* (Linn.).

——t_1 not deeply excised 11

11. f_1 with a low tubercle on lower surface (Fig. 24), t_1 straight; 2-2.5 mm: ♂ *Sepsis orthocnemis* Frey.

——Basal 2/3 of f_1 incrassate, t_1 S-shaped; 2-3 mm: ♂ *Sepsis flavimana* Meig. and ♂ *Sepsis biflexuosa* Strobl.

12. Abdominal tergites, mainly the 4th and 5th, with distinct setae near posterior margins as in Fig. 19 13

——Abdominal tergites with setulae only 14

13. 2nd abdominal tergite without a tuft of longer hairs laterally, legs mostly black; 2.5-3.5 mm: ♀ *Sepsis fulgens* Hffmsgg.

——2nd abdominal tergite with a tuft of hairs laterally, legs mostly testaceous 15

14. t_3 with 1 ad in middle, f_2 without a medial seta anteriorly; 3-3.5 mm: ♀ *Sepsis cynipsea* (Linn.).

——t_3 without ad: ♀♀ *Sepsis orthocnemis* Frey, *Sepsis flavimana* Meig., and *Sepsis biflexuosa* Strobl.

15. Humeral bristles present, sternopleurae pollinose all over, legs partly testaceous; 2nd tarsus of male as in Fig. 26; 3-4 mm: *Themira annulipes* (Meig.).

——Humeral bristles absent 16

16. Jowls broader than 3rd antennal article; body 4-4.5 mm: 17

——Jowls not broader than 3rd antennal article, thorax black, shining; body 2-3.5 mm: *Themira lucida* Staeger.

17. Antennae black, arista incrassate basally, r_{4+5} and m_{1+2} nearly parallel apically; f_1 of male as in Fig. 27: *Themira putris* (Linn.).

——Antennae rufous, arista slightly incrassate basally, r_{4+5} and m_{1+2} strongly converging apically; f_1 of male as in Fig. 28: *Themira nigricornis* (Meig.).

Piophilidae

1. Dorsum of thorax feebly shining, with distinct coarse sculpture and 3 rows of setulae. Mesopleurae hairy. Male, 3.5-4 mm; female, 4-4.5 mm: *Piophila casei* (Linn.).

——Thorax mostly strongly shining and smooth or only sparsely hairy and feebly wrinkled, always without 3 rows of setulae; 2-4 mm. 2

2. h and prs invariably absent 3

——h and prs invariably present (Fig. 30) 6

3. Mesopleura bare, head entirely black; 2.5-4 mm: *Piophila foveolata* Meig.

——Mesopleura with scattered short setae, head often partly yellowish or reddish 4

4. cx_1 entirely black; 3-4 mm: *Piophila nigricornis* Meig.

——cx_1 partly or entirely yellowish 5

5. Frons yellowish anteriorly, broadly black posteriorly, jowls about as broad as 3rd antennal article, p_2 and p_3 invariably more or less infuscated; 2.5-3 mm: *Piophila varipes* Meig.

——Frons entirely yellowish, only narrowly black on ocellar triangle and around the eyes. Jowls 1.5 times as broad as 3rd antennal article, p_2 and p_3 often entirely yellow, anterior tarsi of female entirely black, those of male yellowish with blackish metatarsus as a rule; 2.5-3 mm: *Piophila nigrimana* Meig.

6. Only 1 pair of dc present (Fig. 30), mesopleura bare, shining, frons and jowls mostly yellowish to reddish; 2-3.5 mm: *Piophila vulgaris* Fall.

——4 pairs of dc present, fore-tarsi distinctly broader than middle and hind-tarsi (Fig. 32); 2-2.5 mm: *Piophila latipes* Meig.

Sphaeroceridae

1. Vein m does not reach the margin of wing, anal cell and hind-basal cell absent (Fig. 37) 12

——Vein m reaches the margin of wing distinctly, anal cell and hind-basal cell present (Figs. 33, 34) 2

2. tp meeting cu_1 in about 2/3 of length of the latter, i.e., far from margin of wing. t_3 with ventral apical spine curved (*Sphaerocera*) 3

——tp meeting cu_1 in close proximity to margin of wing (Fig. 34) 5

3. Oral margin distinctly protruded beyond a level of profrons, scutellum with 2 marginal teeth, t_3 strongly incrassate; 3 mm: *Sphaerocera curvipes* Latr.

——Oral margin deep behind the level of profrons. Scutellum with 6-16 marginal teeth 4

4. Vein m straight or slightly curved backward, thorax dull (sculptured) with glossy longitudinal vittae, scutellum with 6-10 teeth; 1 mm: *Sphaerocera pusilla* (Fall.).

——Vein m distinctly curved forward, thorax uniformly dull, scutellum with 14-16 teeth; 2-3 mm: *Sphaerocera pallidiventris* (Meig.).

5. t_3 without ventral apical spine. Body and legs black, knees testaceous; 2.5-3 mm: *Copromyza atra* (Meig.).

——t_3 with a ventral apical spine 6

6. Postocular setulae in several irregular rows, t_2 with 4-5 strong ad 7

——Postocular setulae in 1 row 9

7. t_3 with 1 conspicuous submedial av 8

——t_3 without a distinct submedial av. Suture between sternopleura and mesopleura with 2-3 arched hirsute bristles; 4-5 mm: *Copromyza* (*Stratioborborus*) *nitida* (Meig.).

8. Thorax dull, with many rows of ac. ta and tp not infuscated, mesopleura entirely dull; 2-4 mm: *Copromyza* (*Crumomyia*) *nigra* (Meig.).

——Thorax glossy, with 4 rows of ac; ta and tp more or less infuscated, mesopleurae glossy; 3 mm: *Copromyza* (*Crumomyia*) *glabrifrons* (Meig.).

9. t_3 with 1 av about in middle; scutellum, besides normal setae with only a very short pubescence 10

——t_3 without a conspicuous av, scutellum with long setulae among 4 marginal setae (Fig. 36) 11

10. 1 pair of dc, body feebly glossy, upper pleurae glossy. Frons with a reddish margin anteriorly: *Copromyza* (*Borborillus*) *hispanica* Duda.

——3 pairs of dc, body black, shining. Frontal triangle dull; 1.5-2 mm: *Copromyza* (*Borborillus*) *costalis* Zett.

11. Lower 2/3 of mesopleura glossy, non-pollinose, 5th sternite of male very long, pointed, and medially excised posteriorly. Body brownish-black, glossy; 4 mm: *Copromyza equina* Fall.

——Lower 1/2 of mesopleura non-pollinose, 5th sternite of male broadly rounded, not excised; 4-4.5 mm: *Copromyza similis* (Collin).

12. Anal vein slightly S-shaped, not broken (Fig. 37) 13

——Anal vein straight at base, broken in an obtuse angle in middle and with apical part convex with respect to anal margin of wing (Fig. 38) 17

13. t_2 with a strong ventral preapical seta and without a ventral apical seta. Costa basally long haired, mt_2 with 1 ventral basal seta, oral margin not protruded, body black, legs partly brown; 2-2.5 mm: *Leptocera fontinalis* (Fall.).

——t_2 without a strong ventral preapical seta. Presutural ac absent, 3rd antennal segment rounded, much broader than 2nd segment, arista inserted rather far from its apical part (subg. *Limosina*) 14

14. Junction of ta and tp conspicuously shorter than tp. Face dirty yellow, cx_1 yellowish; 1.5-1.7 mm: *Leptocera* (*Limosina*) *heteroneura* (Halid.).

——Junction of ta and tp distinctly longer than tp 15

15. Costa reaching far beyond the fork of r_{4+5}. Frons dull black, smooth around the ocelli, body black, with feeble grey pollinosity; 1-1.5 mm: *Leptocera* (*Limosina*) *flavipes* (Meig.).

——Costa reaching slightly beyond the fork of r_{4+5} 16

16. Frons entirely yellow. Thorax and abdomen black, legs yellow, only partly brown; 1.2-1.5 mm: *Leptocera* (*Limosina*) *ochripes* (Meig.).

——Frons more or less brown to black, legs reddish-brown to black, knees brown. t_1 of male strongly incrassate, fore- and hind-tarsi strongly enlarged; 1.5-1.8 mm: *Leptocera* (*Limosina*) *clunipes* (Meig.).

17. Scutellum densely hairy dorsally 18

——Scutellum bare dorsally, ta-tp only 1/3 as long as tp; 0.5-0.8 mm: *Leptocera* (*Elachisoma*) *aterrima* (Halid.).

18. 2nd costal section longer than 3rd section 19

——2nd costal section shorter than 3rd section or both of them equal in length 21

19. r_{4+5} feebly curved toward anterior margin of wing all along its length. Wings intensely brown, thorax reddish to dark brown; 1.2-1.5 mm: *Leptocera* (*Coproica*) *ferruginata* (Stenh.).
——r_{4+5} practically straight from ta up to wing margin 20

20. ta-tp longer than tp. Male with abnormality of venation: tp very oblique and posterior margin of wing with a notch bearing long hairs. Legs and body blackish; 1.2 mm: *Leptocera* (*Coproica*) *acutangula* (Zett.).
——ta-tp as long as tp. Venation of male normal; 1.2 mm: *Leptocera* (*Coproica*) *vagans* (Halid.).

21. r_{4+5} all along its length feebly curved toward the anterior margin of the wing. Jowls strongly enlarged posteriorly. Frons reddish-brown anteriorly as a rule; 0.8-1 mm: *Leptocera hirtula* (Rond.).
——r_{4+5} straight. Jowls only feebly enlarged posteriorly. Frons mostly entirely black; 1-1.2 mm: *Leptocera* (*Coproica*) *lugubris* (Halid.).

Milichiidae

1. Costa ends near the fork of r_{4+5}, proboscis with short labellae. Body 1-2 mm long. Wing as in Fig. 39: *Meoneura* Rond. (*M. obscurella* [Fall.], *M. vagans* [Fall.] and others).
——Costa ends near the mouth of m_{1+2}, proboscis long, with thin labellae 2

2. Thorax and abdomen black, shining, non-pollinose; 2-3 mm: *Madiza glabra* Fall.
——Thorax and frontal vitta dull or pollinose; 1-2 mm: the genera *Leptometopa* Becker and *Desmometopa* Loew.

Drosophilidae (See also *Drosophila* key on page 117.)

1. Acrostical setulae in 6 rows, r_{2+3} apically distinctly curved forward, abdomen brown with enlarged brownish-black maculae; 3-4 mm: *Drosophila funebris* (Fabr.).
——Acrostical setulae in 8 rows, r_{2+3} apically nearly straight, abdomen fulvous with brownish-black marginal maculae; about 2 mm: *Drosophila melanogaster* Meig.

Helomyzidae

1. Anal vein never reaching margin of wing (Fig. 43). Humeral bristle absent, body usually yellow to pale ferrugineous; 5-9 mm: *Suillia*, e.g., *S. pallida* (Fall.), *S. rufa* (Fall.), *S. lurida* (Meig.).
——Anal vein reaching margin of wing 2

2. Prosternum bare. Vein r_1 meeting costa opposite to ta (Fig. 44). t_2 with several apical setae, lunula exposed, 1+3 dc. A row of minute setulae, in front of sternopleural seta; scutellum and end of abdomen partly rufous; 3.5-4 mm: *Neoleria inscripta* (Meig.).
——Prosternum with 1 pair of long setae. Humeral bristle absent, pteropleurae partly haired, t_2 of male with curved apical spines: *Scoliocentra* Loew.

SCATOPHAGIDAE (Treated elsewhere in book under *Anthomyiidae.*)

1. Arista distinctly haired, 3rd antennal segment totally black, environs of ta distinctly infuscated. Body and legs of male with dense golden pubescence; 6-10 mm: *Scatophaga stercoraria* (Linn.).

——Arista nearly bare, 3rd antennal segment rufous, both transverse veins infuscated, postabdomen orange; 4-8 mm: *Scatophaga furcata* (Say).

ANTHOMYIIDAE

1. Propleura with hairs, thorax pale grey and with 4 or 5 contrasting large spots or bands on dorsum 2

——Propleura without hairs, thorax marked otherwise 3

2. 2 black maculae behind the suture (Fig. 45); 5-6 mm: *Anthomyia plurinotata* Brullé.

——3 black maculae behind the suture (Fig. 46); 5-7 mm: *Anthomyia pluvialis* (Linn.).

3. Each tergite with 3 triangular maculae; 3-4 mm: *Calythea albicincta* (Fall.).

——Pattern of abdomen less contrasting, as a rule only a dark medial vitta or pearly irregular maculae 4

4. Thorax with 4 dark vittae, the inner ones narrow and the outer ones broad, lower calyptral scale distinctly protruded beyond margin of upper. Base of wing yellow, tibiae fulvous; 6-11 mm: *Hydrophoria conica* (Wied.).

——Different combination of characters 5

5. Arista plumose or long pubescent 6

——Arista short plumose or bare 9

6. Cross-vein clouded, arista long pubescent (Fig. 48); 5-6 mm: *Hylemya* (*Craspedochaeta*) *pullula* (Zett.).

——Cross-vein not clouded, arista long plumose 7

7. Legs black or largely fuscous 8

——Legs more or less testaceous, frons reddish anteriorly; 7-9.5 mm: *Hylemya* (*Hylemya*) *strigosa* Fabr.

8. Hind-tibia with a short distinctive apical pv, distance between rows of acr equal to that between acr and dc, acr with interstitial microsetae. Antenna as in Fig. 49; 4-4.5 mm: *Hylemya* (*Hylemyza*) *lasciva* (Zett.).

——Hind-tibia without apical pv, rows of acr approximated, without interstitial microsetae; 5-6.5 mm: *Hylemya* (*Hylemya*) *variata* (Fall.).

9. Oral margin distinctly protruded beyond a level with tip of profrons (Fig. 52) (*Hylemya*) 10

——Oral margin not distinctly protruded beyond a level with tip of profrons 13

10. Between notopleural bristles there is a group of setulae (Fig. 50); frons above antennae broadly orange (not invariably so in males),

acr hairlike beyond the suture, in 2-4 irregular rows (Fig. 51); 4-6 mm: *Hylemya radicum* (Linn.). (*H. radicum* subsp. *montalpina* Gregor and Povolný is larger and darker, with microsetae of body 1/3 longer than those of the nominate form.)

——No hairs between notopleural bristles 11

11. acr behind the suture 1/2 as long as dc but strong and in 2 rows, abdomen of male cylindrical, slightly compressed apically 12

——acr behind the suture hairlike and irregularly situated, abdomen of male more or less flattened. Body blackish-grey; 4-5 mm: *Hylemya caerulescens* Strobl.

12. Body black, feebly pollinose, with dull blue luster; 3.5-4.5 mm: *Hylemya grisella* Rond.

——Body with very dense yellowish-grey pollinosity; acr as in Fig. 53; 3-5 mm: *Hylemya cinerella* (Fall.).

13. Legs black, hind-tibia of male with a continuous series of pv setulae from basal to preapical region (Fig. 55); frons of female orange anteriorly. Head as in Fig. 54: *Delia platura* (Meig.).

——Tibiae rufous or fulvous. Head as in Fig. 56: *Pegomya socia* (Fall.).

Muscidae

1. Anal vein 2a curved around the apex of vein cu_{1b}+la (Fig. 70) which ends in middle of the distance between transverse vein z and margin of wing. Abdomen non-maculate, except in *Fannia leucosticta* (Meig.). Female: frontal vitta invariably without cruciate bristles, parafrontals broad and convex toward the middle of frons (Fig. 69). Body length exceptionally more than 7 mm: Fanniinae.

——Anal vein 2a not curved as above; if so (*Azelia*, Mydaeinae), then abdomen with maculae (Figs. 57, 58), t_3 without pd, frontal vitta of male without cruciate bristles 2

2. Lower scale of calyptrae enlarged mesad so as to impinge on base of scutellum, caudal margin transverse (Fig. 96). Vein m_{1+2} curved forward at apical region (Figs. 93, 113) (except *Polietes*, Fig. 99): Muscinae.

——Lower calyptral scale not enlarged mesad so as to impinge on base of scutellum, caudal margin usually semicircular. Vein m_{1+2} straight or feebly curved, exceptionally distinctly curved forward (gen. *Muscina*, *Myospila*, *Graphomya*) 3

3. Hind-tibia with 1 or more conspicuous posterodorsal bristles ("calcar") as long or longer than anterodorsal bristles (exceptions in genus *Hydrotaea*): Phaoniinae.

——Hind-tibia without posterodorsal bristles, or if present they are not longer than anterodorsal bristles: Mydaeinae.

Mydaeinae

1. Sternopleural bristles typically arranged in the form of an equilateral triangle, or nearly so. Front of male like that of female. Predatory, asynanthropic species: *Coenosia*.

——st not arranged in an equilateral triangle; if 3 st are present, they are 1:2. Front of male distinctly narrower than that of female 2

2. Proclinate ors and cruciate bristles present in female and as minute setulae in male. Hind-coxae with caudal setulae on inner margin. Abdomen maculate (Fig. 57). Body 2.5-5 mm: (*Azelia*) 3

——Proclinate ors and cruciate bristles absent, hind-coxae without caudal setulae. Abdomen maculate or not. Body normally much longer than 4 mm 4

3. Male: hind-tibia with an extensive series of ad bristles, mid-tibia with preapical ad (Fig. 59); 3-5 mm: *Azelia macquarti* (Staeger).

——Male: hind-tibia bare on pv surface and with 1 or 2 av on preapical region (Fig. 60), mid-tibia with no mid-post bristle; 3-4 mm: *Azelia triquetra* (Wied.).

4. Prealar bristle and cruciate bristles absent, arista hairs not longer than 1/2 the width of 3rd antennal segment, m vein strongly curved toward r_{4+5}, abdomen many-colored (Fig. 61); 5.5-6.5 mm: *Gymnodia impedita* Pand.

——No such combination of characters 5

5. Wings with setulae on upper surface of basal node to veins r_{2+3} and r_{4+5} 6

——Upper surface of basal node to above veins bare 13

6. Veins m_{1+2} curved forward in apical region (Fig. 63) 7

——Veins m_{1+2} not curved forward in apical region (*Mydaea*) 10

7. Caudal margin of eyes emarginate at middle. Body with marked black maculation (Fig. 62); 6-9 mm: *Graphomya maculata* (Scop.).

——Caudal margin of eyes not emarginate (*Myospila*) 8

8. Longest hairs on arista 1.5-2 times as long as the width of 3rd antennal segment, t_2 with 2 post and 1 ad. Male: frontal vitta 0.19-0.26 mm. Female: paired blotches on abdomen present; 5-9 mm: *Myospila meditabunda meditabunda* (Fabr.).

——Longest hairs on arista about as long as the width of 3rd antennal segment 9

9. Male: frontal vitta 0.08-0.15 mm, basis of wing slightly infumated. Female: abdomen without paired blotches, t_1 without post, t_2 without ad; vein m_{1+2} as in Fig. 63; 5.5-10 mm: *Myospila hennigi* Gregor and Povolný.

——Male: frontal vitta 0.20-0.26 mm, basis of wing distinctly fuscous. Female: abdomen with more or less distinct paired blotches, t_1 with 1-2 post, t_2 with 1-3 ad; 6-10 mm: *Myospila meditabunda alpina* (Hendel).

10. Scutellum more or less fulvous or reddish; 7-9 mm: *Mydaea scutellaris* R.-D.

——Scutellum blackish 11

11. Arista long pubescent, wings in male infuscated basad; 7-9 mm: *Mydaea deserta* (Zett.).

——Arista plumose, hairs longer than 1/2 width of 3rd antennal segment 12

12. Legs blackish, tibiae at most rufous; 7-8 mm: *Mydaea ancilla* (Meig.) (and additional species).

——Legs more or less fulvous; 7.5-10.5 mm: *Mydaea urbana* (Meig.) (and additional species).

13. Prealar bristle absent or setulose, abdomen without paired marks, arista plumose, legs weakly and sparsely bristled (Fig. 64), head evenly flattish across both frons and eyes, parafacials largely obscured from view by anterior margin of eye when viewed from profile. Frontal triangle of female extended to anterior part of interfrontalis (*Hebecnema*) 14

——Species not having the above combination of characters (*Helina*).
(Some 80 species in the Palearctic region; determination of females for the most part very difficult.) Key to males 15

14. Eyes haired, sparsely so in female. Male abdomen subovate, densely pearlaceous-grey, in female speckly and more or less pruinescent; 3-5 mm: *Hebecnema umbratica* (Meig.).

——Eyes bare. Abdomen in male subconical, thinly pruinescent, in female mainly glossy and devoid of pruinescence; 3.5-5 mm: *Hebecnema affinis* Malloch.

15. Legs blackish, tibia at most partially rufous 16

——At least t_2 and t_3 fulvous, 3 dc 18

16. Hypopleura below metathoracic stigma with many fine hairs (Fig. 66) 17

——Hypopleura bare, pra short and weak, eyes bare, t_2 with 1-2 ad; 5-7.5 mm: *Helina duplicata* (Meig.).

17. Scutellum finely haired on sides and entire lower side. f_3 below posteriorly with long preapical setae; 6-8.5 mm: *Quadrularia laetifica* (R.-D.).

——Scutellum bare on lower side, f_3 below posteriorly with long setae for nearly entire length; 6-8.5 mm: *Quadrularia annosa* (Zett.).

18. Prealar bristle long and stout. Abdomen with 2 pairs of distinct maculae, legs nearly entirely testaceous; 5-8 mm: *Helina depuncta* (Fall.).

——pra reduced. Abdomen translucent yellowish, thorax and abdomen with distinct dark maculae; 6-7 mm: *Helina punctata* (R.-D.).

Fanniinae

1. Arista plumose, male with both pairs of orbital bristles, no head dimorphism. Legs yellowish, abdomen ferruginous to testaceous; 4-6 mm: *Platycoenosia mikii* Strobl.

——Arista bare or pubescent, male without lower orbital bristles, head dimorphism present (*Fannia*) 2

2. Abdomen yellowish transparent basally 3

——Abdomen not yellowish basally 4

3. Propleuron setulose, normally 1 av on t_3. Thorax glossy, blackish. Hypopygium with numerous long fine bristles at cercal plate; 5-6 mm: *Fannia difficilis* (Stein.).

——Propleuron bare, t_3 with 2-3 av. Body distinctly pollinose, thorax with vittae. Hypopygium with only short bristles at sur-stylus (Fig. 68); 5-7 mm: *Fannia canicularis* (Linn.).

4. t_2 and t_3 reddish to yellow. Frontal vitta of male as broad as either parafrontal; 5.5-7 mm: *Fannia fuscula* (Fall.).
——t_2 and t_3 entirely fuscous or blackish 5

5. Abdomen with a triangular-sided vitta and dark lateral maculations (Fig. 71); 3-3.5 mm: *Fannia leucosticta* (Meig.).
——Abdomen unicolor or with a triangular or parallel-sided vitta 6

6. Males 7
——Females 15

7. cx_2 with hooklike spines on lower part (Fig. 78) 8
——cx_2 without hooklike spines 10

8. cx_2 with 3 hooklike spines, t_1 without a cluster apically (Fig. 78); 6-7 mm: ♂ *Fannia scalaris* (Fabr.).
——cx_2 with only 1 (lower) spine hooklike. Base of t_1 testaceous, apex with a distinct posterior cluster of flattened bristles (Fig. 72) 9

9. t_3 with a row of pv bristles; 5.5-7.5 mm: ♂ *Fannia manicata* (Meig.).
——t_3 without a complete row of pv; 4-5.5 mm: ♂ *Fannia monilis* (Hal.).

10. mt_2 with a ventral crest or thorn of fused bristles at the base (Fig. 73). t_2 incrassate and slightly curved; 4.5-5.5 mm: ♂ *Fannia armata* (Meig.).
——mt_2 without any crest or spine 11

11. Hind-coxae haired posteroventrally 12
——Hind-coxae bare posteroventrally 14

12. t_3 with 5 or more av and several distinct ad and pv. Prealar bristle little differentiated from the thoracic setulae. Thorax with distinct vittae; 6.5-7.5 mm: ♂ *Fannia incisurata* (Zett.).
——t_3 with only 1-2 av, prealar bristle distinctly stronger than the thoracic setulae 13

13. t_3 with 1-3 pv on basal two-thirds. Squamae dark yellow or brown; (t_2 as in Fig. 74); 4.5-5.5 mm: ♂ *Fannia mutica* (Zett.).
——t_3 without bristles on posterior surface. t_2 1/3 as long as tibial diameter, pv on f_3 erect and 2/3 as long as femoral diameter on basal half, reduced to prostrate setulae on apical 3rd, t_3 with 1-2 av, 5 ad. Hypopygium (Fig. 68) very similar to that of *F. canicularis*; 4.5-6.5 mm: ♂ *Fannia subpubescens* Collin.

14. Parafacials very narrow, hardly visible in lateral view. Hypopygium spherical; 3.5-4 mm: ♂ *Fannia parva* Stein.
——Parafacials broader, distinctly visible in lateral view. Hypopygium flattened; 4-4.5 mm: ♂ *Fannia serena* (Fall.).

15. Hind-coxae haired posteroventrally 16
——Hind-coxae bare posteroventrally 21

16. t_2 ventrally with 1 distinct bristle (av), occiput with 2 oval fields of silvery hairs; 4.5-5.5 mm: ♀ *Fannia mutica* (Zett.).
——t_2 medially without a ventral bristle, occiput without silvery blotch 17

17. Parafacials with 1 or more minute hairs situated below a level with basal margin of 3rd antennal segment. t_3 with several ad, body weakly pollinose, thorax nearly non-vittate; 4.5-6.5 mm: ♀ *Fannia subpubescens* Collin.

——Parafacials not haired below a level with basal margin of 3rd antennal segment 18

18. t_1 with 1 fine ad, thorax grey, not glossy 19

——t_1 absent, thorax black, subglossy, basis of t_1 distinctly reddish 20

19. Scutellum completely setulose, basis of f_2 ventrally without a long bristle; 6-7.5 mm: ♀ *Fannia incisurata* (Zett.).

——Scutellum bare basally, 1 long ventral bristle present on basis of f_2 (Fig. 77); 5.5-7 mm: ♀ *Fannia scalaris* (Fabr.).

20. Body 4-5.5 mm, t_3 with a nearly complete row of ad setae: ♀ *Fannia monilis* (Hal.).

——Body 5.5-7.5 mm, t_3 with a partial series of ad setae on proximal half: ♀ *Fannia manicata* (Meig.).

21. Parafrontals shiny; 4-4.5 mm: ♀ *Fannia serena* (Fall.).

——Parafrontals evenly pollinose; 3.5-4 mm: ♀ *Fannia parva* Stein.

PHAONIINAE

1. Hind-coxae with setulae on caudal surface (Fig. 79). Legs, palpi and basis of antennae fulvous, abdomen largely fulvous with a dark medial stripe; 7-8 mm: *Lasiops simplex* (Wied.).

——Hind-coxae without setulae on caudal surface 2

2. t_3 with several pd of equal length and strength standing over one another, subgenal sclerite of head upwardly extended, invading and thereby restricting the dimension of the bare cheek or genal sclerite. Body black, thorax with brown pollinosity (striped in the female), abdomen with grey pollinosity; 6-7 mm: *Trichopticoides decolor* (Fall.).

——t_3 with a single (or without) strong pd on distal half ("calcar") 3

3. st 1:1, arista bare or pubescent 4

——st 1:2 as a rule; if 1:1, then arista pubescent 25

4. Dorsum of thorax entirely black and glossy, eyes nearly as high or higher than length of fore-tibia, female with frontal triangle extended to anterior half of interfrontalia (Fig. 80) (*Ophyra*) 5

——Body more or less pubescent and pollinose as a rule, frontal triangle of female normally reaching the middle of front only, male fore-femur notched or excavated on preapical region of ventral surface (*Hydrotaea*) 6

5. Lower squama brown, t_3 of male as in Fig. 81; 4-7.5 mm: *Ophyra leucostoma* (Wied.).

——Lower squama white or yellowish, t_3 of male as in Fig. 82; 4-5.5 mm: *Ophyra capensis* (Wied.).

6. Males 7

——Females 17

7. f_3 basally spined on ventral surface (Fig. 83), eyes with numerous hairs; 4.5-6 mm: ♂ *Hydrotaea occulta* (Meig.).
——f_3 not spined as above, eyes mostly bare 8

8. Eyes plumose 9
——Eyes bare 10

9. t_3 as in Fig. 84; 5-6 mm: ♂ *Hydrotaea penicillata* (Rond.).
——t_3 as in Fig. 85; 5.5-8 mm: ♂ *Hydrotaea cyrtoneurina* (Zett.).

10. Abdomen more or less yellowish basally 11
——Abdomen entirely dark 12

11. t_3 as in Fig. 86; 6-7.5 mm: ♂ *Hydrotaea pellucens* Porch.
——t_3 as in Fig. 87; 6-7 mm: ♂ *Hydrotaea meridionalis* Porch.

12. f_3 below in middle with 1 stout blunt seta (Fig. 90); 4.5-6 mm: ♂ *Hydrotaea armipes* (Fall.).
——f_3 in middle without such seta 13

13. Orbitae distinctly separated by a black frontal vitta, ori reaching ocellar triangle 14
——Orbitae touching in narrowest place, the row of ori ends far before ocellar triangle 15

14. f_1 below basally with several short thorns, t_3 with 2-3 (-4) av, frontal vitta as broad or broader than the distance between outer margins of both upper ocelli; 6-8 mm: ♂ *Hydrotaea dentipes* (Fabr.).
——f_1 below basally without short thorns, t_3 with 6-8 av, frontal vitta distinctly narrower than the distance between outer margins of upper ocelli; 7.5-9.5 mm: ♂ *Hydrotaea similis* Meade.

15. f_3 straight, f_1 as in Fig. 91; 4-6 mm: ♂ *Hydrotaea meteorica* (Linn.).
——f_3 curved 16

16. mt_2 below with a coarse brushlike pubescence, t_3 as in Fig. 88; 5.5-7 mm: ♂ *Hydrotaea irritans* (Fall.).
——mt_2 below with fine sparse hairs, t_3 as in Fig. 89; 5.5-7 mm: ♂ *Hydrotaea pandellei* Stein.

17. Abdomen basally more or less transparently rufous or fulvous; 6-7.5 mm: ♀ *Hydrotaea pellucens* Porch. or ♀ *Hydrotaea meridionalis* Porch.
——Abdomen dark including the base 18

18. Halteres yellowish-white; 5-7 mm: ♀ *Hydrotaea irritans* (Fall.) or ♀ *Hydrotaea penicillata* (Rond.).
——Knobs of halteres more or less brownish to black apically 19

19. t_2 without ad 20
——t_2 with 1 ad 24

20. Parafrontalia and parafacialis silvery pollinose except for a black glossy blotch opposite the antennae; 4.5-6 mm: ♀ *Hydrotaea occulta* (Meig.).
——Parafacialia and parafrontalia uniformly pollinose even opposite the antennae 21

21. t_3 without pd; 4-5 mm: ♀ *Hydrotaea armipes* (Fall.).
t_3 with 1 pd 22

22. Distinct presutural acr absent; 5.5-7 mm: ♀ *Hydrotaea pandellei* Stein.

——Presutural acr present 23

23. t_3 with 1 av bristle, acr distinctly shorter than dc; 4-6 mm: ♀ *Hydrotaea meteorica* (Linn.).

——t_3 with 4-5 av bristles, acr nearly as long and stout as dc; 5.5-8 mm: ♀ *Hydrotaea cyrtoneurina* (Zett.).

24. t_3 with 5-6 av; 7.5-9.5 mm: ♀ *Hydrotaea similis* Meade.

——t_3 with 2 (1)-3 av; 6-8 mm: ♀ *Hydrotaea dentipes* (Fabr.).

25. Vein m_{1+2} curved forward in apical region, scutellum invariably more or less ferruginously translucent on apex (*Muscina*) 26

——Vein m_{1+2} straight 29

26. Legs more or less rufous or fulvous. Wing as in Fig. 93; 6-8.5 mm: *Muscina stabulans* (Fall.).

——Legs blackish 27

27. Palpi fulvous, vein r_{4+5} slightly curved upward (Fig. 94); 6.5-9 mm: *Muscina assimilis* (Fall.).

——Palpi black, vein r_{4+5} strongly curved (Fig. 95) 28

28. Lower squama as in Fig. 96. Abdomen with abundant greyish pollinosity; 7-9 mm: *Muscina pabulorum* (Fall.).

——Lower squama as in Fig. 97. Abdomen more or less glossy; 7-10 mm: *Muscina pascuorum* (Meig.).

29. Cheeks with 1 or 2 strong upcurved buccal bristles; female with anterior pair of paraorbital bristles strong and proclinate (Fig. 98). Body black, with grey pollinosity, abdomen with a broad black median vitta; 6-8 mm: *Dendrophaonia querceti* (Bouché).

——Cheeks without notably strong upcurved buccal bristles, female with anterior pair of paraorbital bristles not strong or absent (*Phaonia*) 31

30. Body mostly fulvous; 5-7.5 mm: *Phaonia pallida* (Fall.).

——Body mostly dark: *Phaonia viarum* R.-D., *P. errans* (Meig.), *P. variegata* (Meig.), *P. basalis* (Zett.), and others.

Muscinae

1. Proboscis adapted for piercing, not retractile (Fig. 115). Arista with unequal pilosity: long haired above and partly haired or bare below. (Stomoxydini) 24

——Proboscis with developed labellae, retractile. Arista with equal pilosity above and below 2

2. Vein m_{1+2} straight up to the margin of the wing (Fig. 99) (*Polietes*) 3

——Vein m_{1+2} arched or angled 4

3. Sternopleural bristles 1:1, halteres black, m_{1+2} as in Fig. 99; 7-12 mm: *Polietes lardaria* (Fabr.).

——st 1:2 (-4), halteres yellow, the terminal section of m longer than the preceding one; 5-7.5 mm: *Polietes albolineata* (Fall.).

4. Vein m_{1+2} rounded at its bend (Fig. 100) 5
——Vein m_{1+2} angularly rounded at its bend (Fig. 113) 22

5. Basal region of wings and both calyptral scales densely yellow. Mesonotum, scutellum, and abdomen with very fine vestiture (*Mesembrina*) 6
——No such combination of characters 7

6. Thorax and abdomen uniformly black; 9-13 mm: *Mesembrina meridiana* (Linn.).
——Thorax fulvous anteriorly, abdomen with yellowish pilosity posteriorly; 12-18 mm: *Mesembrina mystacea* (Linn.).

7. t_2 without pv bristle (*Morellia*) 8
——t_2 with a distinct pv bristle 10

8. Prosternum with marginal hairs; 5-8 mm: *Morellia simplex* (Loew).
——Prosternum bare 9

9. Male: t_2 as in Fig. 101. Female: next to the series of long and stout pv there is no accessory series of hairs (Fig. 102). Abdomen less pollinose on posterior margins of tergites, more or less blue glossy: 6-10 mm: *Morellia podagrica* (Loew).
——Male: t_2 as in Fig. 103. Female: long accessory hairs present next to the series of long pv (Fig. 104). Abdomen with distinct pearly maculae and normally with a brownish-green luster; 5-9 mm: *Morellia hortorum* (Fall.).

10. r_{4+5} on lower side of wing haired up to ta, stem vein distinctly haired below. Eyes mostly long and densely haired (*Dasyphora*) 11
——r_{4+5} on lower side of wing bare, stem vein bare or nearly so. Eyes bare or indistinctly setulose (*Pyrellia*) 21

11. ad on t_2 situated in middle between pv and end of tibia (Fig. 105). Eyes almost bare. Thorax with blue metallic luster and feeble silvery striation; 5.5-9 mm: *Pyrellia cyanicolor* Zett.
——ad and pv on t_2 on the same level. Eyes distinctly haired 12

12. Males 13
——Females 18

13. mt_3 with a basal tuft of hairs situated on a tubercle (Fig. 106); 8-10.5 mm: ♂ *Dasyphora penicillata* (Egger), syn. *D. versicolor* (Meig.).
——mt_3 normal 14

14. Mesonotum without acr in front of the suture 15
——Mesonotum with 1-2 strong acr in front of the suture 17

15. Mesonotum in front of the suture only with microsetae of equal length between the dc (Fig. 107). Body metallic and glossy but distinctly pollinose; 7.5-9.5 mm: ♂ *Dasyphora pratorum* (Meig.).
——Mesonotum in front of the suture between the dc with 2 series of shorter macrosetae standing on inner margins of dark longitudinal vittae (Fig. 108). Body metallic and glossy, feebly pollinose 16

16. Frontal vitta as broad as 3rd antennal segment. Wings slightly infumated, squama white or yellowish; 7.5-8 mm: ♂ *Dasyphora eriophthalma* (Macq.).

——Frontal vitta 1/2-1/3 as broad as 3rd antennal segment. Squama and wings at base and anterior margin slightly infuscated; 5.5-7.5 mm: ♂ *Dasyphora cyanella* (Meig.).

17. In front of the suture there are 1-2 pairs of acr, marginal bristle of 3rd tergite nearly as long as 4th tergite. Body rather dark, bronze olive green in color; 7-9 mm: ⚥ *Dasyphora albofasciata* (Macq.), syn. *D. saltuum* Rond.

——In front of the suture there is invariably one 1 pair of acr, marginal bristle on 3rd tergite at most 1/2 as long as 4th tergite. Body bronze green or golden green, abdomen with very dense yellowish-grey or golden grey pollinosity; 6.5-8 mm: ⚥ *Dasyphora asiatica* Zimin.

18. No distinct acr in front of the suture 19

——1-2 pairs of distinct acr in front of the suture 17

19. Mesonotum in front of the suture between the dc only with microsetae of equal length, abdomen densely pollinose: ♀ *Dasyphora pratorum* (Meig.), body mostly with bluish-green luster, 7.5-9.5 mm; ♀ *Dasyphora penicillata* (Egger), body mostly with greenish-blue or blue luster; 8-10.5 mm. (Reliable separation difficult).

——Mesonotum in front of the suture between the dc with 2 series of rather short macrosetae standing on inner margins of dark longitudinal vittae. Body metallic, glossy, feebly pollinose 20

20. t_2 above ad with 1-2 short setae; 7.5-8 mm: ♀ *Dasyphora eriophthalma* (Macq.).

——t_2 above ad without setae; 5.5-7.5 mm: ♀ *Dasyphora cyanella* (Meig.).

21. Anterior spiracles brown, body totally non-pollinose, metallic golden green to bluish-green; 4-7 mm: *Pyrellia cadaverina* (Linn.).

——Anterior spiracles white or pale yellow, mesonotum with feeble silvery medial vitta; 4.5-6.5 mm: *Pyrellia ignita* R.-D.

22. Body intensely golden green, glossy, without pollinosity. Mid-tibia with mid-ventral bristle (*Orthellia*) 23

——Body with more or less distinct grey pollinosity, frequently partly fulvous. Mid-tibia without mid-ventral bristle (*Musca*, see key to adults, p. 110).

23. Mesonotum with 3 pairs of postsutural dc, presutural acr present; 5-8 mm: *Orthellia caesarion* (Meig.).

——Mesonotum with 4 pairs of postsutural dc, presutural acr absent; 6-9 mm: *Orthellia cornicina* (Fabr.).

24. Only 1 (posterior) sternopleural bristle present, proboscis and palpi as in Fig. 115; 5-7 mm: *Stomoxys calcitrans* (Linn.).

——2 sternopleural bristles present, palpi about as long as proboscis 25

25. Arista below with 1 or more setae (Fig. 116), notopleura haired; 5-6 mm: *Lyperosiops stimulans* (Meig.).

——Arista hairy only above, notopleura bare 26

26. Anal vein short, ending before 1/3 of the distance from the margin of wing. Hind-tarsi of male normal (Fig. 117); 2.5-4 mm: *Haematobia titillans* (Bezzi).

——Anal vein ends 2/3 of the distance from the margin of wing. First 3 segments of hind-tarsi of male flattened and lobate, with longer hairs (Fig. 118); 2.7-4.5 mm: *Haematobia irritans* (Linn.).

CALLIPHORIDAE

1. Base of the radius before the humeral cross vein ciliated posteriorly above (Fig. 119) (Chrysomyinae) 2

——Base of the radius bare posteriorly above (exception *Pollenia atramentaria* [Meig.]). 8

2. Hind-coxae pilose posteriorly, lower part of head yellow to reddish orange 3

——Hind-coxae bare posteriorly, lower part of head predominantly dark 6

3. Lower squama pilose above. Body compact, head large, distinctly concave posteriorly. Body metallic, with more or less uniform but feeble pollinosity 4

——Lower squama bare above. Head not distinctly concave posteriorly. Thorax with dense pollinosity and with 3 dark bands; abdomen with conspicuous shifting pattern of pollinosity (non-palearctic) 5

4. Anterior spiracles white, stigmatic bristle (Fig. 120) absent normally; body metallic green, posterior margins of abdominal segments dark; 6-12 mm: *Chrysomya albiceps* (Wied.).

——Anterior spiracles brown to black, stigmatic bristle invariably present. Body metallic bluish-green; 8-12 mm: *Chrysomya megacephala* (Fabr.).

5. Parafrontal with black hair anteriorly, outside frontal row of bristles. Body deep bluish-black with partial green or purple luster; 8-10 mm: *Cochliomyia hominivorax* (Cocquerel).

——Parafrontal with pale hair anteriorly, outside frontal row of bristles. Body normally deep metallic green; 6-9 mm: *Cochliomyia macellaria* (Fabr.).

6. Arista long-haired 7

——Arista short, pubescent. Thorax flattened, mesothoracic spiracle much enlarged. Body black with bluish luster; 6-9 mm: *Boreellus atriceps* (Zett.).

7. Mesonotum flattened on disc, preacrostical bristles reduced. Body black with bluish or greenish-blue luster, front of male broad; 8-12 mm: *Protophormia terraenovae* (R.-D.).

——Mesonotum convex, preacrostical bristles well developed. Body dark green metallic, yellowish-green or purple in places. Eyes of male subcontiguous; 8-10 mm: *Phormia regina* (Meig.).

8. Non-metallic species. Propleura bare, thorax with long wavy golden or brown hair in addition to the black setae and setulose hairs (the wavy hair is easily rubbed off on dorsum). (A taxonomically difficult group, particularly females) (*Pollenia*) 9

——Metallic species, propleura pilose 13

9. Abdomen glossy black, without any conspicuous pollinosity 10

——Abdomen with conspicuous pollinosity, which is adpressed in different directions so as to create shifting pattern 11

10. Stem vein finely haired, R_5 normally petiolate (Fig. 121). Basicosta black; 8-11 mm: *Pollenia atramentaria* (Meig.).

——Stem vein bare, R_5 normally open, not petiolate. Basicosta black; 4-12 mm: *Pollenia vespillo* (Fabr.).

11. Basicosta pale to dull testaceous, t_2 normally with 2 pv; 5-12 mm: *Pollenia rudis* (Fabr.).

——Basicosta dark brown or black 12

12. Jowls with adpressed black hairs, male with very long iv; 9-14 mm: *Pollenia vera* Jacentk.

——Jowls hirsute with pale hairs, male with short iv; 5-10 mm: *Pollenia varia* (Meig.) and *Pollenia intermedia* Macq.

13. Head conspicuously golden yellow, front wide, protruding. 1 pair of acr in front of the scutellum as a rule; 9-15 mm: *Cynomya mortuorum* (Linn.).

——No such combination of characters 14

14. Body metallic blue, shining, with pale pollinosity forming stripes on thorax and more or less distinct mother-of-pearl blotches on abdomen. Lower squama normally haired (*Calliphora*) 15

——Body metallic green or blue, without distinct pollinosity. Lower squama bare (*Lucilia, Phaenicia*) 19

15. Basicosta yellow to yellow-brown, anterior thoracic spiracle orange. Bucca reddish except for posterior third of occipital part; 5-12 mm: *Calliphora vicina* R.-D.

——Basicosta black 16

16. Lower calyptra more or less broadly infuscate (less intensely in female). Male: cerci and paralobi covered with fine setulae only 17

——Lower calyptra greyish-white, with the part adjacent to the scutellum somewhat infuscate. Male: cerci and paralobi covered with stout bristles, lobes of 5th sternite very large and erect 18

17. Hair on lower part of jowls and of occiput fulvous to golden; 8-14 mm: *Calliphora vomitoria* (Linn.).

——Hair on jowls and occiput black. Anterior thoracic spiracle strongly infuscate: *Calliphora uralensis* Villen.

18. Scutellum with 4-5 pairs of strong marginals, lateral seta well developed; 7-13 mm: *Calliphora subalpina* (Ringd.).

——Scutellum with only 3 pairs of strong marginals, lateral seta absent; 8-11 mm: *Calliphora alpina* (Zett.).

19. Basicosta black or brown 20

——Basicosta yellowish, at most with infumated margin 24

20. 3rd tergite with conspicuous marginales (Fig. 122). Subcostal sclerite with microscopic pile only 21

——3rd tergite without outstanding marginal setae. Subcostal sclerite in addition to the microscopic pile with some blackish setulae near apex 22

21. 3 post acr, palpi fuscous brown to black; 5-10 mm: *Bufolucilia silvarum* (Meig.).

——As a rule with only posterior 2 post acr. Palpi greyish-brown, with dull light brown basal half; 5-10 mm: *Bufolucilia bufonivora* Moniez.

22. Male: hypopyg large, protruding, glossy bluish-green. Inferior forceps bifid at apex (Fig. 124), eyes separated by about 1/2 the width of the 3rd antennal segment. Female: 6th tergite feebly convex, its posterior margin in middle with only 1 or 2 pairs of small setae (Fig. 123). 6-11 mm: *Lucilia caesar* (Linn.).

——Male: hypopyg small, inconspicuous, black or green, paralobi never bifid at apex. Female: 6th tergite flattened, with a complete row of long marginal bristles 23

23. Male: eyes separated by almost the width of 3rd antennal segment. Inferior forceps with a slight apical knob which is curved forward (Fig. 125). Female: underside of arista with 9-12 rays, 3rd antennal segment less long than width of interfrontalia plus 1 parafrontale, 2.5 times as long as wide; 5-9 mm: *Lucilia illustris* (Meig.).

——Male: eyes separated by about 1/2 the width of the 3rd antennal segment. Inferior forceps parallel-sided with the apex broadly rounded (Fig. 126). Female: underside of arista with 18-22 rays; 6-11 mm: *Lucilia ampullacea* Villen.

24. 3rd tergite with 2 or more long lateral setae, front of male 1/3 as wide in the narrowest place as the diameter of eye; 5-9 mm: *Lucilia regalis* (Meig.).

——3rd tergite without conspicuous marginal setae 25

25. Mid-tibia with 1 ad seta only 26

——Mid-tibia with 2 strong ad setae 27

26. Posterior part of humeral callus with 2-4 setulae (Fig. 127). Abdomen usually strongly coppery. Male: abdominal sternites long-haired, 2 pairs of ocellar bristles, front about 1/5 head width. Female: bucca less than 1/3 eye height; 5-10 mm: *Phaenicia cuprina* (Wied.).

——Posterior part of humeral callus with 6-8 setulae (Fig. 128). Abdomen usually bright green. Male: abdominal sternite with ordinary pile, accessory pair of ocellars absent, front about 1/8 head width. Female: bucca almost 2/5 eye height; 5-10 mm: *Phaenicia sericata* (Meig.).

27. Male: eyes separated by not much more than width of 3rd antennal segment, abdominal sternites of male and female normally haired; 5-11 mm: *Lucilia richardsi* Collin.

——Male: eyes separated by 1/3 eye width, abdominal sternites of male with dense tufts of long hairs (Fig. 130), those of female with strong setae; 7-10 mm: *Lucilia pilosiventris* Kramer.

SARCOPHAGIDAE

1. Abdominal tergites with irregular pearly maculae 2

——Abdominal tergites with symmetrical pattern of dark, dull, oval maculae on sides and medial stripe 9

2. 3 strong postsutural dc 3

——4 or more postsutural dc, or at least the series distinctly spaced for 4 or more 5

3. 2nd genital segment and hypopygium largely red, frontal bristles straight or very slightly curved outward at front end; 4-8.5 mm: *Ravinia striata* Fabr.

——2nd genital segment entirely black, frontal bristles opposite the antennae diverging outward (Fig. 132) 4

4. Male: 1st genital segment without marginal setae, hypopygium as in Fig. 134. Female: frons just over 1/3 head width; 6-13 mm: *Thyrsocnema incisilobata* (Pand.).

——Male: 1st genital segment with distinct marginal setae (Fig. 131), hypopygium as in Fig. 133. Female: frons distinctly broader than eye; 6.5-13 mm: *Sarcophaga melanura* Meig.

5. Prescutellar ac absent, genital segments mostly red. Hypopygium of male as in Fig. 135; 8-14 mm: *Sarcophaga haemorrhoidalis* (Fall.).

——Prescutellar ac present 6

6. 3rd tergite with strong marginals, normally 1 pair present; genital tergite of male with a series of marginal setae 7

——3rd tergite without long marginals, genital tergite of male without a series of marginal setae as a rule 8

7. Male: hypopygium as in Fig. 136, 1st genital segment dusted except on basal part. Female: 8th tergite absent; 8-15 mm: *Sarcophaga subvicina* Rohd.

——Male: hypopygium as in Fig. 137, subsp. *carnaria* Linn. or in Fig. 138, subsp. *lehmanni* Mull. 1st genital segment glossy for the most part. Female: 8th tergite present; 8-16 mm: *Sarcophaga carnaria* (Linn.).

8. Genital segments red, hypopygium as in Fig. 139: *Parasarcophaga barbata* Thom.

——Genital segments black. This group includes a number of *Parasarcophaga* spp., reliably distinguishable only by their hypopygia: *P. albiceps* (Meig.) (Fig. 140), *P. aratrix* Pand. (Fig. 141), *P. teretirostris* Pand. (Fig. 142), *P. tuberosa* Pand. (Fig. 143), *P. jacobsoni* Rohd. (Fig. 144), *P. scoparia* Pand. (Fig. 145), *P. similis* Pand. (Fig. 146). [Many species of *Parasarcophaga*, e.g., *albiceps*, *argyrostoma*, *barbata*, *hirtipes*, and *parkeri*, are placed in *Sarcophaga* by some specialists.]

9. Lateral maculae nearer to anterior margin of tergites, 3-4 notopleural bristles, 3 sternopleural bristles more or less in 1 row (1:1:1); 6-16 mm: *Bellieria maculata* (Meig.).

——Lateral maculae nearer to posterior margin of tergites 10

10. Sternopleural bristles in a position 2 (or 3):1, frons in either sex relatively very broad (Fig. 147); 4.5-9 mm: *Agria latifrons* Fall.

——Sternopleural bristles only 1:1. *Wohlfartia* B.B. (Obligatory causers of myiases.)

The Genus *Musca*

Adrian Pont and Hugh E. Paterson

THE FOLLOWING KEYS are designed to assist in the identification of the more common species of the genus *Musca* in the various zoogeographical regions. A few less common species are included as well when they have been mentioned in the literature on synanthropic flies. A list of the flies mentioned and their more common synonyms is provided as an aid to tracing species which have been referred to in the literature under names which are now regarded as synonyms.

An attempt was made to keep the keys as simple as possible. Most of the characters used are in common use in the taxonomy of Diptera, but a few less frequently encountered characters are illustrated in Fig. 149. Figure 150 illustrates the two types of male paralobus which occur within *Musca domestica s.l.* Occasional reference is made in the keys to the position of the setae on the tibiae. To follow these, the legs should be imagined to be extended horizontally and at right angles to the thorax. In referring to the abdomen the following convention is accepted: that all the sternites from 1 to 5 are visible and distinct, but that tergite 1 and tergite 2 are fused to form the first visible tergite which is referred to as tergite 1 + 2. Tergites 3 to 5 are discrete.

It will be noticed that *Musca sorbens* Wiedemann is referred to in the keys as "*M. sorbens* Wd. *s.l.*" This is done advisedly because recent unpublished studies have revealed biological discontinuities within *M. sorbens* as recognized by the keys. At least three distinct forms may eventually be recognized.

Musca domestica L. is subdivided into four subspecies: *M. d. domestica*, *M. d. nebulo*, *M. d. calleva*, and *M. d. curviforceps.* This is to some extent provisional since no adequate studies have been made in the Oriental region. The cline in male eye size (and color of the abdomen) which exists within *M. d. domestica* is not, of course, taxonomically recognized.

GENUS: *MUSCA* LINNAEUS, 1758

Musca alpesa Walker, 1849
Musca amita Hennig, 1964
  Musca amica Zimin, 1951, nec Linnaeus
Musca autumnalis De Geer, 1776
  Musca corvina Fabricius, 1781*
  Musca somalorum Bezzi, 1892 (as ssp.)
  Musca ovipara Portschinsky, 1910
  Musca prashadii Patton, 1922
  Musca ugandae Emden, 1939 (as ssp.)
  Musca pseudocorvina Emden, 1939 (as ssp.)
Musca bezzii Patton and Cragg, 1913
Musca conducens Walker, 1860
  Musca lineata Brunetti, 1910
  Musca pulla Bezzi, 1911
Musca convexifrons Thomson, 1869
Musca crassirostris Stein, 1903
  Musca insignis (Austen), 1909*
  Musca modesta de Meijere, 1904
Musca domestica Linnaeus, 1758
ssp. *domestica* s.s.
  nebulo Fabricius, 1794*
  calleva Walker, 1849
  curviforceps Saccà and Rivosecchi, 1955
  Musca analis Macquart, 1843
  Musca australis Macquart, 1843
  Musca antiquissima Walker, 1849
  Musca minor Macquart, 1851
  Musca vicina Macquart, 1851
  Musca determinata Walker, 1852
  Musca cuthbertsoni Patton, 1936
Musca fasciata Stein, 1910
Musca fergusoni Johnston and Bancroft, 1920
Musca gabonensis Macquart, 1855
Musca gibsoni Patton and Cragg, 1913
Musca inferior Stein, 1909
  Musca gurneyi Patton and Cragg, 1912
Musca larvipara Portschinsky, 1910
  Musca vivipara Malloch, 1925, err. typ.
Musca lucidula Loew, 1856
  Musca africana Bezzi, 1892
Musca lusoria Wiedemann, 1824
Musca osiris Wiedemann, 1830
Musca pattoni Austen, 1910
  Musca incerta Patton, 1922

**M. corvina, M. hilli, M. insignis,* and *M. nebulo* are treated as species in the lists in Chapters v and vi which were set up before the latest systematic opinions presented in this section were available (B.G.).

    Musca yerburyi Patton, 1923
Musca planiceps Wiedemann, 1824
    Musca cingalaisina Bigot, 1887
    Musca pollinosa Stein, 1909
Musca sorbens Wiedemann, 1830
    Musca humilis Wiedemann, 1830
    Musca spectanda Wiedemann, 1830
    Musca vetustissima Walker, 1849
    Musca primitiva Walker, 1849
    Musca angustifrons Thomson, 1869
    Musca alba Malloch, 1929
Musca tempestatum (Bezzi), 1908
Musca tempestiva Fallén, 1817
Musca terraereginae Johnston and Bancroft, 1920
    Musca hilli Johnston and Bancroft, 1920, *syn. nov.**
Musca ventrosa Wiedemann, 1830
    Musca pungoana Karsch, 1886
    Musca nigrithorax Stein, 1909
Musca vitripennis Meigen, 1826
Musca xanthomelas Wiedemann, 1824

New World Species

1. Propleural depression haired, rarely individually bare. Suprasquamal ridge bare. Throughout Nearctic and Neotropical regions: *domestica* L.

——Propleural depression bare. Suprasquamal ridge with the tympanic tuft of hairs. North America only: *autumnalis* De Geer.

Palearctic Species

1. Propleural depression haired, rarely individually bare. Suprasquamal ridge bare. Male frons between 0.1 and 0.2 of the total head width: *domestica* L.

——Propleural depression bare 2

2. Eyes distinctly haired in both sexes, sometimes rather inconspicuously so in females. Prothoracic spiracles white 3

——Eyes bare in both sexes 4

3. Male: frons at narrowest point broader than width of 3rd antennal segment; frontal setae in 2 rows alongside the interfrontalia. Female: parafrontalia outside the row of frontal setae bearing setulae that in some places are as long as the frontal setae: *vitripennis* Meigen.

——Male: frons at narrowest point at most as broad as width of 3rd antennal segment; frontal setae in 1 row alongside the interfrontalia. Female: parafrontalia outside the row of frontal setae bearing only short setulae that are all shorter than the frontal setae: *osiris* Wiedemann.

4. Suprasquamal ridge entirely bare 5

——Suprasquamal ridge with at least the tympanic tuft of setulae 7

5. Fore-tibia with a submedian posterior seta. Prothoracic spiracle white: *conducens* Walker.

——Fore-tibia without a submedian posterior seta. Prothoracic spiracle brown 6

6. No presutural dorsocentral setae present, at most a few longer ground-setulae present. Discal cell at least partly devoid of clothing setulae: *lucidula* Loew.

——2 pairs of strong presutural dorsocentral setae present. Discal cell entirely covered with clothing setulae: *tempestiva* Fallén.

7. Suprasquamal ridge with only the tympanic tuft of setulae 8

——Suprasquamal ridge with both tympanic and parasquamal tufts of setulae 9

8. Setulae on the genae abundant, extending up as far as frontal suture. Male: parafacialia at level of the 2nd antennal segment narrower than width of the 3rd antennal segment. Female: with several rows of inclinate frontal setae: *autumnalis* De Geer.

——Setulae on the genae fewer, not extending up as far as the frontal suture. Male: parafacialia at the level of 2nd antennal segment broader than width of 3rd antennal segment. Female: with only a single row of inclinate frontal setae: *amita* Hennig.

9. Male: vein 3 on ventral surface with the setulae confined to the node at base; tergite 1+2 entirely dark, rarely buff at sides or on hind-margin; tergite 5 without 3 dark vittae. Female: interfrontalia at middle 1 to 1.5 times as broad as a parafrontale; 3rd antennal segment at most 2.5 times as long as broad: *larvipara* Portschinsky.

——Male: vein 3 on ventral surface with the setulae usually extending beyond the small cross-vein; tergite 1+2 yellow except for a dark median vitta; tergite 5 with 3 narrow complete dark vittae. Female: interfrontalia at middle twice as broad as a parafrontale; 3rd antennal segment 3 to 3.5 times as long as broad: *convexifrons* Thomson.

Oriental Species

1. Propleural depression haired, rarely individually bare; suprasquamal ridge bare 2

——Propleural depression bare 3

2. Male: frons at its narrowest about 1.5 times the 3rd antennal segment. Female: apical segments of abdomen relatively dark, tergite 4 dark except for buff anterior lateral angles, tergite 5 entirely dark: *domestica* L.

——Male: frons at its narrowest about the same diameter as the anterior ocellus. Female: apical segments of the abdomen paler, either all yellow or at most with a dark midline: *domestica nebulo* F.

3. Mid-tibia with a submedian anteroventral seta. Proboscis enormously dilated, boat-shaped, and adapted for blood sucking. Palpi yellow. Abdomen and thorax densely grey to yellowish-grey dusted on a dark ground: *crassirostris* Stein.

——Mid-tibia without a submedian anteroventral seta. Proboscis normal. Palpi dark 4

4. Eyes distinctly haired in both sexes, sometimes rather inconspicuously so in females. Prothoracic spiracle white. Suprasquamal ridge bare 5

——Eyes bare in both sexes, or, if haired, then suprasquamal ridge with both tympanic and parasquamal tufts of setulae 6

5. Male: frons at narrowest point broader than width of 3rd antennal segment; frontal setae in 2 rows alongside the interfrontalia. Female: parafrontalia outside the row of frontal setae bearing setulae that in some places are as long as the frontal setae: *vitripennis* Meigen.

——Male: frons at narrowest point at most as broad as width of 3rd antennal segment; frontal setae in 1 row alongside the interfrontalia. Female: parafrontalia outside the row of frontal setae bearing only short setulae that are all shorter than the frontal setae: *osiris* Wiedemann.

6. Suprasquamal ridge entirely bare 7

——Suprasquamal ridge with at least the tympanic tuft of setulae 13

7. Fore-tibia with a submedian posterior seta 8

——Fore-tibia without a submedian posterior seta 9

8. Male: tergite 1+2 largely dark, tergites 3 and 4 orange except for a brown median vitta. Female: tergites entirely dark: *conducens* Walker.

——Male and female: tergite 1+2 entirely or largely, tergites 3 and 4 largely, orange-yellow in ground color: *planiceps* Wiedemann.

9. Abdomen entirely orange-yellow in both sexes, sometimes infuscate at apex in pinned specimens due to the dried viscera, without any dark markings or pruinose tessellated pattern. Mesonotum with 4 undusted longitudinal vittae; dusting bluish: *ventrosa* Wiedemann.

——Abdomen with at least a complete or partial dark median vitta, and always with some pruinose tessellated spots 10

10. Prothoracic spiracle white 11

——Prothoracic spiracle brown 12

11. Paramedian thoracic vittae fused to form one broad vitta on each side behind but not in front of the suture. Male: tergite 1+2 dark above, a median dark vitta on tergite 3, and at the base of tergite 4; the vitta expands toward apex of tergite 4 so that the whole of the 5th tergite is dark; rest of abdomen pale in ground color. Female: whole abdomen dark in ground color, nowhere yellow. Hypopleuron with setulae present below spiracle and in lower posterior angle above hind coxa; sternite 1 bare: *sorbens* Wiedemann, *s.l.*

——Paramedian thoracic vittae quite separate in female and more or less separate in male both in front and behind the suture. Male and female: tergite 1+2 largely yellow-orange except along midline: *pattoni* Austen.

12. Without presutural dorsocentral setae, at most a few longer ground setulae present. Discal cell with at least a part devoid of clothing setulae: *lucidula* Loew.

——With 2 pairs of strong presutural dorsocentral setae. Discal cell entirely covered with clothing setulae: *tempestiva* Fallén.

13. Suprasquamal ridge with only the tympanic tuft of setulae 14
——Suprasquamal ridge with both the tympanic and parasquamal tufts of setulae 16

14. Vein 3 with setulae on its ventral surface almost to its apex; stem-vein with 3-7 setulae on posterior side of dorsal surface: *lusoria* Wiedemann.
——Vein 3 with setulae on ventral surface confined to the node at base, not reaching the small cross-vein; stem-vein with only 1-2 setulae on dorsal surface 15

15. Tergite 1+2 orange-yellow except for the basal declivity: *xanthomelas* Wiedemann.
——Tergite 1+2 entirely dark: *autumnalis* De Geer.

16. Lower squama with black setulae on dorsal surface: *inferior* Stein.
——Lower squama entirely bare 17

17. Larger species. Stem-vein with 4-7 setulae on posterior side of dorsal surface. Tergite 1+2 usually (male), or always (female) entirely dark in ground color: *bezzii* Patton and Cragg.
——Smaller species. Stem-vein with 1-3 setulae on posterior side of dorsal surface. Tergite 1+2 mainly or wholly buff (male), or with at least a pair of buff patches (female): *gibsoni* Patton and Cragg.

ETHIOPIAN SPECIES

1. Propleural depression haired, rarely individually bare; suprasquamal ridge bare 2
——Propleural depression bare 3

2. Anterior postsutural dorsocentral seta short, usually less than 1/2 the length of the last. Abdomen largely pale in both sexes, including the 5th tergite which has at most a dark midline. Male: frons width at narrowest point usually less than 0.08 of the greatest head width; paralobus truncated apically (Fig. 150a): *domestica calleva* Walker.
——Anterior postsutural dorsocentral seta longer, usually more than 1/2 the length of the last. Abdomen darker in both sexes; in both the 5th tergite is largely dark. Male: frons width at narrowest point usually more than 0.10 of the greatest head width; paralobus hook-shaped (Fig. 150b): *domestica curviforceps* Saccà and Rivosecchi.

3. Suprasquamal ridge bare 4
——Suprasquamal ridge with at least the tympanic tuft of setulae present 9

4. Fore-tibia with a postero-dorsal seta beyond the middle 5
——Fore-tibia without a submedian postero-dorsal seta 6

5. With 1 weak presutural dorsocentral seta. Labellum of proboscis unusually long. Head profile unusual in *Musca*, with the lower margin straight. Abdominal ground color dark in both sexes: *tempestatum* Bezzi.
——With 2 pairs of well-developed presutural dorsocentral setae. Labellum of proboscis and head profile of the usual *Musca* type. Abdomen

dark in the female but the male has extensive lateral pale areas on tergites 3 and 4: *conducens* Walker.

6. Abdomen entirely orange-yellow in both sexes, sometimes infuscate at apex in pinned specimens due to dried viscera, without any dark markings or pruinose tessellated pattern. Mesonotum with 4 undusted longitudinal vittae; dusting bluish: *ventrosa* Wiedemann.

——Abdomen with at least a complete or partial dark median vitta, and always with some pruinose tessellated spots 7

7. Mid-tibia with a prominent antero-ventral seta near the middle. Proboscis enormously dilated, boat-shaped, and adapted for blood sucking. Palpi yellow. Abdomen and thorax densely grey to yellowish-grey dusted on dark ground: *crassirostris* Stein.

——Mid-tibia without an antero-ventral seta near the middle 8

8. Prothoracic spiracles dark. Abdomen in both sexes dark with a pattern of grey dusting. Male with prominent curled hairs on the 3rd and 4th segments of the hind-tarsus. Small species: *fasciata* Stein.

——Prothoracic spiracle whitish. Male: tergite 1+2 dark above, tergite 3 and base of tergite 4 with a dark median vitta which broadens as it approaches tergite 5 which is all dark above; remainder of abdomen pale. Female: abdomen dark in ground color throughout with grey or yellowish-grey tessellated pattern of pruinosity. Hypopleuron with setulae present below spiracle and in lower posterior angle above hind-coxa. Sternite 1 bare: *sorbens* Wiedemann, *s.l.*

9. Suprasquamal ridge with only the tympanic tuft of setulae present 10

——Suprasquamal ridge with both the tympanic and parasquamal tufts of setulae. 3rd vein with setulae below to well beyond the small cross-vein; basicostal scale at base of wing yellow: *alpesa* Walker.

10. 4 distinct longitudinal vittae on mesonotum 11

——2 more or less distinct vittae on mesonotum due to the fusion, or near fusion, of the paramedian vittae behind the suture. Stem-vein with 3-5 setulae above; 3rd vein usually with 4-5 setulae basad of the small cross-vein on the ventral surface. Abdomen largely yellow in both sexes: *gabonensis* Macquart.

11. Stem-vein with 1-2 setulae above; any setulae on the ventral surface of the 3rd vein confined to basal parts. Abdomen in both sexes extensively yellow: *xanthomelas* Wiedemann.

——Stem-vein with 4-8 setulae above; setulae on ventral surface of 3rd vein extending apically to well beyond the small cross-vein. Abdomen of female largely dark in ground color with pattern of grey dusting; abdomen of male extensively yellow: *lusoria* Wiedemann.

Australian Species

1. Propleural depression haired, rarely individually bare. Suprasquamal ridge bare. Mean width of male frons at narrowest point about 0.1 of the greatest head width: *domestica* L.

——Propleural depression bare 2

2. Suprasquamal ridge with only the tympanic tuft of setulae present. Mesonotum with four longitudinal dark vittae. Abdomen in both sexes with a median dark vitta. Male frons very narrow. A moderately large species: *fergusoni* Johnston and Bancroft.

——Suprasquamal ridge bare 3

3. Both the hypopleuron below the spiracle and the 1st sternite bare. Undusted vittae on mesonotum narrow, each at level of the suture, being half as wide as median dusted vitta and as wide as the dusted vitta laterad of it. Abdomen entirely yellow in both sexes: *ventrosa* Wiedemann.

——Hypopleuron, at least, setulose below the spiracle 4

4. Paramedian thoracic vittae fused to form 1 broad vitta on each side behind but not in front of the suture. Hypopleuron with setulae present below the spiracle and in the lower posterior angle above the hind-coxa. Sternite 1 bare. Male: tergite 1+2 dark above a median dark vitta on tergite 3, and at the base of tergite 4; the vitta expands toward the apex of tergite 4 so that the whole of the 5th tergite is dark; rest of abdomen pale in ground color. Female: whole abdomen dark in ground color: *sorbens* Wiedemann, *s.l.*

——Paramedian thoracic vittae separate behind and in front of the suture. These vittae, at level of the suture, each usually as wide as the median dusted vitta and wider than the dusted vitta laterad of it. Sternite 1 setulose. Abdomen largely pale in both sexes: *terraereginae* Johnston and Bancroft.

The Genus *Drosophila*

Dr. Marshall R. Wheeler

MANY SPECIES OF *Drosophila* utilize damaged or decaying fruits and similar garbage components for food, and because of this they have probably been associated with man for centuries. Some early names, such as "pomace fly" and "vinegar gnat," reflect this relationship. Although these flies are usually considered pests, this food habit allowed the virtual domestication of *Drosophila melanogaster* early in this century, providing an ideal experimental animal for genetic experiments. A few species, with fungivorous larvae, have been considered pests by mushroom growers, but the greatest economic aspect concerns the tremendous populations of *Drosophila* which build up around canning factories. The greatest concentrations are found near factories that produce grape juice and wine, tomatoes and tomato juice, apple juice and cider-vinegar, and orange juice. The fruit and garbage type *Drosophila* species are remarkably attracted to bananas; in the tropics especially but elsewhere on occasion, any of these species might be found within homes and markets due to this attraction. On the other hand, it seems remarkable that only seven species, all of them now cosmopolitan, have become primarily associated with human habitations. Three subgenera are represented, as follows:

Subgenus *Drosophila. D. repleta* Wollaston: this is sometimes a "wild" species in the tropics, but occurs only in buildings in cooler climates where it is commonly seen in dirty toilets. *D. hydei* Sturtevant: this is found around all sorts of spoiling fruit and is often attracted by the odors of beer and wine. *D. immigrans* Sturtevant: this is occasionally a "wild" species, but it is seen more often in buildings around spoiling fruits. *D. funebris* (Fabricius): this species occurs farther north and farther south than any other *Drosophila*; it is found around spoiling fruits and vegetables, especially potatoes. *D. virilis* Sturtevant: this is a "wild" species in Asia but elsewhere it is best known in markets. Both *funebris* and *virilis* larvae have been found at times in spoiled or preserved meat.

Subgenus *Sophophora. D. melanogaster* Meigen: this is the most common kitchen-inhabiting *Drosophila.* A very similar species, *D. simulans* Sturtevant, is also found in houses in tropical areas.

Subgenus *Dorsilopha. D. busckii* Coquillett: this is found more often around spoiled vegetables and seems to be especially attracted to rotting onions and potatoes.

Key to the Commonest Domestic Species of *Drosophila*

1. Mesonotum of thorax dull grayish to brownish, densely spotted, each hair and bristle arising from a dark brown spot 2

——Thorax without such spots, usually unicolorous or with a central striped pattern 3

2. Basal segments (coxae) of front legs much blacker than rest of legs; lateral regions of abdominal segments with darkened areas bearing paler centers: *repleta*

Legs all pale brownish; lateral regions of abdominal segments solidly dark: *hydei*

3. Mesonotum dull yellowish, with distinct black longitudinal stripes, the middle one forked posteriorly: *busckii*

——Mesonotum lacking such obvious stripes, usually unicolorous tan, brown, or black 4

4. Mesonotum tan; abdomen with a pattern of lighter and darker regions 5

——Mesonotum brown to blackish; abdomen usually unicolorous brown to black also 6

5. Mesonotum dull tan; inner side of first femur of both sexes with a row of short, stout, black spinelike bristles; wing veins darker where they meet wing margin: *immigrans*

——Mesonotum semi-shining tan; basal tarsal segment of male with a short, nearly transverse, row of stout black teeth; wing veins not clouded anywhere: *melanogaster*

6. Body color brown; wings lacking darker clouded areas; basal scutellar bristles convergent; male genitalia with a cluster of large black spines, readily seen externally (*funebris*) (Figs. 42, 43)

——Body dark brown to black; posterior cross vein of wing with a narrow darker cloud; basal scutellars divergent; male genitalia lacking the stout spines: *virilis*

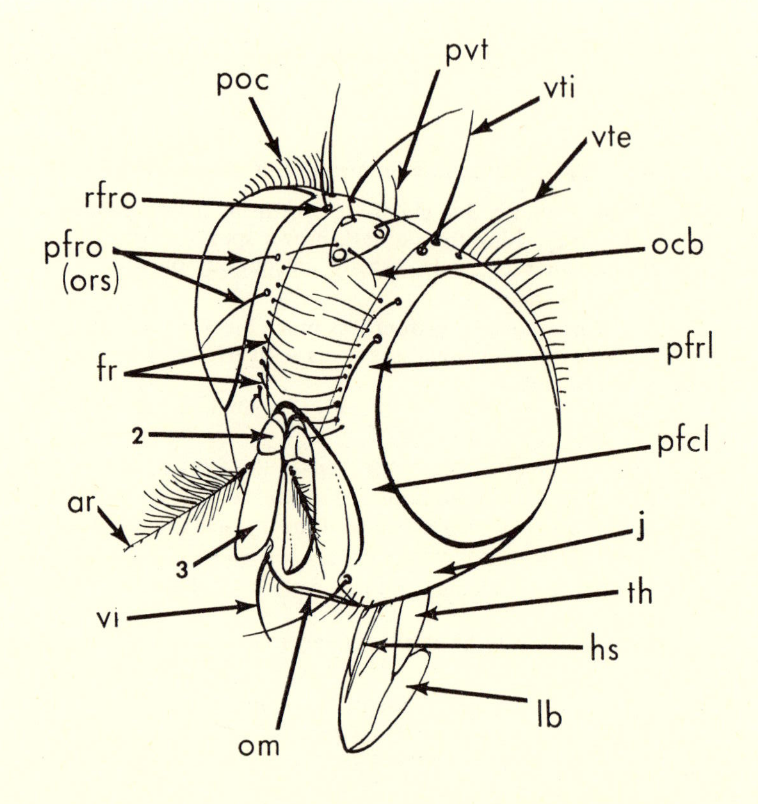

FIG. 1. Fronto-lateral view of muscoid head

ar = arista
fr = frontals
hs = haustellum
j = jowls
lb = labella
ocb = ocular (ocellar) bristles
om = oral margin
pfcl = parafacial
pfrl = parafrontal
pfro (= ors) = proclinate fronto-orbitals
poc = postocular setulae
pvt = postverticals
rfro = reclinate fronto-orbitals
th = theca
vi = vibrissae
vte = external verticals
vti = internal verticals
2, 3 = 2nd, 3rd antennal segments

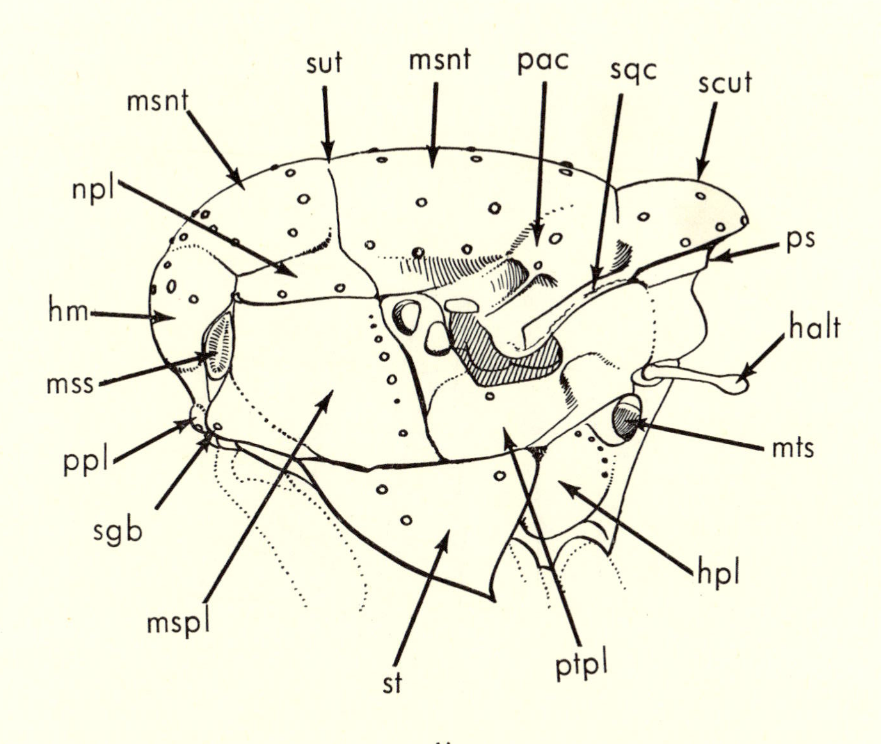

FIG. II. Chaetotaxy of thorax (lateral view)

halt = halter
hm = humerus
hpl = hypopleuron
msnt = mesonotum
mspl = mesopleuron
mss = mesothoracic spiracle (anterior spiracle)
mts = metathoracic spiracle (posterior spiracle)
npl = notopleuron
pac = postalar callus
ppl = propleuron
ps = postscutellum
ptpl = pteropleuron
scut = scutellum
sgb = stigmatic bristle
sqc = suprasquamal carina
st = sternopleuron (position of sternopleuralbristles 2:1)
sut = suture

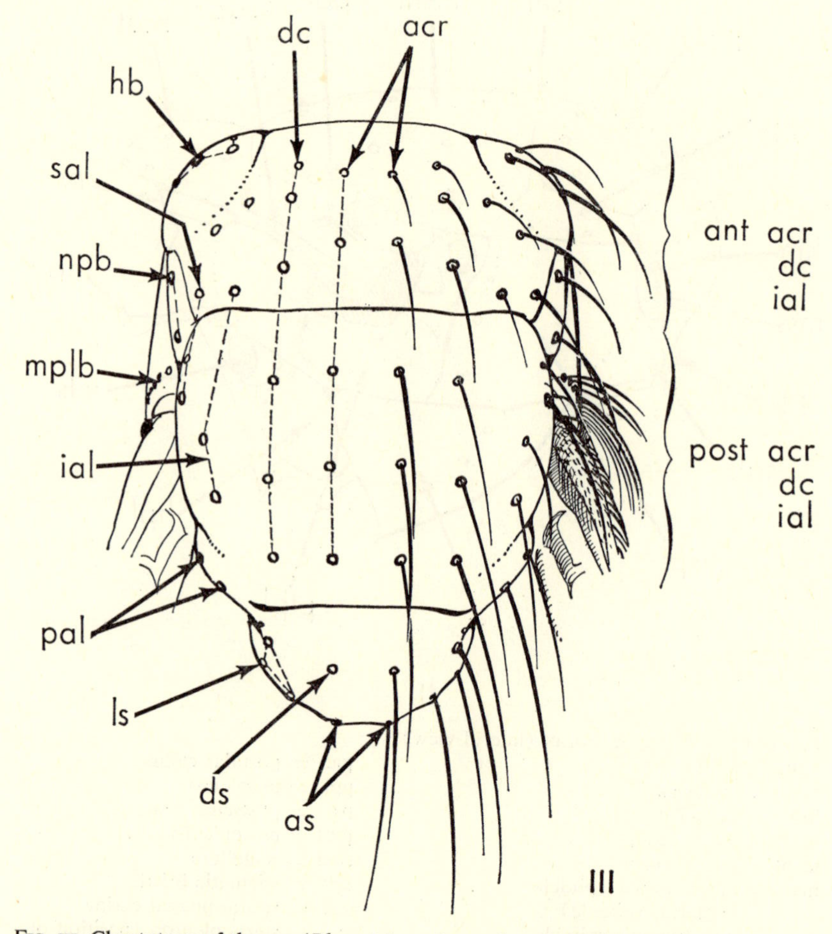

FIG. III. Chaetotaxy of thorax (*Phaenicia sericata*, dorsal view)

acr = acrostical bristles
as = apicoscutellar bristles
ds = discoscutellar bristles
dc = dorsocentral bristles
hb = humeral bristles
ial = intraalar bristles
ls = lateroscutellar bristles
mplb = mesopleural bristles
npb = notopleural bristles
pal = postalar bristles
sal = supraalar bristles

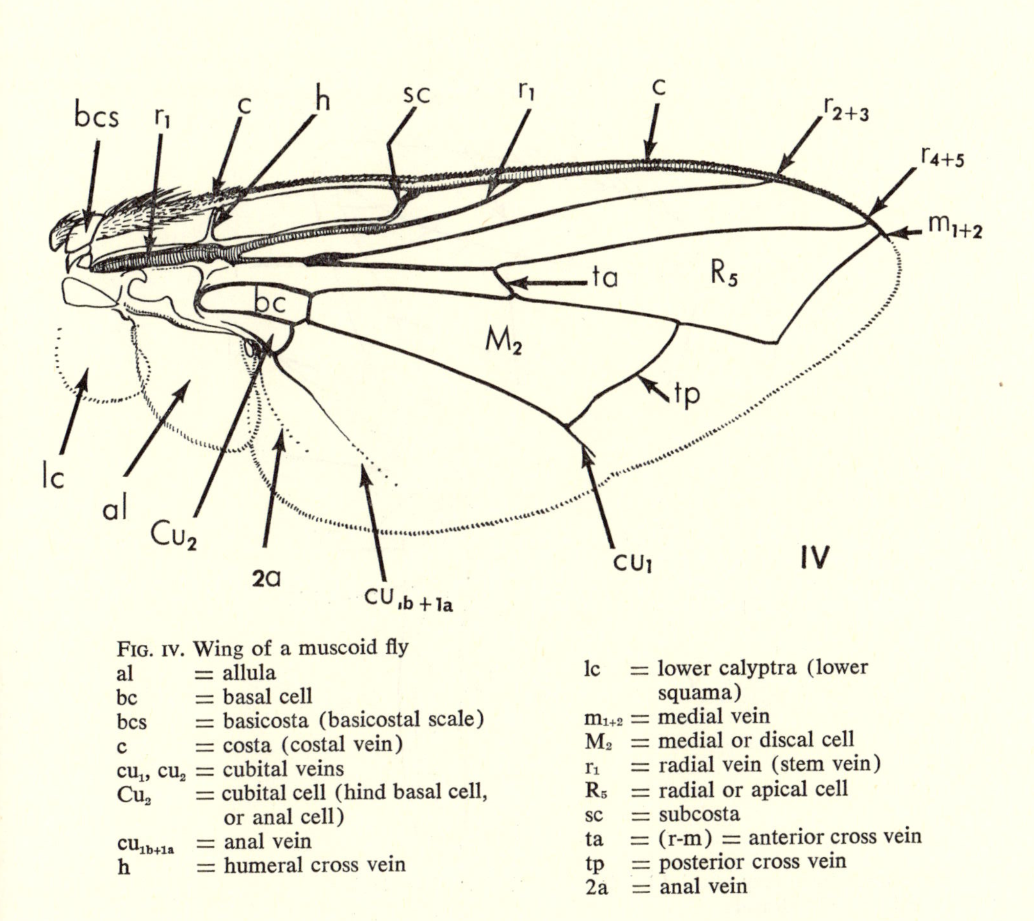

FIG. IV. Wing of a muscoid fly

al = allula
bc = basal cell
bcs = basicosta (basicostal scale)
c = costa (costal vein)
cu_1, cu_2 = cubital veins
Cu_2 = cubital cell (hind basal cell, or anal cell)
cu_{1b+1a} = anal vein
h = humeral cross vein
lc = lower calyptra (lower squama)
m_{1+2} = medial vein
M_2 = medial or discal cell
r_1 = radial vein (stem vein)
R_5 = radial or apical cell
sc = subcosta
ta = (r-m) = anterior cross vein
tp = posterior cross vein
2a = anal vein

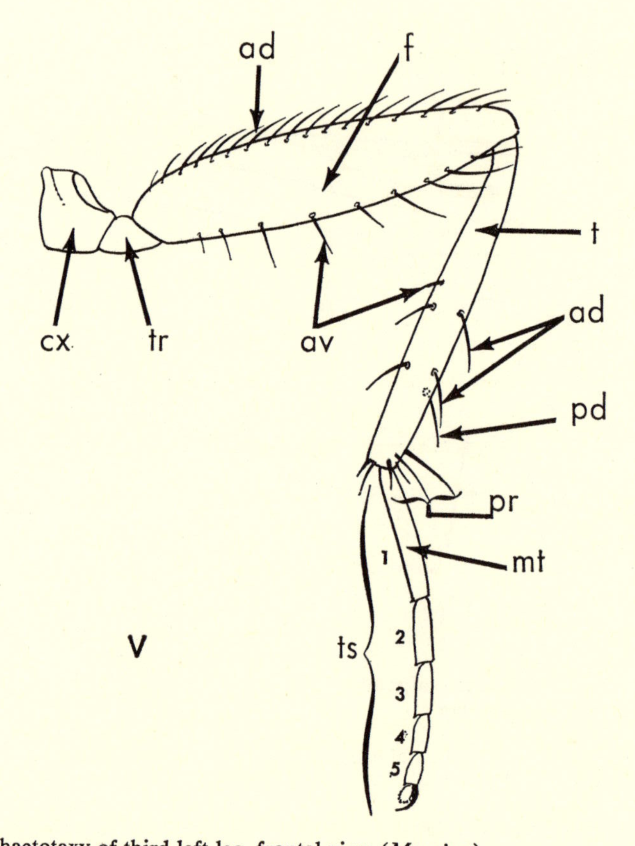

FIG. V. Chaetotaxy of third left leg, frontal view (*Muscina*)

ad = anterodorsals
av = anteroventrals
cx = coxa
f = femur
mt = metatarsus
pd = posterodorsals
pr = preapicals
t = tibia
tr = trochanter
ts = tarsal segments

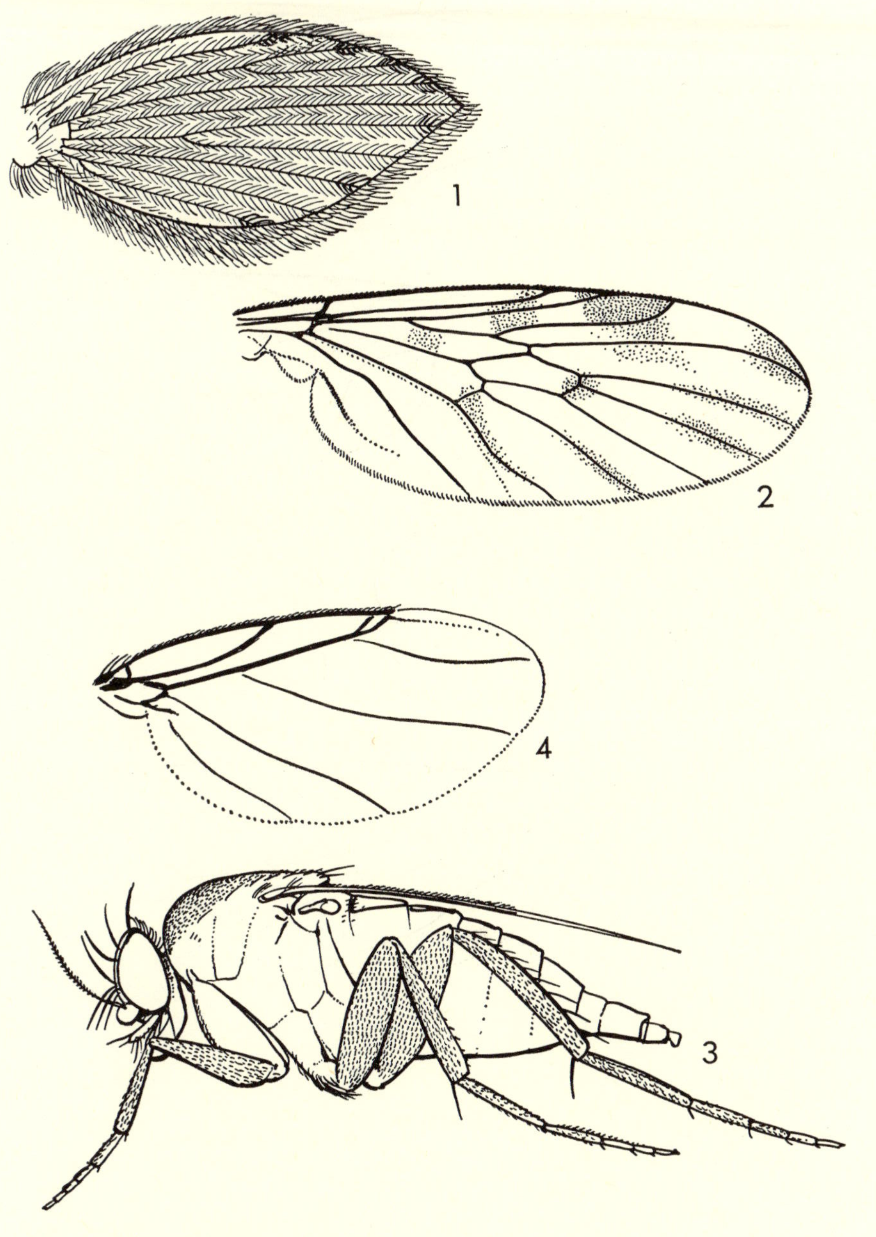

FIG. 1. Wing of the genus *Psychoda*
FIG. 2. Wing of *Syvicola fenestralis* (Scop.)
FIG. 3. Habitus of female *Megaselia* sp.
FIG. 4. Wing of *Megaselia* sp.

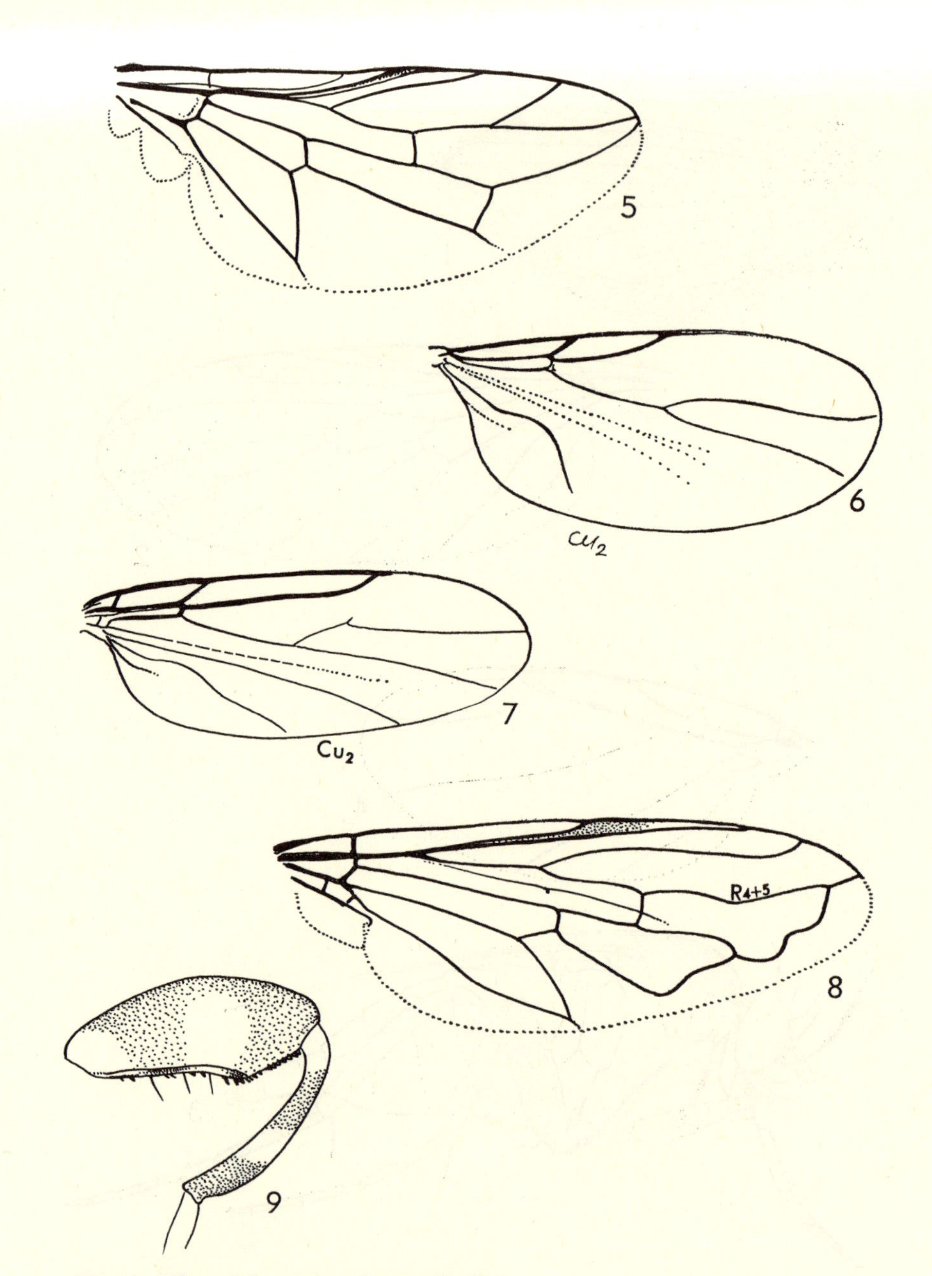

FIG. 5. Wing of *Scenopinus fenestralis* (L.)
FIG. 6. Wing of *Scatopse fuscipes* Meigen
FIG. 7. Wing of *Scatopse notata* (L.)
FIG. 8. Wing of *Xylota pipiens* (L.)
FIG. 9. Hind leg of *Xylota pipiens* (L.)

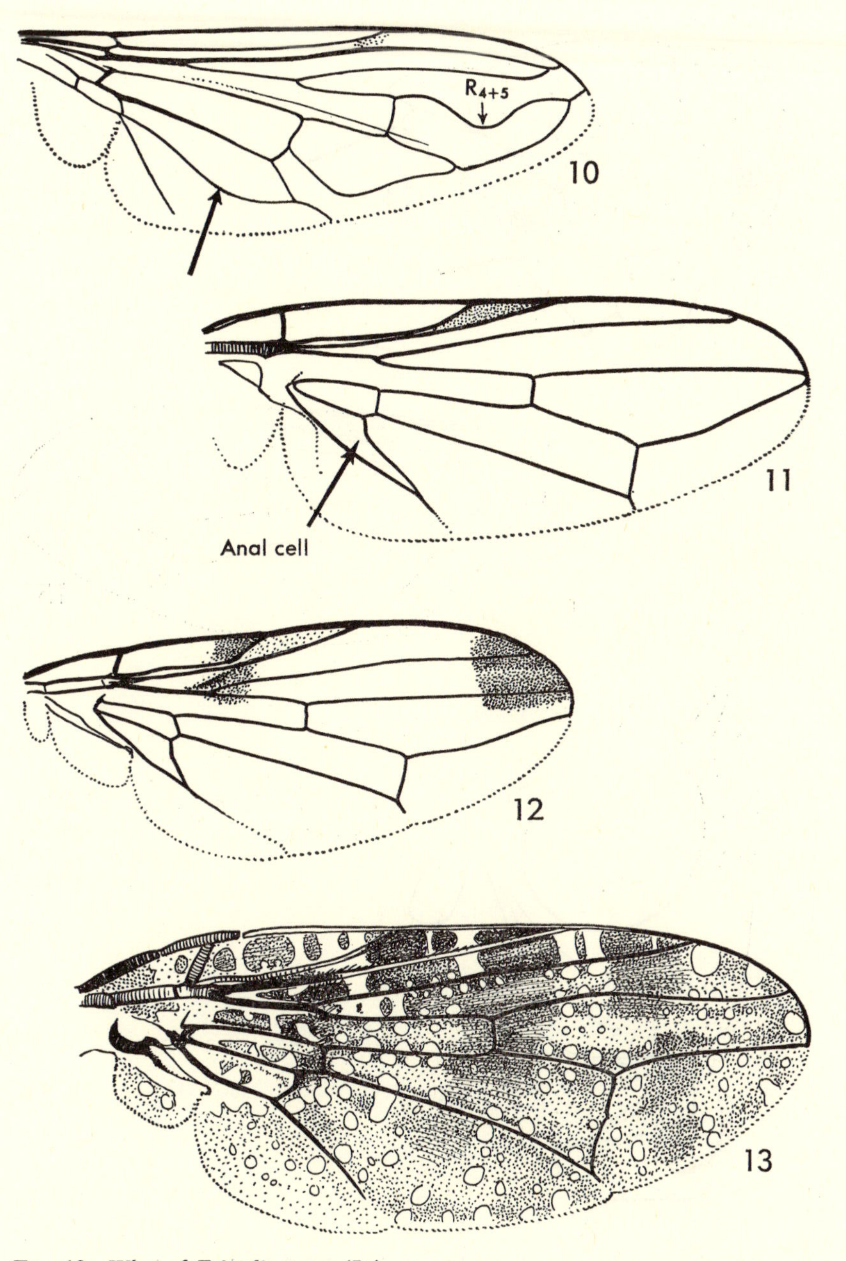

FIG. 10. Wing of *Eristalis tenax* (L.)
FIG. 11. Wing of *Physiphora demandata* (F.)
FIG. 12. Wing of *Euxesta* sp.
FIG. 13. Wing of *Platystoma lugubre* R.-D.

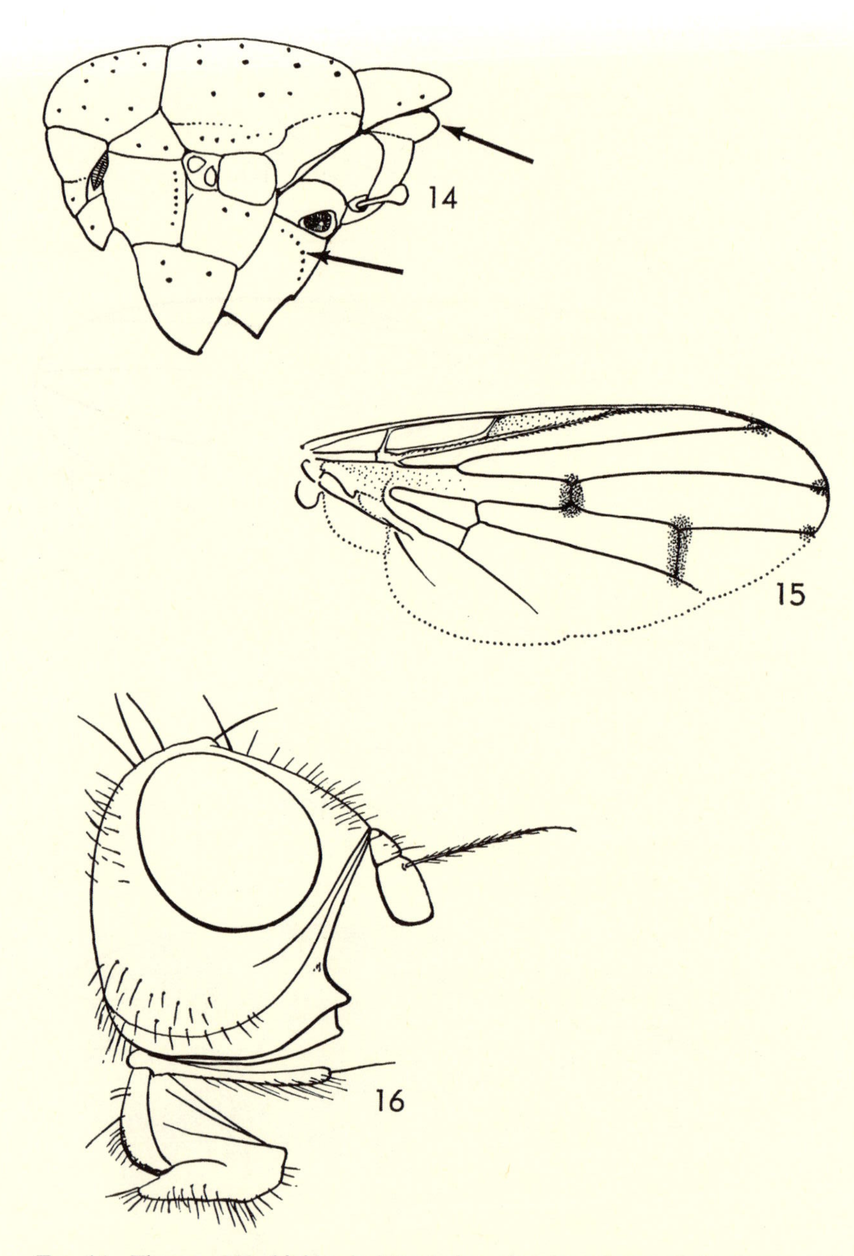

Fig. 14. Thorax of Tachinidae in lateral view, showing scheme of chaetotaxy and postscutellum
Fig. 15. Wing of *Dryomyza anilis* Fall.
Fig. 16. Head of female *Dryomyza flaveola* (Fabr.) (after Séguy)

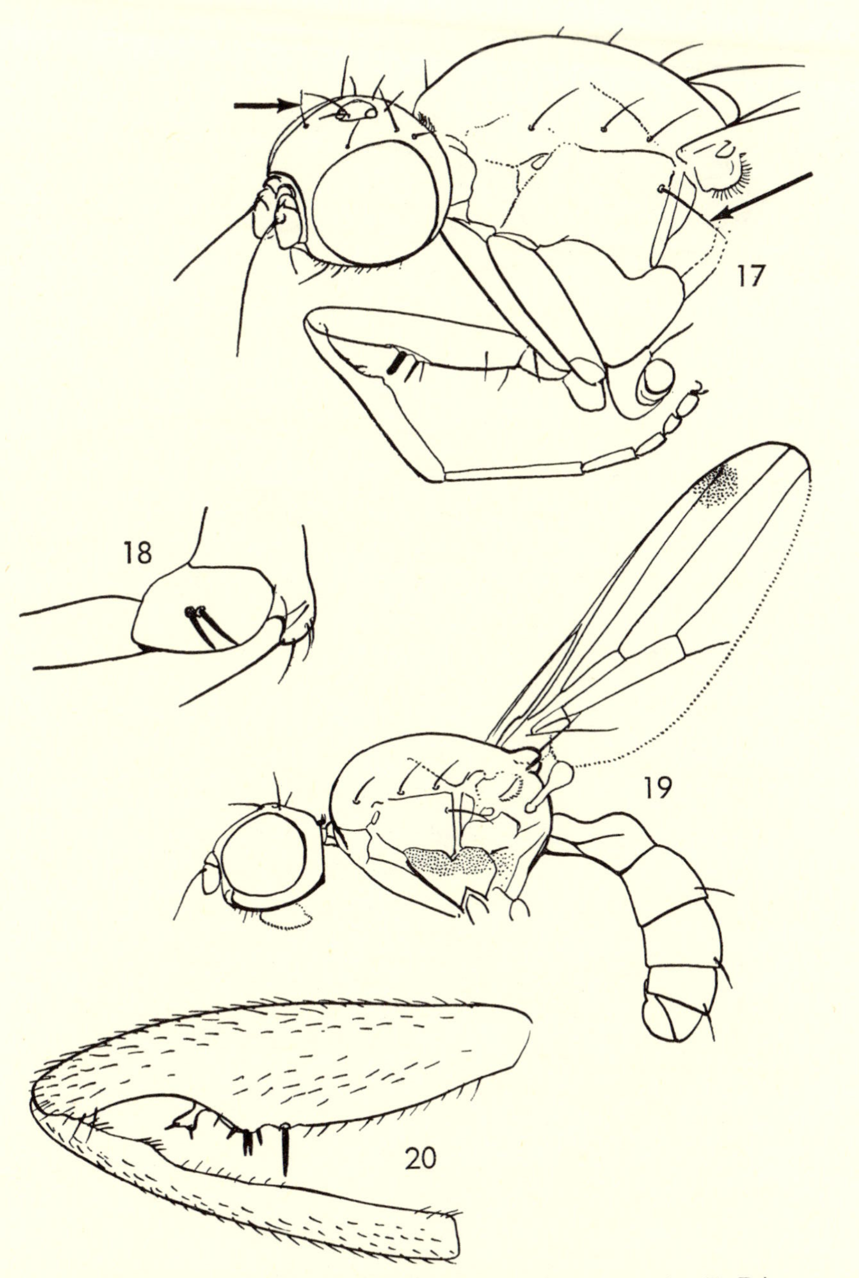

FIG. 17. Head, thorax, and right fore-leg of *Meroplius stercorarius* (R.-D.)
FIG. 18. Third trochanter of *Nemopoda nitidula* (Fall.)
FIG. 19. Body and wings of male *Sepsis thoracica* (R.-D.)
FIG. 20. Fore-femur and tibia of *Sepsis fulgens* Hffmsgg.

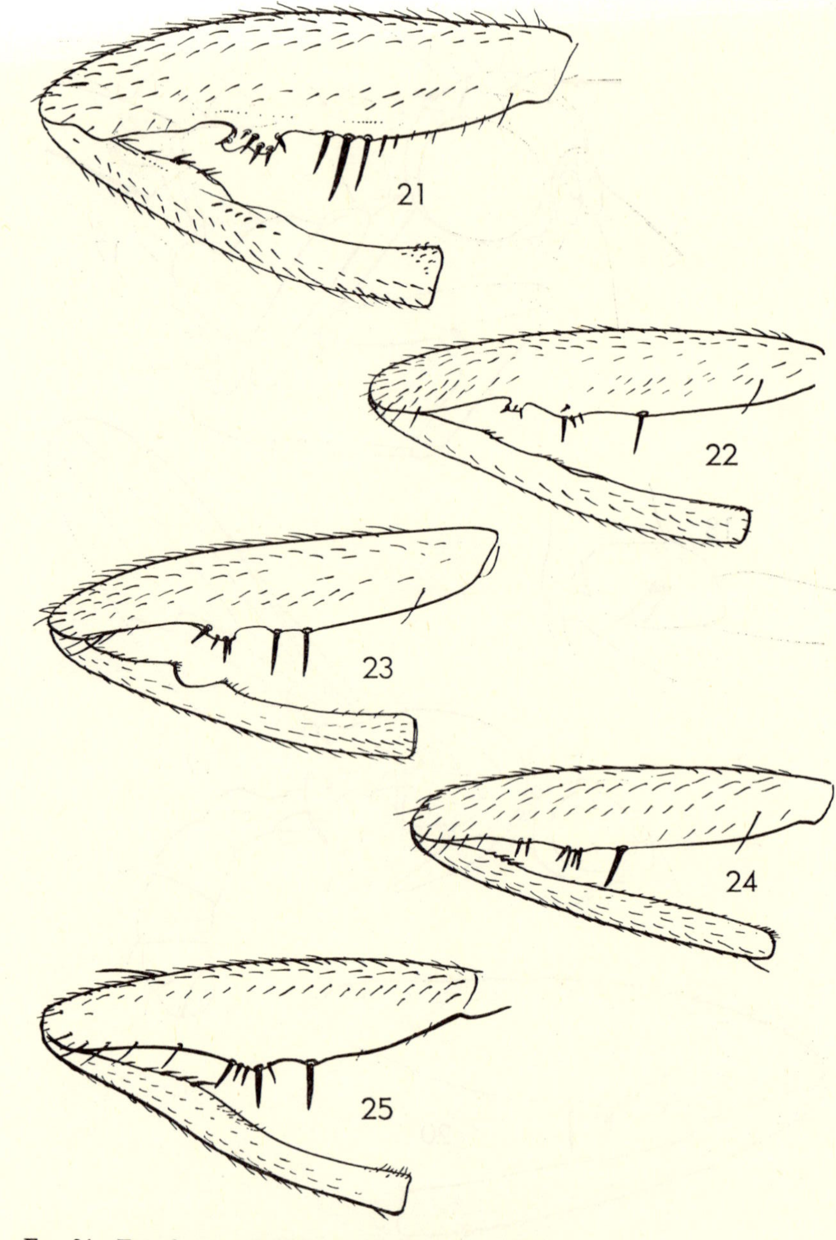

FIG. 21. Fore-femur and tibia of *Sepsis punctum* (Fabr.)
FIG. 22. Fore-femur and tibia of *Sepsis violacea* Meig.
FIG. 23. Fore-femur and tibia of *Sepsis cynipsea* (L.)
FIG. 24. Fore-femur and tibia of *Sepsis orthocnemis* Frey.
FIG. 25. Fore-femur and tibia of *Sepsis flavimana* Meig.

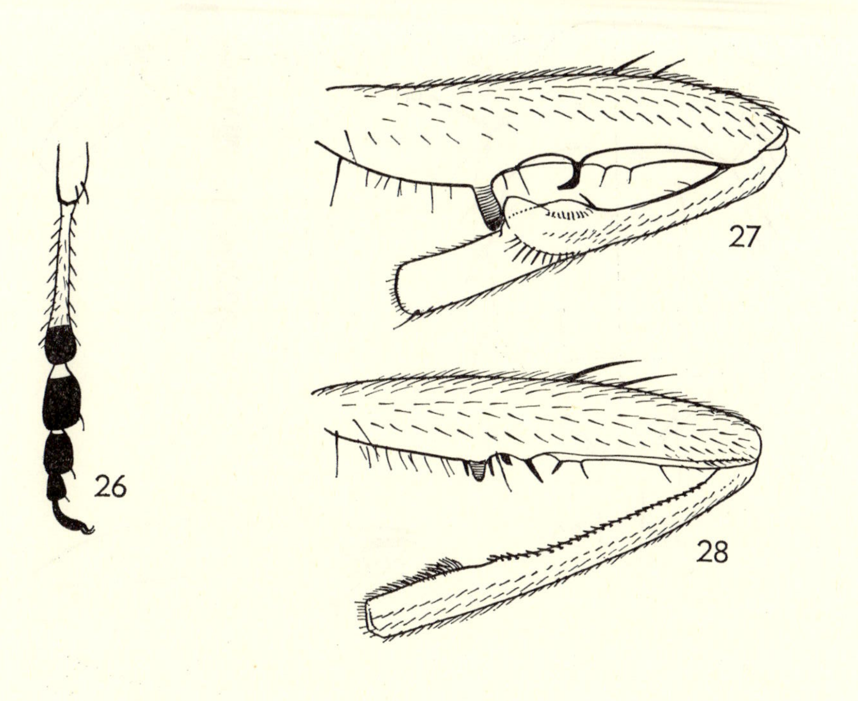

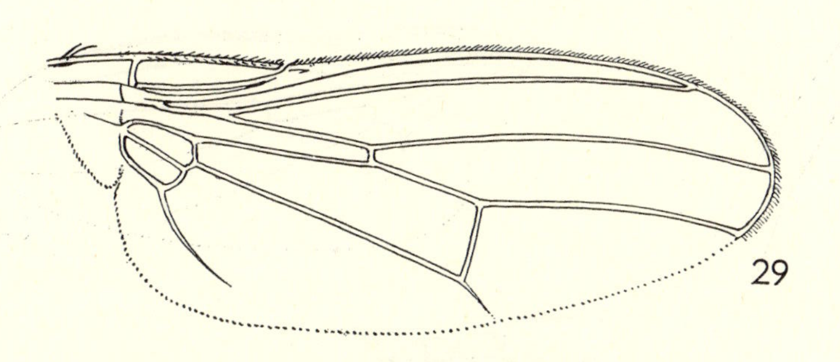

FIG. 26. Middle tarsus of *Themira annulipes* (Meig.)
FIG. 27. Fore-femur and tibia of male *Themira putris* (L.)
FIG. 28. Fore-femur and tibia of male *Themira nigricornis* (Meig.)
FIG. 29. Wing of *Piophila casei* (L.)

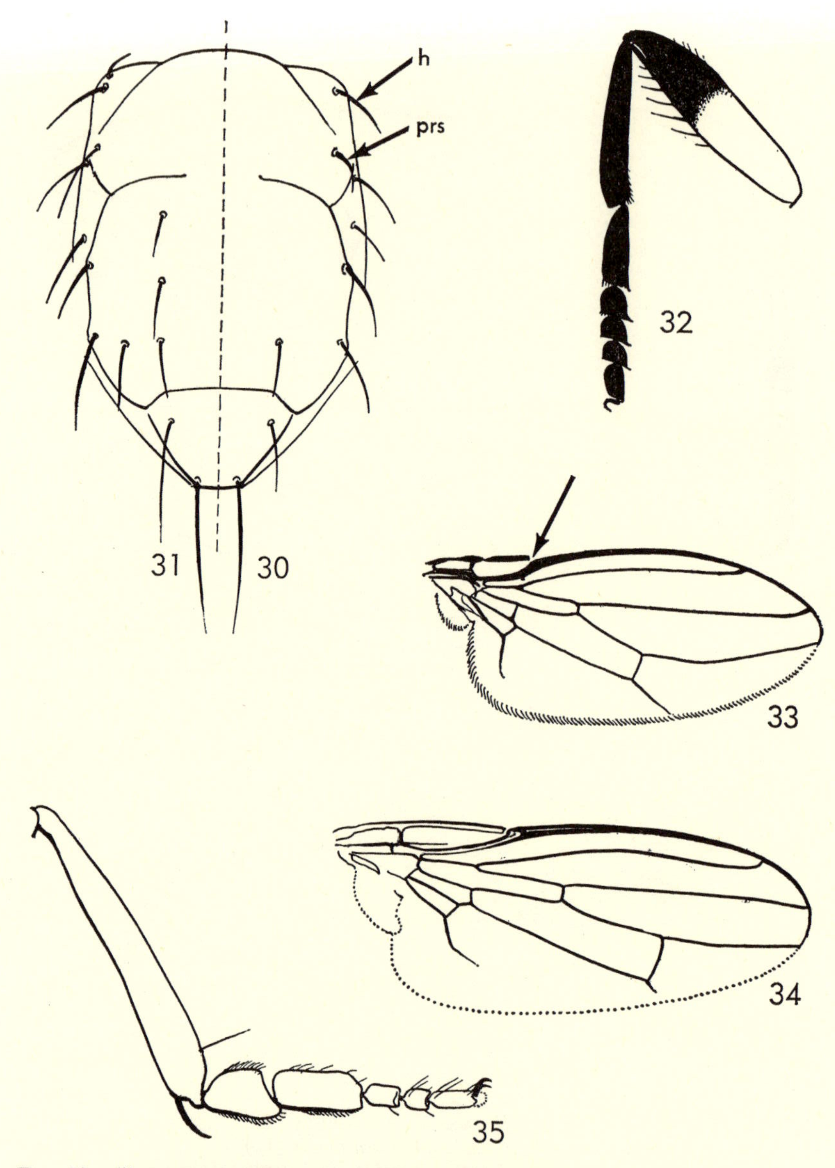

FIG. 30. Chaetotaxy of thorax in *Piophila vulgaris* Fall.
FIG. 31. Chaetotaxy of thorax in *Piophila latipes* Meig.
FIG. 32. Fore-leg of *Piophila latipes* Meig.
FIG. 33. Wing of *Sphaerocera curvipes* Latr.
FIG. 34. Wing of *Copromyza equina* Fall.
FIG. 35. Hind-tibia and tarsus of *Copromyza equina* Fall.

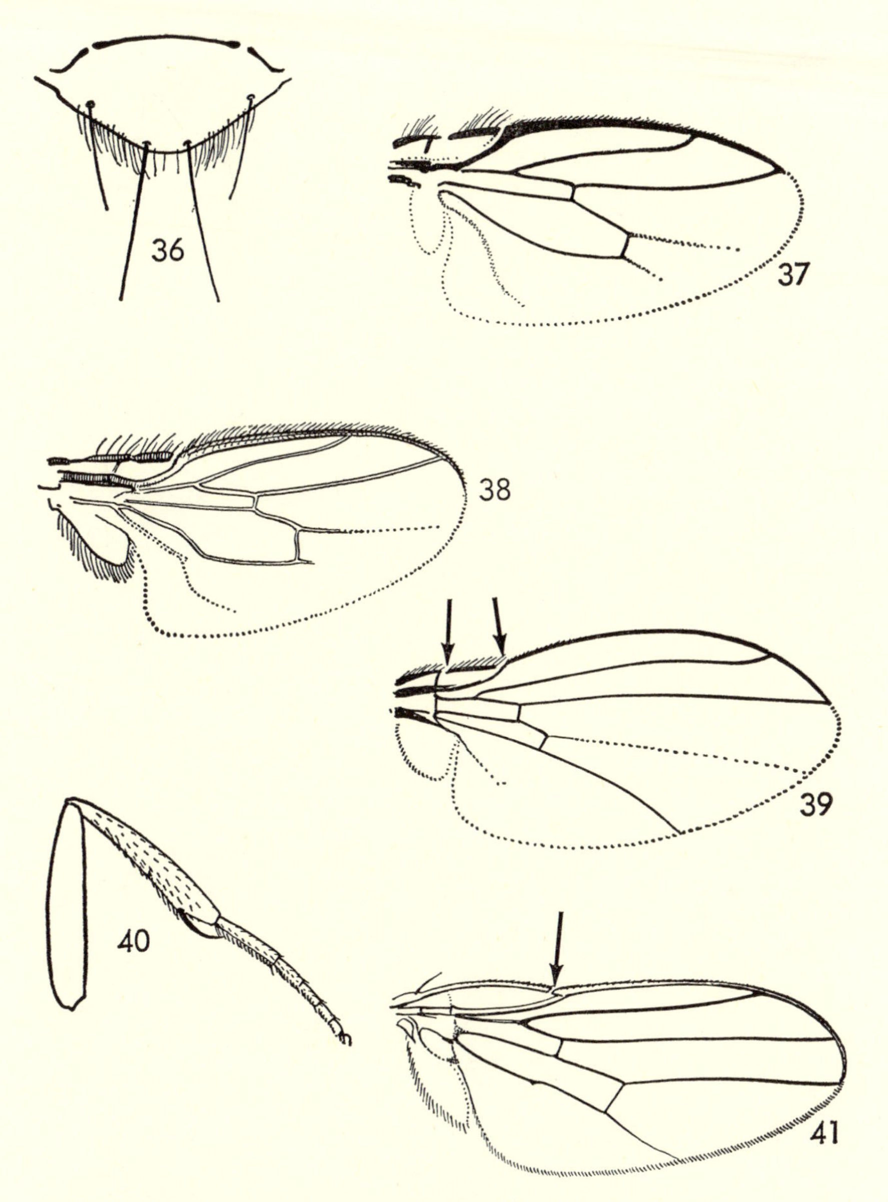

FIG. 36. Scutellum of *Copromyza similis* (Collin)
FIG. 37. Wing of *Leptocera fontinalis* (Fall.) (after Séguy)
FIG. 38. Wing of *Leptocera ferruginata* (Stenh.)
FIG. 39. Wing of *Meoneura vagans* (Fall.)
FIG. 40. Hind-leg of *Hippelates* sp.
FIG. 41. Wing of *Hippelates* sp.

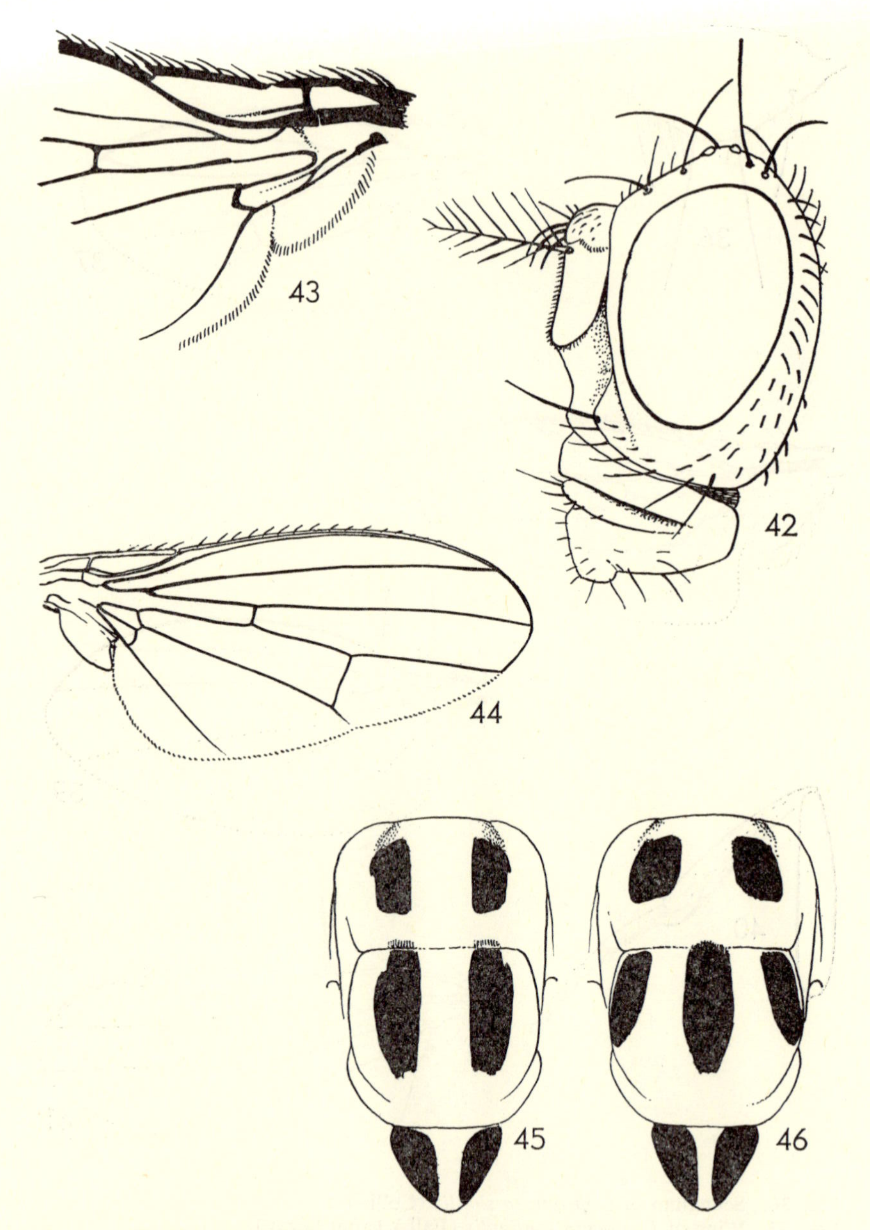

Fig. 42. Head of *Drosophila funebris* (Fabr.) (after Hendel)
Fig. 43. Wing base of *Drosophila funebris* (Fabr.)
Fig. 44. Wing of *Neoleria inscripta* (Meig.)
Fig. 45. Thorax pattern in *Anthomyia plurinotata* Brullé
Fig. 46. Thorax pattern in *Anthomyia pluvialis* (L.)

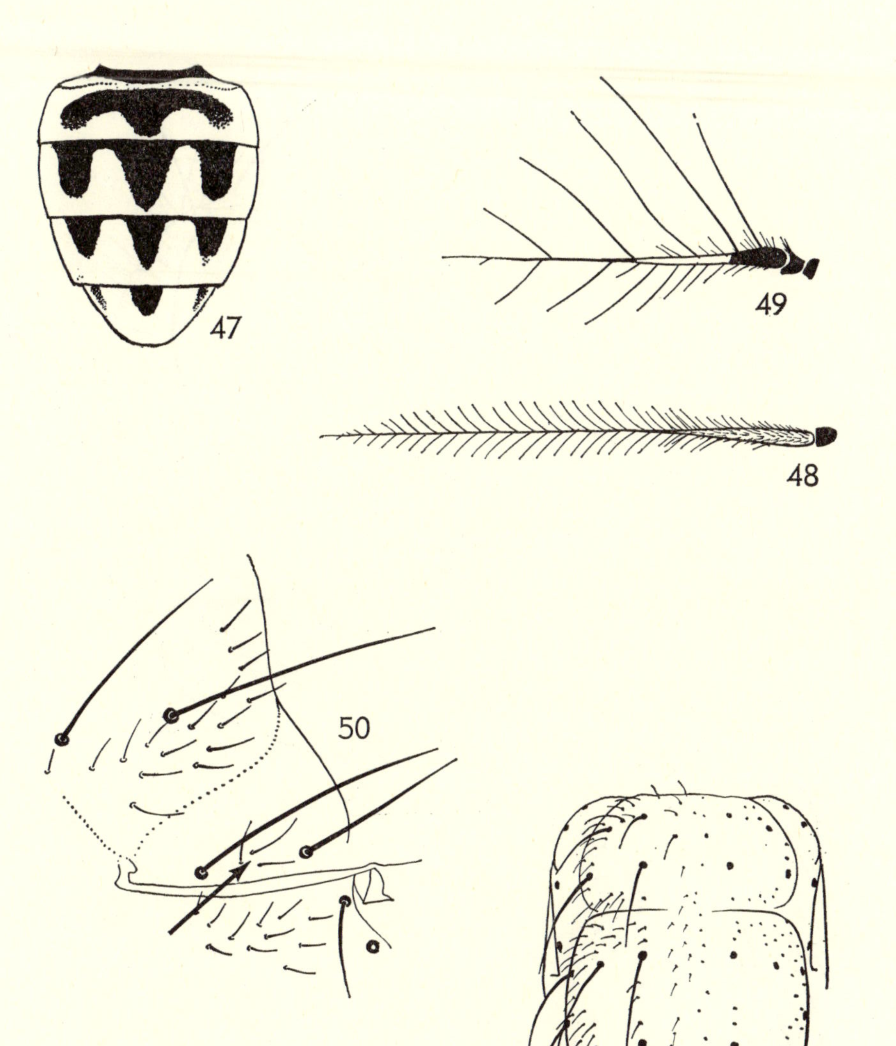

Fig. 47. Pattern of abdomen in female *Calythea albicincta* (Fall.)
Fig. 48. Arista of *Hylemya pullula* (Zett.)
Fig. 49. Arista of *Hylemya lasciva* (Zett.)
Fig. 50. Left notopleura of *Hylemya radicum* (L.)
Fig. 51. Chaetotaxy of thorax in *Hylemya radicum* (L.)

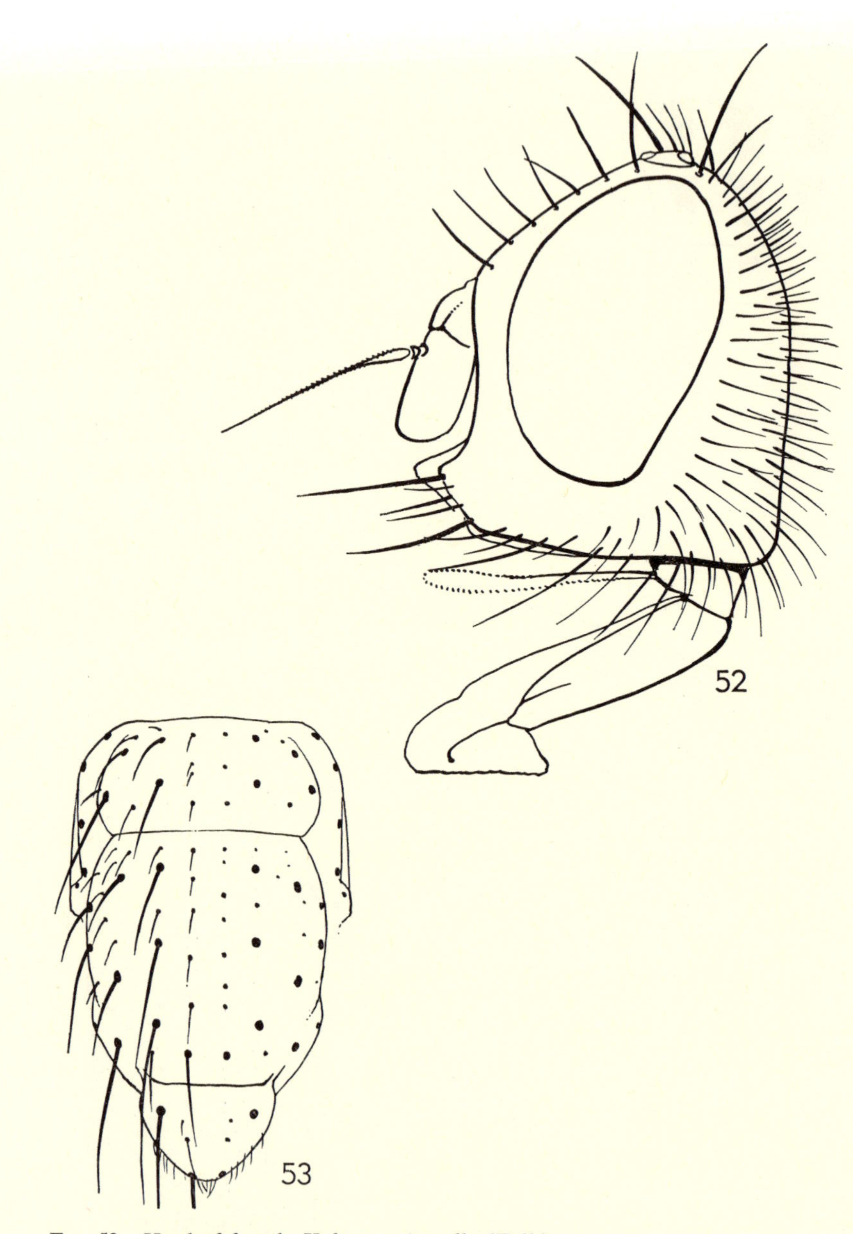

Fig. 52. Head of female *Hylemya cinerella* (Fall.)
Fig. 53. Chaetotaxy of thorax in *Hylemya cinerella* (Fall.)

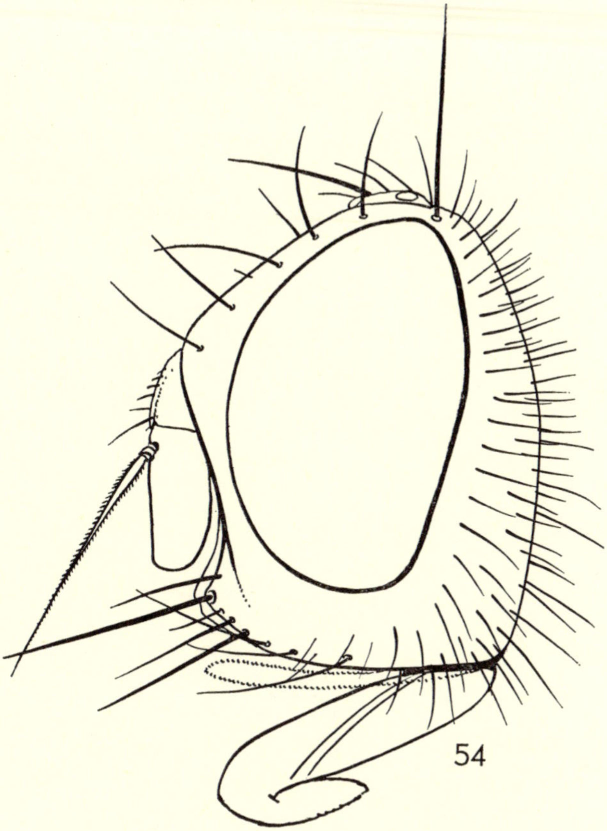

54

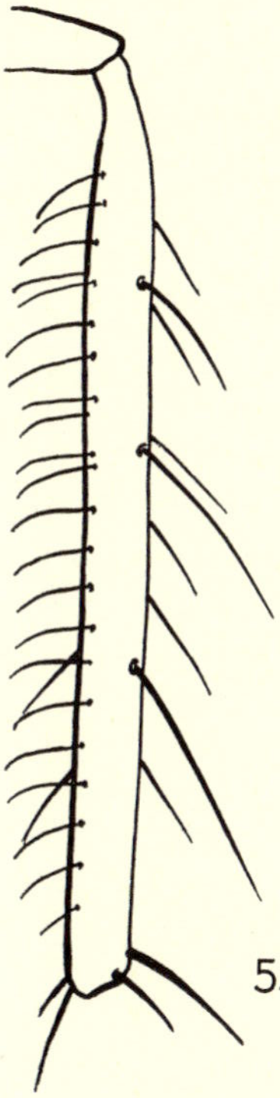

55

FIG. 54. Head of female *Delia platura* (Meig.)
FIG. 55. Hind-tibia of male *Delia platura* (Meig.)

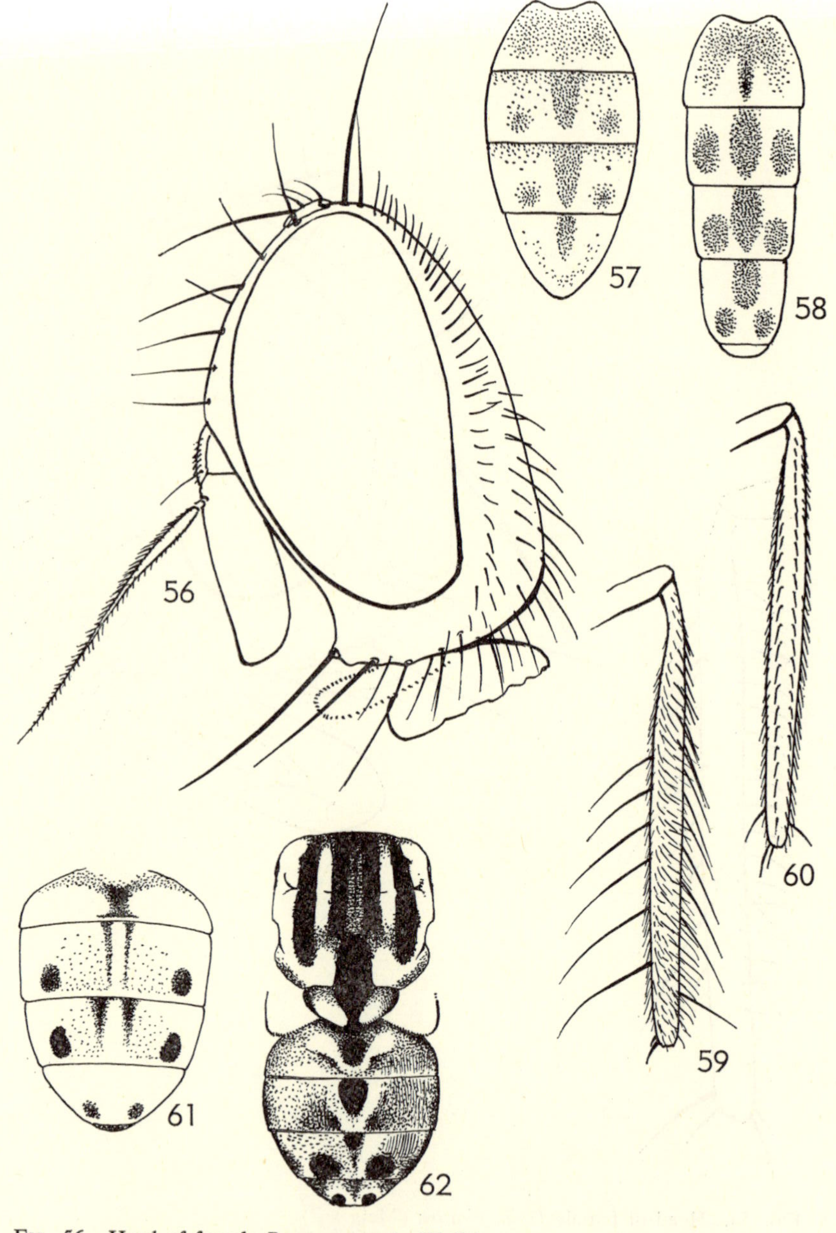

FIG. 56. Head of female *Pegomya socia* (Fall.)
FIG. 57. Schematic pattern of abdomen in female *Azelia macquarti* (Staeger)
FIG. 58. Schematic pattern of abdomen in male *Azelia macquarti* (Staeger)
FIG. 59. Middle tibia of male *Azelia macquarti* (Staeger)
FIG. 60. Middle tibia of male *Azelia triquetra* (Wied.)
FIG. 61. Schematic pattern of abdomen in male *Gymnodia impedita* Pand. (after Séguy)
FIG. 62. Body pattern in *Graphomya maculata* (Scop.)

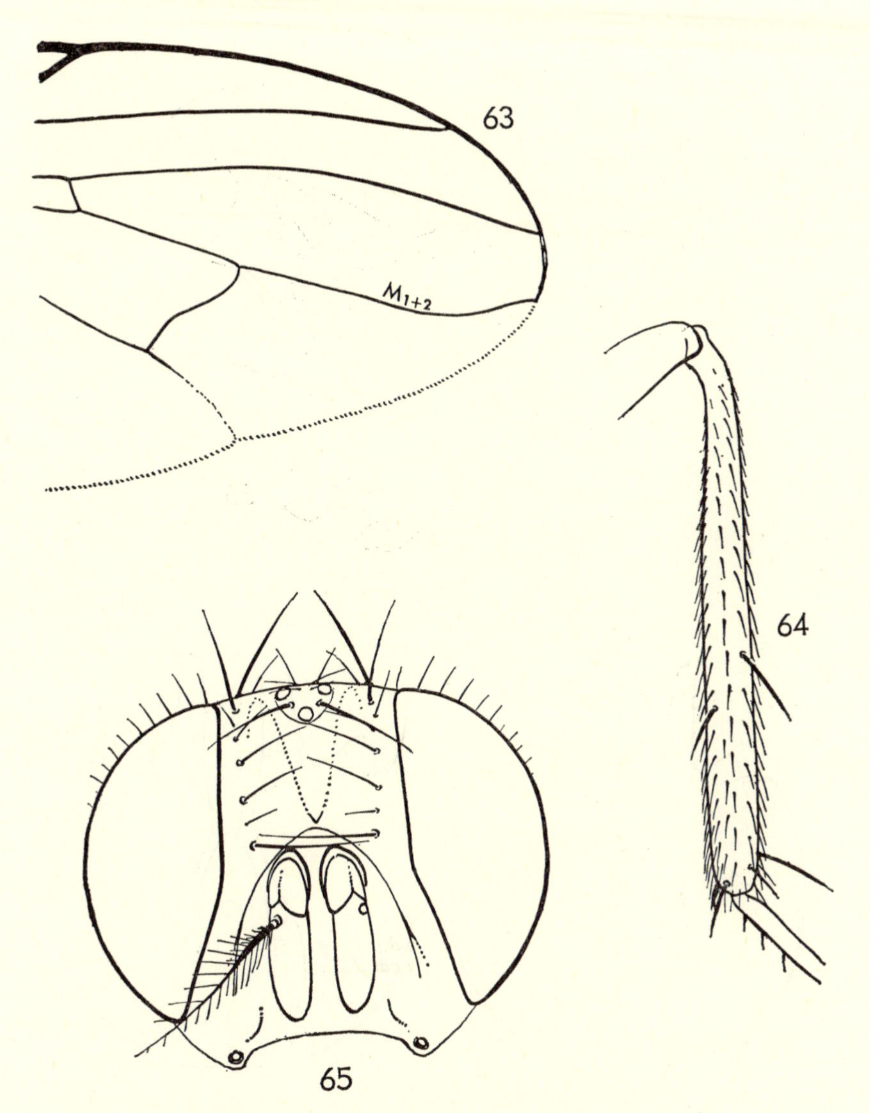

FIG. 63. Wing apex of female *Myospila hennigi* Gregor and Povolný
FIG. 64. Left hind-leg of female *Hebecnema* sp. in frontal view
FIG. 65. Head of female *Hebecnema umbratica* (Meig.) in frontal view

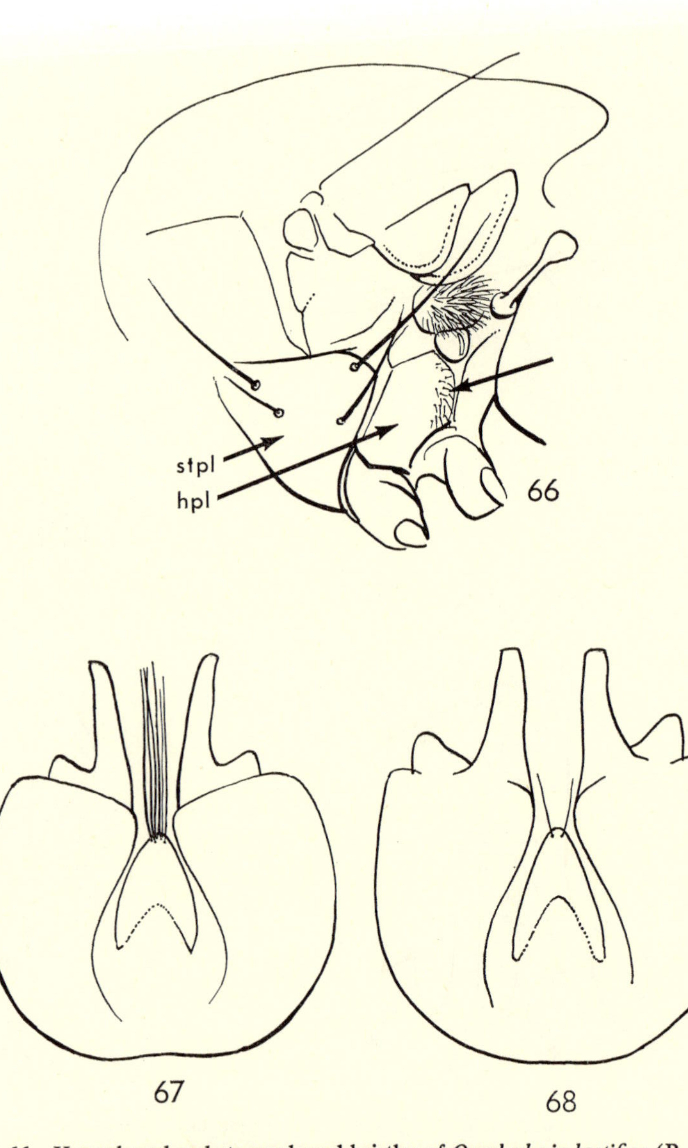

Fig. 66. Hypopleural and sternopleural bristles of *Quadrularia laetifica* (R.-D.)
Fig. 67. Scheme of hypopygium in *Fannia difficilis* Stein
Fig. 68. Scheme of hypopygium in *Fannia canicularis* (L.)

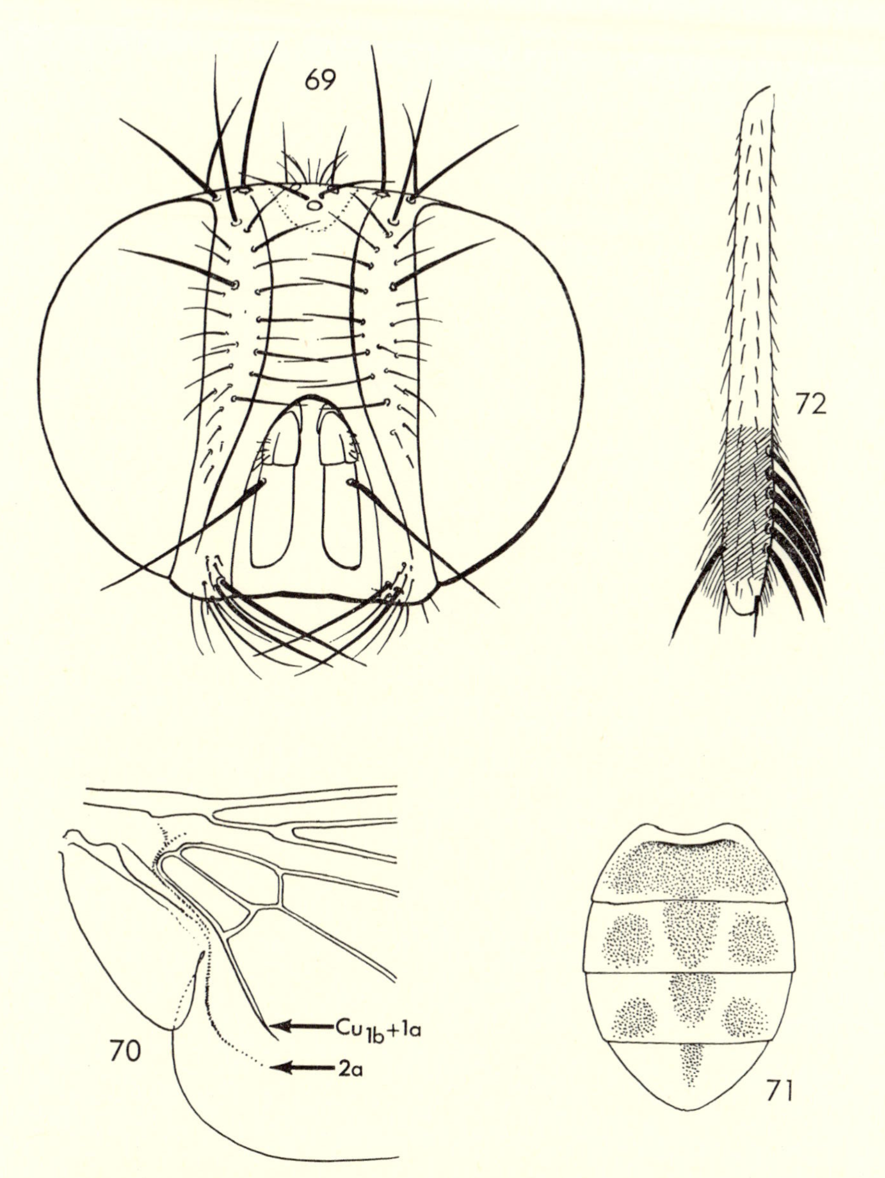

FIG. 69. Head of female *Fannia scalaris* (Fabr.)
FIG. 70. Wing base of *Fannia* sp.
FIG. 71. Schematic pattern of abdomen in female *Fannia leucosticta* (Meig.)
FIG. 72. Fore-tibia of male *Fannia manicata* (Meig.)

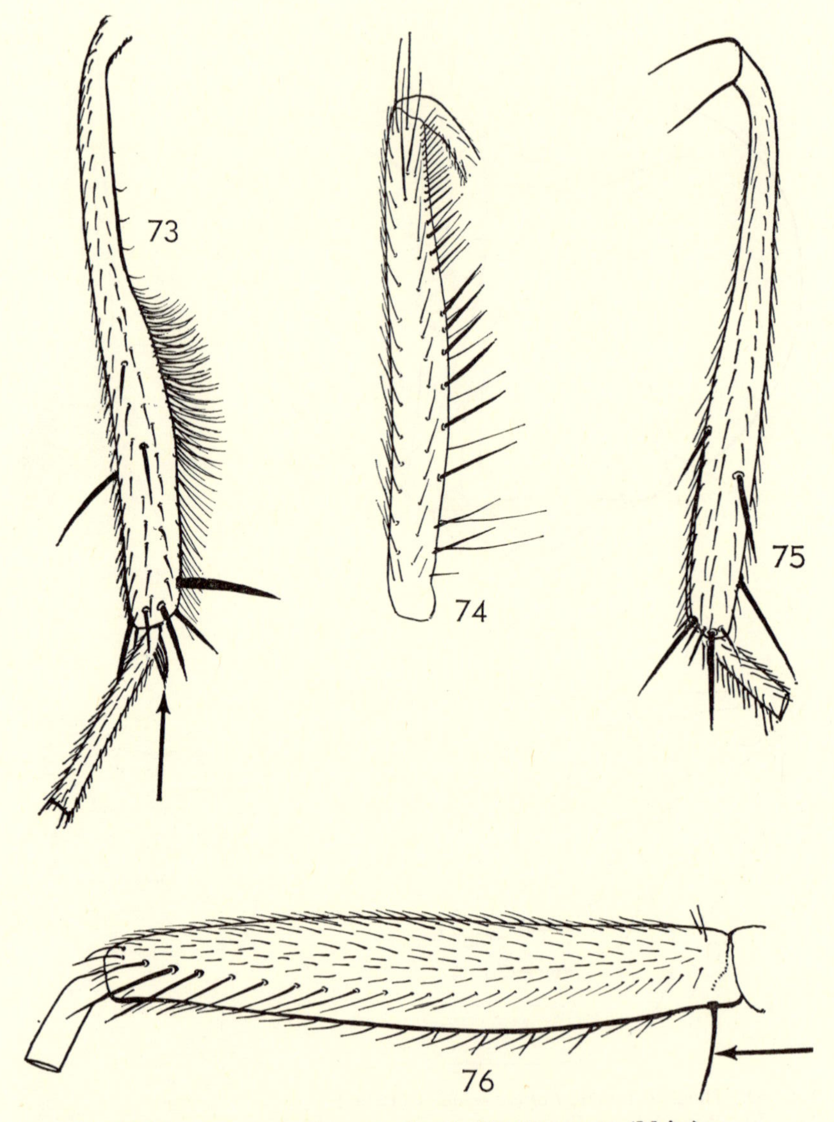

FIG. 73. Middle femur and pretarsus of male *Fannia armata* (Meig.)
FIG. 74. Left middle femur of male *Fannia mutica* (Zett.)
FIG. 75. Left middle tibia of female *Fannia mutica* (Zett.)
FIG. 76. Left middle femur of female *Fannia scalaris* (Fabr.)

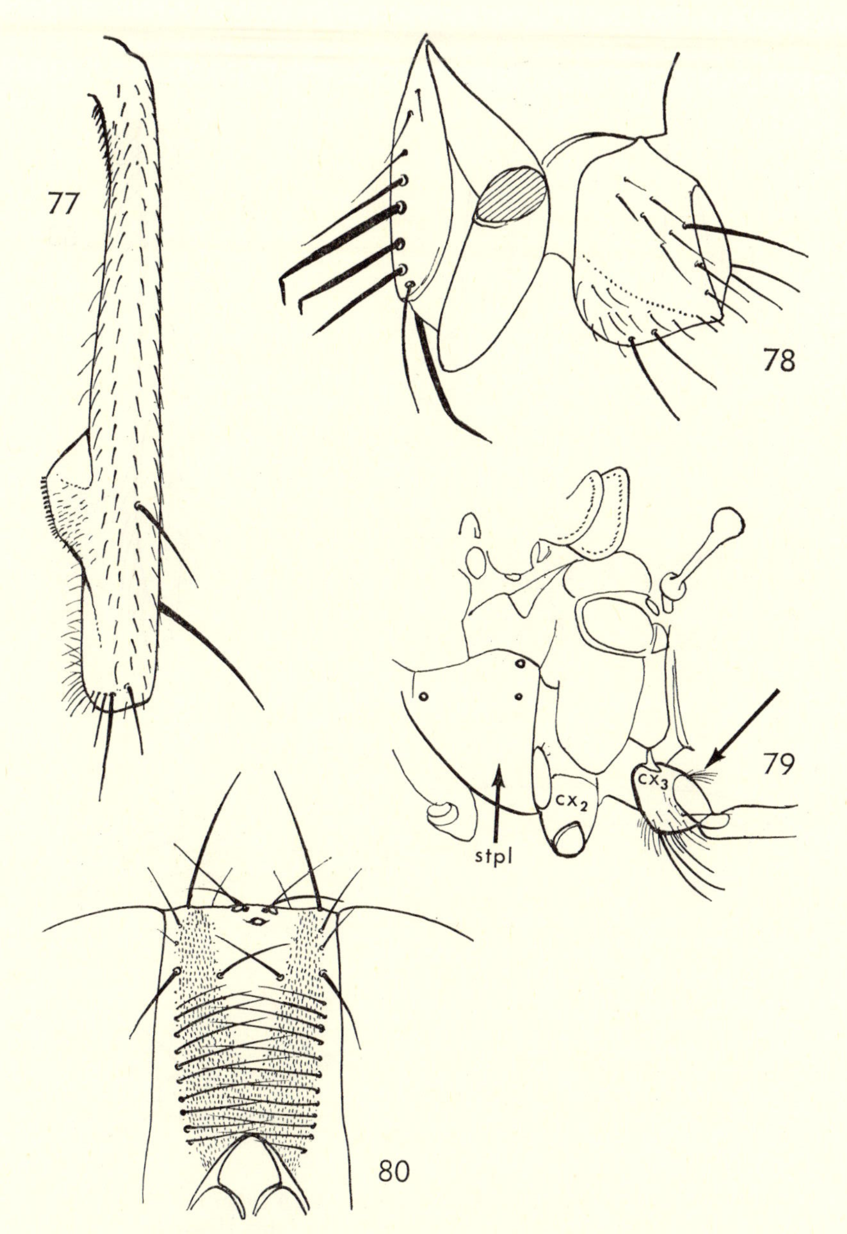

FIG. 77. Left middle tibia of male *Fannia scalaris* (Fabr.)
FIG. 78. Left middle and hind-coxa of male *Fannia scalaris* (Fabr.)
FIG. 79. Left hind-coxa and a part of thorax of *Lasiops simplex* (Wied.)
FIG. 80. Frons of female *Ophyra leucostoma* (Wied.)

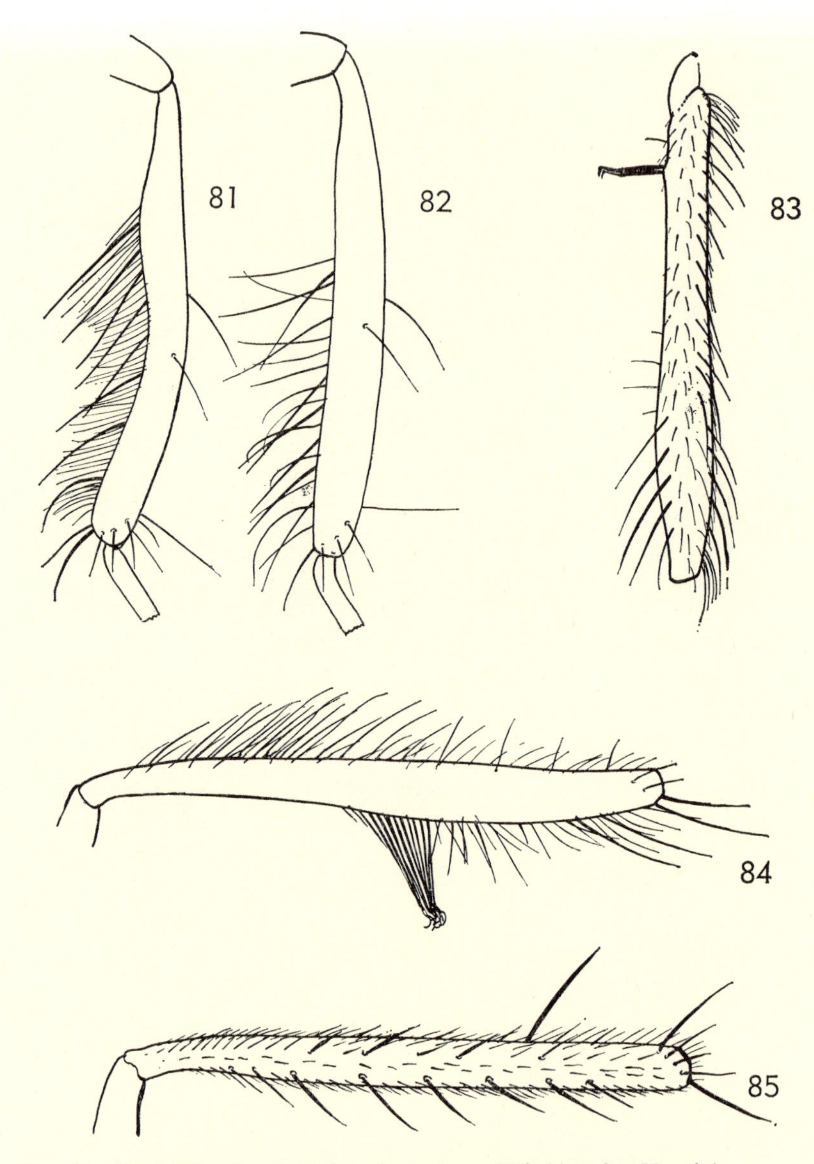

Fig. 81. Hind-tibia of male *Ophyra leucostoma* (Wied.) (after Hennig)
Fig. 82. Hind-tibia of male *Ophyra capensis* (Wied.) (after Hennig)
Fig. 83. Hind-femur of male *Hydrotaea occulta* (Meig.)
Fig. 84. Left hind-tibia of male *Hydrotaea penicillata* (Rond.) (after Hennig)
Fig. 85. Left hind-tibia of male *Hydrotaea cyrtoneurina* (Zett.)

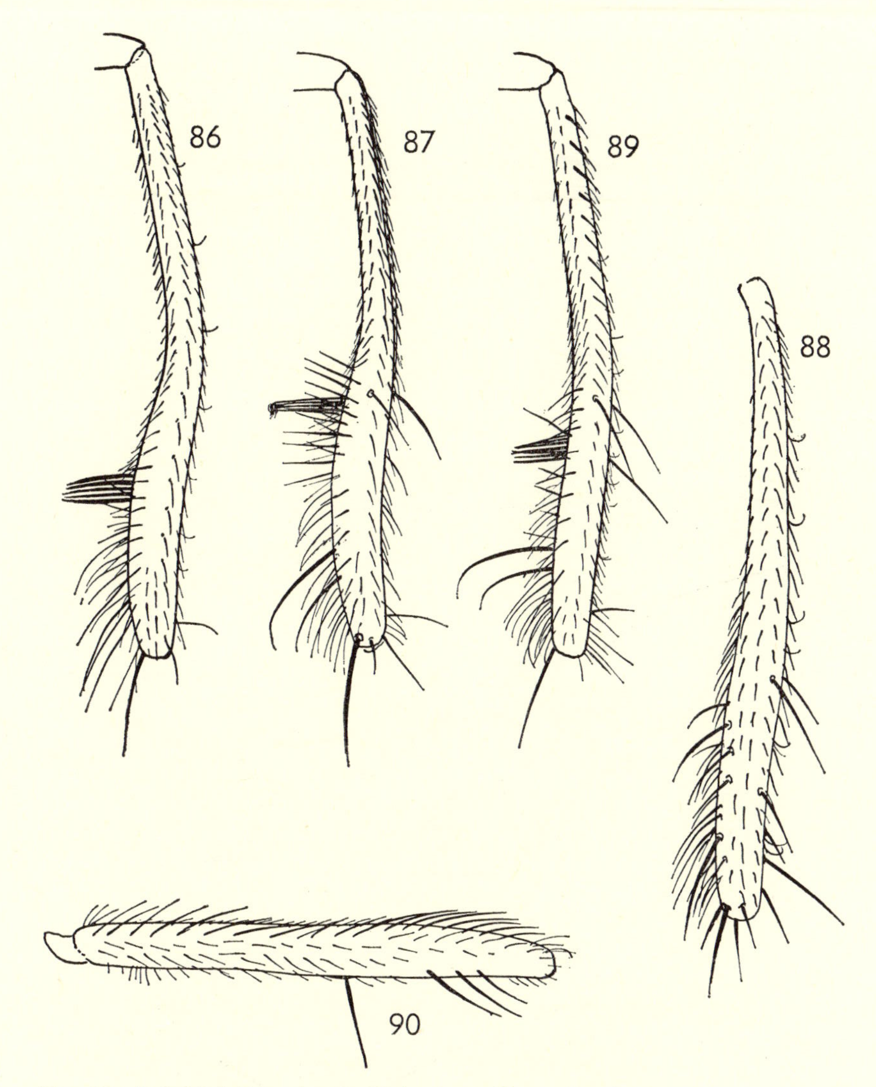

Fig. 86. Left hind-tibia of male *Hydrotaea pellucens* Porch.
Fig. 87. Left hind-tibia of male *Hydrotaea meridionalis* Porch.
Fig. 88. Left hind-tibia of male *Hydrotaea irritans* (Fall.)
Fig. 89. Left hind-tibia of male *Hydrotaea pandellei* Stein
Fig. 90. Left hind-femur of male *Hydrotaea armipes* (Fall.)

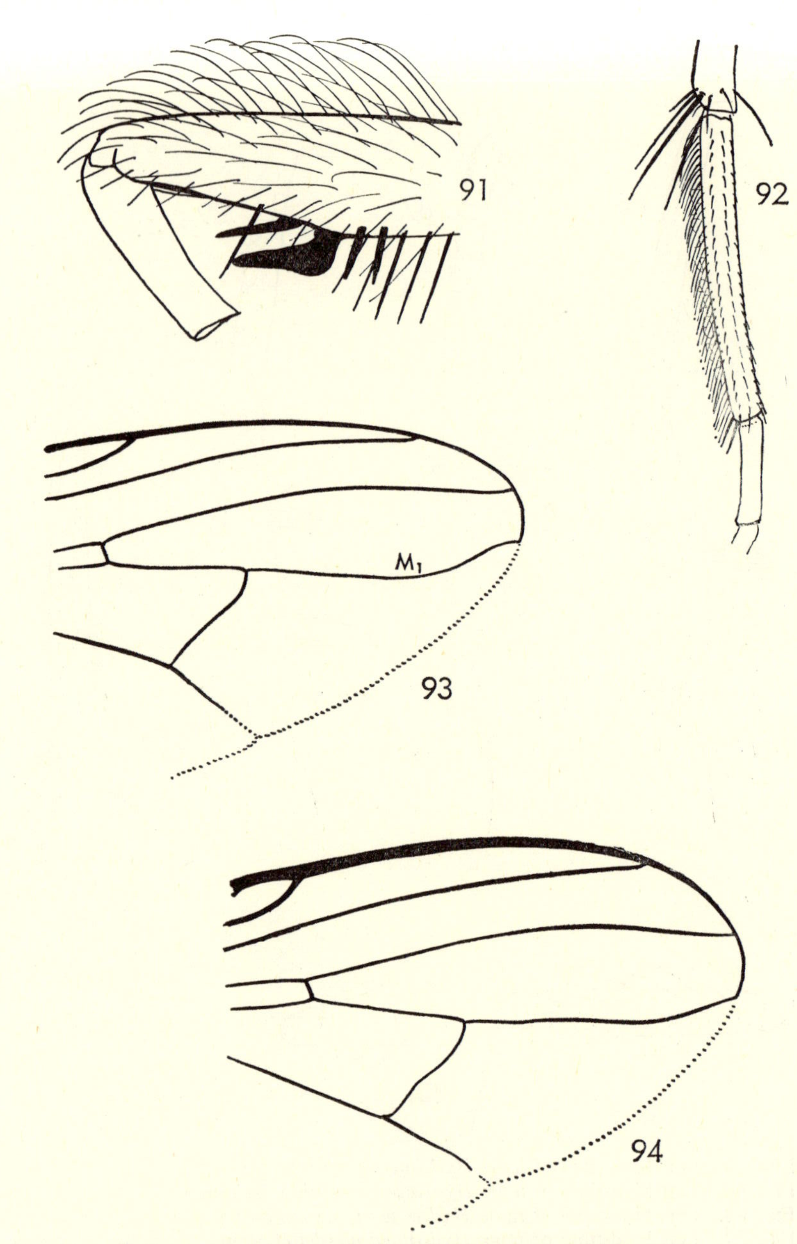

FIG. 91. Distal part of left femur of male *Hydrotaea meteorica* (L.)
FIG. 92. Middle metatarsus of male *Hydrotaea irritans* (Fall.)
FIG. 93. Apex of wing in *Muscina stabulans* (Fall.)
FIG. 94. Apex of wing in *Muscina assimilis* (Fall.)

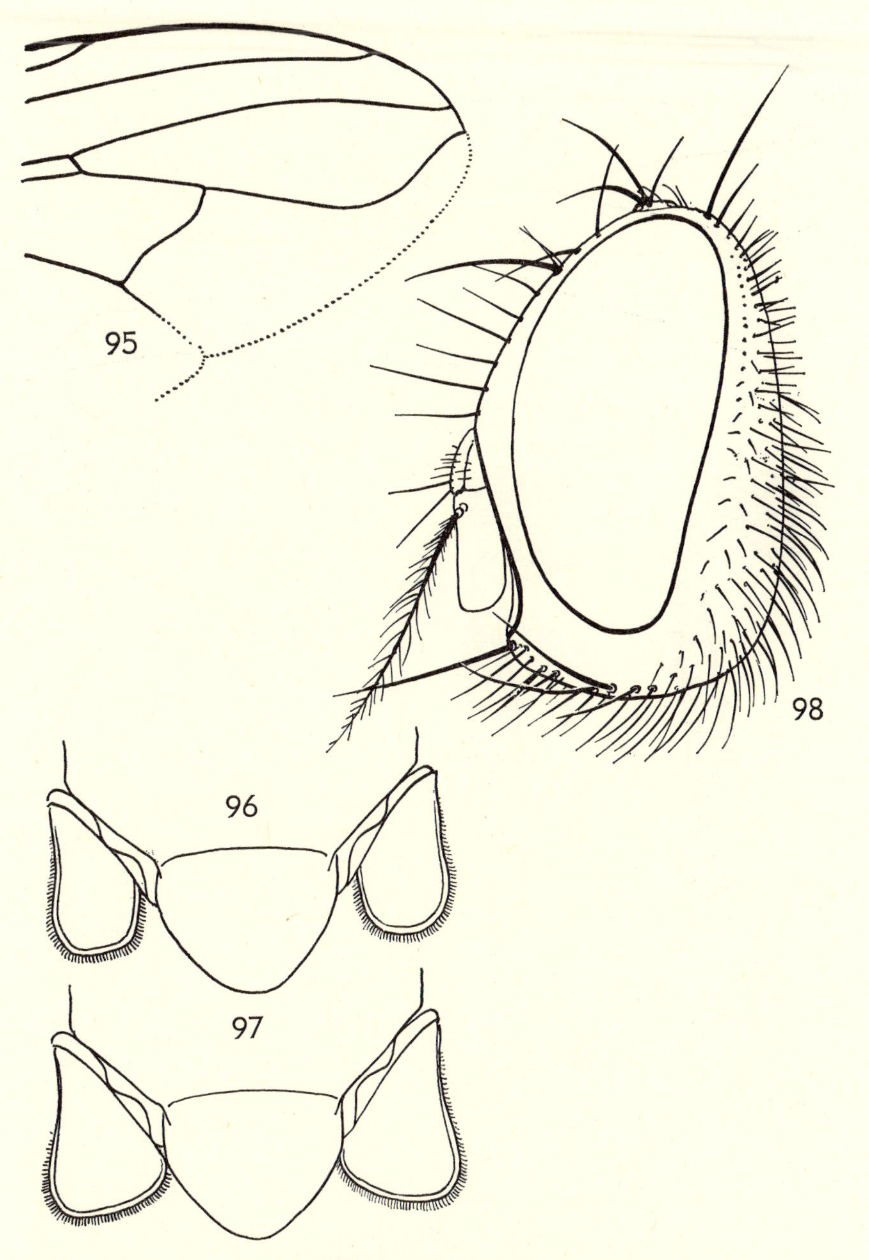

Fig. 95. Apex of wing in *Muscina pabulorum* (Fall.)
Fig. 96. Lower squama of *Muscina pabulorum* (Fall.)
Fig. 97. Lower squama of *Muscina pascuorum* (Meig.)
Fig. 98. Head of female *Dendrophaonia querceti* (Bouché)

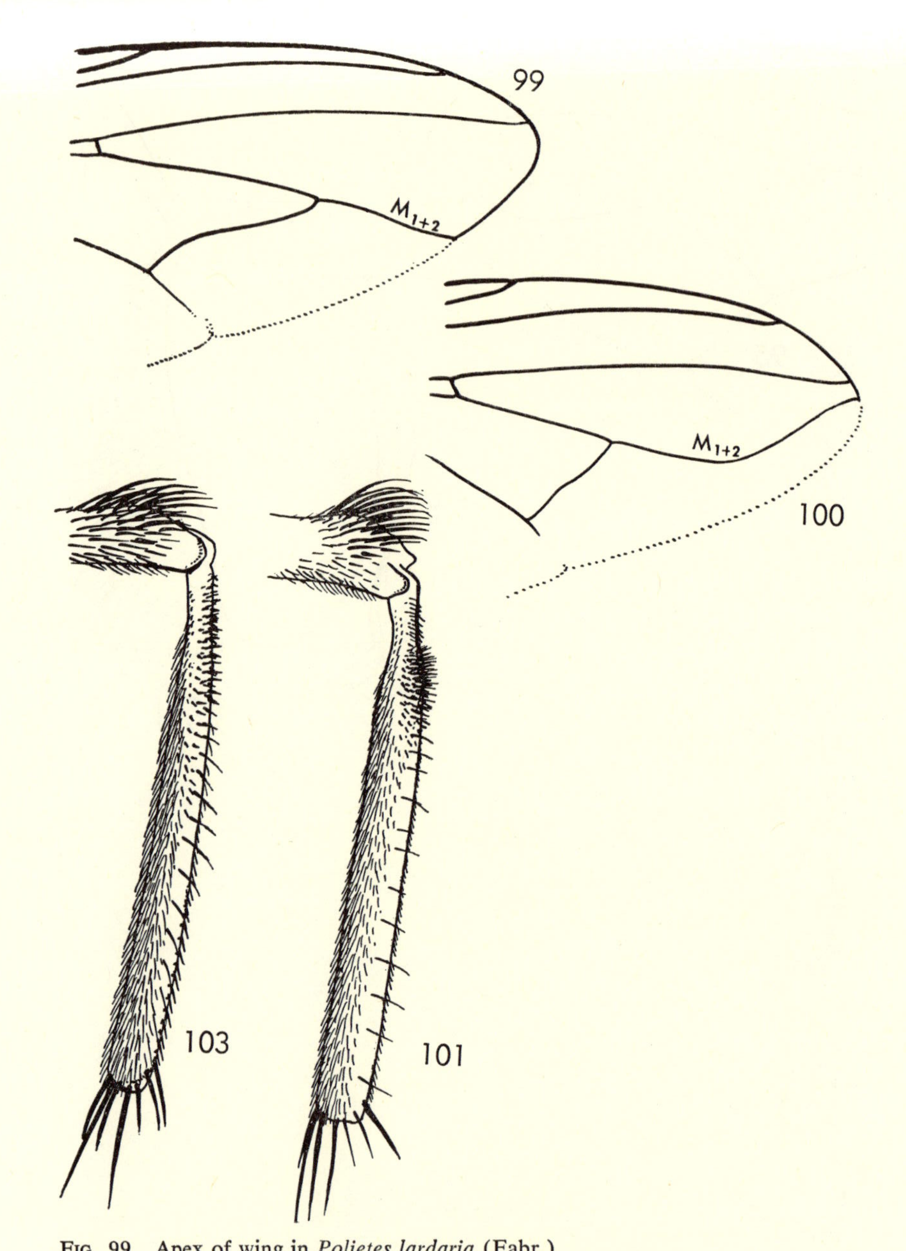

FIG. 99. Apex of wing in *Polietes lardaria* (Fabr.)
FIG. 100. Apex of wing in *Morellia podagrica* (Loew.)
FIG. 101. Left middle tibia of male *Morellia podagrica* (Loew.)
FIG. 103. Left middle tibia of male *Morellia hortorum* (Fall.)

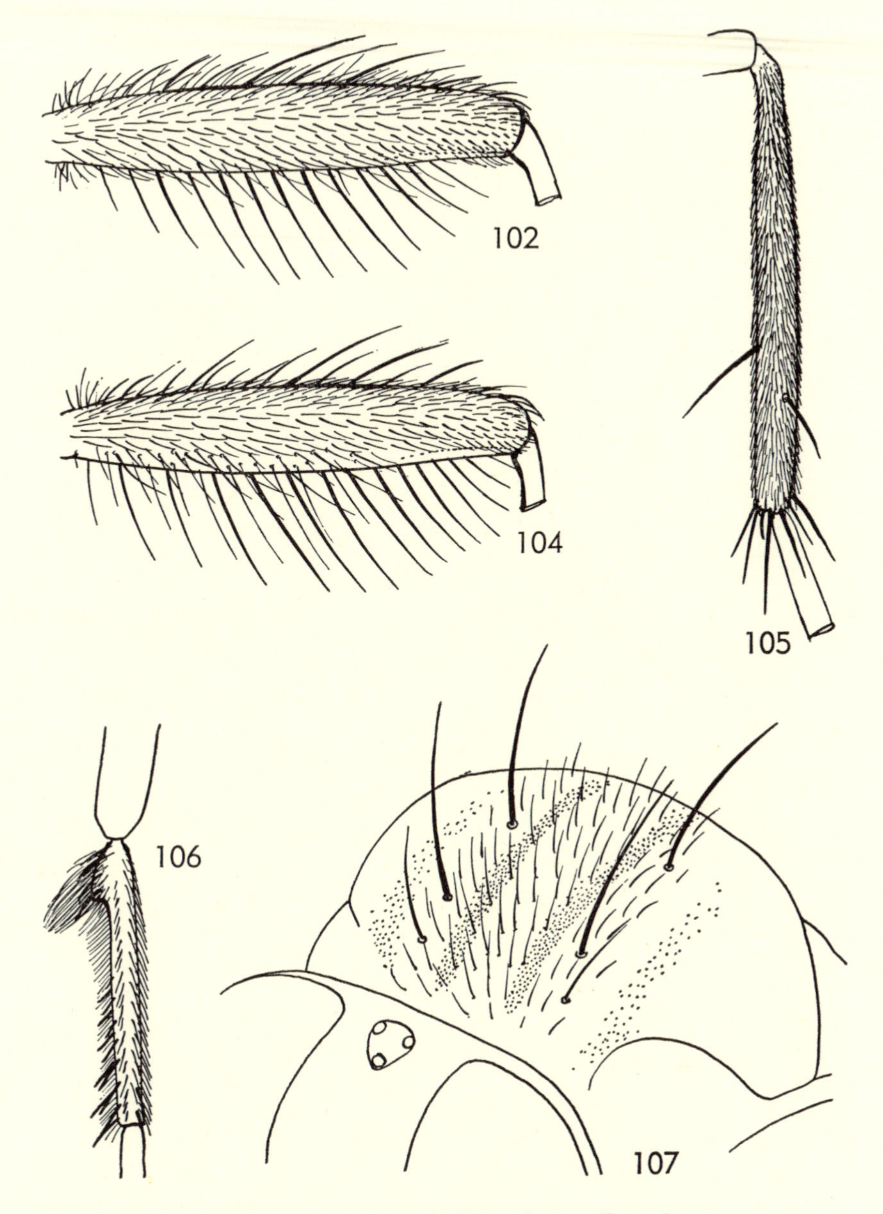

Fig. 102. Left fore-femur of female *Morellia podagrica* (Loew.)
Fig. 104. Left fore-femur of female *Morellia hortorum* (Fall.)
Fig. 105. Left middle tibia of *Pyrellia cyanicolor* (Zett.)
Fig. 106. Hind-metatarsus of male *Dasyphora penicillata* (Egger)
Fig. 107. Chaetotaxy of presutural part of mesonotum in *Dasyphora pratorum* (Meig.)

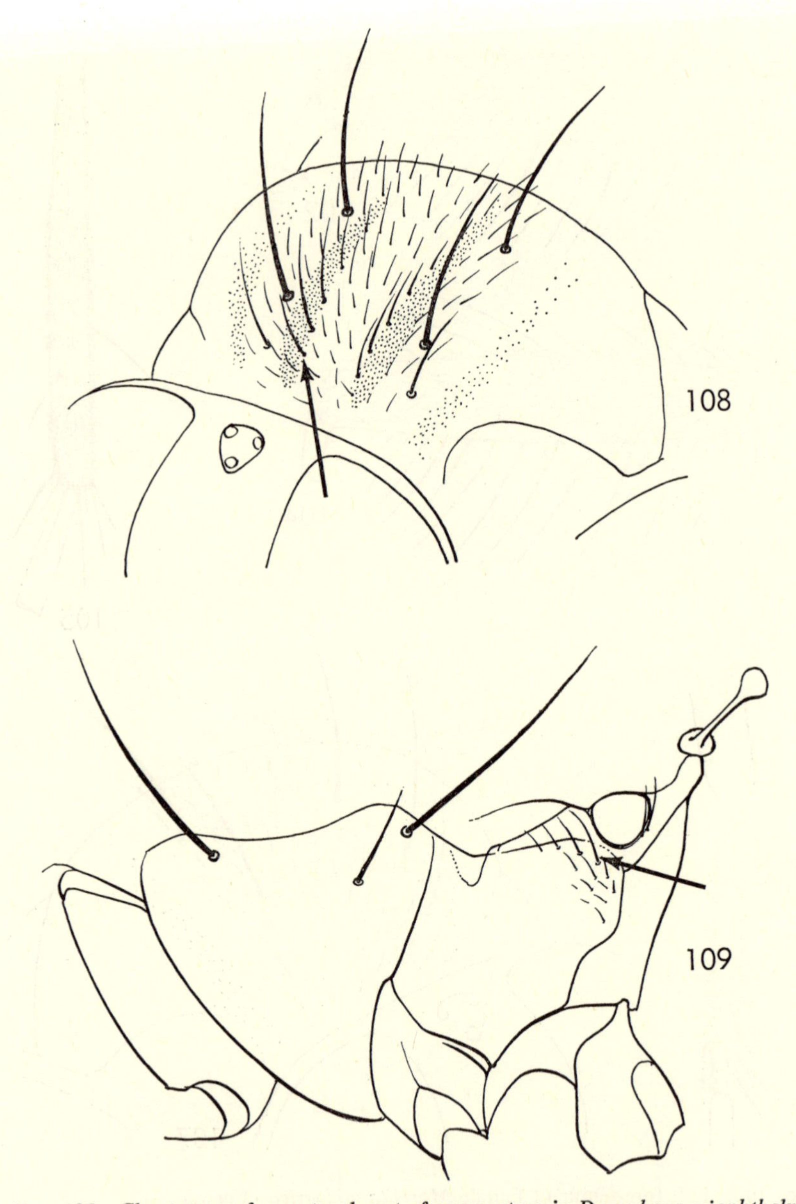

Fig. 108. Chaetotaxy of presutural part of mesonotum in *Dasyphora eriophthalma* (Macq.)

Fig. 109. Sternopleura and hypopleura of *Musca sorbens* Wied.

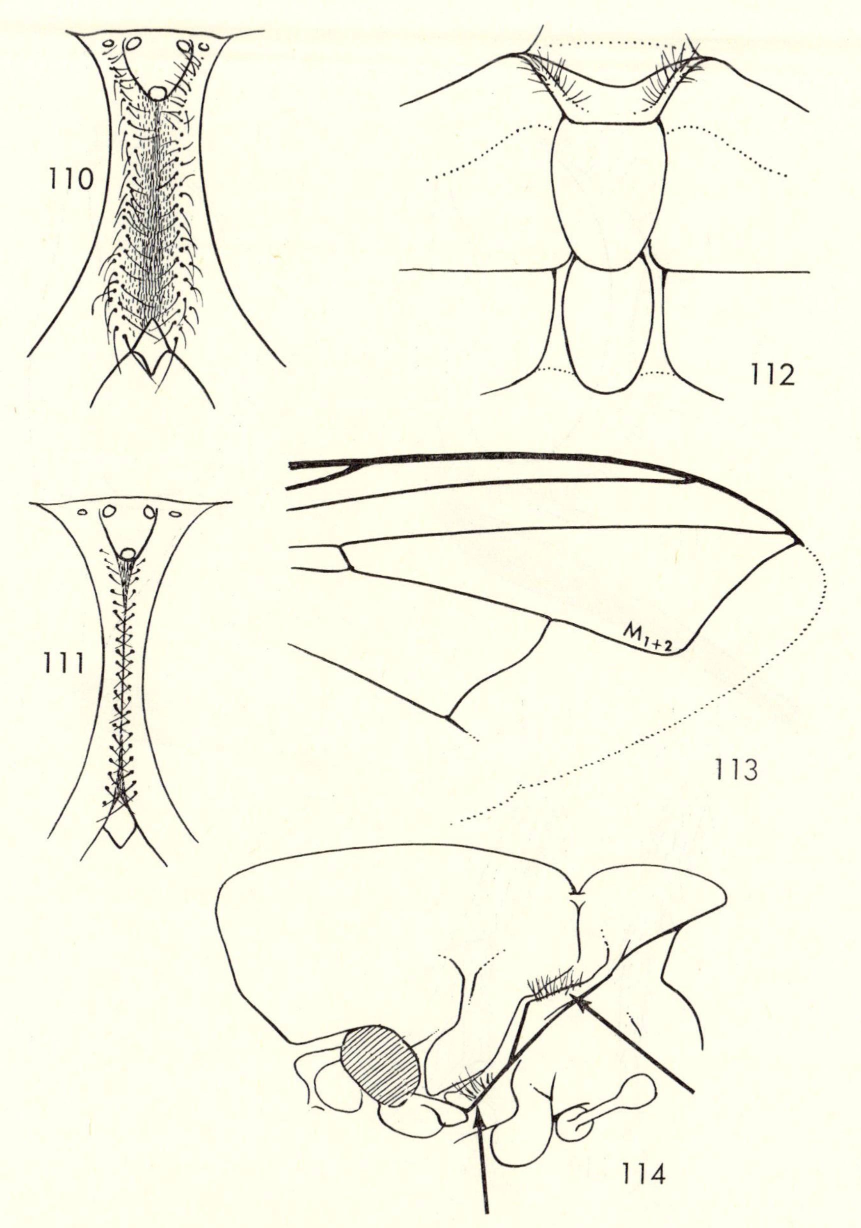

FIG. 110. Frons of male *Musca osiris* Wied.
FIG. 111. Frons of male *Musca vitripennis* Meig.
FIG. 112. First to third abdominal sternite of *Musca autumnalis* De Geer
FIG. 113. Apex of wing of *Musca autumnalis* De Geer
FIG. 114. Suprasquamal carina of *Musca larvipara* Porch.

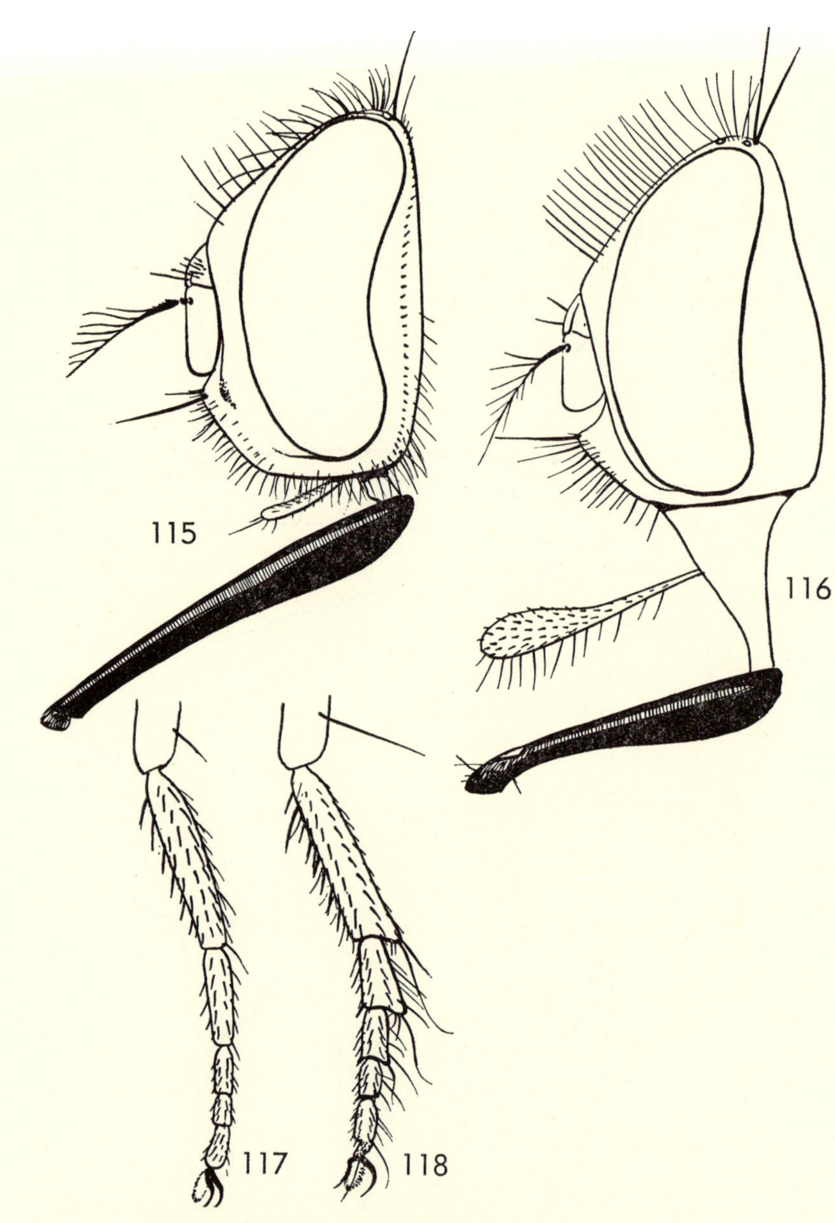

FIG. 115. Head of *Stomoxys calcitrans* (L.)
FIG. 116. Head of *Lyperosiops stimulans* (Meig.) (after Hennig)
FIG. 117. Hind-tarsi of *Haematobia titillans* (Bezzi) (after Hennig)
FIG. 118. Hind-tarsi of *Haematobia irritans* (L.)

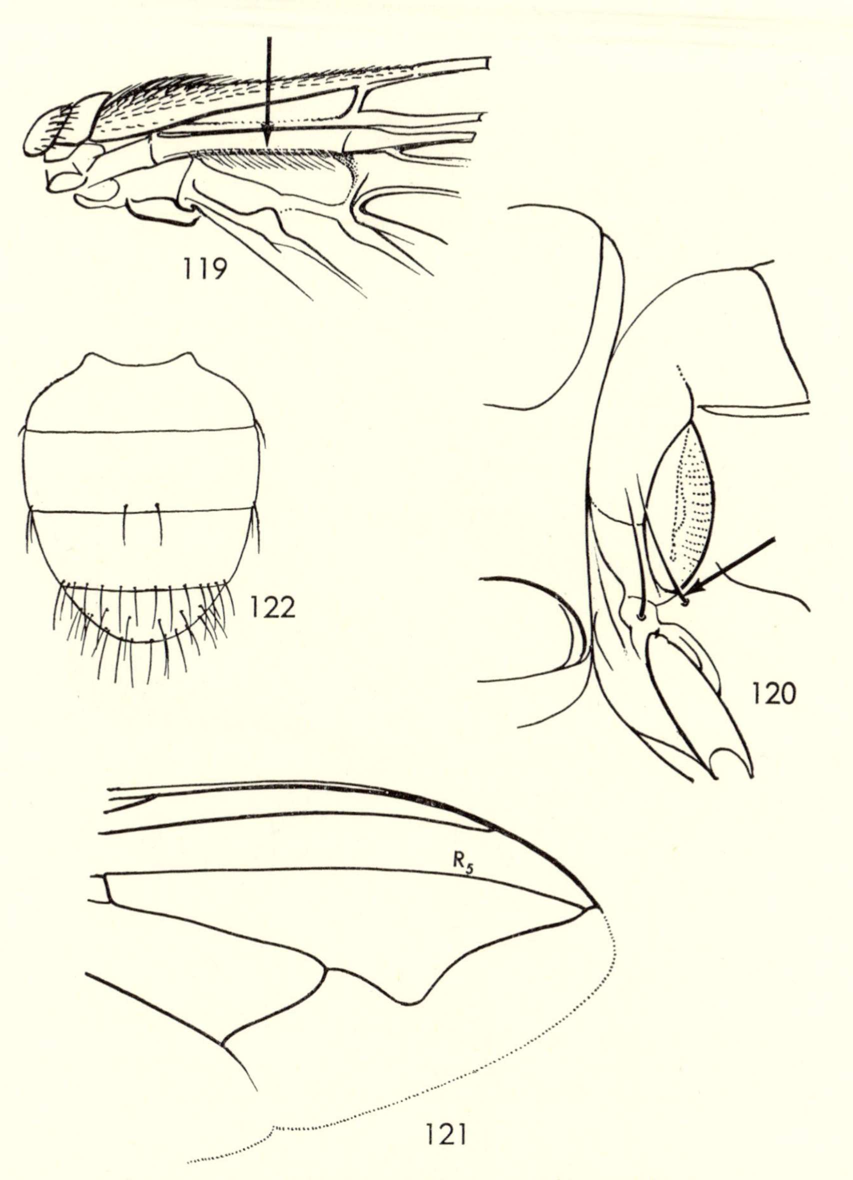

Fig. 119. Upper surface of wing base of *Chrysomya albiceps* (Wied.)
Fig. 120. Area surrounding anterior spiracle of *Chrysomya albiceps* (Wied.)
Fig. 121. Apex of wing of *Pollenia atramentaria* (Meig.)
Fig. 122. Macrochaetae on abdominal tergites of *Bufolucilia silvarum* (Meig.)

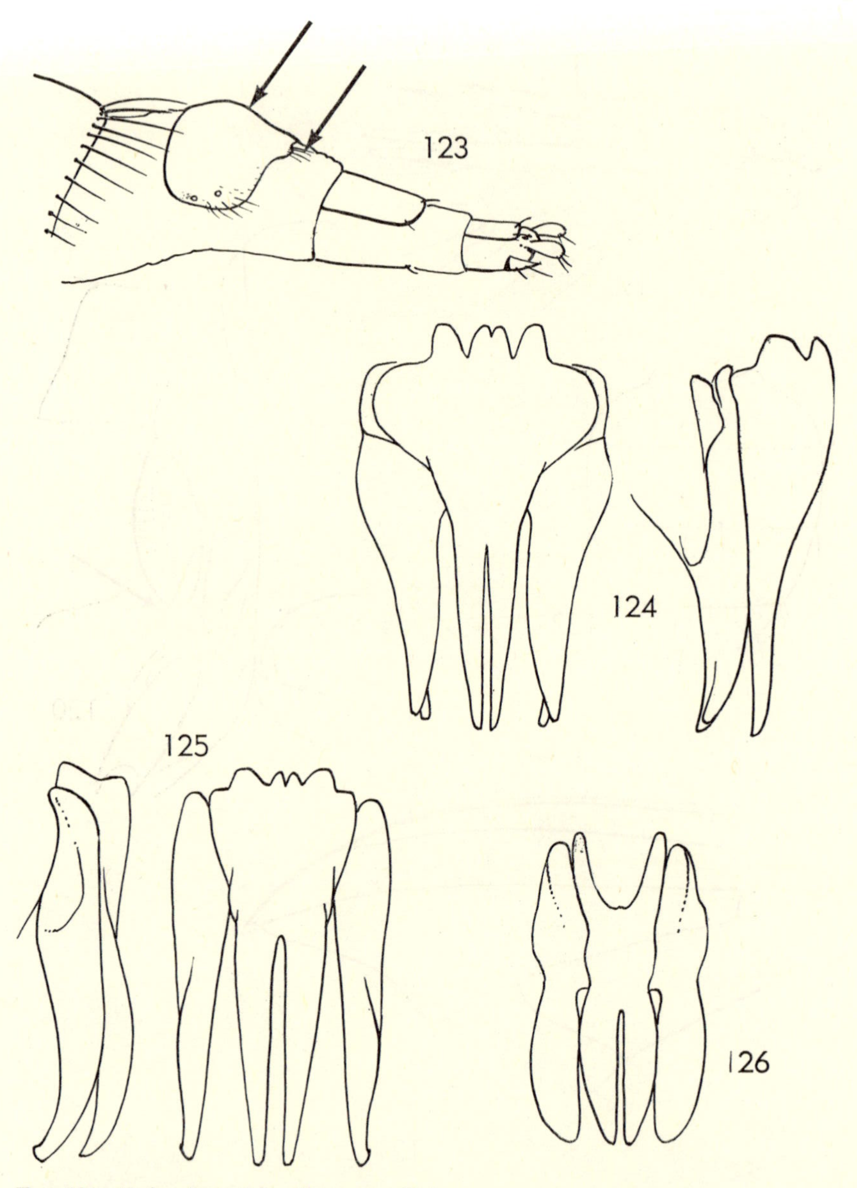

Fig. 123. Ovipositor of *Lucilia caesar* (L.)

Fig. 124. Cerci and surstyli in dorsoventral (left) and lateral (right) view in *Lucilia caesar* (L.)

Fig. 125. Cerci and surstyli in lateral (left) and dorsoventral (right) view in *Lucilia illustris* (Meig.)

Fig. 126. Cerci and surstyli in dorsoventral view in *Lucilia ampullacea* Vill.

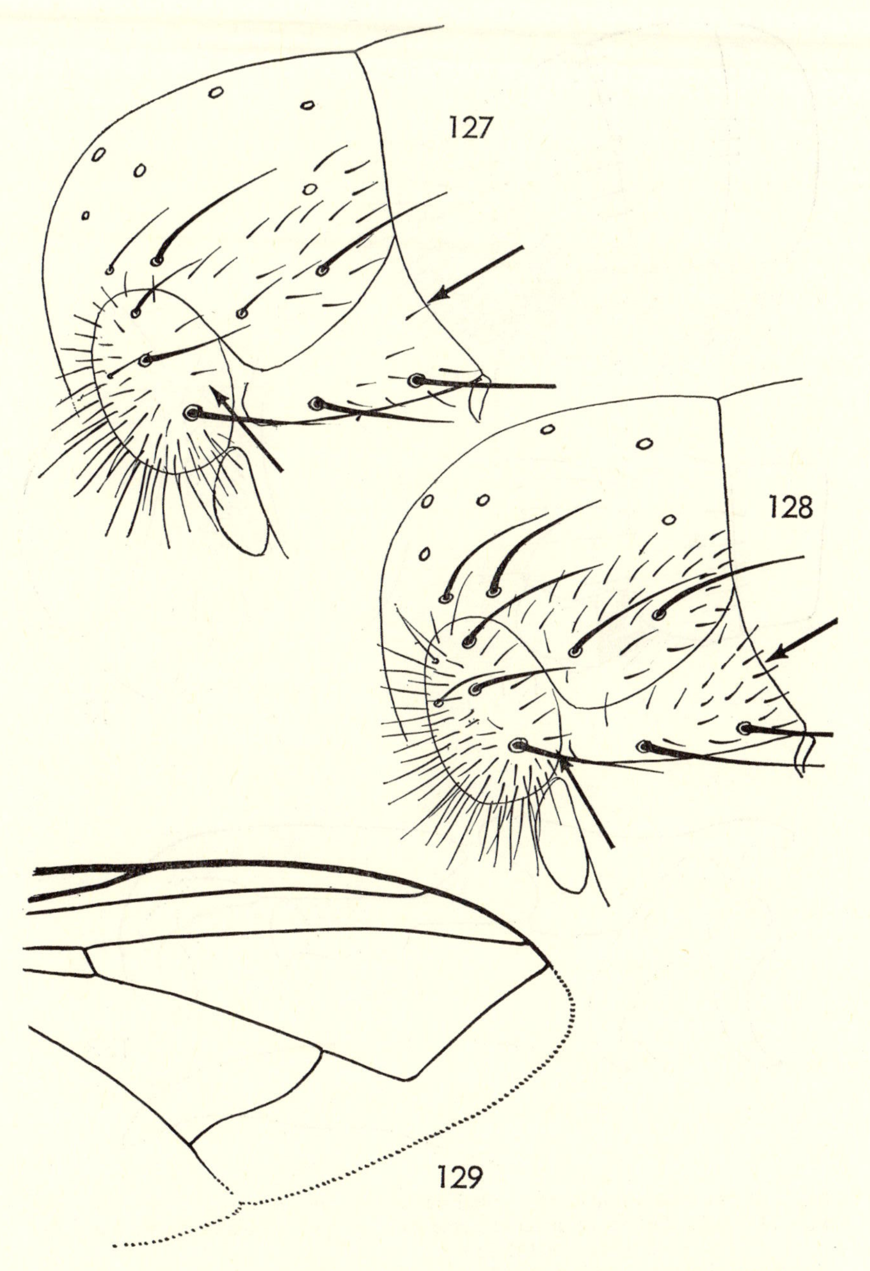

FIG. 127. Microsetae on humerus and notopleura in *Phaenicia cuprina* (Wied.)
FIG. 128. Microsetae on humerus and notopleura in *Phaenicia sericata* (Meig.)
FIG. 129. Apex of wing of *Phaenicia sericata* (Meig.)

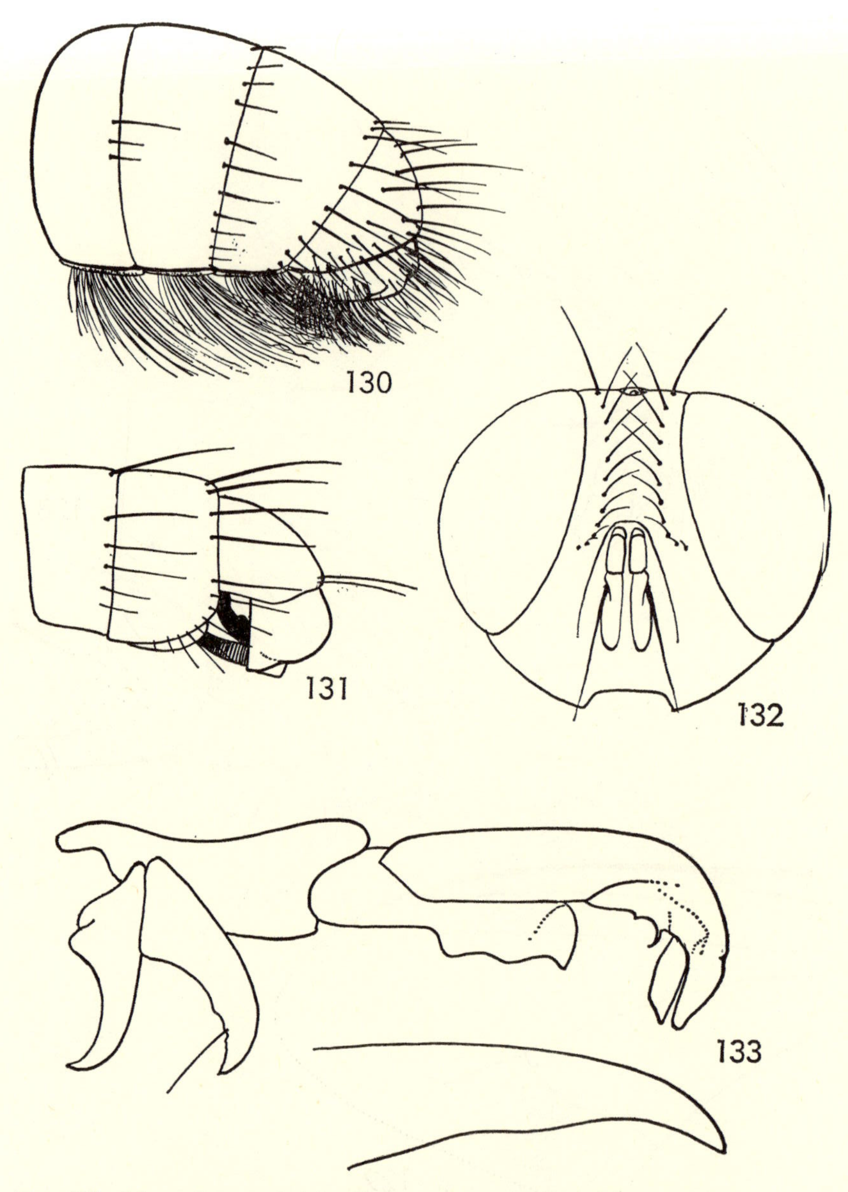

FIG. 130. Macrochaetae of abdominal sternites in *Lucilia pilosiventris* Kram.
FIG. 131. Distribution of macrochaetae on the apex of abdomen in male *Sarcophaga melanura* (Meig.)
FIG. 132. Head of female *Sarcophaga melanura* (Meig.)
FIG. 133. Phallosoma and apex of cerci of *Sarcophaga melanura* (Meig.)

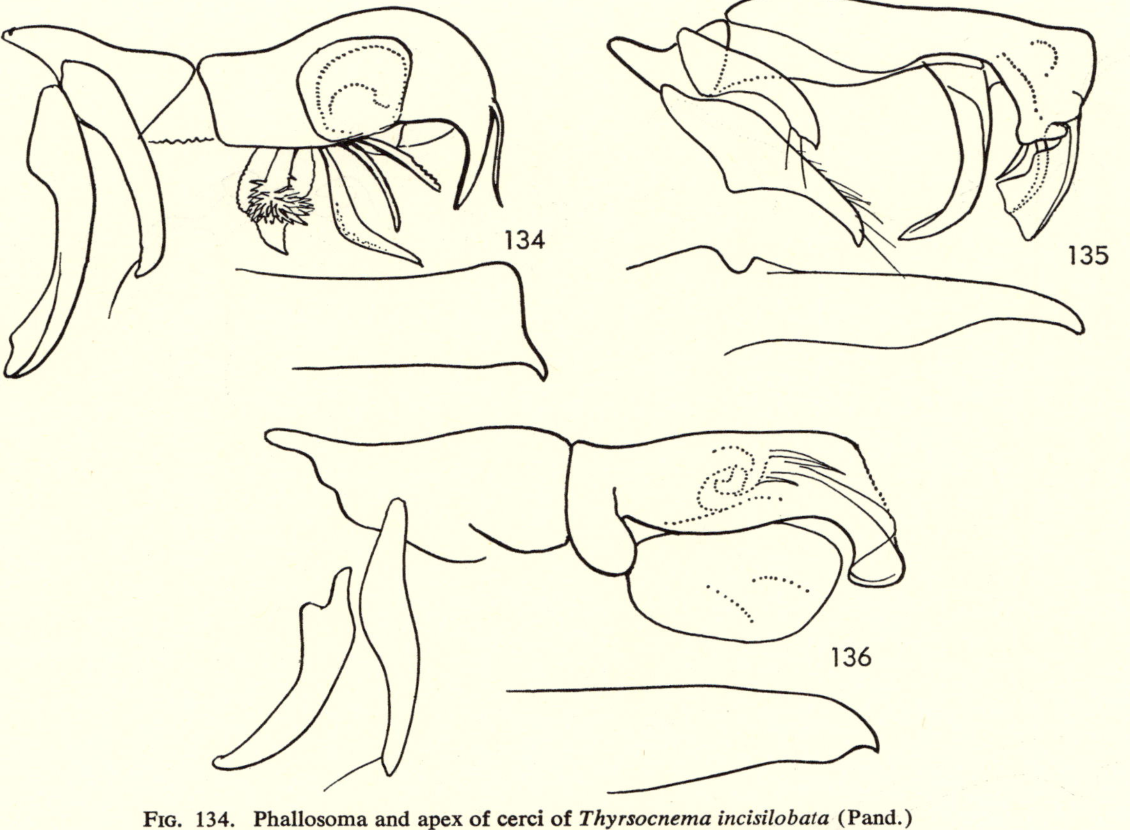

FIG. 134. Phallosoma and apex of cerci of *Thyrsocnema incisilobata* (Pand.)
FIG. 135. Phallosoma and apex of cerci of *Sarcophaga haemorrhoidalis* (Fall.)
FIG. 136. Phallosoma and apex of cerci of *Sarcophaga subvicina* Rohd.

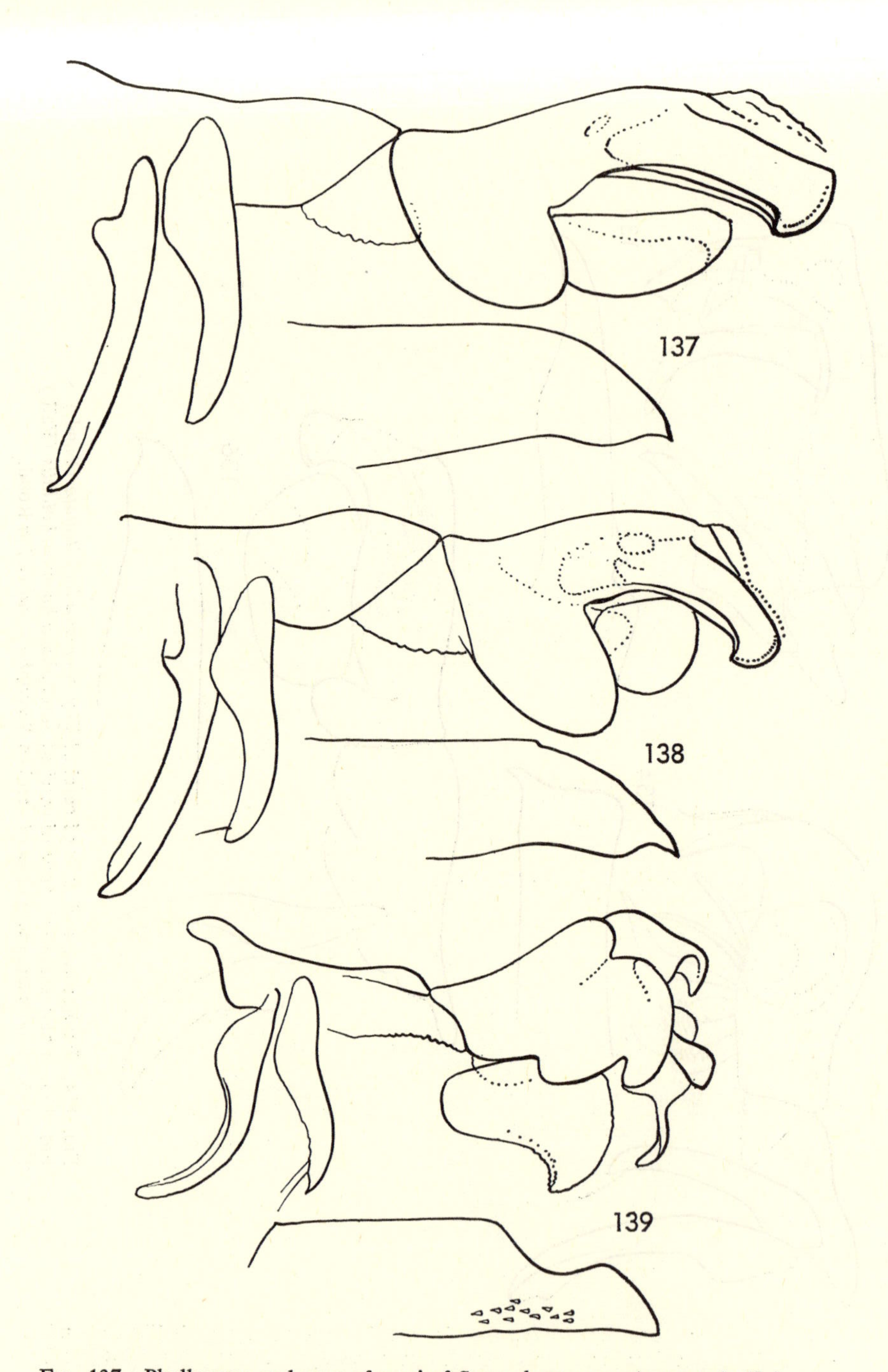

FIG. 137. Phallosoma and apex of cerci of *Sarcophaga carnaria carnaria* (L.)
FIG. 138. Phallosoma and apex of cerci of *Sarcophaga carnaria lehmanni* Müll.
FIG. 139. Phallosoma and apex of cerci of *Parasarcophaga barbata* Thom.

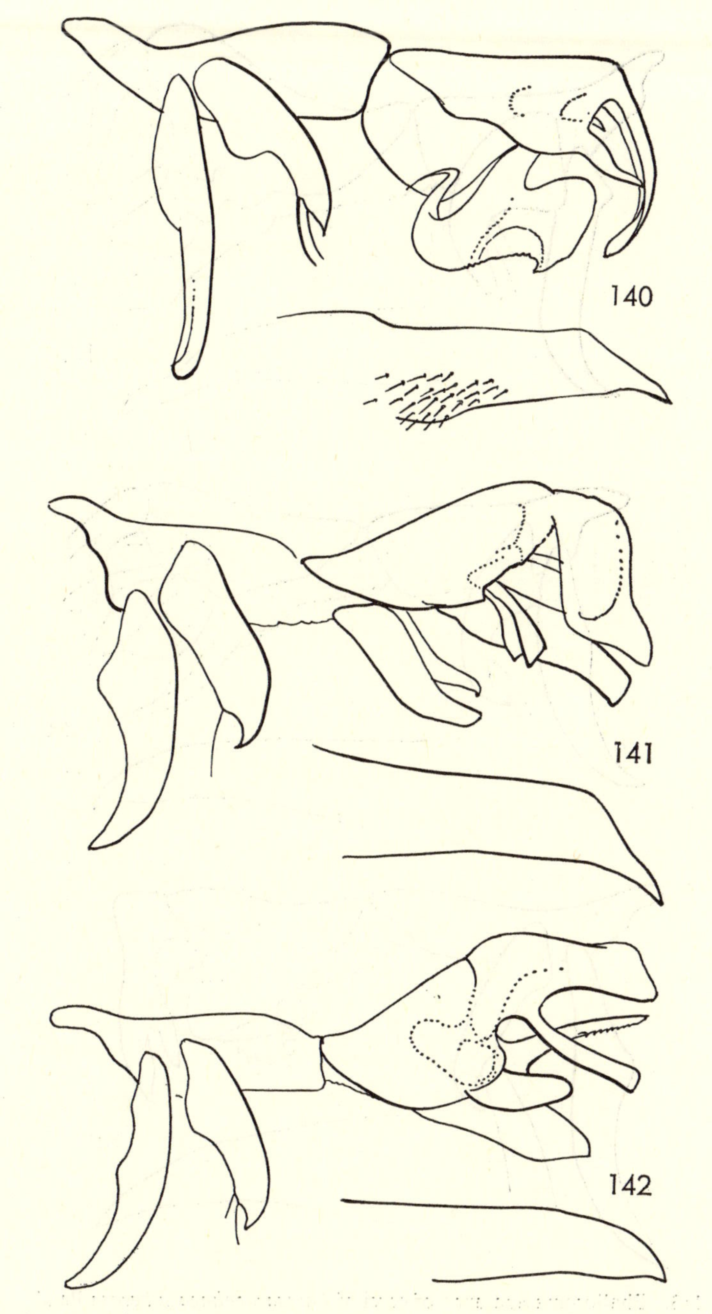

Fig. 140. Phallosoma and apex of cerci of *Parasarcophaga albiceps* (Meig.)
Fig. 141. Phallosoma and apex of cerci of *Parasarcophaga aratrix* Pand.
Fig. 142. Phallosoma and apex of cerci of *Parasarcophaga teretirostris* Pand.

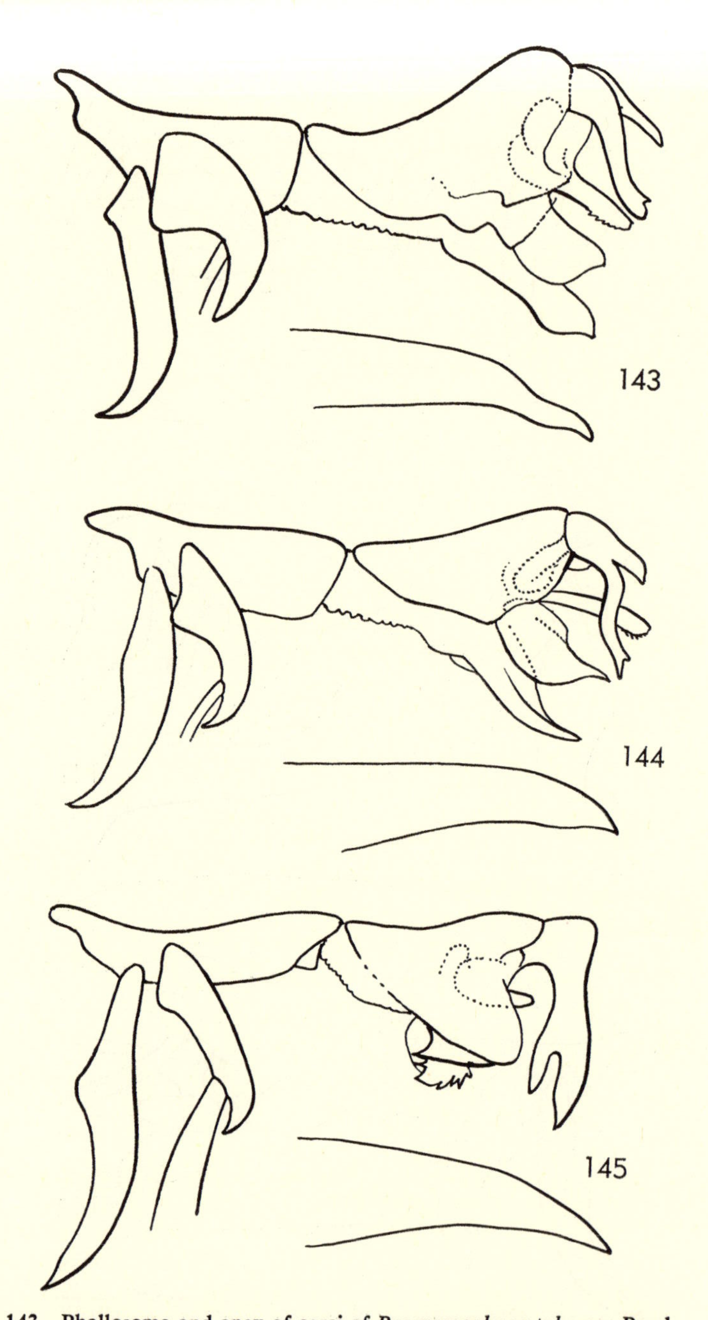

FIG. 143. Phallosoma and apex of cerci of *Parasarcophaga tuberosa* Pand.
FIG. 144. Phallosoma and apex of cerci of *Parasarcophaga jacobsoni* Rohd.
FIG. 145. Phallosoma and apex of cerci of *Parasarcophaga scoparia* Pand.

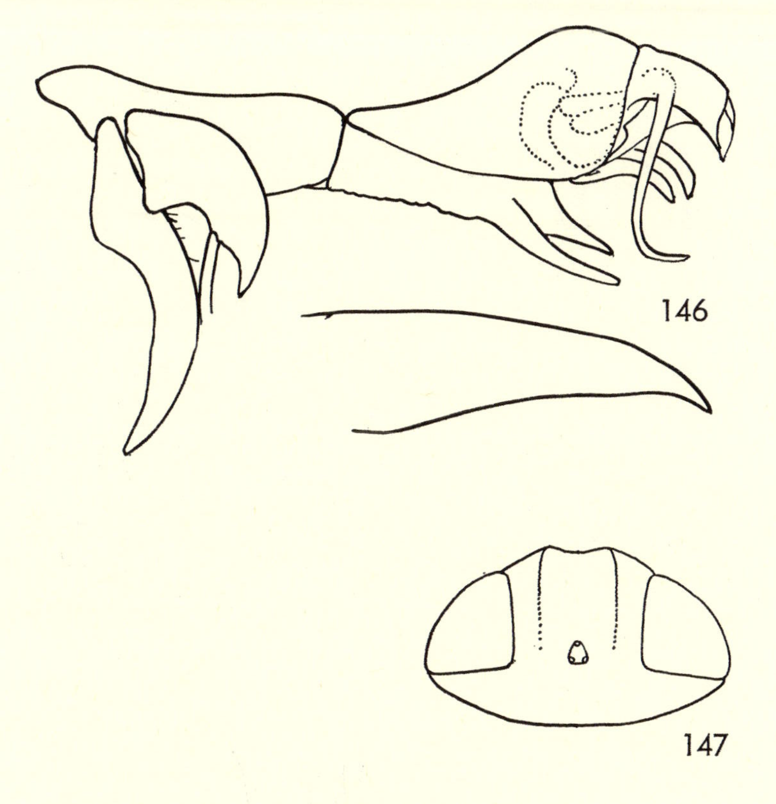

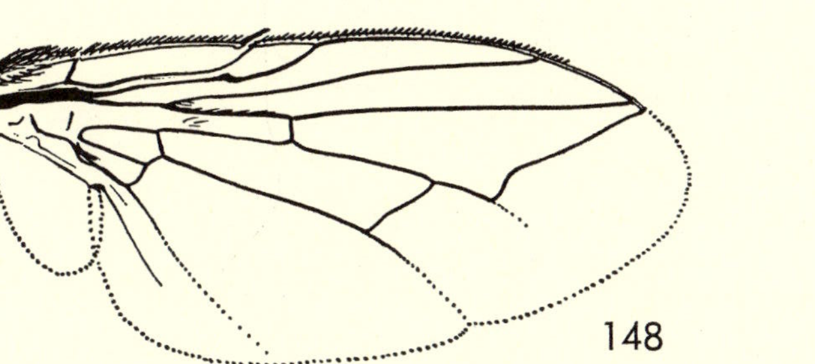

FIG. 146. Phallosoma and apex of cerci of *Parasarcophaga similis* Pand.
FIG. 147. Head of male *Agria latifrons* Fall.
FIG. 148. Wing of *Agria latifrons* Fall.

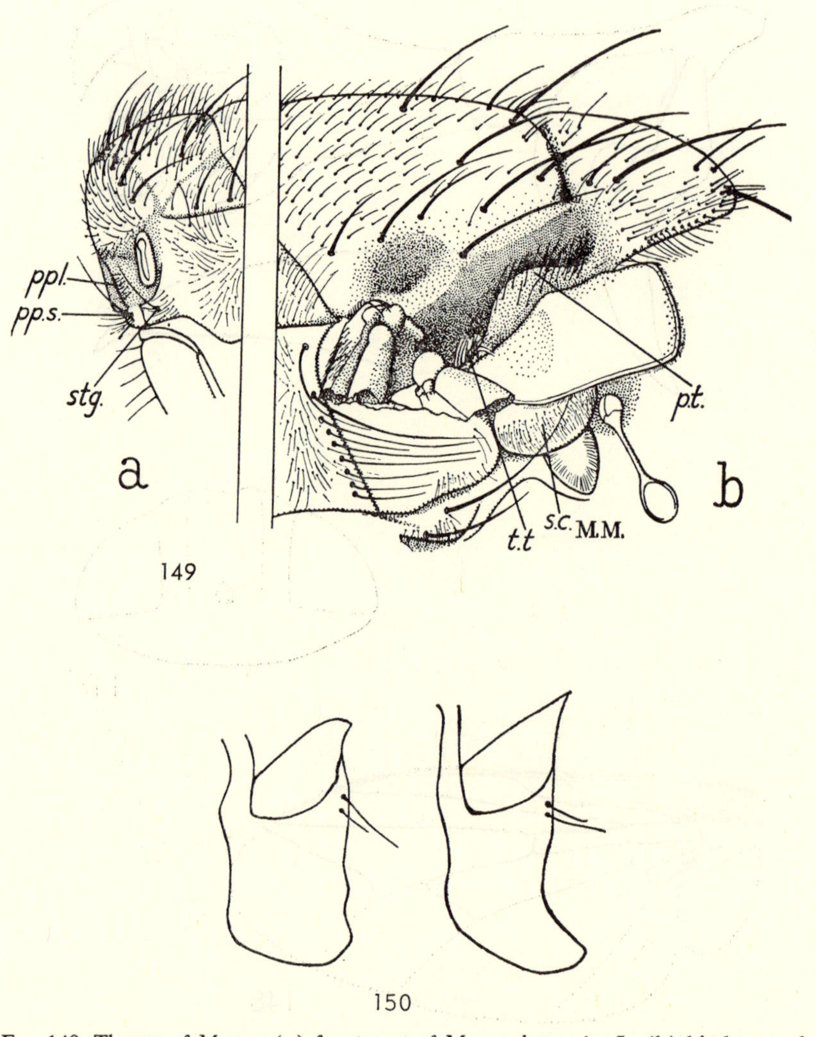

Fig. 149. Thorax of *Musca*. (a) front part of *Musca domestica* L. (b) hind part of *M. natalensis* Villen. *ppl.* propleura. *pp. s.* propleural seta. *stg.* prostigmatal seta. *t.t.* bristles on lower front end (tympanic tuft), and *p.t.* bristles on posterior part of suprasquamal ridge. *s.c.* supraspiracular convexity
Fig. 150. Paralobus shape within *Musca domestica* s. l. (a) *M. d. curviforceps* (b) other subspecies

5

Key to Larvae

Jindra Dušek

INTRODUCTION

THE POORER STATE of our knowledge of the larvae is reflected in the smaller number of species treated in the larval key compared with the adult keys in the preceding chapter. The key includes all well-known larvae of Palaearctic and cosmopolitan species of synanthropic flies. Those species whose larvae have been incompletely described or have not been available to the author are not included in the key; hence, their identification with the aid of this key is not possible. For more intensive study, the reader would do well to consult Hennig's classic "Die Larvenformen der Dipteren."

1. Larvae eucephalous, head capsule strongly sclerotized 2
——Larvae acephalous 5

2. Larvae peripneustic, posterior spiracles on tubular projections (Fig. 1): *Scatopse* Geoff.
——Larvae amphipneustic 3

3. Posterior spiracles on sides of last segment: *Scenopinus fenestralis* (Linn.).
——Posterior spiracles on apex of last segment 4

4. Last segment of body cylindrical, prolonged into a distinct respiratory tubule bearing spiracles apically (Fig. 2): *Psychoda* Latr.
——Last segment of body divided into 5 lobes, spiracles not petiolate (Fig. 3): *Sylvicola fenestralis* (Scop.).

5. Last segment prolonged into a long telescopic respiratory tubule (rat-tailed maggot) (Fig. 4) (Eristalinae) 6
——Last segment of different shape 14

6. Main tracheal branches straight (Fig. 5) (*Eristalis* Latr.) 7
——Main tracheal branches twisted (Fig. 6) (*Helophilus* Meig.) 13

7. Anteriorly directed crochets in short double row just in front of 6th abdominal prolegs 8
——Without crochets in front of 6th abdominal prolegs 9

8. With 7 to 9 larger crochets in posterior row of the above series: *Eristalinus sepulcralis* (Linn.).
——With only 5 larger crochets in posterior row of the above series: *Lathyrophthalmus aeneus* (Scop.).

9. Cuticle with reticulate pattern of brown patches: *Eristalis nemorum* (Linn.).
——Cuticle without reticulate pattern of brown patches 10

10. Primary crochets stout and strongly hooked, about as long as wide, with distal 2/5 strongly sclerotized (Fig. 7), anterior spiracle short with broadly rounded recurved tip (Fig. 8): *Eristalis tenax* (Linn.).

——Primary crochets long and slender, weakly hooked, nearly twice as long as wide; with distal quarter or less, weakly sclerotized (Fig. 9) 11

11. Anterior spiracle when viewed postero-laterally elongate with a narrow triangular tip, dark brownish 12

——Anterior spiracle short with broadly truncate end (Fig. 10), pale brownish: *Eristalis intricarius* (Linn.).

12. Anal papillae with 20 branches (Fig. 11). Anterior spiracle with lower part of facet band running horizontally around the spiracle for half the circumference with the facets directed slightly upward; spiracle abruptly narrowed above this band and curved outward; face of spiracle generally rounded, apex straight: *Eristalis pertinax* (Scop.).

——Anal papillae with 24 branches (Fig. 12). Anterior spiracle (Fig. 13) with the lower part of facet band acutely angled with the upright part; lower band only extending around for about 1/3 of the circumference, facets of lower band directed outward; spiracle gradually tapering to apex above this band and curved slightly backward, face of spiracle flattened, apex slightly hooked: *Eristalis arbustorum* (Linn.).

13. Crochets of 6th abdominal prolegs with crochets on 1 proleg separated from crochets of opposite proleg by a distance greater than length of largest crochet. A conspicuous tuft of long spinules just behind the anus which are as long as the adjacent papillae: *Helophilus pendulus* (Linn.).

——Crochets of 6th abdominal prolegs almost confluent, separated by less than half maximum crochet length. Only few short spinules in tuft just behind anus: *Helophilus hybridus* Loew.

14. Posterior spiracles with respiratory cleft more or less straight, oval, at most singly curved (Figs. 57, 58, 24, 25). Posterior segment mostly with fleshy prolongations or papillae (Figs. 45-47) 16

——Posterior spiracles of different shape 15

15. Posterior spiracles irregular (Fig. 14), situated on a long, unpaired, strongly sclerotized respiratory tubule. Anterior part of larva bluntly rounded, posterior part tapering with 3 pairs of lobes (Fig. 15): *Xylota pipiens* (Linn.).

——Posterior spiracles with respiratory clefts more or less undulated (Figs. 59-65), without or with very short paired stalks. Anterior part of larva pointed, posterior part bluntly rounded, mostly without fleshy prolongations or papillae (Muscinae) 43

16. Mandibular sclerites with rather large accessory sclerites of frequently complicated structure (Fig. 16) (Phaoniinae, Mydaeinae) 73

——Mandibular sclerites simple (Fig. 44) or at most with small accessory sclerites of simple structure (Fig. 43) 17

17. Posterior segment convex or abruptly truncate, without fleshy or spiny projections or papillae 18

——Posterior segment of different shape 20

18. Anterior spiracles fanlike with 12-16 buds 19

——Anterior spiracles dendriform or hornlike, bearing 3-8 buds: Sphaeroceridae.

19. Posterior spiracles nearly sessile, anterior spiracles with 16 buds (Otitidae): *Physiphora demandata* (Fabr.).

——Posterior spiracles on low knoblike sockets, anterior spiracles with 12 buds: Platystomatidae.

20. Posterior segment with at most 4 pairs of fleshy projections or papillae 21

——Posterior segment with more than 4 pairs of projections or papillae 26

21. Larvae conspicuously flattened, body fringed with long projections which are pinnate as a rule (Figs. 34-39), posterior spiracles frequently divided into 3 fingerlike projections (Fig. 42) (Fanniinae) 38

——Larvae cylindrical, projections present at most on posterior segment 22

22. Anterior spiracles fanlike, posterior segment of body not incrassate, apical segments bare without hairs or spines (Figs. 17, 18) (Piophilidae): *Piophila casei* (Linn.).

——Anterior spiracles dendriform (Figs. 19, 20), posterior segment slightly club-shaped, apical segments with hairs, the penultimate and ultimate ones with minute spinules (Figs. 21, 22) (Sepsidae) 23

23. Anterior spiracles much prolonged, with medial axis at least 3 times as long as wide, with more than 20 buds (Fig. 19): *Themira putris* (Linn.).

——Anterior spiracles little or not prolonged, with medial axis at most twice as long as wide, with less than 10 buds as a rule 24

24. Anterior spiracle with more than 10 buds (Fig. 20), respiratory clefts of posterior spiracles broadly ellipsoidal (Fig. 23): *Nemopoda nitidula* (Fall.).

——Anterior spiracles with less than 10 buds. Respiratory clefts of posterior spiracles not shortly ellipsoidal 25

25. Facial mask without salivary furrows, consisting of pectinate chitinous teeth. Respiratory clefts parallel to edges of spiracular area (Fig. 24): *Meroplius stercorarius* (R.-D.).

——Facial mask with salivary furrows, respiratory clefts not parallel to edges of spiracular area (Fig. 25): *Sepsis.*

26. Posterior segment of body with 6 pairs of fleshy projections: Dryomyzidae.

——Posterior segment of body with more than 6 pairs of fleshy projections or papillae 27

27. Posterior segment of body with 7 pairs of fleshy projections or papillae 28

——Posterior segment of body with more than 7 pairs of fleshy projections or papillae 29

28. Anterior spiracles with only 4 buds, posterior spiracles on rather long stalks situated on a cylindrical fleshy projection, the latter slender and long (Figs. 26, 27): *Drosophila fasciata* (Meig.).

——Anterior spiracles with 12-18 buds. Posterior spiracles on rather short, sclerotized stalks, sessile on last segment which bears only minute, flattened, and mostly indistinct papillae: Helomyzidae.

29. Segments 2 to 7 with 6 pairs of papillae each, the largest larvae not longer than 10 mm, posterior spiracles with 4 respiratory clefts (Phoridae): *Megaselia* Rond.

——Segments of body without papillae, full-grown larvae frequently larger than 10 mm, posterior spiracles with 3 respiratory clefts 30

30. Besides 8 pairs of abdominal papillae there is an odd (postanal) papilla between the ventral ones (Fig. 45), the posterior segment covered with dense marked spinules (Scatophagidae): *Scatophaga stercoraria* (Linn.).

——Posterior segment with paired papillae only, the spinulose cover absent or very fine 31

31. Posterior segment bearing 8 pairs of fleshy conical papillae, those of the 6th and 7th pairs approximated or at least nearly on the same level (Fig. 46) (Anthomyiinae) 32

——Posterior segment bearing seven to nine pairs of fleshy conical or cushionlike papillae, those of the 6th and 7th pairs rather apart and the papilla of the 7th pair situated considerably higher than that of the 6th pair (Fig. 47): Calliphoridae, Sarcophagidae 59

32. Anterior spiracles with more than 20 buds 33

——Anterior spiracles with less than 20 buds 34

33. Anterior spiracles with 24-25 buds: *Hylemya strigosa* Fabr.

——Anterior spiracles with around 30 buds: *Hylemya variata* (Fall.).

34. Minute papillae of the 6th and 7th pairs, nearly equal in size, are situated close to one another (Fig. 28) 35

——Papillae of 6th and 7th pairs of different size and apart from one another. Papillae of 5th and 6th pairs large and forming a two-pointed group of papillae. Papillae of the 7th pair minute (Fig. 29): *Delia platura* (Meig.).

35. Anterior spiracles with more than 9 buds 36

——Anterior spiracles with 7-9 buds: *Hylemya cinerella* (Fall.).

36. Segments of body smooth with irregular rows of spinules only at the division lines, only the last segment with a more or less dense cover of minute and subtle spinules 37

——Not only the last segment covered with rather dense and marked spinules. Anterior spiracles with 9-13 buds: *Hylemya radicum* (Linn.).

37. Posterior spiracles on rather long stalks (Fig. 30). Mandibular sclerites with a separate dental sclerite (Fig. 31): *Hylemya pullula* (Zett.).

——Posterior spiracles nearly sessile (Fig. 32). Mandibular sclerites with a long, not separated, ventral projection. (Fig. 33): *Anthomyia pluvialis* (Linn.).

38. Dorso-medial appendages long and filiform, protruding far beyond the division lines, those of the 7th segment reaching far beyond the posterior spiracles (Fig. 34): *Fannia canicularis* (Linn.).

——Dorso-medial appendages much shorter, not filiform, not or slightly protruding beyond the division lines (in the latter case, they are pinnate), those of the 7th segment not reaching beyond the posterior spiracles 39

39. Posterior spiracles on a slightly convex tubercle, not stalked (Fig. 40): *Fannia manicata* (Meig.).

——Posterior spiracles distinctly stalked (Fig. 41) 40

40. Dorso-medial appendages long and pinnate, slightly protruding beyond the division lines. Dorso-lateral appendages of much the same structure (Fig. 35): *Fannia leucosticta* (Meig.).

——Dorso-medial appendages short, not pinnate, bearing short spinules, not protruding beyond the division lines. Dorso-lateral appendages short and inconspicuous 41

41. Marginal appendages of the last abdominal segment as well as the lateral ones of the preceding segments distinctly long pinnate at base, the medial one not shorter than the remaining ones (Figs. 37, 38) 42

——Latero-dorsal appendages not distinctly pinnate at base but of similar structure as in *F. canicularis.* Dorso-lateral appendages totally absent. Of the 3 marginal appendages on either side of the last abdominal segment, the middle one is distinctly shorter than the remaining ones (Fig. 36). Anterior spiracle with only three finger-like buds: *Fannia lineata* Stein.

42. Dorso-medial appendages present in form of minute, nearly invisible tubercles. Latero-dorsal appendages very wide and widely pinnate, the barbs of 2 neighboring appendages on the last segment partly overlapping (Fig. 37): *Fannia incisurata* (Zett.).

——Dorso-medial appendages longer. Latero-dorsal appendages less wide. The barbs of two neighboring appendages on the last segment not distinctly overlapping (Fig. 38): *Fannia scalaris* (Fabr.).

43. Caudal end of body obliquely truncate. Spiracular area separated from the remaining area of the segment by a sharp ledge. Cuticle very strong (firm) and bearing distinct rings of spines at the division lines 44

——Caudal end of body more or less hemispherically convex. Spiracular area not separated from the remaining area by a ledge. Larvae relatively soft-skinned, the rings of spines feebly developed 45

44. Respiratory clefts of posterior spiracles strongly twisted. The upper cleft with some 14, the middle one with 10, and the lower one with 11 twists (Fig. 59): *Morellia hortorum* (Fall.).

——Respiratory clefts of posterior spiracles with few twists. The upper one with some 8, the middle one with 5, and the lower one with 4-5 twists (Fig. 60). Surface of cuticle with deep depressions: *Morellia simplex* (Loew).

45. Cephalo-pharyngeal skeleton with accessory chitinous sclerites below the mandibular ones (Fig. 43) 46

——Cephalo-pharyngeal skeleton without accessory chitinous sclerites below the mandibular ones. Left mandibular sclerite invariably somewhat shorter (Fig. 44) 54

46. Cephalo-pharyngeal skeleton with 2 accessory pairs of sclerites below the mandibular ones 47

——Cephalo-pharyngeal skeleton with only 1 pair of accessory sclerites below the mandibular ones (Fig. 43) 53

47. Respiratory clefts of posterior spiracles S-shaped (Fig. 61). Anterior spiracles with 10-13 buds each. Larvae more than 20 mm in length: *Mesembrina mystacea* (Linn.).

——Respiratory clefts of posterior spiracles much twisted (Fig. 62) 48

48. Anterior spiracles with about 10 appendages. Respiratory clefts of posterior spiracles less strongly twisted 49

——Anterior spiracles with 15-18 appendages each. Respiratory clefts of posterior spiracles much twisted (Fig. 62). Body over 20 mm in length: *Mesembrina meridiana* (Linn.).

49. Besides the 2 pairs of accessory sclerites there are additional chitinous teeth present in the cephalo-pharyngeal region. Anterior spiracles with oblong chitinous sclerite: *Polietes albolineata* (Fall.).

——No additional chitinous teeth in the cephalo-pharyngeal region. Anterior spiracles without a distinct sclerite: *Polietes lardaria* (Fabr.).

50. 7 papillae present in the region of anal plate 51

——At most 5 papillae present in the region of anal plate. Between the dorsal connecting bridge and the anterior ventral prolongations of the anterior connecting bridge, there is a well developed, strongly pigmented chitinous membrane: *Dasyphora cyanella* (Meig.).

51. 1 papilla on either side above the anal plate, the latter reaching at most to middle of the side of body. A pigmented chitinous membrane between the dorsal connecting bridge and the anterior ventral prolongations of the anterior connecting bridge present: *Pyrellia cadaverina* (Linn.).

——No papillae above the anal plate, the latter not reaching beyond middle of the side of body. Pigmented chitinous membrane between the dorsal connecting bridge and the anterior ventral prolongations of the anterior connecting bridge absent 52

52. Anal plate narrow. Papillae in the region of anal plate relatively short: *Orthellia cornicina* (Fabr.).

——Anal plate broad. Papillae in the region of anal plate strongly developed: *Orthellia caesarion* (Meig.).

53. Ventral wing of cephalo-pharyngeal sclerite without cupolalike projection. Dorsal wing very narrow. Arched chitinous ribs at anterior margin of pharyngeal base present. Hypostomal sclerite relatively slender. Respiratory clefts of posterior spiracles S-shaped (Fig. 63) except in *Haematobia irritans* (Fig. 65) 54

——Ventral wing of cephalo-pharyngeal sclerite with a cupolalike projection on upper side. Hypostomal sclerite about as long as high. Anterior connecting bridge relatively wide. Respiratory clefts of posterior spiracles much twisted (Fig. 64) 57

54. Posterior spiracular plates separated by a space at least 1.5 times as wide as 1 plate, the latter of more or less triangular shape. Larvae more than 8 mm in length 55

——Distance of posterior spiracular plates at most as wide as the width of 1 plate, the latter not triangular. Larvae at most 7 mm in length 56

55. A spinose post-anal papilla present (Fig. 48): *Stomoxys calcitrans* (Linn.).

——Post-anal papilla absent: *Lyperosiops stimulans* (Meig.).

56. Posterior spiracular plate approximated, 1/3 to 1/5 of the width of 1 plate apart: *Haematobia irritans* (Linn.).

——Posterior spiracular plates separated by a space 1/2 as wide as the width of 1 spiracular plate: *Haematobia titillans* (Bezzi).

57. The anal plate, surrounded by thorns, reaches beyond the middle of body on either side, more or less rectangular apically. On dorsal side of the last segment there are 3 longitudinal furrows surrounded by thorns: *Musca autumnalis* De Geer.

——Anal plate not reaching the middle of body, rounded apically. Dorsal side of last segment without longitudinal furrows 58

58. Thorns present behind anal plate: *Musca domestica* Linn.

——Thorns behind anal plate absent: *Musca tempestiva* Fall.

59. Sides and ventral side of segments with rather long, conspicuous papillae terminated by a tuft of obtuse spines (Fig. 66): *Chrysomya albiceps* (Wied.).

——Sides and ventral side of segments without any papillae 60

60. Posterior spiracles situated in a deep excavation (Fig. 49). Dorsal wing of cephalo-pharyngeal sclerite parted (Fig. 67) (Sarcophaginae) 70

——Posterior spiracles not situated in an excavation, dorsal wing of cephalo-pharyngeal sclerite simple 61

61. Posterior spiracles with peritreme closed (Fig. 58) 63

——Posterior spiracles with peritreme open (Fig. 57) 62

62. 1st pair of papillae on abdominal segment larger than 1/2 the width of 1 posterior spiracle (Fig. 50). Dorsal thorns near posterior margin of 10th segment present: *Protophormia terraenovae* (R.-D.).

——1st pair of papillae on abdominal segment smaller than 1/2 the width of 1 posterior spiracle. Dorsal thorns near posterior margin of 10th segment absent: *Phormia regina* (Meig.).

63. Peritreme of posterior spiracles feebly pigmented: *Pollenia rudis* (Fabr.).

——Peritreme of posterior spiracles strongly pigmented 64

64. An unpaired sclerite present between mandibular sclerites (Fig. 69) 65

——No unpaired sclerite between mandibular sclerites (Fig. 70) 67

65. A pigmented chitinous sclerite present on posterior end of ventral wings of cephalo-pharyngeal sclerite: *Lucilia ampullacea* Vill.

——No pigmented chitinous sclerite on posterior end of ventral wings of cephalo-pharyngeal sclerite 66

66. Upper, lower, and lateral surface of last segment smooth, without spinules: *Calliphora uralensis* Vill.

——Upper, lower, and lateral surface of last segment covered with spinules: *Calliphora vicina* R.-D.

67. A pigmented chitinous sclerite present on posterior end of ventral wings of cephalo-pharyngeal sclerite: *Lucilia illustris* (Meig.).

——Pharynx without this pigmented area 68

68. Papillae of the 1st pair separated by a distance equal to that between those of 2nd and 3rd pairs (Fig. 51): *Phaenicia sericata* (Meig.).

——Papillae of the 1st pair separated by a distance greater than that between papillae of 1st and 2nd and between those of 2nd and 3rd pairs. The distance between papillae of 1st pair roughly equal to that between papillae of 1st and 3rd pairs (Fig. 52) 69

69. Anterior margins of segments 2-11 with complete ring of spines; posterior margins of segments 8-11 with complete ring of spines: *Lucilia bufonivora* Moniez.

——Dorsal spines on anterior margin of segment 11 absent. Complete posterior rings of spines mostly on segments 9-11 only: *Bufolucilia silvarum* (Meig.).

70. Abdominal papillae of the 6th pair equal in length or longer than those of the 5th pair. Atrium of anterior spiracles strongly enlarged in upper part: *Ravinia striata* Fabr.

——Abdominal papillae of the 6th pair smaller and shorter than those of the 5th pair. Atrium of anterior spiracles hemispherical, not enlarged in upper part 71

71. Second segment without spines on posterior part of upper surface: *Parasarcophaga jacobsoni* Rohd.

——Second segment partly or entirely covered with minute spinules on upper surface 72

72. Abdominal papillae of the 1st pair roughly equal in size to those of the 5th pair (Fig. 53). Last segment coarsely rugose below papillae of 6th and 7th pairs: *Sarcophaga haemorrhoidalis* (Fall.).

——Papillae of the 1st pair smaller than those of the 5th pair. Last segment covered with spinules all over: *Sarcophaga melanura* Meig.

73. Last segment slender, protracted, and tapering as the anterior part of body; it bears 5 pairs of papillae. Ventral side of body with conspicuous parapodia: *Graphomya maculata* (Scop.).

——Last segment bluntly rounded or truncate, invariably much stouter than the anterior part of body 74

74. Lower surface of last segment with conspicuous spinules behind anal papillae (Fig. 54) 75

——Lower surface of last segment without spinules behind anal papillae 79

75. Anal papillae long, smooth, without spinules. Behind the anal papillae there are 4-5 rows of strong as well as minute spinules. Body long and slender, with shining cuticle: *Dendrophaonia querceti* (Bouché).

——Anal papillae covered with spinules to some extent 76

76. Last segment short, rounded. Anterior spiracles with 5-7 buds. Thickenings of anterior margins of abdominal segments with a deep medial longitudinal furrow without spinules. Broad cephalo-pharyngeal sclerite with dorsal wing and ventral wing equal in length, or the former is slightly longer (Fig. 68) 77

——Last segment prolonged, more or less angular in lateral view. Anterior spiracles with 4 buds. Thickenings of anterior margins of segments without a medial longitudinal furrow. Slender cephalo-pharyngeal sclerite with dorsal wing much longer than ventral wing. Cuticle very firm 78

77. Anal plate convex in lateral parts to form structures resembling anal papillae: *Muscina assimilis* (Fall.).

——Anal plate not showing lateral structures resembling papillae (Fig. 55): *Muscina stabulans* (Fall.).

78. Last segment short. Anal plate enlarged toward the end. Conspicuous medial paraanal papillae covered with spinules: *Ophyra leucostoma* (Wied.).

——Last segment long. Anal plate tapering toward the end. Paraanal papillae not protruding conspicuously over the inner and external papillae: *Ophyra capensis* (Wied.).

79. Cuticle of last segment smooth, without marked sculpture 80

——Cuticle of last segment with marked scaled or wrinkled sculpture on lateral and posterior surfaces (apparent at 20 x magnification) 82

80. Unpaired post-anal papilla bears minute spinules (visible at 20 x magnification): *Myospila meditabunda* (Fabr.).

——Unpaired post-anal papilla smooth, without spinules, or absent 81

81. Unpaired post-anal papilla present. 2 posterior thoracic and first abdominal segments without oblique lateral rows of spinules. Posterior spiracles separated by a distance 1/2 as long as diameter of spiracular plate. Hypostomal sclerite roughly as long as wide: *Gymnodia impedita* Pand.

——Unpaired post-anal papilla absent. 2 posterior thoracic and first abdominal segments with transverse lateral rows of spinules. Posterior spiracles separated by a distance equalling 1.5 the diameter of spiracular plate. Hypostomal sclerite 2 to 3 times as long as wide: *Hebecnema fumosa* (Meig.).

82. Posterior spiracles large, separated by a distance as long as the width of spiracle. Posterior surface of last segment with several concentric furrowlike impressions. Near the anal opening there are 7 anal papillae, the medial unpaired one being covered with spinules (Fig. 56): *Hydrotaea dentipes* (Fabr.).

——Posterior spiracles very small, separated by a distance 5-7 times as long as their diameter. Posterior surface of last segment with radial furrows or scales 83

83. Posterior surface of last segment with scaly sculpture passing into radially furrowed sculpture in the sides of the segment. Subanal papillae fusing with paraanal papillae and situated on a common cylindrical projection. Second thoracic segment covered with dense longitudinal furrows all over its surface. Thickenings of anterior margins of middle abdominal segments with spines of two different sizes: *Hydrotaea armipes* (Fall.).

——Posterior surface of last segment with subtle radial furrows. Anal papillae thin. Furrows on posterior part of last segment very numerous; each furrow bifurcated before margin of segment. Second thoracic segment with sparse and deep furrows in middle and in posterior part, the furrows not reaching margin of segment. Thickenings of ventral anterior margins of middle and hind-abdominal segments with spines of nearly equal size: *Hydrotaea occulta* (Meig.).

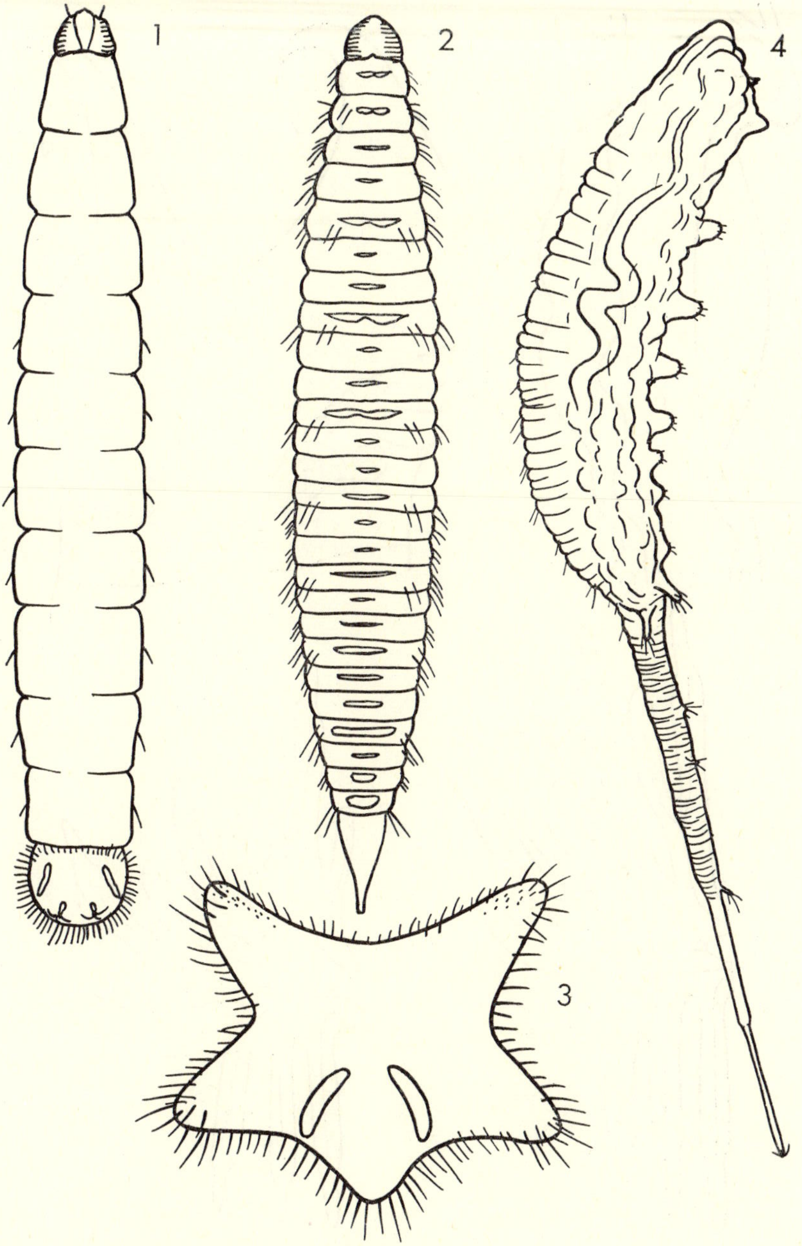

1 2 3 4

FIG. 1. Larva of the genus *Scatopse* Geoffr.
FIG. 2. Larva of the genus *Psychoda* Latr.
FIG. 3. Posterior segment of larva of *Sylvicola fenestralis* (Scop.)
FIG. 4. Larva of *Helophilus pendulus* (L.)

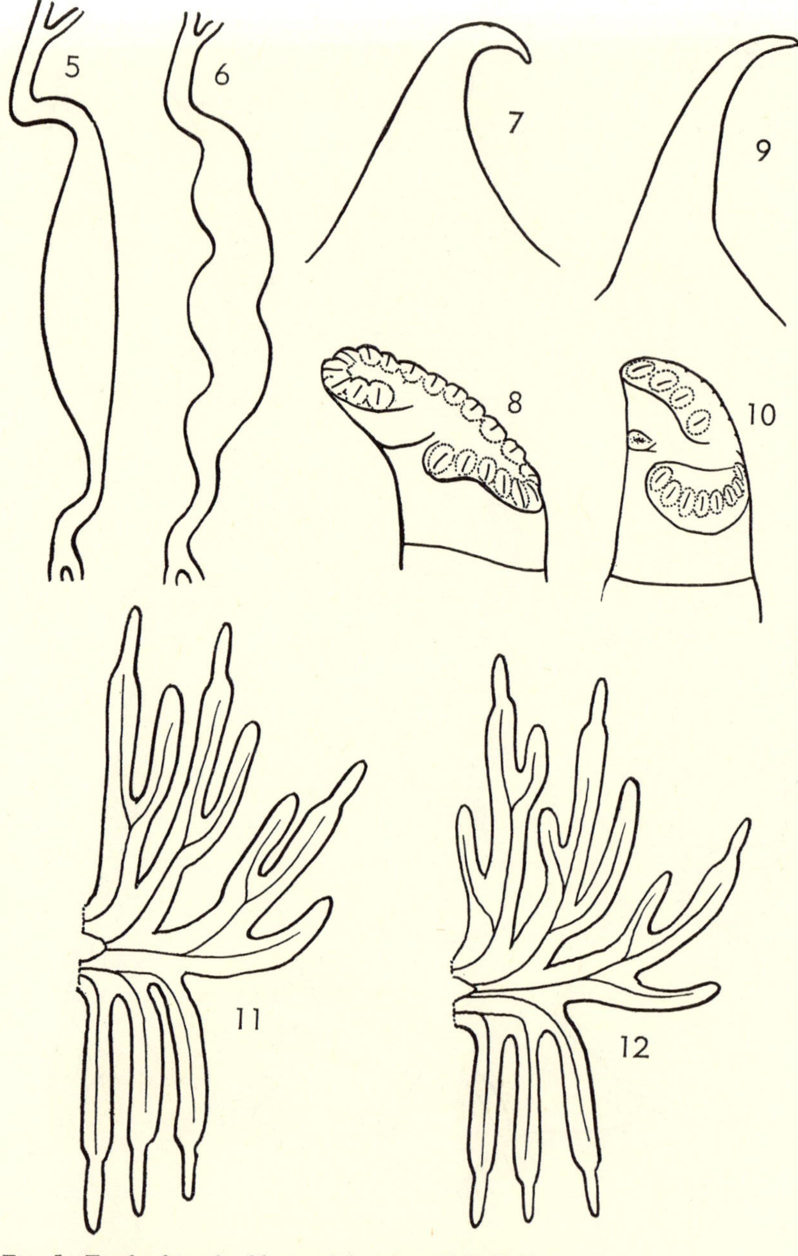

Fig. 5. Tracheal trunk of larva of the genus *Eristalis* Latr.
Fig. 6. Tracheal trunk of larva of the genus *Helophilus* Meig.
Fig. 7. Primary crochets of *Eristalis tenax* (L.)
Fig. 8. Prothoracic spiracle of *Eristalis tenax* (L.)
Fig. 9. Primary crochets of *Eristalis pertinax* (Scop.)
Fig. 10. Prothoracic spiracle of *Eristalis intricarius* (L.)
Fig. 11. Left halves of the evaginated anal papillae of *Eristalis tenax* (L.)
Fig. 12. Left halves of the evaginated anal papillae of *Eristalis arbustorum* (L.)

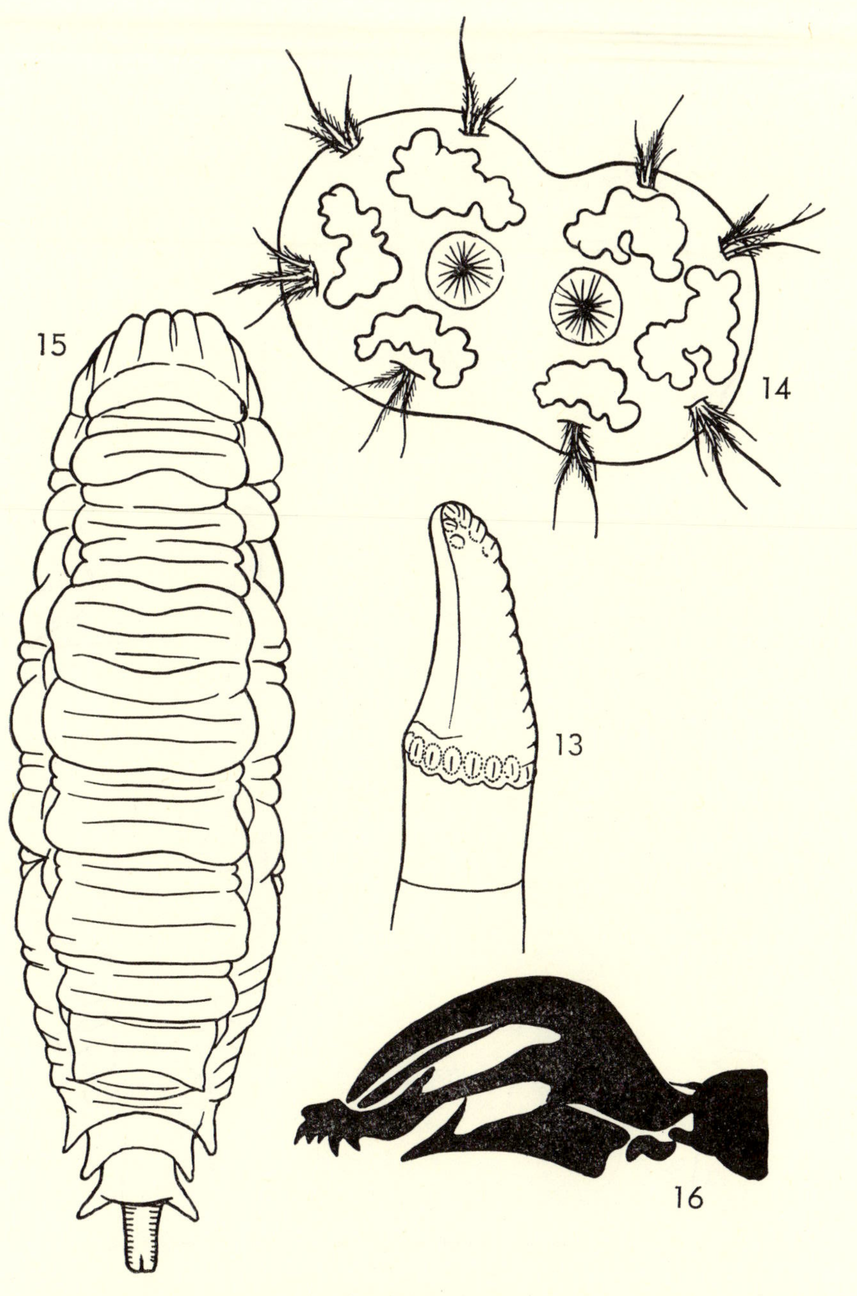

FIG. 13. Prothoracic spiracle of *Eristalis arbustorum* (L.)
FIG. 14. Spiracular plate of *Xylota pipiens* (L.)
FIG. 15. Full-grown larva of *Xylota pipiens* (L.)
FIG. 16. Accessory sclerites of mouth crochets of larvae of the genus *Phormia* R.-D.

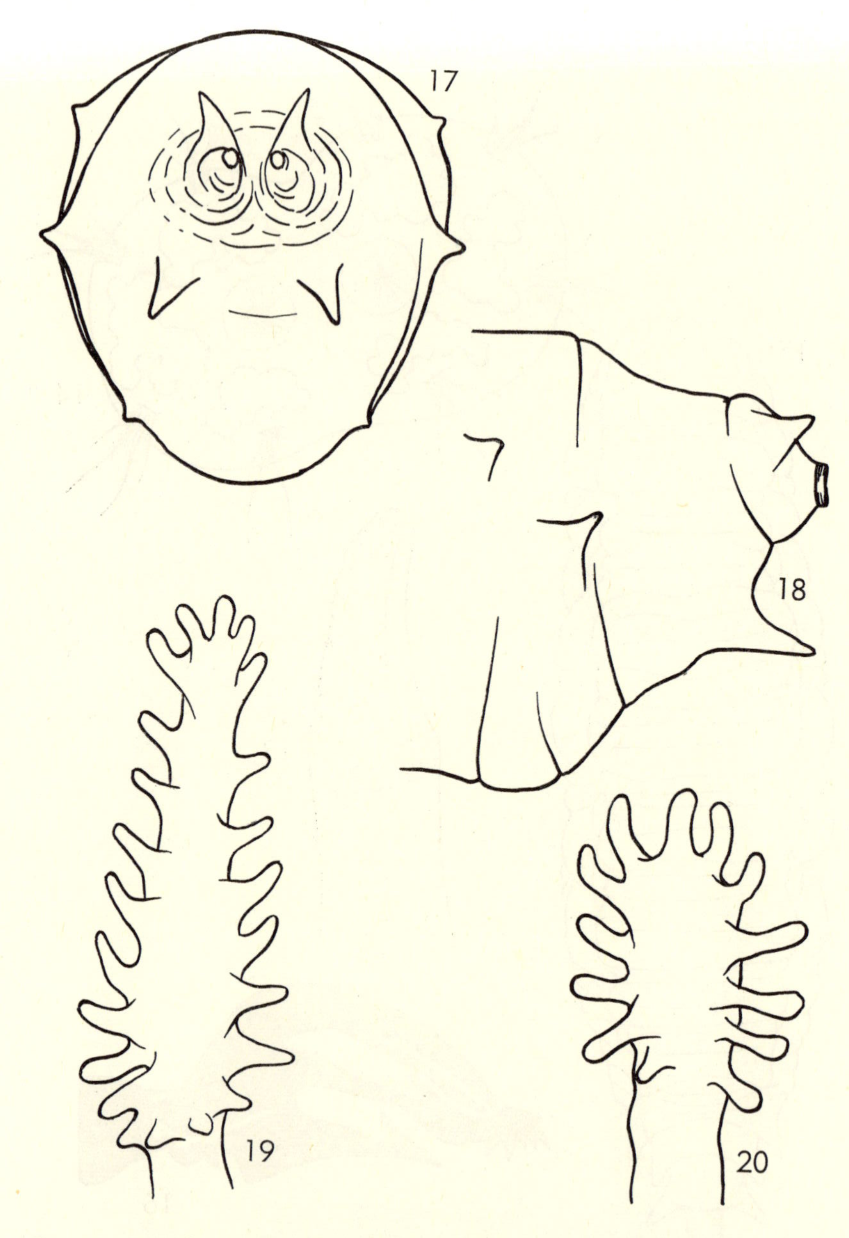

Fig. 17. Posterior segment of larva of *Piophila casei* (L.) in frontal view
Fig. 18. Same as Fig. 17, lateral view
Fig. 19. Anterior spiracle of *Themira putris* (L.)
Fig. 20. Anterior spiracle of *Nemopoda nitidula* (Fall.)

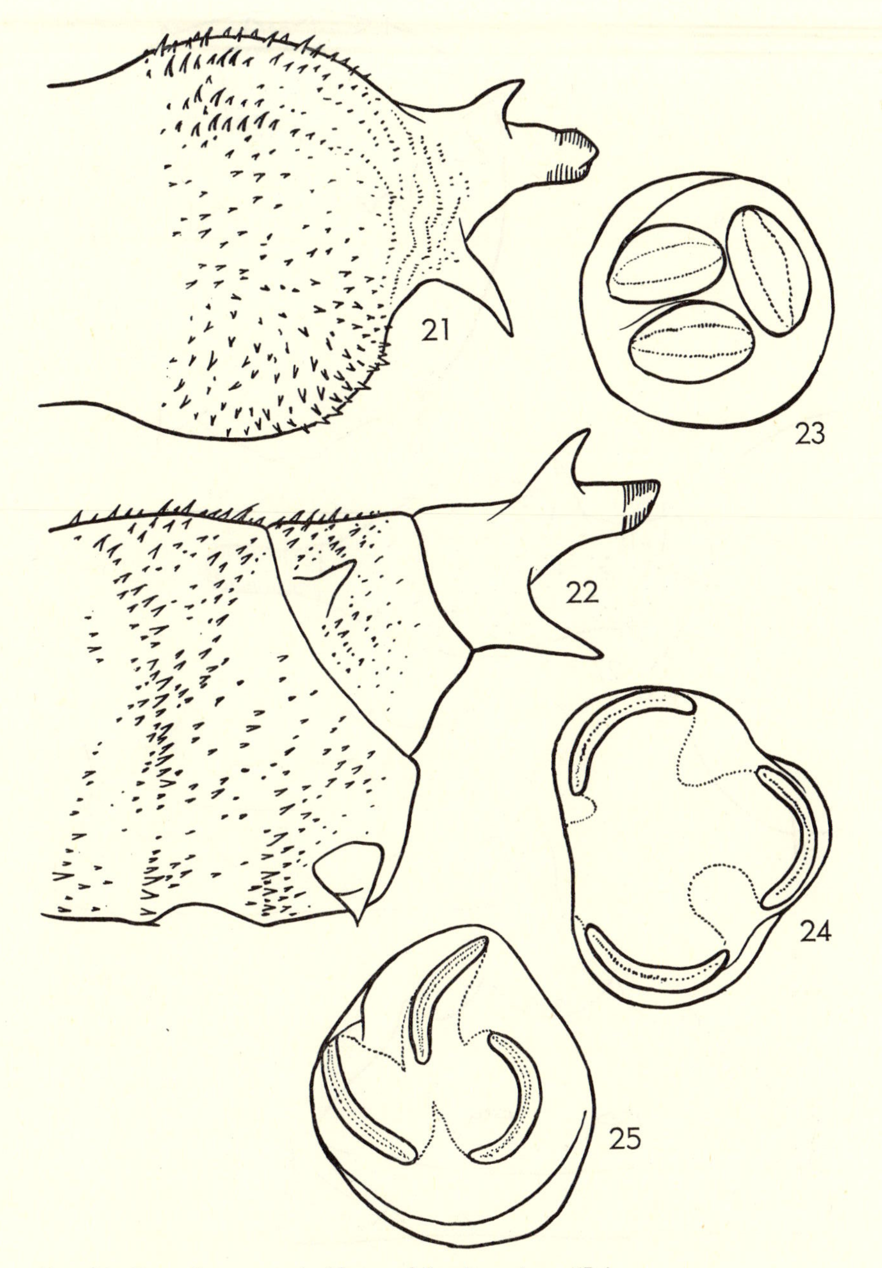

FIG. 21. Posterior segment of larva of *Sepsis cynipsea* (L.)
FIG. 22. Posterior segment of larva of *Nemopoda nitidula* (Fall.)
FIG. 23. Posterior spiracle of *Nemopoda nitidula* (Fall.)
FIG. 24. Posterior spiracle of *Meroplius stercorarius* (R.-D.)
FIG. 25. Posterior spiracle of *Sepsis cynipsea* (L.)

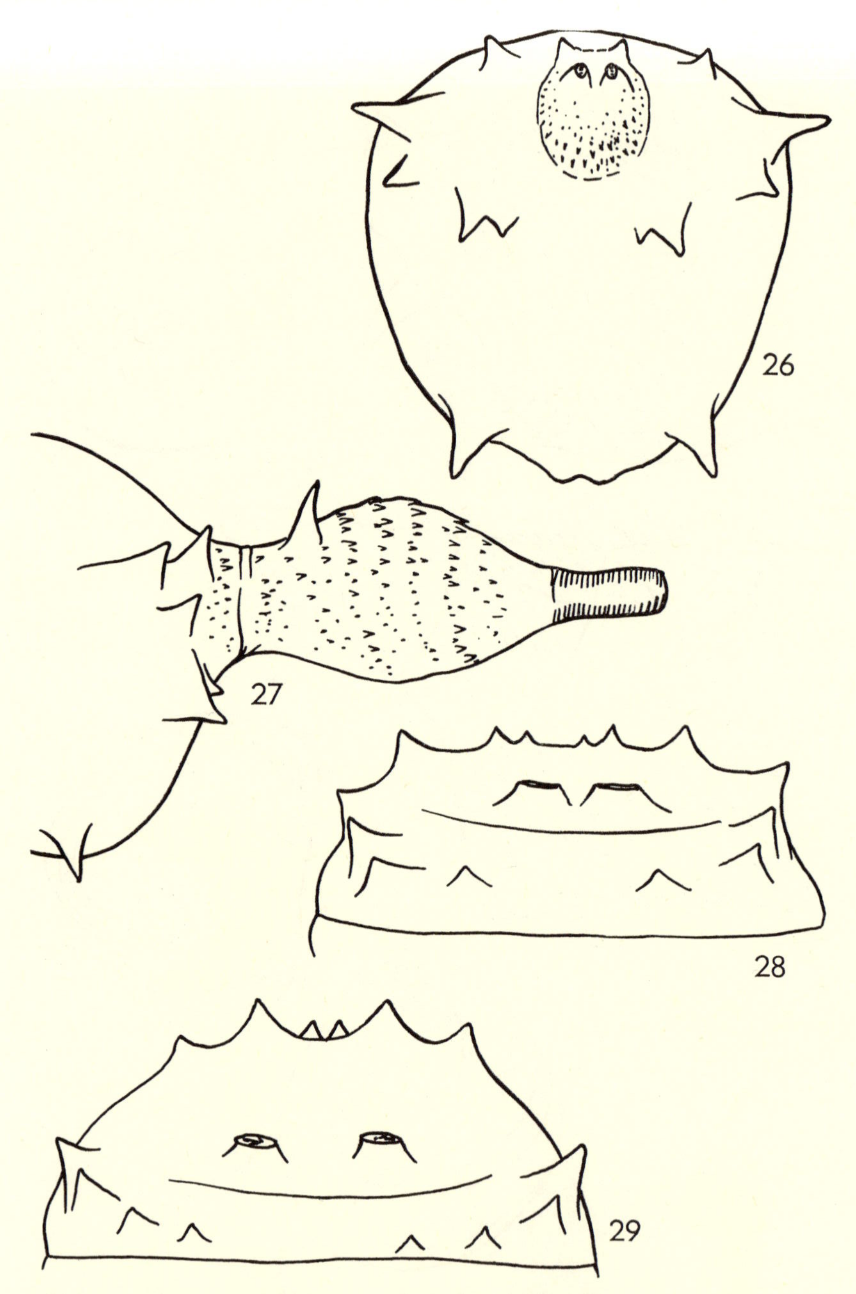

Fig. 26. Posterior segment of larva of *Drosophila fasciata* (Meig.) in frontal view
Fig. 27. Same as Fig. 26, lateral view
Fig. 28. Posterior segment of larva of *Anthomyia pluvialis* (L.) in dorsal view
Fig. 29. Posterior segment of larva of *Delia platura* (Meig.) in dorsal view

FIG. 30. Posterior segment of larva of *Hylemya pullula* (Zett.) in lateral view
FIG. 31. Cephalo-pharyngeal skeleton of larva of *Hylemya pullula* (Zett.)
FIG. 32. Posterior segment of larva of *Anthomyia pluvialis* (L.) in lateral view
FIG. 33. Cephalo-pharyngeal skeleton of *Anthomyia pluvialis* (L.)

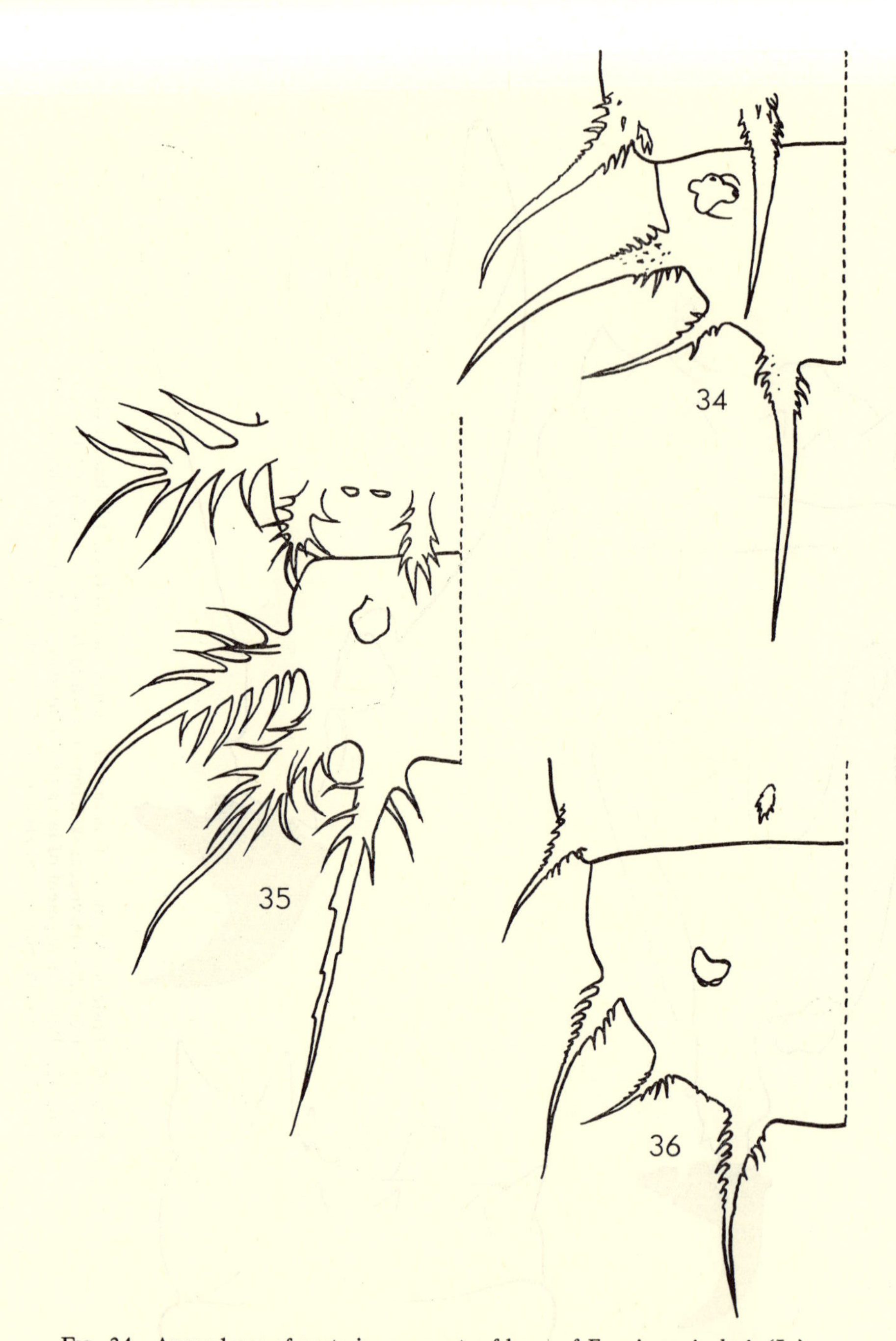

Fig. 34. Appendages of posterior segments of larva of *Fannia canicularis* (L.)
Fig. 35. Appendages of posterior segments of larva of *Fannia leucosticta* (Meig.)
Fig. 36. Appendages of posterior segments of larva of *Fannia lineata* Stein

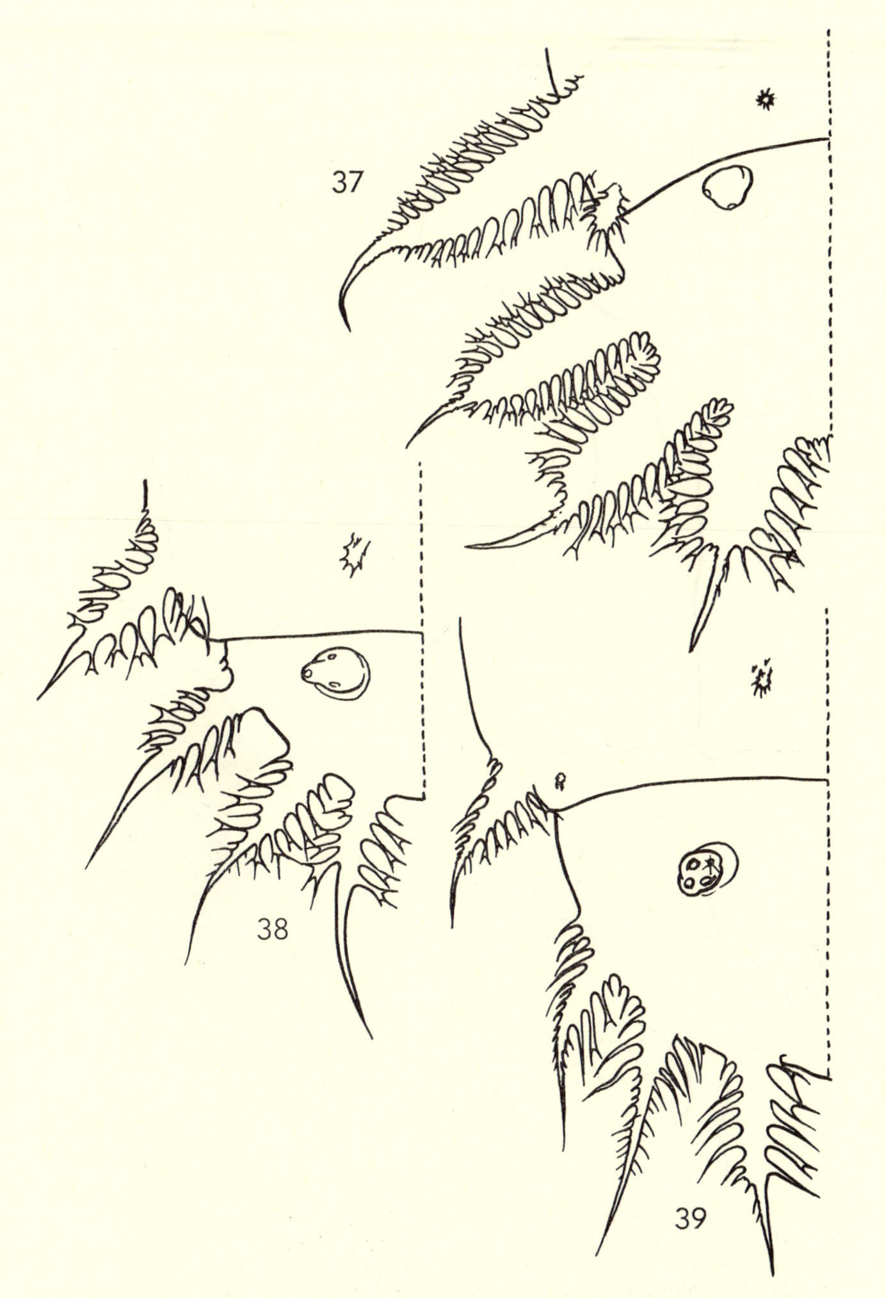

Fig. 37. Appendages of posterior segments of larva of *Fannia incisurata* (Zett.)
Fig. 38. Appendages of posterior segments of larva of *Fannia scalaris* (Fabr.)
Fig. 39. Appendages of posterior segments of larva of *Fannia manicata* (Meig.)

FIG. 40. Lateral view of posterior segment of larva of *Fannia manicata* (Meig.)
FIG. 41. Lateral view of posterior segment of larva of *Fannia scalaris* (Fabr.)
FIG. 42. Posterior spiracle of larva of *Fannia canicularis* (L.)
FIG. 43. Cephalo-pharyngeal skeleton of *Orthellia caesarion* (Meig.)

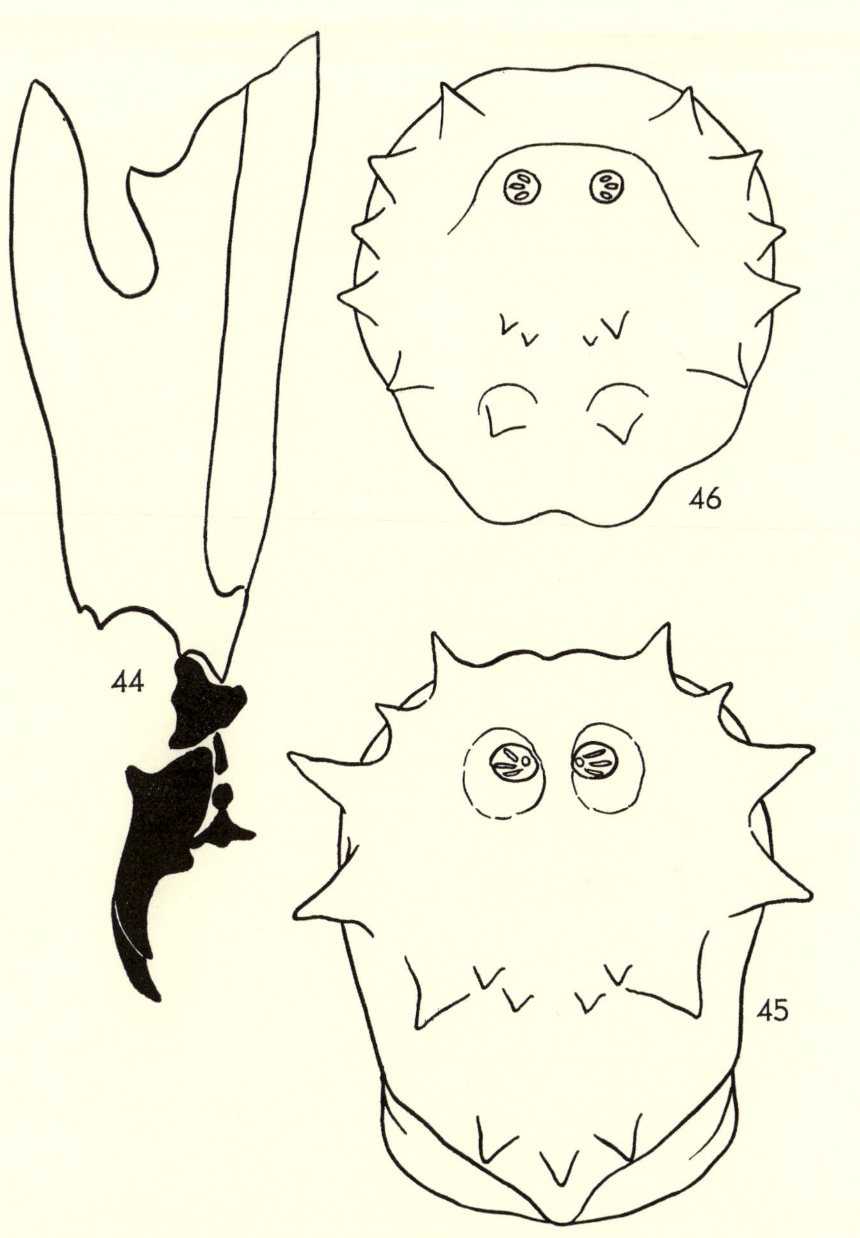

Fig. 44. Cephalo-pharyngeal skeleton of *Musca autumnalis* De Geer
Fig. 45. Frontal view of posterior segment of larva of *Scatophaga stercoraria* (L.)
Fig. 46. Frontal view of posterior segment of larva of *Anthomyia pluvialis* (L.)

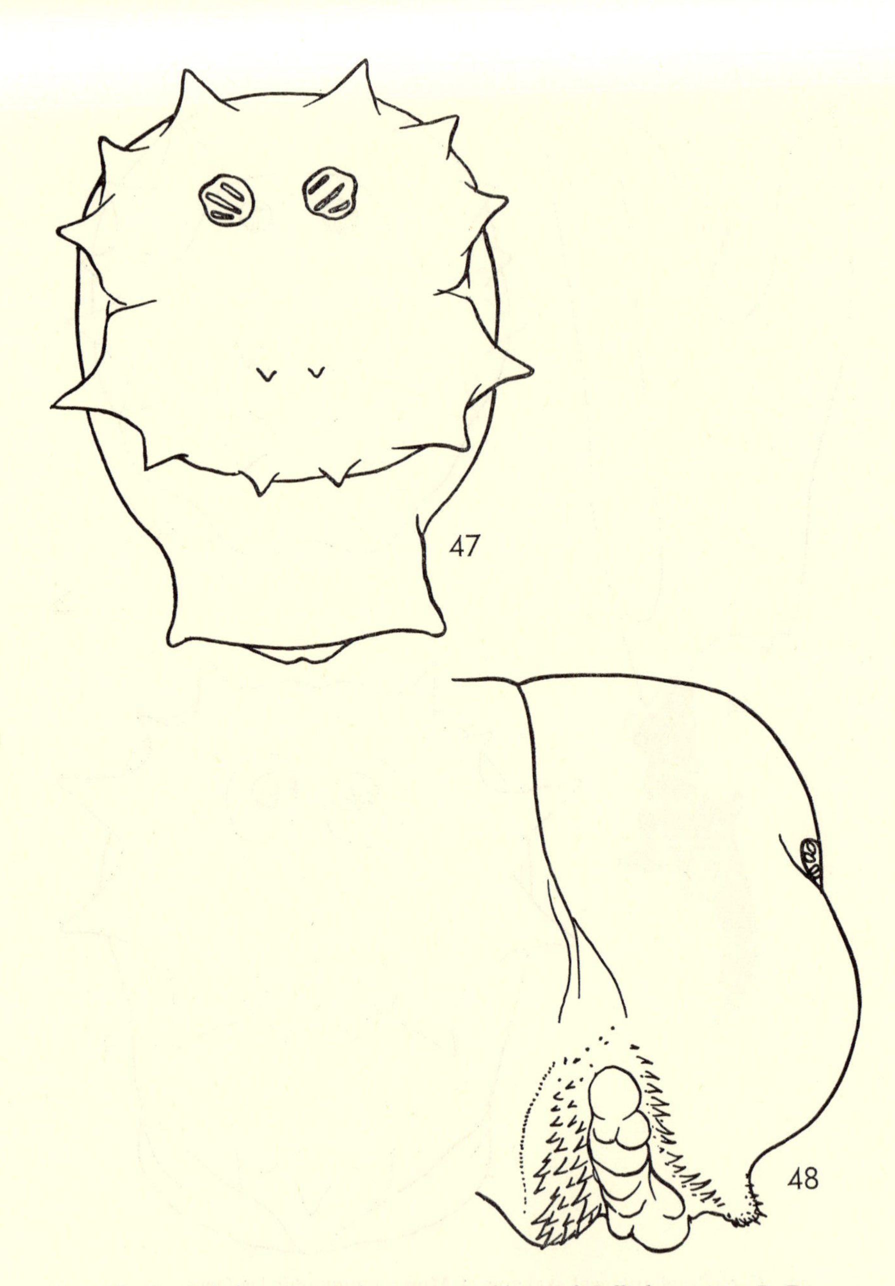

Fig. 47. Frontal view of posterior segment of larva of *Calliphora vicina* R.-D.
Fig. 48. Lateral view of posterior segment of larva of *Stomoxys calcitrans* (L.)

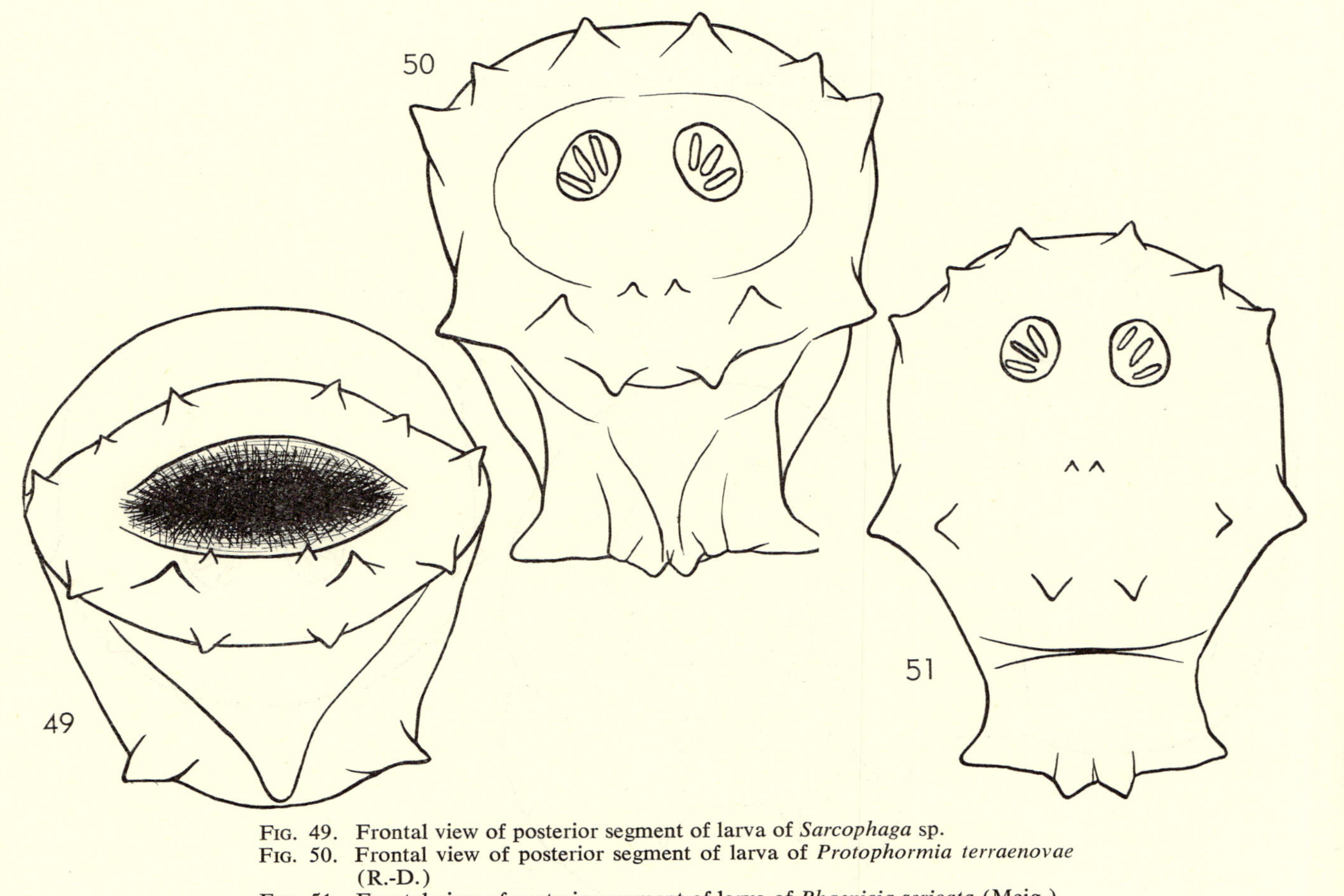

Fig. 49. Frontal view of posterior segment of larva of *Sarcophaga* sp.
Fig. 50. Frontal view of posterior segment of larva of *Protophormia terraenovae* (R.-D.)
Fig. 51. Frontal view of posterior segment of larva of *Phaenicia sericata* (Meig.)

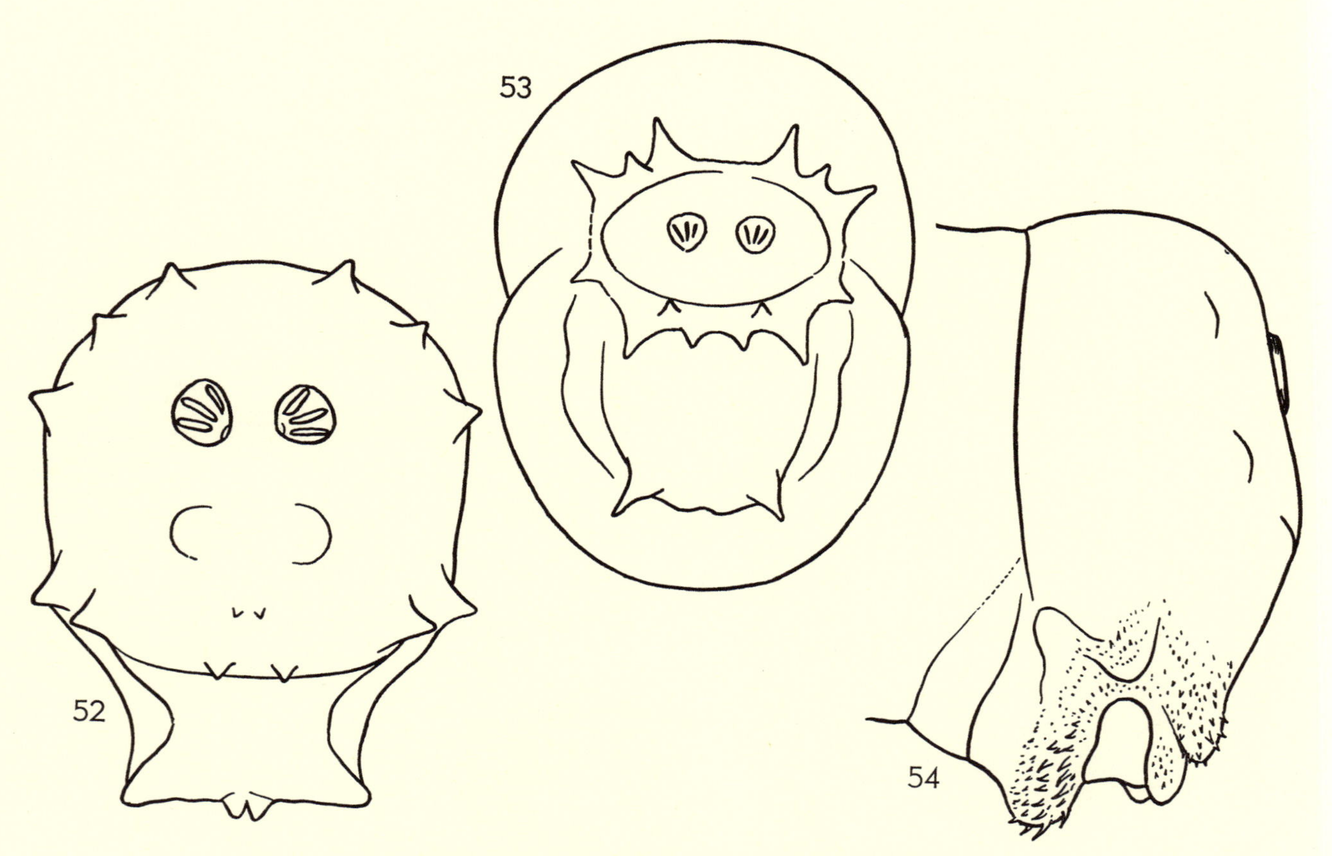

FIG. 52. Frontal view of posterior segment of larva of *Lucilia bufonivora* Moniez
FIG. 53. Frontal view of posterior segment of larva of *Sarcophaga haemorrhoidalis* (Fall.)
FIG. 54. Lateral view of posterior segment of larva of *Muscina stabulans* (Fall.)

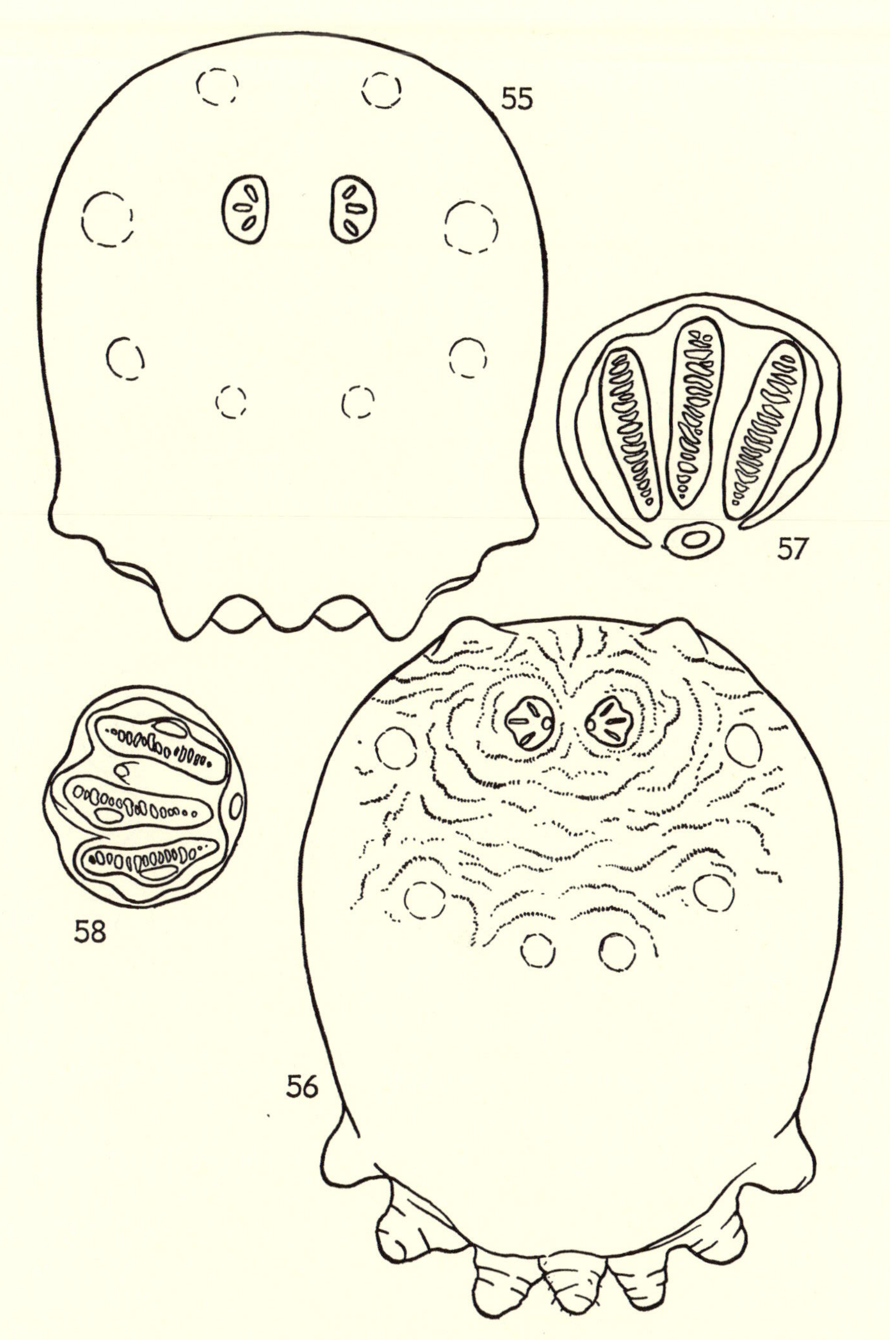

Fig. 55. Frontal view of posterior segment of larva of *Muscina stabulans* (Fall.)
Fig. 56. Frontal view of posterior segment of larva of *Hydrotaea dentipes* (Fabr.)
Fig. 57. Posterior spiracle of *Protophormia terraenovae* (R.-D.)
Fig. 58. Posterior spiracle of *Lucilia ampullacea* Vill.

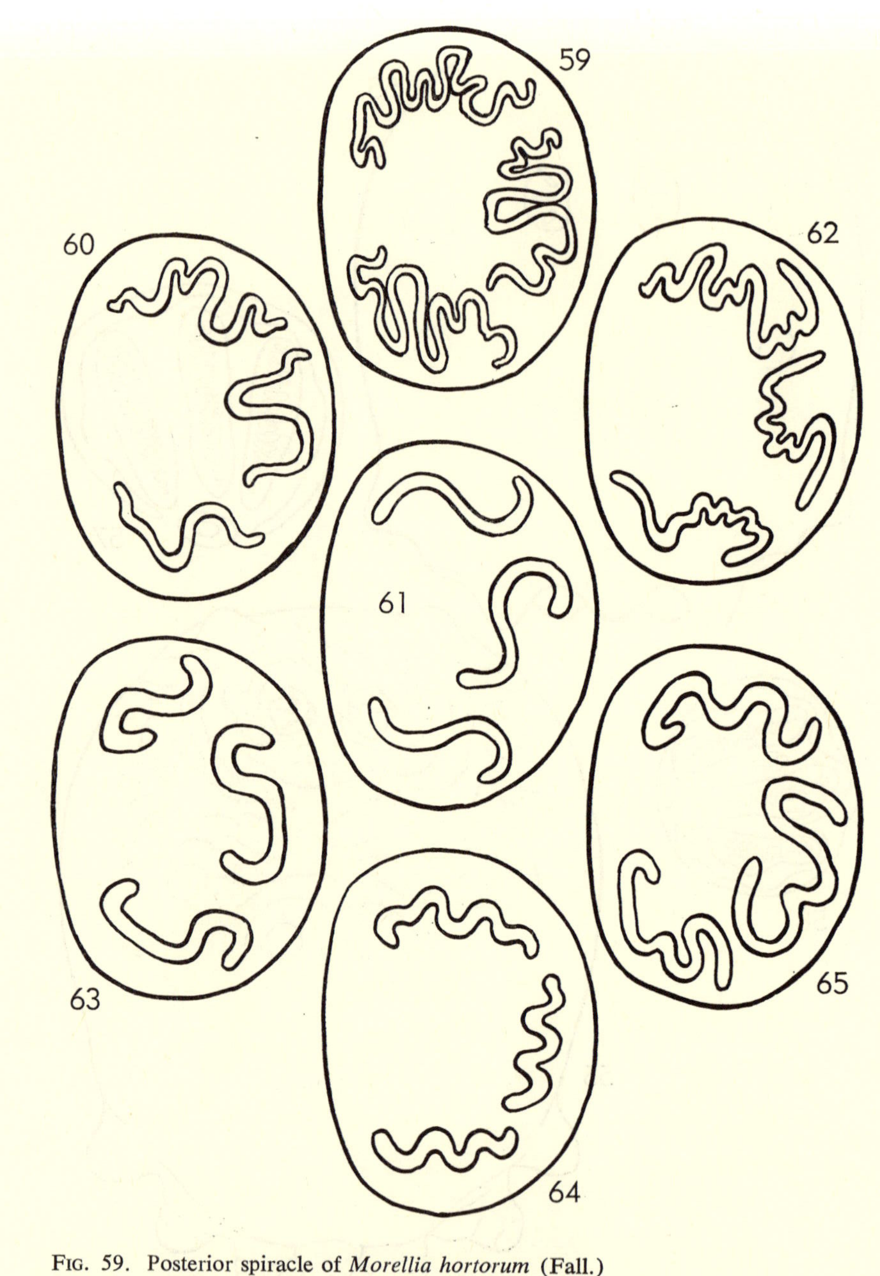

Fig. 59. Posterior spiracle of *Morellia hortorum* (Fall.)
Fig. 60. Posterior spiracle of *Morellia simplex* (Loew)
Fig. 61. Posterior spiracle of *Mesembrina mystacea* (L.)
Fig. 62. Posterior spiracle of *Mesembrina meridiana* (L.)
Fig. 63. Posterior spiracle of *Stomoxys calcitrans* (L.)
Fig. 64. Posterior spiracle of *Musca autumnalis* De Geer
Fig. 65. Posterior spiracle of *Haematobia irritans* (L.)

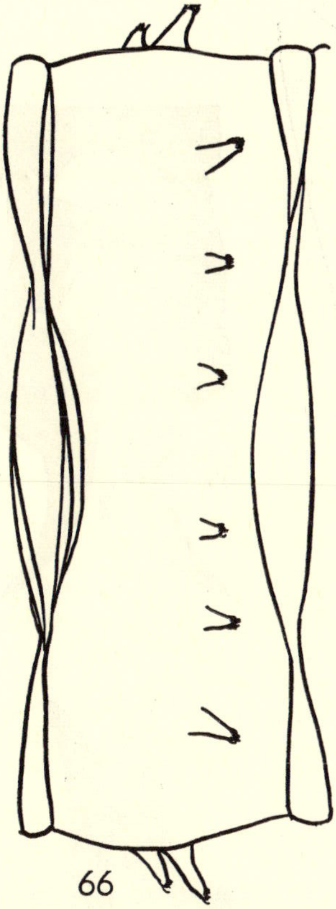

66

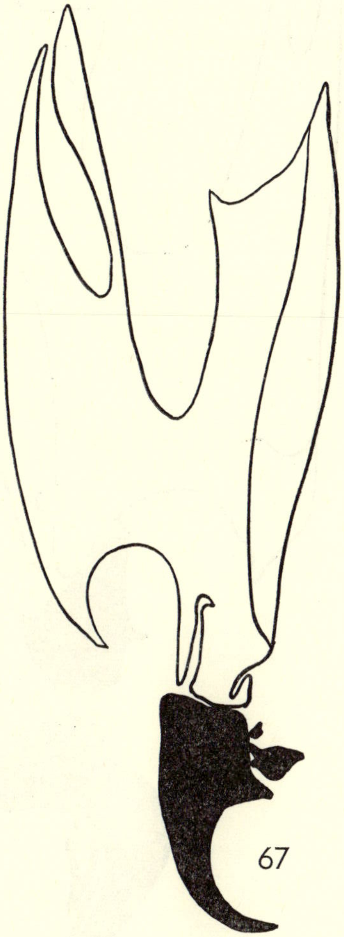

67

FIG. 66. Lateral view of abdominal segment of *Chrysomya albiceps* (Wied.)
FIG. 67. Cephalo-pharyngeal skeleton of *Sarcophaga haemorrhoidalis* (Fall.)

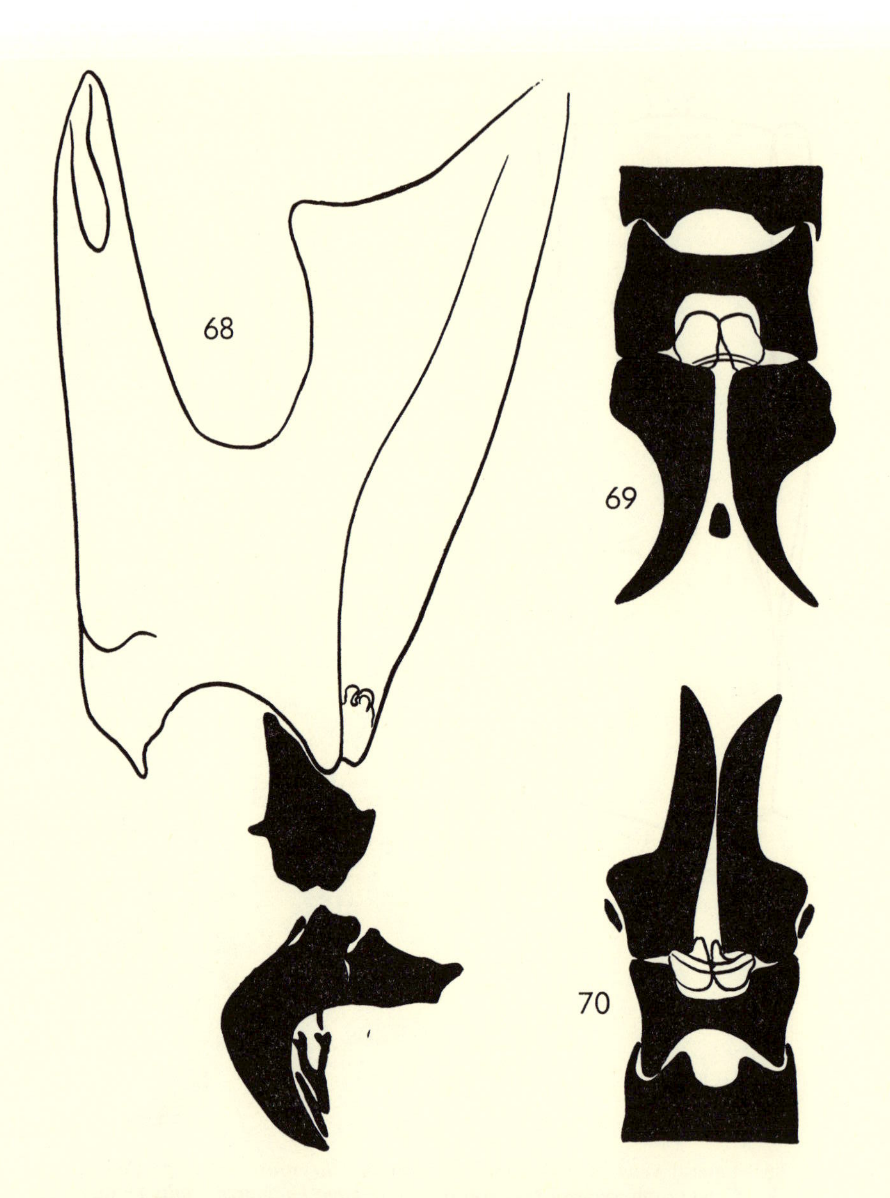

FIG. 68. Cephalo-pharyngeal skeleton of *Muscina stabulans* (Fall.)
FIG. 69. Mandibular sclerites of *Lucilia ampullacea* Vill.
FIG. 70. Mandibular sclerites of *Phaenicia sericata* (Meig.)

6

Fly-Organism Associations with Systematic List of Fly Species

SYSTEMATIC LIST OF FLIES ASSOCIATED WITH OTHER ORGANISMS

FAMILIES: 29; GENERA: 119; SPECIES: 346

FAMILY	GENUS	SPECIES, VARIETIES, AND SUBSPECIES	TOTAL NO. SPECIES
Sciaridae	*Sciara*		
Scatopsidae	*Scatopse*	*fuscipes, notata*	2
Stratiomyidae	*Chorisops*	*tibialis*	1
	Hermetia	*illucens*	1
Rhagionidae	*Rhagio*	*lineola*	1
Scenopinidae	*Scenopinus*	*albicincta, bouvieri, fenestralis, glabrifrons, senilis*	5
Empididae	*Hilara*	*flavipes, litorea*	2
Phoridae	*Phora*		
	Megaselia	*fasciata, ferruginea, iroquoiana, pulicaria, rufipes, scalaris*	6
Syrphidae	*Syrphus*	*albimanus, peltatus*	2
	Melanostoma	*mellinum, scalare*	2
	Platycheirus	*radicum*	1
	Volucella	*obesa*	1
	Xylota	*pipiens*	1
	Eristalis	*aeneus, tenax*	2
Richardiidae	*Richardia*	*viridiventris*	1
Otitidae	*Ceroxys*	*robusta*	1
	Melieria	*crassipennis*	1
	Euxesta	*annonae*	1
	Physiphora	*demandata*	1
Platystomatidae	*Scholastes*		
Coelopidae	*Coelopa*	*frigida, pilipes*	2
Dryomyzidae	*Dryomyza*	*anilis*	1
Sepsidae	*Paleosepsis*		
	Saltella	*sphondylii*	1
	Sepsis	*biflexuosa, cynipsea, nigripes, pilipes, punctum, violacea*	6
Sciomyzidae	*Sciomyza*		
Lauxaniidae	*Lauxania*	*aenea*	1
	Sapromyza	*obesa*	1
Piophilidae	*Piophila*	*casei, nigrimana, pectiniventris*	3

FAMILY	GENUS	SPECIES, VARIETIES, AND SUBSPECIES	TOTAL NO. SPECIES
Lonchaeidae	*Lonchaea*	*chorea*	1
Sphaeroceridae	*Borborus*		
	Sphaerocera	*curvipes*	1
	Copromyza	*atra*	
	Leptocera	*ferruginata, fontinalis, hirtula, humida, moesta, vagans*	6
	Limosina	*punctipennis*	1
Milichiidae	*Desmometopa*	*m-nigrum, tarsalis*	2
	Milichiella	*lacteipennis*	1
Ephydridae	*Teichomyza*	*fusca*	1
Drosophilidae	*Drosophila*	*ananassae, bifasciata, busckii, cellaris, confusa, equinoxialis, fasciata, fenestrarum, ferruginea, funebris, hydei, immigrans, kuntzei, littoralis, melanogaster, melanogaster* (Oregon R.C.), *melanogaster* (strain rho), *nebulosa, obscura, paramelanica, parthenogenetica, paulistorum, phalerata, plurilineata, prosaltans, repleta, robusta, rubrio-striata, silvestris, similis, simulans, subobscura, virilis, willistoni*	32
	Scaptomyza	*graminum*	1
Chloropidae	*Hippelates*	*collusor, currani, illicis, flavipes, peruanus, pusio*	6
	Siphunculina	*funicola*	1
	Oscinella	*frit*	1
	Conioscinella	*mars*	1
Heleomyzidae	*Suillia*	*lineata, ustulata*	2
	Heleomyza	*gigantea, modesta*	2
Anthomyiidae	*Scatophaga*	*furcata, lutaria, stercoraria*	3
	Fucellia	*rufitibia*	1
	Hylemya	*aestiva, antiqua, brassicae, cardui, cinerea, cinerella, coarctata, fugax, nigrimana, platura, radicum, strigosa, urbana, variata*	14
	Pegomya	*finitima, haemorrhoa, hyoscyami, winthemi*	4
	Hydrophoria	*ruralis*	1
	Anthomyia	*lindigii, pluvialis*	2
	Calythea	*albicincta*	1
	Paraprosalpia	*billbergi, silvestris*	2
	Leucophoria	*albiseta*	1
Muscidae	*Coenosia*	*lineatipes, sexnotata, tigrina* var. *leonina, tricolor*	4
	Allognota	*agromyzina*	1
	Atherigona	*orientalis*	1
	Gymnodia	*arcuata*	1
	Limnophora		
	Zenomyia	*oxycera*	1

FAMILY	GENUS	SPECIES, VARIETIES, AND SUBSPECIES	TOTAL NO. SPECIES
	Helina	*anceps, clara, depuncta, duplicata, impuncta, lucorum, quadrum*	7
	Myospila	*meditabunda*	1
	Mydaea	*detrita, pagana, pertusa, tincta, urbana*	5
	Fannia	*aerea, armata, canicularis, corvina, femoralis, incisurata, kowarzi, leucosticta, lineata, manicata, polychaeta, pusio, scalaris, serena, similis, subscalaris*	16
	Hydrotaea	*albipuncta, armipes, australis, dentipes, irritans, meterorica, occulta, parva*	8
	Ophyra	*aenescens, anthrax, calcogaster, leucostoma, nigra*	5
	Lasiops	*simplex, variabilis*	2
	Dendrophaonia	*querceti*	1
	Phaonia	*basalis, cincta, corbetti, errans, erratica, exoleta, goberti, incana, querceti, scutellaris, signata, trimaculata, vagans, variegata*	14
	Muscina	*assimilis, pabulorum, pascuorum, stabulans*	4
	Dasyphora	*asiatica, cyanella, gussakovskii, pratorum*	4
	Polietes	*albolineata, lardaria*	2
	Mesembrina	*meridiana*	1
	Graphomya	*maculata*	1
	Synthesiomyia	*nudiseta*	1
	Neomuscina		
	Philornis		
	Morellia	*hortorum, micans, simplex*	3
	Pyrellia		
	Orthellia	*caerulea, caesarion, cyanea*	3
	Sarcopromusca	*arcuata*	1
	Musca	*amita, autumnalis, autumnalis autumnalis, bezzii, conducens, convexifrons, corvina,* * *crassirostris, domestica, domestica domestica, domestica vicina, fergusoni, gibsoni, hilli, inferior, insignis, larvipara, lusoria, nebulo, osiris, planiceps, sorbens, sorbens sorbens, tempestiva, terraereginae, ventrosa*	22
	Haematobia	*angustifrons, exigua, irritans, titillans*	4
	Lyperosiops	*stimulans*	1
	Stomoxys	*bilineata, boueti, brunnipes, calcitrans, calcitrans* var. *soudanense, nigra*	5

* *M. corvina, hilli,* and *nebulo* are treated as species in this and the following chapters; see also the treatment by Pont and Paterson in Chapter 4.

FAMILY	GENUS	SPECIES, VARIETIES, AND SUBSPECIES	TOTAL NO. SPECIES
Calliphoridae	*Huascaromusca*	*bicolor*	1
	Rhyncomyia	*pictifacies*	1
	Cochliomyia	*hominivorax, macellaria*	2
	Chrysomya	*albiceps, chloropyga, marginalis, megacephala, pinguis, putoria, rufifacies, varipes*	8
	Paralucilia	*fulvipes*	1
	Phormia	*regina, sordida*	2
	Boreellus	*atriceps*	1
	Protophormia	*terraenovae*	1
	Protocalliphora	*azurea*	1
	Bufolucilia	*bufonivora, silvarum*	2
	Lucilia	*argyricephala, caesar, illustris, jeddensis, porphyrina*	5
	Phaenicia	*cluvia, cuprina, eximia, mexicana, pallescens, sericata*	6
	Eucalliphora	*lilaea*	1
	Aldrichina	*grahami*	1
	Calliphora	*clausa, coloradensis, icela, lata, stygia, terraenovae, vicina, vomitoria, vomitoria* var. *dunensis, uralensis*	9
	Neopollenia	*stygia*	1
	Anastellorhina	*augur*	1
	Onesia	*cognata, sepulchralis, townsendi*	3
	Cynomyopsis	*cadaverina*	1
	Cynomya	*mortuorum*	1
	Auchmeromyia	*luteola*	1
	Pollenia	*rudis*	1
Sarcophagidae	*Agria*	*latifrons*	1
	Pseudo-sarcophaga	*affinis*	1
	Wohlfahrtia	*nuba*	1
	Blaesoxipha	*impar, lineata, plinthopyga*	3
	Helicobia	*quadrisetosa*	1
	Oxysarcodexia	*ochripyga, ventricosa*	2
	Ravinia	*derelicta, effrenata, latisetosa, lherminieri, querula, striata, sueta*	7
	Sarcodexia	*pedunculata*	1
	Sarcophaga	*argyrostoma, aurifinis, aurifrons, bullata, carnaria, destructor, dux, dux* var. *tuberosa, frontalis, fuscicauda, haemorrhoidalis, hirtipes, impatiens, melanura, misera, nurus, parkeri, peregrina, ruficornis, sarraceniae, sarracenioides, securifera, tibialis, wiedemanni*	23
Tachinidae	*Tricharaea*	*haemorrhoidalis*	1

CLASSIFICATION OF FLIES	ASSOCIATED ORGANISMS
Suborder: NEMATOCERA	
Superfamily: MYCETOPHILOIDEA	
Family: SCIARIDAE	
Sciara Meigen (Synonym: *Lycoria*)	SCHIZOPHYTA: SCHIZOMYCETES EUBACTERIALES: ENTEROBACTERIACEAE *Shigella* Castellani and Chalmers Boikov, 1932 THALLOPHYTA: PHYCOMYCETES MUCORALES: MUCORACEAE *Mucor* Micheli Baumberger, 1919 THALLOPHYTA: FUNGI IMPERFECTI MONILIALES: MONILIACEAE *Gliocladium* Corda Baumberger, 1919 MONILIALES: TUBERCULARIACEAE *Fusarium* Link Baumberger, 1919
Family: SCATOPSIDAE	
Subfamily: SCATOPSINAE	
Scatopse fuscipes Meigen	ARTHROPODA: ARACHNIDA ACARINA: UNDETERMINED PARASITIDAE Sychevskaîa, 1964a ACARINA: MACROCHELIDAE *Macrocheles muscaedomesticae* (Scopoli) Sychevskaîa, 1964a
Scatopse notata Linnaeus (prey)	ARTHROPODA: INSECTA DIPTERA: ANTHOMYIIDAE *Scatophaga merdaria* (Fabricius) Hewitt, 1914
Suborder: BRACHYCERA	
Superfamily: TABANOIDEA	
Family: STRATIOMYIDAE	
Subfamily: BERIDINAE	
Chorisops tibialis Meigen (prey)	ARTHROPODA: INSECTA HYMENOPTERA: SPHECIDAE *Oxybelus bipunctatus* Olivier Chevalier, 1926
Subfamily: HERMETIINAE	
Hermetia illucens (Linnaeus) (prey)	ARTHROPODA: INSECTA HYMENOPTERA: SPHECIDAE *Stictia denticornis* (Handlirsch) Bodkin, 1917

Fly	Associated organism
Family: RHAGIONIDAE	
Subfamily: RHAGIONINAE	
Rhagio lineola Fabricius (Synonym: *Leptis lineola*)	THALLOPHYTA: PHYCOMYCETES ENTOMOPHTHORALES: ENTOMOPHTHORACEAE *Entomophthora muscivora* Schroeter Rostrup, 1916
Superfamily: ASILOIDEA	
Family: SCENOPINIDAE	
Scenopinus albicincta (Rossi) (Synonym: *Omphrale albicincta*)	CHORDATA: AVES PASSERIFORMES: HIRUNDINIDAE *Hirundo rustica* Linnaeus susbp. *rustica* Séguy, 1929
Scenopinus bouvieri (Séguy) (Synonym: *Omphrale bouvieri*)	CHORDATA: AVES PASSERIFORMES: PLOCEIDAE *Passer domesticus* (Linnaeus) Séguy, 1929
Scenopinus fenestralis (Linnaeus) (Synonym: *Omphrale fenestralis*)	CHORDATA: AVES PASSERIFORMES: HIRUNDINIDAE *Delichon urbica* (Linnaeus) subsp. *urbica* Séguy, 1929 *Hirundo rustica* Linnaeus subsp. *rustica* Séguy, 1929 CHORDATA: MAMMALIA RODENTIA: MURIDAE *Apodemus sylvaticus* Linnaeus Séguy, 1929
Scenopinus glabrifrons Meigen (Synonym: *Omphrale glabrifrons*)	CHORDATA: AVES PASSERIFORMES: STURNIDAE *Sturnus vulgaris* Linnaeus Séguy, 1929
Scenopinus senilis (Fabricius) (Synonym: *Omphrale senilis*)	CHORDATA: AVES PASSERIFORMES: HIRUNDINIDAE *Hirundo rustica* Linnaeus subsp. *rustica* Séguy, 1929 PASSERIFORMES: PLOCEIDAE *Passer domesticus* (Linnaeus) Séguy, 1929
Superfamily: EMPIDOIDEA	
Family: EMPIDIDAE	
Subfamily: EMPIDINAE	
Hilara flavipes Meigen (prey)	ARTHROPODA: INSECTA HYMENOPTERA: SPHECIDAE *Crossocerus quadrimaculatus* (Fabricius) Hamm et al., 1926

Fly	Associated organism
Hilara litorea Fallén (prey)	ARTHROPODA: INSECTA HYMENOPTERA: SPHECIDAE *Crabro leucostoma* (Linnaeus) Hamm et al., 1926 *Crossocerus quadrimaculatus* (Fabricius) Hamm et al., 1926

Suborder: CYCLORRHAPHA
Division: ASCHIZA
Superfamily: PHOROIDEA
Family: PHORIDAE
Subfamily: PHORINAE

Fly	Associated organism
Phora Latreille	ARTHROPODA: INSECTA DIPTERA: MUSCIDAE *Muscina assimilis* (Fallén) Keilin, 1917

Subfamily: METOPININAE

Fly	Associated organism
Megaselia Brues (Synonym: *Aphiochaeta*) (host for parasite)	ARTHROPODA: INSECTA HYMENOPTERA: BRACONIDAE *Aspilota nervosa* Haliday Evans, 1933
Megaselia fasciata (Fallén) (Synonym: *Phora fasciata*) (larval parasite)	ARTHROPODA: INSECTA COLEOPTERA: COCCINELLIDAE *Adonia variegata* Goeze Martelli, 1913 *Coccinella septempunctata* Linnaeus Rondani du Buysson, 1917 *Thea vigintiduopunctata* Linnaeus Martelli, 1913 Lichtenstein, 1920 *Vibidia duodecimguttata* Poda Lichtenstein, 1920 HYMENOPTERA: ENCYRTIDAE *Homalotylus eitelweinii* Ratzeberg du Buysson, 1921
Megaselia ferruginea (Brunetti) (Synonym: *Aphiochaeta ferruginea*)	SCHIZOPHYTA: SCHIZOMYCETES PSEUDOMONADALES: SPIRILLACEAE *Vibrio comma* (Schroeter) Winslow et al. Roberg, 1915
Megaselia iroquoiana (Malloch) (pupa is host for parasite)	ARTHROPODA: INSECTA HYMENOPTERA: PTEROMALIDAE *Spalangia nigroaenea* Curtis Bouček, 1963
Megaselia ?iroquoiana (Malloch) (pupa is host for parasite)	ARTHROPODA: INSECTA HYMENOPTERA: PTEROMALIDAE *Spalangia nigra* Latreille Roberts, 1935

Fly	Associated organism
Megaselia pulicaria (Fallén) (Synonym: *Aphiochaeta pulicaria*) (prey)	ARTHROPODA: INSECTA HYMENOPTERA: SPHECIDAE *Crabro elongatus* (Provancher) Hamm et al., 1926
Megaselia rufipes (Meigen) (Synonym: *Aphiochaeta rufipes*) (larval prey)	ARTHROPODA: INSECTA DIPTERA: MUSCIDAE *Muscina assimilis* (Fallén) Keilin, 1917
	CHORDATA: AVES PASSERIFORMES: PARIDAE *Parus caeruleus* Linnaeus subsp. *caeruleus* Séguy, 1929
Megaselia scalaris (Loew) (Synonym: *Aphiochaeta scalaris*)	ARTHROPODA: INSECTA ORTHOPTERA: LOCUSTIDAE *Schistocerca gregaria* Fo̊rskal Nocedo, 1921
Superfamily: SYRPHOIDEA	
Family: SYRPHIDAE	
Subfamily: SYRPHINAE	
Syrphus Fabricius	THALLOPHYTA: PHYCOMYCETES ENTOMOPHTHORALES: ENTOMOPHTHORACEAE *Entomophthora muscae* Cohn Thaxter, 1888
Syrphus albimanus Fabricius (Synonym: *Platychira albimanus*) (prey)	ARTHROPODA: INSECTA HYMENOPTERA: SPHECIDAE *Crabro cavifrons* Thomson Hamm et al., 1926
Syrphus peltatus (Meigen) (Synonym: *Platychira peltatus*) (prey)	ARTHROPODA: INSECTA HYMENOPTERA: SPHECIDAE *Crabro cavifrons* Thomson Hamm et al., 1926
Melanostoma mellinum (Linnaeus)	THALLOPHYTA: PHYCOMYCETES ENTOMOPHTHORALES: ENTOMOPHTHORACEAE *Entomophthora muscae* Cohn Rostrup, 1916
Melanostoma scalare Fabricius	THALLOPHYTA: PHYCOMYCETES ENTOMOPHTHORALES: ENTOMOPHTHORACEAE *Entomophthora muscae* Cohn Rostrup, 1916

Platycheirus radicum Fabricius (Synonym: *Platychira radicum*) (prey) .... ARTHROPODA: INSECTA
HYMENOPTERA: SPHECIDAE
Ectemnius quadricinctus (Fabricius)
Hamm et al., 1926

Subfamily: MILESIINAE

Volucella obesa (Fabricius) .... SCHIZOPHYTA: SCHIZOMYCETES
EUBACTERIALES: ENTEROBACTERIACEAE
Salmonella Lignières
Alcivar et al., 1946
Salmonella amersfoort Henning
Alcivar et al., 1946
Salmonella anatum (Rettger and Scoville) Bergey et al.
Alcivar et al., 1946
Salmonella ballerup Kauffmann and Møller
Alcivar et al., 1946
Salmonella choleraesuis (Smith) Weldin
Alcivar et al., 1946
Salmonella dublin Kauffmann
Alcivar et al., 1946
Salmonella enteritidis (Gaertner) Castellani and Chalmers
Alcivar et al., 1946
Salmonella florida Cherry, Edwards and Bruner
Alcivar et al., 1946
Salmonella give Kauffmann
Alcivar et al., 1946
Salmonella kentucky Edwards
Alcivar et al., 1946
Salmonella kottbus Schütze et al.
Alcivar et al., 1946
Salmonella meleagridis Bruner and Edwards
Alcivar et al., 1946
Salmonella narashino Kauffmann
Alcivar et al., 1946
Salmonella onderstepoort Henning
Alcivar et al., 1946
Salmonella oranienburg Kauffmann
Alcivar et al., 1946
Salmonella oregon Edwards and Bruner
Alcivar et al., 1946

Fly	Organism
Volucella obesa (cont.)	*Salmonella paratyphi* B Castellani and Chalmers Alcivar et al., 1946 *Salmonella reading* Schütze Alcivar et al., 1946 *Salmonella typhimurium* (Loeffler) Castellani and Chalmers Alcivar et al., 1946 *Salmonella zanzibar* Kauffmann Alcivar et al., 1946 *Shigella alkalescens* (Andrewes) Weldin Alcivar et al., 1946 *Shigella boydii* Ewing, 1, 2, 3, 4, 5, 6, 8, 9 Alcivar et al., 1946 *Shigella ceylonensis* (Weldin) Weldin Alcivar et al., 1946 *Shigella dispar* (Andrewes) Bergey et al. Alcivar et al., 1946 *Shigella flexneri* Castellani and Chalmers, 1a, 2a, Y, 6 Alcivar et al., 1946 ACTINOMYCETALES: MYCOBACTERIACEAE *Mycobacterium phlei* Lehmann and Neumann Currie, 1910
Xylota pipiens (Linnaeus) (Synonym: *Syritta pipiens*)	ARTHROPODA: ARACHNIDA ACARINA: MACROCHELIDAE *Macrocheles muscaedomesticae* (Scopoli) Sychevskaīā, 1964a
Eristalis Latreille	SCHIZOPHYTA: SCHIZOMYCETES EUBACTERIALES: BACILLACEAE *Clostridium botulinum* (van Ermengem) Bergey et al. Bail, 1900 ASCHELMINTHES: NEMATODA ASCARIDIDEA: ASCARIDIDAE *Ascaris lumbricoides* Linnaeus Sukhacheva, 1963
Eristalis aeneus (Scopoli)	PLATYHELMINTHES: CESTOIDEA CYCLOPHYLLIDEA: TAENIIDAE *Taeniarhynchus saginatum* (Goeze) Sychevskaīā et al., 1958

Fly	Associated organism
Eristalis tenax (Linnaeus) (Synonym: *Tubifera tenax*)	SCHIZOPHYTA: SCHIZOMYCETES
	PSEUDOMONADALES: SPIRILLACEAE
	Vibrio comma (Schroeter) Winslow et al.
	Maddox, 1885b
	EUBACTERIALES: MICROCOCCACEAE
	Micrococcus Cohn
	Maddox, 1885b
	ARTHROPODA: INSECTA
	COLEOPTERA: STAPHYLINIDAE
	Aleochara curtula Goeze
	Balduf, 1935
	Philonthus fimetarius Gravenhorst
	Balduf, 1935
	Philonthus politus Linnaeus
	Balduf, 1935
	HYMENOPTERA: CHALCIDIDAE
	Brachymeria vicina Walker
	Nikol'skai͡a, 1960a
	HYMENOPTERA: DIAPRIIDAE
	Diapria conica (Fabricius)
	Sychevskai͡a, 1964b
UNDETERMINED SYRPHIDAE	THALLOPHYTA: PHYCOMYCETES
	ENTOMOPHTHORALES: ENTOMOPHTHORACEAE
	Entomophthora muscae Cohn
	Thaxter, 1888
	Entomophthora ovispora Nowakowski
	Thaxter, 1888
Division: SCHIZOPHORA	
Section: ACALYPTRATAE	
Superfamily: TEPHRITOIDEA	
Family: RICHARDIIDAE	
Richardia viridiventris Wulp	ENTEROVIRUSES
	Poliovirus Types 1 and 3
	Paul et al., 1962
	ECHO virus Types 8, 12, 14
	Paul et al., 1962
Family: OTITIDAE	
Subfamily: OTITINAE	
Ceroxys robusta Loew	ARTHROPODA: ARACHNIDA
	ACARINA: MACROCHELIDAE
	Macrocheles Latreille
	Sychevskai͡a, 1964a
Melieria crassipennis Fabricius (Synonym: *Ceroxys crassipennis*)	PROTOZOA: MASTIGOPHORA
	PROTOMONADIDA: TRYPANOSOMATIDAE
	Herpetomonas ?muscae (Carter)
	Laveran et al., 1920
Subfamily: ULIDIINAE	

Euxesta annonae (Fabricius) (prey) .... ARTHROPODA: INSECTA

DERMAPTERA: LABIDURIDAE

Euborellia annulipes (Lucas)
Illingworth, 1923a

DERMAPTERA: LABIIDAE

Labia pilicornis (Motschulsky)
Illingworth, 1923a

Spingolabis hawaiiensis (Bormans)
Illingworth, 1923a

COLEOPTERA: HYDROPHILIDAE

Cryptopleurum minutum Fabricius
Illingworth, 1923a

Dactylosternum abdominale Fabricius
Illingworth, 1923a

COLEOPTERA: STAPHYLINIDAE

Oxytelus Gravenhorst
Illingworth, 1923a

Philonthus discoideus Gravenhorst
Illingworth, 1923a

Philonthus longicornis Stephens
Illingworth, 1923a

Tachyporus Gravenhorst
Illingworth, 1923a

HYMENOPTERA: FORMICIDAE

Pheidole megacephala (Fabricius)
Illingworth, 1923a

Ponera perkinsi Forel
Illingworth, 1923a

Physiphora Fallén (Synonym: *Chrysomyza*) .... PROTOZOA: MASTIGOPHORA

POLYMASTIGIDA: CHILOMASTIGIDAE

Chilomastix mesnili (Wenyon)
Harris et al., 1946

POLYMASTIGIDA: HEXAMITIDAE

Giardia lamblia Stiles
Harris et al., 1946

TRICHOMONADIDA: TRICHOMONADIDAE

Trichomonas hominis (Davaine)
Harris et al., 1946

PROTOZOA: SARCODINA

AMOEBIDA: ENDAMOEBIDAE

Endolimax nana (Wenyon and O'Connor)
Harris et al., 1946

Entamoeba coli (Grassi)
Harris et al., 1946

Entamoeba histolytica Schaudinn
Harris et al., 1946

Fly	Organism associations
	Iodamoeba bütschlii (Prowazek) Harris et al., 1946 PLATYHELMINTHES: CESTOIDEA CYCLOPHYLLIDEA: HYMENOLEPIDIDAE *Hymenolepis diminuta* (Rudolphi) Blanchard Harris et al., 1946 ASCHELMINTHES: NEMATODA TRICHURIDEA: TRICHURIDAE *Trichuris trichiura* (Linnaeus) Harris et al., 1946 STRONGYLIDEA: ANCYLOSTOMATIDAE *Ancylostoma* (Dubini) Harris et al., 1946 ASCARIDIDEA: ASCARIDIDAE *Ascaris lumbricoides* Linnaeus Harris et al., 1946
Physiphora demandata (Fabricius) (Synonym: *Chloria demandata*)	SCHIZOPHYTA: SCHIZOMYCETES EUBACTERIALES: ENTEROBACTERIACEAE *Escherichia coli* (Migula) Castellani and Chalmers Povolný et al., 1961 *Serratia* Bizio Povolný et al., 1961 *Proteus vulgaris* Hauser Povolný et al., 1961 *Providencia* Kauffmann Povolný et al., 1961 EUBACTERIALES: MICROCOCCACEAE *Staphylococcus aureus* Rosenbach Povolný et al., 1961 EUBACTERIALES: LACTOBACILLACEAE *Diplococcus pneumoniae* Weichselbaum Povolný et al., 1961 EUBACTERIALES: BACILLACEAE *Bacillus* Cohn Povolný et al., 1961 ARTHROPODA: INSECTA HYMENOPTERA: SPHECIDAE *Oxybelus bipunctatus* Olivier Chevalier, 1926 *Oxybelus uniglumis* (Linnaeus) Chevalier, 1926

Fly	Associated organism
Family: PLATYSTOMATIDAE	
Subfamily: PLATYSTOMATINAE	
Scholastes Loew	PROTOZOA: MASTIGOPHORA
	POLYMASTIGIDA: CHILOMASTIGIDAE
	Chilomastix mesnili (Wenyon) Harris et al., 1946
	POLYMASTIGIDA: HEXAMITIDAE
	Giardia lamblia Stiles Harris et al., 1946
	TRICHOMONADIDA: TRICHOMONADIDAE
	Trichomonas hominis (Davaine) Harris et al., 1946
	PROTOZOA: SARCODINA
	AMOEBIDA: ENDAMOEBIDAE
	Endolimax nana (Wenyon and O'Connor) Harris et al., 1946
	Entamoeba coli (Grassi) Harris et al., 1946
	Entamoeba histolytica Schaudinn Harris et al., 1946
	Iodamoeba bütschlii (Prowazek) Harris et al., 1946
	PLATYHELMINTHES: CESTOIDEA
	CYCLOPHYLLIDEA: HYMENOLEPIDIDAE
	Hymenolepis diminuta (Rudolphi) Blanchard Harris et al., 1946
	ASCHELMINTHES: NEMATODA
	TRICHURIDEA: TRICHURIDAE
	Trichuris trichiura (Linnaeus) Harris et al., 1946
	STRONGYLIDEA: ANCYLOSTOMATIDAE
	Ancylostoma (Dubini) Harris et al., 1946
	ASCARIDIDEA: ASCARIDIDAE
	Ascaris lumbricoides Linnaeus Harris, et al., 1946
Superfamily: SCIOMYZOIDEA	
Family: COELOPIDAE	
Coelopa frigida (Fabricius) (Synonym: *Coelopa eximia*) (prey)	ARTHROPODA: INSECTA
	COLEOPTERA: STAPHYLINIDAE
	Aleochara algarum Fauvel Lesne et al., 1922
Coelopa pilipes Haliday (prey)	ARTHROPODA: INSECTA
	COLEOPTERA: STAPHYLINIDAE
	Aleochara algarum Fauvel Lesne et al., 1922

Fly	Associated organism
Family: DRYOMYZIDAE	
Dryomyza anilis Fallén (Synonym: *Neuroctena anilis*)	PROTOZOA: MASTIGOPHORA PROTOMONADIDA: TRYPANOSOMATIDAE *Herpetomonas ?muscarum* (Leidy) Kent Mackinnon, 1910
Family: SEPSIDAE	
Paleosepsis Duda	SCHIZOPHYTA: SCHIZOMYCETES EUBACTERIALES: ENTEROBACTERIACEAE *Salmonella anatum* (Rettger and Scoville) Bergey et al. Greenberg et al., 1963b *Salmonella derby* Warren and Scott Greenberg et al., 1963b
Saltella sphondylii (Schrank) (Synonym: *Pandora scutellaris*)	ARTHROPODA: INSECTA HYMENOPTERA: UNDETERMINED ICHNEUMONIDAE Hammer, 1942
Sepsis Fallén	SCHIZOPHYTA: SCHIZOMYCETES EUBACTERIALES: ENTEROBACTERIACEAE *Serratia marcescens* Bizio Nicholls, 1912 *Salmonella typhi* Warren and Scott Nicholls, 1912 EUBACTERIALES: MICROCOCCACEAE *Staphylococcus aureus* Rosenbach Nicholls, 1912 ARTHROPODA: INSECTA HYMENOPTERA: CYNIPIDAE *Cothonaspis nigricornis* (Kieffer) Sychevskaiā, 1964b *Kleidotoma certa* Belizin Sychevskaiā, 1964b
Sepsis biflexuosa Strobl	SCHIZOPHYTA: SCHIZOMYCETES EUBACTERIALES: ENTEROBACTERIACEAE *Salmonella anatum* (Rettger and Scoville) Bergey et al. Greenberg et al., 1963b *Salmonella derby* Warren and Scott Greenberg et al., 1963b
Sepsis cynipsea (Linnaeus)	ASCHELMINTHES: NEMATODA ENOPLIDA: TETRADONEMATIDAE *Mermithonema entomophilum* Goodey Goodey, 1941

Sepsis cynipsea (cont.)	ARTHROPODA: ARACHNIDA ACARINA: TROMBIDIIDAE *Allothrombium fuliginosum* (Hermann) = [*Leptus coccineus* per Oudemans, A.C., 1929; = *Pediculeus coccineus* Scopoli] Scopoli, 1763 ARTHROPODA: INSECTA HYMENOPTERA: SPHECIDAE *Crabro leucostoma* (Linnaeus) Hamm et al., 1926 DIPTERA: ANTHOMYIIDAE *Scatophaga stercoraria* (Linnaeus) Hobby, 1934b
Sepsis nigripes Meigen (prey)	ARTHROPODA: INSECTA HYMENOPTERA: SPHECIDAE *Crabro leucostoma* (Linnaeus) Hamm et al., 1926
Sepsis pilipes Wulp (Synonym: *Sepsidimorpha pilipes*) (host for parasite)	ARTHROPODA: INSECTA HYMENOPTERA: UNDETERMINED ICHNEUMONIDAE Hammer, 1942
Sepsis punctum (Fabricius)	SCHIZOPHYTA: SCHIZOMYCETES EUBACTERIALES: ENTEROBACTERIACEAE *Escherichia coli* (Migula) Castellani and Chalmers Povolný et al., 1961 *Serratia* Bizio Povolný et al., 1961 *Proteus vulgaris* Hauser Povolný et al., 1961 *Providencia* Kauffmann Povolný et al., 1961 *Salmonella anatum* (Rettger and Scoville) Bergey et al. Greenberg et al., 1963b *Salmonella derby* Warren and Scott Greenberg et al., 1963b EUBACTERIALES: MICROCOCCACEAE *Staphylococcus aureus* Rosenbach Povolný et al., 1961 EUBACTERIALES: LACTOBACILLACEAE *Diplococcus pneumoniae* Weichselbaum Povolný et al., 1961 EUBACTERIALES: BACILLACEAE *Bacillus* Cohn Povolný et al., 1961

Sepsis violacea Meigen (Synonym: *Sepsis ?punctum*)	SCHIZOPHYTA: SCHIZOMYCETES EUBACTERIALES: ENTEROBACTERIACEAE *Escherichia coli* var. *communior* (Topley and Wilson) Yale Beyer, 1925 ARTHROPODA: ARACHNIDA ACARINA: MACROCHELIDAE *Macrocheles* Latreille Sychevskaîa, 1964a ACARINA: ANOETIDAE *Myianoetus* Oudemans Sychevskaîa, 1964a
Family: SCIOMYZIDAE Subfamily: SCIOMYZINAE *Sciomyza* Fallén	THALLOPHYTA: PHYCOMYCETES ENTOMOPHTHORALES: ENTOMOPHTHORACEAE *Entomophthora muscivora* Schroeter Rostrup, 1916
Superfamily: LAUXANIOIDEA Family: LAUXANIIDAE *Lauxania aenea* Fallén	THALLOPHYTA: PHYCOMYCETES ENTOMOPHTHORALES: ENTOMOPHTHORACEAE *Entomophthora muscivora* Schroeter Rostrup, 1916 *Entomophthora richteri* (Bresadola and Staritz) Bubák Bubák, 1903 Lakon, 1915
Sapromyza Fallén	THALLOPHYTA: PHYCOMYCETES ENTOMOPHTHORALES: ENTOMOPHTHORACEAE *Entomophthora ovispora* Nowakowski Thaxter, 1888
Sapromyza obesa Zetterstedt (Synonym: *Sapromyza rorida*)	THALLOPHYTA: PHYCOMYCETES ENTOMOPHTHORALES: ENTOMOPHTHORACEAE *Entomophthora muscivora* Schroeter Rostrup, 1916

Superfamily: PALLOPTEROIDEA
Family: PIOPHILIDAE
Piophila casei (Linnaeus) .... SCHIZOPHYTA: SCHIZOMYCETES

EUBACTERIALES:
ACHROMOBACTERACEAE

Alcaligenes faecalis Castellani and Chalmers
Gregor et al., 1960
Lysenko et al., 1961

Flavobacterium devorans (Zimmermann) Bergey et al.
Gregor et al., 1960

Flavobacterium invisible (Vaughan) Bergey et al.
Gregor et al., 1960

EUBACTERIALES:
ENTEROBACTERIACEAE

Escherichia Castellani and Chalmers
Lysenko et al., 1961

Escherichia coli (Migula) Castellani and Chalmers
Beyer, 1925
Gregor et al., 1960
Povolný et al., 1961

Aerobacter aerogenes (Kruse) Beijerinck
Gregor et al., 1960

Klebsiella Trevisan
Lysenko et al., 1961

Klebsiella cloacae (Jordan) Brisou
Gregor et al., 1960

Cloaca Castellani and Chalmers
Lysenko et al., 1961

Serratia Bizio
Povolný et al., 1961

Serratia marcescens Bizio
Gregor et al., 1960

Proteus inconstans (Ornstein) Shaw and Clarke
Gregor et al., 1960

Proteus mirabilis Hauser
Gregor et al., 1960
Lysenko et al., 1961

Proteus morganii (Winslow, Kligler, and Rothberg) Yale
Gregor et al., 1960
Lysenko et al., 1961

Piophila casei (cont.)	*Proteus rettgeri* (Hadley et al.) Rustigian and Stuart
	Gregor et al., 1960
	Lysenko et al., 1961
	Proteus vulgaris Hauser
	Gregor et al., 1960
	Lysenko et al., 1961
	Povolný et al., 1961
	Providencia Kauffmann
	Povolný et al., 1961
	Citrobacter freundii Werkman and Gillen
	Gregor et al., 1960
	UNDETERMINED ENTEROBACTERIACEAE
	Klebsiella-Cloaca intermed. forms
	Lysenko et al., 1961
	EUBACTERIALES: MICROCOCCACEAE
	Staphylococcus afermentans Shaw, Stitt, and Cowan
	Gregor et al., 1960
	Staphylococcus aureus Rosenbach
	Gregor et al., 1960
	Povolný et al., 1961
	Staphylococcus lactis Shaw, Stitt, and Cowan
	Gregor et al., 1960
	Staphylococcus saprophyticus Fairbrother
	Gregor et al., 1960
	Lysenko et al., 1961
	EUBACTERIALES: LACTOBACILLACEAE
	Diplococcus pneumoniae Weichselbaum
	Povolný et al., 1961
	Streptococcus durans Sherman and Wing
	Gregor et al., 1960
	Lysenko et al., 1961
	Streptococcus faecalis var. *liquefaciens* (Sternberg) Mattick
	Lysenko et al., 1961
	EUBACTERIALES: BACILLACEAE
	Bacillus Cohn
	Povolný et al., 1961
	Bacillus anthracis Cohn
	Legroux et al., 1945
	Bacillus cereus var. *mycoides* (Flügge) Smith et al.
	Gregor et al., 1960

Fly	Associated organism
Piophila casei (cont.)	*Bacillus subtilis* Cohn Lysenko et al., 1961 *Clostridium botulinum* (van Ermengen) Bergey et al. Legroux et al., 1945 ARTHROPODA: INSECTA COLEOPTERA: CLERIDAE *Necrobia rufipes* De Geer Simmons, 1922 HYMENOPTERA: BRACONIDAE *Aphaereta pallipes* Say Beard, 1964 HYMENOPTERA: PTEROMALIDAE *Nasonia vitripennis* (Walker) Roberts, 1935 *Pachycrepoideus vindemmiae* Rondani Simmons, 1922 Crandell, 1939
Piophila nigrimana Meigen	ARTHROPODA: ARACHNIDA ACARINA: MACROCHELIDAE *Macrocheles muscaedomesticae* (Scopoli) Sychevskaîa, 1964a
Piophila pectiniventris Duda	SCHIZOPHYTA: SCHIZOMYCETES EUBACTERIALES: ENTEROBACTERIACEAE *Escherichia coli* (Migula) Castellani and Chalmers Lysenko, 1958 *Proteus mirabilis* Hauser Lysenko, 1958 *Proteus rettgeri* (Hadley et al.) Rustigian and Stuart Lysenko, 1958 *Proteus vulgaris* Hauser Lysenko, 1958
Family: LONCHAEIDAE	
Lonchaea chorea (Fabricius) (Synonym: *Lonchaea vaginalis*)	THALLOPHYTA: PHYCOMYCETES ENTOMOPHTHORALES: ENTOMOPHTHORACEAE *Entomophthora ovispora* Nowakowski Thaxter, 1888
Superfamily: MILICHIOIDEA	
Family: SPHAEROCERIDAE	
Borborus Meigen	PROTOZOA: MASTIGOPHORA PROTOMONADIDA: TRYPANOSOMATIDAE *Herpetomonas muscarum* (Leidy) Kent Patton, 1921a

Fly	Organism
Sphaerocera curvipes Latreille (Synonym: *Sphaerocera subsultans*)	PROTOZOA: MASTIGOPHORA PROTOMONADIDA: TRYPANOSOMATIDAE *Leptomonas legerorum* Chatton Chatton, 1912a
Copromyza Fallén (Synonym: *Cypsela*)	ARTHROPODA: ARACHNIDA ACARINA: PARASITIDAE *Parasitus lunaris* Berlese Sychevskaia͡, 1964a
Copromyza atra (Meigen) (Synonyms: *Borborus ater*, *Olina geniculata*)	SCHIZOPHYTA: SCHIZOMYCETES EUBACTERIALES: ENTEROBACTERIACEAE *Escherichia coli* (Migula) Castellani and Chalmers Povolný et al., 1961 *Proteus vulgaris* Hauser Povolný et al., 1961 EUBACTERIALES: BACILLACEAE *Bacillus* Cohn Povolný et al., 1961 ARTHROPODA: INSECTA HYMENOPTERA: UNDETERMINED ICHNEUMONIDAE Hammer, 1942
Leptocera Macquart (Synonym: *Limosina*)	SCHIZOPHYTA: SCHIZOMYCETES EUBACTERIALES: ENTEROBACTERIACEAE *Salmonella typhi* Warren and Scott Boikov, 1932 *Shigella* Castellani and Chalmers Boikov, 1932 PROTOZOA: MASTIGOPHORA PROTOMONADIDA: TRYPANOSOMATIDAE *Herpetomonas muscarum* (Leidy) Kent Becker, 1922-23
Leptocera ferruginata (Stenhammar) (Synonym: *Leptocera feruginata*)	SCHIZOPHYTA: SCHIZOMYCETES EUBACTERIALES: ACHROMOBACTERIACEAE *Alcaligenes faecalis* Castellani and Chalmers Lysenko, 1958 *Flavobacterium invisible* (Vaughan) Bergey et al. Lysenko, 1958 EUBACTERIALES: ENTEROBACTERIACEAE *Escherichia coli* (Migula) Castellani and Chalmers Lysenko, 1958

Fly	Associated organism
Leptocera ferruginata (cont.)	*Klebsiella cloacae* (Jordan) Brisou Lysenko, 1958 *Proteus inconstans* (Ornstein) Shaw and Clarke Lysenko, 1958 *Proteus mirabilis* Hauser Lysenko, 1958 *Proteus vulgaris* Hauser Lysenko, 1958 *Salmonella derby* Warren and Scott Greenberg et al., 1963b *Salmonella newport* Schütze Greenberg et al., 1963b EUBACTERIALES: MICROCOCCACEAE *Staphylococcus afermentans* Shaw, Stitt, and Cowan Lysenko, 1958 *Staphylococcus lactis* Shaw, Stitt, and Cowan Lysenko, 1958 *Staphylococcus saprophyticus* Fairbrother Lysenko, 1958 EUBACTERIALES: BACILLACEAE *Bacillus subtilis* Cohn Lysenko, 1958
Leptocera fontinalis (Fallén)	ARTHROPODA: ARACHNIDA ACARINA: MACROCHELIDAE *Macrocheles muscaedomesticae* (Scopoli) Sychevskaîa, 1964a
Leptocera hirtula var. *thalhammeri* Strobl (Synonym: *Limosina hirtula* var. *thalhammeri*)	PROTOZOA: MASTIGOPHORA PROTOMONADIDA: TRYPANOSOMATIDAE *Leptomonas* Kent Chatton, 1912a
Leptocera humida Haliday (prey)	ARTHROPODA: INSECTA DIPTERA: MUSCIDAE *Coenosia tigrina* (Zetterstedt) Hobby, 1934b
Leptocera moesta Villeneuve (Synonym: ?*Scotophilella moesta*) (host for parasite)	ARTHROPODA: INSECTA HYMENOPTERA: UNDETERMINED ICHNEUMONIDAE Hammer, 1942
Leptocera vagans (Haliday)	SCHIZOPHYTA: SCHIZOMYCETES EUBACTERIALES: ENTEROBACTERIACEAE *Salmonella derby* Warren and Scott Greenberg et al., 1963b

Fly	Organism
	Salmonella newport Schütze Greenberg et al., 1963b
Limosina punctipennis (Wiedemann)	SCHIZOPHYTA: SCHIZOMYCETES EUBACTERIALES: ENTEROBACTERIACEAE *Escherichia coli* (Migula) Castellani and Chalmers Nicholls, 1912 *Serratia marcescens* Bizio Nicholls, 1912 *Salmonella typhi* Warren and Scott Nicholls, 1912 EUBACTERIALES: MICROCOCCACEAE *Staphylococcus aureus* Rosenbach Nicholls, 1912 ASCHELMINTHES: NEMATODA TRICHURIDEA: TRICHURIDAE *Trichuris trichiura* (Linnaeus) Nicholls, 1912 STRONGYLIDEA: ANCYLOSTOMATIDAE *Necator americanus* (Stiles) Nicholls, 1912 ASCARIDIDEA: ASCARIDIDAE *Ascaris lumbricoides* Linnaeus Nicholls, 1912
UNDETERMINED SPHAEROCERIDAE	SCHIZOPHYTA: SCHIZOMYCETES EUBACTERIALES: ENTEROBACTERIACEAE *Escherichia coli* (Migula) Castellani and Chalmers Povolný et al., 1961 *Serratia* Bizio Povolný et al., 1961 *Proteus vulgaris* Hauser Povolný et al., 1961 *Providencia* Kauffmann Povolný et al., 1961 EUBACTERIALES: MICROCOCCACEAE *Staphylococcus aureus* Rosenbach Povolný et al., 1961 EUBACTERIALES: LACTOBACILLACEAE *Diplococcus pneumoniae* Weichselbaum Povolný et al., 1961 EUBACTERIALES: BACILLACEAE *Bacillus* Cohn Povolný et al., 1961

Fly	Associated organism
Family: MILICHIIDAE Subfamily: MADIZINAE	
Desmometopa m-nigrum (Zetterstedt)	THALLOPHYTA: PHYCOMYCETES MUCORALES: MUCORACEAE *Mucor* Micheli Baumberger, 1919
Desmometopa tarsalis Loew	ARTHROPODA: ARACHNIDA ACARINA: MACROCHELIDAE *Macrocheles muscaedomesticae* (Scopoli) Sychevskaîa, 1964a ACARINA: ANOETIDAE *Myianoetus* Oudemans Sychevskaîa, 1964a
Subfamily: MILICHIINAE	
Milichiella lacteipennis (Loew) (prey)	ARTHROPODA: INSECTA DERMAPTERA: LABIDURIDAE *Euborellia annulipes* (Lucas) Illingworth, 1923a DERMAPTERA: LABIIDAE *Labia pilicornis* (Motschulsky) Illingworth, 1923a *Spingolabis hawaiiensis* (Bormans) Illingworth, 1923a COLEOPTERA: HYDROPHILIDAE *Cryptopleurum minutum* Fabricius Illingworth, 1923a *Dactylosternum abdominale* Fabricius Illingworth, 1923a COLEOPTERA: STAPHYLINIDAE *Oxytelus* Gravenhorst Illingworth, 1923a *Philonthus discoideus* Gravenhorst Illingworth, 1923a *Philonthus longicornis* Stephens Illingworth, 1923a *Tachyporus* Gravenhorst Illingworth, 1923a HYMENOPTERA: FORMICIDAE *Pheidole megacephala* (Fabricius) Illingworth, 1923a *Ponera perkinsi* Forel Illingworth, 1923a
Superfamily: DROSOPHILOIDEA Family: EPHYDRIDAE Subfamily: EPHYDRINAE	
Teichomyza fusca Macquart (Synonym: *Theicomyza fusca*)	PROTOZOA: MASTIGOPHORA PROTOMONADIDA: TRYPANOSOMATIDAE *Herpetomonas muscarum* (Leidy) Kent Léger, 1903

Family: DROSOPHILIDAE
Subfamily: DROSOPHILINAE
Drosophila Fallén .... INSECT VIRUSES
Sigma virus
Vigier, 1961
THALLOPHYTA: FUNGI IMPERFECTI
MONILIALES: MONILIACEAE
Aspergillus (Micheli) Corda
Blochwitz, 1929
PROTOZOA: MASTIGOPHORA
PROTOMONADIDA: TRYPANOSOMATIDAE
Herpetomonas muscarum (Leidy) Kent
Patton, 1921a
PROTOMONADIDA: BODONIDAE
Rhynchoidomonas roubaudi Roubaud
Roubaud, 1912b
PROTOZOA: CNIDOSPORIDIA
MICROSPORIDA: MRAZEKIIDAE
Octosporea muscae-domesticae Flu
Kramer, 1965b
UNDETERMINED MICROSPORIDA
Bell, 1952
Stalker et al., 1963
ASCHELMINTHES: NEMATODA
RHABDITIDA: RHABDITIDAE
Rhabditis pellio (Schneider)
Aubertot, 1923
RHABDITIDA: PANAGROLAIMIDAE
Panagrellus zymosiphilus (Brunold)
Brunold, 1950
ARTHROPODA: ARACHNIDA
ARANEAE: THERIDIIDAE
Latrodectus mactans (Fabricius)
Herms et al., 1935
ACARINA: TROMBICULIDAE
Ascoschoengastia indica (Hirst)
Wharton et al., 1946
Euschoengastia peromysci (Ewing)
Lipovsky, 1954
Eutrombicula alfreddugesi (Oudemans)
Lipovsky, 1954
Eutrombicula splendens (Ewing)
Lipovsky, 1954
Fonsecia gurneyi (Ewing)
Lipovsky, 1954
Hannemania Oudemans
Lipovsky, 1954

Fly	Organism
Drosophila (cont.)	*Kayella lacerta* (Brennan) Lipovsky, 1954 *Neoschoengastia americana* (Hirst) Lipovsky, 1954 *Neotrombicula lipovskyi* (Brennan and Wharton) Lipovsky, 1954 *Neotrombiculoides montanensis* (Brennan) Lipovsky, 1954 *Pseudoschoengastia farneri* Lipovsky Lipovsky, 1954 *Pseudoschoengastia hungerfordi* Lipovsky Lipovsky, 1954 ACARINA: ANOETIDAE *Histiostoma laboratorium* Hughes Hughes, 1950 Perron, 1954a ARTHROPODA: INSECTA HEMIPTERA: PHYMATIDAE *Phymata pennsylvanica americana* Melin Balduf, 1947 HEMIPTERA: REDUVIIDAE *Pristhesancus papuensis* Stål Noble, 1936 *Sinea diadema* (Fabricius) Balduf, 1947 HYMENOPTERA: CYNIPIDAE *Eucoila drosophilae* Kieffer Boche, 1939 HYMENOPTERA: PTEROMALIDAE *Spalangia drosophilae*[1] Ashmead Ashmead, 1887, 1900 Riley et al., 1891 Richardson, 1913a Viereck, 1916
Drosophila ananassae Doleschall	SCHIZOPHYTA: SCHIZOMYCETES EUBACTERIALES: ENTEROBACTERIACEAE *Shigella flexneri* Castellani and Chalmers Gabaldón et al., 1956

[1] Although not specifically mentioned in each entry, it is assumed that all *Spalangia* are pupal parasites.

Host	Associated organism
Drosophila bifasciata Pomini	INSECT VIRUSES "Sex-ratio" virus Magni, 1954 Poulson et al., 1961a
Drosophila busckii Coquillet (host for parasite)	ARTHROPODA: INSECTA HYMENOPTERA: CYNIPIDAE *Pseudeucoila bochei* Weld Jenni, 1951
Drosophila cellaris Linnaeus	THALLOPHYTA: ASCOMYCETES ENDOMYCETALES: ENDOMYCETACEAE *Saccharomyces cerevisiae* Hansen Berlese, 1896 *Saccharomyces pastorianus* Hansen Berlese, 1896 THALLOPHYTA: FUNGI IMPERFECTI MONILIALES: DEMATIACEAE *Torula* Persoon Berlese et al., 1897 CRYPTOCOCCALES: CRYPTOCOCCACEAE *Candida* Berkhout Berlese, 1896 Berlese et al., 1897 *Kloeckera apiculata* (Reess emend Klöker) Janke Berlese, 1896
Drosophila confusa Staeger	PROTOZOA: MASTIGOPHORA PROTOMONADIDA: TRYPANOSOMATIDAE *Herpetomonas roubaudi* (Chatton) Drbohlav Chatton, 1912b *Leptomonas* Kent Chatton et al., 1912 *Leptomonas drosophilae* Chatton and Alilaire Chatton et al., 1908, 1911a, 1911b, 1911c, 1912 *Leptomonas drosophilae* Chatton and Alilaire=[*Trypanosoma drosophilae*] Chatton et al., 1908, 1911a, 1911b, 1912 PROTOZOA: CNIDOSPORIDIA MICROSPORIDA: MRAZEKIIDAE *Octosporea monospora* Chatton and Krempf Chatton et al., 1911d

Drosophila confusa (cont.)	*Octosporea muscae-domesticae* Flu Chatton et al., 1911d
	ASCHELMINTHES: NEMATODA TYLENCHIDA: UNDETERMINED ALLANTONEMATIDAE Welch, 1959
	ARTHROPODA: INSECTA DIPTERA: MUSCIDAE *Muscina assimilis* (Fallén) Keilin, 1917
Drosophila equinoxialis Dobzhansky	INSECT VIRUSES "Sex-ratio" virus Malogolowkin et al., 1960 Poulson et al., 1961a
Drosophila fasciata Dufour	SCHIZOPHYTA: SCHIZOMYCETES SPIROCHAETALES: TREPONEMATACEAE *Treponema* Schaudinn Ikeda, 1965
Drosophila fenestrarum Fallén	SCHIZOPHYTA: SCHIZOMYCETES PSEUDOMONADALES: PSEUDOMONADACEAE *Acetobacter pasteurianus* Beijerinck Henneberg, 1902 *Acetobacter zylinum* (Brown) Henneberg, 1902
	EUBACTERIALES: ENTEROBACTERIACEAE *Serratia marcescens* Bizio Henneberg, 1902
	EUBACTERIALES: LACTOBACILLACEAE *Lactobacillus* Beijerinck Henneberg, 1902
Drosophila ?fenestrarum Fallén	ARTHROPODA: ARACHNIDA ACARINA: ANOETIDAE *Myianoetus muscarum* (Linnaeus) =[*Pediculus muscae fenestrarum* per Goeze;=*Acarus muscarum* per Oudemans, A.C., 1929] Goeze, 1776
Drosophila ferruginea Becker	SCHIZOPHYTA: SCHIZOMYCETES EUBACTERIALES: ENTEROBACTERIACEAE *Shigella dysenteriae* (Shiga) Castellani and Chalmers Beyer, 1925

Host	Associated organism
Drosophila funebris (Fabricius)	SCHIZOPHYTA: SCHIZOMYCETES
	PSEUDOMONADALES: PSEUDOMONADACEAE
	Acetobacter pasteurianus Beijerinck
	Henneberg, 1902
	Acetobacter zylinum (Brown)
	Henneberg, 1902
	EUBACTERIALES: ENTEROBACTERIACEAE
	Serratia marcescens Bizio
	Henneberg, 1902
	Shigella Castellani and Chalmers
	Boikov, 1932
	EUBACTERIALES: LACTOBACILLACEAE
	Lactobacillus Beijerinck
	Henneberg, 1902
	THALLOPHYTA: ASCOMYCETES
	ENDOMYCETALES: ENDOMYCETACEAE
	Coccidiascus legeri Chatton
	Chatton, 1913
	THALLOPHYTA: UNDETERMINED BASIDIOMYCETES
	"Mushrooms"
	Austin, 1933
	PROTOZOA: CNIDOSPORIDIA
	MICROSPORIDA: MRAZEKIIDAE
	Octosporea muscae-domesticae Flu
	Chatton et al., 1911d
	ARTHROPODA: INSECTA
	HYMENOPTERA: BRACONIDAE
	Aspilota concolor Nees
	Austin, 1933
	HYMENOPTERA: CYNIPIDAE
	Pseudeucoila bochei Weld
	Jenni, 1951
	HYMENOPTERA: PTEROMALIDAE
	Pachyneuron vindemmiae Rondani
	Milani, 1947
	HYMENOPTERA: FORMICIDAE
	Iridomyrmex humilis (Mayr)
	Pavan, 1952
Drosophila hydei Sturtevant (host for parasite)	ARTHROPODA: INSECTA
	HYMENOPTERA: CYNIPIDAE
	Pseudeucoila bochei Weld
	Jenni, 1951
Drosophila immigrans Sturtevant (host for parasite)	ARTHROPODA: INSECTA
	HYMENOPTERA: PTEROMALIDAE
	Pachyneuron vindemmiae Rondani
	Milani, 1947

Fly	Associated organisms
Drosophila kuntzei Duda	ASCHELMINTHES: NEMATODA TYLENCHIDA: ALLANTONEMATIDAE *Howardula aoronymphium* Welch Welch, 1959
Drosophila littoralis Meigen	ARTHROPODA: INSECTA HYMENOPTERA: CYNIPIDAE *Pseudeucoila bochei* Weld Jenni, 1951
Drosophila melanogaster Meigen (Synonym: *Drosophila ampelophila*)	VIRUSES ARBOVIRUSES Sindbis virus Herreng, 1967 Ohanessian et al., 1967 ENTEROVIRUSES Poliovirus Type 2, Lansing strain Toomey et al., 1947 OTHER ANIMAL VIRUSES Rous-sarcoma virus (Oncogenic virus) Burdette et al., 1967 INSECT VIRUSES "Sex-ratio" virus Sakaguchi et al., 1960 Counce et al., 1961 Poulson et al., 1961a Rico, 1964 Sigma virus Plus, 1950, 1954 de Lestrange, 1954 Duhamel et al., 1956 L'Héritier, 1957 Bussereau, 1964 Berkaloff et al., 1965 Bregliano, 1965 Seecof, 1965, 1969 Bernard, 1968 Sigma virus, P^- strain Bernard, 1964 SCHIZOPHYTA: SCHIZOMYCETES PSEUDOMONADALES: PSEUDOMONADACEAE *Acetobacter* Beijerinck Baumberger, 1919 *Acetobacter aceti* (Beijerinck) Delcourt et al., 1910 EUBACTERIALES: ENTEROBACTERIACEAE *Erwinia amylovora* (Burrill) Winslow et al. Ark et al., 1936

Drosophila melanogaster (cont.)	*Serratia marcescens* Bizio Nicholls, 1912 *Salmonella typhi* Warren and Scott Nicholls, 1912 EUBACTERIALES: MICROCOCCACEAE *Staphylococcus aureus* Rosenbach Nicholls, 1912 EUBACTERIALES: BACILLACEAE *Bacillus* Cohn Nicholls, 1912 Tatum, 1939 SPIROCHAETALES: TREPONEMATACEAE *Treponema* Schaudinn Ikeda, 1965 THALLOPHYTA: PHYCOMYCETES MUCORALES: MUCORACEAE *Mucor* Micheli Baumberger, 1919 Griffith, 1952 *Rhizopus* Ehrenberg Baumberger, 1919 ENTOMOPHTHORALES: ENTOMOPHTHORACEAE *Entomophthora muscae* Cohn Goldstein, 1927 THALLOPHYTA: ASCOMYCETES ENDOMYCETALES: ENDOMYCETACEAE *Saccharomyces* Hansen Baumberger, 1919 *Saccharomyces cerevisiae* Hansen Baumberger, 1919 *Saccharomyces ?melli* (Fabian and Quinet) Delcourt et al., 1910 *Saccharomyces theobromae* Nicholls, 1912 THALLOPHYTA: BASIDIOMYCETES AGARICALES: AGARICACEAE *Agaricus campestris* Fries Baumberger, 1919 THALLOPHYTA: FUNGI IMPERFECTI MONILIALES: MONILIACEAE *Aspergillus* (Micheli) Corda Baumberger, 1919 *Aspergillus clavatus* Desmazières Griffith, 1952 *Aspergillus niger* van Tieghem Griffith, 1952

Fly	Organism and references
Drosophila melanogaster (cont.)	*Monilia* (Persoon) Saccardo
	Griffith, 1952
	Penicillium camemberti Thom
	Griffith, 1952
	Penicillium glaucum Link
	Baumberger, 1919
	Griffith, 1952
	MONILIALES: DEMATIACEAE
	Alternaria Nees
	Griffith, 1952
	Curvularia Boedijn
	Griffith, 1952
	Hormodendrum Bonorden
	Griffith, 1952
	Macrosporium Fries
	Griffith, 1952
	MONILIALES: TUBERCULARIACEAE
	Fusarium Link
	Griffith, 1952
	PROTOZOA: MASTIGOPHORA
	PROTOMONADIDA: TRYPANOSOMATIDAE
	Leptomonas ampelophilae Chatton and Léger
	Chatton et al., 1911c, 1912
	PROTOZOA: CNIDOSPORIDIA
	MICROSPORIDA: MRAZEKIIDAE
	Octosporea muscae-domesticae Flu
	Chatton et al., 1911d
	UNDETERMINED MICROSPORIDA
	Wolfson et al., 1957
	Stalker et al., 1963
	ASCHELMINTHES: NEMATODA
	TYLENCHIDA: ALLANTONEMATIDAE
	Parasitylenchus diplogenus Welch
	Welch, 1959
	ARTHROPODA: ARACHNIDA
	ACARINA: TROMBICULIDAE
	Ascoschoengastia indica (Hirst)
	Wharton, 1946
	ACARINA: ANOETIDAE
	Histiostoma laboratorium Hughes
	Stolpe, 1938
	Brown, 1965
	Myianoetus muscarum (Linnaeus)
	Greenberg et al., 1960
	ARTHROPODA: INSECTA
	HEMIPTERA: PHYMATIDAE
	Phymata pennsylvanica americana Melin
	Balduf, 1941, 1948

Fly	Associated organisms
Drosophila melanogaster (cont.)	COLEOPTERA: STAPHYLINIDAE *Baryodma ontarionis* Casey Colhoun, 1953 HYMENOPTERA: BRACONIDAE *Asobara tabida* (Nees) Jenni, 1951 Walker, 1963 HYMENOPTERA: CYNIPIDAE *Eucoila* Westwood Jenni, 1947 *Eucoila drosophilae* Kieffer Boche, 1939 *Pseudeucoila bochei*[1] Weld Jenni, 1951 Schlegel-Oprecht, 1953 Nøstvik, 1954 Walker, 1959, 1961, 1963 HYMENOPTERA: CHALCIDIDAE *Dirhinus pachycerus* Masi Roy et al., 1940 HYMENOPTERA: PTEROMALIDAE *Pachyneuron vindemmiae* Rondani Milani, 1947 *Spalangia drosophilae* Ashmead Simmonds, 1944 Bouček, 1963
Drosophila ?melanogaster Meigen	INSECT VIRUSES Sigma virus L'Héritier et al., 1937
Drosophila melanogaster Meigen Oregon R. C.	INSECT VIRUSES Sigma virus Brun et al., 1955a Plus, 1955
Drosophila melanogaster Meigen Strain ρ (rho)	INSECT VIRUSES Sigma virus Brun et al., 1955b
Drosophila nebulosa Sturtevant	INSECT VIRUS "Sex-ratio" virus Counce et al., 1961 Poulson et al., 1961a SCHIZOPHYTA: SCHIZOMYCETES SPIROCHAETALES: TREPONEMATACEAE *Treponema* Schaudinn Poulson et al., 1961b
Drosophila obscura Fallén	ASCHELMINTHES: NEMATODA TYLENCHIDA: ALLANTONEMATIDAE *Parasitylenchus diplogenus* Welch Welch, 1959

[1] Although not specifically mentioned in each entry, it is assumed that all *Pseudeucoila bochei* are parasites.

Fly	Organism
Drosophila paramelanica Patterson	PROTOZOA: CNIDOSPORDIA UNDETERMINED MICROSPORIDA Wolfson et al., 1957
Drosophila parthenogenetica Stalker	PROTOZOA: CNIDOSPORIDIA UNDETERMINED MICROSPORIDA Wolfson et al., 1957
Drosophila paulistorum Dobzhansky	INSECT VIRUS "Sex-ratio" virus Malogolowkin, 1958
Drosophila phalerata Meigen	PROTOZOA: CNIDOSPORIDIA MICROSPORIDA: MRAZEKIIDAE *Octosporea muscae-domesticae* Flu Chatton et al., 1911d ASCHELMINTHES: NEMATODA TYLENCHIDA: ALLANTONEMATIDAE *Howardula aoronymphium* Welch Welch, 1959 ARTHROPODA: INSECTA HYMENOPTERA: CYNIPIDAE *Pseudeucoila bochei* Weld Jenni, 1951
Drosophila plurilineata Villeneuve	PROTOZOA: CNIDOSPORIDIA MICROSPORIDA: MRAZEKIIDAE *Octosporea monospora* Chatton and Krempf Chatton et al., 1911d *Octosporea muscae-domesticae* Flu Chatton et al., 1911d
Drosophila prosaltans Duda	INSECT VIRUS "Sex-ratio" virus Cavalcanti et al., 1957, 1958
Drosophila repleta Wollaston	THALLOPHYTA: PHYCOMYCETES ENTOMOPHTHORALES: ENTOMOPHTHORACEAE *Entomophthora muscae* Cohn Goldstein, 1927
Drosophila robusta Sturtevant	PROTOZOA: CNIDOSPORIDIA UNDETERMINED MICROSPORIDA Stalker et al., 1963
Drosophila rubrio-striata Becker (Synonym: *Drosophila plurilineata*)	PROTOZOA: MASTIGOPHORA PROTOMONADIDA: TRYPANOSOMATIDAE *Herpetomonas rubrio-striatae* Chatton and Léger Chatton et al., 1911c, 1912 *Leptomonas* Kent Chatton et al., 1912

Fly	Associated organism
Drosophila silvestris Basden	ASCHELMINTHES: NEMATODA TYLENCHIDA: ALLANTONEMATIDAE *Parasitylenchus diplogenus* Welch Welch, 1959
Drosophila similis Williston	ARTHROPODA: INSECTA HYMENOPTERA: PTEROMALIDAE *Pachyneuron vindemmiae* Rondani Milani, 1947
Drosophila simulans Sturtevant	ARTHROPODA: INSECTA DIPTERA: DROSOPHILIDAE *Drosophila melanogaster* Meigen Moore, 1952
Drosophila subobscura Collin	ASCHELMINTHES: NEMATODA TYLENCHIDA: ALLANTONEMATIDAE *Parasitylenchus diplogenus* Welch Welch, 1959 ARTHROPODA: INSECTA HYMENOPTERA: CYNIPIDAE *Pseudeucoila bochei* Weld Jenni, 1951
Drosophila virilis Sturtevant	PROTOZOA: MASTIGOPHORA PROTOMONADIDA: TRYPANOSOMATIDAE *Crithidia* Léger (from *Arilus cristatus*) McGhee et al., 1965 *Crithidia* Léger (from *Euryophthalmus davisi*) McGhee et al., 1965 *Crithidia acanthocephali* Hanson and McGhee McGhee et al., 1965 *Crithidia fasciculata* Léger McGhee et al., 1965 *Crithidia oncopelti* (Noguchi and Tilden) sensu M. and A. Lwoff McGhee et al., 1965 ARTHROPODA: INSECTA HYMENOPTERA: FORMICIDAE *Iridomyrmex humilis* (Mayr) Pavan, 1952
Drosophila willistoni Sturtevant	INSECT VIRUS "Sex-ratio" virus Malogolowkin et al., 1957, 1959, 1960 Malogolowkin, 1958 Sakaguchi et al., 1959, 1960, 1961 Poulson et al., 1959, 1961a Counce et al., 1961

Fly	Organism
Drosophila willistoni (cont.)	SCHIZOPHYTA: SCHIZOMYCETES SPIROCHAETALES: TREPONEMATACEAE *Treponema* Schaudinn Poulson et al., 1961b Ikeda, 1965 PROTOZOA: CNIDOSPORIDIA MICROSPORIDA: NOSEMATIDAE *Nosema kingi* Kramer Kramer, 1964b UNDETERMINED MICROSPORIDA Burnett et al., 1962
Scaptomyza Hardy (Synonym: *Buonostoma*)	PROTOZOA: CNIDOSPORIDIA UNDETERMINED MICROSPORIDA Stalker, 1963
Scaptomyza graminum (Fallén)	ARTHROPODA: INSECTA HYMENOPTERA: SPHECIDAE *Crabro elongatus* (Provancher) Hamm et al., 1926 *Crabro palmarius* (Schreber) Hamm et al., 1926
UNDETERMINED DROSOPHILIDAE	SCHIZOPHYTA: SCHIZOMYCETES EUBACTERIALES: ENTEROBACTERIACEAE *Escherichia coli* (Migula) Castellani and Chalmers Povolný et al., 1961 *Serratia* Bizio Povolný et al., 1961 *Proteus vulgaris* Hauser Povolný et al., 1961 *Providencia* Kauffmann Povolný et al., 1961 EUBACTERIALES: MICROCOCCACEAE *Staphylococcus aureus* Rosenbach Povolný et al., 1961 EUBACTERIALES: LACTOBACILLACEAE *Diplococcus pneumoniae* Weichselbaum Povolný et al., 1961 EUBACTERIALES: BACILLACEAE *Bacillus* Cohn Povolný et al., 1961 ASCHELMINTHES: NEMATODA SPIRURIDEA: SPIRURIDAE *Habronema ?megastoma* (Rudolphi) Seurat Crawford, 1926

Superfamily: CHLOROPOIDEA
Family: CHLOROPIDAE
Subfamily: OSCINELLINAE

Hippelates Loew .... SCHIZOPHYTA: SCHIZOMYCETES
EUBACTERIALES: LACTOBACILLACEAE
Streptococcus agalactiae Lehmann and Neumann
Sanders, 1940a
SPIROCHAETALES: TREPONEMATACEAE
Treponema carateum Brumpt
Leon Blanco et al., 1941
Soberón y Parra et al., 1944
ARTHROPODA: INSECTA
HYMENOPTERA: CYNIPIDAE
Hexacola Foerster
Mulla, 1962
Trybliographa Foerster
Legner et al., 1964
HYMENOPTERA: PTEROMALIDAE
Spalangia drosophilae Ashmead
Legner et al., 1964

Hippelates collusor (Townsend) .... ARTHROPODA: INSECTA
HYMENOPTERA: CYNIPIDAE
Trybliographa Foerster
Legner et al., 1969b
HYMENOPTERA: PTEROMALIDAE
Eupteromales hemipterus (Walker)
Legner et al., 1965b
Legner et al., 1966
Eupteromales nidulans (Thomson)
Bay, unpublished
Spalangia[1] *drosophilae* Ashmead
Bay et al., 1963
Bay, unpublished
HYMENOPTERA: DIAPRIIDAE
Phaenopria occidentalis Fouts
Bay et al., 1963
Bay, unpublished
Legner, 1967, 1969

Hippelates currani Aldrich .... SCHIZOPHYTA: SCHIZOMYCETES
EUBACTERIALES: LACTOBACILLACEAE
Streptococcus Rosenbach, Lancefield's Group A
Bassett, 1967
Streptococcus pyogenes Rosenbach
Bassett, 1970

[1] Although not specifically mentioned in each entry, it is assumed that all *Spalangia* are pupal parasites.

Fly	Organism
Hippelates currani (cont.)	SCHIZOPHYTA: SCHIZOMYCETES SPIROCHAETALES: TREPONEMATACEAE *Treponema pertenue* Castellani Kumm et al., 1936
Hippelates illicis Curran	SCHIZOPHYTA: SCHIZOMYCETES SPIROCHAETALES: TREPONEMATACEAE *Treponema pertenue* Castellani Kumm et al., 1936
Hippelates flavipes Loew (Synonyms: *Oscinis pallipes*, *Hippelates pallipes* [a valid species in temperate zones])	SCHIZOPHYTA: SCHIZOMYCETES EUBACTERIALES: MICROCOCCACEAE *Staphylococcus aureus* Rosenbach Nicholls, 1912 Taplin et al., 1966 EUBACTERIALES: LACTOBACILLACEAE *Streptococcus pyogenes* Rosenbach Bassett, 1970 SPIROCHAETALES: TREPONEMATACEAE *Borrelia refringens* (Schaudinn and Hoffmann) Bergey et al. Kumm, 1935 Kumm et al., 1935 *Treponema pertenue* Castellani Kumm, 1935 Kumm et al., 1935, 1936 PROTOZOA: MASTIGOPHORA PROTOMONADIDA: TRYPANOSOMATIDAE *Leptomonas* Kent Kumm et al., 1935 ASCHELMINTHES UNDETERMINED NEMATODA Kumm et al., 1935
Hippelates peruanus Becker	SCHIZOPHYTA: SCHIZOMYCETES EUBACTERIALES: LACTOBACILLACEAE *Streptococcus* Rosenbach, Lancefield's Group A Bassett, 1967 *Streptococcus pyogenes* Rosenbach Bassett, 1970 SPIROCHAETALES: TREPONEMATACEAE *Treponema pertenue* Castellani Kumm et al., 1936
Hippelates pusio Loew	SCHIZOPHYTA: SCHIZOMYCETES SPIROCHAETALES: TREPONEMATACEAE *Treponema pertenue* Castellani Kumm et al., 1936

Fly	Organism / Reference
Hippelates pusio (cont.)	ARTHROPODA: INSECTA
	HYMENOPTERA: CYNIPIDAE
	Trybliographa Foerster
	Legner et al., 1965d
	HYMENOPTERA: PTEROMALIDAE
	Spalangia Latreille
	Legner et al., 1965d
	Spalangia drosophilae Ashmead
	Legner et al., 1965d
	HYMENOPTERA: UNDETERMINED PTEROMALIDAE
	Legner et al., 1965d
	HYMENOPTERA: ENCYRTIDAE
	Ooencyrtus submetallicus (Howard)
	Legner et al., 1965d
	HYMENOPTERA: FORMICIDAE
	Monomorium pharaonis (Linnaeus)
	Legner et al., 1965d
	Solenopsis geminata (Fabricius)
	Legner et al., 1965d
	Tapinoma melanocephalum (Fabricius)
	Legner et al., 1965d
	Tetramorium guineense (Fabricius)
	Legner et al., 1965d
	Wasmannia auropunctata (Roger)
	Legner et al., 1965d
Siphunculina funicola de Meijere	SCHIZOPHYTA: SCHIZOMYCETES
	EUBACTERIALES: MICROCOCCACEAE
	Staphylococcus aureus Rosenbach
	Syddiq, 1938
	EUBACTERIALES: LACTOBACILLACEAE
	Streptococcus Rosenbach (non-hemolytic)
	Syddiq, 1938
	Streptococcus Rosenbach
	Syddiq, 1938
	EUBACTERIALES: CORYNEBACTERIACEAE
	?*Corynebacterium* Lehmann and Neumann
	Syddiq, 1938
	THALLOPHYTA: PHYCOMYCETES
	MUCORALES: MUCORACEAE
	Mucor mucedo (Linnaeus) Brefeld
	Syddiq, 1938

Fly	Associated organism
Siphunculina funicola (cont.)	PROTOZOA: MASTIGOPHORA PROTOMONADIDA: TRYPANOSOMATIDAE *Herpetomonas* Kent Patton, 1921c *Leishmania donovani* Laveran and Mesnil Patton, 1921-22 PROTOMONADIDA: BODONIDAE *Rhynchoidomonas siphunculinae* Patton Patton, 1921c
Oscinella frit (Linnaeus) (host for parasite)	ARTHROPODA: INSECTA HYMENOPTERA: CYNIPIDAE *Hexacola* Foerster Bay et al., 1963 HYMENOPTERA: PTEROMALIDAE *Spalangia drosophilae* Ashmead Bay et al., 1963
Conioscinella mars (Curran) (Synonym: *Oscinella mars*)	SCHIZOPHYTA: SCHIZOMYCETES SPIROCHAETALES: TREPONEMATACEAE *Treponema pertenue* Castellani Kumm et al., 1936
Family: HELEOMYZIDAE	
Subfamily: SUILLIINAE	
Suillia lineata Robineau-Desvoidy (Synonym: *Helomyza lineata*) (prey)	ARTHROPODA: INSECTA DIPTERA: MUSCIDAE *Muscina stabulans* (Fallén) Porchinskiĭ, 1913
Suillia ustulata Meigen (Synonym: *Helomyza ustulata*) (prey)	ARTHROPODA: INSECTA DIPTERA: MUSCIDAE *Muscina stabulans* (Fallén) Porchinskiĭ, 1913
Subfamily: HELEOMYZINAE	
Heleomyza Fallén (Synonym: *Helomyza*) (prey)	ARTHROPODA: INSECTA DIPTERA: MUSCIDAE *Muscina stabulans* (Fallén) Porchinskiĭ, 1913
Heleomyza gigantea Meigen (Synonym: *Helomyza gigantea*) (prey)	ARTHROPODA: INSECTA DIPTERA: MUSCIDAE *Muscina stabulans* (Fallén) Porchinskiĭ, 1913
Heleomyza modesta Meigen (Synonym: *Helomyza modesta*)	SCHIZOPHYTA: SCHIZOMYCETES EUBACTERIALES: ENTEROBACTERIACEAE *Salmonella paratyphi* C Castellani and Chalmers Anonymous, 1952

Fly	Organism
Heleomyza modesta (cont.)	*Salmonella typhi* Warren and Scott Anonymous, 1952
Section: CALYPTRATAE Superfamily: MUSCOIDEA Family: ANTHOMYIIDAE Subfamily: SCATOPHAGINAE	
Scatophaga Meigen	ENTEROVIRUSES Poliovirus Power et al., 1942
Scatophaga furcata (Say) (Synonym: *Scatophaga squalida*)	THALLOPHYTA: PHYCOMYCETES ENTOMOPHTHORALES: ENTOMOPHTHORACEAE *Entomophthora muscae* Cohn Rostrup, 1916
Scatophaga lutaria (Fabricius)	PROTOZOA: MASTIGOPHORA PROTOMONADIDA: TRYPANOSOMATIDAE *Herpetomonas ?muscarum* (Leidy) Kent Mackinnon, 1910
Scatophaga stercoraria (Linnaeus) (Synonyms: *Scopeuma stercoraria, Scopeuma stercorarium*)	SCHIZOPHYTA: SCHIZOMYCETES EUBACTERIALES: ENTEROBACTERIACEAE *Escherichia coli* (Migula) Castellani and Chalmers Povolný et al., 1961 *Serratia* Bizio Povolný et al., 1961 EUBACTERIALES: BACILLACEAE *Bacillus* Cohn Povolný et al., 1961 THALLOPHYTA: PHYCOMYCETES ENTOMOPHTHORALES: ENTOMOPHTHORACEAE *Entomophthora* Fresenius Bail, 1860 *Entomophthora muscae* Cohn Rostrup, 1916 Graham-Smith, 1919 Hammer, 1942 ARTHROPODA: ARACHNIDA ACARINA: LAELAPTIDAE *Holostaspis badius* (Koch) Berlese Weiss, 1915 ARTHROPODA: INSECTA COLLEMBOLA: SMINTHURIDAE *Sminthurus viridis* (Linnaeus) Walters, 1966 COLEOPTERA: STAPHYLINIDAE *Ontholestes tessellatus* Fourcroy Hammer, 1942

Fly	Associated organisms
Scatophaga stercoraria (cont.)	*Philonthus* Curtis Hammer, 1942 HYMENOPTERA: UNDETERMINED ICHNEUMONIDAE Hammer, 1942 HYMENOPTERA: BRACONIDAE *Aphaereta minuta* Nees Sychevskaia͡, 1964b HYMENOPTERA: CYNIPIDAE *Pseudeucoila trichopsila* (Hartig) Sychevskaia͡, 1964b DIPTERA: SPHAEROCERIDAE *Copromyza equina* (Fallén) Cotterell, 1920
Subfamily: FUCELLINAE *Fucellia rufitibia* Stein	SCHIZOPHYTA: SCHIZOMYCETES EUBACTERIALES: ENTEROBACTERIACEAE *Escherichia freundii* (Braak) Yale O'Keefe et al., 1954 *Paracolobactrum aerogenoides* Borman, Stuart, and Wheeler O'Keefe et al., 1954 *Paracolobactrum intermedium* Borman, Stuart, and Wheeler O'Keefe et al., 1954 *Proteus vulgaris* Hauser O'Keefe et al., 1954
Subfamily: ANTHOMYIINAE *Hylemya* Robineau-Desvoidy (Synonyms: *Hylemyia, Phorbia*)	SCHIZOPHYTA: SCHIZOMYCETES PSEUDOMONADALES: SPIRILLACEAE *Vibrio comma* (Schroeter) Winslow et al. Anonymous, 1952 EUBACTERIALES: ENTEROBACTERIACEAE *Salmonella typhi* Warren and Scott Anonymous, 1952 *Shigella dysenteriae* (Shiga) Castellani and Chalmers Anonymous, 1952 EUBACTERIALES: BRUCELLACEAE *Pasteurella pestis* (Lehmann and Neumann) Holland Anonymous, 1952 EUBACTERIALES: BACILLACEAE *Bacillus anthracis* Cohn Anonymous, 1952

Fly	Associated organism
Hylemya (cont.)	ARTHROPODA: INSECTA HYMENOPTERA: SPHECIDAE *Crabro leucostoma* (Linnaeus) Hamm et al., 1926
Hylemya aestiva (Meigen) (Synonym: *Paregle aestiva*)	THALLOPHYTA: PHYCOMYCETES ENTOMOPHTHORALES: ENTOMOPHTHORACEAE *Entomophthora* Fresenius Hammer, 1942
	ARTHROPODA: INSECTA DIPTERA: ANTHOMYIIDAE *Scatophaga stercoraria* (Linnaeus) Hobby, 1934b
Hylemya antiqua (Meigen) (Synonym: *Hylemyia antiqua*)	SCHIZOPHYTA: SCHIZOMYCETES EUBACTERIALES: ENTEROBACTERIACEAE *Escherichia freundii* (Braak) Yale O'Keefe et al., 1954 *Paracolobactrum aerogenoides* Borman, Stuart, and Wheeler O'Keefe et al., 1954 *Proteus vulgaris* Hauser O'Keefe et al., 1954
	ASCHELMINTHES: NEMATODA TYLENCHIDA: CONTORTYLENCHIDAE *Heterotylenchus aberrans* Bovien Stoffolano, 1969
	ARTHROPODA: INSECTA HYMENOPTERA: PTEROMALIDAE *Spalangia rugosicollis* Ashmead Perron, 1954b
Hylemya brassicae Bouché	ARTHROPODA: ARACHNIDA ACARINA: MACROCHELIDAE *Macrocheles merdarius* (Berlese) Chant, 1960
	COLEOPTERA: STAPHYLINIDAE *Baryodma ontarionis* Casey Colhoun, 1953
Hylemya cardui (Meigen) (Synonym: *Hylemyia cardui*)	THALLOPHYTA: PHYCOMYCETES ENTOMOPHTHORALES: ENTOMOPHTHORACEAE *Entomophthora muscae* Cohn Rostrup, 1916
Hylemya cinerea Fallén (Synonym: *Phorbia cinerea*) (pupa is host for parasite)	ARTHROPODA: INSECTA HYMENOPTERA: PTEROMALIDAE *Spalangia erythromera brachyceps* Bouček Bouček, 1963

Hylemya cinerella (Fallén) (Synonym: *Paregle cinerella*) .... SCHIZOPHYTA: SCHIZOMYCETES

EUBACTERIALES:
ENTEROBACTERIACEAE

Escherichia coli (Migula) Castellani and Chalmers
Povolný et al., 1961

Serratia Bizio
Povolný et al., 1961

Proteus vulgaris Hauser
Povolný et al., 1961

Providencia Kauffmann
Povolný et al., 1961

EUBACTERIALES: MICROCOCCACEAE

Staphylococcus aureus Rosenbach
Povolný et al., 1961

Staphylococcus saprophyticus Fairbrother
Lysenko, 1958

EUBACTERIALES: LACTOBACILLACEAE

Diplococcus pneumoniae Weichselbaum
Povolný et al., 1961

Streptococcus durans Sherman and Wing
Lysenko, 1958

Streptococcus faecalis Andrewes and Horder var. *faecalis*
Lysenko, 1958

EUBACTERIALES: BACILLACEAE

Bacillus Cohn
Povolný et al., 1961

Bacillus subtilis Cohn
Lysenko, 1958

ARTHROPODA: ARACHNIDA

ACARINA: MACROCHELIDAE

Macrocheles muscaedomesticae (Scopoli)
Sychevskaîa, 1964a

ACARINA: PYEMOTIDAE

Pygmephorus Kramer
Sychevskaîa, 1964a

ACARINA: ANOETIDAE

Myianoetus Oudemans
Sychevskaîa, 1964a

ARTHROPODA: INSECTA

HYMENOPTERA: SPHECIDAE

Crabro venator (Rohwer)
Kurczewski et al., 1968

Fly	Associated organism
Hylemya coarctata (Fallén) (Synonym: *Hylemyia coarctata*)	THALLOPHYTA: PHYCOMYCETES ENTOMOPHTHORALES: ENTOMOPHTHORACEAE *Entomophthora muscae* Cohn Rostrup, 1916
Hylemya fugax (Meigen) (Synonym: *Hylemyia fugax*) (prey)	ARTHROPODA: INSECTA HYMENOPTERA: SPHECIDAE *Crabro leucostoma* (Linnaeus) Hamm et al., 1926
Hylemya nigrimana (Meigen) (Synonym: *Hylemyia nigrimana*) (prey)	ARTHROPODA: INSECTA HYMENOPTERA: SPHECIDAE *Crabro peltarius* (Schreber) Hamm et al., 1926
Hylemya platura (Meigen) (Synonym: *Phorbia platura*) (pupa is host for parasite)	ARTHROPODA: INSECTA HYMENOPTERA: PTEROMALIDAE *Spalangia erythromera brachyceps* Bouček Bouček, 1963
Hylemya radicum (Linnaeus) (Synonyms: *Anthomyia radicum*, *Egle radicum*)	THALLOPHYTA: PHYCOMYCETES ENTOMOPHTHORALES: ENTOMOPHTHORACEAE *Entomophthora muscae* Cohn Graham-Smith, 1919 ARTHROPODA: INSECTA HYMENOPTERA: SPHECIDAE *Crabro leucostoma* (Linnaeus) Hamm et al., 1926 *Crabro peltarius* (Schreber) Hamm et al., 1926 *Oxybelus bipunctatus* Olivier Chevalier, 1926
Hylemya strigosa (Fabricius) (Synonym: *Hylemyia strigosa*) (prey)	ARTHROPODA: INSECTA HYMENOPTERA: SPHECIDAE *Crabro cribraria* (Linnaeus) Hamm et al., 1926 DIPTERA: MUSCIDAE *Myospila meditabunda* (Fabricius) Porchinskiĭ, 1910
Hylemya urbana Malloch (prey)	ARTHROPODA: INSECTA HYMENOPTERA: SPHECIDAE *Sericophorus sydneyi* Rayment Rayment, 1954
Hylemya variata (Fallén) (Synonym: *Hylemyia variata*) (prey)	ARTHROPODA: INSECTA HYMENOPTERA: SPHECIDAE *Crabro peltarius* (Schreber) Hamm et al., 1926

Fly	Associated organism
Pegomya Robineau-Desvoidy (Synonym: *Pegomyia*)	ARTHROPODA: INSECTA HYMENOPTERA: PTEROMALIDAE *Spalangia erythromera brachyceps* Bouček Bouček, 1963
Pegomya finitima Stein	ARTHROPODA: INSECTA HYMENOPTERA: SPHECIDAE *Crabro advenus* Smith Kurczewski et al., 1968
Pegomya haemorrhoa (Zetterstedt) (Synonym: ?*Pegomyia transversa*)	ARTHROPODA: INSECTA DIPTERA: MUSCIDAE *Phaonia variegata* (Meigen) Keilin, 1917
Pegomya hyoscyami (Panzer) (Synonym: *Pegomyia hyoscyami*)	ARTHROPODA: INSECTA COLEOPTERA: STAPHYLINIDAE *Aleochara curtula* Goeze Kemner, 1926 *Aleochara laevigata* Gyllenhall Kemner, 1926
Pegomya winthemi (Meigen) (Synonym: *Pegomyia winthemi*) (prey)	ARTHROPODA: INSECTA DIPTERA: MUSCIDAE *Phaonia variegata* (Meigen) Keilin, 1917
Hydrophoria ruralis Meigen (Synonym: *Anthomyia maculata*)	PROTOZOA: MASTIGOPHORA PROTOMONADIDA: TRYPANOSOMATIDAE *Leptomonas anthomyia* (Franchini) Wenyon Franchini, 1922
Anthomyia Meigen	THALLOPHYTA: PHYCOMYCETES ENTOMOPHTHORALES: ENTOMOPHTHORACEAE ?*Entomophthora* Fresenius Roubaud, 1911a *Entomophthora muscae* Cohn Thaxter, 1888
	ARTHROPODA: INSECTA HYMENOPTERA: SPHECIDAE *Crabro clypeatus* (Schreber) Hamm et al., 1926 *Crabro peltarius* (Schreber) Hamm et al., 1926
Anthomyia lindigii Schiner (phoresy)	ARTHROPODA: INSECTA DIPTERA: CUTEREBRIDAE *Dermatobia hominis* (Linnaeus) Lutz, 1917

Fly	Organism
Anthomyia pluvialis (Linnaeus) (Synonym: *Anthomyia procellaris*)	SCHIZOPHYTA: SCHIZOMYCETES
	EUBACTERIALES: ENTEROBACTERIACEAE
	Escherichia coli (Migula) Castellani and Chalmers
	Povolný et al., 1961
	Serratia Bizio
	Povolný et al., 1961
	Proteus vulgaris Hauser
	Povolný et al., 1961
	Providencia Kauffmann
	Povolný et al., 1961
	EUBACTERIALES: MICROCOCCACEAE
	Staphylococcus aureus Rosenbach
	Povolný et al., 1961
	EUBACTERIALES: LACTOBACILLACEAE
	Diplococcus pneumoniae Weichselbaum
	Povolný et al., 1961
	EUBACTERIALES: BACILLACEAE
	Bacillus Cohn
	Povolný et al., 1961
	ARTHROPODA: INSECTA
	HYMENOPTERA: SPHECIDAE
	Crabro leucostoma (Linnaeus)
	Hamm et al., 1926
	CHORDATA: AVES
	PASSERIFORMES: PARIDAE
	Aegithalus caudatus (Linnaeus)
	Séguy, 1929
	Parus caeruleus Linnaeus subsp. *caeruleus*
	Séguy, 1929
	PASSERIFORMES: STURNIDAE
	Sturnus vulgaris Linnaeus
	Séguy, 1929
	PASSERIFORMES: FRINGILLIDAE
	Fringilla coelebs Linnaeus
	Séguy, 1929
	PASSERIFORMES: PLOCEIDAE
	Passer domesticus (Linnaeus)
	Séguy, 1929
Calythea albicincta (Fallén)	SCHIZOPHYTA: SCHIZOMYCETES
	EUBACTERIALES: BACILLACEAE
	Bacillus Cohn
	Zmeev, 1943
	ARTHROPODA: ARACHNIDA
	ACARINA: UNDETERMINED PARASITIDAE
	Sychevskaia͡, 1964a

Fly	Associated organism
Paraprosalpia billbergi (Zetterstedt) (Synonym: *Prosalpia billbergi*) (predator)	ARTHROPODA: INSECTA DIPTERA: UNDETERMINED CHIRONOMIDAE Hobby, 1934a
Paraprosalpia silvestris (Fallén) (Synonym: *Prosalpia sylvestris*) (predator)	ARTHROPODA: INSECTA DIPTERA: EMPIDIDAE *Empis grisea* Fallén Hobby, 1934a
Leucophora albiseta (von Roser) (Synonym: *Anthomyia albescens*) (parasite)	ARTHROPODA: INSECTA HYMENOPTERA: SPHECIDAE *Dinetus pictus* (Fabricius) Chevalier, 1923c *Diodontus minutus* (Fabricius) Chevalier, 1923c *Oxybelus quattuordecimnotatus* Jurine Chevalier, 1926
Family: MUSCIDAE	
Subfamily: COENOSIINAE	
Coenosia lineatipes (Zetterstedt)	ARTHROPODA: INSECTA HYMENOPTERA: SPHECIDAE *Crabro leucostoma* (Linnaeus) Hamm et al., 1926
Coenosia sexnotata Meigen (prey)	ARTHROPODA: INSECTA HYMENOPTERA: SPHECIDAE *Crabro leucostoma* (Linnaeus) Hamm et al., 1926
Coenosia tigrina (Fabricius) var. *leonina* (prey)	ARTHROPODA: INSECTA HYMENOPTERA: SPHECIDAE *Crabro advenus* Smith Kurczewski et al., 1968 *Crabro peltarius* (Schreber) Hamm et al., 1926
Coenosia tricolor (Zetterstedt) (prey)	ARTHROPODA: INSECTA HYMENOPTERA: SPHECIDAE *Crabro palmarius* (Schreber) Hamm et al., 1926
Allognota agromyzina (Fallén) (predator)	ANNELIDA: UNDETERMINED OLIGOCHAETA Keilin, 1917 ARTHROPODA: INSECTA DIPTERA: TIPULIDAE *Ula macroptera* (Macquart) Keilin, 1917
Atherigona Rondani	PROTOZOA: MASTIGOPHORA POLYMASTIGIDA: CHILOMASTIGIDAE *Chilomastix mesnili* (Wenyon) Harris et al., 1946

Fly	Organism
	POLYMASTIGIDA: HEXAMITIDAE
	Giardia lamblia Stiles
	Harris et al., 1946
	TRICHOMONADIDA: TRICHOMONADIDAE
	Trichomonas hominis (Davaine)
	Harris et al., 1946
	PROTOZOA: SARCODINA
	AMOEBIDA: ENDAMOEBIDAE
	Endolimax nana (Wenyon and O'Connor)
	Harris et al., 1946
	Entamoeba coli (Grassi)
	Harris et al., 1946
	Entamoeba histolytica Schaudinn
	Harris et al., 1946
	Iodamoeba bütschlii (Prowazek)
	Harris et al., 1946
	PLATYHELMINTHES: CESTOIDEA
	CYCLOPHYLLIDEA: HYMENOLEPIDIDAE
	Hymenolepis diminuta (Rudolphi) Blanchard
	Harris et al., 1946
	ASCHELMINTHES: NEMATODA
	TRICHURIDEA: TRICHURIDAE
	Trichuris trichiura (Linnaeus)
	Harris et al., 1946
	STRONGYLIDEA: ANCYLOSTOMATIDAE
	Ancylostoma (Dubini)
	Harris et al., 1946
	ASCARIDIDEA: ASCARIDIDAE
	Ascaris lumbricoides Linnaeus
	Harris et al., 1946
Atherigona orientalis Schiner (Synonym: *Atherigona excisa*)	ENTEROVIRUSES
	Poliovirus Types 1 and 3
	Paul et al., 1962
	ECHO virus Types 8, 12, 14
	Paul et al., 1962
	SCHIZOPHYTA: SCHIZOMYCETES
	SPIROCHAETALES: TREPONEMATACEAE
	Treponema pertenue Castellani
	Satchell et al., 1953
Subfamily: LIMNOPHORINAE	
Gymnodia arcuata (Stein)	ENTEROVIRUSES
	Poliovirus
	Francis, Jr. et al., 1948

Fly	Associated organism
Limnophora Robineau-Desvoidy	ENTEROVIRUSES Poliovirus Types 1 and 3 Paul et al., 1962 ECHO virus Types 8, 12, 14 Paul et al., 1962 ARTHROPODA: INSECTA DIPTERA: CUTEREBRIDAE *Dermatobia hominis* (Linnaeus) Dunn, 1930 Neel et al., 1955
Xenomyia oxycera Emden	ARTHROPODA: INSECTA DIPTERA: SIMULIIDAE *Simulium damnosum* Theobald Crosskey et al., 1962
Subfamily: MYDAEINAE	
Helina Robineau-Desvoidy (Synonym: *Spilogaster*)	SCHIZOPHYTA: SCHIZOMYCETES EUBACTERIALES: ENTEROBACTERIACEAE *Escherichia coli* (Migula) Castellani and Chalmers Graham-Smith, 1909 *Escherichia coli* var. *acidilactici* (Topley and Wilson) Yale Graham-Smith, 1909 *Escherichia coli* var. *communior* (Topley and Wilson) Yale Graham-Smith, 1909 *Escherichia coli* var. *neapolitana* (Topley and Wilson) Yale Graham-Smith, 1909 *Aerobacter aerogenes* (Kruse) Beijerinck Graham-Smith, 1909
Helina anceps (Zetterstedt) (prey)	ARTHROPODA: INSECTA HYMENOPTERA: SPHECIDAE *Ectemnius quadricinctus* (Fabricius) Hamm et al., 1926
Helina clara (Meigen) (Synonym: *Spilogaster clara*) (prey)	ARTHROPODA: INSECTA HYMENOPTERA: SPHECIDAE *Crossocerus quadrimaculatus* (Fabricius) Hamm et al., 1926
Helina depuncta (Fallén) (Synonym: *Helina tetrastigma*) (prey)	ARTHROPODA: INSECTA HYMENOPTERA: SPHECIDAE *Crabro peltarius* (Schreber) Hamm et al., 1926

Fly	Associated organism
Helina duplicata (Meigen) (Synonyms: *Spilogaster duplicata, Helina communis*) (prey)	ARTHROPODA: INSECTA HYMENOPTERA: SPHECIDAE *Crabro leucostoma* (Linnaeus) Hamm et al., 1926 *Crabro palmarius* (Schreber) Hamm et al., 1926 *Crabro peltarius* (Schreber) Hamm et al., 1926
Helina impuncta (Fallén) (prey)	ARTHROPODA: INSECTA HYMENOPTERA: SPHECIDAE *Crabro palmarius* (Schreber) Hamm et al., 1926 *Crabro peltarius* (Schreber) Hamm et al., 1926 *Ectemnius quadricinctus* (Fabricius) Hamm et al., 1926
Helina lucorum (Fallén) (prey)	ARTHROPODA: INSECTA HYMENOPTERA: SPHECIDAE *Crabro peltarius* (Schreber) Hamm et al., 1926
Helina quadrum (Fabricius) (prey)	ARTHROPODA: INSECTA HYMENOPTERA: SPHECIDAE *Crabro peltarius* (Schreber) Hamm et al., 1926 *Crossocerus quadrimaculatus* (Fabricius) Hamm et al., 1926
Myospila meditabunda (Fabricius) (Synonym: *Myiospila meditabunda*)	ENTEROVIRUSES Poliovirus Francis, Jr. et al., 1948 ARTHROPODA: INSECTA HYMENOPTERA: SPHECIDAE *Crabro venator* (Rohwer) Kurczewski et al., 1968 LEPIDOPTERA: NOCTUIDAE *Macronoctua onusta* Grote Breakey, 1929
Mydaea Robineau-Desvoidy	THALLOPHYTA: ASCOMYCETES HYPOCREALES: HYPOCREACEAE *Cordyceps dipterigena* Berkeley and Broome Petch, 1923 MOLLUSCA: GASTROPODA SIGMURETHRA: ACHATINIDAE *Burtoa nilotica* (Pfeiffer) Rodhain et al., 1916

Fly	Associated organism
Mydaea detrita Zetterstedt (prey)	ARTHROPODA: INSECTA HYMENOPTERA: SPHECIDAE *Ectemnius quadricinctus* (Fabricius) Hamm et al., 1926
Mydaea pagana (Fabricius) (cannibalism)	ARTHROPODA: INSECTA DIPTERA: MUSCIDAE *Mydaea pagana* (Fabricius) Thomson, 1937
Mydaea pertusa Meigen	CHORDATA: AVES PASSERIFORMES: PARIDAE *Parus caeruleus* Linnaeus subsp. *caeruleus* Séguy, 1929
Mydaea tincta (Zetterstedt) (predator?)	ARTHROPODA: INSECTA DIPTERA: MYCETOPHILIDAE *Mycetophila lineola* Meigen Keilin, 1917
Mydaea urbana (Meigen)	SCHIZOPHYTA: SCHIZOMYCETES EUBACTERIALES: ENTEROBACTERIACEAE *Escherichia coli* (Migula) Castellani and Chalmers Povolný et al., 1961 *Serratia* Bizio Povolný et al., 1961 *Proteus morganii* (Winslow, Kligler, and Rothberg) Yale Povolný et al., 1961 *Proteus vulgaris* Hauser Povolný et al., 1961 ARTHROPODA: INSECTA HYMENOPTERA: SPHECIDAE *Ectemnius quadricinctus* (Fabricius) Hamm et al., 1926
Subfamily: FANNIINAE	
Fannia Robineau-Desvoidy	ENTEROVIRUSES Poliovirus Toomey et al., 1941 Trask et al., 1943 Melnick et al., 1945 Poliovirus Types 1 and 3 Riordan et al., 1961 Paul et al., 1962 Coxsackievirus Type A9 (A9), Type B5 (B5) Riordan et al., 1961 ECHO virus Types 1, 5, 11 Riordan et al., 1961

Fly	Organism
Fannia (cont.)	ECHO virus Types 8, 12, 14 Paul et al., 1962 SCHIZOPHYTA: SCHIZOMYCETES PSEUDOMONADALES: SPIRILLACEAE *Vibrio comma* (Schroeter) Winslow et al. Savchenko, 1892a, 1892b EUBACTERIALES: ENTEROBACTERIACEAE *Proteus morganii* (Winslow, Kligler, and Rothberg) Yale Shimizu et al., 1965 PROTOZOA: MASTIGOPHORA POLYMASTIGIDA: HEXAMITIDAE *Giardia intestinalis* (Lambl) Wenyon et al., 1917a PROTOZOA: SARCODINA AMOEBIDA: ENDAMOEBIDAE *Entamoeba coli* (Grassi) Wenyon et al., 1917a *Entamoeba histolytica* Schaudinn Wenyon et al., 1917a ASCHELMINTHES: NEMATODA ASCARIDIDEA: ASCARIDIDAE *Ascaris lumbricoides* Linnaeus Sukhacheva, 1963 ARTHROPODA: INSECTA HYMENOPTERA: BRACONIDAE *Alysia manducator* Panzer Myers, 1927 CHORDATA: AVES GALLIFORMES: PHASIANIDAE *Gallus gallus* Linnaeus Rodriguez et al., 1962b
Fannia aërea (Zetterstedt) (prey)	ARTHROPODA: INSECTA DIPTERA: ASILIDAE *Isopogon brevirostris* (Meigen) Hobby, 1934b
Fannia armata (Meigen) (predator)	ARTHROPODA: INSECTA HYMENOPTERA: SPHECIDAE *Crabro leucostoma* (Linnaeus) Hamm et al., 1926 *Crabro peltarius* (Schreber) Hamm et al., 1926
Fannia canicularis (Linnaeus) (Synonym: *Homalomyia canicularis*)	ENTEROVIRUSES Poliovirus Francis, Jr. et al., 1948

Fannia canicularis (cont.)

SCHIZOPHYTA: SCHIZOMYCETES

EUBACTERIALES:
ACHROMOBACTERIACEAE

Alcaligenes faecalis Castellani and Chalmers
Gregor et al., 1960

Flavobacterium devorans (Zimmermann) Bergey et al.
Gregor et al., 1960

Flavobacterium invisible (Vaughan) Bergey et al.
Gregor et al., 1960

EUBACTERIALES:
ENTEROBACTERIACEAE

Escherichia coli (Migula) Castellani and Chalmers
Yao et al., 1929
Gregor et al., 1960
Povolný et al., 1961

Escherichia coli var. *acidilactici* (Topley and Wilson) Yale
Graham-Smith, 1909
Zmeev, 1943

Escherichia coli var. *communior* (Topley and Wilson) Yale
Graham-Smith, 1909

Escherichia coli var. *neapolitana* (Topley and Wilson) Yale
Graham-Smith, 1909

Escherichia intermedia (Werkman and Gillen) Vaughn and Levine
Zmeev, 1943
O'Keefe et al., 1954

Aerobacter aerogenes (Kruse) Beijerinck
Zmeev, 1943
O'Keefe et al., 1954
Gregor et al., 1960
Povolný et al., 1961

Klebsiella cloacae (Jordan) Brisou
Gregor et al., 1960

Cloaca Castellani and Chalmers
Zmeev, 1943

Paracolobactrum aerogenoides Borman, Stuart, and Wheeler
Zmeev, 1943

Serratia marcescens Bizio
Gregor et al., 1960

Fannia canicularis (cont.)

Proteus inconstans (Ornstein) Shaw and Clarke
  Gregor et al., 1960
Proteus mirabilis Hauser
  O'Keefe et al., 1954
  Gregor et al., 1960
Proteus morganii (Winslow, Kligler, and Rothberg) Yale
  O'Keefe et al., 1954
  Gregor et al., 1960
Proteus rettgeri (Hadley et al.) Rustigian and Stuart
  Gregor et al., 1960
Proteus vulgaris Hauser
  O'Keefe et al., 1954
  Gregor et al., 1960
  Povolný et al., 1961
Salmonella hirschfeldii Weldin
  Yao et al., 1929
Citrobacter freundii Werkman and Gillen
  Zmeev, 1943
  Gregor et al., 1960
Shigella dysenteriae (Shiga) Castellani and Chalmers
  Yao et al., 1929
Shigella flexneri Castellani and Chalmers
  Sychevskaia͡ et al., 1959a
  Arskiĭ et al., 1961
Shigella sonnei (Levine) Weldin
  Arskiĭ et al., 1961

EUBACTERIALES: UNDETERMINED ENTEROBACTERIACEAE

Paraescherichia, unvalidated
  Zmeev, 1943

EUBACTERIALES: MICROCOCCACEAE

Staphylococcus afermentans Shaw, Stitt, and Cowan
  Gregor et al., 1960
Staphylococcus aureus Rosenbach
  Gregor et al., 1960
  Povolný et al., 1961
Staphylococcus lactis Shaw, Stitt, and Cowan
  Gregor et al., 1960
Staphylococcus saprophyticus Fairbrother
  Gregor et al., 1960

Fannia canicularis (cont.)

EUBACTERIALES: LACTOBACILLACEAE
Streptococcus durans Sherman and Wing
Gregor et al., 1960

EUBACTERIALES: BACILLACEAE
Bacillus Cohn
Povolný et al., 1961
Bacillus cereus var. *mycoides* (Flügge) Smith et al.
Gregor et al., 1960
Bacillus thuringiensis Berliner
Eversole et al., 1965

ACTINOMYCETALES: MYCOBACTERIACEAE
Mycobacterium leprae (Hansen) Lehmann and Neumann
Asami, 1934

THALLOPHYTA: PHYCOMYCETES
ENTOMOPHTHORALES: ENTOMOPHTHORACEAE
Entomophthora Fresenius
Graham-Smith, 1916
Entomophthora muscae Cohn
Buchanan, 1913
Graham-Smith, 1919
Wilhelmi, 1919
Tischler, 1950
Steve, 1959

PROTOZOA: MASTIGOPHORA
PROTOMONADIDA: TRYPANOSOMATIDAE
Herpetomonas Kent
Dunkerly, 1911
Herpetomonas muscarum (Leidy) Kent
Dunkerly, 1911
Patton, 1921a
Leptomonas Kent
Dunkerly, 1911

PROTOMONADIDA: BODONIDAE
Bodo Ehrenberg
Dunkerly, 1912

POLYMASTIGIDA: CHILOMASTIGIDAE
Chilomastix mesnili (Wenyon)
Root, 1921

POLYMASTIGIDA: HEXAMITIDAE
Giardia intestinalis (Lambl)
Root, 1921

Fannia canicularis (cont.)

PROTOZOA: SARCODINA
AMOEBIDA: ENDAMOEBIDAE
Endolimax nana (Wenyon and O'Connor)
Root, 1921
Yao et al., 1929
Entamoeba coli (Grassi)
Root, 1921
Yao et al., 1929
Entamoeba histolytica Schaudinn
Root, 1921
Yao et al., 1929

PLATYHELMINTHES: CESTOIDEA
PSEUDOPHYLLIDEA: DIPHYLLOBOTHRIIDAE
Bothriocephalus marginatus Krefft
Nicoll, 1911b

ASCHELMINTHES: NEMATODA
SPIRURIDEA: SPIRURIDAE
Habronema megastoma (Rudolphi) Seurat
Roubaud et al., 1922b
Habronema muscae (Carter)
Roubaud et al., 1922b

ARTHROPODA: ARACHNIDA
ARANEAE: SALTICIDAE
Salticus scenicus (Clerck)
Steve, 1959
ACARINA: MACROCHELIDAE
Glyptholaspis confusa (Foa)
Axtell, 1961a
Macrocheles glaber (Müller)
Filipponi, 1960
Macrocheles muscaedomesticae (Scopoli)
Buchanan, 1916
Steve, 1959
Filipponi, 1960
Axtell, 1961a
Anderson, 1964
Singh et al., 1966
O'Donnell, 1967
Macrocheles perglaber Filipponi and Pegazzano
Filipponi, 1960
Macrocheles subbadius Berlese
Filipponi, 1960
ACARINA: UROPODIDAE
Fuscuropoda Vitzthum
Anderson, 1964

Fannia canicularis (cont.)

Fuscuropoda vegetans (De Geer)
O'Donnell, 1967

ACARINA: ANOETIDAE

Myianoetus muscarum (Linnaeus)
Greenberg et al., 1960

ARTHROPODA: INSECTA

ORTHOPTERA: LOCUSTIDAE

Nomadacris septembasciata Serville
Jack, 1935

Schistocerca gregaria Főrskal
Nocedo, 1921
Régnier, 1931

DERMAPTERA: FORFICULIDAE

Forficula auricularia Linnaeus
Anderson, 1964

LEPIDOPTERA: PYRALIDIDAE

Epischnia incanella Holst
Heeger, 1848 (see Hewitt, 1912)

LEPIDOPTERA: SPHINGIDAE

Acherontia atropos Linnaeus
Keilin, 1917

COLEOPTERA: UNDETERMINED STAPHYLINIDAE
Anderson, 1964

COLEOPTERA: SCARABAEIDAE

Lachnosterna Hope
Davis, 1919

COLEOPTERA: CURCULIONIDAE

Conotrachelus retentus Say
Brooks, 1922

HYMENOPTERA: ICHNEUMONIDAE

Stilpnus anthyomyidiperda Viereck
Legner et al. (pre-publication), 1965a
Legner (pre-publication), 1966
Legner et al., 1966b

HYMENOPTERA: PTEROMALIDAE

Muscidifurax raptor Girault and Sanders
Legner (pre-publication), 1966

Pachycrepoideus vindemmiae Rondani
Steve, 1959

Spalangia[1] Latreille
Sychevskaīā, 1964b

Spalangia cameroni Perkins
Legner (pre-publication), 1966

[1] Although not specifically mentioned in each entry, it is assumed that all *Spalangia* are pupal parasites.

Fannia canicularis (cont.)

Spalangia endius Walker
Legner (pre-publication), 1966
Spalangia nigroaenea Curtis
Legner (pre-publication), 1966
HYMENOPTERA: DIAPRIIDAE
Trichopria Ashmead
Legner (pre-publication), 1966
HYMENOPTERA: FORMICIDAE
Camponotus pennsylvanicus De Geer
Steve, 1959
HYMENOPTERA: SPHECIDAE
Crabro advenus Smith
Kurczewski et al., 1968
Crossocerus quadrimaculatus (Fabricius)
Hamm et al., 1926
Oxybelus uniglumis (Linnaeus)
Chevalier, 1926
HYMENOPTERA: VESPIDAE
Vespa germanica Fabricius
Kühlhorn, 1961
Vespa vulgaris Linnaeus
Kühlhorn, 1961
HYMENOPTERA: APIDAE
Bombus terrestris Linnaeus
Hewitt, 1912
DIPTERA: ANTHOMYIIDAE
Scatophaga stercoraria (Linnaeus)
Hewitt, 1914
Steve, 1959
DIPTERA: MUSCIDAE
Ophyra leucostoma (Wiedemann)
Anderson, 1964
Muscina assimilis (Fallén)
Keilin, 1917
DIPTERA: SARCOPHAGIDAE
Ravinia striata Fabricius
Sychevskaîa, 1954
Sarcophaga haemorrhoidalis (Fallén)
Sychevskaîa, 1954
CHORDATA: AVES
PASSERIFORMES: HIRUNDINIDAE
Hirundo rustica Linnaeus subsp. *rustica*
Séguy, 1929
PASSERIFORMES: TURDIDAE
Luscinia megarhynchos Brehm
Séguy, 1929

Fannia canicularis (cont.)

CHORDATA: MAMMALIA
RODENTIA: UNDETERMINED
GLIRIDAE
Hewitt, 1912

Fannia ?corvina (Verrall) (Synonym: *Homalomyia corvina*) ....

PROTOZOA: MASTIGOPHORA
PROTOMONADIDA: TRYPANOSOMATIDAE
Herpetomonas ?muscarum (Leidy) Kent
Mackinnon, 1910

Fannia femoralis (Stein) ....

ARTHROPODA: INSECTA
HYMENOPTERA: PTEROMALIDAE
Muscidifurax raptor Girault and Sanders
Legner et al. (pre-publication), 1965a
Legner (pre-publication), 1966
Legner et al., 1966c
Nasonia vitripennis[1] (Walker)
Roberts, 1933a
Legner et al. (pre-publication), 1965a
Legner (pre-publication), 1966
Legner et al., 1966b
Spalangia[2] *cameroni* Perkins
Legner et al. (pre-publication), 1965a
Legner (pre-publication), 1966
Legner et al., 1966b
Spalangia endius Walker
Legner et al. (pre-publication), 1965a
Legner (pre-publication), 1966
Legner et al., 1966b
Spalangia nigroaenea Curtis
Legner et al. (pre-publication), 1965a
Legner (pre-publication), 1966
Legner et al., 1966b
Spalangia simplex Perkins
Legner et al. (pre-publication), 1965a
HYMENOPTERA: DIAPRIIDAE
Trichopria Ashmead
Legner (pre-publication), 1966

[1] Although not specifically mentioned in each entry, it is assumed that all *Nasonia vitripennis* are pupal parasites.

[2] Although not specifically mentioned in each entry, it is assumed that all *Spalangia* are pupal parasites.

Fly	Associated organism
Fannia femoralis (cont.)	DIPTERA: MUSCIDAE *Ophyra leucostoma* (Wiedemann) Legner et al. (pre-publication), 1965a
Fannia incisurata (Zetterstedt)	SCHIZOPHYTA: SCHIZOMYCETES EUBACTERIALES: ENTEROBACTERIACEAE *Escherichia coli* (Migula) Castellani and Chalmers Lysenko, 1958 *Proteus inconstans* (Ornstein) Shaw and Clarke Lysenko, 1958 *Proteus rettgeri* (Hadley et al.) Rustigian and Stuart Lysenko, 1958 *Proteus vulgaris* Hauser Lysenko, 1958 *Citrobacter freundii* Werkman and Gillen Lysenko, 1958 ARTHROPODA: INSECTA HYMENOPTERA: ICHNEUMONIDAE *Stilpnus* Gravenhorst Sychevskaia͡, 1964b HYMENOPTERA: SPHECIDAE *Crossocerus quadrimaculatus* (Fabricius) Hamm et al., 1926 CHORDATA: AVES PASSERIFORMES: UNDETERMINED PARIDAE Séguy, 1929
Fannia kowarzi (Verrall) (prey)	ARTHROPODA: INSECTA HYMENOPTERA: SPHECIDAE *Crabro leucostoma* (Linnaeus) Hamm et al., 1926
Fannia leucosticta (Meigen)	SCHIZOPHYTA: SCHIZOMYCETES EUBACTERIALES: ENTEROBACTERIACEAE *Escherichia coli* var. *acidilactici* (Topley and Wilson) Yale Zmeev, 1943 *Escherichia intermedia* (Werkman and Gillen) Vaughn and Levine Zmeev, 1943 *Aerobacter* Beijerinck Zmeev, 1943

Fly	Associated organism
Fannia leucosticta (cont.)	*Aerobacter aerogenes* (Kruse) Beijerinck Zmeev, 1943 *Cloaca* Castellani and Chalmers Zmeev, 1943 *Paracolobactrum aerogenoides* Borman, Stuart, and Wheeler Zmeev, 1943 *Citrobacter freundii* Werkman and Gillen Zmeev, 1943 ARTHROPODA: ARACHNIDA ACARINA: MACROCHELIDAE *Macrocheles muscaedomesticae* (Scopoli) Sychevskaîa, 1964a *Macrocheles plumiventris* Hull Sychevskaîa, 1964a ARTHROPODA: INSECTA HYMENOPTERA: PTEROMALIDAE *Spalangia*[1] Latreille Sychevskaîa, 1964b *Spalangia cameroni* Perkins Bouček, 1963 *Spalangia endius* Walker Bouček, 1963 *Spalangia nigripes* Curtis Bouček, 1963 DIPTERA: SARCOPHAGIDAE *Ravinia striata* Fabricius Sychevskaîa, 1954 *Sarcophaga haemorrhoidalis* (Fallén) Sychevskaîa, 1954
Fannia lineata (Stein)	CHORDATA: AVES PASSERIFORMES: STURNIDAE *Sturnus vulgaris* (Linnaeus) Séguy, 1929
Fannia manicata (Meigen) (prey)	ARTHROPODA: INSECTA HYMENOPTERA: SPHECIDAE *Crabro leucostoma* (Linnaeus) Hamm et al., 1926
Fannia polychaeta Stein (prey)	ARTHROPODA: INSECTA HYMENOPTERA: SPHECIDAE *Crabro leucostoma* (Linnaeus) Hamm et al., 1926 *Crabro pubescens* Shuckard Hamm et al., 1926

[1] Although not specifically mentioned in each entry, it is assumed that all *Spalangia* are pupal parasites.

Fly	Associated organism
Fannia polychaeta (cont.)	*Oxybelus bipunctatus* Olivier Chevalier, 1926
Fannia pusio (Wiedemann) (prey)	ARTHROPODA: INSECTA DERMAPTERA: LABIDURIDAE *Euborellia annulipes* (Lucas) Illingworth, 1923a DERMAPTERA: LABIIDAE *Labia pilicornis* (Motschulsky) Illingworth, 1923a *Spingolabis hawaiiensis* (Bormans) Illingworth, 1923a COLEOPTERA: HYDROPHILIDAE *Cryptopleurum minutum* Fabricius Illingworth, 1923a *Dactylosternum abdominale* Fabricius Illingworth, 1923a COLEOPTERA: STAPHYLINIDAE *Oxytelus* Gravenhorst Illingworth, 1923a *Philonthus discoideus* Gravenhorst Illingworth, 1923a *Philonthus longicornis* Stephens Illingworth, 1923a *Tachyporus Gravenhorst* Illingworth, 1923a HYMENOPTERA: FORMICIDAE *Pheidole megacephala* (Fabricius) Illingworth, 1923a *Ponera perkinsi* Forel Illingworth, 1923a
Fannia scalaris (Fabricius) (Synonym: *Homalomyia scalaris*)	SCHIZOPHYTA: SCHIZOMYCETES EUBACTERIALES: ENTEROBACTERIACEAE *Escherichia* Castellani and Chalmers Zmeev, 1943 *Escherichia coli* (Migula) Castellani and Chalmers Yao et al., 1929 Povolný et al., 1961 *Escherichia coli* var. *acidilactici* (Topley and Wilson) Yale Zmeev, 1943 *Escherichia coli* var. *acidilactici* (Topley and Wilson) Yale= [*Escherichia paraacidilactici*] Zmeev, 1943

Fannia scalaris (cont.)

Escherichia coli var. *communior* (Topley and Wilson) Yale
Zmeev, 1943

Escherichia intermedia (Werkman and Gillen) Vaughn and Levine
Zmeev, 1943
O'Keefe et al., 1954

Aerobacter Beijerinck
Zmeev, 1943

Aerobacter aerogenes (Kruse) Beijerinck
Zmeev, 1943

Cloaca Castellani and Chalmers
Zmeev, 1943

Serratia Bizio
Povolný et al., 1961

Proteus morganii (Winslow, Kligler, and Rothberg) Yale
O'Keefe et al., 1954

Proteus vulgaris Hauser
O'Keefe et al., 1954
Povolný et al., 1961

Providencia Kauffmann
Povolný et al., 1961

Salmonella hirschfeldii Weldin
Yao et al., 1929

Citrobacter freundii Werkman and Gillen
Zmeev, 1943

Citrobacter freundii Werkman and Gillen=[*Escherichia anindolica*]
Zmeev, 1943

Shigella Castellani and Chalmers
Zaidenov, 1961

Shigella dysenteriae (Shiga) Castellani and Chalmers
Yao et al., 1929

Shigella flexneri Castellani and Chalmers
Sychevskaīā et al., 1959a

EUBACTERIALES: MICROCOCCACEAE

Staphylococcus aureus Rosenbach
Povolný et al., 1961

EUBACTERIALES: LACTOBACILLACEAE

Diplococcus pneumoniae Weichselbaum
Povolný et al., 1961

Fannia scalaris (cont.)

EUBACTERIALES: BACILLACEAE
Bacillus Cohn
Povolný et al., 1961

PROTOZOA: MASTIGOPHORA
PROTOMONADIDA: TRYPANOSOMATIDAE
Herpetomonas muscarum (Leidy) Kent
Léger, 1903
Brug, 1915

PROTOZOA: SARCODINA
AMOEBIDA: ENDAMOEBIDAE
Endolimax nana (Wenyon and O'Connor)
Yao et al., 1929
Entamoeba coli (Grassi)
Yao et al., 1929
Entamoeba histolytica Schaudinn
Yao et al., 1929

PROTOZOA: CNIDOSPORIDIA
MICROSPORIDA: NOSEMATIDAE
Thelohania ovata Dunkerly
Dunkerly, 1912
MICROSPORIDA: MRAZEKIIDAE
Octosporea monospora Chatton and Krempf
Brug, 1914

ARTHROPODA: ARACHNIDA
ACARINA: MACROCHELIDAE
Macrocheles Latreille
Sychevskaîa, 1964a
Macrocheles muscaedomesticae (Scopoli)
Sychevskaîa, 1964a
ACARINA: ANOETIDAE
Myianoetus Oudemans
Sychevskaîa, 1964a

ARTHROPODA: INSECTA
HYMENOPTERA: ICHNEUMONIDAE
Stilpnus Gravenhorst
Sychevskaîa, 1964b
HYMENOPTERA: PTEROMALIDAE
Pachycrepoideus vindemmiae Rondani
Steve, 1959
Spalangia[1] Latreille
Sychevskaîa, 1964b

[1] Although not specifically mentioned in each entry, it is assumed that all *Spalangia* are pupal parasites.

Fly	Associated organism
Fannia scalaris (cont.)	*Spalangia cameroni* Perkins Bouček, 1963 *Spalangia endius* Walker Bouček, 1963 *Spalangia nigripes* Curtis Bouček, 1963 HYMENOPTERA: SPHECIDAE *Crabro signatus* Panzer Chevalier, 1923b *Crossocerus quadrimaculatus* (Fabricius) Hamm et al., 1926 *Oxybelus uniglumis* (Linnaeus) Chevalier, 1926 DIPTERA: SARCOPHAGIDAE *Ravinia striata* Fabricius Sychevskaiā, 1954 *Sarcophaga haemorrhoidalis* (Fallén) Sychevskaiā, 1954 CHORDATA: AVES PASSERIFORMES: HIRUNDINIDAE *Delichon urbica* (Linnaeus) subsp. *urbica* Séguy, 1929 *Hirundo rustica* Linnaeus subsp. *rustica* Séguy, 1929 PASSERIFORMES: PLOCEIDAE *Passer domesticus* (Linnaeus) Séguy, 1929
Fannia ?serena (Fallén) (prey)	ARTHROPODA: INSECTA DIPTERA: ANTHOMYIIDAE *Scatophaga stercoraria* (Linnaeus) Hobby, 1934b
Fannia similis Stein (prey)	ARTHROPODA: INSECTA HYMENOPTERA: SPHECIDAE *Crabro leucostoma* (Linnaeus) Hamm et al., 1926
Fannia subscalaris Zimin (competition)	ARTHROPODA: INSECTA DIPTERA: SARCOPHAGIDAE *Ravinia striata* Fabricius Sychevskaiā, 1954 *Sarcophaga haemorrhoidalis* (Fallén) Sychevskaiā, 1954

Subfamily: PHAONIINAE

Fly	Associated organism
Hydrotaea albipuncta (Zetterstedt) (Synonym: *Hydrotaea albipunctata*)	ARTHROPODA: INSECTA HYMENOPTERA: UNDETERMINED ICHNEUMONIDAE Hammer, 1942 HYMENOPTERA: SPHECIDAE *Crabro leucostoma* (Linnaeus) Hamm et al., 1926
Hydrotaea armipes (Fallén)	ARTHROPODA: ARACHNIDA ACARINA: ASCAIDAE *Gamasellus* Berlese Sychevskaîa, 1964a HYMENOPTERA: ANOETIDAE *Myianoetus* Oudemans Sychevskaîa, 1964a ARTHROPODA: INSECTA HYMENOPTERA: SPHECIDAE *Crabro leucostoma* (Linnaeus) Hamm et al., 1926
Hydrotaea australis Malloch	ASCHELMINTHES: NEMATODA FILARIIDAE: DIPETALONEMATIDAE *Oncocerca gibsoni* (Cleland and Johnston) Henry, 1927 ARTHROPODA: INSECTA HYMENOPTERA: PTEROMALIDAE *Spalangia endius* Walker Bouček, 1963 *Spalangia orientalis* Graham Mackerras, 1932 HYMENOPTERA: DIAPRIIDAE *Phaenopria* Ashmead Mackerras, 1932
Hydrotaea dentipes (Fabricius)	SCHIZOPHYTA: SCHIZOMYCETES EUBACTERIALES: ACHROMOBACTERIACEAE *Alcaligenes faecalis* Castellani and Chalmers Lysenko, 1958 Gregor et al., 1960 *Flavobacterium devorans* (Zimmermann) Bergey et al. Lysenko, 1958 Gregor et al., 1960 *Flavobacterium invisible* (Vaughan) Bergey et al. Gregor et al., 1960

Hydrotaea dentipes (cont.)

EUBACTERIALES: ENTEROBACTERIACEAE

Escherichia coli (Migula) Castellani and Chalmers
- Lysenko, 1958
- Gregor et al., 1960
- Povolný et al., 1961

Aerobacter aerogenes (Kruse) Beijerinck
- Gregor et al., 1960

Klebsiella cloacae (Jordan) Brisou
- Gregor et al., 1960

Serratia Bizio
- Povolný et al., 1961

Serratia marcescens Bizio
- Gregor et al., 1960

Proteus inconstans (Ornstein) Shaw and Clarke
- Lysenko, 1958
- Gregor et al., 1960

Proteus mirabilis Hauser
- Lysenko, 1958
- Gregor et al., 1960

Proteus morganii (Winslow, Kligler, and Rothberg) Yale
- Gregor et al., 1960

Proteus rettgeri (Hadley et al.) Rustigian and Stuart
- Lysenko, 1958
- Gregor et al., 1960

Proteus vulgaris Hauser
- Lysenko, 1958
- Gregor et al., 1960
- Povolný et al., 1961

Providencia Kauffmann
- Povolný et al., 1961

Citrobacter freundii Werkman and Gillen
- Gregor et al., 1960

EUBACTERIALES: MICROCOCCACEAE

Staphylococcus afermentans Shaw, Stitt, and Cowan
- Lysenko, 1958
- Gregor et al., 1960

Staphylococcus aureus Rosenbach
- Gregor et al., 1960
- Povolný et al., 1961

Staphylococcus lactis Shaw, Stitt, and Cowan
- Gregor et al., 1960

Hydrotaea dentipes (cont.)

Staphylococcus saprophyticus Fairbrother
Lysenko, 1958
Gregor et al., 1960
EUBACTERIALES: LACTOBACILLACEAE
Diplococcus pneumoniae Weichselbaum
Povolný et al., 1961
Streptococcus durans Sherman and Wing
Lysenko, 1958
Gregor et al., 1960
Streptococcus faecalis Andrewes and Horder
Lysenko, 1958
EUBACTERIALES: BACILLACEAE
Bacillus Cohn
Povolný et al., 1961
Bacillus cereus var. *mycoides* (Flügge) Smith et al.
Gregor et al., 1960
THALLOPHYTA: PHYCOMYCETES
ENTOMOPHTHORALES: ENTOMOPHTHORACEAE
Entomophthora Fresenius
Graham-Smith, 1916
Entomophthora muscae Cohn
Graham-Smith, 1919
ARTHROPODA: ARACHNIDA
ACARINA: LAELAPTIDAE
Holostaspis Kolenati
Graham-Smith, 1916
ARTHROPODA: INSECTA
HYMENOPTERA: ICHNEUMONIDAE
Atractodes gravidus Gravenhorst
Meyers, 1929
HYMENOPTERA: CYNIPIDAE
Kleidotoma Westwood
James, 1928
Kleidotoma marshalli Marshall
James, 1928
HYMENOPTERA: FIGITIDAE
Figites anthomyiarum (Bouché)
James, 1928
Figites striolatus Hartig
Myers, 1929
HYMENOPTERA: PTEROMALIDAE
Spalangia cameroni Perkins
Bouček, 1963

Hydrotaea dentipes (cont.)	HYMENOPTERA: SPHECIDAE *Oxybelus quattuordecimnotatus* Jurine Chevalier, 1926 DIPTERA: ANTHOMYIIDAE *Scatophaga stercoraria* (Linnaeus) Hobby, 1934b DIPTERA: MUSCIDAE *Muscina stabulans* (Fallén) Porchinskiĭ, 1913 *Polietes albolineata* (Fallén) Porchinskiĭ, 1910
Hydrotaea irritans (Fallén)	SCHIZOPHYTA: SCHIZOMYCETES EUBACTERIALES: ACHROMOBACTERACEAE *Alcaligenes faecalis* Castellani and Chalmers Gregor et al., 1960 *Flavobacterium devorans* (Zimmermann) Bergey et al. Gregor et al., 1960 *Flavobacterium invisible* (Vaughan) Bergey et al. Gregor et al., 1960 EUBACTERIALES: ENTEROBACTERIACEAE *Escherichia coli* (Migula) Castellani and Chalmers Gregor et al., 1960 *Aerobacter aerogenes* (Kruse) Beijerinck Gregor et al., 1960 *Klebsiella cloacae* (Jordan) Brisou Gregor et al., 1960 *Serratia marcescens* Bizio Gregor et al., 1960 *Proteus inconstans* (Ornstein) Shaw and Clarke Gregor et al., 1960 *Proteus mirabilis* Hauser Gregor et al., 1960 *Proteus morganii* (Winslow, Kligler, and Rothberg) Yale Gregor et al., 1960 *Proteus rettgeri* (Hadley et al.) Rustigian and Stuart Gregor et al., 1960 *Proteus vulgaris* Hauser Gregor et al., 1960

Hydrotaea irritans (cont.)

Citrobacter freundii Werkman and Gillen
Gregor et al., 1960

EUBACTERIALES: MICROCOCCACEAE

Staphylococcus afermentans Shaw, Stitt, and Cowan
Gregor et al., 1960

Staphylococcus aureus Rosenbach
Gregor et al., 1960

Staphylococcus lactis Shaw, Stitt, and Cowan
Gregor et al., 1960

Staphylococcus saprophyticus Fairbrother
Gregor et al., 1960

EUBACTERIALES: LACTOBACILLACEAE

Streptococcus durans Sherman and Wing
Gregor et al., 1960

EUBACTERIALES: CORYNEBACTERIACEAE

Corynebacterium pyogenes (Glaze) Eberson
Bahr, 1952, 1953

EUBACTERIALES: BACILLACEAE

Bacillus cereus var. *mycoides* (Flügge) Smith et al.
Gregor et al., 1960

ARTHROPODA: ARACHNIDA

ACARINA: MACROCHELIDAE

Macrocheles Latreille
Sychevskaia͡, 1964a

ACARINA: ANOETIDAE

Myianoetus Oudemans
Sychevskaia͡, 1964a

ARTHROPODA: INSECTA

HYMENOPTERA: SPHECIDAE

Crabro leucostoma (Linnaeus)
Hamm et al., 1926

Crabro peltarius (Schreber)
Hamm et al., 1926

Ectemnius quadricinctus (Fabricius)
Hamm et al., 1926

DIPTERA: ASILIDAE

Laphria flava (Linnaeus)
Hobby, 1934b

Hydrotaea ?irritans (Fallén) (prey) .... ARTHROPODA: INSECTA
DIPTERA: ANTHOMYIIDAE
Scatophaga stercoraria (Linnaeus)
Hobby, 1934b

Hydrotaea meteorica (Linnaeus) (prey) .... ASCHELMINTHES: NEMATODA
TYLENCHIDA: CONTORTYLENCHIDAE
Heterotylenchus autumnalis Nickle
Világiová, 1968
ARTHROPODA: INSECTA
HYMENOPTERA: SPHECIDAE
Crabro leucostoma (Linnaeus)
Hamm et al., 1926
Oxybelus bipunctatus Olivier
Chevalier, 1926

Hydrotaea occulta (Meigen) .... SCHIZOPHYTA: SCHIZOMYCETES
EUBACTERIALES: ENTEROBACTERIACEAE
Escherichia coli (Migula) Castellani and Chalmers
Povolný et al., 1961
Serratia Bizio
Povolný et al., 1961
Proteus vulgaris Hauser
Povolný et al., 1961
Providencia Kauffmann
Povolný et al., 1961
EUBACTERIALES: MICROCOCCACEAE
Staphylococcus aureus Rosenbach
Povolný et al., 1961
EUBACTERIALES: LACTOBACILLACEAE
Diplococcus pneumoniae Weichselbaum
Povolný et al., 1961
EUBACTERIALES: BACILLACEAE
Bacillus Cohn
Povolný et al., 1961

Hydrotaea parva Meade (prey) ...... ARTHROPODA: INSECTA
DIPTERA: MUSCIDAE
Coenosia tigrina (Zetterstedt)
Hobby, 1934b

Ophyra Robineau-Desvoidy (Synonym: *?Phyra*) .... ENTEROVIRUSES
Poliovirus
Trask et al., 1943
Poliovirus Types 1, 3
Riordan et al., 1961
Coxsackievirus Types A9 (A9), B5 (B5)
Riordan et al., 1961
ECHO virus Types 1, 5, 11
Riordan et al., 1961

Ophyra (cont.)

SCHIZOPHYTA: SCHIZOMYCETES
EUBACTERIALES:
ENTEROBACTERIACEAE
Proteus rettgeri (Hadley et al.) Rustigian and Stuart
Shimizu et al., 1965
Providencia Kauffmann
Shimizu et al., 1965
Salmonella chester Kauffmann
Bulling et al., 1959
Salmonella typhimurium (Loeffler) Castellani and Chalmers
Bulling et al., 1959
Shigella Castellani and Chalmers
Shimizu et al., 1965
Shigella boydii Ewing 2, 5, 6
Richards et al., 1961
Shigella dysenteriae (Shiga) Castellani and Chalmers 2
Richards et al., 1961
Shigella flexneri Castellani and Chalmers 1a, 2a, 3, 4a, 4b, 6
Richards et al., 1961
Shigella sonnei (Levine) Weldin 1
Richards et al., 1961
ASCHELMINTHES: NEMATODA
ASCARIDIDEA: ASCARIDIDAE
Ascaris lumbricoides Linnaeus
Sukhacheva, 1963
CHORDATA: AVES
GALLIFORMES: PHASIANIDAE
Gallus gallus Linnaeus
Rodriguez et al., 1962b

Ophyra aenescens (Wiedemann) ....

ENTEROVIRUSES
Poliovirus Types 1, 3
Paul et al., 1962
ECHO virus Types 8, 12, 14
Paul et al., 1962
SCHIZOPHYTA: SCHIZOMYCETES
EUBACTERIALES:
ENTEROBACTERIACEAE
Escherichia coli var. *communior* (Topley and Wilson) Yale
Beyer, 1925
Salmonella Lignières
Greenberg et al., 1964
Salmonella anatum (Rettger and Scoville) Bergey et al.
Greenberg et al., 1963b
Salmonella blockley Kauffmann
Greenberg et al., 1964

Fly	Organism
Ophyra aenescens (cont.)	*Salmonella bovis-morbificans* (Basenau) Schütze et al. Greenberg et al., 1963b *Salmonella derby* Warren and Scott Greenberg et al., 1963b *Salmonella new brunswick* Edwards Greenberg et al., 1963b *Shigella* Castellani and Chalmers Greenberg et al., 1964 *Shigella dysenteriae* (Shiga) Beyer, 1925 ARTHROPODA: INSECTA HYMENOPTERA: PTEROMALIDAE *Nasonia vitripennis* (Walker) Roberts, 1933a
Ophyra anthrax (Meigen)	ARTHROPODA: ARACHNIDA ACARINA: PARASITIDAE *Parasitus lunaris* Berlese Sychevskaīa, 1964a ACARINA: MACROCHELIDAE *Macrocheles muscaedomesticae* (Scopoli) Sychevskaīa, 1964a ACARINA: ANOETIDAE *Myianoetus* Oudemans Sychevskaīa, 1964a ARTHROPODA: INSECTA HYMENOPTERA: CYNIPIDAE *Cothonaspis nigricornis* (Kieffer) Sychevskaīa, 1964b HYMENOPTERA: FIGITIDAE *Figites discordis* Belizin Sychevskaīa, 1964b HYMENOPTERA: PTEROMALIDAE *Spalangia*[1] Latreille Sychevskaīa, 1964b *Spalangia cameroni* Perkins Bouček, 1963 *Spalangia stomoxysiae* Girault Bouček, 1963
Ophyra calcogaster Wiedemann	THALLOPHYTA: PHYCOMYCETES MUCORALES: MUCORACEAE *Mucor ambiguus* Vuillemin Usui, 1960
Ophyra leucostoma (Wiedemann)	ENTEROVIRUSES Poliovirus Paul et al., 1941

[1] Although not specifically mentioned in each entry, it is assumed that all *Spalangia* are pupal parasites.

Ophyra leucostoma (cont.)

Sabin et al., 1942
Melnick et al., 1945
Poliovirus Strain 113
Levkovich et al., 1957
Poliovirus Type 2, Lansing strain
Levkovich et al., 1957

SCHIZOPHYTA: SCHIZOMYCETES

EUBACTERIALES:
ACHROMOBACTERIACEAE

Alcaligenes faecalis Castellani and Chalmers
Gregor et al., 1960
Flavobacterium devorans (Zimmermann) Bergey et al.
Gregor et al., 1960
Flavobacterium invisible (Vaughan) Bergey et al.
Gregor et al., 1960

EUBACTERIALES:
ENTEROBACTERIACEAE

Escherichia coli (Migula) Castellani and Chalmers
Gregor et al., 1960
Povolný et al., 1961
Escherichia coli var. *acidilactici* (Topley and Wilson) Yale
Shura-Bura, 1952
Aerobacter aerogenes (Kruse) Beijerinck
Gregor et al., 1960
Klebsiella cloacae (Jordan) Brisou
Gregor et al., 1960
Serratia marcescens Bizio
Gregor et al., 1960
Proteus inconstans (Ornstein) Shaw and Clarke
Lysenko, 1958
Gregor et al., 1960
Proteus mirabilis Hauser
Gregor et al., 1960
Proteus morganii (Winslow, Kligler, and Rothberg) Yale
Gregor et al., 1960
Povolný et al., 1961
Proteus rettgeri (Hadley et al.) Rustigian and Stuart
Lysenko, 1958
Gregor et al., 1960

Ophyra leucostoma (cont.)

Proteus vulgaris Hauser
Gregor et al., 1960
Povolný et al., 1961
Citrobacter freundii Werkman and Gillen
Lysenko, 1958
Gregor et al., 1960
EUBACTERIALES: MICROCOCCACEAE
Staphylococcus afermentans Shaw, Stitt, and Cowan
Gregor et al., 1960
Staphylococcus aureus Rosenbach
Gregor et al., 1960
Staphylococcus lactis Shaw, Stitt, and Cowan
Gregor et al., 1960
Staphylococcus saprophyticus Fairbrother
Gregor et al., 1960
EUBACTERIALES: LACTOBACILLACEAE
Streptococcus Rosenbach
Povolný et al., 1961
Streptococcus durans Sherman and Wing
Gregor et al., 1960
EUBACTERIALES: BACILLACEAE
Bacillus Cohn
Povolný et al., 1961
Bacillus cereus var. *mycoides* (Flügge) Smith et al.
Gregor et al., 1960
ARTHROPODA: ARACHNIDA
PSEUDOSCORPIONIDEA: CHERNETIDAE
Lamprochernes nodosus (Schrank)
Graham-Smith, 1916
ACARINA: ASCAIDAE
Gamasellus Berlese
Sychevskaia͡, 1964a
ACARINA: MACROCHELIDAE
Macrocheles muscaedomesticae (Scopoli)
Sychevskaia͡, 1964a
ACARINA: LAELAPTIDAE
Holostaspis Kolenati
Graham-Smith, 1916
ACARINA: ANOETIDAE
Myianoetus Oudemans
Sychevskaia͡, 1964a

Ophyra leucostoma (cont.)

ARTHROPODA: INSECTA

HYMENOPTERA: ICHNEUMONIDAE

Atractodes ?gravidus Gravenhorst
Sychevskaia͡, 1964b

HYMENOPTERA: BRACONIDAE

Alysia ridibunda Say
Roberts, 1935

HYMENOPTERA: CYNIPIDAE

Eucoila Westwood
Roberts, 1935

HYMENOPTERA: PTEROMALIDAE

Muscidifurax raptor Girault and Sanders
Legner et al. (pre-publication), 1965a
Legner et al., 1966b
Legner (pre-publication), 1966

Nasonia vitripennis (Walker)
Roberts, 1933a
Legner (pre-publication), 1966

Spalangia[1] *cameroni* Perkins
Legner (pre-publication), 1966

Spalangia endius Walker
Legner et al. (pre-publication), 1965a
Legner et al., 1966b
Legner (pre-publication), 1966

Spalangia nigroaenea Curtis
Legner et al. (pre-publication), 1965a
Legner (pre-publication), 1966
Legner et al., 1966b

HYMENOPTERA: SPHECIDAE

Crabro leucostoma (Linnaeus)
Hamm et al., 1926

Crabro peltarius (Schreber)
Hamm et al., 1926

Oxybelus quattuordecimnotatus Jurine
Chevalier, 1926

CHORDATA: AVES

PASSERIFORMES: HIRUNDINIDAE

Delichon urbica (Linnaeus) subsp. *urbica*
Séguy, 1929

Hirundo rustica (Linnaeus) subsp. *rustica*
Séguy, 1929

[1] Although not specifically mentioned in each entry, it is assumed that all *Spalangia* are pupal parasites.

Fly	Associated organisms
Ophyra nigra Wiedemann	ARTHROPODA: INSECTA
	DERMAPTERA: LABIDURIDAE
	Euborellia annulipes (Lucas)
	Illingworth, 1923a
	DERMAPTERA: LABIIDAE
	Labia pilicornis (Motschulsky)
	Illingworth, 1923a
	Spingolabis hawaiiensis (Bormans)
	Illingworth, 1923a
	COLEOPTERA: HYDROPHILIDAE
	Cryptopleurum minutum Fabricius
	Illingworth, 1923a
	Dactylosternum abdominale Fabricius
	Illingworth, 1923a
	COLEOPTERA: STAPHYLINIDAE
	Oxytelus Gravenhorst
	Illingworth, 1923a
	Philonthus discoideus Gravenhorst
	Illingworth, 1923a
	Philonthus longicornis Stephens
	Illingworth, 1923a
	Tachyporus Gravenhorst
	Illingworth, 1923a
	HYMENOPTERA: PTEROMALIDAE
	Nasonia vitripennis (Walker)
	Froggatt, 1919
	Johnston et al., 1921
	HYMENOPTERA: DIAPRIIDAE
	Hemilexomyia abrupta Dodd
	Dodd, 1920
	Johnston et al., 1921
	HYMENOPTERA: FORMICIDAE
	Pheidole megacephala Fabricius
	Illingworth, 1923a
	Ponera perkinsi Forel
	Illingworth, 1923a
Lasiops simplex (Wiedemann) (Synonym: *Alloeostylus simplex*)	SCHIZOPHYTA: SCHIZOMYCETES
	EUBACTERIALES: ENTEROBACTERIACEAE
	Escherichia coli (Migula) Castellani and Chalmers
	Povolný et al., 1961
	Serratia Bizio
	Povolný et al., 1961
	Proteus morganii (Winslow, Kligler, and Rothberg) Yale
	Povolný et al., 1961

Fly	Associated organism
Lasiops variabilis (Robineau-Desvoidy) (Synonym: *Hyetodesia variabilis*)	THALLOPHYTA: PHYCOMYCETES ENTOMOPHTHORALES: ENTOMOPHTHORACEAE *Entomophthora muscae* Cohn Rostrup, 1916
Dendrophaonia querceti (Bouché)	ARTHROPODA: ARACHNIDA ACARINA: MACROCHELIDAE *Macrocheles muscaedomesticae* (Scopoli) Sychevskaia͡, 1964a ARTHROPODA: INSECTA HYMENOPTERA: ICHNEUMONIDAE *Stilpnus* Gravenhorst Sychevskaia͡, 1964b HYMENOPTERA: CHALCIDIDAE *Brachymeria minuta* (Fabricius) Sychevskaia͡, 1966
Phaonia Robineau-Desvoidy (Synonym: *Hyetodesia*) (prey)	ARTHROPODA: INSECTA HYMENOPTERA: SPHECIDAE *Crabro leucostoma* (Linnaeus) Hamm et al., 1926
Phaonia basilis (Zetterstedt) (prey)	ARTHROPODA: INSECTA HYMENOPTERA: SPHECIDAE *Ectemnius quadricinctus* (Fabricius) Hamm et al., 1926
Phaonia cincta (Zetterstedt) (predator)	ARTHROPODA: INSECTA DIPTERA: UNDETERMINED CERATOPOGONIDAE Keilin, 1917 DIPTERA: ANISOPODIDAE *Mycetobia pallipes* Meigen Keilin, 1917 SYRPHIDAE: UNDETERMINED ERISTALINI Keilin, 1917 DIPTERA: AULACIGASTRIDAE *Aulacigaster leucopeza* (Meigen) Keilin, 1917
Phaonia corbetti Malloch (pupa is host for parasite)	ARTHROPODA: INSECTA HYMENOPTERA: PTEROMALIDAE *Spalangia nigroaenea* Curtis Bouček, 1963
Phaonia errans (Meigen) (prey)	ARTHROPODA: INSECTA HYMENOPTERA: SPHECIDAE *Ectemnius quadricinctus* (Fabricius) Hamm et al., 1926

Fly	Associated organism
Phaonia erratica (Fallén) (prey)	ARTHROPODA: INSECTA HYMENOPTERA: SPHECIDAE *Ectemnius quadricinctus* (Fabricius) Hamm et al., 1926
Phaonia exoleta Meigen (Synonym: *Phaonia mirabilis*) (predator)	ARTHROPODA: INSECTA DIPTERA: CULICIDAE *Aedes geniculatus* Olivier Tate, 1935 *Culex pipiens* Linnaeus Tate, 1935
Phaonia goberti (Mik) (predator)	ARTHROPODA: INSECTA DIPTERA: CLUSIIDAE *Clusiodes albimanus* (Meigen) Keilin, 1917
Phaonia incana (Wiedemann)	SCHIZOPHYTA: SCHIZOMYCETES EUBACTERIALES: ENTEROBACTERIACEAE *Escherichia coli* (Migula) Castellani and Chalmers Hoffmann, 1950
Phaonia querceti (Bouché) (Synonym: *Mydaea platyptera*)	SCHIZOPHYTA: SCHIZOMYCETES EUBACTERIALES: ENTEROBACTERIACEAE *Escherichia anindolica* (Chester) Bergey et al. Zmeev, 1943 *Escherichia coli* (Migula) Castellani and Chalmers Zmeev, 1943 *Escherichia coli* var. *acidilactici* (Topley and Wilson) Yale Zmeev, 1943 *Escherichia coli* var. *acidilactici* (Topley and Wilson) Yale= [*Escherichia paraacidilactici* A] Zmeev, 1943 *Escherichia coli* var. *acidilactici* (Topley and Wilson) Yale= [*Escherichia paraacidilactici* B] Zmeev, 1943 *Cloaca* Castellani and Chalmers Zmeev, 1943 EUBACTERIALES: UNDETERMINED ENTEROBACTERIACEAE *Paraescherichia*, unvalidated Zmeev, 1943

Fly	Associated organism
Phaonia querceti (cont.)	ARTHROPODA: INSECTA HYMENOPTERA: PTEROMALIDAE *Spalangia*[1] *cameroni* Perkins Bouček, 1963 *Spalangia ?endius* Walker Bouček, 1963 *Spalangia nigroaenea* Curtis Bouček, 1963 HYMENOPTERA: SPHECIDAE *Crabro signatus* Panzer Chevalier, 1923b, 1926 *Ectemnius quadricinctus* (Fabricius) Hamm et al., 1926 *Oxybelus uniglumis* (Linnaeus) Chevalier, 1926 CHORDATA: AVES PASSERIFORMES: UNDETERMINED PARIDAE Séguy, 1929
Phaonia scutellaris (Fallén)	ARTHROPODA: INSECTA HYMENOPTERA: SPHECIDAE *Ectemnius quadricinctus* (Fabricius) Hamm et al., 1926
Phaonia signata (Meigen)	ARTHROPODA: INSECTA COLEOPTERA: CARABIDAE *Carabus italicus* Dejean Giglio-Tos, 1892 HYMENOPTERA: SPHECIDAE *Ectemnius quadricinctus* (Fabricius) Hamm et al., 1926
Phaonia trimaculata (Bouché) (predator?)	ARTHROPODA: INSECTA DIPTERA: ANTHOMYIIDAE *Hylemya brassicae* (Bouché) Bouché, 1834
Phaonia vagans (Fallén) (prey)	ARTHROPODA: INSECTA HYMENOPTERA: SPHECIDAE *Ectemnius chrysostoma* Lepeletier and Brulle Hamm et al., 1926
Phaonia variegata (Meigen)	ARTHROPODA: INSECTA HYMENOPTERA: SPHECIDAE *Ectemnius quadricinctus* (Fabricius) Hamm et al., 1926

[1] Although not specifically mentioned in each entry, it is assumed that all *Spalangia* are pupal parasites.

Fly	Associated organisms
Phaonia variegata (cont.)	DIPTERA: TRICHOCERIDAE *Trichocera* Meigen Thomson, 1937 HYMENOPTERA: MYCETOPHILIDAE *Mycetophila ornata* Stephens Thomson, 1937 SYRPHIDAE: UNDETERMINED MILESIINAE Keilin, 1917
Muscina Robineau-Desvoidy	ENTEROVIRUSES Poliovirus Trask et al., 1943 Poliovirus Types 1, 3 Riordan et al., 1961 Coxsackievirus Type A9 (A9), B5 (B5) Riordan et al., 1961 ECHO virus Types 1, 5, 11 Riordan et al., 1961 SCHIZOPHYTA: SCHIZOMYCETES EUBACTERIALES: ENTEROBACTERIACEAE *Shigella boydii* Ewing 2, 5, 6 Richards et al., 1961 *Shigella dysenteriae* (Shiga) Castellani and Chalmers 2 Richards et al., 1961 *Shigella flexneri* Castellani and Chalmers 1a, 2a, 3, 4a, 4b, 6 Richards et al., 1961 *Shigella sonnei* (Levine) Weldin 1 Richards et al., 1961 ARTHROPODA: INSECTA HYMENOPTERA: SPHECIDAE *Crabro clypeatus* (Schreber) Hamm et al., 1926 CHORDATA: AVES GALLIFORMES: PHASIANIDAE *Gallus gallus* Linnaeus Rodriguez et al., 1962b
Muscina assimilis (Fallén)	ENTEROVIRUSES Poliovirus Francis, Jr. et al., 1948 PROTOZOA: CNIDOSPORIDIA MICROSPORIDA: NOSEMATIDAE *Thelohania thomsoni* Kramer Kramer, 1961b

Fly	Organism
Muscina assimilis (cont.)	ARTHROPODA: INSECTA
	HYMENOPTERA: SPHECIDAE
	Crabro advenus Smith
	*Kurczewski et al., 1968
	LEPIDOPTERA: NOCTUIDAE
	Macronoctua onusta Grote
	Breakey, 1929
Muscina pabulorum (Fallén)	ARTHROPODA: INSECTA
	LEPIDOPTERA: LYMANTRIIDAE
	Lymantria monacha Linnaeus
	Ratzeburg, 1844
	LEPIDOPTERA: LASIOCAMPIDAE
	Dendrolimus pini Linnaeus
	Ratzeburg, 1844
	Sitowski, 1928
	HYMENOPTERA: SPHECIDAE
	Ectemnius quadricinctus (Fabricius)
	Hamm et al., 1926
	HYMENOPTERA: APIDAE
	Bombus agrorum Fabricius
	Verhoeff, 1891
	DIPTERA: SCIARIDAE
	Sciara militaris Nowicki
	Beling, 1868
	Porchinskiĭ, 1913
Muscina pascuorum (Meigen)	THALLOPHYTA: BASIDIOMYCETES
	AGARICALES: AGARICACEAE
	Agaricus campestris Linnaeus
	von Bremi, 1846
Muscina stabulans (Fallén)	VIRUSES
	ENTEROVIRUSES
	Poliovirus
	Paul et al., 1941
	Sabin et al., 1942
	Melnick et al., 1945
	Francis, Jr. et al., 1948
	Poliovirus Types 1, 3
	Paul et al., 1962
	Poliovirus Type 2, Lansing strain
	Bang et al., 1943
	Theiler's mouse encephalomyelitis virus, GD VII strain
	Bang et al., 1943
	ECHO virus Types 8, 12, 14
	Paul et al., 1962
	OTHER ANIMAL VIRUSES
	Foot and mouth disease virus
	Lebailly, 1924

* Authors: *Musca* [*sic*] *assimilis* (Fallén).

Muscina stabulans (cont.)	SCHIZOPHYTA: SCHIZOMYCETES
	PSEUDOMONADALES:
	PSEUDOMONADACEAE
	Pseudomonas aeruginosa (Schroeter) Migula
	Havlik, 1964
	EUBACTERIALES:
	ACHROMOBACTERACEAE
	Alcaligenes faecalis Castellani and Chalmers
	Gregor et al., 1960
	Havlik, 1964
	Flavobacterium devorans (Zimmermann) Bergey et al.
	Gregor et al., 1960
	Flavobacterium invisible (Vaughan) Bergey et al.
	Gregor et al., 1960
	EUBACTERIALES:
	ENTEROBACTERIACEAE
	Escherichia coli (Migula) Castellani and Chalmers
	Beyer, 1925
	Zmeev, 1943
	Gregor et al., 1960
	Povolný et al., 1961
	Havlik, 1964
	Escherichia freundii (Braak) Yale
	O'Keefe et al., 1954
	Aerobacter Beijerinck
	Zmeev, 1943
	Aerobacter aerogenes (Kruse) Beijerinck
	Gregor et al., 1960
	Havlik, 1964
	Klebsiella cloacae (Jordan) Brisou
	Gregor et al., 1960
	Cloaca Castellani and Chalmers
	Zmeev, 1943
	Paracolobactrum aerogenoides Borman, Stuart, and Wheeler
	O'Keefe et al., 1954
	Serratia Bizio
	Povolný et al., 1961
	Serratia marcescens Bizio
	Gregor et al., 1960
	Havlik, 1964
	Proteus inconstans (Ornstein) Shaw and Clarke
	Gregor et al., 1960

Muscina stabulans (cont.)	*Proteus mirabilis* Hauser Gregor et al., 1960 *Proteus morganii* (Winslow, Kligler, and Rothberg) Yale Gregor et al., 1960 Povolný et al., 1961 Havlik, 1964 *Proteus rettgeri* (Hadley et al.) Rustigian and Stuart Gregor et al., 1960 *Proteus vulgaris* Hauser O'Keefe et al., 1954 Gregor et al., 1960 Povolný et al., 1961 Havlik, 1964 *Salmonella typhi* Warren and Scott Anonymous, 1952 *Citrobacter freundii* Werkman and Gillen Gregor et al., 1960 *Shigella* Castellani and Chalmers Zaidenov, 1961 ?*Shigella* Castellani and Chalmers Sychevskaīa et al., 1959a *Shigella flexneri* Castellani and Chalmers Sychevskaīa et al., 1959a Arskiĭ et al., 1961 *Shigella flexneri* Castellani and Chalmers 1a Sychevskaīa et al., 1959b *Shigella sonnei* (Levine) Weldin Arskiĭ et al., 1961 EUBACTERIALES: BRUCELLACEAE *Brucella abortus* (Schmidt) Meyer and Shaw Ruhland et al., 1941 EUBACTERIALES: MICROCOCCACEAE *Staphylococcus afermentans* Shaw, Stitt, and Cowan Gregor et al., 1960 *Staphylococcus aureus* Rosenbach Gregor et al., 1960 *Staphylococcus lactis* Shaw, Stitt, and Cowan Gregor et al., 1960 *Staphylococcus saprophyticus* Fairbrother Gregor et al., 1960

Muscina stabulans (cont.)

EUBACTERIALES: NEISSERIACEAE
Neisseria catarrhalis (Froschand and Kolle) Holland
Beyer, 1925

EUBACTERIALES: LACTOBACILLACEAE
Streptococcus Rosenbach
Povolný et al., 1961
Streptococcus durans Sherman and Wing
Gregor et al., 1960

EUBACTERIALES: BACILLACEAE
Bacillus Cohn
Povolný et al., 1961
Bacillus cereus var. *mycoides* (Flügge) Smith et al.
Gregor et al., 1960

ACTINOMYCETALES: MYCOBACTERIACEAE
Mycobacterium ?leprae (Hansen) Lehman and Neumann
Honeij et al., 1914

THALLOPHYTA: PHYCOMYCETES
ENTOMOPHTHORALES: ENTOMOPHTHORACEAE
Entomophthora muscae Cohn
Judd, 1955

PROTOZOA: MASTIGOPHORA
PROTOMONADIDA: TRYPANOSOMATIDAE
Herpetomonas muscarum (Leidy) Kent
Becker, 1922-23

PROTOMONADIDA: BODONIDAE
Rhynchoidomonas Patton
Franchini, 1922

PROTOZOA: SPOROZOA
HAPLOSPORIDA: UNKNOWN TAXONOMIC POSITION AND FAMILY
Toxoplasma Nicolle and Manceaux
Kåss, 1954

PLATYHELMINTHES: CESTOIDEA
CYCLOPHYLLIDEA: HYMENOLEPIDIDAE
Hymenolepis nana (V. Siebold) Blanchard
Bogoi͡avlenskiĭ et al., 1928
Sychevskai͡a et al., 1959b

Muscina stabulans (cont.)

CYCLOPHYLLIDEA: TAENIIDAE
Taeniarhynchus saginatum (Goeze)
Bogoi͡avlenskiĭ et al., 1928
Sychevskai͡a et al., 1958

ASCHELMINTHES: NEMATODA
TRICHURIDEA: TRICHURIDAE
Trichuris trichiura (Linnaeus)
Bogoi͡avlenskiĭ et al., 1928

STRONGYLIDEA: ANCYLOSTOMATIDAE
Ancylostoma duodenale (Dubini) Creplin
Bogoi͡avlenskiĭ et al., 1928

STRONGYLIDEA: TRICHOSTRONGYLIDAE
Trichostrongylus colubriformis (Giles)
Bogoi͡avlenskiĭ et al., 1928

OXYURIDEA: OXYURIDAE
Enterobius vermicularis (Linnaeus) Leach
Bogoi͡avlenskiĭ et al., 1928

ASCARIDIDEA: ASCARIDIDAE
Ascaris lumbricoides Linnaeus
Bogoi͡avlenskiĭ et al., 1928

SPIRURIDEA: SPIRURIDAE
Habronema megastoma (Rudolphi) Seurat
Roubaud et al., 1922b
Habronema muscae (Carter)
Roubaud et al., 1922b

ARTHROPODA: ARACHNIDA
ACARINA: MACROCHELIDAE
Macrocheles glaber (Müller)
Sychevskai͡a, 1964a
Macrocheles muscaedomesticae (Scopoli)
Anderson, 1964
Sychevskai͡a, 1964a

ACARINA: GAMASOLAELAPTIDAE
Saintdidieria sexclavatus (Oudemans)
Sychevskai͡a, 1964a

ACARINA: LAELAPTIDAE
Holostaspis Kolenati
Graham-Smith, 1916

ACARINA: UROPODIDAE
Fuscuropoda Vitzthum
Anderson, 1964

Muscina stabulans (cont.)

ACARINA: ANOETIDAE

Myianoetus Oudemans
Sychevskaîa, 1964a

Myianoetus muscarum (Linnaeus)
Berlese, 1881
Greenberg et al., 1960
Greenberg, 1961

ARTHROPODA: INSECTA

ORTHOPTERA: LOCUSTIDAE

Chortophaga viridifasciata (De Geer)
Knutson, 1941

Encoptolophus sordidus (Burmeister) subsp. *costalis*
Knutson, 1941

Nomadacris septemfasciata Serville
Jack, 1935

Schistocerca gregaria Fòrskal
Nocedo, 1921
Régnier, 1931

Xanthippus corallipes (Haldeman) subsp. *patherinis*
Knutson, 1941

DERMAPTERA: FORFICULIDAE

Forficula auricularia Linnaeus
Anderson, 1964

LEPIDOPTERA: OLETHREUTIDAE

Epiblema otiosana Clemens
Decker, 1932

Suleima helianthana Riley
Satterthwait, 1943

LEPIDOPTERA: NOCTUIDAE

Agrotis castanea var. *neglecta* Hubner
von Gercke, 1882

Agrotis segetum (Denis and Schiffermüller)
Herold, 1923

Archanara oblogna Grote
Cole, 1930

Heliothis obsoleta Fabricius
Rodionov, 1927

Laphygma frugiperda (Smith)
Smith, 1921

Leucania unipuncta Hawarth
Satterthwait, 1943

Macronoctua onusta Grote
Breakey, 1929

Papaipema nebris Guenée
Decker, 1931

Muscina stabulans (cont.)

Peridroma margitosa form *saucia* Hubner
Fletcher, 1900

LEPIDOPTERA: LIPARIDAE

Liparis dispar Linnaeus
Levitt, 1935

LEPIDOPTERA: SPHINGIDAE

Acherontia atropos Linnaeus
Keilin, 1917

LEPIDOPTERA: LASIOCAMPIDAE

Dendrolimus pini Linnaeus
Keilin, 1917
Sitowski, 1928

Malacosoma americana Fabricius
Curran, 1942

COLEOPTERA: SCARABAEIDAE

Melolontha melolontha Linnaeus
Kamner, 1928

COLEOPTERA: CERAMBYCIDAE

Dorysthenes forficatus Fabricius
Bouhélier et al., 1936

COLEOPTERA: CHRYSOMELIDAE

Gallerucella luteola Müller
Curran, 1942

COLEOPTERA: CURCULIONIDAE

Pissodes strobi Peck
MacAloney, 1930

Rhodobaenus tredecimpunctatus (Illiger)
Satterthwait, 1943

HYMENOPTERA: DIPRIONIDAE

Diprion pallidus Klug
van der Wulp, 1869

Diprion pini (Linnaeus)
Keilin, 1917

HYMENOPTERA: CHALCIDIDAE

Brachymeria minuta (Fabricius)
Sychevskaîa, 1966

HYMENOPTERA: PTEROMALIDAE

Dibrachys cavus (Walker)
Graham-Smith, 1919

Nasonia vitripennis (Walker)
Séguy, 1929

Spalangia endius Walker
Legner (pre-publication), 1966

Spalangia rugulosa Foerster
Bouček, 1963

Stenomolina muscarum (Linnaeus)
Ratzeburg, 1844

Fly	Associated organism
Muscina stabulans (cont.)	HYMENOPTERA: SPHECIDAE *Ectemnius quadricinctus* (Fabricius) Hamm et al., 1926 DIPTERA: MUSCIDAE *Hydrotaea dentipes* (Fabricius) Porchinskiĭ, 1913 *Ophyra leucostoma* (Wiedemann) Anderson, 1964 *Polietes albolineata* (Fallén) Porchinskiĭ, 1913 CHORDATA: AVES PASSERIFORMES: HIRUNDINIDAE *Delichon urbica* (Linnaeus) subsp. *urbica* Séguy, 1929 *Hirundo rustica* Linnaeus subsp. *rustica* Séguy, 1929 PASSERIFORMES: STURNIDAE *Sturnus vulgaris* Linnaeus Séguy, 1929
Subfamily: MUSCINAE	
Dasyphora asiatica Zimin	PLATYHELMINTHES: CESTOIDEA CYCLOPHYLLIDEA: TAENIIDAE *Taeniarhynchus saginatum* (Goeze) Sychevskaîâ et al., 1958
Dasyphora cyanella (Meigen)	ARTHROPODA: INSECTA DIPTERA: MUSCIDAE *Myospila meditabunda* (Fabricius) Thomson, 1937 *Mydaea urbana* (Meigen) Thomson, 1937 *Mesembrina meridiana* Linnaeus Thomson, 1937
Dasyphora gussakovskii Zimin	ARTHROPODA: ARACHNIDA ACARINA: UNDETERMINED PARASITIDAE Sychevskaîâ, 1964a
Dasyphora pratorum (Meigen)	PROTOZOA: MASTIGOPHORA PROTOMONADIDA: TRYPANOSOMATIDAE *Crithidia lesnei* (Léger) Léger, 1903
Polietes albolineata (Fallén)	ARTHROPODA: INSECTA DIPTERA: SPHAEROCERIDAE *Borborus* Meigen Porchinskiĭ, 1910

Fly	Associated organism
Polietes lardaria Fabricius	ARTHROPODA: INSECTA HYMENOPTERA: SPHECIDAE *Ectemnius quadricinctus* (Fabricius) Hamm et al., 1926
Mesembrina meridiana (Linnaeus)	ARTHROPODA: INSECTA HYMENOPTERA: UNDETERMINED ICHNEUMONIDAE Hammer, 1942 HYMENOPTERA: SPHECIDAE *Ectemnius quadricinctus* (Fabricius) Hamm et al., 1926
Graphomya maculata (Scopoli) (Synonym: *Graphomyia maculata*)	ENTEROVIRUSES Poliovirus Types 1 and 3 Paul et al., 1962 ECHO virus Types 8, 12, 14 Paul et al., 1962 PROTOZOA: MASTIGOPHORA PROTOMONADIDA: TRYPANOSOMATIDAE *Leptomonas mirabilis* Roubaud Franchini, 1922 ARTHROPODA: INSECTA DIPTERA: UNDETERMINED TIPULIDAE Keilin, 1917 DIPTERA: PTYCHOPTERIDAE *Ptychoptera contaminata* (Linnaeus) Keilin, 1917 DIPTERA: UNDETERMINED TABANIDAE Keilin, 1917 DIPTERA: SYRPHIDAE *Eristalis* Latreille Keilin, 1917 UNDETERMINED SCATOPHAGINAE Keilin, 1917
Synthesiomyia nudiseta (Wulp) (Synonyms: *Synthesiomyia brasiliensis, Synthesiomyia brasiliana*)	ENTEROVIRUSES Poliovirus Types 1 and 3 Paul et al., 1962 ECHO virus Types 8, 12, 14 Paul et al., 1962 SCHIZOPHYTA: SCHIZOMYCETES EUBACTERIALES: ENTEROBACTERIACEAE *Escherichia coli* (Migula) Castellani and Chalmers Beyer, 1925

Fly	Associated organism
Synthesiomyia nudiseta (cont.)	*Shigella dysenteriae* (Shiga) Castellani and Chalmers Beyer, 1925 ARTHROPODA: INSECTA HYMENOPTERA: BRACONIDAE *Alysia ridibunda* Say Roberts, 1935 HYMENOPTERA: CYNIPIDAE *Eucoila* Westwood Roberts, 1935 HYMENOPTERA: CHALCIDIDAE *Brachymeria fonscolombei* (Dufour) Roberts, 1933a, 1933b, 1935 HYMENOPTERA: PTEROMALIDAE *Nasonia vitripennis* (Walker) Roberts, 1933a DIPTERA: CUTEREBRIDAE *Dermatobia hominis* (Linnaeus) Lutz, 1917
Neomuscina Townsend	ENTEROVIRUSES Poliovirus Types 1 and 3 Paul et al., 1962 ECHO virus Types 8, 12, 14 Paul et al., 1962
Philornis Meinert	ENTEROVIRUSES Poliovirus Types 1 and 3 Paul et al., 1962 ECHO virus Types 8, 12, 14 Paul et al., 1962
Morellia Robineau-Desvoidy	ARTHROPODA: INSECTA HYMENOPTERA: SPHECIDAE *Ectemnius quadricinctus* (Fabricius) Hamm et al., 1926
Morellia hortorum (Fallén)	ASCHELMINTHES: NEMATODA TYLENCHIDA: CONTORTYLENCHIDAE *Heterotylenchus* Stoffolano and Nickle Stoffolano, 1969 ARTHROPODA: INSECTA HYMENOPTERA: UNDETERMINED ICHNEUMONIDAE Hammer, 1942 HYMENOPTERA: SPHECIDAE *Crabro cribraria* (Linnaeus) Hamm et al., 1926

Fly	Associated organism
Morellia micans (Macquart)	PROTOZOA: MASTIGOPHORA PROTOMONADIDA: TRYPANOSOMATIDAE *Herpetomonas muscarum* (Leidy) Kent Becker, 1924
Morellia simplex (Loew)	ASCHELMINTHES: NEMATODA TYLENCHIDA: CONTORTYLENCHIDAE *Heterotylenchus autumnalis* Nickle Világiová, 1968
Pyrellia Robineau-Desvoidy	PROTOZOA: MASTIGOPHORA PROTOMONADIDA: TRYPANOSOMATIDAE *Leptomonas* Kent Roubaud, 1912a
Orthellia Robineau-Desvoidy (Synonyms: *Pseudopyrellia*, *Cryptolucilia*)	ASCHELMINTHES: NEMATODA SPIRURIDEA: SPIRURIDAE *Habronema megastoma* (Rudolphi) Seurat Johnston, 1920 Johnston et al., 1920a Neveu-Lemaire, 1936 *Habronema microstoma* (Schneider) Johnston, 1920 *Habronema muscae* (Carter) Johnston, 1920 Johnston et al., 1920a Neveu-Lemaire, 1936 ARTHROPODA: INSECTA COLEOPTERA: STAPHYLINIDAE *Baryodma bimaculata* Gravenhorst Lindquist, 1936 HYMENOPTERA: PTEROMALIDAE *Spalangia drosophilae* Ashmead Lindquist, 1936 *Spalangia endius* Walker Bouček, 1963 *Spalangia stomoxysiae* Girault Lindquist, 1936 Peck, 1951
Orthellia caerulea Wiedemann	SCHIZOPHYTA: SCHIZOMYCETES EUBACTERIALES: ENTEROBACTERIACEAE *Escherichia coli* (Migula) Castellani and Chalmers Shimizu et al., 1965

Orthellia caerulea (cont.)

Proteus morganii (Winslow, Kligler, and Rothberg) Yale
Shimizu et al., 1965
Proteus vulgaris Hauser
Shimizu et al., 1965

Orthellia caesarion (Meigen)
(Synonyms: *Cryptolucilia caesarion, Pseudopyrellia cornicina, Musca cornicina, Orthellia cornicina, Cryptolucilia cornicina*)

ASCHELMINTHES: NEMATODA
TYLENCHIDA: CONTORTYLENCHIDAE
Heterotylenchus autumnalis Nickle
Stoffolano, 1970

ARTHROPODA: INSECTA
COLEOPTERA: HYDROPHILIDAE
Sphaeridium scarabaeoides (Linnaeus)
Hammer, 1942

COLEOPTERA: STAPHYLINIDAE
Ontholestes tessellatus Fourcroy
Hammer, 1942
Philonthus Curtis
Hammer, 1942

HYMENOPTERA: UNDETERMINED ICHNEUMONIDAE
Hammer, 1942

HYMENOPTERA: BRACONIDAE
Aphaereta pallipes Say
Blickle, 1961
Benson et al., 1963
Sanders et al., 1966
Houser et al., 1967

HYMENOPTERA: CYNIPIDAE
Cothonaspis nigricornis (Kieffer)
Sychevskaîa, 1964b
Eucoila Westwood
Blickle, 1961
Sanders et al., 1966

HYMENOPTERA: FIGITIDAE
Xyalophora quinquelineata (Say)
Blickle, 1961

HYMENOPTERA: PTEROMALIDAE
Spalangia muscidarum Richardson
Pinkus, 1913

HYMENOPTERA: SPHECIDAE
Crabro venator (Rohwer)
Kurczewski et al., 1968
Bembix spinolae (Lepeletier)
Parker, 1917
Mellinus Fabricius
Hammer, 1942

HYMENOPTERA: VESPIDAE
Vespa Linnaeus
Hammer, 1942

Fly	Associated organisms
Orthellia caesarion (cont.)	DIPTERA: ANTHOMYIIDAE *Scatophaga stercoraria* (Linnaeus) Hewitt, 1914 DIPTERA: MUSCIDAE *Hebecnema umbratica* (Meigen) Thomson, 1937 *Myospila meditabunda* (Fabricius) Porchinskiĭ, 1910 Thomson, 1937 *Mydaea urbana* (Meigen) Thomson, 1937 *Polietes albolineata* (Fallén) Porchinskiĭ, 1910 *Polietes hirticrura* Meade Thomson, 1937 *Mesembrina meridiana* Linnaeus Thomson, 1937
Orthellia cyanea (Fabricius) (Synonym: *Lasiopyrellia cyanea*)	ARTHROPODA: INSECTA HYMENOPTERA: CYNIPIDAE *Bothrochacis stercoraria* Bridwell Bridwell, 1919
Sarcopromusca arcuata Townsend (Synonym: *Sarcophaga terminalis*)	ARTHROPODA: INSECTA DIPTERA: CUTEREBRIDAE *Dermatobia hominis* (Linnaeus) Lins de Almeida, 1933
Musca Linnaeus (Synonym: *Philaematomyia*)	ENTEROVIRUSES Poliovirus Trask et al., 1943 Poliovirus Types 1 and 3 Paul et al., 1962 ECHO virus Types 8, 12, 14 Paul et al., 1962 SCHIZOPHYTA: SCHIZOMYCETES PSEUDOMONADALES: PSEUDOMONADACEAE *Pseudomonas* Migula de la Paz, 1938 PSEUDOMONADALES: SPIRILLACEAE *Vibrio* Müller de la Paz, 1938 *Vibrio comma* (Schroeter) Winslow et al. de la Paz, 1938 Lal et al., 1939 EUBACTERIALES: ACHROMOBACTERACEAE *Alcaligenes* Castellani and Chalmers de la Paz, 1938

Musca (cont.)

EUBACTERIALES:
ENTEROBACTERIACEAE
Escherichia Castellani and Chalmers
de la Paz, 1938
Aerobacter Beijerinck
de la Paz, 1938
Proteus Hauser
de la Paz, 1938
Proteus morganii (Winslow, Kligler, and Rothberg) Yale
de la Paz, 1938
Shimizu et al., 1965
Proteus rettgeri (Hadley et al.) Rustigian and Stuart
de la Paz, 1938
Shimizu et al., 1965
Providencia Kauffmann
Shimizu et al., 1965
Salmonella cholerae-suis (Smith) Weldin
de la Paz, 1938
Salmonella columbensis (Castellani) Castellani and Chalmers
de la Paz, 1938
Salmonella gallinarum (Klein) Bergey
da la Paz, 1938
Salmonella typhi Warren and Scott
de la Paz, 1938
Eberthella belfastiensis (Weldin and Levine) Bergey et al.
de la Paz, 1938
Eberthella dubia (Chester) Bergey et al.
de la Paz, 1938
Eberthella kandiensis (Castellani) Bergey et al.
de la Paz, 1938
Eberthella talavensis (Castellani) Bergey et al.
de la Paz, 1938
Shigella ambigua (Andrewes) Weldin
de la Paz, 1938
Shigella dysenteriae (Shiga) Castellani and Chalmers
de la Paz, 1938

Musca (cont.)

Shigella flexneri Castellani and Chalmers
de la Paz, 1938
Shigella giumai (Castellani) Hauduroy et al.
de la Paz, 1938
Shigella madampensis (Castellani) Weldin
de la Paz, 1938
Shigella pfaffi (Hadley et al.) Weldin
de la Paz, 1938
Shigella sonnei (Levine) Weldin
de la Paz, 1938
EUBACTERIALES: MICROCOCCACEAE
Staphylococcus Rosenbach
de la Paz, 1938
EUBACTERIALES: LACTOBACILLACEAE
Streptococcus Rosenbach
de la Paz, 1938
ACTINOMYCETALES: MYCOBACTERIACEAE
Mycobacterium ?leprae (Hansen) Lehmann and Neumann
Arizumi, 1934
SCHIZOPHYTA: MICROTATOBIOTES
RICKETTSIALES: CHLAMYDIACEAE
Colesiota conjunctivae (Coles) Rake
Mitscherlich, 1941
THALLOPHYTA: PHYCOMYCETES
ENTOMOPHTHORALES: ENTOMOPHTHORACEAE
Entomophthora sphaerosperma Fresenius
Thaxter, 1888
PROTOZOA: MASTIGOPHORA
POLYMASTIGIDA: CHILOMASTIGIDAE
Chilomastix mesnili (Wenyon)
Harris et al., 1946
POLYMASTIGIDA: HEXAMITIDAE
Giardia intestinalis (Lambl)
Wenyon et al., 1917a
Giardia lamblia Stiles
Harris et al., 1946
TRICHOMONADIDA: TRICHOMONADIDAE
Trichomonas hominis (Davaine)
Harris et al., 1946

Musca (cont.)

PROTOZOA: SARCODINA
AMOEBIDA: ENDAMOEBIDAE
Endolimax nana (Wenyon and O'Connor)
Harris et al., 1946
Entamoeba coli (Grassi)
Wenyon et al., 1917a
Harris et al., 1946
Entamoeba histolytica Schaudinn
Wenyon et al., 1917a
Harris et al., 1946
Iodamoeba bütschlii (Prowazek)
Harris et al., 1946

PLATYHELMINTHES: CESTOIDEA
CYCLOPHYLLIDEA: HYMENOLEPIDIDAE
Hymenolepis diminuta (Rudolphi) Blanchard
Harris et al., 1946

ASCHELMINTHES: NEMATODA
TRICHURIDEA: TRICHURIDAE
Trichuris Roederer
Tao, 1936
Trichuris trichiura (Linnaeus)
Harris et al., 1946
STRONGYLIDEA: ANCYLOSTOMATIDAE
Ancylostoma (Dubini)
Tao, 1936
Harris et al., 1946
ASCARIDIDEA: ASCARIDIDAE
Ascaris Linnaeus
Tao, 1936
Ascaris lumbricoides Linnaeus
Harris et al., 1946

ARTHROPODA: INSECTA
HEMIPTERA: PHYMATIDAE
Phymata pennsylvanica americana Melin
Balduf, 1947
HEMIPTERA: REDUVIIDAE
Sinea diadema (Fabricius)
Balduf, 1947
HYMENOPTERA: CHALCIDIDAE
Brachymeria fonscolombiae (Dufour)
Masi, 1916
Stenomalus muscarum Linnaeus
Waterston, 1916

Fly	Associated organisms
Musca (cont.)	HYMENOPTERA: PTEROMALIDAE
	Nasonia vitripennis (Walker)
	Johnston et al., 1921
	Parker et al., 1928
	Spalangia cameroni Perkins
	Bouček, 1963
	Spalangia muscidarum Richardson
	Johnston et al., 1921
	HYMENOPTERA: SPHECIDAE
	Bembix Fabricius
	Peckham et al., 1905
	Bembix lunata Fabricius
	Ramakrishna Ayyar, 1920
	Crabro vagus (Linnaeus)
	Hamm et al., 1926
	Crossocerus lentus (Fox)
	Peckham et al., 1905
	DIPTERA: CUTEREBRIDAE
	Dermatobia hominis (Linnaeus)
	Townsend, 1922
Musca amita Hennig (Synonym: *Musca amica*)	ASCHELMINTHES: NEMATODA
	SPIRURIDEA: THELAZIIDAE
	Thelazia gulosa Railliet and Henry
	Krastin, 1950, 1952
	Thelazia skrjabini Erschov
	Krastin, 1950, 1952
	ARTHROPODA: INSECTA
	HYMENOPTERA: SPHECIDAE
	Oxybelus Latreille
	Pridantseva, 1959
Musca autumnalis De Geer	SCHIZOPHYTA: SCHIZOMYCETES
	EUBACTERIALES: BRUCELLACEAE
	Moraxella bovis (Hauduroy et al.) Murray
	Brown, 1965
	Steve et al., 1965
	EUBACTERIALES: BACILLACEAE
	Bacillus thuringiensis Berliner
	Yendol et al., 1967
	ASCHELMINTHES: NEMATODA
	SPIRURIDEA: THELAZIIDAE
	Thelazia californiensis Price
	Sabrosky, 1959
	Thelazia gulosa Railliet and Henry
	Világiová, 1962
	Thelazia rhodesii (Desmarest)
	Klesov, 1949
	Sabrosky, 1959
	Világiová, 1962

Musca autumnalis (cont.)

TYLENCHIDA: CONTORTYLENCHIDAE

Heterotylenchus Stoffolano and Nickle
Stoffolano et al., 1966

Heterotylenchus autumnalis Nickle
Jones et al., 1967
Nickle, 1967
Stoffolano, 1967, 1969, 1970
Treece et al., 1968
Világiová, 1968

ARTHROPODA: ARACHNIDA

ACARINA: MACROCHELIDAE

Macrocheles muscaedomesticae (Scopoli)
Singh et al., 1966

ACARINA: ANOETIDAE

Myianoetus Oudemans
Sychevskaia͡, 1964

ARTHROPODA: INSECTA

COLEOPTERA: STAPHYLINIDAE

Aleochara bimaculata (Gravenhorst)
Thomas et al., 1968

Aleochara tristis Gravenhorst
Drea, 1966
Jones, 1969

HYMENOPTERA: BRACONIDAE

Aphaereta pallipes Say
Blickle, 1961
Benson et al., 1963
Beard, 1964
Sanders et al., 1966
Houser et al., 1967
Thomas et al., 1968
Turner et al., 1968

HYMENOPTERA: CYNIPIDAE

Eucoila Westwood
Blickle, 1961

Eucoila impatiens (Say)
Thomas et al., 1968
Turner et al., 1968

HYMENOPTERA: FIGITIDAE

Xyalophora quinquelineata (Say)
Blickle, 1961

HYMENOPTERA: PTEROMALIDAE

Nasonia vitripennis (Walker)
Beard, 1964
Hair et al., 1965

Fly	Associated organism
Musca autumnalis (cont.)	*Muscidifurax raptor* Girault Thomas et al., 1968 Turner et al., 1968 *Spalangia nigra* Latreille Turner et al., 1968 HYMENOPTERA: SPHECIDAE *Crabro advenus* Smith Kurczewski et al., 1968 *Mellinus* Fabricius Hammer, 1942 HYMENOPTERA: VESPIDAE *Vespa* Linnaeus Hammer, 1942 DIPTERA: ASILIDAE *Lasiopogon cinctus* (Fabricius) Hobby, 1934b
Musca autumnalis autumnalis De Geer (Synonyms: *Musca ovipara, Musca corvina*)	ARTHROPODA: INSECTA HYMENOPTERA: SPHECIDAE *Ectemnius quadricinctus* (Fabricius) Hamm et al., 1926 DIPTERA: MUSCIDAE *Myospila meditabunda* (Fabricius) Porchinskiĭ, 1910
Musca bezzii Patton and Cragg	SCHIZOPHYTA: SCHIZOMYCETES ACTINOMYCETALES: MYCOBACTERIACEAE *Mycobacterium ?leprae* (Hansen) Lehmann and Neumann de Mello et al., 1926 PROTOZOA: MASTIGOPHORA PROTOMONADIDA: TRYPANOSOMATIDAE *Leptomonas craggi* (Patton) Wenyon Patton, 1921a
Musca conducens Walker	ASCHELMINTHES: NEMATODA FILARIIDEA: STEPHANOFILARIIDAE *Stephanofilaria assamensis* Pande Srivastava et al., 1963 Patnaik, 1965
Musca convexifrons Thomson	ASCHELMINTHES: NEMATODA SPIRURIDEA: THELAZIIDAE *Thelazia rhodesii* (Desmarest) Krastin, 1949a, 1949b

Fly	Associated organisms
Musca convexifrons (cont.)	ARTHROPODA: INSECTA HYMENOPTERA: SPHECIDAE *Sericophorus teliferopodus* Rayment Rayment, 1954 *Sericophorus viridis roddi* Rayment Rayment, 1954
Musca corvina Fabricius	THALLOPHYTA: PHYCOMYCETES ENTOMOPHTHORALES: ENTOMOPHTHORACEAE ?*Entomophthora* Fresenius Roubaud, 1911a *Entomophthora muscae* Cohn Graham-Smith, 1919 PROTOZOA: MASTIGOPHORA PROTOMONADIDA: TRYPANOSOMATIDAE *Herpetomonas muscarum* (Leidy) Kent Roubaud, 1909 ARTHROPODA: ARACHNIDA PSEUDOSCORPIONIDEA: CHERNETIDAE *Lamprochernes nodosus* (Schrank) Graham-Smith, 1916 ARTHROPODA: INSECTA HYMENOPTERA: SPHECIDAE *Crabro leucostoma* (Linnaeus) Hamm et al., 1926 *Crabro peltarius* (Schreber) Hamm et al., 1926 *Nysson* Latreille Froggatt, 1917 *Sericophorus sydneyi* Rayment Rayment, 1954 *Stizus* Latreille Froggatt, 1917 *Stizus turneri* Froggatt Froggatt, 1917 Rayment, 1954 DIPTERA: MUSCIDAE *Polietes albolineata* (Fallén) Porchinskiĭ, 1910
Musca crassirostris Stein [Synonym: *Philaematomyia insignis*]	PROTOZOA: MASTIGOPHORA PROTOMONADIDA: TRYPANOSOMATIDAE *Trypanosoma evansi* (Steel) Nieschulz, 1927

Musca domestica Linnaeus .... VIRUSES

PSITTACOSIS GROUP

Trachoma virus
- Nicolle et al., 1919
- Zardi, 1964

POXVIRUSES

Variola
- Terni, 1908b

Vaccinia
- Terni, 1908b
- Merk, 1910

ENTEROVIRUSES

Poliovirus
- Francis, 1914
- Paul et al., 1941
- Sabin et al., 1942
- Gordon, 1943
- Melnick et al., 1945, 1949
- Francis, Jr. et al., 1948
- Melnick, 1949

Poliovirus non-paralytic
- Melnick et al., 1949

Poliovirus Type 1
- Melnick et al., 1953
- Riordan et al., 1961
- Brygoo et al., 1962
- Downey, 1963

Poliovirus Attenuated Type 1, LSc
- Gudnadóttir, 1960

Poliovirus Texas '48 strain (Type 1)
- Melnick et al., 1952

Poliovirus Type 2, Lansing strain
- Bang et al., 1943
- Rendtorff et al., 1943
- Hurlbut, 1950

Poliovirus Type 3
- Riordan et al., 1961

Theiler's mouse encephalomyelitis virus, GD VII strain
- Bang et al., 1943

Coxsackievirus
- Melnick et al., 1954
- Duca et al., 1958

Coxsackievirus Alaska 5 (A10)
- Melnick et al., 1953

Coxsackievirus Connecticut 5 (B1)
- Melnick et al., 1953

Coxsackievirus Easton 14 (A5)
- Melnick et al., 1953

Musca domestica (cont.)

Coxsackievirus Nancy (B3)
Melnick et al., 1953
Coxsackievirus Texas 1 (A4)
Melnick et al., 1953
Coxsackievirus Texas 12 (new type)
Melnick et al., 1953
Coxsackievirus Texas 13 (B4)
Melnick et al., 1953
Coxsackievirus Texas '48 (new type)
Melnick et al., 1953
Coxsackievirus Type 2 (A2)
Melnick et al., 1953
Coxsackievirus Type 3 (A3)
Melnick et al., 1953
Coxsackievirus Type A9 (A9)
Riordan et al., 1961
Coxsackievirus Type B5 (B5)
Riordan et al., 1961
ECHO virus Types 1, 5, 11
Riordan et al., 1961
ECHO virus Types 6 and 7
Downey, 1963

OTHER ANIMAL VIRUSES

Danysz rat virus
Graham-Smith, 1910
Foot and mouth disease virus
Lebailly, 1924
Kunike, 1927b
Dhennin et al., 1961
Mink enteritis virus
Bouillant et al., 1965
Tortor bovis virus
Sen, 1925-26
Tortor suis virus
Mohler, 1920

INSECT VIRUSES

Borrelina bombycis virus
Smuidsinovicia, 1889

BACTERIOPHAGES

Bacteriophage (of *Staphylococcus muscae*)
Glaser, 1938
Bacteriophage (of *Streptococcus* C55)
Shope, 1927

Musca domestica (cont.)

SCHIZOPHYTA
PSEUDOMONADALES:
PSEUDOMONADACEAE
Pseudomonas Migula
Prado et al., 1955
Pseudomonas aeruginosa (Schroeter) Migula
Bacot, 1911a, 1911b
Graham-Smith, 1911b
Ledingham, 1911
Cox et al., 1912
Duncan, 1926
Ostrolenk et al., 1942a
Peppler, 1944
Vanni, 1946
Stephens, 1963
Havlik, 1964
Pseudomonas eisenbergii Migula
Cao, 1906
Pseudomonas fluorescens Migula
Cao, 1906
Lysenko et al., 1961
Pseudomonas ichthyosmia (Hammer) Bergey et al.
Hawley et al., 1951
Pseudomonas jaegeri Migula
Cao, 1906
PSEUDOMONADALES: SPIRILLACEAE
Vibrio Müller
Smuidsinovicia, 1889
Vibrio comma (Schroeter) Winslow et al.
Chantemesse et al., 1905
Ganon, 1908
Graham-Smith, 1910
Alessandrini, 1912
Duncan, 1926
Shope, 1927
Dishon, 1956
EUBACTERIALES:
ACHROMOBACTERACEAE
Alcaligenes Castellani and Chalmers
Prado et al., 1955
Alcaligenes faecalis Castellani and Chalmers
Hamilton, 1903
Testi, 1909
Lysenko et al., 1961
Havlik, 1964

Musca domestica (cont.)

Achromobacter Bergey, Harrison, Breed, Hammer, and Huntoon
Prado et al., 1955

Flavobacterium Bergey et al.
Lysenko et al., 1961

Flavobacterium devorans (Zimmermann) Bergey et al.
Lysenko, 1958

EUBACTERIALES: ENTEROBACTERIACEAE

Escherichia Castellani and Chalmers=[MacConkey's bacillus No. 1]
Nicoll, 1911a

Escherichia Castellani and Chalmers=[MacConkey's bacillus No. 33a]
Nicoll, 1911a

Escherichia Castellani and Chalmers=[MacConkey's bacillus No. 66]
Nicoll, 1911a

Escherichia Castellani and Chalmers=[MacConkey's bacillus No. 66a]
Nicoll, 1911a

Escherichia Castellani and Chalmers=[MacConkey's bacillus No. 71]
Nicoll, 1911a

Escherichia Castellani and Chalmers=[MacConkey's bacillus No. 101]
Nicoll, 1911a

Escherichia Castellani and Chalmers=[MacConkey's bacillus No. 106]
Nicoll, 1911a

Escherichia Castellani and Chalmers=[MacConkey's bacillus No. 106a]
Nicoll, 1911a

Escherichia Castellani and Chalmers
Prado et al., 1955
Lysenko et al., 1961

Escherichia anaerogenes (Chester) Bergey et al.
Scott, 1917a

Musca domestica (cont.)

Jones, 1941
Hawley et al., 1951
Escherichia coli (Migula) Castellani and Chalmers
Hamilton, 1903
Graham-Smith, 1909
Nicoll, 1911a
Tebbutt, 1913
Scott, 1917a
Reinstorf, 1923
Beyer, 1925
Shope, 1927
Yao et al., 1929
Parisot et al., 1934
Jones, 1941
Ostrolenk et al., 1942a
Vanni, 1946
Emmel, 1949
Hoffmann, 1950
Gerberich, 1951
Hawley et al., 1951
Gabaldón, 1955
Cova Garcia, 1956
Ingram et al., 1956
Radvan, 1956, 1960a, 1960b, 1960c
Coutinho et al., 1957
Povolný et al., 1961
Greenberg et al., 1963a
Havlik, 1964
Escherichia coli (Migula) Castellani and Chalmers 086-B7
Coutinho et al., 1957
Escherichia coli var. *acidilactici* (Topley and Wilson) Yale
Graham-Smith, 1909
Nicoll, 1911a
Cox et al., 1912
Scott, 1917a
Escherichia coli var. *communior* (Topley and Wilson) Yale
Graham-Smith, 1909
Scott, 1917a
Escherichia coli var. *communis* (Escherich) Breed
Nicoll, 1911a
Cox et al., 1912
Buchanan, 1913
Duncan, 1926
Tarasov et al., 1941

Musca domestica (cont.)	*Escherichia coli* var. *neapolitana* (Topley and Wilson) Yale
	Graham-Smith, 1909
	Nicoll, 1911a
	Cox et al., 1912
	Buchanan, 1913
	Aerobacter Beijerinck= [MacConkey's bacillus No. 67]
	Nicoll, 1911a
	Aerobacter Beijerinck= [MacConkey's bacillus No. 75]
	Nicoll, 1911a
	Aerobacter Beijerinck= [MacConkey's bacillus No. 102]
	Nicoll, 1911a
	Aerobacter Beijerinck= ?[MacConkey's bacillus No. 108a]
	Nicoll, 1911a
	Aerobacter aerogenes (Kruse) Beijerinck
	Graham-Smith, 1909
	Cox et al., 1912
	Buchanan, 1913
	Ostrolenk et al., 1942a
	Hawley et al., 1951
	Greenberg, 1959b
	Povolný et al., 1961
	Greenberg et al., 1963a
	Havlik, 1964
	Aerobacter cloacae (Jordan) Bergey et al.
	Nicoll, 1911a
	Povolný et al., 1961
	Aerobacter oxytocum (Trevisan) Bergey et al.
	Nicoll, 1911a
	Klebsiella Trevisan
	Prado et al., 1955
	Lysenko et al., 1961
	Klebsiella cloacae (Jordan) Brisou
	Lysenko, 1958
	Klebsiella pneumoniae (Schroeter) Trevisan
	Buchanan, 1913
	Shope, 1927
	Klebsiella ?pneumoniae (Schroeter) Trevisan
	Testi, 1909
	Cloaca Castellani and Chalmers
	Lysenko et al., 1961

Musca domestica (cont.)	*Paracolobactrum* Borman, Stuart, and Wheeler
	Prado et al., 1955
	Paracolobactrum coliforme Borman, Stuart, and Wheeler
	Jones, 1941
	Radvan, 1960a, 1960b
	Paracolobactrum intermedium Borman, Stuart, and Wheeler
	Prado et al., 1955
	Erwinia amylovora (Burrill) Winslow et al.
	Ark et al., 1936
	Serratia Bizio
	Prado et al., 1955
	Povolný et al., 1961
	Serratia marcescens Bizio
	Graham-Smith, 1910, 1911a, 1911b
	Orton et al., 1910
	Herms, 1911
	Ledingham, 1911
	Duncan, 1926
	Bychkov, 1932
	Peppler, 1944
	Vanni, 1946
	Gerberich, 1951
	Radvan, 1956, 1960a, 1960c
	Stephens, 1963
	Havlik, 1964
	Proteus Hauser
	Hamilton, 1903
	Reinstorf, 1923
	Prado et al., 1955
	Coutinho et al., 1957
	Proteus inconstans (Ornstein) Shaw and Clarke
	Lysenko, 1958
	Proteus mirabilis Hauser
	Cao, 1906
	Greenberg, 1959b
	Lysenko et al., 1961
	Proteus morganii (Winslow, Kligler, and Rothberg) Yale
	Morgan et al., 1909
	Nicoll, 1911a
	Cox et al., 1912
	Buchanan, 1913
	Tebbutt, 1913
	Jones, 1941

Musca domestica (cont.)

Hawley et al., 1951
Radvan, 1956, 1960a
Lysenko, 1958
Lysenko et al., 1961
Povolný et al., 1961
Havlik, 1964

Proteus rettgeri (Hadley et al.) Rustigian and Stuart
Lysenko et al., 1961

Proteus vulgaris Hauser
Ficker, 1903
Cao, 1906
Testi, 1909
Ledingham, 1911
Buchanan, 1913
Scott, 1917a
Shope, 1927
Gerberich, 1951
Radvan, 1956, 1960a, 1960b, 1960c
Greenberg, 1959b
Lysenko et al., 1961
Povolný et al., 1961
Greenberg et al., 1963a
Havlik, 1964

Providencia Kauffmann
Radvan, 1960b, 1960c
Lysenko et al., 1961
Shimizu et al., 1965

Salmonella Lignières
Alcivar et al., 1946
Lindsay et al., 1953
Gabaldón et al., 1956
Bolaños, 1959
Greenberg et al., 1964

Salmonella amersfoort Henning
Alcivar et al., 1946

Salmonella anatum (Rettger and Scoville) Bergey et al.
Alcivar et al., 1946
Bolaños, 1959
Greenberg et al., 1963b

Salmonella ballerup Kauffmann and Møller
Alcivar et al., 1946

Salmonella blockley Kauffmann
Greenberg et al., 1964

Salmonella budapest Kauffmann
Prado et al., 1955

Musca domestica (cont.)	*Salmonella choleraesuis* (Smith) Weldin
	Scott, 1917a
	Alcivar et al., 1946
	Salmonella derby Warren and Scott
	Prado et al., 1955
	Greenberg et al., 1963b
	Tacal et al., 1967
	Salmonella dublin Kauffmann
	Alcivar et al., 1946
	Salmonella enteritidis (Gaertner) Castellani and Chalmers
	Graham-Smith, 1910
	Cox et al., 1912
	Buchanan, 1913
	Ostrolenk et al., 1942a, 1942b
	Peppler, 1944
	Alcivar et al., 1946
	Gross et al., 1953
	Prado et al., 1955
	Greenberg, 1959a, 1959c
	Salmonella florida Cherry, Edwards, and Bruner
	Alcivar et al., 1946
	Salmonella gallinarum (Klein) Bergey
	Gerberich, 1951
	Salmonella give Kauffmann
	Alcivar et al., 1946
	Gross et al., 1953
	Bolaños, 1959
	Salmonella hirschfeldii Weldin
	Yao et al., 1929
	Salmonella kentucky Edwards
	Alcivar et al., 1946
	Greenberg et al., 1963b
	Salmonella kottbus Schütz et al.
	Alcivar et al., 1946
	Salmonella meleagridis Bruner and Edwards
	Alcivar et al., 1946
	Salmonella minnesota Edwards and Bruner
	Prado et al., 1955
	Salmonella narashino Kauffmann
	Alcivar et al., 1946
	Salmonella new brunswick Edwards
	Greenberg, 1963b
	Salmonella newport Schütze
	Bolaños, 1959

Fly	Organism and references
Musca domestica (cont.)	*Salmonella onderstepoort* Henning Alcivar et al., 1946 *Salmonella oranienburg* Kauffmann Alcivar et al., 1946 *Salmonella oregon* Edwards and Bruner Alcivar et al., 1946 *Salmonella panama* Kauffmann Bolaños, 1959 *Salmonella paratyphi* A (Brion and Kayser) Castellani and Chalmers Shope, 1927 Boikov, 1932 Gross et al., 1951, 1953 Prado et al., 1955 *Salmonella paratyphi* B Castellani and Chalmers Nicoll, 1911a Alcivar et al., 1946 Gross et al., 1951, 1953 Dishon, 1956 Radvan, 1956 Greenberg, 1959a *Salmonella pullorum* (Rettger) Bergey Gwatkin et al., 1944 Gerberich, 1951, 1952 *Salmonella reading* Schütze Alcivar et al., 1946 Bolaños, 1959 *Salmonella san diego* Kauffmann Bolaños, 1959 *Salmonella saint paul* Edwards and Bruner Bolaños, 1959 *Salmonella schottmuelleri* (Winslow, Kligler, and Rothberg) Bergey et al. Hawley et al., 1951 Radvan, 1960a, 1960b, 1960c *Salmonella typhi* Warren and Scott Firth et al., 1902 Ficker, 1903 Hamilton, 1903 Buchanan, 1907 Graham-Smith, 1910 Ledingham, 1911 Cochrane, 1912 Thomson, 1912

Musca domestica (cont.)

Krontowski, 1913
Tebbutt, 1913
Teodoro, 1916
Nicoll, 1917a
Wollman, 1921
Duncan, 1926
Shope, 1927
Ara et al., 1932
Boikov, 1932
Jones, 1941
Curbelo et al., 1945
Gross et al., 1951, 1953
Prado et al., 1955
Radvan, 1956, 1960a
Webb et al., 1956
Greenberg, 1959a, 1959c
Greenberg et al., 1963a

Salmonella typhimurium (Loeffler) Castellani and Chalmers
McNeil et al., 1944
Alcivar et al., 1946
Radvan, 1956, 1960a
Greenberg, 1964

Salmonella worthington Edwards and Bruner
Greenberg et al., 1963b

Salmonella zanzibar Kauffmann
Alcivar et al., 1946
Bolaños, 1959

Citrobacter Werkman and Gillen
Lysenko et al., 1961

Citrobacter freundii Werkman and Gillen
Lysenko, 1958
Greenberg, 1959b

Shigella Castellani and Chalmers
Boikov, 1932
Lindsay et al., 1953
Boyd, 1957
Zaidenov, 1961
Greenberg et al., 1964

Shigella alkalescens (Andrewes) Weldin
Alcivar et al., 1946
Gabaldón et al., 1956

Shigella boydii Ewing
Bolaños, 1959

Shigella boydii Ewing 1, 2, 3, 4, 5, 6, 8, 9
Alcivar et al., 1946

Musca domestica (cont.)

Shigella boydii Ewing 2, 5, 6
  Richards et al., 1961
Shigella ceylonensis (Weldin) Weldin
  Alcivar et al., 1946
Shigella dispar (Andrewes) Bergey et al.
  Alcivar et al., 1946
  Gabaldón et al., 1956
Shigella dysenteriae (Shiga) Castellani and Chalmers
  Krontowski, 1913
  Manson-Bahr, 1919
  Paraf, 1920
  Wollman, 1921
  Reinstorf, 1923
  Duncan, 1926
  Yao et al., 1929
  Emmel, 1949
  Hawley et al., 1951
  Boyd, 1957
  Stephens, 1963
Shigella dysenteriae (Shiga) Castellani and Chalmers 2
  Richards et al., 1961
Shigella flexneri Castellani and Chalmers
  Auché, 1906
  Tebbutt, 1913
  Emmel, 1949
  Gross et al., 1953
  Shura-Bura, 1955, 1957
  Gabaldón et al., 1956
  Greenberg, 1959c
  Evtodienko, 1968
Shigella flexneri Castellani and Chalmers=[Pseudodysenterie A6154]
  Reinstorf, 1923
Shigella flexneri Castellani and Chalmers=[Pseudodysenterie D9112]
  Reinstorf, 1923
Shigella flexneri Castellani and Chalmers 1a
  Sychevskaia͡ et al., 1959b
  Richards et al., 1961
Shigella flexneri Castellani and Chalmers 1a, 2a, Y
  Alcivar et al., 1946

Musca domestica (cont.)

Shigella flexneri Castellani and Chalmers 2a
  Greenberg, 1959a
  Richards et al., 1961
Shigella flexneri Castellani and Chalmers 3, 4a, 4b
  Richards et al., 1961
Shigella flexneri Castellani and Chalmers 6
  Kuhns et al., 1944
  Alcivar et al., 1946
  Prado et al., 1955
  Richards et al., 1961
Shigella sonnei (Levine) Weldin
  Gross et al., 1953
  Shura-Bura, 1955
  Gabaldón et al., 1956
  Radvan, 1956, 1960a, 1960b, 1960c
  Evtodienko, 1968
Shigella sonnei (Levine) Weldin 1
  Richards et al., 1961

EUBACTERIALES: UNDETERMINED ENTEROBACTERIACEAE

Coliform
  Ficker, 1903
  Peppler, 1944
Klebsiella-Cloaca intermediate forms
  Lysenko et al., 1961
MacConkey's bacillus No. 7, unvalidated
  Nicoll, 1911a
MacConkey's bacillus No. 36, unvalidated
  Nicoll, 1911a
MacConkey's bacillus No. 36a, unvalidated
  Nicoll, 1911a
MacConkey's bacillus No. 74, unvalidated
  Nicoll, 1911a
MacConkey's bacillus No. 109
  Nicoll, 1911a

EUBACTERIALES: BRUCELLACEAE

Pasteurella multocida (Lehmann and Neumann) Rosenbusch and Merchant
  Scott, 1917a, 1917b

Musca domestica (cont.)

Shope, 1927
Skidmore, 1932

Pasteurella pestis (Lehmann and Neumann) Holland
Nuttall, 1897
Hunter, 1906
Gosio, 1925
Duncan, 1926
Wollman, 1927
Russo, 1930

Pasteurella tularensis (McCoy and Chapin) Bergey et al.
Wayson, 1914

Brucella (Meyer and Shaw) Pribram
Antonov, 1945

Brucella abortus (Schmidt) Meyer and Shaw
Duncan, 1926
Wollman, 1927
Ruhland et al., 1941

Brucella melitensis (Hughes) Meyer and Shaw
Duncan, 1926

Haemophilus influenzae (Lehmann and Neumann) Winslow et al. (Weeks' bacillus)
Nicolle et al., 1919
Wollman, 1927

Moraxella lacunata (Eyre) Lwoff
Wollman, 1927

Cillopasteurella delendae-muscae (Roubaud and Descazeaux) Prévot
Roubaud et al., 1923

EUBACTERIALES: BACTEROIDACEAE

Bacteroides vulgatus Eggerth and Gagnon
Duncan, 1926

EUBACTERIALES: MICROCOCCACEAE

Micrococcus Cohn
Smuidsinovicia, 1889
Prado et al., 1955

Micrococcus Cohn
Coutinho et al., 1957

Staphylococcus Rosenbach (non-pathogenic)
Ewing, 1942

Musca domestica (cont.)

Staphylococcus Rosenbach (pathogenic)
Ewing, 1942
Staphylococcus Rosenbach
Cox et al., 1912
Picado, 1935
Staphylococcus aureus Rosenbach
Buchanan, 1907
Herms, 1911
Reinstorf, 1923
Duncan, 1926
Shope, 1927
Gerberich, 1951
Hawley et al., 1951
Coutinho et al., 1957
Radvan, 1960a, 1960b, 1960c
Greenberg et al., 1963a
Staphylococcus aureus Rosenbach=[*Micrococcus citreus*]
Scott, 1917a
Staphylococcus aureus Rosenbach
Duncan, 1926
Staphylococcus aureus Rosenbach=[*Micrococcus pyogenes* var. *albus*]
Scott, 1917a
Staphylococcus aureus Rosenbach=[*Micrococcus pyogenes* var. *aureus*]
Scott, 1917a
Staphylococcus aureus Rosenbach=[*Staphylococcus albus*]
Povolný et al., 1961
Staphylococcus aureus Rosenbach=[*Staphylococcus pyogenes*]
Povolný et al., 1961
Staphylococcus muscae Glaser
Glaser, 1924
Shope, 1927
Glaser, 1938
Staphylococcus saprophyticus Fairbrother
Lysenko, 1958
Lysenko et al., 1961
Gaffkya tetragena (Gaffky) Trevisan
Scott, 1917a

Musca domestica (cont.)

Sarcina Goodsir
Hamilton, 1903
Cox et al., 1912
Reinstorf, 1923

Sarcina flava de Bary
Lysenko et al., 1961

Sarcina loewenbergi Macé
Galli-Vallerio, 1908

Sarcina lutea Schroeter
Hawley et al., 1951

EUBACTERIALES: NEISSERIACEAE

Neisseria catarrhalis (Froschand and Kolle) Holland
Beyer, 1925

EUBACTERIALES: BREVIBACTERIACEAE

Brevibacterium ammoniagenes (Cooke and Keith) Breed
Hawley et al., 1951

Kurthia zenkeri (Hauser) Bergey et al.
Cao, 1906

Kurthia zopfii (Kurth)
Hamilton, 1903

EUBACTERIALES: LACTOBACILLACEAE

Diplococcus pneumoniae Weichselbaum
Povolný et al., 1961

Diplococcus pneumoniae Weichselbaum I
Shope, 1927

Diplococcus pneumoniae Weichselbaum II
Shope, 1927

Streptococcus Rosenbach
Ledingham, 1911
Cox et al., 1912
Reinstorf, 1923
Ewing, 1942
Radvan, 1960c
Povolný et al., 1961

Streptococcus Rosenbach, C54
Shope, 1927

Streptococcus Rosenbach, C55
Shope, 1927

Streptococcus Rosenbach, 744
Shope, 1927

Musca domestica (cont.)

Streptococcus agalactiae Lehmann and Neumann
- Sanders, 1940b
- Ewing, 1942
- Schumann, 1961

Streptococcus faecalis Andrewes and Horder
- Scott, 1917a
- Duncan, 1926
- Gerberich, 1951
- Lysenko et al., 1961

Streptococcus faecalis var. *liquefaciens* (Sternberg) Mattick
- Lysenko et al., 1961

Streptococcus lactis (Lister) Löhnis
- Greenberg et al., 1963a

Streptococcus mitis Andrewes and Horder
- Radvan, 1960c

Streptococcus pyogenes Rosenbach
- Scott, 1917a
- Wellmann, 1955

EUBACTERIALES: CORYNEBACTERIACEAE

Corynebacterium diphtheriae (Kruse) Lehmann and Neumann
- Graham-Smith, 1919

Listeria monocytogenes (Murray et al.) Pirie
- Radvan, 1956, 1960a

EUBACTERIALES: BACILLACEAE

Bacillus Cohn
- Ledingham, 1911
- Povolný et al., 1961

Bacillus Cohn=["*Bacillus* A," Ledingham]
- Tebbutt, 1913

Bacillus anthracis Cohn
- Raimbert, 1869, 1870
- Buchanan, 1907
- Graham-Smith, 1910, 1911a, 1911b, 1912
- Wollman, 1921
- Duncan, 1926
- Sen et al., 1944
- Radvan, 1956, 1960a, 1960b, 1960c

Musca domestica (cont.)	*Bacillus cereus* Frankland and Frankland
	Hawley et al., 1951
	Briggs, 1960
	Harvey et al., 1960
	Radvan, 1960a, 1960b, 1960c
	Stephens, 1963
	Bacillus cereus var. *mycoides* (Flügge) Smith et al.
	Hamilton, 1903
	Duncan, 1926
	Bacillus coli mutabilis Neisser
	Nicoll, 1911a
	Bacillus gasoformans Eisenberg
	Nicoll, 1911a
	Bacillus gruenthali Morgan
	Nicoll, 1911a
	Bacillus megaterium de Bary
	Hamilton, 1903
	Bacillus megaterium de Bary
	Hawley et al., 1951
	Bacillus mesentericus Trevisan
	Hamilton, 1903
	Duncan, 1926
	Bacillus ?radiciformis Eberbach
	Cao, 1906
	Bacillus subtilis Cohn
	Cao, 1906
	Hamilton, 1903
	Reinstorf, 1923
	Duncan, 1926
	Radvan, 1956, 1960a, 1960c
	Lysenko, 1958
	Lysenko et al., 1961
	Greenberg et al., 1963a
	Bacilllus thuringiensis Berliner
	Hall et al., 1959
	Figueiredo et al., 1960
	Burns et al., 1961
	Borgatti et al., 1963
	Feigin, 1963
	Greenwood, 1964
	Harvey, 1964
	Tonkonozhenko, 1967
	Bacillus thuringiensis var. *sotto* Heimpel and Angus
	Briggs, 1960
	Bacillus thuringiensis var. *thuringiensis* Heipel and Angus
	Briggs, 1960
	Dunn, 1960

Musca domestica (cont.)

Burgerjohn et al., 1965
Gingrich, 1965
Harvey et al., 1965
Bacillus vesiculosus Matzuschita
Nicoll, 1911a
Clostridium botulinum (van Ermengem) Bergey et al.
Bail, 1900
Clostridium chauvoei (Arloing, Cornevin and Thomas) Scott
Sauer, 1908
Clostridium parabotulinum bovis Robinson
Theiler, 1927
Clostridium tetani (Flügge) Bergey et al.
Bail, 1900
ACTINOMYCETALES:
MYCOBACTERIACEAE
Mycobacterium leprae (Hansen) Lehmann and Neumann
Currie, 1910, 1911
Leboeuf, 1912, 1914
Minett, 1912
Asami, 1934
Mycobacterium ?leprae (Hansen) Lehmann and Neumann
Wherry, 1908a, 1908b
Sandes, 1911, 1912
Honeij et al., 1914
Barros et al., 1947
Mycobacterium lepraemurium Marchoux and Sorel
Wherry, 1908a
Marchoux, 1916
Mycobacterium ?lepraemurium Marchoux and Sorel
Wherry, 1908b
Mycobacterium phlei Lehmann and Neumann
Currie, 1910
Mycobacterium tuberculosis (Zopf) Lehmann and Neumann
Hayward, 1904
André, 1906
Buchanan, 1907
Graham-Smith, 1910
Wollman, 1921
Morellini et al., 1947, 1953
Tison, 1950
Morellini, 1952, 1956

Musca domestica (cont.)

ACTINOMYCETALES:
DERMATOPHILACEAE
Dermatophilus congolensis Van Saceghem
Richard et al., 1966

SPIROCHAETALES:
SPIROCHAETACEAE
Spirochaeta Ehrenberg
Wollman, 1927
Spirochaeta acuminata Castellani
Castellani, 1907
Spirochaeta obtusa Castellani
Castellani, 1907

SPIROCHAETALES:
TREPONEMATACEAE
Borrelia berbera (Sergent and Foley) Bergey et al.
Sergent et al., 1910
Wollman et al., 1928
Treponema carateum Brumpt
Soberón y Parra et al., 1944
Treponema pertenue Castellani
Castellani, 1907, 1908
Oho, 1921
Leptospira canicola Okell et al.
Kunert et al., 1952
Schmidtke, 1959
Leptospira grippotyphosa Topley and Wilson
Kunert et al., 1952
Schmidtke, 1959
Leptospira icterohaemorrhagiae (Inada and Ido) Noguchi
Kunert et al., 1952
Schmidtke, 1959

BACTERIA OF UNCERTAIN AFFINITY
Bacterium mathisi Roubaud and Teillard, unvalidated
Roubaud et al., 1935

SCHIZOPHYTA: MICROTATOBIOTES
RICKETTSIALES: RICKETTSIACEAE
Rickettsia da Rocha-Lima
Hertig et al., 1924
Coxiella burneti (Derrick) Philip
Philip, 1948

RICKETTSIALES: CHLAMYDIACEAE
Colesiota conjunctivae (Coles) Rake
Mitscherlich, 1943

Musca domestica (cont.)

THALLOPHYTA: PHYCOMYCETES
MUCORALES: MUCORACEAE
Mucor mucedo (Linnaeus) Brefeld
Reinstorf, 1923
Rhizopus nigricans Ehrenberg
Baumberger, 1919
ENTOMOPHTHORALES: ENTOMOPHTHORACEAE
Entomophthora Fresenius
Graham-Smith, 1916
?*Entomophthora* Fresenius
Roubaud, 1911a
Entomophthora americana Thaxter
Thaxter, 1888
Entomophthora muscae Cohn
Cohn, 1855
White, 1880
Thaxter, 1888
Cavara, 1899
Brefeld, 1908
Buchanan, 1913
Güssow, 1913
Picard, 1914
Bishopp et al., 1915a
Dove, 1916
Lodge, 1916
Graham-Smith, 1919
Lakon, 1919
Friederichs, 1920
Roubaud, 1922
Roubaud et al., 1922b
Schweizer, 1936, 1947
de Salles et al., 1944
Judd, 1955
Baird, 1957
Krenner, 1961

THALLOPHYTA: ASCOMYCETES
ENDOMYCETALES: ENDOMYCETACEAE
Saccharomyces cerevisiae Hansen
Peppler, 1944
GYMNOASCALES: GYMNOASCACEAE
Trichophyton tonsurans Malmsten
=[*T. crateriforme* Sabouraud]
Trichophyton mentagrophytes (Robin) Blanchard var. *granulosum*
Gip et al., 1968

Musca domestica (cont.)

THALLOPHYTA: FUNGI IMPERFECTI
MONILIALES: MONILIACEAE
Aspergillus (Micheli) Corda
Baumberger, 1919
Aspergillus flavus Link
Amonkar et al., 1965
Beard et al., 1965
Beauveria bassiana (Balsamo) Vuillemin
Dresner, 1949, 1950
Steinhaus et al., 1953
Dunn et al., 1963
?*Botrytis* Micheli
Gómez et al., 1888-89
Geotrichum candidum Link
Ficker, 1903
Metarrhizium anisopliae (Metschnikoff) Sorokin
Friederichs, 1919, 1920
Notini et al., 1944
Penicillium Link
Baumberger, 1919
Penicillium glaucum Link
Baumberger, 1919
Reinstorf, 1923
MONILIALES: STILBACEAE
Isaria Persoon ex Fries
Bail, 1860
MONILIALES: BLASTOCYSTIDACEAE (uncertain position)
Blastocystis Alexeieff
Buxton, 1920

PROTOZOA: MASTIGOPHORA
PROTOMONADIDA: TRYPANOSOMATIDAE
Crithidia luciliae (Strickland) Wallace and Clark
Wallace et al., 1959
Herpetomonas Kent
Wenyon, 1911a
Darling, 1912b
Herpetomonas muscarum (Leidy) Kent
Prowazek, 1904
Carter, 1909
Roubaud, 1909
Rosenbusch, 1910
Wenyon, 1911a
Cardamatis, 1912a
Chatton et al., 1913

Fly	Organism / Reference
Musca domestica (cont.)	Franchini et al., 1915
	Root, 1921
	Becker, 1922-3, 1924
	Ross et al., 1924
	Fantham et al., 1927
	Vanni, 1946
	Coutinho et al., 1957
	Wallace et al., 1959
	Kramer, 1961a
	Herpetomonas muscarum (Leidy) Kent=[*Herpetomonas muscae domesticae*]
	Flu, 1911
	Herpetomonas muscarum (Leidy) Kent=[*Leptomonas muscae domesticae*]
	Flu, 1911
	Leishmania tropica (Wright)
	Carter, 1909
	Row, 1911
	Laveran, 1915
	Wollman, 1927
	Wollman et al., 1928
	Leishmania ?tropica (Wright)
	Cardamatis et al., 1911
	Wenyon, 1911b
	Trypanosoma Gruby
	Wenyon, 1911a
	Trypanosoma brucei Plimmer and Bradford
	Vanni, 1946
	Trypanosoma equinum Vages
	Sivori et al., 1902
	Trypanosoma evansi (Steel)
	Musgrave et al., 1903
	Mitzmain, 1912b
	Trypanosoma gambiense Dutton
	Beck, 1910
	Trypanosoma hippicum Darling
	Darling, 1912a, 1912b
	POLYMASTIGIDA: CHILOMASTIGIDAE
	Chilomastix mesnili (Wenyon)
	Root, 1921
	POLYMASTIGIDA: HEXAMITIDAE
	Giardia intestinalis (Lambl)
	Roubaud, 1918
	Buxton, 1920
	Root, 1921
	Jausion et al., 1923
	Aleksander et al., 1935

Host	Organism / Reference
Musca domestica (cont.)	Sieyro, 1942
	Vanni, 1946
	Rendtorff et al., 1954
	TRICHOMONADIDA: TRICHOMONADIDAE
	Trichomonas Donné
	Sieyro, 1942
	Trichomonas hominis (Davaine)
	Hegner, 1928
	Trichomonas ?hominis (Davaine)
	Simitch et al., 1937
	Tritrichomonas foetus (Riedmüller)
	Morgan, 1942
	Holz, 1953
	PROTOZOA: SARCODINA
	AMOEBIDA: AMOEBIDAE
	Hartmannella hyalina Dangeard
	Coutinho et al., 1957
	AMOEBIDA: ENDAMOEBIDAE
	Endolimax nana (Wenyon and O'Connor)
	Root, 1921
	Yao et al., 1929
	Coutinho et al., 1957
	Entamoeba coli (Grassi)
	Roubaud, 1918
	Buxton, 1920
	Root, 1921
	Yao et al., 1929
	Aleksander et al., 1935
	Vanni, 1946
	Roberts, 1947
	Rendtorff et al., 1954
	Coutinho et al., 1957
	Entamoeba histolytica Schaudinn
	Kuenen et al., 1913
	Roubaud, 1918
	Buxton, 1920
	Root, 1921
	Jausion et al., 1923
	Yao et al., 1929
	Aleksander et al., 1935
	Pipkin, 1942, 1949
	Sieyro, 1942
	Vanni, 1946
	Roberts, 1947
	Iodamoeba bütschlii (Prowazek)
	Aleksander et al., 1935
	Coutinho et al., 1957

Musca domestica (cont.)

PROTOZOA: SPOROZOA

COCCIDIDA: EIMERIIDAE

Eimeria Schneider
Coutinho et al., 1957

Eimeria acervalina (Tyzzer)
Roberts, 1947

Eimeria irresidua Kessel and Janciewicz
Metelkin, 1935

Eimeria perforans (Leuckart) Sluiter and Swellengrebel
Metelkin, 1935

HAPLOSPORIDA: UNCERTAIN TAXONOMIC POSITION AND FAMILY

Toxoplasma Nicolle and Manceaux
Kåss, 1954
Schmidtke, 1959

Toxoplasma gondii Nicolle and Manceaux
Kunert et al., 1953
Varela et al., 1961

PROTOZOA: CNIDOSPORIDIA

MICROSPORIDA: NOSEMATIDAE

Nosema apis Zander
Kramer, 1962

Nosema bombycis Nägeli
Smuidsinovicia, 1889
Teodoro, 1926

Nosema kingi Kramer
Kramer, 1964b

MICROSPORIDA: MRAZEKIIDAE

Octosporea muscae-domesticae Flu
Flu, 1911
Porter, 1953
Fantham et al., 1958
Kramer, 1964a, 1965b, 1966

Octosporea ?muscae-domesticae Flu
Cardamatis, 1912b

Spirolugea Léger and Hesse
Fanthom et al., 1958

MICROSPORIDA: TELOMYXIDAE

Telomyxa Léger and Hesse
Fanthom et al., 1958

PLATYHELMINTHES: CESTOIDEA

PSEUDOPHYLLIDEA: DIPHYLLOBOTHRIIDAE

Bothriocephalus marginatus Krefft
Nicoll, 1911b

Musca domestica (cont.)

Diphyllobothrium Cobbold
Pokrovskiĭ et al., 1938
Diphyllobothrium latum (Linnaeus) Lühe
Pod"i͡apol'skai͡a et al., 1934
Aleksander et al., 1935

CYCLOPHYLLIDEA: ANOPLOCEPHALIDAE

Moniezia expansa (Rudolphi)
Nicoll, 1911b

CYCLOPHYLLIDEA: DAVAINEIDAE

Raillietina cesticillus (Molin)
Grassi et al., 1889
Gutberlet, 1916
Ackert, 1919
Neveu-Lemaire, 1936
Raillietina tetragona (Molin)
Gutberlet, 1916
Ackert, 1920
Neveu-Lemaire, 1936

CYCLOPHYLLIDEA: DILEPIDIDAE

Choanotaenia infundibulum (Bloch)
Gutberlet, 1916
Neveu-Lemaire, 1936
Wetzel, 1936
Reid et al., 1937
Dipylidium caninum (Linnaeus) Railliet
Nicoll, 1911b

CYCLOPHYLLIDEA: HYMENOLEPIDIDAE

Hymenolepis Weinland
Pokrovskiĭ et al., 1938
Coutinho et al., 1957
Hymenolepis diminuta (Rudolphi) Blanchard
Nicoll, 1911b
Hymenolepis nana (V. Siebold) Blanchard
Buxton, 1920
Bogoi͡avlenskiĭ et al., 1928
Aleksander et al., 1935
Vampirolepis fraterna (Stiles) Spassky
Joyeux, 1920

CYCLOPHYLLIDAE: TAENIIDAE

Taenia pisiformis (Bloch)
Nicoll, 1911b

Musca domestica (cont.)

Taeniarhynchus saginatum (Goeze)
Buxton, 1920
Bogoi͡avlenskiĭ et al., 1928
Aleksander et al., 1935

ASCHELMINTHES: NEMATODA

RHABDITIDA: RHABDITIDAE

Rhabditis pellio (Schneider)
Menzel, 1924

RHABDIASIDEA: STRONGYLOIDIDAE

Strongyloides stercoralis (Babay)
Buxton, 1920

TRICHURIDEA: TRICHURIDAE

Trichuris trichiura (Linnaeus)
Nicoll, 1911b
Bogoi͡avlenskiĭ et al., 1928
Buxton, 1920
Pod"i͡apol'skai͡a et al., 1934
Aleksander et al., 1935

STRONGYLIDEA: ANCYLOSTOMATIDAE

Ancylostoma (Dubini)
Buxton, 1920

Ancylostoma caninum (Ercolani)
Nicoll, 1911b
Harada, 1953, 1954

Ancylostoma duodenale (Dubini) Creplin
Buxton, 1920
Bogoi͡avlenskiĭ et al., 1928

Necator americanus (Stiles)
Buxton, 1920

STRONGYLIDEA: STRONGYLIDAE

Strongylus equinus Mueller
Nicoll, 1911b

STRONGYLIDEA: SYNGAMIDAE

Syngamus trachea (Montagu)
Clapham, 1939

STRONGYLIDEA: TRICHOSTRONGYLIDAE

Trichostrongylus colubriformis (Giles)
Bogoi͡avlenskiĭ et al., 1928

OXYURIDEA: OXYURIDAE

Enterobius vermicularis (Linnaeus) Leach
Bogoi͡avlenksiĭ et al., 1928
Pod"i͡apol'skai͡a et al., 1934
Aleksander et al., 1935

Musca domestica (cont.)

ASCARIDIDEA: ASCARIDIDAE

Ascaris Linnaeus
- Pod"i͡apol'skai͡a et al., 1934
- Pokrovskiĭ et al., 1938

Ascaris lumbricoides Linnaeus
- Stiles, 1889
- Bogoi͡avlenskiĭ et al., 1928
- Roberts, 1934
- Aleksander et al., 1935
- Sukhacheva, 1963

Parascaris equorum (Goeze)
- Nicoll, 1911b

Toxascaris leonina (Linstow)
- Nicoll, 1911b

SPIRURIDEA: ACUARIIDAE

Dispharynx ?nasuta (Rudolphi)
- Piana, 1896

SPIRURIDEA: SPIRURIDAE

Habronema Diesing
- Generali, ?1886
- Mello et al., 1943b
- Coutinho et al., 1957

Habronema megastoma (Rudolphi) Seurat
- Bull, 1919
- Hill, 1919
- Johnston, 1920
- Johnston et al., 1920a
- Roubaud et al., 1921, 1922a, 1922b
- Descazeaux et al., 1933
- Neveu-Lemaire, 1936

Habronema ?megastoma (Rudolphi) Seurat
- Bull, 1918

Habronema microstoma (Schneider)
- Bull, 1919
- Hill, 1919
- Johnston, 1920
- Roubaud et al., 1922b
- Neveu-Lemaire, 1936

Habronema muscae (Carter)
- Carter, 1861
- Johnston, 1913, 1920
- Ransom, 1913
- van Saceghem, 1917, 1918
- Bull, 1919
- Hill, 1919
- Roubaud et al., 1922a, 1922b

Musca domestica (cont.)

de Margarinos Torres et al., 1923
Descazeaux et al., 1933
Neveu-Lemaire, 1936
Mello et al., 1943a

Habronema ?muscae (Carter)
Piana, 1896

TYLENCHIDA: CONTORTYLENCHIDAE

Heterotylenchus autumnalis Nickle
Stoffolano, 1970

FILARIIDEA: DIPETALONEMATIDAE

Oncocera gibsoni (Cleland and Johnson)
Henry, 1927

FILARIIDEA: FILARIIDAE

?Filaria Mueller
Place, 1915
Iwanoff, 1934

ASCHELMINTHES: UNDETERMINED NEMATODA

Coutinho et al., 1957

MOLLUSCA: GASTROPODA

SIGMURETHRA: HELICIDAE

Cryptomphalus aspersa (Müller)
Séguy, 1934

ARTHROPODA: ARACHNIDA

PSEUDOSCORPIONIDEA: CHERNETIDAE

Lamprochernes nodosus (Schrank)
Graham-Smith, 1916

Toxochernes panzeri (Loch)
Preudhomme de Borre, 1873

PSEUDOSCORPIONIDEA: CHELIFERIDAE

Chelifer Geoffroy
Stainton, 1864
Stevens, 1866
Hagen, 1867
Knab, 1897

ARANEAE: THERIDIIDAE

Latrodectus mactans (Fabricius)
Herms et al., 1935

Latrodectus mactans (Fabricius) *tredecimguttatus*
Bettini, 1965

Steatoda Sundevall
Pavlovskiĭ, 1921

Teutana Simon
Pavlovskiĭ, 1921

Theridium Walckenaer
Pavlovskiĭ, 1921

Musca domestica (cont.)

ARANEAE: ERESIDAE

Stegodyphus mimosarum Pavesi
Steyn, 1959

ARANEAE: ARGIOPIDAE

Epeira diadema Walker
Pavlovskiĭ, 1921

ACARINA: MACROCHELIDAE

Glyptholaspis confusa Foa
Axtell, 1961a, 1961b, 1963a, 1963b
Rodriguez et al., 1962a

Macrocheles Latreille
Axtell, 1961b

Macrocheles glaber (Müller)
Filipponi, 1960
Filipponi et al., 1963a

Macrocheles merdarius (Berlese)
Axtell, 1963b
Filipponi et al., 1963b, 1964

Macrocheles muscaedomesticae (Scopoli)
Dugès, 1834
Berlese, 1882a, 1882b
Leonardi, 1900
Oudemans, 1900, 1904
Ewing, 1913
Pereira et al., 1945, 1947
Filipponi, 1955, 1960, 1964, 1965
Rodriguez et al., 1960, 1962a
Axtell, 1961a, 1961b, 1963a, 1963b
Rodriguez, 1961
Wade et al., 1961
Filipponi et al., 1963a, 1963b, 1964
Wallwork et al., 1963
Anderson, 1964
O'Donnell et al., 1965
Farish et al., 1966
Singh et al., 1966
Kinn, 1966
Willis et al., 1968

Macrocheles peniculatus Berlese
Filipponi et al., 1963a, 1964a, 1964b
Filipponi, 1964

Macrocheles perglaber Filipponi and Pegazzano
Filipponi, 1960

Musca domestica (cont.)

Filipponi et al., 1963a, 1964a
Macrocheles plumiventris Hull
Rodriguez et al., 1960
Macrocheles robustulus (Berlese)
Axtell, 1961b, 1963b
Filipponi et al., 1963a
Macrocheles scutatus (Berlese)
Filipponi et al., 1963a
Macrocheles subbadius Berlese
Filipponi, 1960
Axtell, 1961b, 1963b
Filipponi et al., 1964

ACARINA: LAELAPTIDAE

Holostaspis Kolenati
Graham-Smith, 1916
Holostaspis badius (Koch) Berlese
Zanini, 1930
Holostaspis ?badius (Koch) Berlese= ? [*Acarus muscaedomesticae* per Scopoli; = *Macrocheles badius* per Oudemans, 1929]
Scopoli, 1772
Holostaspis ?badius (Koch) Berlese=?[*Acarus muscaedomesticae* Scopoli per Goeze and per Oudemans, 1929]
Goeze, 1776

ACARINA: UROPODIDAE

Fuscuropoda Vitzthum
Anderson, 1964
Fuscuropoda vegetans (De Geer)
O'Donnell et al., 1965
Willis et al., 1968

ACARINA: TROMBIDIIDAE

Microtrombidium ?muscarum (Linnaeus)
Riley, 1877
Griffith, 1907
Ewing et al., 1918

ACARINA: TROMBICULIDAE

Euschoengastia peromysci (Ewing)
Lipovsky, 1954
Eutrombicula alfreddugesi (Oudemans)
Lipovsky, 1954
Eutrombicula splendens (Ewing)
Lipovsky, 1954
Fonsecia gurneyi (Ewing)
Lipovsky, 1954

Musca domestica (cont.)

Hannemania Oudemans
Lipovsky, 1954
Kayella lacerta (Brennan)
Lipovsky, 1954
Neoschoengastia americana (Hirst)
Lipovsky, 1954
Neotrombicula lipovskyi (Brennan and Wharton)
Lipovsky, 1954
Neotrombiculoides montanensis (Brennan)
Lipovsky, 1954
Pseudoschoengastia farneri Lipovsky
Lipovsky, 1954
Pseudoschoengastia hungerfordi Lipovsky
Lipovsky, 1954

ACARINA: ACARIDAE

Acarus siro Linnaeus
Graham-Smith, 1916

ACARINA: ANOETIDAE

Myianoetus muscarum (Linnaeus) =?[*Pediculus muscarum* per Menzel;=*Acarus muscarum* per Oudemans, 1929]
Menzel et al., 1683
Myianoetus muscarum (Linnaeus) =?[*Acarus rufus muscarum* per De Geer;=*Acarus muscarum* per Oudemans, 1929]
De Geer, 1778
Myianoetus muscarum (Linnaeus)
Greenberg et al., 1960

ACARINA: SARCOPTIDAE

Sarcoptes Latreille
Newsad, 1930

ACARINA: UNDETERMINED FAMILY

Gamasus musci, unvalidated
Buchanan, 1916

ARTHROPODA: MYRIAPODA

SCUTIGEROMORPHA: SCUTIGERIDAE

Scutigera coleoptrata (Linnaeus)
Pavlovskiĭ, 1921

ARTHROPODA: INSECTA

DERMAPTERA: LABIDURIDAE

Euborellia annulipes (Lucas)
Illingworth, 1923a

Musca domestica (cont.)

DERMAPTERA: LABIIDAE
Labia pilicornis (Motschulsky)
Illingworth, 1923a
Spingolabis hawaiiensis (Bormans)
Illingworth, 1923a
DERMAPTERA: FORFICULIDAE
Forficula auricularia Linnaeus
Anderson, 1964
HEMIPTERA: PHYMATIDAE
Phymata pennsylvanica americana Melin
Balduf, 1941, 1948
HEMIPTERA: REDUVIIDAE
Apiomerus pilipes (Fabricius)
Uribe, 1926
LEPIDOPTERA: HYPONOMEUTIDAE
Hyponomeuta padella Linnaeus
Séguy, 1934
LEPIDOPTERA: SPHINGIDAE
Acherontia atropos Linnaeus
Keilin, 1917
COLEOPTERA: HYDROPHILIDAE
Cryptopleurum minutum Fabricius
Illingworth, 1923a
Dactylosternum abdominale Fabricius
Illingworth, 1923a
COLEOPTERA: STAPHYLINIDAE
Aleochara taeniata Erichson
Legner et al., 1966a
White et al., 1966
Baryodma ontarionis Casey
Colhoun, 1953
Creophilus maxillosus (Linnaeus)
Mourier et al., 1968
Oxytelus Gravenhorst
Illingworth, 1923a
Oxytelus sculptus Gravenhorst
Mourier et al., 1968
Ont[h]olestes tessellatus Fourcr.
Mourier et al., 1968
Philonthus discoideus Gravenhorst
Illingworth, 1923a
Philonthus longicornis Stephens
Illingworth, 1923a
Tachyporus Gravenhorst
Illingworth, 1923a
COLEOPTERA: UNDETERMINED STAPHYLINIDAE
Anderson, 1964

Musca domestica (cont.)

COLEOPTERA: HISTERIDAE
Pachylister chinensis Quensel
Simmonds, 1958
Saprinus lugens Erichson
Vogt, 1948

COLEOPTERA: DERMESTIDAE
Dermestes ater De Geer
Vogt, 1948
Dermestes caninus Germar
Vogt, 1948

COLEOPTERA: UNDETERMINED DERMESTIDAE
Packard, 1873

COLEOPTERA: SCARABAEIDAE
Copris incertus Say
Simmonds, 1958

HYMENOPTERA: ICHNEUMONIDAE
Opius nitidulator Nees
Rambousek, 1929
Phygadeuon Gravenhorst
Legner et al., 1966a
Mourier et al., 1968

HYMENOPTERA: BRACONIDAE
Alysia manducator Panzer
Froggatt, 1922
Gurney et al., 1926a
Legner et al., 1966a
Aphaereta pallipes Say
Blickle, 1961
Beard, 1964

HYMENOPTERA: CYNIPIDAE
Eucoila Westwood
Simmonds, 1958
Blickle, 1961
Kleidotoma Westwood
James, 1928
Kleidotoma marshalli Marshall
James, 1928

HYMENOPTERA: FIGITIDAE
Figites Latreille
Legner, 1966
Figites anthomyiarum (Bouché)
James, 1928
Xyalophora quinquelineata (Say)
Blickle, 1961

HYMENOPTERA: CHALCIDIDAE
Dirhinus Dalman
Simmonds, 1940

Musca domestica (cont.)

HYMENOPTERA: PTEROMALIDAE

Muscidifurax raptor[1] Girault and Sanders
- Girault et al., 1910a, 1910b
- Roberts, 1935
- Puerto Rico, U.S.D.A. Experimental Station, 1938
- Bartlett, 1939
- Legner et al. (pre-publication), 1965a
- Legner et al., 1965c, 1966a
- Legner (pre-publication), 1966
- Mourier et al., 1968
- Legner, 1969

Nasonia vitripennis[2] (Walker)
- Girault et al., 1910a
- Roubaud, 1917
- Altson, 1920
- Johnston et al., 1920b, 1921
- Froggatt, 1921, 1922
- Séguy, 1929
- Fujita, 1932
- Smirnov et al., 1933, 1934
- Smirnov, 1934
- Vladimirova et al., 1934
- DeBach et al., 1947
- Edwards, 1954
- Wylie, 1962
- Nagel et al., 1963
- Pimentel, 1963
- Pimentel et al., 1963, 1965a
- Beard, 1964
- Madden, 1964
- Hair et al., 1965
- Madden et al., 1965
- Wylie, 1965a, 1965b, 1965c
- Legner (pre-publication), 1966
- Legner et al., 1966a

Pachycrepoideus vindemmiae Rondani
- Bartlett, 1939
- McCoy, 1963
- Legner et al., 1965c, 1966a

Prospalangia platensis Brèthes
- Brèthes, 1915

[1] Although not specifically mentioned in each entry, it is assumed that all *Muscidifurax raptor* are pupal parasites.

[2] Although not specifically mentioned in each entry, it is assumed that all *Nasonia vitripennis* are pupal parasites.

Musca domestica (cont.)

Spalangia[1] Latreille
- Girault et al., 1910a
- Simmonds, 1922
- Vandenberg, 1930
- Legner et al., 1966a

Spalangia cameroni Perkins
- Legner (pre-publication), 1966
- Legner et al., 1966a, 1969
- Gerling et al., 1968
- Mourier et al., 1968
- Legner, 1969

Spalangia chontalensis (Cameron)
- Legner et al., 1966a

Spalangia drosophilae Ashmead
- Lindquist, 1936

Spalangia endius Walker
- Legner et al. (pre-publication), 1965a
- Legner et al., 1965c, 1966a
- Legner (pre-publication), 1966

Spalangia hirta Haliday
- Richardson, 1913a

Spalangia longepetiolata Bouček
- Legner et al., 1969a

Spalangia muscae Howard
- Howard, 1911
- Girault, 1916
- Peck, 1951

Spalangia muscidarum Richardson
- Pinkus, 1913
- Richardson, 1913a, 1913b
- Johnston et al., 1920b, 1921
- Froggatt, 1922
- Bartlett, 1939
- Simmonds, 1958

Spalangia nigra Latreille
- Richardson, 1913b
- Parker, 1924
- Parker et al., 1928
- Sanders, 1942
- Legner, 1969

Spalangia nigripes Curtis
- Bouček, 1963
- Legner et al., 1969a

Spalangia nigroaenea Curtis
- Legner et al., 1965c, 1966a, 1969a
- Legner (pre-publication), 1966

[1] Although not specifically mentioned in each entry, it is assumed that all *Spalangia* are pupal parasites.

Musca domestica (cont.)

Spalangia platensis (Brèthes)
Bouček, 1963
Spalangia philippinensis Fullaway
Bartlett, 1939
Spalangia simplex Perkins
Legner et al., 1966a
Spalangia stomoxysiae Girault
Lindquist, 1936
Legner et al., 1965c
Sphegigaster Spinola
Legner et al., 1969a
HYMENOPTERA: UNDETERMINED PTEROMALIDAE
Bishopp, 1913a
HYMENOPTERA: ENCYRTIDAE
Tachinaephagus giraulti Johnston and Tiegs
Legner et al., 1966a
HYMENOPTERA: EULOPHIDAE
Syntomosphyrum albiclavus Kerrich
Saunders, 1964
Syntomosphyrum glossinae Waterston
Saunders, 1964
HYMENOPTERA: DIAPRIIDAE
Hemilexomyia abrupta Dodd
Johnston et al., 1921
Trichopria Ashmead
Bartlett, 1939
Legner et al., 1965c, 1966a, 1969a
Legner (pre-publication), 1966
Trichopria commoda Muesebeck
Muesebeck, 1961
HYMENOPTERA: FORMICIDAE
Iridomyrmex humilis (Mayr)
Pavan, 1952
Paratrechina longicornis (Latreille)
Pimentel, 1955
Pheidole megacephala (Fabricius)
Bridwell, 1918
Illingworth, 1923a
Phillips, 1934
Simmonds, 1958
Pheidole subarmata borinquenensis Wheeler
Pimentel, 1955
Pheidologeton affinis (Jerd.)
Pimentel et al., 1969

Musca domestica (cont.)

Ponera perkinsi Forel
Illingworth, 1923a
Solenopsis corticalis Forel
Pimentel, 1955
Solenopsis geminata (Fabricius)
Pimentel, 1955
Tapinoma melanocephalum (Fabricius)
Pimentel, 1955

HYMENOPTERA: SPHECIDAE

Bembix spinolae (Lepeletier)
Parker, 1917
Mellinus arvensis (Linnaeus)
Herold, 1922
Oxybelus uniglumis (Linnaeus)
Chevalier, 1926

HYMENOPTERA: VESPIDAE

Polistes hebraeus (Fabricius)
Jepson, 1915
Vespa germanica Fabricius
Herold, 1922
Kühlhorn, 1961
Vespa vulgaris Linnaeus
Kühlhorn, 1961

DIPTERA: STRATIOMYIDAE

Hermetia illucens Linnaeus
Furman et al., 1959
Vasquez-Gonzales et al., 1962-63

DIPTERA: ANTHOMYIIDAE

Scatophaga stercoraria (Linnaeus)
Hewitt, 1914

DIPTERA: MUSCIDAE

Hydrotaea dentipes (Fabricius)
Porchinskiĭ, 1911
Derbeneva-Ukhova, 1935, 1961
Ophyra leucostoma (Wiedemann)
Anderson, 1964
Anderson et al., 1964
Muscina stabulans (Fallén)
Porchinskiĭ, 1913

DIPTERA: CALLIPHORIDAE

Protophormia terraenovae (Robineau-Desvoidy)
Vladimirova et al., 1938

DIPTERA: CUTEREBRIDAE

Dermatobia hominis (Linnaeus)
Neiva et al., 1917
Neel et al., 1955
Zeledón, 1957

Fly	Associated organism
Musca domestica (cont.)	CHORDATA: REPTILIA SQUAMATA: LACERTIDAE *Lacerta agilis* Linnaeus Antonov, 1945 CHORDATA: AVES GALLIFORMES: PHASIANIDAE *Gallus gallus* Linnaeus Bushnell et al., 1924 Rodriguez et al., 1962b PASSERIFORMES: HIRUNDINIDAE *Delichon urbica* (Linnaeus) subsp. *urbica* Séguy, 1929 *Hirundo rustica* Linnaeus subsp. *rustica* Séguy, 1929
House flies (presumably *Musca domestica* Linnaeus)	VIRUSES ENTEROVIRUSES Poliovirus Flexner et al., 1911 Toomey et al., 1941 OTHER ANIMAL VIRUSES Foot and mouth disease virus Kunike, 1927a SCHIZOPHYTA: SCHIZOMYCETES PSEUDOMONADALES: PSEUDOMONADACEAE *Pseudomonas aeruginosa* (Schroeter) Migula Manning, 1902 PSEUDOMONADALES: SPIRILLACEAE *Vibrio comma* (Schroeter) Winslow et al. Cattani, 1886 Celli et al., 1888 Craig, 1894 Alessandrini et al., 1912 Flu, 1915 CHLAMYDOBACTERIALES: CHLAMYDOBACTERIACEAE *Leptothrix* Kützing Wenyon et al., 1917b EUBACTERIALES: ENTEROBACTERIACEAE *Escherichia coli* (Migula) Castellani and Chalmers Purdy, 1906 Jenkins et al., 1954

House flies (cont.)

Escherichia coli var. *communis* (Escherich) Breed
  Manning, 1902
Serratia marcescens Bizio
  Marpmann, 1884
  Manning, 1902
  Terry, 1912
Proteus morganii (Winslow, Kligler, and Rothberg) Yale
  Nicoll, 1917b
Salmonella pullorum (Rettger) Bergey
  Gwatkin et al., 1952
Salmonella typhi Warren and Scott
  Celli et al., 1888
  Manning, 1902
  Bertarelli, 1909
  Ara, 1933
Shigella dysenteriae (Shiga) Castellani and Chalmers
  Bahr, 1914
Shigella flexneri Castellani and Chalmers
  Auché, 1906
  Bahr, 1914

EUBACTERIALES: UNDETERMINED ENTEROBACTERIACEAE

Coliform
  Geldreich et al., 1963

EUBACTERIALES: MICROCOCCACEAE

Micrococcus Cohn
  Ramirez, 1898
Staphylococcus aureus Rosenbach
  Celli et al., 1888
  Manning, 1902
Staphylococcus muscae Glaser
  Glaser, 1926
Sarcina alba Zimmermann
  Manning, 1902
Sarcina aurantiaca Flügge
  Manning, 1902

EUBACTERIALES: LACTOBACILLACEAE

Streptococcus Rosenbach
  Geldreich et al., 1963
Streptococcus faecalis Andrewes and Horder
  Geldreich et al., 1963

House flies (cont.)

EUBACTERIALES:
CORYNEBACTERIACEAE
Corynebacterium diphtheriae (Kruse) Lehmann and Neumann
Smith, 1898

EUBACTERIALES: BACILLACEAE
Bacillus Cohn
Ramirez, 1898
Graham-Smith, 1913

Bacillus anthracis Cohn
Dalrymple, 1912, 1914
Morris, 1920
Rinonpoli, 1930

Clostridium bifermentans (Weinberg and Séguin) Bergey et al.
Marpmann, 1884

Clostridium putrefaciens (McBryde) Sturges and Drake
Purdy, 1906

ACTINOMYCETALES:
MYCOBACTERIACEAE
Mycobacterium leprae (Hansen) Lehmann and Neumann
Römer, 1906

Mycobacterium tuberculosis (Zopf) Lehmann and Neumann
Celli et al., 1888
Hofmann, 1888
Lord, 1904, 1905
André, 1908

SPIROCHAETALES:
TREPONEMATACEAE
Treponema pertenue Castellani
Yasuyama, 1928

THALLOPHYTA: PHYCOMYCETES
ENTOMOPHTHORALES:
ENTOMOPHTHORACEAE
Entomophthora muscae Cohn
Cohn, 1857
Solms-Laubach, 1870
White et al., 1877
Goldstein, 1927
Yeager, 1939

THALLOPHYTA: FUNGI IMPERFECTI
MONILIALES: MONILIACEAE
Metarrhizium anisopliae (Metschnikoff) Sorokin
Schaerffenberg, 1959

House flies (cont.)

Penicillium Link
Cohn, 1857
MONILIALES: DEMATIACEAE
Torula Persoon
Berlese, 1896
CRYPTOCOCCALES:
CRYPTOCOCCACEAE
Candida Berkhout
Berlese, 1896
PROTOZOA: MASTIGOPHORA
PROTOMONADIDA:
TRYPANOSOMATIDAE
Herpetomonas muscarum (Leidy) Kent
Werner, 1909a, 1911b, 1913
Glaser, 1922
Leishmania tropica (Wright)
Blanc et al., 1921
PROTOMONADIDA: BODONIDAE
Bodo Ehrenberg
Flu, 1915
Attimonelli, 1940b
Bodo Ehrenberg=[*Prowazekia*]
Attimonelli, 1940b
POLYMASTIGIDA: TETRAMITIDAE
Tetramitus Perty
Attimonelli, 1940b
POLYMASTIGIDA: HEXAMITIDAE
Giardia Kunstler
Frye et al., 1932
Attimonelli, 1940b
Giardia intestinalis (Lambl)
Wenyon et al., 1917a, 1917b
TRICHOMONADIDA:
TRICHOMONADIDAE
Trichomonas Donné
Wenyon et al., 1917b
Attimonelli, 1940b
Akatov, 1955
PROTOZOA: SARCODINA
AMOEBIDA: ENDAMOEBIDAE
Endolimax nana (Wenyon and O'Connor)
Werner, 1909b
Frye et al., 1932
Entamoeba coli (Grassi)
Wenyon et al., 1917a, 1917b
Frye et al., 1932
Entamoeba histolytica Schaudinn
Werner, 1909b

House flies (cont.)

Wenyon et al., 1917a, 1917b
Rogers, 1929
Frye et al., 1932

PROTOZOA: SPOROZOA
COCCIDIDA: EIMERIIDAE
Eimeria Schneider
Wenyon et al., 1917a

PROTOZOA: CILIATA
GYMNOSTOMATIDA: HOLOPHRYIDAE
Holophrya Ehrenberg
Attimonelli 1940b

PLATYHELMINTHES: TREMATODA
DIGENEA: HETEROPHYIDAE
Heterophyes heterophyes
(V. Siebold) Stiles and Hassall
Wenyon et al., 1917a

DIGENEA: SCHISTOSOMATIDAE
Schistosoma Weinland
Wenyon et al., 1917a

PLATYHELMINTHES: UNDETERMINED TREMATODA
Trematode operculated ova
Wenyon et al., 1917b
Trematode operculated ova (30 μ)
Wenyon et al., 1917b
Trematode operculated ova (35 x 20 μ)
Wenyon et al., 1917b

PLATYHELMINTHES: CESTOIDEA
CYCLOPHYLLIDEA: DAVAINEIDAE
Davainea nana (Fuhrmann)
Calandruccio, 1906

CYCLOPHYLLIDEA: TAENIIDAE
Taeniarhynchus saginatum (Goeze)
Wenyon et al., 1917a

PLATYHELMINTHES: UNDETERMINED CESTOIDEA
Cestode ova
Wenyon et al., 1917b

ASCHELMINTHES: NEMATODA
TRICHURIDEA: TRICHURIDAE
Trichuris trichiura (Linnaeus)
Wenyon et al., 1917a

STRONGYLIDEA: ANCYLOSTOMATIDAE
Ancylostoma duodenale (Dubini)
Creplin
Wenyon et al., 1917a

House flies (cont.)

SPIRURIDEA: SPIRURIDAE
Habronema muscae (Carter)
Leidy, 1874
Ransom, 1911
ARTHROPODA: ARACHNIDA
UNDETERMINED
PSEUDOSCORPIONIDEA
Manning, 1902
ARANEAE: DICTYNIDAE
Coenothele gregalis Simon
Diguet, 1909
ACARINA: MACROCHELIDAE
Macrocheles muscaedomesticae (Scopoli)
Schröck, 1686
Axtell, 1964
Macrocheles subbadius Berlese
Axtell, 1964
ACARINA: LAELAPTIDAE
Alliphis halleri (Canestrini) = ?[*Pediculus* or *Pulex muscarum* per Heister; = *Gamasus halleri* per Oudemans, 1926]
Heister, 1727
Holostaspis badius (Koch) Berlese = ?[*Acarus muscaedomesticae* per Scopoli; = *Macrocheles badius* per Oudemans, 1929]
Scopoli, 1772
ACARINA: TROMBIDIIDAE
Atomus parasiticus (De Geer) = ?[Small red mite per Goeze; = *Acarus parasiticus* per Oudemans, 1929]
Goeze, 1774
Atomus parasiticus (De Geer)
De Geer, 1778
Murray, 1877
ACARINA: ANOETIDAE
Myianoetus muscarum (Linnaeus) = ?[*Tinea muscarum* and *Teredo muscarum* per Gahrliep; = *Acarus muscarum* per Oudemans, 1926]
Gahrliep, 1696
Myianoetus muscarum (Linnaeus) = ?[insect per Winterschmidt; = *Acarus muscarum* per Oudemans, 1929]

House flies (cont.)

Winterschmidt, 1765

Myianoetus muscarum (Linnaeus) = ?[very small mites per De Geer; = *Acarus muscarum* per Oudemans, 1929]
De Geer, 1771, v. 2

Myianoetus muscarum (Linnaeus) = ?[*Acarus coleoptratorum* per Schrank; = *Acarus muscarum* per Oudemans, 1929]
Schrank, 1776

Myianoetus muscarum (Linnaeus) = ?[*Acarus muscorum* per Fabricius; = *Acarus muscarum* per Oudemans, 1929]
Fabricius, 1780

Myianoetus muscarum (Linnaeus)
Mohr, 1786
Walckenaer, 1802

ARTHROPODA: INSECTA

HYMENOPTERA: TENTHREDINIDAE

Tenthredo pectoralis Norton
Venables, 1914

HYMENOPTERA: CYNIPIDAE

Eucoila Westwood
Simmonds, 1940

HYMENOPTERA: CHALCIDIDAE

Dirhinus luzonensis Rohwer
Dresner, 1954

HYMENOPTERA: PTEROMALIDAE

Spalangia[1] *cameroni* Perkins
Simmonds, 1929

Spalangia muscidarum Richardson
Simmonds, 1940

Spalangia philippinensis Fullaway
Fullaway, 1917a

HYMENOPTERA: FORMICIDAE

Pheidole megacephala (Fabricius)
Illingworth, 1915

DIPTERA: ASILIDAE

Efferia aestuans (Linnaeus)
Bromley, 1946

Efferia pogonias (Wiedemann)
Bromley, 1946

Neoitamus flavofemoratus (Hine)
Bromley, 1946

[1] Although not specifically mentioned in each entry, it is assumed that all *Spalangia* are pupal parasites.

Fly	Organism and reference
House flies (cont.)	*Proctacanthus philadelphicus* Macquart
	Bromley, 1946
	Regasilus notatus (Wiedemann)
	Bromley, 1946
	Regasilus sadyates (Walker)
	Bromley, 1946
Musca domestica domestica Linnaeus	ENTEROVIRUSES
	Poliovirus strain 113
	Levkovich et al., 1957
	Poliovirus Type 2, Lansing strain
	Levkovich et al., 1957
Musca domestica vicina Macquart (Synonyms: *Musca vicina*, *Musca vitrina*)	ENTEROVIRUSES
	Poliovirus Type 1
	Brygoo et al., 1962
	SCHIZOPHYTA: SCHIZOMYCETES
	PSEUDOMONADALES: PSEUDOMONADACEAE
	Pseudomonas Migula
	Floyd et al., 1953
	EUBACTERIALES: ACHROMOBACTERACEAE
	Alcaligenes Castellani and Chalmers
	Floyd et al., 1953
	Alcaligenes faecalis Castellani and Chalmers
	Shterngol'd, 1949
	EUBACTERIALES: ENTEROBACTERIACEAE
	Escherichia coli (Migula) Castellani and Chalmers
	Floyd et al., 1953
	Silverman et al., 1953
	Levinson, 1960
	Paracolobactrum Borman, Stuart, and Wheeler
	Shterngol'd, 1949
	Proteus Hauser
	Shterngol'd, 1949
	Floyd et al., 1953
	Proteus vulgaris Hauser
	Shterngol'd, 1949
	Providencia Kauffmann
	Floyd et al., 1953
	Salmonella typhi Warren and Scott
	Floyd et al., 1953
	Salmonella typhimurium (Loeffler) Castellani and Chalmers
	Floyd et al., 1953
	Brygoo et al., 1962

Host	Organism / Reference
Musca domestica vicina (cont.)	*Bethesda* Edwards, West, and Bruner
	Floyd et al., 1953
	?*Shigella* Castellani and Chalmers
	Sychevskaia͡ et al., 1959a
	Shigella dysenteriae Castellani and Chalmers
	Shterngol'd, 1949
	Shigella dysenteriae (Shiga) Castellani and Chalmers 1
	Sychevskaia͡ et al., 1959a
	Shigella dysenteriae (Shiga) Castellani and Chalmers 2
	Floyd et al., 1953
	Sychevskaia͡ et al., 1959a
	Shigella dysenteriae (Shiga) Castellani and Chalmers 3-7
	Floyd et al., 1953
	Shigella flexneri Castellani and Chalmers
	Shterngol'd, 1949
	Sukhova, 1954
	Shura-Bura, 1955
	Sychevskaia͡ et al., 1959a
	Arskiĭ et al., 1961
	Shigella flexneri Castellani and Chalmers = [*Shigella paradysentariae*]
	Shterngol'd, 1949
	Shigella flexneri Castellani and Chalmers, 1a
	Sychevskaia͡ et al., 1959a, 1959b
	Shigella flexneri Castellani and Chalmers, 2a, 5, X
	Floyd et al., 1953
	Shigella flexneri Castellani and Chalmers, 6
	Sychevskaia͡ et al., 1959b
	Shigella flexneri Castellani and Chalmers, Y
	Shterngol'd, 1949
	Shigella sonnei (Levine) Weldin
	Floyd et al., 1953
	Arskiĭ et al., 1961
	EUBACTERIALES: UNDETERMINED ENTEROBACTERIACEAE
	Coliform
	McGuire et al., 1957

Musca domestica vicina (cont.)	EUBACTERIALES: BRUCELLACEAE
	Hemophilus influenzae (Lehmann and Neumann) Winslow et al.
	McGuire et al., 1957
	EUBACTERIALES: MICROCOCCACEAE
	Staphylococcus Rosenbach (beta-hemolytic)
	McGuire et al., 1957
	Staphylococcus Rosenbach (non-hemolytic)
	McGuire et al., 1957
	Staphylococcus Rosenbach (pathogenic)
	Brygoo et al., 1962
	Gaffkya Trevisan
	McGuire et al., 1957
	Sarcina Goodsir
	Silverman et al., 1953
	McGuire et al., 1957
	EUBACTERIALES: NEISSERIACEAE
	Neisseria gonorrhoeae Trevisan
	McGuire, 1957
	EUBACTERIALES: LACTOBACILLACEAE
	Streptococcus Rosenbach (alpha-hemolytic)
	McGuire et al., 1957
	Streptococcus Rosenbach (beta-hemolytic)
	McGuire et al., 1957
	Streptococcus Rosenbach (non-hemolytic)
	McGuire et al., 1957
	Lactobacillus Beijerinck
	Silverman et al., 1953
	EUBACTERIALES: CORYNEBACTERIACEAE
	Corynebacterium Lehmann and Neumann
	McGuire et al., 1957
	EUBACTERIALES: BACILLACEAE
	Bacillus Cohn
	McGuire et al., 1957
	Bacillus anthracis Cohn
	Anonymous, 1952
	Bacillus subtilis Cohn
	Silverman et al., 1953
	Clostridium Prazmowski
	McGuire et al., 1957

Musca domestica vicina (cont.)	SPIROCHAETALES: TREPONEMATACEAE *Treponema pertenue* Castellani Satchell et al., 1953 THALLOPHYTA: FUNGI IMPERFECTI MONILIALES: MONILIACEAE *Aspergillus flavus* Link Usui, 1960 PROTOZOA: MASTIGOPHORA PROTOMONADIDA: TRYPANOSOMATIDAE *Herpetomonas* Kent Chang, 1940 *Herpetomonas muscarum* (Leidy) Kent Pletneva, 1937 POLYMASTIGIDA: CHILOMASTIGIDAE *Chilomastix mesnili* (Wenyon) Chang, 1940 POLYMASTIGIDA: HEXAMITIDAE *Giardia intestinalis* (Lambl) Pletneva, 1937 TRICHOMONADIDA: TRICHOMONADIDAE *Trichomonas hominis* (Davaine) Chang, 1940 PROTOZOA: SARCODINA AMOEBIDA: ENDAMOEBIDAE *Endolimax nana* (Wenyon and O'Connor) Chang, 1940 *Entamoeba coli* (Grassi) Pletneva, 1937 Chang, 1940 *Entamoeba histolytica* Schaudinn Pletneva, 1937 Chang, 1940 *Iodamoeba butschli* (Prowazek) Chang, 1940 PLATYHELMINTHES: CESTOIDEA CYCLOPHYLLIDEA: HYMENOLEPIDIDAE *Hymenolepis nana* (V. Siebold) Blanchard Pletneva, 1937 Sychevskaia͡ et al., 1959b CYCLOPHYLLIDEA: TAENIIDAE *Taeniarhynchus saginatum* (Goeze) Zmeev, 1936 Sychevskaia͡ et al., 1958

Musca domestica vicina (cont.)

CYCLOPHYLLIDEA: UNDETERMINED TAENIIDAE
Sychevskaîa et al., 1959b

SPECIES OF UNCERTAIN AFFINITY
Bassus laetatorius, unvalidated
Sychevskaîa, 1964b

ASCHELMINTHES: NEMATODA

TRICHURIDEA: TRICHURIDAE
Trichuris trichiura (Linnaeus)
Chang, 1940

STRONGYLIDEA: ANCYLOSTOMATIDAE
Ancylostoma (Dubini)
Chang, 1940

OXYURIDEA: OXYURIDAE
Enterobius vermicularis (Linnaeus) Leach
Pletneva, 1937

ASCARIDIDEA: ASCARIDIDAE
Ascaris lumbricoides Linnaeus
Chang, 1940
Sychevskaîa et al., 1958

TYLENCHIDA: ALLANTONEMATIDAE
Allantonema muscae (Roy and Mukherjee)
Roy et al., 1937a
Allantonema stricklandi (Roy and Mukherjee)
Roy et al., 1937b

ARTHROPODA: ARACHNIDA

ACARINA: UNDETERMINED PARASITIDAE
Sychevskaîa, 1964a

ACARINA: MACROCHELIDAE
Macrocheles Latreille
DeCoursey et al., 1956
Macrocheles glaber (Müller)
Sychevskaîa, 1964a
Macrocheles muscaedomesticae (Scopoli)
Sychevskaîa, 1964a

ACARINA: GAMASOLAELAPTIDAE
Digamasellus Berlese
Sychevskaîa, 1964a

ARTHROPODA: INSECTA

HYMENOPTERA: CHALCIDIDAE
Brachymeria minuta (Fabricius)
Sychevskaîa, 1966
Dirhinus pachycerus Masi
Roy et al., 1940

Host	Associated organism
Musca domestica vicina (cont.)	HYMENOPTERA: PTEROMALIDAE
	Spalangia[1] Latreille
	Roy et al., 1939
	Sychevskai͡a, 1964b
	Spalangia cameroni Perkins
	Bouček, 1963
	Spalangia endius Walker
	Bouček, 1963
	Spalangia nigripes Curtis
	Bouček, 1963
	Spalangia nigroaenea Curtis
	Sychevskai͡a, 1964b
	Spalangia stomoxysiae Girault
	Sychevskai͡a, 1964b
Musca domestica ?vicina Macquart	ARTHROPODA: INSECTA
	HYMENOPTERA: CYNIPIDAE
	Cothonaspis nigricornis (Kieffer)
	Sychevskai͡a, 1964b
Musca fergusoni Johnston and Bancroft	ASCHELMINTHES: NEMATODA
	SPIRURIDEA: SPIRURIDAE
	Habronema megastoma (Rudolphi) Seurat
	Johnston, 1920
	Johnston et al., 1920a
	Neveu-Lemaire, 1936
	Habronema microstoma (Schneider)
	Johnston, 1920
	Habronema muscae (Carter)
	Johnston, 1920
	Johnston et al., 1920a
	Neveu-Lemaire, 1936
	ARTHROPODA: INSECTA
	HYMENOPTERA: PTEROMALIDAE
	Spalangia nigroaenea Curtis
	Bouček, 1963
Musca gibsoni Patton and Cragg	ARTHROPODA: INSECTA
	HYMENOPTERA: PTEROMALIDAE
	Spalangia muscophaga Girault
	Bouček, 1963
Musca hilli Johnston and Bancroft	ASCHELMINTHES: NEMATODA
	SPIRURIDEA: SPIRURIDAE
	Habronema megastoma (Rudolphi) Seurat
	Johnston, 1920
	Johnston et al., 1920a

[1] Although not specifically mentioned in each entry, it is assumed that all *Spalangia* are pupal parasites.

Fly	Organism
Musca hilli (cont.)	*Habronema microstoma* (Schneider) Johnston, 1920 *Habronema muscae* (Carter) Johnston, 1920 Johnston et al., 1920a ARTHROPODA: INSECTA HYMENOPTERA: PTEROMALIDAE *Nasonia vitripennis* (Walker) Johnston et al., 1920b *Spalangia muscidarum* Richardson Johnston et al., 1920b *Spalangia nigroaenea* Curtis Bouček, 1963
Musca inferior Stein	SCHIZOPHYTA: SCHIZOMYCETES EUBACTERIALES: BRUCELLACEAE *Pasteurella multocida* (Lehmann and Neumann) Rosenbusch and Merchant Nieschulz et al., 1929 EUBACTERIALES: BACILLACEAE *Bacillus anthracis* Cohn Nieschulz, 1928a PROTOZOA: MASTIGOPHORA PROTOMONADIDA: TRYPANOSOMATIDAE *Trypanosoma evansi* (Steel) Nieschulz, 1927 ARTHROPODA: INSECTA HYMENOPTERA: CHALCIDIDAE *Dirhinus pachycerus* Masi Roy et al., 1939
Musca insignis (Austen) (Synonym: *Philaematomyia insignis*)	PROTOZOA: MASTIGOPHORA PROTOMONADIDA: TRYPANOSOMATIDAE *Trypanosoma evansi* (Steel) Fletcher, 1916
Musca larvipara Porchinskiĭ	SCHIZOPHYTA: SCHIZOMYCETES EUBACTERIALES: ENTEROBACTERIACEAE *Escherichia coli* var. *acidilactici* (Topley and Wilson) Yale Zmeev, 1943 *Escherichia coli* var. *acidilactici* (Topley and Wilson) Yale = [*Escherichia paraacidilactici* A] Zmeev, 1943 *Escherichia coli* var. *acidilactici* (Topley and Wilson) Yale

Musca larvipara (cont.)	= [*Escherichia paraacidilactici* B] Zmeev, 1943 ?*Shigella* Castellani and Chalmers Sychevskaîa et al., 1959a ASCHELMINTHES: NEMATODA SPIRURIDEA: THELAZIIDAE *Thelazia gulosa* Raillet and Henry Klesov, 1951 Világiová, 1962 *Thelazia rhodesii* (Desmarest) Klesov, 1949, 1951 Világiová, 1962 Tukhmanyants et al., 1963 TYLENCHIDA: CONTORTYLENCHIDAE *Heterotylenchus autumnalis* Nickle Világiová, 1968 ARTHROPODA: ARACHNIDA ACARINA: UNDETERMINED PARASITIDAE Sychevskaîa, 1964a ACARINA: GAMASOLAELAPTIDAE *Saintdidieria sexclavatus* (Oudemans) Sychevskaîa, 1964a ACARINA: TROMBIBIIDAE *Atomus inexpectatus* (Oudemans) Sychevskaîa, 1964a
Musca lusoria Wiedemann	ASCHELMINTHES: NEMATODA SPIRURIDEA: SPIRURIDAE *Habronema megastoma* (Rudolphi) Seurat Neveu-Lemaire, 1936 *Habronema muscae* (Carter) Neveu-Lemaire, 1936 ARTHROPODA: INSECTA HYMENOPTERA: BRACONIDAE *Idiasta lusoriae* (Bridwell) Bridwell, 1919 HYMENOPTERA: CYNIPIDAE *Bothrochacis stercoraria* Bridwell Bridwell, 1919
Musca nebulo Fabricius	SCHIZOPHYTA: PSEUDOMONADALES: PSEUDOMONADACEAE *Pseudomonas septica* Bergey et al. Amonkar, 1967

Fly	Associated organism
Musca nebulo (cont.)	PROTOZOA: MASTIGOPHORA PROTOMONADIDA: TRYPANOSOMATIDAE *Herpetomonas muscarum* (Leidy) Kent Patton, 1910a Patton, 1912a *Herpetomonas muscarum* (Leidy) Kent = [*Herpetomonas muscaedomesticae*] Patton, 1921a *Herpetomonas muscarum* (Leidy) Kent = [*Herpetomonas sarcophagae*] Patton, 1921a *Leishmania donovani* Laveran and Mesnil Patton, 1921-22 *Leptomonas mirabilis* Roubaud Patton, 1921a PROTOMONADIDA: BODONIDAE *Rhynchoidomonas luciliae* (Patton) Patton Patton, 1910c ARTHROPODA: INSECTA HYMENOPTERA: CHALCIDIDAE *Dirhinus pachycerus* Masi Roy et al., 1940 HYMENOPTERA: EULOPHIDAE *Syntomosphyrum glossinae* Waterston Lamborn, 1925
Musca osiris Wiedemann	SCHIZOPHYTA: SCHIZOMYCETES EUBACTERIALES: ENTEROBACTERIACEAE ?*Shigella* Castellani and Chalmers Sychevskaia͡ et al., 1959a
Musca planiceps Wiedemann	THALLOPHYTA: ASCOMYCETES LABOULBENIALES: LABOULBENIACEAE *Stygmatomyces baeri* (Knoch) Peyritsch Senior-White et al., 1945 ARTHROPODA: INSECTA COLEOPTERA: STAPHYLINIDAE *Aleochara trivialis* Kraatz Senior-White et al., 1945

Fly	Organisms
Musca sorbens Wiedemann (Synonyms: *Musca humilis*, *Musca spectanda*)	VIRUSES PSITACOSIS GROUP Trachoma virus Gear et al., 1962 SCHIZOPHYTA: SCHIZOMYCETES EUBACTERIALES: ENTEROBACTERIACEAE ?*Shigella* Castellani and Chalmers Sychevskaiā et al., 1959a *Shigella dysenteriae* (Shiga) Castellani and Chalmers 1 Sychevskaiā et al., 1959a *Shigella flexneri* Castellani and Chalmers Sychevskaiā et al., 1959a EUBACTERIALES: UNDETERMINED ENTEROBACTERIACEAE Coliform McGuire et al., 1957 EUBACTERIALES: BRUCELLACEAE *Haemophilus influenzae* (Lehmann and Neumann) Winslow et al. McGuire et al., 1957 *Haemophilus ?influenzae* (Lehmann and Neumann) Winslow et al. Sukhova, 1953 EUBACTERIALES: MICROCOCCACEAE *Staphylococcus* Rosenbach (beta-hemolytic) McGuire et al., 1957 *Staphylococcus* Rosenbach (non-hemolytic) McGuire et al., 1957 *Gaffkya* Trevisan McGuire et al., 1957 *Sarcina* Goodsir McGuire et al., 1957 EUBACTERIALES: NEISSERIACEAE *Neisseria gonorrhoeae* Trevisan McGuire et al., 1957 EUBACTERIALES: LACTOBACILLACEAE *Streptococcus* Rosenbach (alpha-hemolytic) McGuire et al., 1957 *Streptococcus* Rosenbach (beta-hemolytic) McGuire et al., 1957 *Streptococcus* Rosenbach (non-hemolytic) McGuire et al., 1957

Musca sorbens (cont.)

EUBACTERIALES:
CORYNEBACTERIACEAE
Corynebacterium Lehmann and Neumann
McGuire et al., 1957
EUBACTERIALES: BACILLACEAE
Bacillus Cohn
McGuire et al., 1957
Clostridium Prazmowski
McGuire et al., 1957
ACTINOMYCETALES:
MYCOBACTERIACEAE
Mycobacterium leprae (Hansen) Lehmann and Neumann = [*Bacillus leprae*]
Lamborn, 1935, 1937
Mycobacterium ?leprae (Hansen) Lehmann and Neumann = [acid-fast bacillus]
Lamborn, 1938
Mycobacterium tuberculosis (Zopf) Lehmann and Neumann
Lamborn, 1938, 1939
SPIROCHAETALES:
TREPONEMATACEAE
Treponema pertenue Castellani
Thomson et al., 1934
Lamborn, 1936a, 1936b
Satchell et al., 1953
PROTOZOA: MASTIGOPHORA
PROTOMONADIDA:
TRYPANOSOMATIDAE
Herpetomonas Kent
Chang, 1940
Herpetomonas muscarum (Leidy) Kent
Patton, 1921a
Leishmania donovani Laveran and Mesnil
Patton, 1921-22
Thomson et al., 1934
Lamborn, 1935, 1955
Leishmania donovani Laveran and Mesnil = [*Leishmania infantum*]
Thomson et al., 1934
Lamborn, 1955
Leishmania tropica (Wright)
Thomson et al., 1934
Lamborn, 1935, 1955

Musca sorbens (cont.)

Trypanosoma brucei Plimmer and Bradford
Lamborn, 1934
Thomson et al., 1934
Trypanosoma congolense Broden
Lamborn, 1934
Trypanosoma rhodesiense Stephens and Fantham
Lamborn, 1936a
Lamborn et al., 1936
Trypanosoma suis Ochmann
Lamborn, 1934
POLYMASTIGIDA: CHILOMASTIGIDAE
Chilomastix mesnili (Wenyon)
Chang, 1940
POLYMASTIGIDA: HEXAMITIDAE
Giardia lamblia Stiles
Chang, 1940
TRICHOMONADIDA: TRICHOMONADIDAE
Trichomonas hominis (Davaine)
Chang, 1940
PROTOZOA: SARCODINA
AMOEBIDA: ENDAMOEBIDAE
Endolimax nana (Wenyon and O'Connor)
Chang, 1940
Entamoeba coli (Grassi)
Chang, 1940
Entamoeba histolytica Schaudinn
Chang, 1940
Iodamoeba bütschlii (Prowazek)
Chang, 1940
PROTOZOA: CNIDOSPORIDIA
MICROSPORIDA: MRAZEKIIDAE
Octosporea muscae-domesticae Flu
Kramer, 1965b
UNDETERMINED MICROSPORIDA
Lamborn, 1935, 1955
PLATYHELMINTHES: CESTOIDEA
CYCLOPHYLLIDEA: TAENIIDAE
Taeniarhynchus saginatum (Goeze)
Sychevskai͡a et al., 1958
ASCHELMINTHES: NEMATODA
TRICHURIDEA: TRICHURIDAE
Trichuris trichiura (Linnaeus)
Chang, 1940
STRONGYLIDEA: ANCYLOSTOMATIDAE
Ancylostoma (Dubini)
Chang, 1940

Musca sorbens (cont.)

ASCARIDIDEA: ASCARIDIDAE
Ascaris lumbricoides Linnaeus
Chang, 1940

SPIRURIDEA: SPIRURIDAE
Habronema megastoma (Rudolphi) Seurat
Neveu-Lemaire, 1936
Habronema muscae (Carter)
Neveu-Lemaire, 1936

TYLENCHIDA: CONTORTYLENCHIDAE
Heterotylenchus Stoffolano and Nickle
Hughes et al., 1969

ARTHROPODA: ARACHNIDA
ACARINA: MACROCHELIDAE
Macrocheles Latreille
DeCoursey et al., 1956

ARTHROPODA: INSECTA
HYMENOPTERA: PTEROMALIDAE
Spalangia[1] *endius* Walker
Bouček, 1963
Spalangia nigroaenea Curtis
Bouček, 1963

Musca sorbens sorbens Wiedemann (Synonym: *Musca vetustissima*) ....

SCHIZOPHYTA: SCHIZOMYCETES
EUBACTERIALES: BACILLACEAE
Bacillus anthracis Cohn
Cleland, 1912, 1913

ASCHELMINTHES: NEMATODA
SPIRURIDEA: SPIRURIDAE
Habronema Diesing
Johnston, 1912
Tryon, 1914
Habronema megastoma (Rudolphi) Seurat
Johnston, 1920
Johnston et al., 1920a
Neveu-Lemaire, 1936
Habronema microstoma (Schneider)
Johnston, 1920
Habronema ?muscae (Carter)
Johnston, 1913
Habronema muscae (Carter)
Johnston, 1920
Johnston et al., 1920a
Neveu-Lemaire, 1936

[1] Although not specifically mentioned in each entry, it is assumed that all *Spalangia* are pupal parasites.

Musca sorbens sorbens (cont.)	FILARIIDEA: DIPETALONEMATIDAE *Oncocerca gibsoni* (Cleland and Johnston) Henry, 1927 FILARIIDEA: FILARIIDAE ?*Filaria* Mueller Place, 1915 ARTHROPODA: INSECTA HYMENOPTERA: PTEROMALIDAE *Nasonia vitripennis* (Walker) Johnston et al., 1920b *Spalangia muscidarum* Richardson Johnston et al., 1920b *Spalangia nigroaenea* Curtis Bouček, 1963 HYMENOPTERA: SPHECIDAE *Sericophorus sydneyi* Rayment Rayment, 1954 *Sericophorus viridis roddi* Rayment Rayment, 1954
Musca tempestiva Fallén	PROTOZOA: MASTIGOPHORA PROTOMONADIDA: TRYPANOSOMATIDAE *Trypanosoma brucei* Plimmer and Bradford Lamborn, 1934 *Trypanosoma congolense* Broden Lamborn, 1934 *Trypanosoma suis* Ochmann Lamborn, 1934 ASCHELMINTHES: NEMATODA TYLENCHIDA: CONTORTYLENCHIDAE *Heterotylenchus autumnalis* Nickle Világiová, 1968 ARTHROPODA: INSECTA HYMENOPTERA: SPHECIDAE *Oxybelus* Latreille Pridantseva, 1959
Musca terraereginae .... Johnston and Bancroft	ASCHELMINTHES: NEMATODA SPIRURIDEA: SPIRURIDAE *Habronema megastoma* (Rudolphi) Seurat Johnston et al., 1920a Johnston, 1920 Neveu-Lemaire, 1936 *Habronema microstoma* (Schneider) Johnston, 1920

Host	Associated organisms
Musca terraereginae (cont.)	*Habronema muscae* (Carter) Johnston et al., 1920a Johnston, 1920 Neveu-Lemaire, 1936 ARTHROPODA: INSECTA HYMENOPTERA: PTEROMALIDAE *Nasonia vitripennis* (Walker) Johnston et al., 1920b *Spalangia muscidarum* Richardson Johnston et al., 1920b *Spalangia nigroaenea* Curtis Bouček, 1963
Musca ventrosa Wiedemann	ASCHELMINTHES: NEMATODA SPIRURIDEA: SPIRURIDAE *Habronema megastoma* (Rudolphi) Seurat Neveu-Lemaire, 1936 *Habronema muscae* (Carter) Neveu-Lemaire, 1936
Subfamily: STOMOXYINAE	
Haematobia Lepeletier and Serville (Synonyms: ?*Lyperosia Bdellolarynx*)	PROTOZOA: MASTIGOPHORA PROTOMONADIDA: TRYPANOSOMATIDAE *Trypanosoma* Gruby Krishna Iyer et al., 1935 *Trypanosoma evansi* (Steel) Mitzmain, 1912a ARTHROPODA: INSECTA ANOPLURA: HAEMATOPINIDAE *Haematopinus tuberculatus* (Burmeister) Mitzmain, 1912a COLEOPTERA: STAPHYLINIDAE *Aleochara handschini* Scheerpeltz Scheerpeltz, 1934 *Aleochara windredi* Scheerpeltz Scheerpeltz, 1934 *Philonthus politus* Linnaeus Fullaway, 1926 HYMENOPTERA: PTEROMALIDAE *Muscidifurax raptor* Girault and Sanders Fullaway, 1917b *Pachycrepoideus vindemmiae* Rondani Fullaway, 1917b *Spalangia*[1] *cameroni* Perkins Fullaway, 1917b

[1] Although not specifically mentioned in each entry, it is assumed that all *Spalangia* are pupal parasites.

Fly	Organism
Haematobia (cont.)	*Spalangia philippinensis* Fullaway Fullaway, 1917b HYMENOPTERA: SPHECIDAE *Bembix lunata* Fabricius Ramakrishna Ayyar, 1920
Haematobia angustifrons Malloch (Synonym: *Bdellolarynx latifrons*)	PROTOZOA: MASTIGOPHORA PROTOMONADIDA: TRYPANOSOMATIDAE *Trypanosoma brucei* Plimmer and Bradford Lamborn, 1934 *Trypanosoma congolense* Broden Lamborn, 1934
Haematobia exigua (de Meijere) (Synonyms: *Lyperosia exigua*, *Siphona exigua*)	SCHIZOPHYTA: SCHIZOMYCETES EUBACTERIALES: BRUCELLACEAE *Pasteurella boviseptica* (Kruse) Holland Nieschulz et al., 1929 EUBACTERIALES: BACILLACEAE *Bacillus anthracis* Cohn Nieschulz, 1928a PROTOZOA: MASTIGOPHORA PROTOMONADIDA: TRYPANOSOMATIDAE *Trypanosoma evansi* (Steel) Schat, 1909 Nieschulz, 1927 ASCHELMINTHES: NEMATODA SPIRURIDEA: SPIRURIDAE *Habronema microstoma* (Schneider) Neveu-Lemaire, 1936 ARTHROPODA: INSECTA COLEOPTERA: HISTERIDAE *Pachylister chinensis* Quens. Bornemissza, 1968 COLEOPTERA: STAPHYLINIDAE *Aleochara handschini* Scheerpeltz Handschin, 1934a *Aleochara windredi* Scheerpeltz Handschin, 1934a *Oxytelus ocularis* Fauvel Handschin, 1932 HYMENOPTERA: PTEROMALIDAE *Pachycrepoideus vindemmiae* Rondani Handschin, 1934a

Fly	Organism
Haematobia exigua (cont.)	*Spalangia*[1] *cameroni* Perkins Bouček, 1963 *Spalangia endius* Walker Bouček, 1963 *Spalangia nigroaenea* Curtis Bouček, 1963 *Spalangia orientalis* Graham Handschin, 1932, 1934a, 1934b *Spalangia sundaica* Graham Handschin, 1932, 1934a, 1934b Lever, 1936 HYMENOPTERA: ENCYRTIDAE *Cerchysius lyperosae* Ferrière Handschin, 1934a *Tachinaephagus giraulti* Johnston and Tiegs Handschin, 1934a HYMENOPTERA: EULOPHIDAE *Trichospilus pupivora* Ferrière Handschin, 1934a HYMENOPTERA: DIAPRIIDE *Phaenopria* Ashmead Handschin, 1932 *Phaenopria fimicola* Ferrière Handschin, 1934a *Trichopria* Ashmead Handschin, 1932 HYMENOPTERA: FORMICIDAE *Pheidole oceanica* Mayr Lever, 1936
Haematobia ?exigua (de Meijere)	PROTOZOA: MASTIGOPHORA PROTOMONADIDA: TRYPANOSOMATIDAE *Trypanosoma evansi* (Steel) Schat, 1903
Haematobia irritans (Linnaeus) (Synonyms: *Lyperosia irritans, Siphona irritans, Haematobia serrata*)	ENTEROVIRUSES Poliovirus Francis, 1914 SCHIZOPHYTA: SCHIZOMYCETES EUBACTERIALES: ACHROMOBACTERACEAE *Alcaligenes* Castellani and Chalmers Stirrat et al., 1955 *Achromobacter* Bergey, Harrison, Breed, Hammer, and Huntoon Stirrat et al., 1955

[1] Although not specifically mentioned in each entry, it is assumed that all *Spalangia* are pupal parasites.

Haematobia irritans (cont.)

EUBACTERIALES: ENTEROBACTERIACEAE

Escherichia coli (Migula) Castellani and Chalmers
Stirrat et al., 1955

Escherichia freundii (Braak) Yale
Stirrat et al., 1955

Escherichia intermedia (Werkman and Gillen) Vaughn and Levine
Stirrat et al., 1955

Aerobacter aerogenes (Kruse) Beijerinck
Stirrat et al., 1955

Aerobacter cloacae (Jordan) Bergey et al.
Stirrat et al., 1955

Proteus rettgeri (Hadley et al.) Rustigian and Stuart
Stirrat et al., 1955

EUBACTERIALES: MICROCOCCACEAE

Micrococcus candidus Cohn
Stirrat et al., 1955

Micrococcus caseolyticus Evans
Stirrat et al., 1955

Staphylococcus epidermidis (Winslow and Winslow) Evans
Stirrat et al., 1955

Staphylococcus flavus (Trevisan) Wood
Stirrat et al., 1955

EUBACTERIALES: LACTOBACILLACEAE

Streptococcus Rosenbach
Stirrat et al., 1955

EUBACTERIALES: BACILLACEAE

Bacillus Cohn
Stirrat et al., 1955

Bacillus anthracis Cohn
Morris, 1918, 1920

Bacillus cereus var. *mycoides* (Flügge) Smith et al.
Stirrat et al., 1955

Bacillus subtilis Cohn
Stirrat et al., 1955

Bacillus subtilis var. *niger* Smith et al.
Stirrat et al., 1955

Bacillus thuringiensis var. *thuringiensis* Heimpel and Angus
Gingrich, 1965

Haematobia irritans (cont.)

THALLOPHYTA: FUNGI IMPERFECTI
MONILIALES: MONILIACEAE
Aspergillus (Micheli) Corda
Stirrat et al., 1955
Geotrichum Link
Stirrat et al., 1955
Oidium Link
Stirrat et al., 1955
Penicillium Link
Stirrat et al., 1955
Trichothecium Link
Stirrat et al., 1955
EUBACTERIALES: DEMATIACEAE
Cladosporium Link
Stirrat et al., 1955
Pullalaria Berkhout
Stirrat et al., 1955

PROTOZOA: MASTIGOPHORA
PROTOMONADIDA: TRYPANOSOMATIDAE
Trypanosoma congolense Broden
Jack, 1917

ARTHROPODA: ARACHNIDA
ACARINA: ACARIDAE
Acarus Linnaeus
Wilhelmi, 1917a

ARTHROPODA: INSECTA
ORTHOPTERA: GRYLLIDAE
Nemobius fasciatus (De Geer)
Bourne et al., 1967
COLEOPTERA: STAPHYLINIDAE
Aleochara Gravenhorst
Wolcott, 1922
HYMENOPTERA: UNDETERMINED ICHNEUMONIDAE
Hammer, 1942
HYMENOPTERA: FIGITIDAE
Neralsia bifoveolata (Cresson)
Wolcott, 1922
HYMENOPTERA: PTEROMALIDAE
Spalangia[1] Latreille
Wolcott, 1922
Spalangia cameroni Perkins
Bouček, 1963
Spalangia drosophilae Ashmead
Lindquist, 1936

[1] Although not specifically mentioned in each entry, it is assumed that all *Spalangia* are pupal parasites.

Host	Associated organisms and references
Haematobia irritans (cont.)	Bartlett, 1939
	Bouček, 1963
	Spalangia endius Walker
	Bouček, 1963
	Spalangia haematobiae Ashmead
	Riley et al., 1891
	Ashmead, 1894, 1900
	von Dalla Torre, 1898
	Schmiedeknecht, 1909
	Viereck, 1909, 1916
	Richardson, 1913b
	Bartlett, 1939
	Bouček, 1963
	Spalangia muscidarum Richardson
	Pinkus, 1913
	Spalangia nigra Latreille
	Bouček, 1963
	Spalangia nigroaenea Curtis
	Bouček, 1963
	Spalangia philippinensis Fullaway
	Bartlett, 1939
	Spalangia stomoxysiae Girault
	Lindquist, 1936
	Peck, 1951
	DIPTERA: MUSCIDAE
	Hydrotaea dentipes (Fabricius)
	Wilhelmi, 1917a
Haematobia titillans (Bezzi) (Synonym: *Lyperosia titillans*)	ASCHELMINTHES: NEMATODA
	SPIRURIDEA: ACUARIIDAE
	Parabronema skrjabini Rassowska
	Ivashkin, 1959
	ARTHROPODA: ARACHNIDA
	ACARINA: UNDETERMINED ASCAIDAE
	Pridantseva, 1959
	ACARINA: UNDETERMINED PARASITIDAE
	Pridantseva, 1959
	ACARINA: UNDETERMINED TARSONEMIDAE
	Pridantseva, 1959
	ACARINA: UNDETERMINED TROMBICULIDAE
	Pridantseva, 1959
	ARTHROPODA: INSECTA
	COLEOPTERA: UNDETERMINED CARABIDAE
	Pridantseva, 1959
	COLEOPTERA: UNDETERMINED STAPHYLINIDAE
	Pridantseva, 1959

Haematobia titillans (cont.)

COLEOPTERA: UNDETERMINED
HISTERIDAE
Pridantseva, 1959

HYMENOPTERA: SPHECIDAE
Oxybelus Latreille
Pridantseva, 1959

Lyperosiops stimulans (Meigen) (Synonym: *Haematobia stimulans*) ....

ARTHROPODA: INSECTA
HYMENOPTERA: SPHECIDAE
Crabro alpinus Imhof
Vergne, 1931
Mellinus Fabricius
Hammer, 1942

HYMENOPTERA: VESPIDAE
Vespa Linnaeus
Hammer, 1942

DIPTERA: MUSCIDAE
Myospila meditabunda (Fabricius)
Thomson, 1937
Mydaea urbana (Meigen)
Thomson, 1937
Polietes albolineata (Fallén)
Porchinskiĭ, 1910
Mesembrina meridiana Linnaeus
Thomson, 1937

Stomoxys Geoffroy ....

POXVIRUSES
Vaccinia
Terni, 1909
Variola
Terni, 1909

ENTEROVIRUSES
Poliovirus
Paul et al., 1941
Trask et al., 1943

SCHIZOPHYTA: SCHIZOMYCETES
EUBACTERIALES:
CORYNEBACTERIACEAE
Erysipelothrix insidiosa (Trevisan) Langford and Hansen
Megnin, 1874, 1875

EUBACTERIALES: BACILLACEAE
Bacillus anthracis Cohn
Megnin, 1874
Terni, 1908a

PROTOZOA: MASTIGOPHORA
PROTOMONADIDA:
TRYPANOSOMATIDAE
Herpetomonas Kent
Wenyon, 1911a

Stomoxys (cont.)

Herpetomonas muscarum (Leidy) Kent
- Patton et al., 1909
- Wenyon, 1911b

Trypanosoma Gruby
- Nabarro et al., 1905
- Sergent et al., 1905, 1922a
- Wenyon, 1908

Trypanosoma congolense Broden
- van Saceghem, 1922

Trypanosoma ?congolense Broden
- Jowett, 1911

Trypanosoma dimorphon Laveran and Mesnil
- Nabarro et al., 1905

Trypanosoma evansi (Steel)
- Penning, 1904
- Nabarro et al., 1905
- Fraser, 1909
- Donatien et al., 1922, 1923
- Broudin et al., 1926

Trypanosoma gambiense Dutton
- Greig et al., 1905
- Gray, 1907
- Dutton et al., 1908
- Minchin, 1908
- Beck, 1910

Trypanosoma pecorum Bruce, Hamerton, Bateman, and Mackie
- van Saceghem, 1921, 1922b

Trypanosoma rhodesiense Stephens and Fantham
- Duke, 1934

Trypanosoma vivax Ziemann
- Bouffard, 1907a, 1907b

ASCHELMINTHES: NEMATODA

FILARIIDEA: FILARIIDAE

?Filaria Mueller
- Wenyon, 1911a

Setaria cervi (Rudolphi)
- Noe, 1903
- Neveu-Lemaire, 1936

ARTHROPODA: ARACHNIDA

PSEUDOSCORPIONIDEA: CHERNETIDAE

Pselaphochernes scorpioides (Hermann)
- Graham-Smith, 1916

Fly	Organism
Stomoxys (cont.)	HYMENOPTERA: PTEROMALIDAE *Spalangia sundaica* Graham Handschin, 1932, 1934a HYMENOPTERA: SPHECIDAE *Bembix lunata* Fabricius Ramakrishna Ayyar, 1920 *Crabro clypeatus* (Schreber) Hamm et al., 1926 *Oxybelus* Latreille Roubaud, 1911a Barotte, 1925 DIPTERA: TABANIDAE *Tabanus striatus* Fabricius Mitzmain, 1913 CHORDATA: AVES PASSERIFORMES: UNDETERMINED STURNIDAE Fletcher, 1920
Stomoxys bilineata Grünberg (Synonym: *Stomoxys sexvittata*)	PROTOZOA: MASTIGOPHORA PROTOMONADIDA: TRYPANOSOMATIDAE *Trypanosoma evansi* (Steel) Bouet et al., 1912 *Trypanosoma evansi* (Steel) = [*Trypanosoma soudanense*] Bouet et al., 1912 *Trypanosoma gambiense* Dutton Roubaud, 1909
Stomoxys boueti Roubaud	PROTOZOA: MASTIGOPHORA PROTOMONADIDA: TRYPANOSOMATIDAE *Trypanosoma brucei* Plimmer and Bradford Bouet et al., 1912 *Trypanosoma vivax* Ziemann Bouet et al., 1912
Stomoxys brunnipes Grünberg	THALLOPHYTA: PHYCOMYCETES ENTOMOPHTHORALES: ENTOMOPHTHORACEAE ?*Entomophthora* Fresenius Roubaud, 1911a PROTOZOA: MASTIGOPHORA PROTOMONADIDA: TRYPANOSOMATIDAE *Trypanosoma evansi* (Steel) Nieschulz, 1930
Stomoxys calcitrans (Linnaeus)	POXVIRUSES Vaccinia Terni, 1908b

Stomoxys calcitrans (cont.)

Borreliota avium
  Schuberg et al., 1912
  Bos, 1932
ARBOVIRUSES
Yellow fever virus
  Hoskins, 1933
ENTEROVIRUSES
Poliovirus
  Anderson et al., 1912
  Rosenau, 1912a
  Rosenau et al., 1912b
  Boudreau et al., 1914
  Francis, 1914
OTHER ANIMAL VIRUSES
Equine infectious anemia virus
  Flocken et al.
  Howard, 1917
  Scott, 1917, 1920, 1922
  Lührs, 1919
  Stein et al., 1942
Foot and mouth disease virus
  Kunike, 1927a, 1927b
  Schmit-Jensen, 1927
  Wilhelmi, 1927
Southwest African horse death virus (Windhuher)
  Schuberg et al., 1912
Tortor suis virus
  Mohler, 1920
Vesicular stomatitis virus
  Ferris et al., 1955
SCHIZOPHYTA: SCHIZOMYCETES
PSEUDOMONADALES: PSEUDOMONADACEAE
Pseudomonas aeruginosa (Schroeter) Migula
  Duncan, 1926
PSEUDOMONADALES: SPIRILLACEAE
Vibrio comma (Schroeter) Winslow et al.
  Duncan, 1926
EUBACTERIALES: ENTEROBACTERIACEAE
Escherichia coli (Migula) Castellani and Chalmers
  Povolný et al., 1961
  Love et al., 1965
Escherichia coli var. *communior* (Topley and Wilson) Yale
  Graham-Smith, 1909

Stomoxys calcitrans (cont.)	*Escherichia coli* var. *communis* (Escherich) Breed Duncan, 1926 *Escherichia coli* var. *neapolitana* (Topley and Wilson) Yale Graham-Smith, 1909 *Escherichia freundii* (Braak) Yale Love et al., 1965 *Escherichia intermedia* (Werkman and Gillen) Vaughn and Levine Love et al., 1965 *Aerobacter aerogenes* (Kruse) Beijerinck Love et al., 1965 *Aerobacter cloacae* (Jordan) Bergey et al. Love et al., 1965 *Serratia* Bizio Povolný et al., 1961 *Serratia marcescens* Bizio Duncan, 1926 *Proteus* Hauser Testi, 1909 *Proteus vulgaris* Hauser Povolný et al., 1961 *Salmonella* Lignières Greenberg et al., 1964 *Salmonella blockley* Kauffmann Greenberg et al., 1964 *Salmonella paratyphi* B Castellani and Chalmers Birk, 1932 *Salmonella typhi* Warren and Scott Duncan, 1926 *Shigella* Castellani and Chalmers Greenberg et al., 1964 *Shigella dysenteriae* (Shiga) Castellani and Chalmers Duncan, 1926 EUBACTERIALES: BRUCELLACEAE *Pasteurella multocida* (Lehmann and Neumann) Rosenbusch and Merchant Nieschulz et al., 1929 *Pasteurella pestis* (Lehmann and Neumann) Holland Duncan, 1926 *Pasteurella tularensis* (McCoy and Chapin) Bergey et al. Wayson, 1914

Stomoxys calcitrans (cont.)

Somov et al., 1937
Olsufiev, 1940
Romanova, 1947
Bozhenko et al., 1948
Brucella abortus (Schmidt)
Meyer and Shaw
Duncan, 1926
Ruhland et al., 1941
Wellmann, 1950-1, 1959
Brucella melitensis (Hughes)
Meyer and Shaw
Duncan, 1926
Eyre et al., 1907
Wellmann, 1950-1, 1959
Brucella suis Huddleson
Wellmann, 1950-1, 1959
Cillopasteurella delendae-muscae (Roubaud and Descazeaux) Prevot
Roubaud et al., 1923

EUBACTERIALES: BACTEROIDACEAE
Bacteroides vulgatus
Duncan, 1926

EUBACTERIALES: MICROCOCCACEAE
Staphylococcus Rosenbach (pathogenic)
Brygoo et al., 1962a
Staphylococcus aureus Rosenbach
Duncan, 1926
Staphylococcus aureus Rosenbach = [*Staphylococcus albus*]
Duncan, 1926

EUBACTERIALES: LACTOBACILLACEAE
Diplococcus pneumoniae Weichselbaum
Povolný et al., 1961
Streptococcus Rosenbach
Schuberg et al., 1913, 1914
Love et al., 1965
Streptococcus durans Sherman and Wing
Duncan, 1926
Streptococcus faecium Orla-Jensen
Love et al., 1965

EUBACTERIALES: CORYNEBACTERIACEAE
Erysipelothrix insidiosa (Trevisan) Langford and Hansen

Stomoxys calcitrans (cont.)

Wellmann, 1948-9a, 1948-9b, 1949, 1950, 1959
Tolstîak, 1956

EUBACTERIALES: BACILLACEAE

Bacillus anthracis Cohn
Schuberg et al., 1912, 1913, 1914
Mitzmain, 1914a, 1914c
Duncan, 1926
Nieschulz, 1928a, 1928b
Sen et al., 1944

Bacillus cereus var. *mycoides* (Flügge) et al.
Duncan, 1926

Bacillus mesentericus Trevisan
Duncan, 1926

Bacillus subtilis Cohn
Duncan, 1926

Bacillus thuringiensis var. *thuringiensis* Heimpel and Angus
Gingrich, 1965

Clostridium botulinum (van Ermengem) Bergey et al.
Bail, 1900

ACTINOMYCETALES: MYCOBACTERIACEAE

Mycobacterium ?leprae (Hansen) Lehmann and Neumann
Honeij et al., 1914

ACTINOMYCETALES: DERMATOPHILACEAE

Dermatophilus congolensis van Saceghem
Richard et al., 1966

SPIROCHAETALES: SPIROCHAETACEAE

Spirochaeta stomoxyae Jegen
Jegen, 1924

SPIROCHAETALES: TREPONEMATACEAE

Borrelia anserina (Sakharoff) Bergey et al.
Schuberg et al., 1909

Borrelia berbera (Sergent and Foley) Bergey et al.
Wollman et al., 1928

Borrelia recurrentis (Lebert) Bergey et al.
Schuberg et al., 1909, 1912
Fränkel, 1912

Leptospira canicola Okell et al.
Schmidtke, 1959

Stomoxys calcitrans (cont.)	*Leptospira grippotyphosa* Topley and Wilson
	Schmidtke, 1959
	Leptospira icterohaemorrhagiae (Inada and Ido) Noguchi
	Reiter, 1917
	Uhlenhuth et al., 1917
	Schmidtke, 1959
	SCHIZOPHYTA: MICROTATOBIOTES
	RICKETTSIALES: RICKETTSIACEAE
	Rickettsia da Rocha-Lima
	Hertig et al., 1924
	RICKETTSIALES: CHLAMYDIACEAE
	Colesiota conjunctivae (Coles) Rake
	Mitscherlich, 1943
	RICKETTSIALES: ANAPLASMATACEAE
	Anaplasma marginale Theiler
	Taylor, 1935
	Anonymous, 1936
	THALLOPHYTA: PHYCOMYCETES
	ENTOMOPHTHORALES: ENTOMOPHTHORACEAE
	?*Entomophthora* Fresenius
	Roubaud, 1911a
	Entomophthora muscae Cohn
	Surcouf, 1923
	PROTOZOA: MASTIGOPHORA
	PROTOMONADIDA: TRYPANOSOMATIDAE
	Crithidia Leger
	Taylor, 1930
	Crithidia haematopotae Jegen
	Jegen, 1924
	Herpetomonas Kent
	Patton, 1912b
	Leishmania mexicana Biagi
	Lainson et al., 1965
	Leishmania tropica (Wright)
	Berberian, 1938, 1939
	Leptomonas stomoxyae Jegen
	Jegen, 1924
	Trypanosoma Gruby
	Martini, 1903
	Nieschulz, 1928b
	Trypanosoma brucei Plimmer and Bradford
	Martin et al., 1908
	Roubaud, 1909

Stomoxys calcitrans (cont.)

Schuberg et al., 1909
Taylor, 1930
Trypanosoma congolense Broden
Roubaud, 1909
Carmichael, 1934
Trypanosoma equinum Vages
Sivori et al., 1902
Lignières, 1903
Trypanosoma equiperdum Doflein
Sieber et al., 1908
Schuberg et al., 1909
Trypanosoma evansi (Steel)
Schat, 1903
Leese, 1909
Baldrey, 1911
Bouet et al., 1912
Mitzmain, 1912b
Sergent et al., 1922b
Nieschulz, 1927, 1928b, 1929a, 1930
Dieben, 1928
Trypanosoma ?evansi (Steel)
Curry, 1902a
Trypanosoma gambiense Dutton
Roubaud, 1909
Schuberg et al., 1909a
Duke, 1913
Trypanosoma rhodesiense Stephens and Fantham
Lamborn, 1933
Trypanosoma vivax Ziemann
Bouet et al., 1912

PROTOMONADIDA: BODONIDAE
Bodo Ehrenberg
Duke, 1913

PROTOZOA: SARCODINA
AMOEBIDA: ENDAMOEBIDAE
Entamoeba histolytica Schaudinn
Jausion et al., 1923

PROTOZOA: SPOROZOA
COCCIDIDA: EIMERIDAE
Eimeria irresidua Kessel and Janciewicz
Metelkin, 1935
Eimeria perforans (Leuckart) Sluiter and Swellengrebel
Metelkin, 1935

HAPLOSPORIDA: UNCERTAIN FAMILY
Toxoplasma Nicolle and Manceaux
van Thiel et al., 1953

Stomoxys calcitrans (cont.)

Schmidtke, 1959

Toxoplasma gondii Nicolle and Manceaux
Blanc et al., 1950
Kunert et al., 1953
Laarman, 1956, 1957

PLATYHELMINTHES: CESTOIDEA

CYCLOPHYLLIDEA: DILEPIDIDAE

Choanotaenia infundibulum (Bloch)
Gutberlet, 1916

CYCLOPHYLLIDEA: HYMENOLEPIDIDAE

Echinolepis carioca (Magalhaes) Spassky and Spasskaja
Gutberlet, 1920
Neveu-Lemaire, 1936

ASCHELMINTHES: NEMATODA

RHABDITIDA: RHABDITIDAE

Rhabditis axei (Cobbold)
Hague, 1963

SPIRURIDEA: SPIRURIDAE

Habronema megastoma (Rudolphi) Seurat
Bull, 1919
Hill, 1919

Habronema microstoma (Schneider)
Bull, 1919
Hill, 1919
Johnston, 1920
Johnston et al., 1920a
Roubaud et al., 1922a, 1922b
Neveu-Lemaire, 1936

Habronema muscae (Carter)
Johnston, 1913
Bull, 1919
Hill, 1919

SPIRURIDEA: FILARIIDAE

?*Filaria* Mueller
Place, 1915
Iwanoff, 1934

Setaria cervi (Rudolphi)
von Linstow, 1875

ARTHROPODA: ARACHNIDA

PSEUDOSCORPIONIDEA: CHERNETIDAE

Lamprochernes nodosus (Schrank)
Graham-Smith, 1916

Stomoxys calcitrans (cont.)

ACARINA: PARASITIDAE
Eulaelaps stabularis Koch
Bouvier et al., 1944
Parasitus coleoptratorum (Linnaeus)
Wilhelmi, 1917b

ACARINA: MACROCHELIDAE
Macrocheles glaber (Müller)
Filipponi, 1960
Macrocheles muscaedomesticae (Scopoli)
Filipponi, 1960
Axtell, 1964
Kinn, 1966
Macrocheles perglaber Filipponi and Pegazzano
Filipponi, 1960
Macrocheles subbadius Berlese
Filipponi, 1960
Axtell, 1964

ACARINA: LAELAPTIDAE
Holostaspis marginatus (Hermann)
Wilhelmi, 1917b

ACARINA: TROMBIDIIDAE
Microtrombidium striaticeps (Oudemans)
Ewing, 1919

ACARINA: ANOETIDAE
Myianoetus muscarum (Linnaeus)
Greenberg et al., 1960

ARTHROPODA: INSECTA

ORTHOPTERA: LOCUSTIDAE
Schistocerca paranensis Burmeister
Lahille, 1907

DERMAPTERA: LABIDURIDAE
Euborellia annulipes (Lucas)
Illingworth, 1923a

DERMAPTERA: LABIIDAE
Labia pilicornis (Motschulsky)
Illingworth, 1923a
Spingolabis hawaiiensis (Bormans)
Illingworth, 1923a

ODONATA: COENAGRIONIDAE
Agria fumipennis (Burmeister)
Wright, 1945
Enallagma durum Hagen
Wright, 1945
Ischnura ramburii Selys
Wright, 1945

Stomoxys calcitrans (cont.)

ODONATA: LIBELLULIDAE
Celithemis amanda Hagen
Wright, 1945
Erythemis simplicicollis Say
Wright, 1945
Libellula auripennis Burmeister
Wright, 1945
Libellula pulchella Drury
Wright, 1945
Libellula vibrans Fabricius
Wright, 1945
Pachydiplax longipennis Burmeister
Wright, 1945
Pantala flavescens Fabricius
Wright, 1945
Tramea carolina Linnaeus
Wright, 1945

ODONATA: AESCHNIDAE
Anax junius Drury
Wright, 1945
Coryphaeschna ingens Rambur
Wright, 1945

UNDETERMINED ODONATA
Brues, 1946

LEPIDOPTERA: LASIOCAMPIDAE
Dendrolimus pini Linnaeus
Sitowski, 1928

COLEOPTERA: HYDROPHILIDAE
Cryptopleurum minutum Fabricius
Illingworth, 1923a
Dactylosternum abdominale Fabricius
Illingworth, 1923a

COLEOPTERA: STAPHYLINIDAE
Oxytelus Gravenhorst
Illingworth, 1923a
Philonthus discoideus Gravenhorst
Illingworth, 1923a
Philonthus longicornis Stephens
Illingworth, 1923a
Tachyporus Gravenhorst
Illingworth, 1923a

COLEOPTERA: UNDETERMINED HISTERIDAE
Bishopp, 1913a

HYMENOPTERA: CHALCIDIDAE
Dirhinus pachycerus Masi
Roy et al., 1940

Stomoxys calcitrans (cont.)

HYMENOPTERA: UNDETERMINED CHALCIDIDAE
Surcouf, 1923

HYMENOPTERA: PTEROMALIDAE

Muscidifurax raptor Girault and Saunders
Bartlett, 1939
Legner (pre-publication), 1966

Nasonia vitripennis (Walker)
Séguy, 1929

Pachycrepoideus vindemmiae Rondani
Bartlett, 1939

Prospalangia platensis Brèthes
Brèthes, 1915

Spalangia[1] Latreille
Roy et al., 1939

Spalangia cameroni Perkins
Legner (pre-publication), 1966
Legner et al., 1969a

Spalangia endius Walker
Legner (pre-publication), 1966

Spalangia longepetiolata Bouček
Legner et al., 1969a

Spalangia muscae Howard
Bishopp, 1913a
Girault, 1921

Spalangia muscidarum Richardson
Pinkus, 1913
Richardson, 1913b
Johnston et al., 1920b, 1921
Bartlett, 1939

Spalangia nigra Latreille
Bishopp, 1913b, 1939
Bouček, 1963

Spalangia nigripes Curtis
Legner et al., 1969a

Spalangia nigroaenea Curtis
Legner (pre-publication), 1966
Legner et al., 1969a

Spalangia philippinensis Fullaway
Bartlett, 1939

Spalangia platensis (Brèthes)
Bouček, 1963

Spalangia rugosicollis Ashmead
Girault, 1920
Peck, 1951

[1] Although not specifically mentioned in each entry, it is assumed that all *Spalangia* are pupal parasites.

Stomoxys calcitrans (cont.)

Spalangia stomoxysiae Girault
Girault, 1916, 1920
Peck, 1951
Sphegigaster Spinola
Legner et al., 1969a
HYMENOPTERA: UNDETERMINED PTEROMALIDAE
Pinkus, 1913
Bishopp, 1913a
HYMENOPTERA: DIAPRIIDAE
Trichopria Ashmead
Bartlett, 1939
Legner et al., 1969a
HYMENOPTERA: FORMICIDAE
Pheidole megacephala (Fabricius)
Illingworth, 1923a
Ponera perkinsi Forel
Illingworth, 1923a
HYMENOPTERA: SPHECIDAE
Bembix oculata Latreille
Fabre, 1914
Crabro cribraria (Linnaeus)
Hamm et al., 1926
Stictia denticornis (Handlirsch)
Bodkin, 1917
HYMENOPTERA: VESPIDAE
Vespa germanica Fabricius
Kühlhorn, 1961
Vespa vulgaris Linnaeus
Kühlhorn, 1961
DIPTERA: UNDETERMINED CULICIDAE
Surcouf, 1923
DIPTERA: UNDETERMINED CHIRONOMIDAE
Surcouf, 1923
DIPTERA: UNDETERMINED ASILIDAE
Bishopp, 1913a
DIPTERA: ANTHOMYIIDAE
Scatophaga stercoraria (Linnaeus)
Hewitt, 1914
DIPTERA: MUSCIDAE
Hydrotaea dentipes (Fabricius)
Derbeneva-Ukhova, 1961
DIPTERA: CUTEREBRIDAE
Dermatobia hominis (Linnaeus)
Neiva et al., 1917
Neel et al., 1955
Zeledón, 1957

Stomoxys calcitrans (cont.)

CHORDATA: AMPHIBIA

OPISTHOCOELA: DISCOGLOSSIDAE

Discoglossus pictus Otth
Surcouf, 1923

PROCOELA: BUFONIDAE

Bufo mauritanicus Schlegel
Surcouf, 1923

Bufo viridis Laurenti
Surcouf, 1923

PROCOELA: HYLIDAE

Hyla viridis Laurenti var. *grisea*
Surcouf, 1923

DIPLASTIOCOELA: RANIDAE

Rana ridibunda Pallas
Surcouf, 1923

CHORDATA: REPTILIA

SQUAMATA: AGAMIDAE

Agama agilis Olivier
Surcouf, 1923

SQUAMATA: SCINCIDAE

Chalcides tridactylus Gray
Surcouf, 1923

SQUAMATA: LACERTIDAE

Acanthodactylus boskianus (Lichtenstein)
Surcouf, 1923

CHORDATA: AVES

GALLIFORMES: PHASIANIDAE

Gallus gallus Linnaeus
Bushnell et al., 1924

PASSERIFORMES: HIRUNDINIDAE

Hirundo rustica Linnaeus subsp. *rustica*
Séguy, 1929

Stomoxys ?calcitrans (Linnaeus) .... PROTOZOA: MASTIGOPHORA

PROTOMONADIDA: TRYPANOSOMATIDAE

Herpetomonas Kent
Gray, 1906

Trypanosoma brucei Plimmer and Bradford
Hall, 1927

Trypanosoma congolense Broden
Hall, 1927

Trypanosoma evansi (Steel)
Musgrave et al., 1903
Schat, 1909

Trypanosoma gambiense Dutton
Gray, 1906

Fly	Organism
Stomoxys ?calcitrans (cont.)	*Trypanosoma vivax* Ziemann Hall, 1927
Stomoxys calcitrans (Linnaeus) var. *soudanense*	PROTOZOA: MASTIGOPHORA PROTOMONADIDA: TRYPANOSOMATIDAE *Trypanosoma brucei* Plimmer and Bradford Bouet et al., 1912
Stomoxys nigra Macquart (Synonyms: *Stomoxys bouvieri, Stomoxys glauca*)	THALLOPHYTA: PHYCOMYCETES ENTOMOPHTHORALES: ENTOMOPHTHORACEAE *?Entomophthora* Fresenius Roubaud, 1911a PROTOZOA: MASTIGOPHORA PROTOMONADIDA: TRYPANOSOMATIDAE *Trypanosoma* Gruby Edington et al., 1907 *Trypanosoma brucei* Plimmer and Bradford Martin et al., 1908 Roubaud, 1909 Bouet et al., 1912 *Trypanosoma congolense* Broden Roubaud, 1909 Carmichael, 1934 *Trypanosoma evansi* (Steel) Bouet et al., 1912 Moutia, 1928 *Trypanosoma gambiense* Dutton Roubaud, 1909 Duke, 1913 *Trypanosoma rhodesiense* Stephens and Fantham Lamborn, 1933 *Trypanosoma vivax* Ziemann Bouet et al., 1912 PROTOMONADIDA: BODONIDAE *Bodo* Ehrenberg Duke, 1913
UNDETERMINED MUSCIDAE	SCHIZOPHYTA: SCHIZOMYCETES EUBACTERIALES: ENTEROBACTERIACEAE *Salmonella chester* Kauffmann Bulling et al., 1959 *Salmonella kottbus* Schütze et al. Bulling et al., 1959 *Salmonella typhimurium* (Loeffler) Castellani and Chalmers Bulling et al., 1959

Fly	Organism
UNDETERMINED MUSCIDAE (cont.)	EUBACTERIALES: BACILLACEAE *Clostridium chauvoei* (Arloing, Cornevin, and Thomas) Scott Sauer, 1908 ACTINOMYCETALES: MYCOBACTERIACEAE *Mycobacterium leprae* (Hansen) Lehmann and Neumann Lamborn, 1936a ARTHROPODA: INSECTA COLEOPTERA: STAPHYLINIDAE *Aleochara handschini* Scheerpeltz Handschin, 1934a HYMENOPTERA: PTEROMALIDAE *Pachycrepoideus vindemmiae* Rondani Handschin, 1934a *Spalangia*[1] *orientalis* Graham Handschin, 1932, 1934b *Spalangia sundaica* Graham Handschin, 1932, 1934a, 1934b HYMENOPTERA: ENCYRTIDAE *Cerchysius lyperosae* Ferrière Handschin, 1934a *Tachinaephagus giraulti* Johnston and Tiegs Handschin, 1934a HYMENOPTERA: EULOPHIDAE *Trichospilus pupivora* Ferrière Handschin, 1934a HYMENOPTERA: DIAPRIIDAE *Phaenopria* Ashmead Handschin, 1934a HYMENOPTERA: FORMICIDAE *Monomorium pharaonis* (Linnaeus) Handschin, 1934a DIPTERA: CUTEREBRIDAE *Dermatobia hominis* (Linnaeus) Neel et al., 1955 ENTEROVIRUSES
Superfamily: OESTROIDEA Family: CALLIPHORIDAE Subfamily: MESEMBRINELLINAE	
Huascaromusca bicolor (Fabricius)	Poliovirus Types 1 and 3 Paul et al., 1962 ECHO virus Types 8, 12, 14 Paul et al., 1962

[1] Although not specifically mentioned in each entry, it is assumed that all *Spalangia* are pupal parasites.

Fly	Associated organism
Subfamily: RHINIINAE	
Rhyncomyia pictifacies Bigot	ARTHROPODA: INSECTA
	DIPTERA: BOMBYLIIDAE
	Thyridanthrax abruptus Loew
	McDonald, 1958
Subfamily: CHRYSOMYINAE	
Cochliomyia Townsend (Synonyms: *Compsomyia, Callitroga*)	ASCHELMINTHES: NEMATODA
	TRICHURIDEA: TRICHURIDAE
	Trichuris Roederer
	Tao, 1936
	STRONGYLIDEA: ANCYLOSTOMATIDAE
	Ancylostoma (Dubini)
	Tao, 1936
	ASCARIDIDEA: ASCARIDIDAE
	Ascaris Linnaeus
	Tao, 1936
	ARTHROPODA: INSECTA
	HYMENOPTERA: FORMICIDAE
	Tapinoma melanocephalum (Fabricius)
	Pimentel, 1955
Cochliomyia hominivorax (Coquerel) (Synonym: *Cochliomyia americana*)	ARTHROPODA: INSECTA
	HYMENOPTERA: FORMICIDAE
	Crematogaster lineolata (Say)
	Lindquist, 1942
	Eciton Latreille
	Strong, 1938
	Forelius foetidus (Buckley)
	Lindquist, 1942
	Labidus coecus (Latreille)
	Lindquist, 1942
	Pheidole Westwood
	Strong, 1938
	Lindquist, 1942
	Pheidole morrisii var. *impexa* Wheeler
	Lindquist, 1942
	Pogonomyrmex barbatus var. *fuscatus* Emery
	Lindquist, 1942
	Pogonomyrmex barbatus var. *molefaciens* (Buckley)
	Lindquist, 1942
	Solenopsis geminata (Fabricius)
	Lindquist, 1942
	ENTEROVIRUSES
Cochliomyia macellaria (Fabricius) (Synonyms: *Callitroga*	Poliovirus
	Melnick et al., 1945
	Francis, Jr. et al., 1948

macellaria, Chrysomyia macellaria, Lucillia macellaria)

Poliovirus Type 1
Melnick et al., 1953
Paul et al., 1962
Poliovirus Type 3
Paul et al., 1962
Coxsackievirus Alaska 5 (A10)
Melnick et al., 1953
Coxsackievirus Texas 1 (A4)
Melnick et al., 1953
Coxsackievirus Texas 15 (A7)
Melnick et al., 1953
Coxsackievirus Type 3 (A3)
Melnick et al., 1953
ECHO virus Types 8, 12, 14
Paul et al., 1962

SCHIZOPHYTA: SCHIZOMYCETES
EUBACTERIALES:
ENTEROBACTERIACEAE
Escherichia coli (Migula) Castellani and Chalmers
Beyer, 1925
Salmonella Lignières
Alcivar et al., 1946
Greenberg et al., 1964
Salmonella alachua Kauffmann
Greenberg et al., 1963b
Salmonella amersfoort Henning
Alcivar, 1946
Salmonella anatum (Rettger and Scoville) Bergey et al.
Alcivar et al., 1946
Greenberg et al., 1963b
Salmonella ballerup Kauffmann and Møller
Alcivar et al., 1946
Salmonella cholerae-suis (Smith) Weldin
Alcivar et al., 1946
Salmonella derby Warren and Scott
Greenberg et al., 1963b
Salmonella dublin Kauffmann
Alcivar et al., 1946
Salmonella enteritidis (Gaertner) Castellani and Chalmers
Alcivar et al., 1946
Salmonella florida Cherry, Edwards, and Bruner
Alcivar et al., 1946

Cochliomyia macellaria (cont.)	*Salmonella give* Kauffmann Alcivar et al., 1946 Greenberg et al., 1963b *Salmonella kentucky* Edwards Alcivar et al., 1946 *Salmonella kottbus* Schütze et al. Alcivar et al., 1946 *Salmonella meleagridis* Bruner and Edwards Alcivar et al., 1946 Greenberg et al., 1963b *Salmonella narashino* Kauffmann Alcivar et al., 1946 *Salmonella new brunswick* Edwards Greenberg, 1963b *Salmonella newport* Schütze Greenberg et al., 1963b *Salmonella onderstepoort* Henning Alcivar et al., 1946 *Salmonella oranienburg* Kauffmann Alcivar et al., 1946 *Salmonella oregon* Edwards and Bruner Alcivar et al., 1946 *Salmonella panama* Kauffmann Greenberg et al., 1963b *Salmonella paratyphi* B Castellani and Chalmers Alcivar et al., 1946 *Salmonella reading* Schütze Alcivar et al., 1946 *Salmonella typhi* Warren and Scott Beyer, 1925 *Salmonella typhimurium* (Loeffler) Castellani and Chalmers Alcivar et al., 1946 *Salmonella zanzibar* Kauffmann Alcivar et al., 1946 *Shigella alkalescens* (Andrewes) Weldin Alcivar et al., 1946 *Shigella boydii* Ewing 1, 2, 3, 4, 5, 6, 8, 9 Alcivar et al., 1946 *Shigella ceylonensis* (Weldin) Weldin Alcivar et al., 1946 *Shigella dispar* (Andrewes) Bergey et al. Alcivar et al., 1946

Cochliomyia macellaria (cont.)

Shigella flexneri Castellani and Chalmers 1a, 2a, Y, 6
Alcivar et al., 1946

EUBACTERIALES: BRUCELLACEAE
Pasteurella pestis (Lehmann and Neumann) Holland
Gosio, 1925

PROTOZOA: MASTIGOPHORA

PROTOMONADIDA: TRYPANOSOMATIDAE
Herpetomonas muscarum (Leidy) Kent
Becker, 1922-3, 1924

POLYMASTIGIDA: CHILOMASTIGIDAE
Chilomastix mesnili (Wenyon)
Root, 1921

POLYMASTIGIDA: HEXAMITIDAE
Giardia intestinalis (Lambl)
Root, 1921

PROTOZOA: SARCODINA

AMOEBIDA: ENDAMOEBIDAE
Endolimax nana (Wenyon and O'Connor)
Root, 1921
Entamoeba coli (Grassi)
Root, 1921
Entamoeba histolytica Schaudinn
Root, 1921
Pipkin, 1942, 1949

PROTOZOA: CNIDOSPORIDIA

MICROSPORIDA: MRAZEKIIDAE
Octosporea muscae-domesticae Flu
Kramer, 1964a, 1965b

ARTHROPODA: INSECTA

HYMENOPTERA: BRACONIDAE
Alysia ridibunda Say
Roberts, 1935

HYMENOPTERA: CYNIPIDAE
Eucoila Westwood
Roberts, 1935

HYMENOPTERA: FIGITIDAE
Neralsia Cameron
Roberts, 1935

HYMENOPTERA: CHALCIDIDAE
Brachymeria fonscolombei (Dufour)
Bishopp, 1929-30
Marlatt, 1931
Roberts, 1933a, 1933b, 1935

Fly	Associated organism
Cochliomyia macellaria (cont.)	HYMENOPTERA: PTEROMALIDAE *Muscidifurax raptor* Girault and Sanders Girault et al., 1910b *Nasonia vitripennis* (Walker) Girault et al., 1910a Marlatt, 1931 Roberts, 1933a, 1935 HYMENOPTERA: FORMICIDAE *Paratrechina longicornis* (Latreille) Pimentel, 1955 *Solenopsis geminata* (Fabricius) Pimentel, 1955 DIPTERA: UNDETERMINED CULICINAE Zepeda, 1913 DIPTERA: CUTEREBRIDAE *Dermatobia hominis* (Linnaeus) Lins de Almeida, 1933 Neel et al., 1955
Chrysomya Robineau-Desvoidy (Synonyms: *Chrysomyia*, *Pycnosoma*)	SCHIZOPHYTA: SCHIZOMYCETES PSEUDOMONADALES: PSEUDOMONADACEAE *Pseudomonas* Migula de la Paz, 1938 PSEUDOMONADALES: SPIRILLACEAE *Vibrio* Müller de la Paz, 1938 EUBACTERIALES: ACHROMOBACTERACEAE *Alcaligenes* Castellani and Chalmers de la Paz, 1938 EUBACTERIALES: ENTEROBACTERIACEAE *Escherichia* Castellani and Chalmers de la Paz, 1938 *Aerobacter* Beijerinck de la Paz, 1938 *Proteus* Hauser de la Paz, 1938 *Proteus morganii* (Winslow, Kligler, and Rothberg) Yale de la Paz, 1938 *Salmonella cholerae-suis* (Smith) Weldin de la Paz, 1938

Chrysomya (cont.)	*Salmonella columbensis* (Castellani) Castellani and Chalmers de la Paz, 1938 *Salmonella typhi* Warren and Scott de la Paz, 1938 *Eberthella belfastiensis* (Weldin and Levine) Bergey et al. de la Paz, 1938 *Shigella ambigua* (Andrewes) Weldin de la Paz, 1938 *Shigella boydii* Ewing 2, 5, 6 Richards et al., 1961 *Shigella dysenteriae* (Shiga) Castellani and Chalmers de la Paz, 1938 *Shigella dysenteriae* (Shiga) Castellani and Chalmers 2 Richards et al., 1961 *Shigella flexneri* Castellani and Chalmers de la Paz, 1938 *Shigella flexneri* Castellani and Chalmers 1a, 2a, 3, 4a, 4b, 6 Richards et al., 1961 *Shigella giumai* (Castellani) Hauduroy et al. de la Paz, 1938 *Shigella sonnei* (Levine) Weldin de la Paz, 1938 *Shigella sonnei* (Levine) Weldin 1 Richards et al., 1961 EUBACTERIALES: MICROCOCCACEAE *Staphylococcus* Rosenbach de la Paz, 1938 EUBACTERIALES: LACTOBACILLACEAE *Streptococcus* Rosenbach de la Paz, 1938 PROTOZOA: MASTIGOPHORA PROTOMONADIDA: TRYPANOSOMATIDAE *Leptomonas soudanensis* Roubaud Roubaud, 1911d PROTOZOA: SARCODINA AMOEBIDA: ENDAMOEBIDAE *Entamoeba histolytica* Schaudinn Anonymous ("Korean authors et al.")

Chrysomya (cont.)

ARTHROPODA: INSECTA
HYMENOPTERA: PTEROMALIDAE
Nasonia vitripennis (Walker)
Roubaud, 1917

Chrysomya albiceps (Wiedemann) ....

SCHIZOPHYTA: SCHIZOMYCETES
EUBACTERIALES: BACILLACEAE
Clostridium parabotulinum bovis Robinson
Theiler, 1927

PLATYHELMINTHES: CESTOIDEA
CYCLOPHYLLIDEA: TAENIIDAE
Taeniarhynchus saginatum (Goeze)
Round, 1961

ARTHROPODA: INSECTA
HYMENOPTERA: BRACONIDAE
Alysia manducator Panzer
Smit, 1929

HYMENOPTERA: PTEROMALIDAE
Nasonia vitripennis[1] (Walker)
Anonymous, 1925
Smit, 1929
Ullyett, 1950

Spalangia endius Walker
Bouček, 1963

HYMENOPTERA: EULOPHIDAE
Syntomosphyrum glossinae Waterston
Nash, 1933

Chrysomya chloropyga (Wiedemann) ....

SCHIZOPHYTA: SCHIZOMYCETES
EUBACTERIALES: BACILLACEAE
Clostridium parabotulinum bovis Robinson
Theiler, 1927

PLATYHELMINTHES: CESTOIDEA
CYCLOPHYLLIDEA: TAENIIDAE
Taeniarhynchus saginatum (Goeze)
Round, 1961

ARTHROPODA: INSECTA
HYMENOPTERA: PTEROMALIDAE
Nasonia vitripennis (Walker)
Ullyett, 1950

HYMENOPTERA: EUPELMIDAE
Mesocomys pulchriceps Cameron
Anonymous, 1925

[1] Although not specifically mentioned in each entry, it is assumed that all *Nasonia vitripennis* are pupal parasites.

Chrysomya chloropyga (cont.)	HYMENOPTERA: EULOPHIDAE *Syntomosphyrum glossinae* Waterston Nash, 1933
Chrysomya marginalis (Wiedemann) (Synonyms: *Chrysomyia marginatum, Pycnosoma marginale*)	SCHIZOPHYTA: SCHIZOMYCETES EUBACTERIALES: BACILLACEAE *Clostridium parabotulinum bovis* Robinson Theiler, 1927 PROTOZOA: MASTIGOPHORA PROTOMONADIDA: TRYPANOSOMATIDAE *Herpetomonas muscarum* (Leidy) Kent Roubaud, 1908b *Leptomonas mirabilis* Roubaud Roubaud, 1908b, 1911c ARTHROPODA: INSECTA HYMENOPTERA: EULOPHIDAE *Syntomosphyrum glossinae* Waterston Nash, 1933
Chrysomya megacephala (Fabricius) (Synonyms: *Calliphora incisuralis, Pycnosoma dux, Chrysomyia dux*)	ENTEROVIRUSES Poliovirus Type 2 Asahina et al., 1963 SCHIZOPHYTA: SCHIZOMYCETES EUBACTERIALES: ENTEROBACTERIACEAE *Escherichia coli* (Migula) Castellani and Chalmers Chow, 1940 *Escherichia coli* (Migula) Castellani and Chalmers Shimizu et al., 1965 *Proteus mirabilis* Hauser Shimizu et al., 1965 *Proteus morganii* (Winslow, Kligler, and Rothberg) Yale Shimizu et al., 1965 *Proteus rettgeri* (Hadley et al.) Rustigian and Stuart Shimizu et al., 1965 *Proteus vulgaris* Hauser Shimizu et al., 1965 *Providencia* Kauffmann Shimizu et al., 1965 *Salmonella* Lignières Shimizu et al., 1965 *Salmonella typhi* Warren and Scott Chow, 1940

Chrysomya megacephala (cont.)

Shigella Castellani and Chalmers
Shimizu et al., 1965
How

Shigella dysenteriae (Shiga) Castellani and Chalmers
Chow, 1940

PROTOZOA: MASTIGOPHORA

PROTOMONADIDA: TRYPANOSOMATIDAE

Herpetomonas Kent
Chang, 1940

Leptomonas mirabilis Roubaud
Patton, 1921a

POLYMASTIGIDA: CHILOMASTIGIDAE

Chilomastix mesnili (Wenyon)
Chang, 1940
Harris et al., 1946

POLYMASTIGIDA: HEXAMITIDAE

Giardia lamblia Stiles
Chang, 1940
Harris et al., 1946

TRICHOMONADIDA: TRICHOMONADIDAE

Trichomonas hominis (Davaine)
Chang, 1940
Harris et al., 1946

PROTOZOA: SARCODINA

AMOEBIDA: ENDAMOEBIDAE

Endolimax nana (Wenyon and O'Connor)
Chang, 1940
Harris et al., 1946

Entamoeba coli (Grassi)
Chang, 1940
Harris et al., 1946

Entamoeba histolytica Schaudinn
Chang, 1940, 1945
Harris et al., 1946

Iodamoeba bütschlii (Prowazek)
Chang, 1940
Harris et al., 1946

PLATYHELMINTHES: CESTOIDEA

CYCLOPHYLLIDEA: HYMENOLEPIDIDAE

Hymenolepis diminuta (Rudolphi) Blanchard
Harris et al., 1946

Fly	Associated organism
Chrysomya megacephala (cont.)	ASCHELMINTHES: NEMATODA
	TRICHURIDEA: TRICHURIDAE
	Trichuris Roederer
	Chang, 1945
	Trichuris trichiura (Linnaeus)
	Chang, 1940
	Harris et al., 1946
	STRONGYLIDEA: ANCYLOSTOMATIDAE
	Ancylostoma (Dubini)
	Chang, 1940, 1945
	Harris et al., 1946
	ASCARIDIDEA: ASCARIDIDAE
	Ascaris Linnaeus
	Chang, 1945
	Ascaris lumbricoides Linnaeus
	Chang, 1940
	Harris et al., 1946
	ARTHROPODA: INSECTA
	HYMENOPTERA: CHALCIDIDAE
	Brachymeria fulvitarsis (Cameron)
	Roy et al., 1939
	Dirhinus pachycerus Masi
	Roy et al., 1939
	HYMENOPTERA: PTEROMALIDAE
	Spalangia[1] Latreille
	Roy et al., 1939
	Spalangia endius Walker
	Bouček, 1963
	Spalangia muscidarum Richardson
	Johnston et al., 1920b, 1921
	Spalangia nigroaenea Curtis
	Bouček, 1963
	HYMENOPTERA: ENCYRTIDAE
	Australencyrtus giraulti Johnston and Tiegs
	Johnston et al., 1921
Chrysomya pinguis (Walker)	SCHIZOPHYTA: SCHIZOMYCETES
	EUBACTERIALES: ENTEROBACTERIACEAE
	Escherichia coli (Migula) Castellani and Chalmers
	Shimizu et al., 1965
	Proteus mirabilis Hauser
	Shimizu et al., 1965
	Proteus rettgeri (Hadley et al.) Rustigian and Stuart
	Shimizu et al., 1965

[1] Although not specifically mentioned in each entry, it is assumed that all *Spalangia* are pupal parasites.

Fly	Organism / Reference
Chrysomya pinguis (cont.)	*Proteus vulgaris* Hauser
	Shimizu et al., 1965
	Providencia Kauffmann
	Shimizu et al., 1965
Chrysomya putoria (Wiedemann) (Synonyms: *Chrysomyia putoria, Chrysomyia putorium, Pycnosoma putorium*)	ENTEROVIRUSES
	Poliovirus Types 1 and 3
	Brygoo et al., 1962
	Coxsackievirus
	Brygoo et al., 1962
	UNDETERMINED ENTEROVIRUSES
	Brygoo et al., 1962
	SCHIZOPHYTA: SCHIZOMYCETES
	EUBACTERIALES: ENTEROBACTERIACEAE
	Escherichia coli (Migula) Castellani and Chalmers 26 B6
	Brygoo et al., 1962
	Escherichia coli (Migula) Castellani and Chalmers III B4
	Brygoo et al., 1962
	Salmonella cholerae-suis (Smith) Weldin
	Brygoo et al., 1962
	Salmonella give Kauffmann
	Brygoo et al., 1962
	Salmonella typhimurium (Loeffler) Castellani and Chalmers
	Brygoo et al., 1962
	Shigella flexneri Castellani and Chalmers
	Brygoo et al., 1962
	EUBACTERIALES: MICROCOCCACEAE
	Staphylococcus Rosenbach (pathogenic)
	Brygoo et al., 1962
	THALLOPHYTA: PHYCOMYCETES
	ENTOMOPHTHORALES: ENTOMOPHTHORACEAE
	?*Entomophthora* Fresenius
	Roubaud, 1911a
	PROTOZOA: MASTIGOPHORA
	PROTOMONADIDA: TRYPANOSOMATIDAE
	Herpetomonas Kent
	Roubaud, 1912a
	Herpetomonas muscarum (Leidy) Kent
	Roubaud, 1908b, 1909
	Leptomonas Kent
	Roubaud, 1909

Fly	Organism / References
Chrysomya putoria (cont.)	*Leptomonas mirabilis* Roubaud Roubaud, 1908b, 1909, 1912a *Leptomonas pycnosomae* Roubaud Roubaud, 1909 *Leptomonas soudanensis* Roubaud Roubaud, 1912a
	ARTHROPODA: INSECTA HYMENOPTERA: EULOPHIDAE *Syntomosphyrum glossinae* Waterston Lamborn, 1925 Nash, 1933
Chrysomya rufifacies (Macquart) (Synonyms: *Chrysomyia albiceps, Pycnosoma rufifacies*)	SCHIZOPHYTA: SCHIZOMYCETES EUBACTERIALES: BACILLACEAE *Bacillus* Cohn Zmeev, 1943
	ACTINOMYCETALES: MYCOBACTERIACEAE *Mycobacterium tuberculosis* (Zopf) Lehmann and Neumann Morellini et al., 1947
	PROTOZOA: MASTIGOPHORA PROTOMONADIDA: TRYPANOSOMATIDAE *Leptomonas mirabilis* Roubaud Patton, 1921a
	PLATYHELMINTHES: CESTOIDEA CYCLOPHYLLIDEA: TAENIIDAE *Taeniarhynchus saginatum* (Goeze) Sychevskaîa et al., 1958
	ARTHROPODA: INSECTA HYMENOPTERA: BRACONIDAE *Alysia manducator* Panzer Miller, 1927 Newman, 1928 Morgan, 1929
	HYMENOPTERA: CHALCIDIDAE *Dirhinus sarcophagae* Froggatt Johnston et al., 1921
	HYMENOPTERA: PTEROMALIDAE *Nasonia vitripennis*[1] (Walker) Froggatt et al., 1917 Froggatt, 1919, 1921, 1922 Johnston et al., 1921 Gurney et al., 1926a

[1] Although not specifically mentioned in each entry, it is assumed that all *Nasonia vitripennis* are pupal parasites.

Fly	Associated organisms
Chrysomya rufifacies (cont.)	*Spalangia*[1] *muscidarum* Richardson Johnston et al., 1920b, 1921 *Spalangia nigroaenea* Curtis Bouček, 1963 HYMENOPTERA: ENCYRTIDAE *Australencyrtus giraulti* Johnston and Tiegs Johnston et al., 1921 *Stenoterys fulvoventralis* Dodd Newman et al., 1930 DIPTERA: CALLIPHORIDAE *Chrysomya villeneuvei* Patton Patton, 1922
Chrysomya varipes (Macquart) (Synonyms: *Microcalliphora varipes, Pycnosoma varipes*)	ARTHROPODA: INSECTA HYMENOPTERA: CHALCIDIDAE *Dirhinus sarcophagae* Froggatt Johnston et al., 1921 HYMENOPTERA: PTEROMALIDAE *Nasonia vitripennis*[2] (Walker) Froggatt et al., 1917 Froggatt, 1919 Johnston et al., 1921 *Spalangia muscidarum* Richardson Johnston et al., 1920b, 1921 *Spalangia nigroaenea* Curtis Bouček, 1963 HYMENOPTERA: ENCYRTIDAE *Australencyrtus giraulti* Johnston and Tiegs Johnston et al., 1921 HYMENOPTERA: SPHECIDAE *Nysson* Latreille Froggatt, 1917 *Stizus* Latreille Froggatt, 1917 *Stizus turneri* Froggatt Froggatt, 1917
Paralucilia fulvipes (Macquart) (Synonyms: ?*Calliphora tibialis*)	ARTHROPODA: INSECTA HYMENOPTERA: SPHECIDAE *Sericophorus teliferopodus* Rayment Rayment, 1954

[1] Although not specifically mentioned in each entry, it is assumed that all *Spalangia* are pupal parasites.

[2] Although not specifically mentioned in each entry, it is assumed that all *Nasonia vitripennis* are pupal parasites.

Fly	Organism / Reference
Phormia Robineau-Desvoidy	ENTEROVIRUSES
	Poliovirus
	Trask et al., 1943
	Coxsackievirus
	Melnick et al., 1954
	SCHIZOPHYTA: SCHIZOMYCETES
	EUBACTERIALES: BACILLACEAE
	Clostridium perfringens (Veillon and Zuber) Holland
	Green, 1953
Phormia regina (Meigen)	VIRUSES
	ARBOVIRUSES
	Eastern equine encephalomyelitis
	Bourke, 1964
	ENTEROVIRUSES
	Poliovirus
	Paul et al., 1941
	Sabin et al., 1942
	Melnick et al., 1945
	Francis Jr. et al., 1948
	Melnick, 1949
	Poliovirus Delaware '48 strain
	Melnick et al., 1952
	Poliovirus Type 1
	Riordan et al., 1961
	Downey, 1963
	Poliovirus Attenuated Type 1, LSc
	Gudnadóttir, 1960
	Gudnadóttir et al., 1960
	Poliovirus Mahoney Type 1
	Gudnadóttir, 1960
	Gudnadóttir et al., 1960
	Poliovirus New York City '44 strain (Type 1)
	Melnick et al., 1947, 1952
	Poliovirus Philippine Islands '45 strain (Type 1)
	Melnick et al., 1952
	Poliovirus Texas '48 strain (Type 1)
	Melnick et al., 1952
	Poliovirus W-S '48 strain (Type 1)
	Melnick et al., 1952
	Poliovirus Type 2, Lansing strain
	Melnick et al., 1947, 1952
	Poliovirus Y-SK strain, murine adapted, Type 2
	Melnick et al., 1947, 1952
	Poliovirus Type 3
	Riordan et al., 1961

Phormia regina (cont.)	Theiler's mouse encephalomyelitis virus, TO strain Melnick et al., 1947, 1952 Coxsackievirus Texas 1 (A4) Melnick et al., 1952 Coxsackievirus Texas '48 (new type) Melnick et al., 1952 Coxsackievirus Type A9 (A9) Riordan et al., 1961 Coxsackievirus Type B5 (B5) Riordan et al., 1961 Coxsackievirus W-S '48 (B1) Melnick et al., 1952 ECHO virus Types 1, 5, 11 Riordan et al., 1961 ECHO virus Types 6 and 7 Downey, 1963 SCHIZOPHYTA: SCHIZOMYCETES PSEUDOMONADALES: PSEUDOMONADACEAE *Pseudomonas aeruginosa* (Schroeter) Migula Greenberg, 1962 EUBACTERIALES: ENTEROBACTERIACEAE *Escherichia coli* (Migula) Castellani and Chalmers Beyer, 1925 Greenberg, 1962 *Escherichia freundii* (Braak) Yale O'Keefe et al., 1954 *Aerobacter* Beijerinck Knuckles, 1959 *Proteus morganii* (Winslow, Kligler, and Rothberg) Yale Knuckles, 1959 *Proteus vulgaris* Hauser Maseritz, 1934 O'Keefe et al., 1954 Greenberg, 1962 *Salmonella blockley* Kauffmann Greenberg et al., 1964 *Salmonella derby* Warren and Scott Greenberg et al., 1963b *Salmonella give* Kauffmann Greenberg et al., 1963b *Salmonella muenchen* Kauffmann Greenberg et al., 1963b

Phormia regina (cont.)

Salmonella new brunswick Edwards
Greenberg et al., 1963b
Salmonella newport Schütze
Greenberg et al., 1963b
Salmonella schottmuelleri (Winslow, Kligler, and Rothberg) Bergey et al.
Knuckles, 1959
Salmonella typhi Warren and Scott
Maseritz, 1934
Greenberg, 1962
Salmonella typhimurium (Loeffler) Castellani and Chalmers
Knuckles, 1959
Shigella Castellani and Chalmers
Greenberg et al., 1964
Shigella flexneri Castellani and Chalmers
Greenberg, 1962

EUBACTERIALES: MICROCOCCACEAE

Staphylococcus Rosenbach
Picado, 1935
Staphylococcus aureus Rosenbach
Maseritz, 1934
Greenberg, 1962
Staphylococcus flavus (Trevisan) Wood
Guyénot, 1907
Maseritz, 1934
Sarcina subflava Ravenel
Beyer, 1925

EUBACTERIALES: LACTOBACILLACEAE

Diplococcus pneumoniae Weichselbaum
Maseritz, 1934
Streptococcus durans Sherman and Wing
Greenberg, 1962

EUBACTERIALES: BACILLACEAE

Bacillus subtilis Cohn
Greenberg, 1962

PROTOZOA: MASTIGOPHORA

PROTOMONADIDA: TRYPANOSOMATIDAE

Herpetomonas muscarum (Leidy) Kent
Becker, 1922-3, 1924

Phormia regina (cont.)

PROTOZOA: SARCODINA
AMOEBIDA: ENDAMOEBIDAE
Entamoeba histolytica Schaudinn
Pipkin, 1942, 1949
PROTOZOA: CNIDOSPORIDIA
MICROSPORIDA: NOSEMATIDAE
Nosema apis Zander
Kramer, 1962
Nosema kingi Kramer
Kramer, 1964b
MICROSPORIDA: MRAZEKIIDAE
Octosporea muscae-domesticae Flu
Kramer, 1964a, 1965a, 1965b, 1965c, 1966
ARTHROPODA: ARACHNIDA
ACARINA: MACROCHELIDAE
Macrocheles muscaedomesticae (Scopoli)
Kinn, 1966
ARTHROPODA: INSECTA
HYMENOPTERA: BRACONIDAE
Aphaereta pallipes Say
Beard, 1964
HYMENOPTERA: CHALCIDIDAE
Brachymeria fonscolombei (Dufour)
Marlatt, 1931
Roberts, 1933b, 1935
HYMENOPTERA: PTEROMALIDAE
Muscidifurax raptor Girault and Sanders
Girault et al., 1910b
Legner (pre-publication), 1966
Nasonia vitripennis[1] (Walker)
Girault et al., 1910a
Marlatt, 1931
Roberts, 1935
Legner (pre-publication), 1966
Nasonia ?vitripennis (Walker)
Bishopp et al., 1915b
CHORDATA: AVES
PASSERIFORMES: HIRUNDINIDAE
Delichon urbica (Linnaeus) subsp. *urbica*
Séguy, 1929
Hirundo rustica Linnaeus subsp. *rustica*
Séguy, 1929

[1] Although not specifically mentioned in each entry, it is assumed that all *Nasonia vitripennis* are pupal parasites.

Fly	Associated organism
Phormia sordida Zetterstedt	ARTHROPODA: INSECTA HYMENOPTERA: PTEROMALIDAE *Nasonia vitripennis* (Walker) Roubaud, 1917
Boreellus atriceps (Zetterstedt) (Synonym: *Phormia coerulea*)	ARTHROPODA: INSECTA DIPTERA: MUSCIDAE *Myospila meditabunda* (Fabricius) Porchinskiĭ, 1910
Protophormia Townsend	ENTEROVIRUSES Poliovirus Trask et al., 1943
Protophormia terraenovae (Robineau-Desvoidy) (Synonyms: *Phormia terraenovae*, *Phormia groenlandica*)	ENTEROVIRUSES Poliovirus Sabin et al., 1942 Poliovirus Strain 113 Levkovich et al., 1957 Poliovirus Type 2, Lansing strain Levkovich et al., 1957 SCHIZOPHYTA: SCHIZOMYCETES PSEUDOMONADALES: PSEUDOMONADACEAE *Pseudomonas aeruginosa* (Schroeter) Migula Havlik, 1964 *Pseudomonas fluorescens* Migula Lysenko et al., 1961 EUBACTERIALES: ACHROMOBACTERACEAE *Alcaligenes faecalis* Castellani and Chalmers Lysenko, 1958 Gregor et al., 1960 Lysenko et al., 1961 Havlik, 1964 *Flavobacterium* Bergey et al. Lysenko et al., 1961 *Flavobacterium devorans* (Zimmermann) Bergey et al. Gregor et al., 1960 *Flavobacterium invisible* (Vaughan) Bergey et al. Gregor et al., 1960 EUBACTERIALES: ENTEROBACTERIACEAE *Escherichia* Castellani and Chalmers Lysenko et al., 1961 *Escherichia coli* (Migula) Castellani and Chalmers Pavillard et al., 1957

Protophormia terraenovae (cont.)	Lysenko, 1958
	Gregor et al., 1960
	Radvan, 1960a, 1960b, 1960c
	Havlik, 1964
	Aerobacter aerogenes (Kruse) Beijerinck
	Gregor et al., 1960
	Havlik, 1964
	Klebsiella Trevisan
	Lysenko et al., 1961
	Klebsiella cloacae (Jordan) Brisou
	Gregor et al., 1960
	Paracolobactrum coliforme Borman, Stuart, and Wheeler
	Radvan, 1960a, 1960b
	Serratia marcescens Bizio
	Radvan, 1960c
	Gregor et al., 1960
	Havlik, 1964
	Proteus inconstans (Ornstein) Shaw and Clarke
	Lysenko, 1958
	Gregor et al., 1960
	Proteus mirabilis Hauser
	Lysenko, 1958
	Gregor et al., 1960
	Lysenko et al., 1961
	Shimizu et al., 1965
	Proteus morganii (Winslow, Kligler, and Rothberg) Yale
	Lysenko, 1958
	Gregor et al., 1960
	Lysenko et al., 1961
	Havlik, 1964
	Shimizu et al., 1965
	Proteus rettgeri (Hadley et al.) Rustigian and Stuart
	Lysenko, 1958
	Gregor et al., 1960
	Lysenko et al., 1961
	Shimizu et al., 1965
	Proteus vulgaris Hauser
	Pavillard et al., 1957
	Lysenko, 1958
	Gregor et al., 1960
	Radvan, 1960a, 1960b, 1960c
	Lysenko et al., 1961
	Havlik, 1964
	Shimizu et al., 1965

Protophormia terraenovae (cont.)	*Providencia* Kauffmann
	Radvan, 1960a, 1960b, 1960c
	Lysenko et al., 1961
	Shimizu et al., 1965
	Salmonella schottmuelleri (Winslow, Kligler, and Rothberg) Bergey et al.
	Radvan, 1960a, 1960b, 1960c
	Salmonella typhi Warren and Scott
	Webb et al., 1956
	Salmonella typhimurium (Loeffler) Castellani and Chalmers
	Pavillard et al., 1957
	Ojala et al., 1966
	Citrobacter freundii Werkman and Gillen
	Gregor et al., 1960
	Shigella Castellani and Chalmers
	Zaidenov, 1961
	Shigella sonnei (Levine) Weldin
	Radvan, 1960a, 1960b, 1960c
	EUBACTERIALES: UNDETERMINED ENTEROBACTERIACEAE
	Klebsiella-Cloaca intermed. forms
	Lysenko et al., 1961
	EUBACTERIALES: MICROCOCCACEAE
	Staphylococcus afermentans Shaw, Stitt, and Cowan
	Gregor et al., 1960
	Staphylococcus aureus Rosenbach
	Pavillard et al., 1957
	Gregor et al., 1960
	Radvan, 1960a, 1960c
	Staphylococcus lactis Shaw, Stitt, and Cowan
	Lysenko, 1958
	Gregor et al., 1960
	Lysenko et al., 1961
	Staphylococcus saprophyticus Fairbrother
	Gregor et al., 1960
	Lysenko et al., 1961
	Sarcina flava de Bary
	Lysenko et al., 1961
	EUBACTERIALES: LACTOBACILLACEAE
	Diplococcus pneumoniae Weichselbaum I
	Pavillard et al., 1957

Protophormia terraenovae (cont.)

Streptococcus Rosenbach
Radvan, 1960a, 1960c
Streptococcus agalactiae Lehmann and Neumann Group B
Pavillard et al., 1957
Streptococcus durans Sherman and Wing
Gregor et al., 1960
Lysenko et al., 1961
Streptococcus faecalis var. *liquefaciens* (Sternberg) Mattick
Lysenko et al., 1961
Streptococcus mitis Andrewes and Horder
Pavillard et al., 1957
Radvan, 1960c
Streptococcus pyogenes Rosenbach Group A
Pavillard et al., 1957

EUBACTERIALES: CORYNEBACTERIACEAE

Listeria monocytogenes (Murray et al.) Pirie
Radvan, 1960a

EUBACTERIALES: BACILLACEAE

Bacillus anthracis Cohn
Radvan, 1960a, 1960b, 1960c
Bacillus cereus Frankland and Frankland
Radvan, 1960b, 1960c
Bacillus cereus var. *mycoides* (Flügge) Smith et al.
Gregor et al., 1960
Bacillus subtilis Cohn
Pavillard et al., 1957
Lysenko, 1958
Radvan, 1960a, 1960c
Lysenko et al., 1961
Clostridium perfringens (Veillon and Zuber) Holland
Pavillard et al., 1957

THALLOPHYTA: FUNGI IMPERFECTI CRYPTOCOCCALES: CRYPTOCOCCACEAE

Candida albicans (Robin) Berkhout
Pavillard et al., 1957

Fly	Associated organisms
Protophormia terraenovae (cont.)	PROTOZOA: SPOROZOA COCCIDIDA: EIMERIIDAE *Eimeria irresidua* Kessel and Janciewicz Metelkin, 1935 *Eimeria perforans* (Leuckart) Sluiter and Swellengrebel Metelkin, 1935 ASCHELMINTHES: NEMATODA ASCARIDIDEA: ASCARIDIDAE *Ascaris lumbricoides* Linnaeus Sukhacheva, 1963 ARTHROPODA: INSECTA HYMENOPTERA: BRACONIDAE *Alysia manducator* Panzer Altson, 1920 HYMENOPTERA: PTEROMALIDAE *Dibrachys cavus* (Walker) Graham-Smith, 1919 *Nasonia vitripennis*[1] (Walker) Altson, 1920 Smirnov, 1934 Smirnov et al., 1934 Vladimirova et al., 1934
Protocalliphora Hough	ENTEROVIRUSES Poliovirus Paul et al., 1941 Trask et al., 1943
Protocalliphora azurea (Fallén)	ARTHROPODA: INSECTA HYMENOPTERA: SPHECIDAE *Ectemnius quadricinctus* (Fabricius) Hamm et al., 1926
Bufolucilia silvarum (Meigen)	ENTEROVIRUSES Poliovirus Melnick et al., 1945
Lucilia Robineau-Desvoidy	ENTEROVIRUSES Poliovirus Paul et al., 1941 SCHIZOPHYTA: SCHIZOMYCETES EUBACTERIALES: ENTEROBACTERIACEAE *Salmonella chester* Kauffmann Bulling et al., 1959 *Salmonella kottbus* Schütze et al. Bulling et al., 1959

[1] Although not specifically mentioned in each entry, it is assumed that all *Nasonia vitripennis* are pupal parasites.

Lucilia (cont.)

Salmonella typhimurium (Loeffler) Castellani and Chalmers
Bulling et al., 1959

EUBACTERIALES: BRUCELLACEAE

Pasteurella pestis (Lehmann and Neumann) Holland
Lang, 1940

Brucella abortus (Schmidt) Meyer and Shaw
Ruhland et al., 1941

EUBACTERIALES: BACILLACEAE

Clostridium botulinum (Van Ermengem) Bergey et al.
Bail, 1900

Clostridium perfringens (Veillon and Zuber) Holland
Green, 1953

ACTINOMYCETALES: MYCOBACTERIACEAE

Mycobacterium leprae (Hansen) Lehmann and Neumann
Currie, 1910
Leboeuf, 1912

Mycobacterium ?leprae (Hansen) Lehmann and Neumann
Honeij et al., 1914
Arizumi, 1934

PROTOZOA: MASTIGOPHORA

PROTOMONADIDA: TRYPANOSOMATIDAE

Crithidia Léger
Roubaud, 1908a

Herpetomonas Kent
Roubaud, 1908a

Herpetomonas muscarum (Leidy) Kent
Drbohlov, 1925

Herpetomonas ?muscarum (Leidy) Kent
Strickland, 1911

Leptomonas mesnili Roubaud
Roubaud, 1908a

PROTOMONADIDA: BODONIDAE

Rhynchoidomonas luciliae (Patton) Patton
Alexeieff, 1911

POLYMASTIGIDA: CHILOMASTIGIDAE

Chilomastix mesnili (Wenyon)
Harris et al., 1946

Lucilia (cont.)

PROTOMONADIDA: HEXAMITIDAE

Giardia intestinalis (Lambl)
Wenyon et al., 1917a

Giardia lamblia Stiles
Harris et al., 1946

TRICHOMONADIDA: TRICHOMONADIDAE

Trichomonas hominis (Davaine)
Harris et al., 1946

PROTOZOA: SARCODINA

AMOEBIDA: ENDAMOEBIDAE

Endolimax nana (Wenyon and O'Connor)
Harris et al., 1946

Entamoeba coli (Grassi)
Wenyon et al., 1917a
Harris et al., 1946

Entamoeba histolytica Schaudinn
Wenyon et al., 1917a
Harris et al., 1946
Anonymous ("Korean authors et al.")

Iodamoeba bütschlii (Prowazek)
Harris et al., 1946

PLATYHELMINTHES: CESTOIDEA

PSEUDOPHYLLIDEA: DIPHYLLOBOTHRIIDAE

Diphyllobothrium Cobbold
Pokrovskiĭ et al., 1938

PSEUDOPHYLLIDEA: HYMENOLEPIDIDAE

Hymenolepis diminuta (Rudolphi) Blanchard
Harris et al., 1946

PSEUDOPHYLLIDEA: TAENIIDAE

Echinococcus granulosus (Batsch) Rudolphi
Perez-Fontana et al., 1961

ASCHELMINTHES: NEMATODA

TRICHURIDEA: TRICHURIDAE

Trichuris Roederer
Tao, 1936
Perez-Fontana et al., 1961

Trichuris trichiura (Linnaeus)
Harris et al., 1946

STRONGYLIDEA: ANCYLOSTOMATIDAE

Ancylostoma (Dubini)
Tao, 1936
Harris et al., 1946

Lucilia (cont.)

OXYURIDEA: OXYURIDAE

Enterobius vermicularis (Linnaeus) Leach
Pokrovskiĭ et al., 1938

ASCARIDIDEA: ASCARIDIDAE

Ascaris Linnaeus
Tao, 1936

Ascaris lumbricoides Linnaeus
Harris et al., 1946

Ascaris ?lumbricoides Linnaeus
Perez-Fontana et al., 1961

ARTHROPODA: INSECTA

HYMENOPTERA: ICHNEUMONIDAE

Atractodes gravidus Gravenhorst
Myers, 1929

HYMENOPTERA: BRACONIDAE

Alysia fossulata Provancher
Roberts, 1935

Alysia manducator Panzer
Myers, 1927

HYMENOPTERA: CHALCIDIDAE

Brachymeria fonscolombei (Dufour)
Masi, 1916
Parker, 1923

Dirhinus sarcophagae Froggatt
Johnston et al., 1921

HYMENOPTERA: PTEROMALIDAE

Nasonia Ashmead
Hardy, 1924

Nasonia vitripennis[1] (Walker)
Roubaud, 1917
Johnston et al., 1921
Hardy, 1925
Parker et al., 1928
Myers, 1929
Roberts, 1935

Spalangia[2] *muscidarum* Richardson
Johnston et al., 1921

Spalangia nigripes Curtis
Bouček, 1963

Spalangia nigroaenea Curtis
Bouček, 1963

HYMENOPTERA: ENCYRTIDAE

Australencyrtus giraulti Johnston and Tiegs
Johnston et al., 1921

[1] Although not specifically mentioned in each entry, it is assumed that all *Nasonia vitripennis* are pupal parasites.

[2] Although not specifically mentioned in each entry, it is assumed that all *Spalangia* are pupal parasites.

Fly	Associated organism
Lucilia (cont.)	*Tachinaephagus* Ashmead Hardy, 1924 HYMENOPTERA: EULOPHIDAE *Syntomosphyrum albiclavus* Kerrich Saunders, 1964 *Syntomosphyrum glossinae* Waterston Saunders, 1964 HYMENOPTERA: DIAPRIIDAE *Paraspilomicrus* Johnston and Tiegs Hardy, 1924 *Paraspilomicrus froggatti* Johnston and Tiegs Johnston et al., 1921 HYMENOPTERA: UNDETERMINED FORMICIDAE Pavlovskiĭ, 1921 DIPTERA: MUSCIDAE *Muscina pabulorum* (Fallén) Thomson, 1937
Lucilia argyricephala Macquart (Synonym: ?*Lucilia serenissima*)	SCHIZOPHYTA: SCHIZOMYCETES ACTINOMYCETALES: MYCOBACTERIACEAE *Mycobacterium leprae* (Hansen) Lehmann and Neumann Asami, 1934 PROTOZOA: MASTIGOPHORA PROTOMONADIDA: TRYPANOSOMATIDAE *Herpetomonas muscarum* (Leidy) Kent Patton, 1912a *Leptomonas mirabilis* Roubaud Patton, 1912a PROTOMONADIDA: BODONIDAE *Rhynchoidomonas luciliae* (Patton) Patton Patton, 1910b ARTHROPODA: INSECTA HYMENOPTERA: SPHECIDAE *Sericophorus teliferopodus* Rayment Rayment, 1954 *Sericophorus viridis roddi* Rayment Rayment, 1954
Lucilia bufonivora Moniez	ARTHROPODA: INSECTA HYMENOPTERA: PTEROMALIDAE *Nasonia vitripennis* (Walker) Cousin, 1933

Fly	Organism
Lucilia caesar (Linnaeus) .... (North American *Lucilia caesar* are considered by most authorities to be *Lucilia illustris*)	ENTEROVIRUSES
	Poliovirus Type 2, Lansing strain
	Bang et al., 1943
	Theiler's mouse encephalomyelitis virus, GD VII strain
	Bang et al., 1943
	SCHIZOPHYTA: SCHIZOMYCETES
	PSEUDOMONADALES: PSEUDOMONADACEAE
	Pseudomonas aeruginosa (Schroeter) Migula
	Muzzarelli, 1925
	Weil et al., 1933
	Pseudomonas eisenbergii Migula
	Cao, 1906
	Pseudomonas fluorescens Migula
	Cao, 1906
	Pseudomonas jaegeri Migula
	Cao, 1906
	PSEUDOMONADALES: SPIRILLACEAE
	Vibrio comma (Schroeter) Winslow et al.
	Alessandrini, 1912
	EUBACTERIALES: ACHROMOBACTERACEAE
	Alcaligenes faecalis Castellani and Chalmers
	Weil et al., 1933
	Gregor et al., 1960
	Flavobacterium devorans (Zimmermann) Bergey et al.
	Gregor et al., 1960
	Flavobacterium invisible (Vaughan) Bergey et al.
	Gregor et al., 1960
	EUBACTERIALES: ENTEROBACTERIACEAE
	Escherichia coli (Migula) Castellani and Chalmers
	Beyer, 1925
	Yao et al., 1929
	Gregor et al., 1960
	Escherichia coli var. *communior* (Topley and Wilson) Yale
	Beyer, 1925
	Weil et al., 1933
	Aerobacter aerogenes (Kruse) Beijerinck
	Gregor et al., 1960

Lucilia caesar (cont.)

Klebsiella cloacae (Jordan) Brisou
Gregor et al., 1960
Serratia marcescens Bizio
Gregor et al., 1960
Proteus inconstans (Ornstein) Shaw and Clarke
Gregor et al., 1960
Proteus mirabilis Hauser
Cao, 1906
Gregor et al., 1960
Proteus morganii (Winslow, Kligler, and Rothberg) Yale
Gregor et al., 1960
Proteus rettgeri (Hadley et al.) Rustigian and Stuart
Gregor et al., 1960
Proteus vulgaris Hauser
Cao, 1906
Weil et al., 1933
Gregor et al., 1960
Salmonella enteritidis (Gaertner) Castellani and Chalmers
Graham-Smith, 1912
Salmonella hirschfeldii Weldin
Yao et al., 1929
Salmonella typhi Warren and Scott
Krontowski, 1913
Teodoro, 1916
Wollman, 1921
Citrobacter freundii Werkman and Gillen
Gregor et al., 1960
Shigella dysenteriae (Shiga) Castellani and Chalmers
Krontowski, 1913
Wollman, 1921
Beyer, 1925
Yao et al., 1929

EUBACTERIALES: BRUCELLACEAE

Pasteurella pestis (Lehmann and Neumann) Holland
Russo, 1930
Brucella abortus
Wollman, 1927

EUBACTERIALES: MICROCOCCACEAE

Staphylococcus afermentans Shaw, Stitt, and Cowan
Gregor et al., 1960

Lucilia caesar (cont.)	*Staphylococcus aureus* Rosenbach Beyer, 1925 Gregor et al., 1960 *Staphylococcus aureus* Rosenbach = [*Micrococcus pyogenes* var. *albus*] Weil et al., 1933 *Staphylococcus aureus* Rosenbach = [*Micrococcus pyogenes* var. *aureus*] Weil et al., 1933 *Staphylococcus epidermidis* (Winslow and Winslow) Evans Weil et al., 1933 *Staphylococcus flavus* (Trevisan) Wood Guyénot, 1907 *Staphylococcus lactis* Shaw, Stitt, and Cowan Gregor et al., 1960 *Staphylococcus saprophyticus* Fairbrother Gregor et al., 1960 *Gaffkya tetragena* (Gaffky) Trevisan Weil et al., 1933 EUBACTERIALES: BREVIBACTERIACEAE *Kurthia zenkeri* (Hauser) Bergey et al. Cao, 1906 EUBACTERIALES: LACTOBACILLACEAE *Streptococcus* Rosenbach (alpha-hemolytic) Weil et al., 1933 *Streptococcus durans* Sherman and Wing Gregor et al., 1960 *Streptococcus pyogenes* Rosenbach Weil et al., 1933 EUBACTERIALES: CORYNEBACTERIACEAE ?*Corynebacterium* Lehmann and Neumann Weil et al., 1933 EUBACTERIALES: BACILLACEAE *Bacillus anthracis* Cohn Graham-Smith, 1912

Lucilia caesar (cont.)

Bacillus cereus var. *mycoides* (Flügge) Smith et al.
Gregor et al., 1960
Bacillus megaterium de Bary
Muzzarelli, 1925
Bacillus radiciformis Eberbach
Cao, 1906
Bacillus subtilis Cohn
Cao, 1906
Clostridium botulinum (van Ermengem) Bergey et al.
Saunders, 1921
Clostridium botulinum Type C Spray
Saunders, 1921
Bengtson, 1922

ACTINOMYCETALES: MYCOBACTERIACEAE

Mycobacterium lepraemurium Marchoux and Sorel
Wherry, 1908a
Mycobacterium ?lepraemurium Marchoux and Sorel
Wherry, 1908b
Mycobacterium paratuberculosis Bergey et al.
Muzzarelli, 1925
Mycobacterium tuberculosis (Zopf) Lehmann and Neumann
Hayward, 1904

SPIROCHAETALES: TREPONEMATACEAE

Borrelia berbera (Sergent and Foley) Bergey et al.
Wollman et al., 1928

THALLOPHYTA: PHYCOMYCETES

MUCORALES: MUCORACEAE

Rhizopus nigricans Ehrenberg
Baumberger, 1919

ENTOMOPHTHORALES: ENTOMOPHTHORACEAE

Entomophthora muscae Cohn
Thaxter, 1888
Graham-Smith, 1919

THALLOPHYTA: FUNGI IMPERFECTI

MONILIALES: MONILIACEAE

Blastomyces Costantin and Rolland
Muzzarelli, 1925
Penicillium glaucum Link
Baumberger, 1919

Lucilia caesar (cont.)

FUNGI OF UNCERTAIN AFFINITY
Streptotricea, unvalidated
Muzzarelli, 1925

PROTOZOA: MASTIGOPHORA
PROTOMONADIDA:
TRYPANOSOMATIDAE
Herpetomonas Kent
Chang, 1940
Leishmania tropica (Wright)
Wollman et al., 1928

POLYMASTIGIDA: CHILOMASTIGIDAE
Chilomastix mesnili (Wenyon)
Root, 1921
Chang, 1940

POLYMASTIGIDA: HEXAMITIDAE
Giardia intestinalis (Lambl)
Root, 1921
Giardia lamblia Stiles
Chang, 1940

TRICHOMONADIDA:
TRICHOMONADIDAE
Trichomonas hominis (Davaine)
Chang, 1940

PROTOZOA: SARCODINA
AMOEBIDA: ENDAMOEBIDAE
Endolimax nana (Wenyon and O'Connor)
Root, 1921
Yao et al., 1929
Chang, 1940
Entamoeba coli (Grassi)
Root, 1921
Yao et al., 1929
Chang, 1940
Entamoeba histolytica Schaudinn
Root, 1921
Yao et al., 1929
Chang, 1940
Iodamoeba bütschlii (Prowazek)
Chang, 1940

PROTOZOA: SPOROZOA
COCCIDIDA: EIMERIIDAE
Eimeria irresidua Kessel and Janciewicz
Metelkin, 1935
Eimeria perforans (Leuckart) Sluiter and Swellengrebel
Metelkin, 1935

Lucilia caesar (cont.)

ASCHELMINTHES: NEMATODA

TRICHURIDEA: TRICHURIDAE

Trichuris trichiura (Linnaeus)
Chang, 1940

STRONGYLIDEA: ANCYLOSTOMATIDAE

Ancylostoma (Dubini)
Chang, 1940

ASCARIDIDEA: ASCARIDIDAE

Ascaris lumbricoides Linnaeus
Chang, 1940

ARTHROPODA: INSECTA

COLEOPTERA: STAPHYLINIDAE

Aleochara curtula Goeze
Kemner, 1926

Creophilus maxillosus Linnaeus
Porchinskiĭ, 1885

Emus hirtus Linnaeus
Porchinskiĭ, 1885

Ontholestes murinus Linnaeus
Porchinskiĭ, 1885

Philonthus politus Linnaeus
Porchinskiĭ, 1885

HYMENOPTERA: BRACONIDAE

Alysia manducator Panzer
Altson, 1920

Opius nitidulator Nees
Rambousek, 1929

HYMENOPTERA: CYNIPIDAE

Kleidotoma Westwood
James, 1928

Kleidotoma marshalli Marshall
James, 1928

HYMENOPTERA: FIGITIDAE

Figites anthomyiarum (Bouché)
James, 1928

HYMENOPTERA: CHALCIDIDAE

Brachymeria obtusata Foerster
Nikol'skai͡a, 1960a

HYMENOPTERA: PTEROMALIDAE

Habrocytus obscurus Dalman
Voukassovitch, 1924b

Nasonia vitripennis[1] (Walker)
Altson, 1920
Smirnov, 1934
Smirnov et al., 1934
Vladimirova et al., 1934

Nasonia ?vitripennis (Walker)
Bishopp et al., 1915b

[1] Although not specifically mentioned in each entry, it is assumed that all *Nasonia vitripennis* are pupal parasites.

Fly	Associated organisms
Lucilia caesar (cont.)	HYMENOPTERA: SPHECIDAE *Crabro cavifrons* Thomson Hamm et al., 1926 *Crabro vagus* (Linnaeus) Hamm et al., 1926 *Ectemnius quadricinctus* (Fabricius) Hamm et al., 1926
Lucilia illustris (Meigen)	ENTEROVIRUSES Poliovirus Melnick et al., 1945 Poliovirus Type 1 Downey, 1963 ECHO virus Types 6 and 7 Downey, 1963 SCHIZOPHYTA: SCHIZOMYCETES EUBACTERIALES: ENTEROBACTERIACEAE *Salmonella typhimurium* (Loeffler) Castellani and Chalmers Nuorteva, 1965 Ojala et al., 1966 CHORDATA: AVES PASSERIFORMES: UNDETERMINED PLOCEIDAE Séguy, 1929
Lucilia ?jeddensis Bigot	PROTOZOA: MASTIGOPHORA PROTOMONADIDA: TRYPANOSOMATIDAE *Herpetomonas muscarum* (Leidy) Kent Yamasaki, 1927
Lucilia porphyrina (Walker) (Synonym: *Lucilia craggii*)	PROTOZOA: MASTIGOPHORA PROTOMONADIDA: TRYPANOSOMATIDAE *Herpetomonas muscarum* (Leidy) Kent Patton, 1921a *Leptomonas mirabilis* Roubaud Patton, 1921a
Phaenicia Robineau-Desvoidy	ENTEROVIRUSES Poliovirus Trask et al., 1943 Poliovirus Types 1 and 3 Paul et al., 1962 Coxsackievirus Melnick (personal communication from Curnen), 1950

Fly	Organism
Phaenicia (cont.)	Melnick et al., 1954 ECHO virus Types 8, 12, 14 Paul et al., 1962 SCHIZOPHYTA: SCHIZOMYCETES EUBACTERIALES: ENTEROBACTERIACEAE *Shigella boydii* Ewing 2, 5, 6 Richards et al., 1961 *Shigella dysenteriae* (Shiga) Castellani and Chalmers 2 Richards et al., 1961 *Shigella flexneri* Castellani and Chalmers 1a, 2a, 3, 4a, 4b, 6 Richards et al., 1961 *Shigella sonnei* (Levine) Weldin 1 Richards et al., 1961 ARTHROPODA: INSECTA HYMENOPTERA: FORMICIDAE *Paratrechina longicornis* (Latreille) Pimentel, 1955 *Solenopsis corticalis* Forel Pimentel, 1955 *Solenopsis geminata* (Fabricius) Pimentel, 1955
Phaenicia cluvia Walker (Synonym: ?*Lucilia pilatei*)	PROTOZOA: MASTIGOPHORA PROTOMONADIDA: TRYPANOSOMATIDAE *Leptomonas mesnili* Roubaud Roubaud, 1909
Phaenicia cuprina Wiedemann (Synonym: *Lucilia cuprina*)	SCHIZOPHYTA: SCHIZOMYCETES EUBACTERIALES: ENTEROBACTERIACEAE *Escherichia coli* (Migula) Castellani and Chalmers Shimizu et al., 1965 *Proteus mirabilis* Hauser Shimizu et al., 1965 *Proteus morganii* (Winslow, Kligler, and Rothberg) Yale Shimizu et al., 1965 *Proteus rettgeri* (Hadley et al.) Rustigian and Stuart Shimizu et al., 1965 *Proteus vulgaris* Hauser Shimizu et al., 1965 *Providencia* Kauffmann Shimizu et al., 1965

Fly	Organism
Phaenicia cuprina (cont.)	PROTOZOA: CNIDOSPORIDA MICROSPORIDA: NOSEMATIDAE *Nosema kingi* Kramer Kramer, 1964b ASCHELMINTHES: NEMATODA ASCARIDIDEA: ASCARIDIDAE *Ascaris lumbricoides* Linnaeus Roberts, 1934 ARTHROPODA: INSECTA HYMENOPTERA: DIAPRIIDAE *Hemilexomyia abrupta* Dodd Dodd, 1920
Phaenicia eximia (Wiedemann) (Synonyms: *Lucilia eximia, Lucilia hirtiforceps, Lucilia ruficornis*)	SCHIZOPHYTA: SCHIZOMYCETES EUBACTERIALES: MICROCOCCACEAE *Staphylococcus* Rosenbach Picado, 1935 ARTHROPODA: INSECTA HYMENOPTERA: CHALCIDIDAE *Brachymeria fonscolombei* (Dufour) Roberts, 1933a HYMENOPTERA: PTEROMALIDAE *Nasonia vitripennis* (Walker) Roberts, 1933a DIPTERA: CUTEREBRIDAE *Dermatobia hominis* (Linnaeus) Blanchard, 1899
Phaenicia mexicana (Macquart) (Synonym: *Lucilia unicolor*)	THALLOPHYTA: PHYCOMYCETES ENTOMOPHTHORALES: ENTOMOPHTHORACEAE *Entomophthora americana* Thaxter Dresner, 1949 ARTHROPODA: INSECTA HYMENOPTERA: BRACONIDAE *Alysia ridibunda* Say Roberts, 1935 HYMENOPTERA: CHALCIDIDAE *Brachymeria fonscolombei* (Dufour) Roberts, 1933a, 1933b, 1935 HYMENOPTERA: PTEROMALIDAE *Nasonia vitripennis* (Walker) Roberts, 1933a
Phaenicia pallescens (Shannon) (Synonym: *Lucilia pallescens*)	ENTEROVIRUSES Poliovirus Melnick et al., 1949 Poliovirus non-paralytic Melnick et al., 1949 Poliovirus Type 1 Melnick et al., 1953

Fly	Organism
Phaenicia pallescens (cont.)	Coxsackievirus Easton 2 (A1) Melnick et al., 1953 Coxsackievirus Easton 10 (A8) Melnick et al., 1953 Coxsackievirus Easton 14 (A5) Melnick et al., 1953 Coxsackievirus Texas 1 (A4) Melnick et al., 1953 Coxsackievirus Texas 12 (new type) Melnick et al., 1953 Coxsackievirus Texas 14 (new type) Melnick et al., 1953 Coxsackievirus Type 3 (A3) Melnick et al., 1953 PROTOZOA: SARCODINA AMOEBIDA: ENDAMOEBIDAE *Entamoeba histolytica* Schaudinn Pipkin, 1942, 1949 ARTHROPODA: INSECTA COLEOPTERA: HISTERIDAE *Saprinus lugens* Erichson Vogt, 1948 COLEOPTERA: DERMESTIDAE *Dermestes ater* De Geer Vogt, 1948 *Dermestes caninus* Germar Vogt, 1948
Phaenicia sericata (Meigen) (Synonyms: *Lucilia sericata, Lucilia latifrons, Lucilia flavipennis*)	VIRUSES ENTEROVIRUSES Poliovirus Sabin et al., 1942 Melnick et al., 1945, 1949 Francis, Jr. et al., 1948 Melnick, 1949 Poliovirus non-paralytic Melnick et al., 1949 Poliovirus Type 1 Melnick et al., 1953 Riordan et al., 1961 Downey, 1963 Poliovirus Attenuated Type 1, LSc Dave et al., 1965 Poliovirus Philippine Islands '45 strain (Type 1) Melnick et al., 1952 Poliovirus Texas '48 strain (Type 1) Melnick et al., 1952 Poliovirus Type 2, Lansing strain Melnick et al., 1947, 1952

Phaenicia sericata (cont.)

Poliovirus Y-SK strain, murine adapted, Type 2
Melnick et al., 1947, 1952
Poliovirus Type 3
Riordan et al., 1961
Poliovirus Attenuated Type 3, Leon
Dave et al., 1965
Poliovirus North Carolina '44 strain (Type 3)
Melnick et al., 1952
Theiler's mouse encephalomyelitis virus, TO strain
Melnick et al., 1947, 1952
Coxsackievirus Easton 2 (A1)
Melnick et al., 1953
Coxsackievirus Easton 10 (A8)
Melnick et al., 1953
Coxsackievirus Easton 14 (A5)
Melnick et al., 1953
Coxsackievirus Texas 1 (A4)
Melnick et al., 1953
Coxsackievirus Texas 12 (new type)
Melnick et al., 1953
Coxsackievirus Texas 14 (new type)
Melnick et al., 1953
Coxsackievirus Texas '48 (new type)
Melnick et al., 1952
Coxsackievirus Type 3 (A3)
Melnick et al., 1953
Coxsackievirus Type A9 (A9)
Riordan et al., 1961
Coxsackievirus Type B5 (B5)
Riordan et al., 1961
ECHO virus Types 1, 5, 11
Riordan et al., 1961
ECHO virus Types 6 and 7
Downey, 1963

OTHER ANIMAL VIRUSES
Foot and mouth disease virus
Dhennin et al., 1961

SCHIZOPHYTA: SCHIZOMYCETES
PSEUDOMONADALES: PSEUDOMONADACEAE
Pseudomonas aeruginosa (Schroeter) Migula
Weil et al., 1933
Greenberg, 1962
Havlik, 1964

Phaenicia sericata (cont.)	*Pseudomonas fluorescens* Migula Lysenko et al., 1961 EUBACTERIALES: ACHROMOBACTERACEAE *Alcaligenes faecalis* Castellani and Chalmers Michelbacher et al., 1932 Weil et al., 1933 Gregor et al., 1960 Havlik, 1964 *Flavobacterium* Bergey et al. Lysenko et al., 1961 *Flavobacterium devorans* (Zimmermann) Bergey et al. Gregor et al., 1960 *Flavobacterium invisible* (Vaughan) Bergey et al. Gregor et al., 1960 EUBACTERIALES: ENTEROBACTERIACEAE *Escherichia* Castellani and Chalmers Lysenko et al., 1961 *Escherichia coli* (Migula) Castellani and Chalmers Michelbacher et al., 1932 Hobson, 1932b, 1935 Zmeev, 1943 Gregor et al., 1960 Povolný et al., 1961 Greenberg, 1962 Havlik, 1964 *Escherichia coli* var. *acidilactici* (Topley and Wilson) Yale =? [*Escherichia paraacidilactici* A] Zmeev, 1943 *Escherichia coli* var. *acidilactici* (Topley and Wilson) Yale =? [*Escherichia paraacidilactici* B] Zmeev, 1943 *Escherichia coli* var. *communior* (Topley and Wilson) Yale Weil et al., 1933 *Aerobacter* Beijerinck Zmeev, 1943 *Aerobacter aerogenes* (Kruse) Beijerinck Gregor et al., 1960 Havlik, 1964

Phaenicia sericata (cont.)

Aerobacter cloacae (Jordan) Bergey et al.
  Michelbacher et al., 1932
Klebsiella Trevisan
  Lysenko et al., 1961
Klebsiella cloacae (Jordan) Brisou
  Gregor et al., 1960
Cloaca Castellani and Chalmers
  Zmeev, 1943
  Lysenko et al., 1961
Paracolobactrum aerogenoides Borman, Stuart, and Wheeler
  Zmeev, 1943
Erwinia amylovora (Burrill) Winslow et al.
  Ark et al., 1936
Serratia Bizio
  Povolný et al., 1961
Serratia marcescens Bizio
  Michelbacher et al., 1932
  Gregor et al., 1960
  Havlik, 1964
Proteus inconstans (Ornstein) Shaw and Clarke
  Gregor et al., 1960
Proteus mirabilis Hauser
  Gregor et al., 1960
  Lysenko et al., 1961
Proteus morganii (Winslow, Kligler, and Rothberg) Yale
  Gregor et al., 1960
  Lysenko et al., 1961
  Havlik, 1964
  Shimizu et al., 1965
Proteus rettgeri (Hadley et al.) Rustigian and Stuart
  Gregor et al., 1960
  Lysenko et al., 1961
  Shimizu et al., 1965
Proteus vulgaris Hauser
  Hobson, 1932a, 1935
  Michelbacher et al., 1932
  Weil et al., 1933
  Simmons, 1935a, 1935b
  O'Keefe et al., 1954
  Sukhova, 1954
  Gregor et al., 1960
  Lysenko et al., 1961
  Povolný et al., 1961
  Greenberg, 1962
  Havlik, 1964

Phaenicia sericata (cont.)	*Providencia* Kauffmann Lysenko et al., 1961 Shimizu et al., 1965 *Salmonella* Lignières Alcivar et al., 1946 Greenberg et al., 1964 Shimizu et al., 1965 *Salmonella amersfoort* Henning Alcivar et al., 1946 *Salmonella anatum* (Rettger and Scoville) Bergey et al. Alcivar et al., 1946 *Salmonella ballerup* Kauffmann and Moller Alcivar et al., 1946 *Salmonella blockley* Kauffmann Greenberg et al., 1964 *Salmonella cholerae-suis* (Smith) Weldin Alcivar et al., 1946 *Salmonella dublin* Kauffmann Alcivar et al., 1946 *Salmonella enteritidis* (Gaertner) Castellani and Chalmers Alcivar et al., 1946 Gross et al., 1953 *Salmonella florida* Cherry, Edwards, and Bruner Alcivar et al., 1946 *Salmonella give* Kauffmann Alcivar et al., 1946 Gross et al., 1953 *Salmonella kentucky* Edwards Alcivar et al., 1946 *Salmonella kottbus* Schütze et al. Alcivar et al., 1946 *Salmonella meleagridis* Bruner and Edwards Alcivar et al., 1946 *Salmonella narashino* Kauffmann Alcivar et al., 1946 *Salmonella new brunswick* Edwards Greenberg et al., 1963b *Salmonella onderstepoort* Henning Alcivar et al., 1946 *Salmonella oranienburg* Kauffmann Alcivar et al., 1946 *Salmonella oregon* Edwards and Bruner Alcivar et al., 1946

Phaenicia sericata (cont.)	*Salmonella paratyphi* A (Brion and Kayser) Castellani and Chalmers Gross et al., 1953 *Salmonella paratyphi* B Castellani and Chalmers Alcivar et al., 1946 Gross et al., 1953 *Salmonella reading* Schütze Alcivar et al., 1946 *Salmonella typhi* Warren and Scott Simmons, 1935a, 1935b Anonymous, 1952 Gross et al., 1953 Sukhova, 1954 Greenberg, 1962 *Salmonella typhimurium* (Loeffler) Castellani and Chalmers Alcivar et al., 1946 Ojala et al., 1966 *Salmonella zanzibar* Kauffmann Alcivar et al., 1946 *Citrobacter freundii* Werkman and Gillen Gregor et al., 1960 *Shigella alkalescens* (Andrewes) Weldin Alcivar et al., 1946 *Shigella boydii* Ewing 1, 2, 3, 4, 5, 6, 8, 9 Alcivar et al., 1946 *Shigella ceylonensis* (Weldin) Weldin Alcivar et al., 1946 *Shigella dispar* (Andrewes) Bergey et al. Alcivar et al., 1946 *Shigella dysenteriae* (Shiga) Castellani and Chalmers 2 Sychevskaı͡a et al., 1959a *Shigella flexneri* Castellani and Chalmers Gross et al., 1953 Arskiĭ et al., 1961 Greenberg, 1962 *Shigella flexneri* Castellani and Chalmers 1a Sychevskaı͡a et al., 1959a *Shigella flexneri* Castellani and Chalmers 1a, 2a, Y, 6 Alcivar et al., 1946

Phaenicia sericata (cont.)

Shigella sonnei (Levine) Weldin
Gross et al., 1953
Arskiĭ et al., 1961

EUBACTERIALES: UNDETERMINED ENTEROBACTERIACEAE

Klebsiella-Cloaca intermed. forms
Lysenko et al., 1961

Paraescherichia, unvalidated
Zmeev, 1943

EUBACTERIALES: MICROCOCCACEAE

Micrococcus rushmorei Brown
Brown, 1927

Staphylococcus afermentans Shaw, Stitt, and Cowan
Gregor et al., 1960

Staphylococcus aureus Rosenbach
Robinson et al., 1933, 1934
Simmons, 1935a, 1935b
Gregor et al., 1960
Greenberg, 1962

Staphylococcus aureus Rosenbach =[*Micrococcus pyogenes* var. *albus*]
Michelbacher et al., 1932

Staphylococcus aureus Rosenbach =[*Micrococcus pyogenes* var. *albus*]
Weil et al., 1933

Staphylococcus aureus Rosenbach =[*Micrococcus pyogenes* var. *aureus*]
Weil et al., 1933

Staphylococcus aureus Rosenbach =[*Staphylococcus pyogenes*]
Greenberg, 1962

Staphylococcus epidermidis (Winslow and Winslow) Evans
Weil et al., 1933

Staphylococcus lactis Shaw, Stitt, and Cowan
Gregor et al., 1960
Lysenko et al., 1961

Staphylococcus saprophyticus Fairbrother
Gregor et al., 1960
Lysenko et al., 1961

Gaffkya tetragena (Gaffky) Trevisan
Weil et al., 1933

Phaenicia sericata (cont.)

Sarcina Goodsir
Michelbacher et al., 1932
Sarcina flava de Bary
Lysenko et al., 1961
EUBACTERIALES: NEISSERIACEAE
Neisseria luciliarum Brown
Brown, 1927
EUBACTERIALES: LACTOBACILLACEAE
Diplococcus pneumoniae Weichselbaum
Gregor et al., 1960
Povolný et al., 1961
Streptococcus Rosenbach (alpha-hemolytic)
Weil et al., 1932
Streptococcus Rosenbach (beta-hemolytic)
Robinson et al., 1934
Streptococcus durans Sherman and Wing
Simmons, 1935a, 1935b
Gregor et al., 1960
Lysenko et al., 1961
Povolný et al., 1961
Greenberg, 1962
Streptococcus faecalis var. *liquefaciens* (Sternberg) Mattick
Lysenko et al., 1961
Streptococcus mitis Andrewes and Horder
Simmons, 1935a, 1935b
Streptococcus pyogenes Rosenbach
Robinson et al., 1933
Weil et al., 1933
Simmons, 1935a, 1935b
EUBACTERIALES: CORYNEBACTERIACEAE
?*Corynebacterium* Lehmann and Neumann
Weil et al., 1933
EUBACTERIALES: BACILLACEAE
Bacillus Cohn
Povolný et al., 1961
Bacillus cereus var. *mycoides* (Flügge) Smith et al.
Gregor et al., 1960
Bacillus lutzae Brown
Brown, 1927
Bacillus subtilis Cohn
Michelbacher et al., 1932

Phaenicia sericata (cont.)

Lysenko et al., 1961
Greenberg, 1962

Bacillus thuringiensis Berliner
Greenwood, 1964

Clostridium botulinum Type C Spray
Bengtson, 1922

Clostridium perfringens (Veillon and Zuber) Holland
Michelbacher et al., 1932
Simmons, 1935a, 1935b

ACTINOMYCETALES: MYCOBACTERIACEAE

Mycobacterium tuberculosis (Zopf) Lehmann and Neumann
Morellini et al., 1947

Leptospira canicola Okell et al.
Kunert et al., 1952
Schmidtke, 1959

Leptospira grippotyphosa Topley and Wilson
Kunert et al., 1952
Schmidtke, 1959

Leptospira icterohaemorrhagiae (Inada and Ido) Noguchi
Kunert et al., 1952
Schmidtke, 1959

Bacterium mathisi Roubaud and Treillard, unvalidated
Roubaud et al., 1935

THALLOPHYTA: PHYCOMYCETES

ENTOMOPHTHORALES: ENTOMOPHTHORACEAE

Entomophthora muscae Cohn
Dove, 1916

PROTOZOA: MASTIGOPHORA

PROTOMONADIDA: TRYPANOSOMATIDAE

Crithidia Léger
Thomson, 1933

Crithidia luciliae (Strickland) Wallace and Clark
Wallace et al., 1959

Herpetomonas Kent
Chang, 1940

Herpetomonas muscarum (Leidy) Kent
Becker, 1922-3, 1924
Drbohlav, 1926
Fantham et al., 1927

Phaenicia sericata (cont.)

Thomson, 1933
Bellosillo, 1937
Wallace et al., 1959

Leptomonas mesnili Roubaud
Roubaud, 1908a, 1909, 1911c

PROTOMONADIDA: BODONIDAE

Rhynchoidomonas intestinalis (Roubaud) Alexieff
Roubaud, 1911b

POLYMASTIGIDA: CHILOMASTIGIDAE

Chilomastix mesnili (Wenyon)
Chang, 1940

POLYMASTIGIDA: HEXAMITIDAE

Giardia lamblia Stiles
Chang, 1940

TRICHOMONADIDA: TRICHOMONADIDAE

Trichomonas hominis (Davaine)
Hegner, 1928
Chang, 1940

PROTOZOA: SARCODINA

AMOEBIDA: ENDAMOEBIDAE

Endolimax nana (Wenyon and O'Connor)
Chang, 1940

Entamoeba coli (Grassi)
Chang, 1940

Entamoeba histolytica Schaudinn
Chang, 1940, 1945

Iodamoeba bütschlii (Prowazek)
Chang, 1940

PROTOZOA: SPOROZOA

HAPLOSPORIDA: UNCERTAIN POSITION

Toxoplasma Nicolle and Manceaux
Schmidtke, 1959

Toxoplasma gondii Nicolle and Manceaux
Kunert et al., 1953

PROTOZOA: CNIDOSPORIDIA

MICROSPORIDA: MRAZEKIIDAE

Octosporea muscae-domesticae Flu
Kramer, 1964a, 1965b

ASCHELMINTHES: NEMATODA

TRICHURIDEA: TRICHURIDAE

Trichuris Roederer
Chang, 1945

Trichuris trichiura (Linnaeus)
Chang, 1940

Phaenicia sericata (cont.)

STRONGYLIDEA: ANCYLOSTOMATIDAE

Ancylostoma (Dubini)
Chang, 1940, 1945

STRONGYLIDEA: SYNGAMIDAE

Syngamus trachea (Montagu)
Clapham, 1939

ASCARIDIDEA: ASCARIDIDAE

Ascaris Linnaeus
Chang, 1945

Ascaris lumbricoides Linnaeus
Chang, 1940

ARTHROPODA: ARACHNIDA

ACARINA: MACROCHELIDAE

Macrocheles glaber (Müller)
Sychevskai͡a, 1964a

Macrocheles muscaedomesticae (Scopoli)
Sychevskai͡a, 1964a

ACARINA: GAMASOLAELAPTIDAE

Digamasellus Berlese
Sychevskai͡a, 1964a

ACARINA: ANOETIDAE

Myianoetus muscarum (Linnaeus)
Greenberg et al., 1960

ARTHROPODA: INSECTA

DERMAPTERA: LABIDURIDAE

Euborellia annulipes (Lucas)
Illingworth, 1923a

DERMAPTERA: LABIIDAE

Labia pilcornis (Motschulsky)
Illingworth, 1923a

Spingolabis hawaiiensis (Bormans)
Illingworth, 1923a

COLEOPTERA: HYDROPHILIDAE

Cryptopleurum minutum Fabricius
Illingworth, 1923a

Dactylosternum abdominale Fabricius
Illingworth, 1923a

COLEOPTERA: SILPHIDAE

Nicrophorus nigritus Mannerheim
Illingworth, 1927

COLEOPTERA: UNDETERMINED SILPHIDAE
Illingworth, 1927

COLEOPTERA: STAPHYLINIDAE

Creophilus maxillosus var. *villosus* Gravenhorst
Illingworth, 1927

Fly	Associated organism / Reference
Phaenicia sericata (cont.)	*Oxytelus* Gravenhorst
	Illingworth, 1923a
	Philonthus discoideus Gravenhorst
	Illingworth, 1923a
	Philonthus longicornis Stephens
	Illingworth, 1923a
	Tachyporus Gravenhorst
	Illingworth, 1923a
	COLEOPTERA: HISTERIDAE
	Saprinus lugens Erichson
	Illingworth, 1927
	Xerosaprinus fimbriatus (LeConte)
	Illingworth, 1927
	Xerosaprinus lubricus (LeConte)
	Illingworth, 1927
	COLEOPTERA: UNDETERMINED HISTERIDAE
	Holdaway, 1930
	COLEOPTERA: UNDETERMINED DERMESTIDAE
	Holdaway, 1930
	COLEOPTERA: OSTOMATIDAE
	Temnochila virescens (Fabricius)
	Struble, 1942
	HYMENOPTERA: BRACONIDAE
	Alysia manducator Panzer
	Altson, 1920
	Gurney et al., 1926a
	Miller, 1927
	Newman, 1928
	Morgan, 1929
	Smit, 1929
	Holdaway et al., 1930, 1932
	Tillyard, 1930
	Salt, 1932
	Evans, 1933
	Alysia ridibunda Say
	Roberts, 1935
	Aphaereta minuta Nees
	Holdaway et al., 1930
	Evans, 1933
	Aphaereta pallipes Say
	Beard, 1964
	Aspilota nervosa Haliday
	Evans, 1933
	HYMENOPTERA: CYNIPIDAE
	Eucoila Westwood
	Roberts, 1935a
	Kleidotoma Westwood
	James, 1928

Phaenicia sericata (cont.)	*Kleidotoma marshalli* Marshall James, 1928 *Pseudeucoila trichopsila* (Hartig) Sychevskaîa, 1964b HYMENOPTERA: FIGITIDAE *Figites anthomyiarum* (Bouché) James, 1928 HYMENOPTERA: CHALCIDIDAE *Brachymeria fonscolombei* (Dufour) Roberts, 1933b, 1935 *Brachymeria minuta* (Fabricius) Sychevskaîa, 1966 *Dirhinus sarcophagae* Froggatt Johnston et al., 1921 HYMENOPTERA: PTEROMALIDAE ?*Nasonia* Ashmead Feng, 1933 *Nasonia vitripennis*[1] (Walker) Froggatt, 1919 Altson, 1920 Johnston et al., 1921 Anonymous, 1925 Smit, 1929 Cousin, 1933 Evans, 1933 Ullyett, 1950 Pimentel et al., 1963 *Nasonia ?vitripennis* (Walker) Bishopp et al., 1915b *Spalangia nigra* Latreille Roberts, 1935 *Spalangia nigroaenea* Curtis Bouček, 1963 HYMENOPTERA: ENCYRTIDAE *Australencyrtus giraulti* Johnston and Tiegs Johnston et al., 1921 HYMENOPTERA: FORMICIDAE *Pheidole megacephala* (Fabricius) Illingworth, 1923a *Ponera perkinsi* Forel Illingworth, 1923a HYMENOPTERA: SPHECIDAE *Crabro advenus* Smith Kurczewski et al., 1968 *Bembix olivata* Dahlbom Ullyett et al., 1941

[1] Although not specifically mentioned in each entry, it is assumed that all *Nasonia vitripennis* are pupal parasites.

Fly	Associated organisms
Phaenicia sericata (cont.)	*Ectemnius quadricinctus* (Fabricius) Hamm et al., 1926 DIPTERA: MUSCIDAE *Musca domestica* Linnaeus Pimentel et al., 1965b DIPTERA: CALLIPHORIDAE *Chrysomya* Robineau-Desvoidy Salt, 1932 *Chrysomya rufifacies* (Macquart) Holdaway, 1930 Mackerras, 1930 *Phaenicia cuprina* Wiedemann Mackerras, 1933 *Calliphora* Robineau-Desvoidy Salt, 1932 DIPTERA: SARCOPHAGIDAE *Sarcophaga* Meigen Salt, 1932 DIPTERA: UNDETERMINED SARCOPHAGIDAE Holdaway, 1930
Eucalliphora lilaea (Walker) (Synonym: *Calliphora latifrons*)	SCHIZOPHYTA: SCHIZOMYCETES EUBACTERIALES: ENTEROBACTERIACEAE *Salmonella typhi* Warren and Scott Gwatkin et al., 1938 EUBACTERIALES: BRUCELLACEAE *Brucella abortus* (Schmidt) Meyer and Shaw Gwatkin et al., 1938 EUBACTERIALES: MICROCOCCACEAE *Staphylococcus aureus* Rosenbach Gwatkin et al., 1938 EUBACTERIALES: LACTOBACILLACEAE *Streptococcus agalactiae* Lehmann and Neumann Gwatkin et al., 1938
Aldrichina grahami (Aldrich)	SCHIZOPHYTA: SCHIZOMYCETES EUBACTERIALES: ENTEROBACTERIACEAE *Escherichia coli* (Migula) Castellani and Chalmers Shimizu et al., 1965 *Proteus morganii* (Winslow, Kligler, and Rothberg) Yale Shimizu et al., 1965 *Proteus rettgeri* (Hadley et al.) Rustigian and Stuart Shimizu et al., 1965

Calliphora Robineau-Desvoidy ....

ENTEROVIRUSES

Poliovirus
  Trask et al., 1943

Poliovirus Type 1
  Riordan et al., 1961
  Downey, 1963

Poliovirus Type 3
  Riordan et al., 1961

Coxsackievirus Type A9 (A9)
  Riordan et al., 1961

Coxsackievirus Type B5 (B5)
  Riordan et al., 1961

ECHO virus Types 1, 5, 11
  Riordan et al., 1961

ECHO virus Types 6 and 7
  Downey, 1963

SCHIZOPHYTA: SCHIZOMYCETES

EUBACTERIALES: ENTEROBACTERIACEAE

Salmonella chester Kauffmann
  Bulling et al., 1959

Salmonella derby Warren and Scott
  Tacal et al., 1967

Salmonella kottbus Schütze et al.
  Bulling et al., 1959

Salmonella typhimurium (Loeffler) Castellani and Chalmers
  Bulling et al., 1959

EUBACTERIALES: BRUCELLACEAE

Brucella abortus (Schmidt) Meyer and Shaw
  Ruhland et al., 1941

EUBACTERIALES: BACILLACEAE

Bacillus Cohn
  Zmeev, 1943

Clostridium perfringens (Veillon and Zuber)
  Green, 1953

THALLOPHYTA: ASCOMYCETES

GYMNOASCALES: GYMNOASCACEAE

Trichophyton mentagrophytes (Robin) Blanchard var. *granulosum*
  Gip et al., 1968

THALLOPHYTA: PHYCOMYCETES

ENTOMOPHTHORALES: ENTOMOPHTHORACEAE

Entomophthora calliphorae Giard
  Thaxter, 1888

Calliphora (cont.)

PROTOZOA: MASTIGOPHORA
PROTOMONADIDA:
TRYPANOSOMATIDAE
Leptomonas gracilis (Léger) Wenyon
Alexeieff, 1911
Leptomonas pyrrhocoris (Zotta) Zotta
Zotta, 1921
POLYMASTIGIDA: HEXAMITIDAE
Giardia intestinalis (Lambl)
Wenyon et al., 1917a, 1917b
PROTOZOA: SARCODINA
AMOEBIDA: ENDAMOEBIDAE
Entamoeba coli (Grassi)
Wenyon et al., 1917a, 1917b
Entamoeba histolytica Schaudinn
Wenyon et al., 1917a, 1917b
PLATYHELMINTHES: CESTOIDEA
CYCLOPHYLLIDEA: TAENIIDAE
Echinococcus granulosus (Batsch) Rudolphi
Perez-Fontana et al., 1961
ASCHELMINTHES: NEMATODA
TRICHURIDEA: TRICHURIDAE
Trichuris Roederer
Perez-Fontana et al., 1961
STRONGYLIDEA:
ANCYLOSTOMATIDAE
Ancylostoma caninum (Ercolani)
Harada, 1953, 1954
ASCARIDIDEA: ASCARIDIDAE
Ascaris lumbricoides Linnaeus
Perez-Fontana et al., 1961
ARTHROPODA: INSECTA
ODONATA: AESCHNIDAE
Aeschna cyanea (Müller)
Freeman, 1934
HYMENOPTERA: BRACONIDAE
Alysia manducator Panzer
Myers, 1927
Morgan, 1929
Laing, 1937
HYMENOPTERA: PTEROMALIDAE
Nasonia vitripennis[1] (Walker)
Parker et al., 1928
Roberts, 1935
Edwards, 1954

[1] Although not specifically mentioned in each entry, it is assumed that all *Nasonia vitripennis* are pupal parasites.

Fly	Associated organism
Calliphora (cont.)	HYMENOPTERA: EULOPHIDAE *Syntomosphyrum albiclavus* Kerrich Saunders, 1964 *Syntomosphyrum glossinae* Waterston Saunders, 1964 HYMENOPTERA: SPHECIDAE *Crabro vagus* (Linnaeus) Hamm et al., 1926 DIPTERA: MUSCIDAE *Muscina pabulorum* (Fallén) Thomson, 1937
Calliphora clausa Macquart (Synonym: *Onesia accepta*)	ANNELIDA: OLIGOCHAETA OPISTHOPORA: LUMBRICIDAE *Allolobophora caliginosa* (Savigny) subsp. *trapezoides* Fuller, 1933 OPISTHOPORA: MEGASCOLECIDAE *Microscolex dubius* (Fletcher) Fuller, 1933 ARTHROPODA: INSECTA HEMIPTERA: APHIDAE *Myzus persicae* (Sulzer) Fuller, 1933
Calliphora coloradensis Hough	PROTOZOA: MASTIGOPHORA PROTOMONADIDA: TRYPANOSOMATIDAE *Herpetomonas muscarum* (Leidy) Kent Swingle, 1911 ARTHROPODA: INSECTA HYMENOPTERA: BRACONIDAE *Alysia ridibunda* Say Roberts, 1935 HYMENOPTERA: CHALCIDIDAE *Brachymeria fonscolombei* (Dufour) Roberts, 1933b, 1935
Calliphora icela (Walker)	ARTHROPODA: INSECTA HYMENOPTERA: BRACONIDAE *Alysia manducator* Panzer Miller, 1927
Calliphora lata Coquillet	SCHIZOPHYTA: SCHIZOMYCETES EUBACTERIALES: ENTEROBACTERIACEAE *Proteus mirabilis* Hauser Shimizu et al., 1965 *Proteus morganii* (Winslow, Kligler, and Rothberg) Yale Shimizu et al., 1965

Fly	Associated organisms
Calliphora lata (cont.)	*Proteus rettgeri* (Hadley et al.) Rustigian and Stuart Shimizu et al., 1965 ACTINOMYCETALES: MYCOBACTERIACEAE *Mycobacterium leprae* (Hansen) Lehmann and Neumann Asami, 1934
Calliphora stygia (Fabricius) (Synonyms: *Calliphora villosa, Pollenia stygia*)	ARTHROPODA: INSECTA HYMENOPTERA: CHALCIDIDAE *Dirhinus sarcophagae* Froggatt Froggatt, 1921, 1922 HYMENOPTERA: UNDETERMINED CHALCIDIDAE Froggatt, 1914b HYMENOPTERA: PTEROMALIDAE *Nasonia vitripennis*[1] (Walker) Froggatt et al., 1914 Froggatt, 1919 Johnston et al., 1921 *Spalangia muscidarum* Richardson Johnston et al., 1921 HYMENOPTERA: DIAPRIIDAE *Hemilexomyia abrupta*[2] Dodd Dodd, 1920 Froggatt, 1921, 1922 Johnston et al., 1921 DIPTERA: CALLIPHORIDAE *Chrysomya rufifacies* (Macquart) Mackerras, 1930
Calliphora terraenovae Macquart (Synonym: *Calliphora vomitoria nigribarda*)	SCHIZOPHYTA: SCHIZOMYCETES EUBACTERIALES: ENTEROBACTERIACEAE *Salmonella typhi* Warren and Scott Gwatkin et al., 1938 EUBACTERIALES: BRUCELLACEAE *Brucella abortus* (Schmidt) Meyer and Shaw Gwatkin et al., 1938 EUBACTERIALES: MICROCOCCACEAE *Staphylococcus aureus* Rosenberg Gwatkin et al., 1938 EUBACTERIALES: LACTOBACILLACEAE *Streptococcus agalactiae* Lehmann and Neumann Gwatkin et al., 1938

[1] Although not specifically mentioned in each entry, it is assumed that all *Nasonia vitripennis* are pupal parasites.

[2] Although not specifically mentioned in each entry, it is assumed that all *Hemilexomyia abrupta* are parasites.

Fly	Organisms
Calliphora vicina Robineau-Desvoidy (Synonyms: *Calliphora erythrocephala, Calliphora rufifacies*)	ENTEROVIRUSES
	Poliovirus
	Sabin et al., 1942
	Poliovirus Type 2, Lansing strain
	Bang et al., 1943
	Theiler's mouse encephalomyelitis virus, GD VII strain
	Bang et al., 1943
	SCHIZOPHYTA: SCHIZOMYCETES
	PSEUDOMONADALES: PSEUDOMONADACEAE
	Pseudomonas aeruginosa (Schroeter) Migula
	Graham-Smith, 1911a, 1911b
	Balzam, 1937
	Greenberg, 1962
	Havlik, 1964
	PSEUDOMONADALES: SPIRILLACEAE
	Vibrio comma (Schroeter) Winslow et al.
	Graham-Smith, 1911a, 1911b
	Alessandrini, 1912
	EUBACTERIALES: ACHROMOBACTERACEAE
	Alcaligenes faecalis Castellani and Chalmers
	Gregor et al., 1960
	Povolný et al., 1961
	Havlik, 1964
	Flavobacterium devorans (Zimmermann) Bergey et al.
	Gregor et al., 1960
	Flavobacterium invisible (Vaughan) Bergey et al.
	Gregor et al., 1960
	EUBACTERIALES: ENTEROBACTERIACEAE
	Escherichia coli (Migula) Castellani and Chalmers
	Yao et al., 1929
	Parisot et al., 1934
	Balzam, 1937
	Hoffmann, 1950
	Gregor et al., 1960
	Povolný et al., 1961
	Greenberg, 1962
	Greenberg et al., 1963c
	Havlik, 1964

Calliphora vicina (cont.)	*Escherichia coli* (Migula) Castellani and Chalmers 25
	Greenberg et al., 1963c
	Escherichia coli var. *acidilactici* (Topley and Wilson) Yale
	Shura-Bura, 1952
	Escherichia coli var. *communior* (Topley and Wilson) Yale
	Graham-Smith, 1909
	Escherichia coli var. *neapolitana* (Topley and Wilson) Yale
	Graham-Smith, 1909
	Aerobacter aerogenes (Kruse) Beijerinck
	Graham-Smith, 1909
	Gregor et al., 1960
	Greenberg, 1962
	Havlik, 1964
	Aerobacter cloacae (Jordan) Bergey et al.
	Povolný et al., 1961
	Klebsiella cloacae (Jordan) Brisou
	Gregor et al., 1960
	Serratia marcescens Bizio
	Graham-Smith, 1911a, 1911b
	Balzam, 1937
	Gregor et al., 1960
	Havlik, 1964
	Proteus inconstans (Ornstein) Shaw and Clarke
	Gregor et al., 1960
	Proteus mirabilis Hauser
	Gregor et al., 1960
	Greenberg et al., 1963c
	Greenberg, 1966
	Proteus morganii (Winslow, Kligler, and Rothberg) Yale
	Gregor et al., 1960
	Povolný et al., 1961
	Havlik, 1964
	Proteus rettgeri (Hadley et al.) Rustigian and Stuart
	Gregor et al., 1960
	Proteus vulgaris Hauser
	Gregor et al., 1960
	Povolný et al., 1961
	Greenberg, 1962
	Havlik, 1964

Fly	Organism
Calliphora vicina (cont.)	*Salmonella enteritidis* (Gaertner) Castellani and Chalmers Graham-Smith, 1911a, 1911b, 1912 Gross et al., 1953 Greenberg, 1966 *Salmonella give* Kauffmann Gross et al., 1953 *Salmonella hirschfeldii* Weldin Yao et al., 1929 *Salmonella paratyphi* A (Brion and Kayser) Castellani and Chalmers Gross et al., 1953 *Salmonella paratyphi* B Castellani and Chalmers Gross et al., 1953 *Salmonella schottmuelleri* (Winslow, Kligler, and Rothberg) Bergey et al. Greenberg, 1962 *Salmonella typhi* Warren and Scott Graham-Smith, 1911a, 1911b Teodoro, 1916 Gwatkin et al., 1938 Gross et al., 1953 *Salmonella typhimurium* (Loeffler) Castellani and Chalmers Greenberg et al., 1963c *Citrobacter freundii* Werkman and Gillen Gregor et al., 1960 ?*Shigella* Castellani and Chalmers Sychevskaia͡ et al., 1959a *Shigella dysenteriae* (Shiga) Castellani and Chalmers Yao et al., 1929 *Shigella dysenteriae* (Shiga) Castellani and Chalmers 2 Sychevskaia͡ et al., 1959a *Shigella flexneri* Castellani and Chalmers Gross et al., 1953 Sychevskaia͡ et al., 1959a Greenberg, 1962 *Shigella flexneri* Castellani and Chalmers la Sychevskaia͡ et al., 1959a *Shigella sonnei* (Levine) Weldin Gross et al., 1953 Greenberg, 1962

Calliphora vicina (cont.)	EUBACTERIALES: BRUCELLACEAE
	Brucella abortus (Schmidt) Meyer and Shaw
	Gwatkin et al., 1938
	Haemophilis influenzae (Lehmann and Neumann) Winslow et al.
	Greenberg, 1962
	EUBACTERIALES: MICROCOCCACEAE
	Staphylococcus afermentans Shaw, Stitt, and Cowan
	Gregor et al., 1960
	Staphylococcus aureus Rosenbach
	Gwatkin et al., 1938
	Gregor et al., 1960
	Greenberg, 1962
	Staphylococcus lactis Shaw, Stitt, and Cowan
	Gregor et al., 1960
	Staphylococcus saprophyticus Fairbrother
	Gregor et al., 1960
	Sarcina loewenbergi Macé
	Galli-Vallerio, 1908
	EUBACTERIALES: LACTOBACILLACEAE
	Diplococcus pneumoniae Weichselbaum
	Povolný et al., 1961
	Streptococcus Rosenbach
	Povolný et al., 1961
	Streptococcus agalactiae Lehmann and Neumann
	Gwatkin et al., 1938
	Streptococcus faecalis Andrewes and Horder
	Gregor et al., 1960
	Streptococcus lactis (Lister) Löhnis
	Greenberg, 1962
	EUBACTERIALES: BACILLACEAE
	Bacillus anthracis Cohn
	Graham-Smith, 1911a, 1911b, 1912
	Sen et al., 1944
	Bacillus cereus var. *mycoides* (Flügge) Smith et al.
	Gregor et al., 1960
	Bacillus subtilis Cohn
	Balzam, 1937
	Greenberg, 1962

Calliphora vicina (cont.)

ACTINOMYCETALES:
MYCOBACTERIACEAE
Mycobacterium tuberculosis (Zopf) Lehmann and Neumann
Morellini et al., 1947
SPIROCHAETALES:
TREPONEMATACEAE
Leptospira canicola Okell et al.
Kunert et al., 1952
Schmidtke, 1959
Leptospira grippotyphosa Topley and Wilson
Kunert et al., 1952
Schmidtke, 1959
Leptospira icterohaemorrhagiae (Inada and Ido) Noguchi
Kunert et al., 1952
Schmidtke, 1959

THALLOPHYTA: PHYCOMYCETES
ENTOMOPHTHORALES:
ENTOMOPHTHORACEAE
Entomophthora Fresenius
Graham-Smith, 1916

THALLOPHYTA: ASCOMYCETES
ENDOMYCETALES:
ENDOMYCETACEAE
Saccharomyces cerevisae Hansen
Berlese, 1896
Saccharomyces pastorianus Hansen
Berlese, 1896

THALLOPHYTA: FUNGI IMPERFECTI
MONILIALES: MONILIACEAE
Beauveria bassiana (Balsamo) Vuillemin
Rostrup, 1916
CRYPTOCOCCALES:
CRYPTOCOCCACEAE
Candida Berkhout
Berlese, 1896
Candida mycoderma (Reess)
Berlese, 1896
Kloeckera apiculata (Reess emend Klöcker) Janke
Berlese, 1896

PROTOZOA: MASTIGOPHORA
PROTOMONADIDA:
TRYPANOSOMATIDAE
Herpetomonas muscarum (Leidy) Kent
Alexeieff, 1911

Calliphora vicina (cont.)

Laveran et al., 1920
Root, 1921
Becker, 1924

Herpetomonas muscarum (Leidy) Kent=[*Crithidia calliphorae*]
Becker, 1922-3

Herpetomonas muscarum (Leidy) Kent=[*Herpetomonas muscaedomesticae*]
Becker, 1922-3

PROTOMONADIDA: BODONIDAE

Rhynchoidomonas luciliae (Patton) Patton
Alexeieff, 1911

POLYMASTIGIDA: CHILOMASTIGIDAE

Chilomastix mesnili (Wenyon)
Root, 1921

POLYMASTIGIDA: HEXAMITIDAE

Giardia intestinalis (Lambl)
Root, 1921

PROTOZOA: SARCODINA

AMOEBIDA: ENDAMOEBIDAE

Endolimax nana (Wenyon and O'Connor)
Root, 1921
Yao et al., 1929

Entamoeba coli (Grassi)
Root, 1921
Yao et al., 1929

Entamoeba histolytica Schaudinn
Root, 1921
Yao et al., 1929

PROTOZOA: SPOROZOA

COCCIDIDA: EIMERIIDAE

Eimeria irresidua Kessel and Janciewicz
Metelkin, 1935

Eimeria perforans (Leuckart) Sluiter and Swellengrebel
Metelkin, 1935

HAPLOSPORIDA: UNCERTAIN POSITION

Toxoplasma Nicolle and Manceaux
van Thiel, 1949
Schmidtke, 1959

Toxoplasma gondii Nicolle and Manceaux
Kunert et al., 1953
Schmidtke, 1955

Calliphora vicina (cont.)

PROTOZOA: CNIDOSPORIDIA
MICROSPORIDA: NOSEMATIDAE
Nosema apis Zander
Fantham et al., 1913
Kramer, 1962
MICROSPORIDA: MRAZEKIIDAE
Octosporea muscae-domesticae Flu
Fantham et al., 1958
Spiroglugea Léger and Hesse
Porter, 1953
Fantham et al., 1958
Toxoglugea Léger and Hesse
Porter, 1953
Fantham et al., 1958
MICROSPORIDA: TELOMYXIDAE
Telomyxa Léger and Hesse
Porter, 1953
Fantham et al., 1958

PLATYHELMINTHES: CESTOIDEA
PSEUDOPHYLLIDEA: DIPHYLLOBOTHRIIDAE
Bothriocephalus marginatus Krefft
Nicoll, 1911b
Diphyllobothrium latum (Linnaeus) Lühe
Pod"i͡apol'skai͡a et al., 1934
CYCLOPHYLLIDEA: TAENIIDAE
Taeniarhynchus saginatum (Goeze)
Sychevskai͡a et al., 1958

ASCHELMINTHES: NEMATODA
TRICHURIDEA: TRICHURIDAE
Trichuris trichiura (Linnaeus)
Pod"i͡apol'skai͡a et al., 1934
OXYURIDEA: OXYURIDAE
Enterobius vermicularis (Linnaeus) Leach
Pod"i͡apol'skai͡a et al., 1934
ASCARIDIDEA: ASCARIDIDAE
Ascaris Linnaeus
Pod"i͡apol'skai͡a et al., 1934
Ascaris lumbricoides Linnaeus
Sukhacheva, 1963

ARTHROPODA: INSECTA
LEPIDOPTERA: SPHINGIDAE
Acherontia atropos Linnaeus
Keilin (reviewing Mik, 1890), 1917
COLEOPTERA: STAPHYLINIDAE
Aleochara curtula Goeze
Fuldner, 1964

Calliphora vicina (cont.)

Creophilus erythrocephalus Fabricius
  Froggatt, 1914a

HYMENOPTERA: ICHNEUMONIDAE

Atractodes bicolor Gravenhorst
  Graham-Smith, 1919

Nemeritis canescens (Gravenhorst)
  Salt, 1957

Phygadeuon speculator Thomson
  Graham-Smith, 1919

HYMENOPTERA: BRACONIDAE

Alysia manducator Panzer
  MacDougall, 1909
  Graham-Smith, 1916, 1919
  Altson, 1920
  Gurney et al., 1926a
  Miller, 1927
  Morgan, 1929
  Holdaway et al., 1932
  Evans, 1933
  Caudri, 1941

Aphaereta cephalotes Haliday
  Graham-Smith, 1916

Aphaereta minuta Nees
  Evans, 1933

HYMENOPTERA: CYNIPIDAE

Kleidotoma Westwood
  James, 1928

Kleidotoma marshalli Marshall
  James, 1928

HYMENOPTERA: FIGITIDAE

Figites anthomyiarum (Bouché)
  James, 1928

HYMENOPTERA: CHALCIDIDAE

Brachymeria minuta (Fabricius)
  Sychevskaîa, 1964b, 1966

HYMENOPTERA: UNDETERMINED CHALCIDIDAE
  Froggatt, 1914a, 1914b

HYMENOPTERA: PTEROMALIDAE

Dibrachys affinis Masi
  Voukassovitch, 1924a

Nasonia vitripennis[1] (Walker)
  Froggatt et al., 1914
  Graham-Smith, 1916
  Roubaud, 1917
  Altson, 1920
  Johnston et al., 1920b, 1921

[1] Although not specifically mentioned in each entry, it is assumed that all *Nasonia vitripennis* are pupal parasites.

Fly	Associated organism
Calliphora vicina (cont.)	Séguy, 1929 Cousin, 1933 Smirnov et al., 1933, 1934 Smirnov, 1934 Vladimirova et al., 1934 Velthuis et al., 1965 *Crabro cavifrons* Thomson Hamm et al., 1926 *Crabro cribraria* (Linnaeus) Hamm et al., 1926 *Crabro peltarius* (Schreber) Hamm et al., 1926 *Ectemnius quadricinctus* (Fabricius) Hamm et al., 1926 *Gorytes* Latreille Froggatt, 1914a DIPTERA: ANTHOMYIIDAE *Scatophaga stercoraria* (Linnaeus) Hewitt, 1914 CHORDATA: AVES PASSERIFORMES: HIRUNDINIDAE *Delichon urbica* (Linnaeus) subsp. *urbica* Séguy, 1929 *Hirundo rustica* Linnaeus subsp. *rustica* Séguy, 1929
Calliphora vomitoria (Linnaeus) (Synonym: *Musca vomitoria*)	VIRUSES ENTEROVIRUSES Poliovirus Noguchi et al., 1917 INSECT VIRUSES Tipula iridescent virus Smith et al., 1961 SCHIZOPHYTA: SCHIZOMYCETES PSEUDOMONADALES: PSEUDOMONADACEAE *Pseudomonas aeruginosa* (Schroeter) Migula Muzzarelli, 1925 *Pseudomonas eisenbergii* Migula Cao, 1906 *Pseudomonas fluorescens* Migula Cao, 1906 *Pseudomonas jaegeri* Migula Cao, 1906

Calliphora vomitoria (cont.)

PSEUDOMONADALES: SPIRILLACEAE
Vibrio comma (Schroeter) Winslow et al.
Maddox, 1885a, 1885b
Alessandrini, 1912
EUBACTERIALES: ACHROMOBACTERACEAE
Alcaligenes faecalis Castellani and Chalmers
Gregor et al., 1960
Flavobacterium devorans (Zimmermann) Bergey et al.
Gregor et al., 1960
Flavobacterium invisible (Vaughan) Bergey et al.
Gregor et al., 1960
EUBACTERIALES: ENTEROBACTERIACEAE
Escherichia coli (Migula) Castellani and Chalmers
Wollman, 1911
Yao et al., 1929
Hoffmann, 1950
Gregor et al., 1960
Povolný et al., 1961
Escherichia coli var. *communior* (Topley and Wilson) Yale
Graham-Smith, 1909
Escherichia coli var. *neapolitana* (Topley and Wilson) Yale
Graham-Smith, 1909
Aerobacter aerogenes (Kruse) Beijerinck
Graham-Smith, 1909
Gregor et al., 1960
Klebsiella cloacae (Jordan) Brisou
Gregor et al., 1960
Paracolobactrum aerogenoides Borman, Stuart, and Wheeler
O'Keefe et al., 1954
Serratia marcescens Bizio
Gregor et al., 1960
Proteus inconstans (Ornstein) Shaw and Clarke
Gregor et al., 1960
Proteus mirabilis Hauser
Cao, 1906
Gregor et al., 1960
Proteus morganii (Winslow, Kligler, and Rothberg) Yale
Gregor et al., 1960

Calliphora vomitoria (cont.)

Proteus rettgeri (Hadley et al.) Rustigian and Stuart
  Gregor et al., 1960
Proteus vulgaris Hauser
  Cao, 1906
  Wollman, 1911
  Gregor et al., 1960
  Povolný et al., 1961
Salmonella cholerae-suis (Smith) Weldin
  Buchanan, 1907
Salmonella hirschfeldii Weldin
  Yao et al., 1929
Salmonella typhi Warren and Scott
  Wollman, 1921
Citrobacter freundii Werkman and Gillen
  Gregor et al., 1960
Shigella Castellani and Chalmers
  Zaidenov, 1961
Shigella dysenteriae (Shiga) Castellani and Chalmers
  Wollman, 1921
  Yao et al., 1929
Shigella flexneri Castellani and Chalmers
  Beyer, 1925

EUBACTERIALES: UNDETERMINED ENTEROBACTERIACEAE

Coliforms
  Testi, 1909

EUBACTERIALES: BRUCELLACEAE

Pasteurella pestis (Lehmann and Neumann) Holland
  Gosio, 1925
  Russo, 1930
Cillopasteurella delendae-muscae (Roubaud and Descazeaux) Prévot
  Roubaud et al., 1923

EUBACTERIALES: MICROCOCCACEAE

Micrococcus Cohn
  Maddox, 1885b
  Bogdanow, 1906, 1908
Staphylococcus afermentans Shaw, Stitt, and Cowan
  Gregor et al., 1960
Staphylococcus aureus Rosenbach
  Wollman, 1911
  Gregor et al., 1960

Calliphora vomitoria (cont.)	*Staphylococcus lactis* Shaw, Stitt, and Cowan Gregor et al., 1960 *Staphylococcus saprophyticus* Fairbrother Gregor et al., 1960 *Gaffkya tetragena* (Gaffky) Trevisan Testi, 1909 EUBACTERIALES: BREVIBACTERIACEAE *Kurthia zenkeri* (Hauser) Bergey et al. Cao, 1906 EUBACTERIALES: LACTOBACILLACEAE *Streptococcus faecalis* Andrewes and Horder Gregor et al., 1960 EUBACTERIALES: BACILLACEAE *Bacillus* Cohn Maddox, 1885a *Bacillus ?anthracis* Cohn Maddox, 1885a *Bacillus anthracis* Cohn Raimbert, 1870 Buchanan, 1907 *Bacillus cereus* var. *mycoides* (Flügge) Smith, Gordon, and Clark Gregor et al., 1960 *Bacillus megaterium* de Bary Muzzarelli, 1925 *Bacillus radiciformis* Eberbach Cao, 1906 *Bacillus subtilis* Cohn Cao, 1906 *Bacillus ?subtilis* Cohn Maddox, 1885a ACTINOMYCETALES: MYCOBACTERIACEAE *Mycobacterium lepraemurium* Marchoux and Sorel Wherry, 1908a *Mycobacterium ?lepraemurium* Marchoux and Sorel Wherry, 1908b *Mycobacterium paratuberculosis* Bergey et al. Muzzarelli, 1925

Calliphora vomitoria (cont.)	*Mycobacterium tuberculosis* (Zopf) Lehmann and Neumann Wollman, 1921 THALLOPHYTA: PHYCOMYCETES ENTOMOPHTHORALES: ENTOMOPHTHORACEAE *Entomophthora americana* Thaxter Rozsypal, 1957 *Entomophthora calliphorae* Giard Giard, 1888 *Entomophthora muscae* Cohn Thaxter, 1888 *Zoophthora vomitoriae* (Rozsypal) Rozsypal, 1966 THALLOPHYTA: FUNGI IMPERFECTI MONILIALES: MONILIACEAE *Blastomyces* Costantin and Rolland Muzzarelli, 1925 ?*Penicillium* Link Maddox, 1885b FUNGI OF UNCERTAIN AFFINITY *Streptotricea* Muzzarelli, 1925 PROTOZOA: SARCODINA AMOEBIDA: ENDAMOEBIDAE *Endolimax nana* (Wenyon and O'Connor) Yao et al., 1929 *Entamoeba coli* (Grassi) Yao et al., 1929 *Entamoeba histolytica* Schaudinn Jausion et al., 1923 Yao et al., 1929 PROTOZOA: CNIDOSPORIDIA MICROSPORIDA: MRAZEKIIDAE *Octosporea muscae-domesticae* Flu Fantham et al., 1958 Kramer, 1965b *Spiroglugea* Léger and Hesse Porter, 1953 Fantham et al., 1958 *Toxoglugea* Léger and Hesse Porter, 1953 Fantham et al., 1958 MICROSPORIDA: TELOMYXIDAE *Telomyxa* Léger and Hesse Porter, 1953 Fantham et al., 1958

Fly	Organism
Calliphora vomitoria (cont.)	ARTHROPODA: ARACHNIDA PSEUDOSCORPIONIDEA: CHELIFERIDAE *Chelifer* Geoffroy Donovan, 1797 ACARINA: PARASITIDAE *Parasitus* Latreille Graham-Smith, 1916 ACARINA: MACROCHELIDAE *Macrocheles muscaedomesticae* (Scopoli) Trojan, 1908 ARTHROPODA: INSECTA HYMENOPTERA: ICHNEUMONIDAE *Theronia atalantae* (Poda) Fahringer, 1922 HYMENOPTERA: BRACONIDAE *Alysia manducator* Panzer Altson, 1920 Holdaway et al., 1932 *Aphaereta minuta* Nees Evans, 1933 *Opius nitidulator* Nees Rambousek, 1929 HYMENOPTERA: PTEROMALIDAE *Dibrachys affinis* Masi Voukassovitch, 1924a *Habrocytus obscurus* Dalman Voukassovitch, 1924b *Nasonia vitripennis* (Walker) Altson, 1920 HYMENOPTERA: SPHECIDAE *Crabro cribraria* (Linnaeus) Hamm et al., 1926 *Crabro vagus* (Linnaeus) Hamm et al., 1926 *Ectemnius quadricinctus* (Fabricius) Hamm et al., 1926
Calliphora vomitoria **var.** ***dunensis*** **Giard**	THALLOPHYTA: PHYCOMYCETES ENTOMOPHTHORALES: ENTOMOPHTHORACEAE *Entomophthora calliphorae* Giard Giard, 1879
Calliphora uralensis **Villeneuve**	ENTEROVIRUSES Poliovirus strain 113 Levkovich et al., 1957 Poliovirus Type 2, Lansing strain Levkovich et al., 1957

Calliphora uralensis (cont.)	SCHIZOPHYTA: SCHIZOMYCETES EUBACTERIALES: ENTEROBACTERIACEAE *Shigella flexneri* Castellani and Chalmers Sukhova, 1950 *Shigella ?flexneri* Castellani and Chalmers Sukhova, 1950
Neopollenia stygia (Fabricius) (Synonym: *Calliphora villosa*)	ARTHROPODA: INSECTA HYMENOPTERA: PTEROMALIDAE *Nasonia vitripennis* (Walker) Johnston et al., 1920b *Spalangia muscidarum* Richardson Johnston et al., 1920b *Spalangia nigroaenea* Curtis Bouček, 1963 HYMENOPTERA: SPHECIDAE *Sericophorus teliferopodus* Rayment Rayment, 1954
Anastellorhina augur (Fabricius) (Synonyms: *Calliphora oceanicae*, *Calliphora oceaniae*, *Paracalliphora augur*, *Calliphora augur*)	ASCHELMINTHES: NEMATODA SPIRURIDEA: SPIRURIDAE *Habronema* Diesing Johnston, 1920 Johnston et al., 1920a ARTHROPODA: INSECTA HYMENOPTERA: BRACONIDAE *Alysia manducator* Panzer Gurney et al., 1926a HYMENOPTERA: CHALCIDIDAE *Brachymeria calliphorae* (Froggatt) Froggatt, 1916 Johnson et al., 1921 HYMENOPTERA: PTEROMALIDAE *Nasonia vitripennis*[1] (Walker) Froggatt et al., 1914 Froggatt, 1919 Johnston et al., 1920b, 1921 *Spalangia muscidarum* Richardson Johnston et al., 1920b, 1921 *Spalangia nigroaenea* Curtis Bouček, 1963 HYMENOPTERA: ENCYRTIDAE *Australencyrtus giraulti* Johnston and Tiegs Johnston et al., 1921

[1] Although not specifically mentioned in each entry, it is assumed that all *Nasonia vitripennis* are pupal parasites.

Fly	Organism
Anastellorhina augur (cont.)	DIPTERA: CALLIPHORIDAE *Chrysomya rufifacies* (Macquart) Mackerras, 1930
Onesia cognata Meigen (Synonym: *Onesia coerulea*)	ARTHROPODA: INSECTA HYMENOPTERA: SPHECIDAE *Crabro cribraria* (Linnaeus) Hamm et al., 1926
Onesia sepulchralis Meigen	ARTHROPODA: INSECTA HYMENOPTERA: SPHECIDAE *Crabro peltarius* (Schreber) Hamm et al., 1926 *Ectemnius quadricinctus* (Fabricius) Hamm et al., 1926 *Oxybelus quattuordecimnotatus* Jurine Chevalier, 1926
Onesia townsendi Hall (Synonym: *Onesia aculeata*)	ARTHROPODA: INSECTA HYMENOPTERA: SPHECIDAE *Ectemnius quadricinctus* (Fabricius) Hamm et al., 1926
Cynomyopsis cadaverina (Robineau-Desvoidy) (Synonym: *Cynomya cadaverina*)	ENTEROVIRUSES Poliovirus Sabin et al., 1942 Melnick, 1949 SCHIZOPHYTA: SCHIZOMYCETES PSEUDOMONADALES: PSEUDOMONADACEAE *Pseudomonas aeruginosa* (Schroeter) Migula Greenberg, 1962 EUBACTERIALES: ENTEROBACTERIACEAE *Escherichia coli* (Migula) Castellani and Chalmers Greenberg, 1962 *Proteus vulgaris* Hauser O'Keefe et al., 1954 Greenberg, 1962 *Salmonella typhi* Warren and Scott Gwatkin et al., 1938 Webb et al., 1956 Greenberg, 1962 *Shigella flexneri* Castellani and Chalmers Greenberg, 1962 EUBACTERIALES: BRUCELLACEAE *Brucella abortus* (Schmidt) Meyer and Shaw Gwatkin et al., 1938

Cynomyopsis cadaverina (cont.)	EUBACTERIALES: MICROCOCCACEAE *Staphylococcus aureus* Rosenbach Gwatkin et al., 1938 Greenberg, 1962 EUBACTERIALES: LACTOBACILLACEAE *Streptococcus agalactiae* Lehmann and Neumann Gwatkin et al., 1938 *Streptococcus durans* Sherman and Wing Greenberg, 1962 EUBACTERIALES: BACILLACEAE *Bacillus subtilis* Cohn Greenberg, 1962 PROTOZOA: MASTIGOPHORA PROTOMONADIDA: TRYPANOSOMATIDAE *Herpetomonas ?muscarum* (Leidy) Kent Wallace et al., 1964 *Leptomonas mirabilis* Roubaud Wallace et al., 1964 TRICHOMONADIDA: TRICHOMONADIDAE *Trichomonas hominis* (Davaine) Hegner, 1928 ARTHROPODA: INSECTA HYMENOPTERA: PTEROMALIDAE *Nasonia vitripennis* (Walker) Girault et al., 1910a
Cynomya mortuorum (Linnaeus) (Synonyms: *Cynomyia mortuorum, Sarcophaga mortuorum*)	SCHIZOPHYTA: SCHIZOMYCETES EUBACTERIALES: ENTEROBACTERIACEAE *Salmonella typhi* Warren and Scott Krontowski, 1913 *Shigella dysenteriae* (Shiga) Castellani and Chalmers Krontowski, 1913 PROTOZOA: SPOROZOA COCCIDIDA: EIMERIIDAE *Eimeria irresidua* Kessel and Janciewicz Metelkin, 1935 *Eimeria perforans* (Leuckart) Sluiter and Swellengrebel Metelkin, 1935
Auchmeromyia luteola (Fabricius)	PROTOZOA: MASTIGOPHORA PROTOMONADIDA: TRYPANOSOMATIDAE *Herpetomonas caulleryi* (Roubaud) Wenyon Roubaud, 1911c

Fly	Associated organism
Auchmeromyia luteola (cont.)	ARTHROPODA: INSECTA HYMENOPTERA: SPHECIDAE *Bembix olivata* Dahlbom Roubaud, 1913
UNDETERMINED LUCILIINI	SCHIZOPHYTA: SCHIZOMYCETES EUBACTERIALES: ENTEROBACTERIACEAE *Escherichia coli* (Migula) Castellani and Chalmers Shimizu et al., 1965 *Proteus morganii* (Winslow, Kligler, and Rothberg) Yale Shimizu et al., 1965 *Proteus rettgeri* (Hadley et al.) Rustigian and Stuart Shimizu et al., 1965 *Proteus vulgaris* Hauser Shimizu et al., 1965 *Providencia* Kauffmann Shimizu et al., 1965 *Salmonella* Lignières Shimizu et al., 1965 *Shigella* Castellani and Chalmers Shimizu et al., 1965
UNDETERMINED CALLIPHORINI	SCHIZOPHYTA: SCHIZOMYCETES EUBACTERIALES: ENTEROBACTERIACEAE *Proteus morganii* (Winslow, Kligler, and Rothberg) Yale Shimizu et al., 1965 *Proteus rettgeri* (Hadley et al.) Rustigian and Stuart Shimizu et al., 1965 *Providencia* Kauffmann Shimizu et al., 1965
UNDETERMINED CALLIPHORINAE	PROTOZOA: MASTIGOPHORA PROTOMONADIDA: TRYPANOSOMATIDAE *Herpetomonas muscarum* (Leidy) Kent Patton, 1921a
Subfamily: POLLENIINAE	
Pollenia Robineau-Desvoidy	ARTHROPODA: INSECTA HYMENOPTERA: SPHECIDAE *Crabro vagus* (Linnaeus) Hamm et al., 1926

Pollenia rudis (Fabricius) .... THALLOPHYTA: PHYCOMYCETES

ENTOMOPHTHORALES: ENTOMOPHTHORACEAE

Entomophthora ?americana Thaxter
Judd, 1955

PROTOZOA: MASTIGOPHORA

PROTOMONADIDA: TRYPANOSOMATIDAE

Herpetomonas muscarum (Leidy) Kent
Léger, 1903

PROTOZOA: CNIDOSPORIDIA

MICROSPORIDA: MRAZEKIIDAE

Octosporea muscae-domesticae Flu
Kramer, 1964a, 1965b

PLATYHELMINTHES: CESTOIDEA

CYCLOPHYLLIDEA: TAENIIDAE

Taeniarhynchus saginatum (Goeze)
Sychevskaîâ et al., 1958

ANNELIDA: OLIGOCHAETA

OPISTHOPORA: LUMBRICIDAE

Helodrilus caliginosus (Savigny)
Pimentel et al., 1960

Helodrilus chloroticus (Savigny)
Keilin, 1909, 1911
Webb et al., 1916
Garrison, 1924
DeCoursey, 1927

Helodrilus foetidus (Savigny)
Pimentel et al., 1960

Helodrilus roseus (Savigny)
DeCoursey, 1927
Pimentel et al., 1960

Lumbricus herculeus Savigny
Barnes, 1924

Lumbricus rubellus Hoffmeister
Pimentel et al., 1960

Lumbricus terrestris Linnaeus
Pimentel et al., 1960

Octolasium lacteum (Orley)
Pimentel et al., 1960

OPISTHOPORA: MEGASCOLECIDAE

Diplocardia communis Garman
Pimentel et al., 1960

UNDETERMINED ANNELIDA
Blakitnaîâ, 1962

Pollenia rudis (cont.)

ARTHROPODA: INSECTA
HYMENOPTERA: SPHECIDAE
Crabro advenus Smith
Kurczewski et al., 1968
Crabro cribraria (Linnaeus)
Hamm et al., 1926
Crabro leucostoma (Linnaeus)
Hamm et al., 1926
Crabro peltarius (Schreber)
Hamm et al., 1926
Crabro serripes Panzer
Hamm et al., 1926
Ectemnius quadricinctus (Fabricius)
Hamm et al., 1926
Mellinus arvensis (Linnaeus)
Chevalier, 1923a
Oxybelus uniglumis (Linnaeus)
Chevalier, 1926
DIPTERA: ANTHOMYIIDAE
Scatophaga stercoraria (Linnaeus)
Hewitt, 1914

UNDETERMINED CALLIPHORIDAE ....

ENTEROVIRUSES
Poliovirus
Toomey et al., 1941
SCHIZOPHYTA: SCHIZOMYCETES
PSEUDOMONADALES: PSEUDOMONADACEAE
Pseudomonas aeruginosa (Schroeter) Migula
Holdaway, 1932
PSEUDOMONADALES: SPIRILLACEAE
Vibrio comma (Schroeter) Winslow et al.
Savchenko, 1892a, 1892b
EUBACTERIALES: ENTEROBACTERIACEAE
Salmonella pullorum (Rettger) Bergey
Gwatkin et al., 1952
Salmonella typhi Warren and Scott
Firth et al., 1902
EUBACTERIALES: BACILLACEAE
Bacillus anthracis Cohn
Morris, 1920
Clostridium perfringens (Veillon and Zuber) Holland
Baer, 1929
West, 1953

UNDETERMINED
CALLIPHORIDAE (cont.)

SPIROCHAETALES: TREPONEMATACEAE
Treponema pertenue Castellani
Yasuyama, 1928

THALLOPHYTA: PHYCOMYCETES
ENTOMOPHTHORALES: ENTOMOPHTHORACEAE
Entomophthora muscae Cohn
White et al., 1877

ARTHROPODA: INSECTA
HEMIPTERA: REDUVIIDAE
Pristhesancus papuensis Stål
Noble, 1936

COLEOPTERA: HISTERIDAE
Saprinus cyaneus Fabricius
Gurney et al., 1926b

HYMENOPTERA: BRACONIDAE
Alysia manducator Panzer
Graham-Smith, 1916
Aphaereta auripes (Provancher)
Roberts, 1935

HYMENOPTERA: PTEROMALIDAE
Nasonia vitripennis (Walker)
Gurney et al., 1926b

HYMENOPTERA: DIAPRIIDAE
Trichopria hirticollis (Ashmead)
Roberts, 1935

HYMENOPTERA: FORMICIDAE
Labidus coecus (Latreille)
Lindquist, 1942
Pheidole megacephala (Fabricius)
Phillips, 1934

CHORDATA: AVES
PASSERIFORMES: MUSCICAPIDAE
Leucocirca leucophrys (Latham)
Froggatt et al., 1917

PASSERIFORMES: MELIPHAGIDAE
Meliphaga penicillata Gould
Froggatt et al., 1917
Myzantha melanocephala Latham
Froggatt et al., 1917

Family: SARCOPHAGIDAE
Subfamily: MILTOGRAMMINAE
Agria latifrons Fallén .... ARTHROPODA: ARACHNIDA
ACARINA: MACROCHELIDAE
Macrocheles muscaedomesticae (Scopoli)
Sychevskaia͡, 1964a

Pseudosarcophaga affinis Fallén (Synonyms: *Sarcophaga affinis, Agria affinis*)	THALLOPHYTA: PHYCOMYCETES ENTOMOPHTHORALES: ENTOMOPHTHORACEAE *Entomophthora muscae* Cohn Baird, 1957 ARTHROPODA: INSECTA LEPIDOPTERA: LIPARIDAE *Liparis dispar* Linnaeus Sitowski, 1928 LEPIDOPTERA: LYMANTRIIDAE *Leucoma salicis* Linnaeus Sitowski, 1928 *Lymantria monacha* Linnaeus Sitowski, 1928 LEPIDOPTERA: LASIOCAMPIDAE *Dendrolimus pini* Linnaeus Sitowski, 1928 COLEOPTERA: STAPHYLINIDAE *Baryodma ontarionis* Casey Colhoun, 1953 HYMENOPTERA: BRACONIDAE *Aphaereta pallipes* Say Barlow, 1964 HYMENOPTERA: CHALCIDIDAE *Brachymeria minuta* (Fabricius) Nikol'skaîa, 1960a
Wohlfahrtia nuba Wiedemann	ARTHROPODA: INSECTA HYMENOPTERA: CHALCIDIDAE *Dirhinoides wohlfahrtiae* Ferrière Nikol'skaîa, 1960b
Subfamily: SARCOPHAGINAE	
Blaesoxipha impar (Aldrich) (Synonym: *Sarcophaga impar*)	ARTHROPODA: INSECTA HYMENOPTERA: CHALCIDIDAE *Brachymeria fonscolombei* (Dufour) Roberts, 1933b, 1935 HYMENOPTERA: PTEROMALIDAE *Spalangia endius* Walker Bouček, 1963 *Spalangia stomoxysiae* Girault Lindquist, 1936 Peck, 1951
Blaesoxipha lineata Fallén	ARTHROPODA: INSECTA ORTHOPTERA: LOCUSTIDAE *Dociostaurus maroccanus* (Thunberg) Baranov, 1924
Blaesoxipha plinthopyga (Wiedemann)	ENTEROVIRUSES Poliovirus Types 1 and 3 Paul et al., 1962

(Synonyms: *Sarcophaga plinthopyas*, *Sarcophaga plinthopyga*)

ECHO virus Types 8, 12, 14
Paul et al., 1962

ARTHROPODA: INSECTA

COLEOPTERA: SILPHIDAE

Nicrophorus nigritus Mannerheim
Illingworth, 1927

COLEOPTERA: UNDETERMINED SILPHIDAE

Illingworth, 1927

COLEOPTERA: HISTERIDAE

Saprinus lugens Erichson
Illingworth, 1927

Xerosaprinus fimbriatus (LeConte)
Illingworth, 1927

Xerosaprinus lubricus (LeConte)
Illingworth, 1927

COLEOPTERA: STAPHYLINIDAE

Creophilus maxillosus var. *villosus* Gravenhorst
Illingworth, 1927

HYMENOPTERA: BRACONIDAE

Alysia fossulata Provancher
Roberts, 1935

Alysia ridibunda Say
Roberts, 1935

HYMENOPTERA: CYNIPIDAE

Eucoila Westwood
Roberts, 1935

HYMENOPTERA: FIGITIDAE

Neralsia Cameron
Roberts, 1935

Neralsia armata (Say)
Roberts, 1935

HYMENOPTERA: CHALCIDIDAE

Brachymeria fonscolombei (Dufour)
Roberts, 1933b, 1935

Dirhinus texanus (Ashmead)
Roberts, 1935

HYMENOPTERA: PTEROMALIDAE

Nasonia vitripennis (Walker)
Roberts, 1935

HYMENOPTERA: DIAPRIIDAE

Trichopria hirticollis (Ashmead)
Roberts, 1935

DIPTERA: CUTEREBRIDAE

Dermatobia hominis (Linnaeus)
Blanchard, 1899

Fly	Associated organism
Helicobia quadrisetosa Coquillett	ARTHROPODA: INSECTA HYMENOPTERA: PTEROMALIDAE *Spalangia muscidarum* Richardson Pinkus, 1913
Oxysarcodexia ochripyga (Wulp) (Synonym: *Sarcophaga orchripyga*)	ENTEROVIRUSES Poliovirus Types 1 and 3 Paul et al., 1962 ECHO virus Types 8, 12, 14 Paul et al., 1962
Oxysarcodexia ventricosa (Wulp) (Synonym: *Sarcophaga ventricosa*)	ARTHROPODA: INSECTA HYMENOPTERA: BRACONIDAE *Aphaereta pallipes* Say Sanders et al., 1966
Ravinia derelicta (Walker) (Synonym: *Sarcophaga derelicta*)	ARTHROPODA: INSECTA HYMENOPTERA: BRACONIDAE *Aphaereta pallipes* Say Benson et al., 1963
Ravinia effrenata (Walker) (Synonym: *Sarcophaga effrenata*)	ARTHROPODA: INSECTA HYMENOPTERA: PTEROMALIDAE *Spalangia endius* Walker Bouček, 1963 *Spalangia stomoxysiae* Girault Lindquist, 1936 Peck, 1951
Ravinia latisetosa Parker (Synonym: *Sarcophaga latisetosa*)	ARTHROPODA: INSECTA HYMENOPTERA: BRACONIDAE *Aphaereta pallipes* Say Blickle, 1961 HYMENOPTERA: CYNIPIDAE *Eucoila* Westwood Blickle, 1961 HYMENOPTERA: FIGITIDAE *Xyalophora quinquelineata* (Say) Blickle, 1961
Ravinia lherminieri (Robineau-Desvoidy) (Synonyms: *Sarcophaga pallinervis, Sarcophaga lherminiera*)	SCHIZOPHYTA: SCHIZOMYCETES ACTINOMYCETALES: MYCOBACTERIACEAE *Mycobacterium phlei* Lehmann and Neumann Currie, 1910 ASCHELMINTHES: NEMATODA TYLENCHIDA: CONTORTYLENCHIDAE *Heterotylenchus autumnalis* Nickle Stoffolano, 1970 ARTHROPODA: INSECTA HYMENOPTERA: BRACONIDAE *Aphaereta pallipes* Say Blickle, 1961

Fly	Associated organism
Ravinia lherminieri (cont.)	HYMENOPTERA: CYNIPIDAE *Eucoila* Westwood Blickle, 1961 HYMENOPTERA: FIGITIDAE *Xyalophora quinquelineata* (Say) Blickle, 1961
Ravinia querula (Walker) .... (Synonym: *Sarcophaga querula*)	ARTHROPODA: INSECTA HYMENOPTERA: BRACONIDAE *Aphaereta pallipes* Say Blickle, 1961 Sanders et al., 1966 Houser et al., 1967 HYMENOPTERA: CYNIPIDAE *Eucoila* Westwood Blickle, 1961 Sanders et al., 1966 *Eucoila impatiens* (Say) Turner et al., 1968 HYMENOPTERA: FIGITIDAE *Xyalophora quinquelineata* (Say) Blickle, 1961 HYMENOPTERA: PTEROMALIDAE *Muscidifurax raptor* Girault Turner et al., 1968 *Spalangia nigra* Latreille Turner et al., 1968
Ravinia striata Fabricius	SCHIZOPHYTA: SCHIZOMYCETES EUBACTERIALES: ENTEROBACTERIACEAE *Shigella flexneri* Castellani and Chalmers 1a Sychevskaîa et al., 1959b ARTHROPODA: ARACHNIDA ACARINA: MACROCHELIDAE *Macrocheles muscaedomesticae* (Scopoli) Sychevskaîa, 1964a ARTHROPODA: INSECTA HYMENOPTERA: BRACONIDAE *Aphaereta minuta* Nees Sychevskaîa, 1964b HYMENOPTERA: CYNIPIDAE *Pseudeucoila trichopsila* (Hartig) Sychevskaîa, 1964b, 1966 HYMENOPTERA: FIGITIDAE *Figites discordis* Belizin Sychevskaîa, 1964b *Figites scutellaris* (Rossi) Sychevskaîa, 1964b

Host	Associated organism

Ravinia striata (cont.)	*Figites striolatus* Hartig
	  Sychevskai͡a, 1964b
	HYMENOPTERA: CHALCIDIDAE
	Brachymeria minuta (Fabricius)
	  Sychevskai͡a, 1964b, 1966
	Dirhinoides vlasovi Nikolśkai͡a
	  Sychevskai͡a, 1964b
	Euchalcidia blanda Nikolśkai͡a
	  Nikolśkai͡a, 1960b
	HYMENOPTERA: PTEROMALIDAE
	Nasonia vitripennis (Walker)
	  Sychevskai͡a, 1964b
	Spalangia[1] *cameroni* Perkins
	  Bouček, 1963
	Spalangia endius Walker
	  Bouček, 1963
	Spalangia nigroaenea Curtis
	  Bouček, 1963
	HYMENOPTERA: DIAPRIIDAE
	Diapria conica (Fabricius)
	  Sychevskai͡a, 1964b

Ravinia sueta (Wulp) (Synonym: *Sarcophaga sueta*)	ARTHROPODA: INSECTA
	HYMENOPTERA: PTEROMALIDAE
	Spalangia endius Walker
	  Bouček, 1963
	Spalangia stomoxysiae Girault
	  Lindquist, 1936
	  Peck, 1951

Sarcodexia pedunculata (Hall) (Synonym: *Sarcophaga pedunculata*)	ARTHROPODA: INSECTA HYMENOPTERA: CYNIPIDAE *Eucoila* Westwood Roberts, 1935

Sarcophaga Meigen	ENTEROVIRUSES
	Poliovirus
	  Trask et al., 1943
	  Melnick et al., 1945, 1949
	  Francis, Jr. et al., 1948
	  Melnick, 1949
	Poliovirus non-paralytic
	  Melnick et al., 1949
	Poliovirus Types 1 and 3
	  Riordan et al., 1961
	  Paul et al., 1962
	Coxsackievirus Type A9 (A9)
	  Riordan et al., 1961
	Coxsackievirus Type B5 (B5)
	  Riordan et al., 1961

[1] Although not specifically mentioned in each entry, it is assumed that all *Spalangia* are pupal parasites.

Sarcophaga (cont.)	ECHO virus Types 1, 5, 11 Riordan et al., 1961 ECHO virus Types 8, 12, 14 Paul et al., 1962 SCHIZOPHYTA: SCHIZOMYCETES PSEUDOMONADALES: SPIRILLACEAE *Vibrio comma* (Schroeter) Winslow et al. Pavan, 1949 EUBACTERIALES: ACHROMOBACTERACEAE *Alcaligenes* Castellani and Chalmers de la Paz, 1938 EUBACTERIALES: ENTEROBACTERIACEAE *Escherichia* Castellani and Chalmers de la Paz, 1938 *Escherichia coli* (Migula) Castellani and Chalmers Pavan, 1949 *Aerobacter* Beijerinck de la Paz, 1938 *Serratia marcescens* Bizio Nicholls, 1912 *Proteus* Hauser de la Paz, 1938 *Proteus morganii* (Winslow, Kligler, and Rothberg) Yale de la Paz, 1938 *Salmonella* Lignières Alcivar et al., 1946 *Salmonella amersfoort* Henning Alcivar et al., 1946 *Salmonella anatum* (Rettger and Scoville) Bergey et al. Alcivar et al., 1946 *Salmonella ballerup* Kauffmann and Møller Alcivar et al., 1946 *Salmonella cholerae-suis* (Smith) Weldin Alcivar et al., 1946 *Salmonella columbensis* (Castellani) Castellani and Chalmers de la Paz, 1938 *Salmonella dublin* Kauffmann Alcivar et al., 1946

Sarcophaga (cont.)

Salmonella enteritidis (Gaertner) Castellani and Chalmers
  Alcivar et al., 1946

Salmonella florida Cherry, Edwards, and Bruner
  Alcivar et al., 1946

Salmonella give Kauffmann
  Alcivar et al., 1946

Salmonella kentucky Edwards
  Alcivar et al., 1946

Salmonella kottbus Schütze et al.
  Alcivar et al., 1946

Salmonella meleagridis Bruner and Edwards
  Alcivar et al., 1946

Salmonella narashino Kauffmann
  Alcivar et al., 1946

Salmonella onderstepoort Henning
  Alcivar et al., 1946

Salmonella oranienburg Kauffmann
  Alcivar et al., 1946

Salmonella oregon Edwards and Bruner
  Alcivar et al., 1946

Salmonella paratyphi B Castellani and Chalmers
  Alcivar et al., 1946
  Pavan, 1949

Salmonella reading Schütze
  Alcivar et al., 1946

Salmonella typhi Warren and Scott
  Nicholls, 1912
  Pavan, 1949

Salmonella typhimurium (Loeffler) Castellani and Chalmers
  Alcivar et al., 1946

Salmonella zanzibar Kauffmann
  Alcivar et al., 1946

Shigella alkalescens (Andrewes) Weldin
  Alcivar et al., 1946

Shigella ambigua (Andrewes) Weldin
  de la Paz, 1938

Shigella boydii Ewing 1, 2, 3, 4, 5, 6, 8, 9
  Alcivar et al., 1946

Shigella boydii Ewing 2, 5, 6
  Richards et al., 1961

Sarcophaga (cont.)

Shigella ceylonensis (Weldin) Weldin
Alcivar et al., 1946
Shigella dispar (Andrewes) Bergey et al.
Alcivar et al., 1946
Shigella dysenteriae (Shiga) Castellani and Chalmers
de la Paz, 1938
Shigella dysenteriae (Shiga) Castellani and Chalmers 2
Richards et al., 1961
Shigella flexneri Castellani and Chalmers
de la Paz, 1938
Pavan, 1949
Shigella flexneri Castellani and Chalmers 1a
Richards et al., 1961
Shigella flexneri Castellani and Chalmers 1a, 2a, Y, 6
Alcivar et al., 1946
Shigella flexneri Castellani and Chalmers 2a, 3, 4a, 4b, 6
Richards et al., 1961
Shigella sonnei (Levine) Weldin 1
Richards et al., 1961

EUBACTERIALES: BRUCELLACEAE

Brucella melitensis (Hughes) Meyer and Shaw
Pavan, 1949

EUBACTERIALES: MICROCOCCACEAE

Staphylococcus Rosenbach
Picado, 1935
de la Paz, 1938
Staphylococcus aureus Rosenbach
Nicholls, 1912
Pavan, 1949

EUBACTERIALES: BACILLACEAE

Bacillus Cohn
Nicholls, 1912
Bacillus anthracis Cohn
Sen et al., 1944
Bacillus bifermentans Weinbery and Séguin
Picado, 1935
Clostridium botulinum (van Ermengem) Bergey et al.
Bail, 1900

Sarcophaga (cont.)

Clostridium perfringens (Veillon and Zuber) Holland
Picado, 1935

ACTINOMYCETALES: MYCOBACTERIACEAE

Mycobacterium phlei Lehmann and Neumann
Pavan, 1949

THALLOPHYTA: PHYCOMYCETES

ENTOMOPHTHORALES: ENTOMOPHTHORACEAE

?*Entomophthora* Fresenius
Roubaud, 1911a

PROTOZOA: MASTIGOPHORA

PROTOMONADIDA: TRYPANOSOMATIDAE

Herpetomonas Kent
Chang, 1940

Herpetomonas muscarum (Leidy) Kent
Patton, 1921a

PROTOMONADIDA: BODONIDAE

Rhynchoidomonas Patton
Franchini, 1922

POLYMASTIGIDA: CHILOMASTIGIDAE

Chilomastix mesnili (Wenyon)
Chang, 1940
Harris et al., 1946

POLYMASTIGIDA: HEXAMITIDAE

Giardia intestinalis (Lambl)
Wenyon et al., 1917b

Giardia lamblia Stiles
Chang, 1940
Harris et al., 1946

TRICHOMONADIDA: TRICHOMONADIDAE

Trichomonas hominis (Davaine)
Chang, 1940
Harris et al., 1946

PROTOZOA: SARCODINA

AMOEBIDA: ENDAMOEBIDAE

Endolimax nana (Wenyon and O'Connor)
Chang, 1940
Harris et al., 1946

Entamoeba coli (Grassi)
Wenyon et al., 1917b
Chang, 1940
Harris et al., 1946

Sarcophaga (cont.)

Entamoeba histolytica Schaudinn
Wenyon et al., 1917b
Anonymous ("Korean authors et al.")
Chang, 1940, 1945
Sieyro, 1942
Harris et al., 1946
Iodamoeba bütschlii (Prowazek)
Chang, 1940
Harris et al., 1946

PLATYHELMINTHES: CESTOIDEA
CYCLOPHYLLIDEA: HYMENOLEPIDIDAE
Hymenolepis diminuta (Rudolphi) Blanchard
Harris et al., 1946
CYCLOPHYLLIDEA: TAENIIDAE
Taeniarhynchus saginatum (Goeze)
Round, 1961

ASCHELMINTHES: NEMATODA
TRICHURIDEA: TRICHURIDAE
Trichuris Roederer
Chang, 1945
Trichuris trichiura (Linnaeus)
Chang, 1940
Harris et al., 1946
STRONGYLIDEA: ANCYLOSTOMATIDAE
Ancylostoma (Dubini)
Chang, 1940, 1945
Harris et al., 1946
OXYURIDEA: OXYURIDAE
Enterobius vermicularis (Linnaeus) Leach
Pokrovskiĭ et al., 1938
ASCARIDIDEA: ASCARIDIDAE
Ascaris Linnaeus
Chang, 1945
Ascaris lumbricoides Linnaeus
Chang, 1940
Harris et al., 1946

ARTHROPODA: INSECTA
COLEOPTERA: STAPHYLINIDAE
Baryodma bimaculata Gravenhorst
Lindquist, 1936
HYMENOPTERA: ICHNEUMONIDAE
Atractodes mallyi Bridwell
Bridwell, 1919
Atractodes muiri Bridwell
Bridwell, 1919

Sarcophaga (cont.)	HYMENOPTERA: BRACONIDAE
	Alysia manducator Panzer
	Holdaway et al., 1932
	Aphaereta sarcophaga Gahan
	Bridwell, 1919
	HYMENOPTERA: CYNIPIDAE
	Eucoila Westwood
	Sanders et al., 1966
	HYMENOPTERA: FIGITIDAE
	Neralsia Cameron
	Roberts, 1935
	HYMENOPTERA: CHALCIDIDAE
	Brachymeria fonscolombei (Dufour)
	Masi, 1916
	Parker, 1923
	Roberts, 1933a
	HYMENOPTERA: PTEROMALIDAE
	Nasonia Ashmead
	Feng, 1933
	Nasonia vitripennis[1] (Walker)
	Girault et al., 1910a
	Roubaud, 1917
	Parker et al., 1928
	Fujita, 1932
	Roberts, 1933a, 1935
	Nasonia ?vitripennis (Walker)
	Bishopp et al., 1915b
	Spalangia[2] Latreille
	Roy et al., 1939
	Spalangia drosophilae Ashmead
	Lindquist, 1936
	Spalangia muscidarum Richardson
	Johnston et al., 1920b
	Spalangia nigra Latreille
	Parker et al., 1928
	Spalangia stomoxysiae Girault
	Lindquist, 1936
	HYMENOPTERA: UNDETERMINED PTEROMALIDAE
	Sanders et al., 1966
	HYMENOPTERA: ENCYRTIDAE
	Australencyrtus giraulti Johnston and Tiegs
	Johnston et al., 1921

[1] Although not specifically mentioned in each entry, it is assumed that all *Nasonia vitripennis* are pupal parasites.

[2] Although not specifically mentioned in each entry, it is assumed that all *Spalangia* are pupal parasites.

Fly	Associated organism
Sarcophaga (cont.)	HYMENOPTERA: EULOPHIDAE *Syntomosphyrum albiclavus* Kerrich Saunders, 1964 *Syntomosphyrum glossinae* Waterston Lamborn, 1925 Saunders, 1964 HYMENOPTERA: FORMICIDAE *Paratrechina longicornis* (Latreille) Pimentel, 1955 *Solenopsis geminata* (Fabricius) Pimentel, 1955 HYMENOPTERA: SPHECIDAE *Bembix spinolae* (Lepeletier) Parker, 1917 *Crabro cavifrons* Thomson Hamm et al., 1926 *Crabro vagus* (Linnaeus) Hamm et al., 1926 *Ectemnius quadricinctus* (Fabricius) Hamm et al., 1926 HYMENOPTERA: MUTILLIDAE *Mutilla glossinae* Turner Lamborn, 1925 DIPTERA: CALLIPHORIDAE *Chrysomya villeneuvei* Patton Patton, 1922
Sarcophaga argyrostoma (Robineau-Desvoidy) (Synonyms: *Sarcophaga falculata, Sarcophaga barbata, Parasarcophaga argyrostoma*)	SCHIZOPHYTA: SCHIZOMYCETES EUBACTERIALES: ENTEROBACTERIACEAE *Escherichia coli* (Migula) Castellani and Chalmers Beyer, 1925 EUBACTERIALES: NEISSERIACEAE *Neisseria catarrhalis* (Froschand and Kolle) Holland Beyer, 1925 ACTINOMYCETALES: MYCOBACTERIACEAE *Mycobacterium leprae* (Hansen) Lehmann and Neumann Currie, 1910 *Mycobacterium phlei* Lehmann and Neumann Currie, 1910

Fly	Organism
Sarcophaga argyrostoma (cont.)	*Mycobacterium tuberculosis* (Zopf) Lehmann and Neumann Morellini et al., 1947, 1949 Levinsky et al., 1948
	ARTHROPODA: INSECTA HYMENOPTERA: CYNIPIDAE *Pseudeucoila trichopsila* (Hartig) Sychevskaia͡, 1964b HYMENOPTERA: CHALCIDIDAE *Brachymeria minuta* (Fabricius) Sychevskaia͡, 1964b DIPTERA: CALLIPHORIDAE *Chrysomya rufifacies* (Macquart) Illingworth, 1923b
Sarcophaga aurifinis Walker	SCHIZOPHYTA: SCHIZOMYCETES EUBACTERIALES: ENTEROBACTERIACEAE *Serratia marcescens* Bizio Nicholls, 1912 *Salmonella typhi* Warren and Scott Nicholls, 1912 EUBACTERIALES: MICROCOCCACEAE *Staphylococcus aureus* Rosenbach Nicholls, 1912 EUBACTERIALES: BACILLACEAE *Bacillus* Cohn Nicholls, 1912
Sarcophaga aurifrons Macquart	ARTHROPODA: INSECTA HYMENOPTERA: CHALCIDIDAE *Dirhinus sarcophagae* Froggatt Froggatt, 1919, 1922 HYMENOPTERA: PTEROMALIDAE *Nasonia vitripennis* (Walker) Froggatt, 1919 *Spalangia muscidarum* Richardson Johnston et al., 1921
Sarcophaga bullata Parker	ENTEROVIRUSES Poliovirus Type 2, Lansing strain Melnick et al., 1947 Poliovirus Y-SK strain, murine adapted, Type 2 Melnick et al., 1947 Theiler's mouse encephalomyelitis virus, TO strain Melnick et al., 1947, 1952 SCHIZOPHYTA: SCHIZOMYCETES EUBACTERIALES: BACILLACEAE *Bacillus thuringiensis* Berliner McConnell et al., 1959

Sarcophaga bullata (cont.)

PROTOZOA: MASTIGOPHORA
PROTOMONADIDA:
TRYPANOSOMATIDAE
Herpetomonas muscarum (Leidy) Kent
Becker, 1922-3, 1924
ARTHROPODA: INSECTA
HYMENOPTERA: BRACONIDAE
Aphaereta pallipes Say
Beard, 1964
HYMENOPTERA: CYNIPIDAE
Eucoila Westwood
Roberts, 1935
HYMENOPTERA: FIGITIDAE
Neralsia armata (Say)
Roberts, 1935

Sarcophaga carnaria (Linnaeus) ....

SCHIZOPHYTA: SCHIZOMYCETES
PSEUDOMONADALES:
PSEUDOMONADACEAE
Pseudomonas aeruginosa (Schroeter) Migula
Muzzarelli, 1925
Petragnani, 1925
Pseudomonas eisenbergii Migula
Cao, 1906
Pseudomonas fluorescens Migula
Cao, 1906
Pseudomonas jaegeri Migula
Cao, 1906
PSEUDOMONADALES: SPIRILLACEAE
Vibrio comma (Schroeter) Winslow et al.
Alessandrini, 1912
Petragnani, 1925
EUBACTERIALES:
ENTEROBACTERIACEAE
Escherichia coli (Migula) Castellani and Chalmers
Testi, 1909
Petragnani, 1925
Yao et al., 1929
Hoffmann, 1950
Povolný et al., 1961
Serratia Bizio
Povolný et al., 1961
Proteus mirabilis Hauser
Cao, 1906
Proteus vulgaris Hauser
Cao, 1906
Testi, 1909

Sarcophaga carnaria (cont.)

EUBACTERIALES: ENTEROBACTERIACEAE

Salmonella hirschfeldii Weldin
Yao et al., 1929

Salmonella typhi Warren and Scott
Krontowski, 1913
Teodoro, 1916
Petragnani, 1925

Shigella dysenteriae (Shiga) Castellani and Chalmers
Krontowski, 1913
Yao et al., 1929

EUBACTERIALES: BRUCELLACEAE

Pasteurella pestis (Lehmann and Neumann) Holland
Petragnani, 1925
Russo, 1930

EUBACTERIALES: MICROCOCCACEAE

Staphylococcus aureus Rosenbach
Testi, 1909

EUBACTERIALES: BREVIBACTERIACEAE

Kurthis zenkeri (Hauser) Bergey et al.
Cao, 1906

EUBACTERIALES: LACTOBACILLACEAE

Diplococcus pneumoniae Weichselbaum
Povolný et al., 1961

EUBACTERIALES: BACILLACEAE

Bacillus anthracis Cohn
Petragnani, 1925

Bacillus megaterium de Bary
Muzzarelli, 1925

Bacillus radiciformis Eberbach
Cao, 1906

Bacillus subtilis Cohn
Cao, 1906

ACTINOMYCETALES: MYCOBACTERIACEAE

Mycobacterium paratuberculosis Bergey et al.
Muzzarelli, 1925

Mycobacterium tuberculosis (Zopf) Lehmann and Neumann
Petragnani, 1925

BACTERIA OF UNCERTAIN AFFINITY

Bacterium mathisi Roubaud and Treillard, unvalidated
Roubaud et al., 1935

Sarcophaga carnaria (cont.)

THALLOPHYTA: PHYCOMYCETES
ENTOMOPHTHORALES: ENTOMOPHTHORACEAE
Entomophthora muscae Cohn
Graham-Smith, 1919

THALLOPHYTA: ASCOMYCETES
ENDOMYCETES: ENDOMYCETACEAE
Saccharomyces cerevisiae Hansen
Berlese, 1896

THALLOPHYTA: FUNGI IMPERFECTI
MONILIALES: MONILIACEAE
Blastomyces Costantin and Rolland
Muzzarelli, 1925
CRYPTOCOCCALES: CRYPTOCOCCACEAE
Candida Berkhout
Berlese, 1896
Kloeckera apiculata (Reess emend Klöcker) Janke
Berlese, 1896

FUNGI OF UNCERTAIN AFFINITY
Streptotricea, unvalidated
Muzzarelli, 1925

PROTOZOA: MASTIGOPHORA
PROTOMONADIDA: TRYPANOSOMATIDAE
Herpetomonas Kent
Franchini, 1922
Leishmania Ross
Franchini, 1922

PROTOZOA: SARCODINA
AMOEBIDA: ENDAMOEBIDAE
Endolimax nana (Wenyon and O'Connor)
Yao et al., 1929
Entamoeba coli (Grassi)
Yao et al., 1929
Entamoeba histolytica Schaudinn
Yao et al., 1929

ARTHROPODA: INSECTA
COLEOPTERA: HISTERIDAE
Saprinus detersus (Illiger)
Pavlovskiĭ, 1921
Saprinus politus (Brahm) or *S. aenus* (Fabricius)
Pavlovskiĭ, 1921
HYMENOPTERA: CYNIPIDAE
Kleidotoma Westwood
James, 1928

Fly	Associated organism
Sarcophaga carnaria (cont.)	*Kleidotoma marshalli* Marshall
	James, 1928
	HYMENOPTERA: FIGITIDAE
	Figites anthomyiarum (Bouché)
	James, 1928
	HYMENOPTERA: CHALCIDIDAE
	Brachymeria fonscolombei (Dufour)
	Stefani, 1889
	Roberts, 1935
	Nikol'skaia͡, 1960b
	HYMENOPTERA: PTEROMALIDAE
	Nasonia vitripennis (Walker)
	Cousin, 1933
	HYMENOPTERA: SPHECIDAE
	Ectemnius quadricinctus (Fabricius)
	Hamm et al., 1926
Sarcophaga destructor Malloch	ARTHROPODA: INSECTA
	ORTHOPTERA: LOCUSTIDAE
	Locusta migratoria Linnaeus
	Wood, 1933
Sarcophaga dux Thompson	ARTHROPODA: INSECTA
	HYMENOPTERA: CHALCIDIDAE
	Brachymeria argentifrons (Ashmead)
	Roy et al., 1939
	HYMENOPTERA: PTEROMALIDAE
	Spalangia nigroaenea Curtis
	Bouček, 1963
Sarcophaga dux Thompson var. *tuberosa*	ARTHROPODA: INSECTA
	HYMENOPTERA: CHALCIDIDAE
	Brachymeria argentifrons (Ashmead)
	Roy et al., 1939
	Dirhinus pachycerus Masi
	Roy et al., 1939
Sarcophaga frontalis Doleschall	ARTHROPODA: INSECTA
	HYMENOPTERA: PTEROMALIDAE
	Spalangia nigroaenea Curtis
	Bouček, 1963
Sarcophaga ?frontalis Doleschall	ARTHROPODA: INSECTA
	HYMENOPTERA: PTEROMALIDAE
	Spalangia muscidarum Richardson
	Johnston et al., 1920b
Sarcophaga fuscicauda Böttcher	ARTHROPODA: INSECTA
	DERMAPTERA: LABIDURIDAE
	Euborellia annulipes (Lucas)
	Illingworth, 1923a

Sarcophaga fuscicauda (cont.)

DERMAPTERA: LABIIDAE

Labia pilicornis (Motschulsky)
Illingworth, 1923a

Spingolabis hawaiiensis (Bormans)
Illingworth, 1923a

COLEOPTERA: HYDROPHILIDAE

Cryptopleurum minutum Fabricius
Illingworth, 1923a

Dactylosternum abdominale Fabricius
Illingworth, 1923a

COLEOPTERA: STAPHYLINIDAE

Oxytelus Gravenhorst
Illingworth, 1923a

Philonthus discoideus Gravenhorst
Illingworth, 1923a

Philonthus longicornis Stephens
Illingworth, 1923a

Tachyporus Gravenhorst
Illingworth, 1923a

HYMENOPTERA: CYNIPIDAE

Eucoila impatiens (Say)
Illingworth, 1923a

HYMENOPTERA: FORMICIDAE

Pheidole megacephala (Fabricius)
Illingworth, 1923a

Ponera perkinsi Forel
Illingworth, 1923a

Sarcophaga haemorrhoidalis (Fallén) (Synonyms: *Coprosarcophaga haemorrhoidalis, Bercaea haemorrhoidalis*) ....

ENTEROVIRUSES

Poliovirus
Paul et al., 1941

Poliovirus Type 2, Lansing strain
Bang et al., 1943

Theiler's mouse encephalomyelitis virus, GD VII strain
Bang et al., 1943

SCHIZOPHYTA: SCHIZOMYCETES

EUBACTERIALES: ENTEROBACTERIACEAE

Escherichia coli (Migula) Castellani and Chalmers
Zmeev, 1943
Povolný et al., 1961

Escherichia coli var. *acidilactici* (Topley and Wilson) = [*Escherichia paraacidilactici* A]
Zmeev, 1943

Sarcophaga haemorrhoidalis (cont.)	*Escherichia coli* var. *acidilactici* (Topley and Wilson) = [*Escherichia paraacidilactici* B] Zmeev, 1943 *Escherichia coli* var. *communior* (Topley and Wilson) Yale Zmeev, 1943 *Escherichia intermedia* (Werkman and Gillen) Vaughn and Gillen Zmeev, 1943 *Aerobacter aerogenes* (Kruse) Beijerinck Zmeev, 1943 *Cloaca* Castellani and Chalmers Zmeev, 1943 *Serratia* Bizio Povolný et al., 1961 *Proteus morganii* (Winslow, Kligler, and Rothberg) Yale Povolný et al., 1961 *Proteus rettgeri* (Hadley et al.) Rustigian and Stuart Povolný et al., 1961 *Proteus vulgaris* Hauser Sukhova, 1954 Povolný et al., 1961 *Salmonella enteritidis* (Gaertner) Castellani and Chalmers Gross et al., 1953 *Salmonella give* Kauffmann Gross et al., 1953 *Salmonella paratyphi* A (Brion and Kayser) Castellani and Chalmers Gross et al., 1953 *Salmonella paratyphi* B Castellani and Chalmers Gross et al., 1953 *Salmonella typhi* Warren and Scott Gross et al., 1953 *Citrobacter freundii* Werkman and Gillen Zmeev, 1943 ?*Shigella* Castellani and Chalmers Sychevskai͡a et al., 1959a *Shigella dysenteriae* (Shiga) Castellani and Chalmers 2 Sychevskai͡a et al., 1959a

Fly	Organism / Reference
Sarcophaga haemorrhoidalis (cont.)	*Shigella flexneri* Castellani and Chalmers Gross et al., 1953 Arskiĭ et al., 1961 *Shigella flexneri* Castellani and Chalmers 1a, 6 Sychevskaia͡ et al., 1959a *Shigella sonnei* (Levine) Weldin Gross et al., 1953 Arskiĭ et al., 1961 EUBACTERIALES: LACTOBACILLACEAE *Diplococcus pneumoniae* Weichselbaum Povolný et al., 1961 *Streptococcus* Rosenbach Povolný et al., 1961 *Streptococcus lactis* (Lister) Löhnis Zmeev, 1943 EUBACTERIALES: BACILLACEAE *Bacillus* Cohn Povolný et al., 1961 *Bacillus megaterium* de Bary Zmeev, 1943 PROTOZOA: MASTIGOPHORA PROTOMONADIDA: TRYPANOSOMATIDAE *Herpetomonas muscarum* (Leidy) Kent Laveran et al., 1920 PLATYHELMINTHES: CESTOIDEA CYCLOPHYLLIDEA: TAENIIDAE *Taeniarhynchus saginatum* (Goeze) Sychevskaia͡ et al., 1958 ASCHELMINTHES: NEMATODA ASCARIDIDEA: ASCARIDIDAE *Ascaris lumbricoides* Linnaeus Sychevskaia͡ et al., 1958 ARTHROPODA: INSECTA ORTHOPTERA: LOCUSTIDAE *Schistocerca gregaria* Fŏrskal Régnier, 1931 HYMENOPTERA: BRACONIDAE *Aphaereta minuta* Nees Sychevskaia͡, 1964b HYMENOPTERA: CYNIPIDAE *Pseudeucoila trichopsila* (Hartig) Sychevskaia͡, 1964b, 1966

Fly	Organism
Sarcophaga haemorrhoidalis (cont.)	HYMENOPTERA: FIGITIDAE *Figites discordis* Belizin Sychevskaîa, 1964b HYMENOPTERA: CHALCIDIDAE *Brachymeria fonscolombei* (Dufour) von Dalla Torre, 1898 Roberts, 1935 *Brachymeria minuta* (Fabricius) Nikol'skaîa, 1960a Sychevskaîa, 1964b, 1966 *Euchalcidia blanda* Nikol'skaîa Nikol'skaîa, 1960a Sychevskaîa, 1964b HYMENOPTERA: PTEROMALIDAE *Nasonia vitripennis* (Walker) Anonymous, 1925 Smit, 1929 *Spalangia*[1] Latreille Sychevskaîa, 1964b *Spalangia cameroni* Perkins Bouček, 1963 *Spalangia endius* Walker Bouček, 1963 HYMENOPTERA: EULOPHIDAE *Syntomosphyrum glossinae* Waterston Nash, 1933
Sarcophaga hirtipes Wiedemann (Synonym: *Parasarcophaga hirtipes*)	SCHIZOPHYTA: SCHIZOMYCETES EUBACTERIALES: ENTEROBACTERIACEAE *Escherichia anaerogenes* (Chester) Bergey et al. Zmeev, 1943 *Escherichia anindolica* (Chester) Bergey et al. Zmeev, 1943 *Escherichia coli* (Migula) Castellani and Chalmers Zmeev, 1943 *Escherichia coli* var. *acidilactici* (Topley and Wilson) Yale Zmeev, 1943 *Escherichia coli* var. *communior* (Topley and Wilson) Yale Zmeev, 1943 *Aerobacter* Beijerinck Zmeev, 1943

[1] Although not specifically mentioned in each entry, it is assumed that all *Spalangia* are pupal parasites.

Fly	Associated organism
Sarcophaga hirtipes (cont.)	*Cloaca* Castellani and Chalmers Zmeev, 1943
	ARTHROPODA: INSECTA HYMENOPTERA: CHALCIDIDAE *Brachymeria minuta* (Fabricius) Nikol'skaia͡, 1960a Sychevskaia͡, 1964b
Sarcophaga impatiens (Walker)	ARTHROPODA: INSECTA HYMENOPTERA: CHALCIDIDAE *Dirhinus sarcophagae* Froggatt Johnston et al., 1921
	HYMENOPTERA: PTEROMALIDAE *Spalangia nigroaenea* Curtis Bouček, 1963
Sarcophaga melanura Meigen (Synonym: *Bellieria melanura*)	SCHIZOPHYTA: SCHIZOMYCETES EUBACTERIALES: ENTEROBACTERIACEAE *Escherichia coli* (Migula) Castellani and Chalmers Povolný et al., 1961 *Serratia* Bizio Povolný et al., 1961
	EUBACTERIALES: BACILLACEAE *Bacillus* Cohn Povolný et al., 1961
	PROTOZOA: MASTIGOPHORA PROTOMONADIDA: TRYPANOSOMATIDAE *Herpetomonas* Kent Franchini, 1922 *Leishmania* Ross Franchini, 1922 *Leptomonas mirabilis* Roubaud Franchini, 1922
	ASCHELMINTHES: NEMATODA SPIRURIDEA: SPIRURIDAE *Habronema microstoma* (Schneider) Neveu-Lemaire, 1936
	ARTHROPODA: INSECTA HYMENOPTERA: BRACONIDAE *Aphaereta cephalotes* Haliday Graham-Smith, 1916 *Aphaereta minuta* Nees Sychevskaia͡, 1964b
	HYMENOPTERA: CYNIPIDAE *Pseudeucoila trichopsila* (Hartig) Sychevskaia͡, 1964b

Sarcophaga melanura (cont.)	HYMENOPTERA: FIGITIDAE *Figites scutellaris* (Rossi) Sychevskaia͡, 1964b HYMENOPTERA: CHALCIDIDAE *Brachymeria minuta* (Fabricius) Sychevskaia͡, 1964b CHORDATA: AVES PASSERIFORMES: HIRUNDINIDAE *Hirundo rustica* Linnaeus subsp. *rustica* Séguy, 1929
Sarcophaga misera Walker	PROTOZOA: SARCODINA AMOEBIDA: ENDAMOEBIDAE *Entamoeba histolytica* Schaudinn Pipkin, 1942, 1949 ASCHELMINTHES: NEMATODA SPIRURIDEA: SPIRURIDAE *Habronema megastoma* (Rudolphi) Seurat Johnston, 1920 Johnston et al., 1920a *Habronema microstoma* (Schneider) Johnston, 1920 *Habronema muscae* (Carter) Johnston, 1920 Johnston et al., 1920a Neveu-Lemaire, 1936 ARTHROPODA: INSECTA HYMENOPTERA: PTEROMALIDAE *Nasonia vitripennis* (Walker) Johnston et al., 1921 *Spalangia muscidarum* Richardson Johnston et al., 1920b, 1921
Sarcophaga nurus Rondani	PROTOZOA: MASTIGOPHORA PROTOMONADIDA: TRYPANOSOMATIDAE *Herpetomonas* Kent Franchini, 1922 *Herpetomonas muscarum* (Leidy) Kent Roubaud, 1909 *Leishmania* Ross Franchini, 1922
Sarcophaga parkeri (Rohdendorf) (Synonym: *Parasarcophaga parkeri*)	ARTHROPODA: INSECTA HYMENOPTERA: CYNIPIDAE *Pseudeucoila trichopsila* (Hartig) Sychevskaia͡, 1964b

Fly	Organism
Sarcophaga parkeri (cont.)	HYMENOPTERA: FIGITIDAE *Figites sarcophagarum* Belizin Sychevskaia͡, 1964b HYMENOPTERA: CHALCIDIDAE *Brachymeria minuta* (Fabricius) Nikol'skaia͡, 1960a HYMENOPTERA: PTEROMALIDAE *Nasonia vitripennis* (Walker) Sychevskaia͡, 1964b
Sarcophaga peregrina Robineau-Desvoidy (Synonyms: *Boettcherisca peregrina*, *Sarcophaga irrequieta*)	SCHIZOPHYTA: SCHIZOMYCETES EUBACTERIALES: ENTEROBACTERIACEAE *Proteus morganii* (Winslow, Kligler, and Rothberg) Yale Shimizu et al., 1965 *Proteus rettgeri* (Hadley et al.) Rustigian and Stuart Shimizu et al., 1965 *Salmonella pullorum* (Rettger) Bergey Kaneko et al., 1960 ARTHROPODA: INSECTA HYMENOPTERA: ICHNEUMONIDAE *Mesoleptus laevigatus* (Gravenhorst) Suenaga et al., 1963 HYMENOPTERA: CHALCIDIDAE *Brachymeria fonscolombei* (Dufour) Suenaga et al., 1963 HYMENOPTERA: PTEROMALIDAE *Nasonia vitripennis* (Walker) Johnston et al., 1921
Sarcophaga ruficornis Fabricius	ARTHROPODA: INSECTA HYMENOPTERA: CHALCIDIDAE *Brachymeria fulvitarsis* (Cameron) Roy et al., 1939 *Dirhinus pachycerus* Masi Roy et al., 1939
Sarcophaga sarraceniae Riley	SCHIZOPHYTA: SCHIZOMYCETES EUBACTERIALES: ENTEROBACTERIACEAE *Proteus* Hauser Beyer, 1925 *Shigella flexneri* Castellani and Chalmers Beyer, 1925

	PROTOZOA: MASTIGOPHORA PROTOMONADIDA: TRYPANOSOMATIDAE *Leptomonas lineata* (Swingle) Swingle, 1911
Sarcophaga sarraceniodes .... Aldrich (Synonym: *Sarcophaga misera* var. *sarracenioides*)	PROTOZOA: SARCODINA AMOEBIDA: ENDAMOEBIDAE *Entamoeba histolytica* Schaudinn Pipkin, 1949 ARTHROPODA: INSECTA HYMENOPTERA: BRACONIDAE *Alysia ridibunda* Say Roberts, 1935 HYMENOPTERA: CYNIPIDAE *Eucoila* Westwood Roberts, 1935
Sarcophaga scoparia .... (Pand.)	ARTHROPODA: INSECTA HYMENOPTERA: SPHECIDAE *Crabro advenus* Smith Kurczewski et al., 1968
Sarcophaga securifera .... Villeneuve	PROTOZOA: MASTIGOPHORA PROTOMONADIDA: TRYPANOSOMATIDAE *Herpetomonas muscarum* (Leidy) Kent Becker, 1922-23, 1924
Sarcophaga tibialis .... Macquart	PLATYHELMINTHES: CESTOIDEA CYCLOPHYLLIDEA: TAENIIDAE *Echinococcus granulosus* (Batsch) Rudolphi Heinz et al., 1955
Sarcophaga wiedemanni ...... Aldrich (Synonym: ?*Sarcophaga chrysostoma*)	ARTHROPODA: INSECTA DIPTERA: CUTEREBRIDAE *Dermatobia hominis* (Linnaeus) Blanchard, 1899
Tricharaea Thomson .... (Synonym: *Sarcophagula*)	ENTEROVIRUSES Poliovirus Melnick et al., 1949 Poliovirus non-paralytic Melnick et al., 1949 Poliovirus Type 1 Melnick et al., 1953 Paul et al., 1962 Poliovirus Type 3 Paul et al., 1962 Coxsackievirus Connecticut 5 (B1) Melnick et al., 1953 Coxsackievirus Easton 10 (A8) Melnick et al., 1953

Tricharaea (cont.)

Coxsackievirus Easton 14 (A5)
Melnick et al., 1953
Coxsackievirus Nancy (B3)
Melnick et al., 1953
Coxsackievirus Texas 1 (A4)
Melnick et al., 1953
Coxsackievirus Texas 12 (new type)
Melnick et al., 1953
Coxsackievirus Texas 14 (new type)
Melnick et al., 1953
Coxsackievirus Type 2 (A2)
Melnick et al., 1953
Coxsackievirus Type 3 (A3)
Melnick et al., 1953
ECHO virus Types 8, 12, 14
Paul et al., 1962

SCHIZOPHYTA: SCHIZOMYCETES
EUBACTERIALES: ENTEROBACTERIACEAE
Serratia marcescens Bizio
Nicholls, 1912
Salmonella typhi Warren and Scott
Nicholls, 1912
EUBACTERIALES: MICROCOCCACEAE
Staphylococcus aureus Rosenbach
Nicholls, 1912
EUBACTERIALES: BACILLACEAE
Bacillus Cohn
Nicholls, 1912

ARTHROPODA: INSECTA
HYMENOPTERA: PTEROMALIDAE
Spalangia philippinensis Fullaway
Bartlett, 1939

UNDETERMINED SARCOPHAGIDAE ....

SCHIZOPHYTA: SCHIZOMYCETES
EUBACTERIALES: ENTEROBACTERIACEAE
Escherichia coli (Migula) Castellani and Chalmers
Shimizu et al., 1965
Proteus mirabilis Hauser
Shimizu et al., 1965
Proteus morganii (Winslow, Kligler, and Rothberg) Yale
Shimizu et al., 1965
Proteus rettgeri (Hadley et al.) Rustigian and Stuart
Shimizu et al., 1965

UNDETERMINED SARCOPHAGIDAE (cont.)	*Proteus vulgaris* Hauser Shimizu et al., 1965 *Providencia* Kauffmann Shimizu et al., 1965 ARTHROPODA: INSECTA HYMENOPTERA: CYNIPIDAE *Eucoila impatiens* (Say) Illingworth, 1923a HYMENOPTERA: CHALCIDIDAE *Brachymeria* Westwood Dowden, 1935 *Dirhinoides vlasovi* Nikol'skaîa Nikol'skaîa, 1960a *Dirhinus* Dalman Simmonds, 1940 HYMENOPTERA: PTEROMALIDAE *Spalangia cameroni* Perkins Simmonds, 1929
Family: TACHINIDAE Subfamily: TACHININAE *Linnaeymya haemorrhoidalis* Fallén (Synonym: *Linnaemyia haemorrhoidalis*)	ARTHROPODA: INSECTA HYMENOPTERA: CHALCIDIDAE *Brachymeria fonscolombei* (Dufour) Nikol'skaîa, 1960b
UNDETERMINED TACHINIDAE	SCHIZOPHYTA: SCHIZOMYCETES ACTINOMYCETALES: MYCOBACTERIACEAE *Mycobacterium leprae* (Hansen) Lehmann and Neumann de Souza-Araujo, 1944 THALLOPHYTA: PHYCOMYCETES ENTOMOPHTHORALES: ENTOMOPHTHORACEAE *Entomophthora muscae* Cohn de Souza-Araujo, 1944

The following associations involving competition between flies were not included in the main body of associations.

Drosophila melanogaster Meigen	ARTHROPODA: INSECTA DIPTERA: DROSOPHILIDAE *Drosophila melanogaster* Meigen L'Héritier et al., 1933, 1934 Miller, 1964 Cannon, 1966 *Drosophila simulans* Sturtevant Miller, 1964

Musca domestica Linnaeus	DIPTERA: MUSCIDAE *Musca domestica* Linnaeus Matsuo, 1962 Sokal et al., 1963 Sullivan et al., 1963, 1965 Bhalla et al., 1964 Enon, 1965
Chrysomya albiceps (Wiedemann)	DIPTERA: CALLIPHORIDAE *Chrysomya chloropyga* (Wiedemann) Ullyet, 1950 *Phaenicia sericata* (Meigen) Ullyet, 1950
Chrysomya chloropyga (Wiedemann)	DIPTERA: CALLIPHORIDAE *Phaenicia sericata* (Meigen) Ullyet, 1950[1]
Phaenicia sericata (Meigen)	DIPTERA: CALLIPHORIDAE *Phaenicia sericata* (Meigen) Salt, 1930
Sarcophaga peregrina Robineau-Desvoidy	DIPTERA: SARCOPHAGIDAE *Sarcophaga peregrina* Robineau-Desvoidy Matsuo, 1962

[1] Also, competition between three species: *C. albiceps*, *C. chloropyga*, and *P. sericata*.

7

Organism-Fly Associations with Systematic Lists of Organisms

SYSTEMATIC LIST OF ORGANISMS ASSOCIATED WITH FLIES

VIRUSES	GROUPS	SUBGROUPS	TYPES AND STRAINS
	7	27	40
PHYLUM	FAMILIES	GENERA	SPECIES
Schizophyta	20	52	178
Thallophyta	13	33	31
Protozoa	14	25	64
Platyhelminthes	8	16	17
Aschelminthes	19	26	33
Mollusca	2	2	2
Annelida	2	6	11
Arthropoda			
Arachnida	20	41	48
Myriapoda	1	1	1
Insecta	76	214	319
Chordata			
Amphibia	4	4	5
Reptilia	3	4	4
Aves	9	12	12
Mammalia	2	1	1
Total	193	437	726

VIRUSES

GROUP	SUBGROUP	TYPES AND STRAINS	TOTAL NO. SPECIES
Psittacosis	Trachoma		1
Poxviruses	Variola		1
	Vaccinia		1
	Borreliota		1
Arborviruses	Eastern equine encephalo-myelitis		1
	Sindbis virus		1
	Yellow fever virus		1

GROUP	SUBGROUP	TYPES AND STRAINS	TOTAL NO. SPECIES
Enteroviruses	Poliovirus	Non-paralytic, Strain 113, Delaware '48 strain, Type 1, Attenuated Type 1, LSc., Mahoney Type 1, New York City '44 strain (Type 1), Philippine Islands '45 strain (Type 1), Texas '48 strain (Type 1), W-S '48 strain (Type 1), Type 2, Y-SK strain, murine adapted, Type 2, Type 3, Attenuated Type 3, Leon, North Carolina '44 strain (Type 3)	15
	Theiler's mouse encephalomyelitis virus, GD VII		1
	Theiler's mouse encephalomyelitis virus, TO strain		1
	Coxsackievirus	Alaska 5(A10), Connecticut 5(B1), Easton 2(A1), Easton 10(A8), Easton 14(A5), Nancy (B3), Texas 1(A4), Texas 12(new type), Texas 13(B4), Texas 14(new type), Texas 15(A7), Texas '48 (new type), Type 2(A2), Type 3(A3), Type A9(A9), Type B5(B5), W-S '48(B1)	17
	ECHO virus	Types 1, 5, 6, 7, 8, 11, 12, 14	8
Other Animal Viruses	Danysz rat virus		1
	Equine infectious anemia virus		1
	Foot and mouth disease virus		1
	Mink enteritis virus		1
	Rous-sarcoma virus (oncogenic)		1

VIRUSES

GROUP	SUBGROUP	TYPES AND STRAINS	TOTAL NO. SPECIES
	Southwest African horse death virus (Windhuker)		1
	Tipula iridescent virus		1
	Tortor bovis virus		1
	Tortor suis virus		1
	Vesicular stomatitis virus		1
Insect Viruses	Borrelina bombycis virus		1
	Sex-ratio virus		1
	Sigma virus	P^- strain	1
Bacteriophages	*Staphylococcus muscae*		1
	Streptococcus C55		1

SCHIZOPHYTA

FAMILY	GENUS	SPECIES, TYPES, AND VARIETIES	TOTAL NO. SPECIES
Pseudo-monadaceae	*Pseudomonas*	*aeruginosa, eisenbergii, fluorescens, ichthyosmia, jaegeri*	5
	Acetobacter	*aceti, pasteurianus, xylinum*	3
Spirillaceae	*Vibrio*	*comma*	1
Chlamydo-bacteriaceae	*Leptothrix*		
Achromo-bacteraceae	*Alcaligenes*	*faecalis*	1
	Achromobacter		
	Flavobacterium	*devorans, invisible*	2

SCHIZOPHYTA

FAMILY	GENUS	SPECIES, TYPES, AND VARIETIES	TOTAL NO. SPECIES
Entero-bacteriaceae	*Escherichia*	*anaerogenes, anindolica, coli, coli* 25, *coli* 26-B6, *coli* 086-B7, *coli* 111B4, *coli* var. *acidilactici, coli* var. *communior, coli* var. *communis, coli* var. *neapolitana, freundii, intermedia*	5
	Aerobacter	*aerogenes, cloacae, oxytocum*	3
	Klebsiella	*cloacae, pneumoniae*	2
	Cloaca		
	Paracolo-bactrum	*aerogenoides, coliforme, intermedium*	3
	Erwinia	*amylovora*	1
	Serratia	*marcesens*	1
	Proteus	*inconstans, mirabilis, morganii, rettgeri, vulgaris*	5
	Providencia		
	Salmonella	*alachua, amersfoort, anatum, ballerup, blockley, bovis-morbificans, budapest, chester, cholerae-suis, columbensis, derby, dublin, enteritidis, florida, gallinarum, give, hirschfeldii, kentucky, kottbus, meleagridis, minnesota, muenchen, narashino, new brunswick, newport, onderstepoort, oranienburg, oregon, panama, paratyphi A, paratyphi B, paratyphi C, pullorum, reading, saintpaul, sandiego, schottmuelleri, typhi, typhimurium, worthington, zanzibar*	41
	Eberthella	*belfastiensis, dubia, kandiensis, telavensis*	4
	Citrobacter	*freundii*	1
	Bethesda		
	Shigella	*alkalescens, ambigua, boydii, boydii* 1, 2, 3, 4, 5, 6, *ceylonensis, dispar, dysenteriae, dysenteriae* 1, 2, 3, 7, *flexneri, flexneri* 1a, 2a, 3, 4a, 4b, 5, 6, X, Y, *giumai, madampensis, pfaffi, sonnei, sonnei* 1	12
Brucellaceae	*Pasteurella*	*boviseptica, multocida, pestis, tularensis*	4
	Brucella	*abortus, melitensis, suis*	3
	Haemophilus	*influenzae*	1
	Moraxella	*bovis, lacunata*	2

SCHIZOPHYTA

FAMILY	GENUS	SPECIES, TYPES, AND VARIETIES	TOTAL NO. SPECIES
	Cillo-pasteurella	*delendae-muscae*	1
Bacteroidaceae	*Bacteroides*	*vulgatus*	1
Micrococcaceae	*Micrococcus*	*candidus, caseolyticus, rushmorei*	3
	Staphylococcus[1]	*afermentans, aureus, epidermidis, flavus, lactis, muscae, saprophyticus*	7
	Gaffkya	*tetragena*	1
	Sarcina	*alba, aurantiaca, flava, loewenbergi, lutea, subflava*	6
Neisseriaceae	*Neisseria*	*catarrhalis, gonorrhoeae, luciliarum*	3
Brevibacteriaceae	*Brevibacterium*	*ammoniagenes*	1
	Kurthia	*zenkeri, zopfii*	2
Lactobacillaceae	*Diplococcus*	*pneumoniae* I, II	1
	Streptococcus[2]	*agalactiae, agalactiae* (group B), *durans, faecalis, faecalis* var. *faecalis, faecalis* var. *liquefaciens, faecium, lactis, mitis, pyogenes*	7
	Lactobacillus		
Corynebacteriaceae	*Corynebacterium*	*diphtheriae, pyogenes*	2
	Listeria	*monocytogenes*	1
	Erysipelothrix	*insidiosa*	1
Bacillaceae	*Bacillus*	*anthracis, bifermentans, cereus, cereus* var. *mycoides, coli mutabilis, gasoformans, gruenthali, lutzae, megaterium, mesentericus, radiciformis, subtilis, subtilis* var. *niger, thuringiensis, thuringiensis* var. *sotto, thuringiensis* var. *thuringiensis, vesiculosus*	13
	Clostridium	*bifermentans, botulinum, botulinum* Type C, *chauvoei, parabotulinum bovis, perfringens, putrefaciens, tetani*	7
Mycobacteriaceae	*Mycobacterium*	*leprae, lepraemurium, paratuberculosis, phlei, tuberculosis*	5
Dermatophilaceae	*Dermatophilus*	*congolensis*	1

[1] *Staphylococcus* = unspecified types distinguished only by hemolytic and pathogenic characteristics are not included here.

[2] *Streptococcus* = unspecified types distinguished only by hemolytic and capsular features are not included here.

SCHIZOPHYTA

FAMILY	GENUS	SPECIES, TYPES, AND VARIETIES	TOTAL NO. SPECIES
Spirochaetaceae	*Spirochaeta*	*acuminata, obtusa, stomoxyae*	3
Treponemataceae	*Borrelia*	*anserina, berbera, refringens*	3
	Treponema	*carateum, pertenue, recurrentis*	3
	Leptospira	*canicola, grippotyphosa, icterohaemorrhagiae*	3
Rickettsiaceae	*Rickettsia*		
	Coxiella	*burneti*	1
Chlamydiaceae	*Colesiota*	*conjunctivae*	1
Anaplasmataceae	*Anaplasma*	*marginale*	1

THALLOPHYTA

FAMILY	GENUS	SPECIES	TOTAL NO. SPECIES
Mucoraceae	*Mucor*	*ambiguus, mucedo*	2
Gymnoascaceae	*Rhizopus*	*nigricans*	1
	Trichophyton	*mentagrophytes* (Robin) Blanchard var. *granulosum, tonsurans*	2
Entomophthoraceae	*Entomophthora*	*americana, calliphorae, muscae, muscivora, ovispora, richteri, sphaerosperma*	7
Laboulbeniaceae	*Stygmatomyces*	*baeri*	1
Endomycetaceae	*Coccidiascus*	*legeri*	1
	Saccharomyces	*cerevisiae, ?melli, pastorianus, theobromae*	4
Hypocreaceae	*Cordyceps*	*dipterigena*	1
Agaricaceae	*Agaricus*	*campestris*	1
Moniliaceae	*Aspergillus*	*clavatus, flavus, niger*	3
	Beauveria	*bassiana*	1
	Blastomyces		
	Botrytis		
	Geotrichum	*candidum*	1
	Gliocladium		
	Metarrhizium	*anisopliae*	1
	Monilia		
	Oidium		
	Penicillium	*camemberti, glaucum*	2
	Trichothecium		

THALLOPHYTA

FAMILY	GENUS	SPECIES	TOTAL NO. SPECIES
Dematiaceae	*Altenaria*		
	Cladosporium		
	Curvularia		
	Hormo-dendrum		
	Macrosporium		
	Pullalaria		
	Torula		
Stilbaceae	*Isaria*		
Tuberculariaceae	*Fusarium*		
Blastocystidaceae	*Blastocystis*		
Cryptococcaceae	*Candida*	*albicans, mycoderma*	2
	Kloeckera	*apiculata*	1
Uncertain Affinity	*Streptotricea*		

PROTOZOA

FAMILY	GENUS	SPECIES	TOTAL NO. SPECIES
Trypanosomatidae	*Crithidia*	*acanthocephali, fasciculata, haematopotae, lesnei, luciliae, oncopelti*	6
	Herpetomonas	*caulleryi, ?muscae, muscarum, roubaudi, rubrio-striatae*	5
	Leishmania	*donovani, mexicana, tropica*	3
	Leptomonas	*ampelophilae, anthomyia, craggi, drosophilae, gracilis, legerorum, lineata, mesnili, mirabilis, pycnosomae, pyrrhocoris, soudanenis, stomoxyae*	13
	Trypanosoma	*brucei, congolense, dimorphon, equinum, equiperdum, evansi, gambiense, hippicum, pecorum, rhodesiense, suis, vivax*	12
Bodonidae	***Bodo***		
	Rhynchoido-monas	*intestinalis, luciliae, roubaudi, siphunculinae*	4
Tetramitidae	*Tetramitus*		
Chilomastigidae	*Chilomastix*	*mesnili*	1
Hexamitidae	*Giardia*	*intestinalis, lamblia*	2

PROTOZOA

FAMILY	GENUS	SPECIES	TOTAL NO. SPECIES
Trichomonadidae	*Trichomonas*	*hominis*	1
	Tritrichomonas	*foetus*	1
Amoebidae	*Hartmannella*	*hyalina*	1
Endamoebidae	*Endolimax*	*nana*	1
	Entamoeba	*coli, histolytica*	2
	Iodamoeba	*bütschlii*	1
Eimeriidae	*Eimeria*	*acervalina, irresidua, perforans*	3
Unknown Family	*Toxoplasma*	*gondii*	1
Nosematidae	*Nosema*	*apis, bombycis, kingi*	3
	Thelohania	*ovata, thomsoni*	2
Mrazekiidae	*Octosporea*	*monospora, muscae-domesticae*	2
	Spiroglugea		
	Toxoglugea		
Telomyxidae	*Telomyxa*		
Holophryidae	*Holophrya*		

PLATYHELMINTHES

FAMILY	GENUS	SPECIES	TOTAL NO. SPECIES
Heterophyidae	*Heterophyes*	*heterophyes*	1
Schistosomatidae	*Schistosoma*		
Diphyllobothriidae	*Bothriocephalus*	*marginatus*	1
	Diphyllobothrium	*latum*	1
Anoplocephalidae	*Moniezia*	*expansa*	1
Davaineidae	*Davainea*	*nana*	1
	Raillietina	*cesticillus, tetragona*	2
Dilepididae	*Choanotaenia*	*infundibulum*	1
	Dipylidium	*caninum*	1
Hymenolepididae	*Echinolepis*	*carioca*	1
	Hymenolepis	*diminuta, nana*	2
	Vampirolepis	*fraterna*	1
Taeniidae	*Echinococcus*	*granulosus*	1
	Taenia	*pisiformis*	1
	Taeniarhynchus	*saginatum*	1
Uncertain Affinity	*Bassus*	*laetatorius*	1

ASCHELMINTHES

FAMILY	GENUS	SPECIES	TOTAL NO. SPECIES
Tetradonematidae	*Mermithonema*	*entomophilum*	1
Rhabditidae	*Rhabditis*	*axei, pellio*	2
Panagrolaimidae	*Panagrellus*	*zymosiphilus*	1
Strongyloididae	*Strongyloides*	*stercoralis*	1
Trichuridae	*Trichuris*	*trichiura*	1
Ancylostomatidae	*Ancylostoma*	*caninum, duodenale*	2
	Necator	*americanus*	1
Strongylidae	*Strongylus*	*equinus*	1
Syngamidae	*Syngamus*	*trachea*	1
Trichostrongylidae	*Tricho-strongylus*	*colubriformis*	1
Oxyuridae	*Enterobius*	*vermicularis*	1
Ascarididae	*Ascaris*	*lumbricoides*	1
	Parascaris	*equorum*	1
	Toxascaris	*leonina*	1
Acuariidae	*Dispharynx*	*?nasuta*	1
	Parabronema	*skrjabini*	1
Spiruridae	*Habronema*	*megastoma, microstoma, muscae*	3
Thelaziidae	*Thelazia*	*californiensis, gulosa, rhodesii, skrjabini*	4
Dipetalonematidae	*Oncocerca*	*gibsoni*	1
Filariidae	*Filaria*		
	Setaria	*cervi*	1
Stephanofilariidae	*Stephano-filaria*	*assamensis*	1
Allantonematidae	*Allantonema*	*muscae, stricklandi*	2
	Howardula	*aoronymphium*	1
	Parasity-lenchus	*diplogenus*	1
Contortylenchidae	*Heteroty-lenchus*	*aberrans*	1

MOLLUSCA

FAMILY	GENUS	SPECIES	TOTAL NO. SPECIES
Achatinidae	*Burtoa*	*nilotica*	1
Helicidae	*Cryptomphalus*	*aspersa*	1

ANNELIDA

FAMILY	GENUS	SPECIES AND SUBSPECIES	TOTAL NO. SPECIES
Lumbricidae	*Allolobophora*	*caliginosa* subsp. *trapezoides*	1
	Helodrilus	*caliginosus, chloroticus, foetidus, roseus*	4
	Lumbricus	*herculeus, rubellus, terrestris*	3
	Octolasium	*lacteum*	1
Megascolecidae	*Diplocardia*	*communis*	1
	Microscolex	*dubius*	1

ARTHROPODA—ARACHNIDA

FAMILY	GENUS	SPECIES, VARIETIES, AND SUBSPECIES	TOTAL NO. SPECIES
Chernetidae	*Lamprochernes*	*nodosus*	1
	Pselapho-chernes	*scorpioides*	1
	Toxochernes	*panzeri*	1
Cheliferidae	*Chelifer*		
Dictynidae	*Coenothele*	*gregalis*	1
Theridiidae	*Lactrodectus*	*mactans, mactans* subsp. *tredecimguttatus*	2
	Steatoda		
	Teutana		
	Theridium		
Eresidae	*Stegodyphus*	*mimosarum*	1
Argiopidae	*Epeira*	*diadema*	1
Salticidae	*Salticus*	*scenicus*	1
Ascaidae	*Gamasellus*		
Parasitidae	*Eulaelaps*	*stabularis*	1
	Parasitus	*coleoptratorum, lunaris*	2

ARTHROPODA—ARACHNIDA

FAMILY	GENUS	SPECIES, VARIETIES, AND SUBSPECIES	TOTAL NO. SPECIES
Macrochelidae	*Glyptholaspis*	*confusa*	1
	Macrocheles	*glaber, merdarius, muscaedomesticae, peniculatus, perglaber, plumiventris, robustulus, scutatus, subbadius*	9
			1
Gamasolaelaptidae	*Digamasellus*		
	Saintdidieria	*sexclavatus*	1
Laelaptidae	*Alliphis*	*halleri*	1
	Holostaspis	*badius, marginatus*	2
Uropodidae	*Fuscuropoda*	*vegetans*	1
Pyemotidae	*Pygmephorus*		
Tarsonemidae			
Trombidiidae	*Allothrombium*	*fuliginosum*	1
	Atomus	*inexpectatus, parasiticus*	2
	Micro-trombidium	*?muscarum, striaticeps*	2
Trombiculidae	*Ascoschoen-gastia*	*indica*	1
	Euschoengastia	*peromysci*	1
	Eutrombicula	*alfreddugesi, splendens*	2
	Fonsecia	*gurneyi*	1
	Hannemania		
	Kayella	*lacerta*	1
	Neoschoen-gastia	*americana*	1
	Neotrombicula	*lipovskyi*	1
	Neotrom-biculoides	*montanensis*	1
	Pseudoschoen-gastia	*farneri, hungerfordi*	2
Acaridae	*Acarus*	*siro*	1
Anoetidae	*Histiostoma*	*laboratorium*	1
	Myianoetus	*muscarum*	1
Sarcoptidae	*Sarcoptes*		
Undetermined Family	*Gamasus*	*musci*	1

ARTHROPODA—MYRIAPODA

FAMILY	GENUS	SPECIES, VARIETIES, AND SUBSPECIES	TOTAL NO. SPECIES
Scutigeridae	*Scutigera*	*coleoptrata*	1

ARTHROPODA—INSECTA

FAMILY	GENUS	SPECIES, VARIETIES, AND SUBSPECIES	TOTAL NO. SPECIES
Sminthuridae	*Sminthurus viridis*		1
Gryllidae	*Nemobius*	*fasciatus*	1
Locustidae	*Chortophaga*	*viridifasciata*	1
	Dociostaurus	*maroccanus*	1
	Encoptolophus	*sordidus* subsp. *costalis*	1
	Locusta	*migratoria*	1
	Nomadacris	*septemfasciata*	1
	Schistocerca	*gregaria, paranensis*	2
	Xanthippus	*corallipes*	1
Labiduridae	*Euborellia*	*annulipes*	1
Labiidae	*Labia*	*minor, pilicornis*	2
	Spingolabis	*hawaiiensis*	1
Forficulidae	*Forficula*	*auricularia*	1
Haematopinidae	*Haematopinus*	*tuberculatus*	1
Coenagrionidae	*Agria*	*fumipennis*	1
	Enallagma	*durum*	1
	Ischnura	*ramburii*	1
Libellulidae	*Celithemis*	*amanda*	1
	Erythemis	*simplicicollis*	1
	Libellula	*auripennis, pulchella, vibrans*	3
	Pachydiplax	*longipennis*	1
	Pantala	*flavescens*	1
	Tramea	*carolina*	1
Aeschnidae	*Aeshna*	*cyanea*	1
	Anax	*junius*	1
	Coryphaeschna	*ingens*	1
Phymatidae	*Phymata*	*pennsylvanica*	1
Reduviidae	*Apiomerus*	*pilipes*	1
	Pristhesancus	*papuensis*	1
	Sinea	*diadema*	1

ARTHROPODA—INSECTA

FAMILY	GENUS	SPECIES, VARIETIES, AND SUBSPECIES	TOTAL NO. SPECIES
Aphidae	*Myzus*	*persicae*	1
Hyponomeutidae	*Hyponomeuta*	*padella*	1
Olethreutidae	*Epiblema*	*otiosana*	1
	Suleima	*helianthana*	1
Pyralididae	*Epischnia*	*incanella*	1
Noctuidae	*Agrotis*	*castanea* var. *neglecta*, *segetum*	2
	Archanara	*oblonga*	1
	Heliothis	*obsoleta*	1
	Laphygma	*frugiperda*	1
	Leucania	*unipuncta*	1
	Macronoctua	*onusta*	1
	Papaipema	*nebris*	1
	Peridroma	*margitosa* form *saucia*	1
Liparidae	*Liparis*	*dispar*	1
Lymantriidae	*Leucoma*	*salicis*	1
	Lymantria	*monacha*	1
Sphingidae	*Acherontia*	*atropos*	1
Lasiocampidae	*Dendrolimus*	*pini*	1
	Malacosoma	*americana*	1
Carabidae	*Carabus*	*italicus*	1
Hydrophilidae	*Cryptopleurum*	*minutum*	1
	Dactylosternum	*abdominale*	1
	Sphaeridium	*scarabaeoides*	1
Silphidae	*Nicrophorus*	*nigritus*	1
Staphylinidae	*Aleochara*	*algarum*, *bimaculata*, *curtula*, *handschini*, *laevigata*, *taeniata*, *tristis*, *trivialis*, *windredi*	9
	Baryodma	*bimaculata*, *ontarionis*	2
	Creophilus	*erythrocephalus*, *maxillosus*, *maxillosus* var. *villosus*	2
	Emus	*hirtus*	1
	Ontholestes	*murinus*, *tessellatus*	2
	Oxytelus	*ocularis*, *sculptus*	2
	Philonthus	*discoideus*, *fimetarius*, *longicornis*, *politus*	4
	Tachyporus		

FAMILY	GENUS	SPECIES, VARIETIES, AND SUBSPECIES	TOTAL NO. SPECIES
Histeridae	*Pachylister*	*chinensis*	1
	Saprinus	*cyaneus, detersus, lugens, politus* or *aenus*	4
	Xerosaprinus	*fimbriatus, lubricus*	2
Cleridae	*Necrobia*	*rufipes*	1
Dermestidae	*Dermestes*	*ater, caninus*	2
Ostomatidae	*Temnochila*	*virescens*	1
Coccinellidae	*Adonia*	*variegata*	1
	Coccinella	*septempunctata*	1
	Thea	*vigintiduopunctata*	1
	Vibidia	*duodecimguttata*	1
Scarabaeidae	*Copris*	*incertus*	1
	Lachnosterna		
	Melolontha	*melolontha*	1
Cerambycidae	*Dorysthenes*	*forficatus*	1
Chrysomelidae	*Gallerucella*	*luteola*	1
Curculionidae	*Conotrachelus*	*retentus*	1
	Pissodes	*strobi*	1
	Rhodobaenus	*tredecimpunctatus*	1
Tenthredinidae	*Tenthredo*	*pectoralis*	1
Diprionidae	*Diprion*	*pallidus, pini*	2
Ichneumonidae	*Atractodes*	*bicolor, gravidus, mallyi, miuri*	4
	Mesoleptus	*laevigatus*	1
	Nemeritis	*canescens*	1
	Opius	*nitidulator*	1
	Phygadeuon	*speculator*	1
	Stilpnus	*anthomyidiperda*	1
	Theronia	*atalantae*	1
Braconidae	*Alysia*	*fossulata, manducator, ridibunda*	3
	Aphaereta	*auripes, cephalotes, minuta, pallipes, sarcophaga*	5
	Asobara	*tabida*	1
	Aspilota	*concolor, nervosa*	2
	Idiasta	*lusoriae*	1
	Opius	*nitidulator*	1
Cynipidae	*Bothrochacis*	*stercoraria*	1
	Cothonaspis	*nigricornis*	1

ARTHROPODA—INSECTA

FAMILY	GENUS	SPECIES, VARIETIES, AND SUBSPECIES	TOTAL NO. SPECIES
	Eucoila	*drosophilae, impatiens*	2
	Hexacola		
	Kleidotoma	*certa, marshalli*	2
	Pseudeucoila	*bochei, trichopsila*	2
	Trybliographa		
Figitidae	*Figites*	*anthomyiarum, discordis, sarcophagarum, scutellaris, striolatus*	5
	Neralsia	*armata, bifoveolata*	2
	Xyalophora	*quinquelineata*	1
Chalcididae	*Brachymeria*	*argentifrons, calliphorae, fulvitarsis, fonscolombei, minuta, obtusata, ontarionis, vicina*	8
	Dirhinoides	*vlasovi, wohlfahrtiae*	2
	Dirhinus	*luzonensis, pachycerus, sarcophagae, texanus*	4
	Euchalcidia	*blanda*	1
	Stenomalus	*muscarum*	1
Pteromalidae	*Dibrachys*	*affinis, cavus*	2
	Eupteromales	*hemipterus, nidulans*	2
	Habrocytus	*obscurus*	1
	Muscidifurax	*raptor*	1
	Nasonia	*vitripennis*	1
	Pachy-crepoideus	*vindemmiae*	1
	Pachyneuron	*vindemmiae*	1
	Prospalangia	*platensis*	1
	Spalangia	*cameroni, chontalensis, drosophilae, endius, erythromera brachyceps, haematobiae, hirta, longepetiolata, muscae, muscidarum, muscophaga, nigra, nigripes, nigroaenea, orientalis, philippinensis, platensis, rugosicollis, rugulosa, simplex, stomoxysiae, sundaica*	22
	Sphegigaster		
	Stenomolina	*muscarum*	1
Encyrtidae	*Australencyrtus*	*giraulti*	1
	Cerchysius	*lyperosae*	1
	Homalotylus	*eitelweinii*	1
	Ooencyrtus	*submetallicus*	1
	Stenoterys	*fulvoventralis*	1
	Tachinae-phagus	*giraulti*	1

FAMILY	GENUS	SPECIES, VARIETIES, AND SUBSPECIES	TOTAL NO. SPECIES
Eupelmidae	*Mesocomys*	*pulchriceps*	1
Eulophidae	*Syntomosphyrum*	*albiclavus, glossinae*	2
	Trichospilus	*pupivora*	1
Diapriidae	*Diapria*	*conica*	1
	Hemilexomyia	*abrupta*	1
	Paraspilomicrus	*froggatti*	1
	Phaenopria	*fimicola, occidentalis*	2
	Trichopria	*commoda, hirticollis*	2
Formicidae	*Camponotus*	*pennsylvanicus*	1
	Crematogaster	*lineolata*	1
	Eciton		
	Forelius	*foetidus*	1
	Iridomyrmex	*humilis*	1
	Labidus	*coecus*	1
	Monomorium	*pharaonis*	1
	Paratrechina	*longicornis*	1
	Pheidole	*megacephala, morrisii* var. *impexa, subarmata borinquenensis, oceanica*	4
	Pogonomyrmex	*barbatus* var. *fuscatus, barbatus* var. *molefaciens*	1
	Pheidologeton	*affinis*	1
	Ponera	*perkinsi*	1
	Solenopsis	*corticalis, geminata*	2
	Tapinoma	*melanocephalum*	1
	Tetramorium	*quineense*	1
	Wasmannia	*auropunctata*	1
Sphecidae	*Bembix*	*lunata, oculata, olivata, spinolae*	4
	Crabro	*advenus, alpinus, cavifrons, clypeatus, cribraria, elongatus, leucostoma, palmarius, peltarius, pubescens, serripes, signatus, vagus*	13
	Crossocerus	*lentus, quadrimaculatus*	2
	Dinetus	*pictus*	1
	Diodontus	*minutus*	1
	Ectemnius	*chrysostomus, quadricinctus*	2
	Gorytes		
	Mellinus	*arvensis*	1
	Nysson		
	Oxybelus	*bipunctatus, quattuordecimnotatus, uniglumis*	3

ARTHROPODA—INSECTA

FAMILY	GENUS	SPECIES, VARIETIES, AND SUBSPECIES	TOTAL NO. SPECIES
	Sericophorus	*sydneyi, teliferopodus, viridis roddi*	3
	Stictia	*denticornis*	1
	Stizus	*turneri*	1
Mutillidae	*Mutilla*	*glossinae*	1
Vespidae	*Polistes*	*hebraeus*	1
	Vespa	*germanica, vulgaris*	2
Apidae	*Bombus*	*agrorum, terrestris*	2
Trichoceridae	*Trichocera*		
Tipulidae	*Ula*	*macroptera*	1
Ptychopteridae	*Ptychoptera*	*contaminata*	1
Culicidae	*Aedes*	*geniculatus*	1
	Culex	*pipiens*	1
Ceratopogonidae			
Chironomidae			
Simuliidae	*Simulium*	*damnosum*	1
Anisopodidae	*Mycetobia*	*pallipes*	1
Mycetophilidae	*Mycetophila*	*lineola, ornata*	2
Sciaridae	*Sciara*	*militaris*	1
Stratiomyidae	*Hermetia*	*illucens*	1
Tabanidae	*Tabanus*	*striatus*	1
Asilidae	*Isopogon*	*brevirostris*	1
	Lasiopogon	*cinctus*	1
	Laphria	*flava*	1
	Efferia	*aestuans, pogonias*	2
	Neoitamus	*flavofemoratus*	1
	Proctacanthus	*philadelphicus*	1
	Regasilus	*notatus, sadyates*	2
Bombyliidae	*Thyridanthrax*	*abruptus*	1
Empididae	*Empis*	*grisea*	1
Syrphidae	*Eristalis*		
Sphaeroceridae	*Borborus*		
	Copromyza	*equina*	1
Drosophilidae	*Drosophila*	*melanogaster*	1
Clusiidae	*Clusiodes*	*albimanus*	1

ARTHROPODA—INSECTA

FAMILY	GENUS	SPECIES, VARIETIES, AND SUBSPECIES	TOTAL NO. SPECIES
Aulacigastridae	*Aulacigaster*	*leucopeza*	1
Anthomyiidae	*Scatophaga*	*merdaria, stercoraria*	2
	Hylemya	*brassicae*	1
Muscidae	*Coenosia*	*tigrina*	1
	Hebecnema	*umbratica*	1
	Myospila	*meditabunda*	1
	Mydaea	*urbana, pagana*	2
	Hydrotaea	*dentipes*	1
	Ophyra	*leucostoma*	1
	Phaonia	*variegata*	1
	Muscina	*assimilis, pabulorum, stabulans*	3
	Polietes	*albolineata, hirticrura*	2
	Mesembrina	*meridiana*	1
	Musca	*domestica*	1
Calliphoridae	*Chrysomya*	*rufifacies, villeneuvei*	2
	Protophormia	*terraenovae*	1
	Phaenicia	*cuprina*	1
	Calliphora		
Sarcophagidae	*Ravinia*	*striata*	1
	Sarcophaga	*haemorrhoidalis*	1
Cuterebridae	*Dermatobia*	*hominis*	1

CHORDATA—AMPHIBIA

FAMILY	GENUS	SPECIES, VARIETIES, AND SUBSPECIES	TOTAL NO. SPECIES
Discoglossidae	*Discoglossus*	*pictus*	1
Bufonidae	*Bufo*	*mauritanicus, viridis*	2
Hylidae	*Hyla*	*viridis* var. *grisea*	1
Ranidae	*Rana*	*ridibunda*	1

CHORDATA—REPTILIA

FAMILY	GENUS	SPECIES, VARIETIES, AND SUBSPECIES	TOTAL NO. SPECIES
Agamidae	*Agama*	*agilis*	1
Scincidae	*Chalcides*	*tridactylus*	1
Lacertidae	*Acanthodactylus*	*boskianus*	1
	Lacerta	*agilis*	1

CHORDATA—AVES

FAMILY	GENUS	SPECIES, VARIETIES, AND SUBSPECIES	TOTAL NO. SPECIES
Phasianidae	*Gallus*	*gallus*	1
Hirundinidae	*Delichon*	*urbica* subsp. *urbica*	1
	Hirundo	*rustica* subsp. *rustica*	1
Paridae	*Aegithalus*	*caudatus*	1
	Parus	*caeruleus* subsp. *caeruleus*	1
Turdidae	*Luscinia*	*megarhynchos*	1
Muscicapidae	*Leucocira*	*leucophrys*	1
Sturnidae	*Sturnus*	*vulgaris*	1
Meliphagidae	*Melliphaga*	*penicillata*	1
	Myzantha	*melanocephala*	1
Fringillidae	*Fringilla*	*coelebs*	1
Ploceidae	*Passer*	*domesticus*	1

CHORDATA—MAMMALIA

FAMILY	GENUS	SPECIES, VARIETIES, AND SUBSPECIES	TOTAL NO. SPECIES
Muridae	*Apodemus*	*sylvaticus*	1
Gliridae			

CLASSIFICATION OF ORGANISMS	*ASSOCIATED FLIES*
VIRUSES	
PSITTACOSIS GROUP	MUSCIDAE: MUSCINAE
Trachoma virus	*Musca domestica* Linnaeus Nicolle et al., 1919 Zardi, 1964 *Musca sorbens Wiedemann* Gear et al., 1962
POXVIRUSES	
Variola	MUSCIDAE: MUSCINAE *Musca domestica* Linnaeus Terni, 1908b MUSCIDAE: STOMOXYINAE *Stomoxys* Geoffroy Terni, 1909
Vaccinia	MUSCIDAE: MUSCINAE *Musca domestica* Linnaeus Terni, 1908b Merk, 1910 MUSCIDAE: STOMOXYINAE *Stomoxys* Geoffroy Terni, 1909 *Stomoxys calcitrans* (Linnaeus) Terni, 1908b
Borreliota avium (Synonym: fowlpox virus)	MUSCIDAE: STOMOXYINAE *Stomoxys calcitrans* (Linnaeus) Schuberg et al., 1912 Bos, 1932
ARBOVIRUSES	
Eastern equine encephalomyelitis (Synonym: Eastern encephalitis virus)	CALLIPHORIDAE: CHRYSOMYINAE *Phormia regina* (Meigen) Bourke, 1964
Sindbus virus	DROSOPHILIDAE: DROSOPHILINAE *Drosophila melanogaster* Meigen Herring, 1967 Ohanessian et al., 1967
Yellow fever virus	MUSCIDAE: STOMOXYINAE *Stomoxys calcitrans* (Linnaeus) Hoskins, 1933
ENTEROVIRUSES	
Poliovirus	ANTHOMYIIDAE: SCATOPHAGINAE *Scatophaga* Meigen Power et al., 1942 MUSCIDAE: LIMNOPHORINAE *Gymnodia arcuata* (Stein) Francis, Jr. et al., 1948

Poliovirus (cont.)

MUSCIDAE: MYDAEINAE

Myospila meditabunda (Fabricius)
Francis, Jr., et al., 1948

MUSCIDAE: FANNIINAE

Fannia Robineau-Desvoidy
Toomey et al., 1941
Trask et al., 1943
Melnick et al., 1945

Fannia canicularis (Linnaeus)
Francis, Jr. et al., 1948

MUSCIDAE: PHAONIINAE

Ophyra Robineau-Desvoidy
Trask et al., 1943

Ophyra leucostoma (Wiedemann)
Paul et al., 1941
Sabin et al., 1942
Melnick et al., 1945

Muscina Robineau-Desvoidy
Trask et al., 1943

Muscina assimilis (Fallén)
Francis, Jr. et al., 1948

Muscina stabulans (Fallén)
Paul et al., 1941
Sabin et al., 1942
Melnick et al., 1945
Francis, Jr. et al., 1948

MUSCIDAE: MUSCINAE

Musca Linnaeus
Trask et al., 1943

Musca domestica Linnaeus
Francis, 1914
Paul et al., 1941
Sabin et al., 1942
Gordon, 1943
Melnick et al., 1945, 1949
Francis, Jr. et al., 1948
Melnick, 1949

House fly (presumably *Musca domestica* Linnaeus)
Flexner, 1911
Toomey et al., 1941

MUSCIDAE: STOMOXYINAE

Haematobia irritans (Linnaeus)
Francis, 1914

Stomoxys Geoffroy
Paul et al., 1941
Trask et al., 1943

Stomoxys calcitrans (Linnaeus)
Anderson et al., 1912
Rosenau, 1912a

Organism	Fly / References
Poliovirus (cont.)	Rosenau et al., 1912b
	Boudreau et al., 1914
	Francis, 1914
	CALLIPHORIDAE: CHRYSOMYINAE
	Cochliomyia macellaria (Fabricius)
	Melnick et al., 1945
	Francis, Jr. et al., 1948
	Phormia Robineau-Desvoidy
	Trask et al., 1943
	Phormia regina (Meigen)
	Paul et al., 1941
	Sabin et al., 1942
	Melnick et al., 1945
	Francis, Jr. et al., 1948
	Melnick, 1949
	Protophormia Townsend
	Trask et al., 1943
	Protophormia terraenovae (Robineau-Desvoidy)
	Sabin et al., 1942
	Protocalliphora Hough
	Paul et al., 1941
	Trask et al., 1943
	CALLIPHORIDAE: CALLIPHORINAE
	Bufolucilia silvarum (Meigen)
	Melnick et al., 1945
	Lucilia Robineau-Desvoidy
	Paul et al., 1941
	Lucilia illustris (Meigen)
	Melnick et al., 1945
	Phaenicia Robineau-Desvoidy
	Trask et al., 1943
	Phaenicia pallescens (Shannon)
	Melnick et al., 1949
	Phaenicia sericata (Meigen)
	Sabin et al., 1942
	Melnick et al., 1945, 1949
	Francis, Jr. et al., 1948
	Melnick, 1949
	Calliphora Robineau-Desvoidy
	Trask et al., 1943
	Calliphora vicina Robineau-Desvoidy
	Sabin et al., 1942
	Calliphora vomitoria (Linnaeus)
	Noguchi et al., 1917
	Cynomyopsis cadaverina (Robineau-Desvoidy)
	Sabin et al., 1942
	Melnick, 1949

Organism	Fly / Reference
Poliovirus (cont.)	UNDETERMINED CALLIPHORIDAE
	Blow flies
	Toomey et al., 1941
	SARCOPHAGIDAE: SARCOPHAGINAE
	Sarcophaga Meigen
	Trask et al., 1943
	Melnick et al., 1945, 1949
	Francis, Jr. et al., 1948
	Melnick, 1949
	Sarcophaga haemorrhoidalis (Fallén)
	Paul et al., 1941
	Tricharaea Thomson
	Melnick et al., 1949
Poliovirus non-paralytic	MUSCIDAE: MUSCINAE
	Musca domestica Linnaeus
	Melnick et al., 1949
	CALLIPHORIDAE: CALLIPHORINAE
	Phaenicia pallescens (Shannon)
	Melnick et al., 1949
	Phaenicia sericata (Meigen)
	Melnick et al., 1949
	SARCOPHAGIDAE: SARCOPHAGINAE
	Sarcophaga Meigen
	Melnick et al., 1949
	Tricharaea Thomson
	Melnick et al., 1949
Poliovirus Strain 113	MUSCIDAE: PHAONIINAE
	Ophyra leucostoma (Wiedemann)
	Levkovich et al., 1957
	MUSCIDAE: MUSCINAE
	Musca domestica domestica Linnaeus
	Levkovich et al., 1957
	CALLIPHORIDAE: CHRYSOMYINAE
	Protophormia terraenovae (Robineau-Desvoidy)
	Levkovich et al., 1957
	CALLIPHORIDAE: CALLIPHORINAE
	Calliphora uralensis Villeneuve
	Levkovich et al., 1957
Poliovirus Delaware '48 strain	CALLIPHORIDAE: CHRYSOMYINAE
	Phormia regina (Meigen)
	Melnick et al., 1952
Poliovirus Type 1	RICHARDIIDAE
	Richardia viridiventris Wulp
	Paul et al., 1962
	MUSCIDAE: COENESIINAE
	Atherigona orientalis Schiner
	Paul et al., 1962

Poliovirus Type 1 (cont.)

MUSCIDAE: LIMNOPHORINAE

Limnophora Robineau-Desvoidy
Paul et al., 1962

MUSCIDAE: FANNIINAE

Fannia Robineau-Desvoidy
Riordan et al., 1961
Paul et al., 1962

MUSCIDAE: PHAONIINAE

Ophyra Robineau-Desvoidy
Riordan et al., 1961

Ophyra aenescens (Wiedemann)
Paul et al., 1962

Muscina Robineau-Desvoidy
Riordan et al., 1961

Muscina stabulans (Fallén)
Paul et al., 1962

MUSCIDAE: MUSCINAE

Graphomya maculata (Scopoli)
Paul et al., 1962

Synthesiomyia nudiseta (Wulp)
Paul et al., 1962

Neomuscina Townsend
Paul et al., 1962

Philornis Meinert
Paul et al., 1962

Musca Linnaeus
Paul et al., 1962

Musca domestica Linnaeus
Melnick et al., 1953
Riordan et al., 1961
Downey, 1963

Musca domestica vicina Macquart
Brygoo et al., 1962

CALLIPHORIDAE: MESEMBRINELLINAE

Huascaromusca bicolor (Fabricius)
Paul et al., 1962

CALLIPHORIDAE: CHRYSOMYINAE

Cochliomyia macellaria (Fabricius)
Melnick et al., 1953
Paul et al., 1962

Chrysomya putoria (Wiedemann)
Brygoo et al., 1962

Phormia regina (Meigen)
Riordan et al., 1961
Downey, 1963

CALLIPHORIDAE: CALLIPHORINAE

Lucilia illustris (Meigen)
Downey, 1963

Phaenicia Robineau-Desvoidy
Paul et al., 1962

Organism	Fly / Reference
Poliovirus Type 1 (cont.)	*Phaenicia pallescens* (Shannon) Melnick et al., 1953 *Phaenicia sericata* (Meigen) Melnick et al., 1953 Riordan et al., 1961 Downey, 1963 *Calliphora* Robineau-Desvoidy Riordan et al., 1961 Downey, 1963 SARCOPHAGIDAE: SARCOPHAGINAE *Blaesoxipha plinthopyga* (Wiedemann) Paul et al., 1962 *Oxysarcodexia ochripyga* (Wulp) Paul et al., 1962 *Sarcophaga* Meigen Riordan et al., 1961 Paul et al., 1962 *Tricharaea* Thomson Melnick et al., 1953 Paul et al., 1962
Poliovirus Attenuated Type 1, LSc	MUSCIDAE: MUSCINAE *Musca domestica* Linnaeus Gudnadóttir, 1960 CALLIPHORIDAE: CHRYSOMYINAE *Phormia regina* (Meigen) Gudnadóttir, 1960 Gudnadóttir et al., 1960 CALLIPHORIDAE: CALLIPHORINAE *Phaenicia sericata* (Meigen) Davé et al., 1965
Poliovirus Mahoney Type 1	CALLIPHORIDAE: CHRYSOMYINAE *Phormia regina* (Meigen) Gudnadóttir, 1960 Gudnadóttir et al., 1960
Poliovirus New York City '44 strain (Type 1)	CALLIPHORIDAE: CHRYSOMYINAE *Phormia regina* (Meigen) Melnick et al., 1947, 1952
Poliovirus Philippine Islands '45 strain (Type 1)	CALLIPHORIDAE: CHRYSOMYINAE *Phormia regina* (Meigen) Melnick et al., 1952 CALLIPHORIDAE: CALLIPHORINAE *Phaenicia sericata* (Meigen) Melnick et al., 1952
Poliovirus Texas '48 strain (Type 1)	MUSCIDAE: MUSCINAE *Musca domestica* Linnaeus Melnick et al., 1952 CALLIPHORIDAE: CHRYSOMYINAE *Phormia regina* (Meigen) Melnick et al., 1952

Organism	Fly association
Poliovirus Texas '48 strain (cont.)	CALLIPHORIDAE: CALLIPHORINAE *Phaenicia sericata* (Meigen) Melnick et al., 1952
Poliovirus W-S '48 strain (Type 1)	CALLIPHORIDAE: CHRYSOMYINAE *Phormia regina* (Meigen) Melnick et al., 1952
Poliovirus Type 2	CALLIPHORIDAE: CHRYSOMYINAE *Chrysomya megacephala* (Fabricius) Asahina et al., 1963
Poliovirus Type 2 Lansing strain	DROSOPHILIDAE: DROSOPHILINAE *Drosophila melanogaster* Meigen Toomey et al., 1947 MUSCIDAE: PHAONIINAE *Ophyra leucostoma* (Wiedemann) Levkovich et al., 1957 *Muscina stabulans* (Fallén) Bang et al., 1943 MUSCIDAE: MUSCINAE *Musca domestica* Linnaeus Bang et al., 1943 Rendtorff et al., 1943 Hurlbut, 1950 *Musca domestica domestica* Linnaeus Levkovich et al., 1957 CALLIPHORIDAE: CHRYSOMYINAE *Phormia regina* (Meigen) Melnick et al., 1947, 1952 *Protophormia terraenovae* (Robineau-Desvoidy) Levkovich et al., 1957 CALLIPHORIDAE: CALLIPHORINAE *Lucilia caesar* (Linnaeus) Bang et al., 1943 *Phaenicia sericata* (Meigen) Melnick et al., 1947, 1952 *Calliphora vicina* Robineau-Desvoidy Bang et al., 1943 *Calliphora uralensis* Villeneuve Levkovich et al., 1957 SARCOPHAGIDAE: SARCOPHAGINAE *Sarcophaga bullata* Parker Melnick et al., 1947 *Sarcophaga haemorrhoidalis* (Fallén) Bang et al., 1943
Poliovirus Y-SK strain murine adapted, Type 2	CALLIPHORIDAE: CHRYSOMYINAE *Phormia regina* (Meigen) Melnick et al., 1947, 1952

CALLIPHORIDAE: CALLIPHORINAE
Phaenicia sericata (Meigen)
Melnick et al., 1947, 1952

SARCOPHAGIDAE: SARCOPHAGINAE
Sarcophaga bullata Parker
Melnick et al., 1947

Poliovirus Type 3 .... RICHARDIIDAE
Richardia viridiventris Wulp
Paul et al., 1962

MUSCIDAE: COENOSIINAE
Atherigona orientalis Schiner
Paul et al., 1962

MUSCIDAE: LIMNOPHORINAE
Limnophora Robineau-Desvoidy
Paul et al., 1962

MUSCIDAE: FANNIINAE
Fannia Robineau-Desvoidy
Riordan et al., 1961
Paul et al., 1962

MUSCIDAE: PHAONIINAE
Ophyra Robineau-Desvoidy
Riordan et al., 1961
Ophyra aenescens (Wiedemann)
Paul et al., 1962
Muscina Robineau-Desvoidy
Riordan et al., 1961
Muscina stabulans (Fallén)
Paul et al., 1962

MUSCIDAE: MUSCINAE
Graphomya maculata (Scopoli)
Paul et al., 1962
Synthesiomyia nudiseta (Wulp)
Paul et al., 1962
Neomuscina Townsend
Paul et al., 1962
Philornis Meinert
Paul et al., 1962
Musca Linnaeus
Paul et al., 1962
Musca domestica Linnaeus
Riordan et al., 1961

CALLIPHORIDAE: MESEMBRINELLINAE
Huascaromusca bicolor (Fabricius)
Paul et al., 1962

CALLIPHORIDAE: CHRYSOMYINAE
Cochliomyia macellaria (Fabricius)
Paul et al., 1962
Chrysomya putoria (Wiedemann)
Brygoo et al., 1962

Poliovirus Type 3 (cont.)	*Phormia regina* (Meigen) Riordan et al., 1961 CALLIPHORIDAE: CALLIPHORINAE *Phaenicia* Robineau-Desvoidy Paul et al., 1962 *Phaenicia sericata* (Meigen) Riordan et al., 1961 *Calliphora* Robineau-Desvoidy Riordan et al., 1961 SARCOPHAGIDAE: SARCOPHAGINAE *Blaesoxipha plinthopyga* (Wiedemann) Paul et al., 1962 *Oxysarcodexia ochripyga* (Wulp) Paul et al., 1962 *Sarcophaga* Meigen Riordan et al., 1961 Paul et al., 1962 *Tricharaea* Thomson Paul et al., 1962
Poliovirus Attenuated Type 3, Leon	CALLIPHORIDAE: CALLIPHORINAE *Phaenicia sericata* (Meigen) Davé et al., 1965
Poliovirus North Carolina '44 strain (Type 3)	CALLIPHORIDAE: CALLIPHORINAE *Phaenicia sericata* (Meigen) Melnick et al., 1947, 1952
Theiler's mouse encephalomyelitis virus, GD VII (Synonym: mouse "poliomyelitis" virus)	MUSCIDAE: PHAONIINAE *Muscina stabulans* (Fallén) Bang et al., 1943 MUSCIDAE: MUSCINAE *Musca domestica* Linnaeus Bang et al., 1943 CALLIPHORIDAE: CALLIPHORINAE *Lucilia caesar* (Linnaeus) Bang et al., 1943 *Calliphora vicina* Robineau-Desvoidy Bank et al., 1943 SARCOPHAGIDAE: SARCOPHAGINAE *Sarcophaga haemorrhoidalis* (Fallén) Bang et al., 1943
Theiler's mouse encephalomyelitis virus, TO strain	CALLIPHORIDAE: CHRYSOMYINAE *Phormia regina* (Meigen) Melnick et al., 1947, 1952 CALLIPHORIDAE: CALLIPHORINAE *Phaenicia sericata* (Meigen) Melnick et al., 1947, 1952 SARCOPHAGIDAE: SARCOPHAGINAE *Sarcophaga bullata* Parker Melnick et al., 1947, 1952

Organism	Fly association
Coxsackievirus	MUSCIDAE: MUSCINAE *Musca domestica* Linnaeus Melnick et al., 1954 Duca et al., 1958 CALLIPHORIDAE: CHRYSOMYINAE *Chrysomya putoria* (Wiedemann) Brygoo et al., 1962 *Phormia* Robineau-Desvoidy Melnick et al., 1954 CALLIPHORIDAE: CALLIPHORINAE *Phaenicia* Robineau-Desvoidy Melnick (personal communication from Curnen), 1950 Melnick et al., 1954
Coxsackievirus Alaska 5 (A10)	MUSCIDAE: MUSCINAE *Musca domestica* Linnaeus Melnick et al., 1953 CALLIPHORIDAE: CHRYSOMYINAE *Cochliomyia macellaria* Fabricius Melnick et al., 1953
Coxsackievirus Connecticut 5 (B1)	MUSCIDAE: MUSCINAE *Musca domestica* Linnaeus Melnick et al., 1953 SARCOPHAGIDAE: SARCOPHAGINAE *Tricharaea* Thomson Melnick et al., 1953
Coxsackievirus Easton 2 (A1)	CALLIPHORIDAE: CALLIPHORINAE *Phaenicia pallescens* (Shannon) Melnick et al., 1953 *Phaenicia sericata* (Meigen) Melnick et al., 1953
Coxsackievirus Easton 10 (A8)	CALLIPHORIDAE: CALLIPHORINAE *Phaenicia pallescens* (Shannon) Melnick et al., 1953 *Phaenicia sericata* (Meigen) Melnick et al., 1953 SARCOPHAGIDAE: SARCOPHAGINAE *Tricharaea* Thomson Melnick et al., 1953
Coxsackievirus Easton 14 (A5)	MUSCIDAE: MUSCINAE *Musca domestica* Linnaeus Melnick et al., 1953 CALLIPHORIDAE: CALLIPHORINAE *Phaenicia pallescens* (Shannon) Melnick et al., 1953 *Phaenicia sericata* (Meigen) Melnick et al., 1953 SARCOPHAGIDAE: SARCOPHAGINAE *Tricharaea* Thomson Melnick et al., 1953

Coxsackievirus Nancy (B3)	MUSCIDAE: MUSCINAE
	Musca domestica Linnaeus
	Melnick et al., 1953
	SARCOPHAGIDAE: SARCOPHAGINAE
	Tricharaea Thomson
	Melnick et al., 1953
Coxsackievirus Texas 1 (A4)	MUSCIDAE: MUSCINAE
	Musca domestica Linnaeus
	Melnick et al., 1953
	CALLIPHORIDAE: CHRYSOMYINAE
	Cochliomyia macellaria Fabricius
	Melnick et al., 1953
	Phormia regina (Meigen)
	Melnick et al., 1952
	CALLIPHORIDAE: CALLIPHORINAE
	Phaenicia pallescens (Shannon)
	Melnick et al., 1953
	Phaenicia sericata (Meigen)
	Melnick et al., 1953
	SARCOPHAGIDAE: SARCOPHAGINAE
	Tricharaea Thomson
	Melnick et al., 1953
Coxsackievirus Texas 12 .... (new type)	MUSCIDAE: MUSCINAE
	Musca domestica Linnaeus
	Melnick et al., 1953
	CALLIPHORIDAE: CALLIPHORINAE
	Phaenicia pallescens (Shannon)
	Melnick et al., 1953
	Phaenicia sericata (Meigen)
	Melnick et al., 1953
	SARCOPHAGIDAE: SARCOPHAGINAE
	Tricharaea Thomson
	Melnick et al., 1953
Coxsackievirus Texas 13 (B4)	MUSCIDAE: MUSCINAE
	Musca domestica Linnaeus
	Melnick et al., 1953
Coxsackievirus Texas 14 .... (new type)	CALLIPHORIDAE: CALLIPHORINAE
	Phaenicia pallescens (Shannon)
	Melnick et al., 1953
	Phaenicia sericata (Meigen)
	Melnick et al., 1953
	SARCOPHAGIDAE: SARCOPHAGINAE
	Tricharaea Thomson
	Melnick et al., 1953
Coxsackievirus Texas 15 (A7)	CALLIPHORIDAE: CHRYSOMYINAE
	Cochliomyia macellaria Fabricius
	Melnick et al., 1953
Coxsackievirus Texas '48 .... (new type)	MUSCIDAE: MUSCINAE
	Musca domestica Linnaeus
	Melnick et al., 1952

CALLIPHORIDAE: CHRYSOMYINAE
Phormia regina (Meigen)
Melnick et al., 1952
CALLIPHORIDAE: CALLIPHORINAE
Phaenicia sericata (Meigen)
Melnick et al., 1952

Coxsackievirus Type 2 (A2) .... MUSCIDAE: MUSCINAE
Musca domestica Linnaeus
Melnick et al., 1953
SARCOPHAGIDAE: SARCOPHAGINAE
Tricharaea Thomson
Melnick et al., 1953

Coxsackievirus Type 3 (A3) .... MUSCIDAE: MUSCINAE
Musca domestica Linnaeus
Melnick et al., 1953
CALLIPHORIDAE: CHRYSOMYINAE
Cochliomyia macellaria Fabricius
Melnick et al., 1953
CALLIPHORIDAE: CALLIPHORINAE
Phaenicia pallescens (Shannon)
Melnick et al., 1953
Phaenicia sericata (Meigen)
Melnick et al., 1953
SARCOPHAGIDAE: SARCOPHAGINAE
Tricharaea Thomson
Melnick et al., 1953

Coxsackievirus Type A9 (A9) .... MUSCIDAE: FANNIINAE
Fannia Robineau-Desvoidy
Riordan et al., 1961
MUSCIDAE: PHAONIINAE
Ophyra Robineau-Desvoidy
Riordan et al., 1961
Muscina Robineau-Desvoidy
Riordan et al., 1961
MUSCIDAE: MUSCINAE
Musca domestica Linnaeus
Riordan et al., 1961
CALLIPHORIDAE: CHRYSOMYINAE
Phormia regina (Meigen)
Riordan et al., 1961
CALLIPHORIDAE: CALLIPHORINAE
Phaenicia sericata (Meigen)
Riordan et al., 1961
Calliphora Robineau-Desvoidy
Riordan et al., 1961
SARCOPHAGIDAE: SARCOPHAGINAE
Sarcophaga Meigen
Riordan et al., 1961

Coxsackievirus Type B5 (B5) .... MUSCIDAE: FANNIINAE
Fannia Robineau-Desvoidy
Riordan et al., 1961

MUSCIDAE: PHAONIINAE
Ophyra Robineau-Desvoidy
Riordan et al., 1961
Muscina Robineau-Desvoidy
Riordan et al., 1961

MUSCIDAE: MUSCINAE
Musca domestica Linnaeus
Riordan et al., 1961

CALLIPHORIDAE: CHRYSOMYINAE
Phormia regina (Meigen)
Riordan et al., 1961

CALLIPHORIDAE: CALLIPHORINAE
Phaenicia sericata (Meigen)
Riordan et al., 1961
Calliphora Robineau-Desvoidy
Riordan et al., 1961

SARCOPHAGIDAE: SARCOPHAGINAE
Sarcophaga Meigen
Riordan et al., 1961

Coxsackievirus W-S '48 (B1) ...... CALLIPHORIDAE: CHRYSOMYINAE
Phormia regina (Meigen)
Melnick et al., 1952

ECHO VIRUS

ECHO virus Type 1 .... MUSCIDAE: FANNIINAE
Fannia Robineau-Desvoidy
Riordan et al., 1961

MUSCIDAE: PHAONIINAE
Ophyra Robineau-Desvoidy
Riordan et al., 1961
Muscina Robineau-Desvoidy
Riordan et al., 1961

MUSCIDAE: MUSCINAE
Musca domestica Linnaeus
Riordan et al., 1961

CALLIPHORIDAE: CHRYSOMYINAE
Phormia regina (Meigen)
Riordan et al., 1961

CALLIPHORIDAE: CALLIPHORINAE
Phaenicia sericata (Meigen)
Riordan et al., 1961
Calliphora Robineau-Desvoidy
Riordan et al., 1961

SARCOPHAGIDAE: SARCOPHAGINAE
Sarcophaga Meigen
Riordan et al., 1961

Organism	Fly association
ECHO virus Type 5	MUSCIDAE: FANNIINAE *Fannia* Robineau-Desvoidy Riordan et al., 1961 MUSCIDAE: PHAONIINAE *Ophyra* Robineau-Desvoidy Riordan et al., 1961 *Muscina* Robineau-Desvoidy Riordan et al., 1961 MUSCIDAE: MUSCINAE *Musca domestica* Linnaeus Riordan et al., 1961 CALLIPHORIDAE: CHRYSOMYINAE *Phormia regina* (Meigen) Riordan et al., 1961 CALLIPHORIDAE: CALLIPHORINAE *Phaenicia sericata* (Meigen) Riordan et al., 1961 *Calliphora* Robineau-Desvoidy Riordan et al., 1961 SARCOPHAGIDAE: SARCOPHAGINAE *Sarcophaga* Meigen Riordan et al., 1961
ECHO virus Type 6	MUSCIDAE: MUSCINAE *Musca domestica* Linnaeus Downey, 1963 CALLIPHORIDAE: CHRYSOMYINAE *Phormia regina* (Meigen) Downey, 1963 CALLIPHORIDAE: CALLIPHORINAE *Lucilia illustris* (Meigen) Downey, 1963 *Phaenicia sericata* (Meigen) Downey, 1963 *Calliphora* Robineau-Desvoidy Downey, 1963
ECHO virus Type 7	MUSCIDAE: MUSCINAE *Musca domestica* Linnaeus Downey, 1963 CALLIPHORIDAE: CHRYSOMYINAE *Phormia regina* (Meigen) Downey, 1963 CALLIPHORIDAE: CALLIPHORINAE *Lucilia illustris* (Meigen) Downey, 1963 *Phaenicia sericata* (Meigen) Downey, 1963 *Calliphora* Robineau-Desvoidy Downey, 1963

ECHO virus Type 8 .... RICHARDIIDAE

Richardia viridiventris Wulp
Paul et al., 1962

MUSCIDAE: COENOSIINAE

Atherigona orientalis Schiner
Paul et al., 1962

MUSCIDAE: LIMNOPHORINAE

Limnophora Robineau-Desvoidy
Paul et al., 1962

MUSCIDAE: FANNIINAE

Fannia Robineau-Desvoidy
Paul et al., 1962

MUSCIDAE: PHAONIINAE

Ophyra aenescens (Wiedemann)
Paul et al., 1962

Muscina stabulans (Fallén)
Paul et al., 1962

MUSCIDAE: MUSCINAE

Graphomya maculata (Scopoli)
Paul et al., 1962

Synthesiomyia nudiseta (Wulp)
Paul et al., 1962

Neomuscina Townsend
Paul et al., 1962

Philornis Meinert
Paul et al., 1962

Musca Linnaeus
Paul et al., 1962

CALLIPHORIDAE: MESEMBRINELLINAE

Huascaromusca bicolor (Fabricius)
Paul et al., 1962

CALLIPHORIDAE: CHRYSOMYINAE

Cochliomyia macellaria (Fabricius)
Paul et al., 1962

CALLIPHORIDAE: CALLIPHORINAE

Phaenicia Robineau-Desvoidy
Paul et al., 1962

SARCOPHAGIDAE: SARCOPHAGINAE

Blaesoxipha plinthopyga (Wiedemann)
Paul et al., 1962

Oxysarcodexia ochripyga (Wulp)
Paul et al., 1962

Sarcophaga Meigen
Paul et al., 1962

Tricharaea Thomson
Paul et al., 1962

ECHO virus Type 11 .... MUSCIDAE: FANNIINAE

Fannia Robineau-Desvoidy
Riordan et al., 1961

MUSCIDAE: PHAONIINAE
Ophyra Robineau-Desvoidy
Riordan et al., 1961
Muscina Robineau-Desvoidy
Riordan et al., 1961
MUSCIDAE: MUSCINAE
Musca domestica Linnaeus
Riordan et al., 1961
CALLIPHORIDAE: CHRYSOMYINAE
Phormia regina (Meigen)
Riordan et al., 1961
CALLIPHORIDAE: CALLIPHORINAE
Phaenicia sericata (Meigen)
Riordan et al., 1961
Calliphora Robineau-Desvoidy
Riordan et al., 1961
SARCOPHAGIDAE: SARCOPHAGINAE
Sarcophaga Meigen
Riordan et al., 1961

ECHO virus Type 12 ....

RICHARDIIDAE
Richardia viridiventris Wulp
Paul et al., 1962
MUSCIDAE: COENOSIINAE
Atherigona orientalis Schiner
Paul et al., 1962
MUSCIDAE: LIMNOPHORINAE
Limnophora Robineau-Desvoidy
Paul et al., 1962
MUSCIDAE: FANNIINAE
Fannia Robineau-Desvoidy
Paul et al., 1962
MUSCIDAE: PHAONIINAE
Ophyra aenescens (Wiedemann)
Paul et al., 1962
Muscina stabulans (Fallén)
Paul et al., 1962
MUSCIDAE: MUSCINAE
Graphomya maculata (Scopoli)
Paul et al., 1962
Synthesiomyia nudiseta (Wulp)
Paul et al., 1962
Neomuscina Townsend
Paul et al., 1962
Philornis Meinert
Paul et al., 1962
Musca Linnaeus
Paul et al., 1962
CALLIPHORIDAE: MESEMBRINELLINAE
Huascaromusca bicolor (Fabricius)
Paul et al., 1962

ECHO virus Type 12 (cont.)	CALLIPHORIDAE: CHRYSOMYINAE
	Cochliomyia macellaria (Fabricius)
	Paul et al., 1962
	CALLIPHORIDAE: CALLIPHORINAE
	Phaenicia Robineau-Desvoidy
	Paul et al., 1962
	SARCOPHAGIDAE: SARCOPHAGINAE
	Blaesoxipha plinthopyga (Wiedemann)
	Paul et al., 1962
	Oxysarcodexia ochripyga (Wulp)
	Paul et al., 1962
	Sarcophaga Meigen
	Paul et al., 1962
	Tricharaea Thomson
	Paul et al., 1962
ECHO virus Type 14	RICHARDIIDAE
	Richardia viridiventris Wulp
	Paul et al., 1962
	MUSCIDAE: COENOSIINAE
	Atherigona orientalis Schiner
	Paul et al., 1962
	MUSCIDAE: LIMNOPHORINAE
	Limnophora Robineau-Desvoidy
	Paul et al., 1962
	MUSCIDAE: FANNIINAE
	Fannia Robineau-Desvoidy
	Paul et al., 1962
	MUSCIDAE: PHAONIINAE
	Ophyra aenescens (Wiedemann)
	Paul et al., 1962
	Muscina stabulans (Fallén)
	Paul et al., 1962
	MUSCIDAE: MUSCINAE
	Graphomya maculata (Scopoli)
	Paul et al., 1962
	Synthesiomyia nudiseta (Wulp)
	Paul et al., 1962
	Neomuscina Townsend
	Paul et al., 1962
	Philornis Meinert
	Paul et al., 1962
	Musca Linnaeus
	Paul et al., 1962
	CALLIPHORIDAE: MESEMBRINELLINAE
	Huascaromusca bicolor (Fabricius)
	Paul et al., 1962
	CALLIPHORIDAE: CHRYSOMYINAE
	Cochliomyia macellaria (Fabricius)
	Paul et al., 1962

ECHO virus Type 14 (cont.)	CALLIPHORIDAE: CALLIPHORINAE
	Phaenicia Robineau-Desvoidy
	Paul et al., 1962
	SARCOPHAGIDAE: SARCOPHAGINAE
	Blaesoxipha plinthopyga (Wiedemann)
	Paul et al., 1962
	Oxysarcodexia ochripyga (Wulp)
	Paul et al., 1962
	Sarcophaga Meigen
	Paul et al., 1962
	Tricharaea Thomson
	Paul et al., 1962
UNDETERMINED ENTEROVIRUSES	CALLIPHORIDAE: CHRYSOMYINAE
	Chrysomya putoria (Wiedemann)
	Brygoo et al., 1962
OTHER ANIMAL VIRUSES	
Danysz rat virus	MUSCIDAE: MUSCINAE
	Musca domestica Linnaeus
	Graham-Smith, 1910
Equine infectious anemia virus	MUSCIDAE: STOMOXYINAE
	Stomoxys calcitrans (Linnaeus)
	Flocken et al.
	Howard, 1917
	Scott, 1917, 1920, 1922
	Lührs, 1919
	Stein et al., 1942
Foot and mouth disease virus	MUSCIDAE: PHAONIINAE
	Muscina stabulans (Fallén)
	Lebailly, 1924
	MUSCIDAE: MUSCINAE
	Musca domestica Linnaeus
	Lebailly, 1924
	Kunike, 1927b
	Dhennin et al., 1961
	House fly (presumably *Musca domestica* Linnaeus)
	Kunike, 1927a
	MUSCIDAE: STOMOXYINAE
	Stomoxys calcitrans (Linnaeus)
	Kunike, 1927a, 1927b
	Schmit-Jensen, 1927
	Wilhelmi, 1927
	CALLIPHORIDAE: CALLIPHORINAE
	Phaenicia sericata (Meigen)
	Dhennin et al., 1961
Mink enteritis virus	MUSCIDAE: MUSCINAE
	Musca domestica Linnaeus
	Bouillant et al., 1965

Organism	Fly / Reference
Rous-sarcoma virus (Oncogenic)	DROSOPHILIDAE: DROSOPHILINAE *Drosophila melanogaster* Meigen Burdette et al., 1967
Southwest African horse death virus (Windhuker)	MUSCIDAE: STOMOXYINAE *Stomoxys calcitrans* (Linnaeus) Schuberg et al., 1912
Tipula iridescent virus	CALLIPHORIDAE: CALLIPHORINAE *Calliphora vomitoria* (Linnaeus) Smith et al., 1961
Tortor bovis	MUSCIDAE: MUSCINAE *Musca domestica* Linnaeus Sen, 1925-26
Tortor suis (Synonym: hog cholera virus)	MUSCIDAE: MUSCINAE *Musca domestica* Linnaeus Mohler, 1920 MUSCIDAE: STOMOXYINAE *Stomoxys calcitrans* (Linnaeus) Mohler, 1920
Vesicular stomatitis virus	MUSCIDAE: STOMOXYINAE *Stomoxys calcitrans* (Linnaeus) Ferris et al., 1955
INSECT VIRUSES	
Borrelina bombycis (Synonym: Polyhedral granules of silkworm jaundice)	MUSCIDAE: MUSCINAE *Musca domestica* Linnaeus Smuidsinovicia, 1889
Sex-ratio virus	DROSOPHILIDAE: DROSOPHILINAE *Drosophila bifasciata* Pomini Magni, 1954 Poulson et al., 1961a *Drosophila equinoxialis* Dobzhansky Malogolowkin et al., 1960 Poulson et al., 1961a *Drosophila melanogaster* Meigen Sakaguchi et al., 1960 Counce et al., 1961 Poulson et al., 1961a Rico, 1964 *Drosophila nebulosa* Sturtevant Counce et al., 1961 Poulson et al., 1961a *Drosophila paulistorum* Dobzhansky Malogolowkin, 1958 *Drosophila prosaltans* Duda Cavalcanti et al., 1957, 1958 *Drosophila willistoni* Sturtevant Malogolowkin et al., 1957, 1959, 1960 Malogolowkin, 1958 Sakaguchi et al., 1959, 1960, 1961

Organism	Fly host and references
	Poulson et al., 1959, 1961a Counce et al., 1961
Sigma virus	DROSOPHILIDAE: DROSOPHILINAE *Drosophila* Fallén Vigier, 1961 *Drosophila melanogaster* Meigen Plus, 1950, 1954 de Lestrange, 1954 Duhamel et al., 1956 L'Héritier, 1957 Bussereau, 1964 Berkaloff et al., 1965 Bregliano, 1965 Seecof, 1965, 1969 Bernard, 1968 *Drosophila ?melanogaster* Meigen L'Héritier et al., 1937 *Drosophila melanogaster* Meigen, Oregon, R. C. Brun et al., 1955a Plus, 1955 *Drosophila melanogaster* Meigen, strain ρ (rho) Brun et al., 1955b
Sigma virus, P^- strain	DROSOPHILIDAE: DROSOPHILINAE *Drosophila melanogaster* Meigen Bernard, 1964
BACTERIOPHAGES	
Bacteriophage (of *Staphylococcus muscae*)	MUSCIDAE: MUSCINAE *Musca domestica* Linnaeus Glaser, 1938
Bacteriophage (of *Streptococcus* C 55)	MUSCIDAE: MUSCINAE *Musca domestica* Linnaeus Shope, 1927
PSEUDOMONADACEAE	
Pseudomonas Migula	MUSCIDAE: MUSCINAE *Musca* Linnaeus de la Paz, 1938 *Musca domestica* Linnaeus Prado et al., 1955 *Musca domestica vicina* Macquart Floyd et al., 1953 CALLIPHORIDAE: CHRYSOMYINAE *Chrysomya* Robineau-Desvoidy de la Paz, 1938
Pseudomonas aeruginosa (Schroeter) Migula (Synonyms: *Bacillus pyocyaneus*, *Pseudomonas pyocyaneus*)	MUSCIDAE: PHAONIINAE *Muscina stabulans* (Fallén) Havlik, 1964

Pseudomonas aeruginosa (cont.)

MUSCIDAE: MUSCINAE

Musca domestica Linnaeus
Bacot, 1911a, 1911b
Graham-Smith, 1911b
Ledingham, 1911
Cox, 1912
Duncan, 1926
Ostrolenk et al., 1942a
Peppler, 1944
Vanni, 1946
Stephens, 1963
Havlik, 1964

House fly (presumably *Musca domestica* Linnaeus)
Manning, 1902

MUSCIDAE: STOMOXYINAE

Stomoxys calcitrans (Linnaeus)
Duncan, 1926

CALLIPHORIDAE: CHRYSOMYINAE

Phormia regina (Meigen)
Greenberg, 1962

Protophormia terraenovae (Robineau-Desvoidy)
Havlik, 1964

CALLIPHORIDAE: CALLIPHORINAE

Lucilia caesar (Linnaeus)
Muzzarelli, 1925
Weil et al., 1933

Phaenicia sericata (Meigen)
Weil et al., 1933
Greenberg, 1962
Havlik, 1964

Calliphora vicina Robineau-Desvoidy
Graham-Smith, 1911a, 1911b
Balzam, 1937
Greenberg, 1962
Havlik, 1964

Calliphora vomitoria (Linnaeus)
Muzzarelli, 1925

Cynomyopsis cadaverina (Robineau-Desvoidy)
Greenberg, 1962

UNDETERMINED CALLIPHORIDAE

Blow flies
Holdaway, 1932

SARCOPHAGIDAE: SARCOPHAGINAE

Sarcophaga carnaria (Linnaeus)
Muzzarelli, 1925
Petragnani, 1925

Organism	Fly / Reference
Pseudomonas eisenbergii Migula (Synonym: *Bacillus fluorescens non-liquefaciens*)	MUSCIDAE: MUSCINAE *Musca domestica* Linnaeus Cao, 1906 CALLIPHORIDAE: CALLIPHORINAE *Lucilia caesar* (Linnaeus) Cao, 1906 *Calliphora vomitoria* (Linnaeus) Cao, 1906 SARCOPHAGIDAE: SARCOPHAGINAE *Sarcophaga carnaria* (Linnaeus) Cao, 1906
Pseudomonas fluorescens Migula (Synonym: *Bacillus fluorescens liquefaciens*)	MUSCIDAE: MUSCINAE *Musca domestica* Linnaeus Cao, 1906 Lysenko et al., 1961 CALLIPHORIDAE: CHRYSOMYINAE *Protophormia terraenovae* (Robineau-Desvoidy) Lysenko et al., 1961 CALLIPHORIDAE: CALLIPHORINAE *Lucilia caesar* (Linnaeus) Cao, 1906 *Phaenicia sericata* (Meigen) Lysenko et al., 1961 *Calliphora vomitoria* (Linnaeus) Cao, 1906 SARCOPHAGIDAE: SARCOPHAGINAE *Sarcophaga carnaria* (Linnaeus) Cao, 1906
Pseudomonas ichthyosmia (Hammer) Bergey et al.	MUSCIDAE: MUSCINAE *Musca domestica* Linnaeus Hawley et al., 1951
Pseudomonas jaegeri Migula (Synonym: *Bacillus proteus fluorescens*)	MUSCIDAE: MUSCINAE *Musca domestica* Linnaeus Cao, 1906 CALLIPHORIDAE: CALLIPHORINAE *Lucilia caesar* (Linnaeus) Cao, 1906 *Calliphora vomitoria* (Linnaeus) Cao, 1906 SARCOPHAGIDAE: SARCOPHAGINAE *Sarcophaga carnaria* (Linnaeus) Cao, 1906
Pseudomonas septica Bergey et al.	MUSCIDAE: MUSCINAE *Musca nebulo* Fabricius Amonkar, 1967
Acetobacter Beijerinck (Synonym: acetic acid bacillus)	DROSOPHILIDAE: DROSOPHILINAE *Drosophila melanogaster* Meigen Baumberger, 1919

Acetobacter aceti (Beijerinck) (Synonym: *Bacillus aceti*)	DROSOPHILIDAE: DROSOPHILINAE *Drosophila melanogaster* Meigen Delcourt et al., 1910
Acetobacter pasteurianus Beijerinck (Synonym: *Bacterium pasteurianum*)	DROSOPHILIDAE: DROSOPHILINAE *Drosophila fenestrarum* Fallén Henneberg, 1902 *Drosophila funebris* (Fabricius) Henneberg, 1902
Acetobacter xylinum (Brown) (Synonym: *Bacterium xylinum*)	DROSOPHILIDAE: DROSOPHILINAE *Drosophila fenestrarum* Fallén Henneberg, 1902 *Drosophila funebris* (Fabricius) Henneberg, 1902
SPIRILLACEAE	
Vibrio Müller	MUSCIDAE: MUSCINAE *Musca* Linnaeus de la Paz, 1938 *Musca domestica* Linnaeus Smuidsinovicia, 1889 CALLIPHORIDAE: CHRYSOMYINAE *Chrysomya* Robineau-Desvoidy de la Paz, 1938
Vibrio comma (Schroeter) Winslow et al. (Synonyms: *Vibrio cholerae*, Cholera, comma bacillus)	PHORIDAE: METOPININAE *Megaselia ferruginea* (Brunetti) Roberg, 1915 SYRPHIDAE: MILESIINAE *Eristalis tenax* (Linnaeus) Maddox, 1885b ANTHOMYIIDAE: ANTHOMYIINAE *Hylemya* Robineau-Desvoidy Anonymous, 1952 MUSCIDAE: FANNIINAE *Fannia* Robineau-Desvoidy Savchenko, 1892a, 1892b MUSCIDAE: MUSCINAE *Musca* Linnaeus de la Paz, 1938 Lal et al., 1939 *Musca domestica* Linnaeus Chantemesse et al., 1905 Ganon, 1908 Graham-Smith, 1910 Alessandrini, 1912 Duncan, 1926 Shope, 1927 Dishon, 1956 House fly (presumably *Musca domestica* Linnaeus) Cattani, 1886 Celli et al., 1888

Craig, 1894
Alessandrini et al., 1912
Flu, 1915

MUSCIDAE: STOMOXYINAE
Stomoxys calcitrans (Linnaeus)
Duncan, 1926

CALLIPHORIDAE: CALLIPHORINAE
Lucilia caesar (Linnaeus)
Alessandrini, 1912
Calliphora vicina Robineau-Desvoidy
Graham-Smith, 1911a, 1911b
Alessandrini, 1912
Calliphora vomitoria (Linnaeus)
Maddox, 1885a, 1885b
Alessandrini, 1912

UNDETERMINED CALLIPHORIDAE
Blue bottle flies
Savchenko, 1892a, 1892b

SARCOPHAGIDAE: SARCOPHAGINAE
Sarcophaga Meigen
Pavan, 1949
Sarcophaga carnaria (Linnaeus)
Alessandrini, 1912
Petragnani, 1925

CHLAMYDOBACTERIACEAE

Leptothrix Kützing .... MUSCIDAE: MUSCINAE
House fly (presumably *Musca domestica* Linnaeus)
Wenyon et al., 1917b

ACHROMOBACTERACEAE

Alcaligenes Castellani and Chalmers (Synonym: *Alkaligenes*) .... MUSCIDAE: MUSCINAE
Musca Linnaeus
de la Paz, 1938
Musca domestica Linnaeus
Prado et al., 1955
Musca domestica vicina Macquart
Floyd et al., 1953

MUSCIDAE: STOMOXYINAE
Haematobia irritans (Linnaeus)
Stirrat et al., 1955

CALLIPHORIDAE: CHRYSOMYINAE
Chrysomya Robineau-Desvoidy
de la Paz, 1938

SARCOPHAGIDAE: SARCOPHAGINAE
Sarcophaga Meigen
de la Paz, 1938

Alcaligenes faecalis Castellani and Chalmers ...... PIOPHILIDAE
Piophila casei (Linnaeus)
Gregor et al., 1960
Lysenko et al., 1961

Alcaligenes faecalis (cont.)

SPHAEROCERIDAE
Leptocera ferruginata (Stenhammar)
Lysenko, 1958
MUSCIDAE: FANNIINAE
Fannia canicularis (Linnaeus)
Gregor et al., 1960
MUSCIDAE: PHAONIINAE
Hydrotaea dentipes (Fabricius)
Lysenko, 1958
Gregor et al., 1960
Hydrotaea irritans (Fallén)
Gregor et al., 1960
Ophyra leucostoma (Wiedemann)
Gregor et al., 1960
Muscina stabulans (Fallén)
Gregor et al., 1960
Havlik, 1964
MUSCIDAE: MUSCINAE
Musca domestica Linnaeus
Hamilton, 1903
Testi, 1909
Lysenko et al., 1961
Havlik, 1964
Musca domestica vicina Macquart
Shterngol'd, 1949
CALLIPHORIDAE: CHRYSOMYINAE
Protophormia terraenovae (Robineau-Desvoidy)
Lysenko, 1958
Gregor et al., 1960
Lysenko et al., 1961
Havlik, 1964
CALLIPHORIDAE: CALLIPHORINAE
Lucilia caesar (Linnaeus)
Weil et al., 1933
Gregor et al., 1960
Phaenicia sericata (Meigen)
Michelbacher et al., 1932
Weil et al., 1933
Gregor et al., 1960
Havlik, 1964
Calliphora vicina Robineau-Desvoidy
Gregor et al., 1960
Povolný et al., 1961
Havlik, 1964
Calliphora vomitoria (Linnaeus)
Gregor et al., 1960

Achromobacter Bergey, Harrison, Breed, Hammer, and Huntoon ....

MUSCIDAE: MUSCINAE
Musca domestica Linnaeus
Prado et al., 1955

Organism	Fly
Achromobacter (cont.)	MUSCIDAE: STOMOXYINAE *Haematobia irritans* (Linnaeus) Stirrat et al., 1955
Flavobacterium Bergey et al.	MUSCIDAE: MUSCINAE *Musca domestica* Linnaeus Lysenko et al., 1961 CALLIPHORIDAE: CHRYSOMYINAE *Protophormia terraenovae* (Robineau-Desvoidy) Lysenko et al., 1961 CALLIPHORIDAE: CALLIPHORINAE *Phaenicia sericata* (Meigen) Lysenko et al., 1961
Flavobacterium devorans (Zimmermann) Bergey et al.	PIOPHILIDAE *Piophila casei* (Linnaeus) Gregor et al., 1960 MUSCIDAE: FANNIINAE *Fannia canicularis* (Linnaeus) Gregor et al., 1960 MUSCIDAE: PHAONIINAE *Hydrotaea dentipes* (Fabricius) Lysenko, 1958 Gregor et al., 1960 *Hydrotaea irritans* (Fallén) Gregor et al., 1960 *Ophyra leucostoma* (Wiedemann) Gregor et al., 1960 *Muscina stabulans* (Fallén) Gregor et al., 1960 MUSCIDAE: MUSCINAE *Musca domestica* Linnaeus Lysenko, 1958 CALLIPHORIDAE: CHRYSOMYINAE *Protophormia terraenovae* (Robineau-Desvoidy) Gregor et al., 1960 CALLIPHORIDAE: CALLIPHORINAE *Lucilia caesar* (Linnaeus) Gregor et al., 1960 *Phaenicia sericata* (Meigen) Gregor et al., 1960 *Calliphora vicina* Robineau-Desvoidy Gregor et al., 1960 *Calliphora vomitoria* (Linnaeus) Gregor et al., 1960
Flavobacterium invisible (Vaughan) Bergey et al.	PIOPHILIDAE *Piophila casei* (Linnaeus) Gregor et al., 1960

Organism	Fly / Reference
Flavobacterium invisible (cont.)	SPHAEROCERIDAE *Leptocera ferruginata* (Stenhammar) Lysenko, 1958 MUSCIDAE: FANNIINAE *Fannia canicularis* (Linnaeus) Gregor et al., 1960 MUSCIDAE: PHAONIINAE *Hydrotaea dentipes* (Fabricius) Gregor et al., 1960 *Hydrotaea irritans* (Fallén) Gregor et al., 1960 *Ophyra leucostoma* (Wiedemann) Gregor et al., 1960 *Muscina stabulans* (Fallén) Gregor et al., 1960 CALLIPHORIDAE: CHRYSOMYINAE *Protophormia terraenovae* (Robineau-Desvoidy) Gregor et al., 1960 CALLIPHORIDAE: CALLIPHORINAE *Lucilia caesar* (Linnaeus) Gregor et al., 1960 *Phaenicia sericata* (Meigen) Gregor et al., 1960 *Calliphora vicina* Robineau-Desvoidy Gregor et al., 1960 *Calliphora vomitoria* (Linnaeus) Gregor et al., 1960
ENTEROBACTERIACEAE *Escherichia* Castellani and Chalmers (Synonyms: MacConkey's bacillus No. 1, MacConkey's bacillus No. 33a, MacConkey's bacillus No. 66, MacConkey's bacillus No. 66a, MacConkey's bacillus No. 71, MacConkey's bacillus No. 101, MacConkey's bacillus No. 106, MacConkey's bacillus No. 106a)	PIOPHILIDAE *Piophila casei* (Linnaeus) Lysenko et al., 1961 MUSCIDAE: FANNIINAE *Fannia scalaris* (Fabricius) Zmeev, 1943 MUSCIDAE: MUSCINAE *Musca* Linnaeus de la Paz, 1938 *Musca domestica* Linnaeus Nicoll, 1911a Prado et al., 1955 Lysenko et al., 1961 CALLIPHORIDAE: CHRYSOMYINAE *Chrysomya* Robineau-Desvoidy de la Paz, 1938 *Protophormia terraenovae* (Robineau-Desvoidy) Lysenko et al., 1961 CALLIPHORIDAE: CALLIPHORINAE *Phaenicia sericata* (Meigen) Lysenko et al., 1961

Organism	Fly association
Escherichia (cont.)	SARCOPHAGIDAE: SARCOPHAGINAE *Sarcophaga* Meigen de la Paz, 1938
Escherichia anaerogenes (Chester) Bergey et al. (Synonym: *Bacterium coli anaerogenes*)	MUSCIDAE: MUSCINAE *Musca domestica* Linnaeus Scott, 1917a Jones, 1941 Hawley et al., 1951 SARCOPHAGIDAE: SARCOPHAGINAE *Sarcophaga hirtipes* Wiedemann Zmeev, 1943
Escherichia anindolica (Chester) Bergey et al.	MUSCIDAE: PHAONIINAE *Phaonia querceti* (Bouché) Zmeev, 1943 SARCOPHAGIDAE: SARCOPHAGINAE *Sarcophaga hirtipes* Wiedemann Zmeev, 1943
Escherichia coli (Migula) Castellani and Chalmers (Synonyms: *Bacterium coli,* MacConkey's bacillus No. 35, *B. schafferi, B. coli, Bacillus coli*)	OTITIDAE: ULIDIINAE *Physiphora demandata* (Fabricius) Povolný et al., 1961 SEPSIDAE *Sepsis punctum* (Fabricius) Povolný et al., 1961 PIOPHILIDAE *Piophila casei* (Linnaeus) Beyer, 1925 Gregor et al., 1960 Povolný et al., 1961 *Piophila pectiniventris* Duda Lysenko, 1958 SPHAEROCERIDAE *Copromyza atra* (Meigen) Povolný et al., 1961 *Leptocera ferruginata* (Stenhammar) Lysenko, 1958 *Limosina punctipennis* (Wiedemann) Nicholls, 1912 UNDETERMINED SPHAEROCERIDAE Povolný et al., 1961 UNDETERMINED DROSOPHILIDAE Povolný et al., 1961 ANTHOMYIIDAE: SCATOPHAGINAE *Scatophaga stercoraria* (Linnaeus) Povolný et al., 1961 ANTHOMYIIDAE: ANTHOMYIINAE *Hylemya cinerella* (Fallén) Povolný et al., 1961 *Anthomyia pluvialis* (Linnaeus) Povolný et al., 1961

Escherichia coli (cont.)

MUSCIDAE: MYDAEINAE

Helina Robineau-Desvoidy
  Graham-Smith, 1909
Mydaea urbana (Meigen)
  Povolný et al., 1961

MUSCIDAE: FANNIINAE

Fannia canicularis (Linnaeus)
  Yao et al., 1929
  Gregor et al., 1960
  Povolný et al., 1961
Fannia incisurata (Zetterstedt)
  Lysenko, 1958
Fannia scalaris (Fabricius)
  Yao et al., 1929
  Povolný et al., 1961

MUSCIDAE: PHAONIINAE

Hydrotaea dentipes (Fabricius)
  Lysenko, 1958
  Gregor et al., 1960
  Povolný et al., 1961
Hydrotaea irritans (Fallén)
  Gregor et al., 1960
Hydrotaea occulta (Meigen)
  Povolný et al., 1961
Ophyra leucostoma (Wiedemann)
  Gregor et al., 1960
  Povolný et al., 1961
Lasiops simplex (Wiedemann)
  Povolný et al., 1961
Phaonia incana (Wiedemann)
  Hoffmann, 1950
Phaonia querceti (Bouché)
  Zmeev, 1943
Muscina stabulans (Fallén)
  Beyer, 1925
  Zmeev, 1943
  Gregor et al., 1960
  Povolný et al., 1961
  Havlik, 1964

MUSCIDAE: MUSCINAE

Synthesiomyia nudiseta (Wulp)
  Beyer, 1925
Orthellia caerulea Wiedemann
  Shimizu et al., 1965
Musca domestica Linnaeus
  Hamilton, 1903
  Graham-Smith, 1909
  Nicoll, 1911a
  Tebbutt, 1913
  Scott, 1917a

Escherichia coli (cont.)

Reinstorf, 1923
Beyer, 1925
Shope, 1927
Yao et al., 1929
Parisot et al., 1934
Jones, 1941
Ostrolenk et al., 1942a
Vanni, 1946
Emmel, 1949
Hoffmann, 1950
Hawley et al., 1951
Gerberich, 1951
Gabaldón, 1955
Cova Garcia, 1956
Ingram et al., 1956
Radvan, 1956
Coutinho et al., 1957
Radvan, 1960a, 1960b, 1960c
Povolný et al., 1961
Greenberg et al., 1963a
Havlik, 1964

House fly (presumably *Musca domestica* Linnaeus)
Purdy, 1906
Jenkins et al., 1954

Musca domestica vicina Macquart
Floyd et al., 1953
Silverman et al., 1953
Levinson, 1960

MUSCIDAE: STOMOXYINAE

Haemotobia irritans (Linnaeus)
Stirrat et al., 1955

Stomoxys calcitrans (Linnaeus)
Povolný et al., 1961
Love et al., 1965

CALLIPHORIDAE: CHRYSOMYINAE

Cochliomyia macellaria (Fabricius)
Beyer, 1925

Chrysomya megacephala (Fabricius)
Chow, 1940
Shimizu et al., 1965

Chrysomya pinguis (Walker)
Shimizu et al., 1965

Phormia regina (Meigen)
Beyer, 1925
Greenberg, 1962

Protophormia terraenovae (Robineau-Desvoidy)
Pavillard et al., 1957
Lysenko, 1958

Escherichia coli (cont.)

Gregor et al., 1960
Radvan, 1960a, 1960b, 1960c
Havlik, 1964

CALLIPHORIDAE: CALLIPHORINAE

Lucilia caesar (Linnaeus)
Beyer, 1925
Yao et al., 1929
Gregor et al., 1960

Phaenicia cuprina Wiedemann
Shimizu et al., 1965

Phaenicia sericata (Meigen)
Michelbacher et al., 1932
Hobson, 1932b, 1935
Zmeev, 1943
Gregor et al., 1960
Povolný et al., 1961
Greenberg, 1962
Havlik, 1964

Aldrichina grahami (Aldrich)
Shimizu et al., 1965

Calliphora vicina Robineau-Desvoidy
Yao et al., 1929
Parisot et al., 1934
Balzam, 1937
Hoffmann, 1950
Gregor et al., 1960
Povolný et al., 1961
Greenberg, 1962
Greenberg et al., 1963c
Havlik, 1964

Calliphora vomitoria (Linnaeus)
Wollman, 1911
Yao et al., 1929
Hoffmann, 1950
Gregor et al., 1960
Povolný et al., 1961

Cynomyopsis cadaverina (Robineau-Desvoidy)
Greenberg, 1962

UNDETERMINED LUCILIINI
Shimizu et al., 1965

SARCOPHAGIDAE: SARCOPHAGINAE

Sarcophaga Meigen
Pavan, 1949

Sarcophaga argyrostoma (Robineau-Desvoidy)
Beyer, 1925

Sarcophaga carnaria (Linnaeus)
Testi, 1909
Petragnani, 1925

Organism	Fly associations
Escherichia coli (cont.)	Yao et al., 1929 Hoffmann, 1950 Povolný et al., 1961 *Sarcophaga haemorrhoidalis* (Fallén) Zmeev, 1943 Povolný et al., 1961 *Sarcophaga hirtipes* Wiedemann Zmeev, 1943 *Sarcophaga melanura* Meigen Povolný et al., 1961 UNDETERMINED SARCOPHAGIDAE Shimizu et al., 1965
Escherichia coli (Migula) Castellani and Chalmers 25	CALLIPHORIDAE: CALLIPHORINAE *Calliphora vicina* Robineau-Desvoidy Greenberg et al., 1963c
Escherichia coli (Migula) Castellani and Chalmers 26 B6	CALLIPHORIDAE: CHRYSOMYINAE *Chrysomya putoria* (Wiedemann) Brygoo et al., 1962
Escherichia coli (Migula) Castellani and Chalmers 086-B7	MUSCIDAE: MUSCINAE *Musca domestica* Linnaeus Coutinho et al., 1957
Escherichia coli (Migula) Castellani and Chalmers 111 B4	CALLIPHORIDAE: CHRYSOMYINAE *Chrysomya putoria* (Wiedemann) Brygoo et al., 1962
Escherichia coli var. *acidilactici* (Topley and Wilson) Yale (Synonyms: *Escherichia paraacidilactici* A, *Escherichia paraacidilactici* B, MacConkey's bacillus No. 2, *B. acidi lactici*, *Escherichia paraacidilactici*)	MUSCIDAE: MYDAEINAE *Helina* Robineau-Desvoidy Graham-Smith, 1909 MUSCIDAE: FANNIINAE *Fannia canicularis* (Linnaeus) Graham-Smith, 1909 Zmeev, 1943 *Fannia leucosticta* (Meigen) Zmeev, 1943 *Fannia scalaris* (Fabricius) Zmeev, 1943 MUSCIDAE: PHAONIINAE *Ophyra leucostoma* (Wiedemann) Shura Bura, 1952 *Phaonia querceti* (Bouché) Zmeev, 1943 MUSCIDAE: MUSCINAE *Musca domestica* Linnaeus Graham-Smith, 1909 Nicoll, 1911a Cox et al., 1912 Scott, 1917a *Musca larvipara* Porchinskiĭ Zmeev, 1943

Organism	Fly / Reference
Escherichia coli var. *acidilactici* (cont.)	CALLIPHORIDAE: CALLIPHORINAE *Phaenicia sericata* (Meigen) Zmeev, 1943 *Calliphora vicina* Robineau-Desvoidy Shura Bura, 1952 SARCOPHAGIDAE: SARCOPHAGINAE *Sarcophaga haemorrhoidalis* (Fallén) Zmeev, 1943 *Sarcophaga hirtipes* Wiedemann Zmeev, 1943
Escherichia coli var. *communior* (Topley and Wilson) Yale (Synonym: *Escherichia communior*)	SEPSIDAE *Sepsis violacea* Meigen Beyer, 1925 *Helina* Robineau-Desvoidy Graham-Smith, 1909 MUSCIDAE: FANNIINAE *Fannia canicularis* (Linnaeus) Graham-Smith, 1909 *Fannia scalaris* (Fabricius) Zmeev, 1943 MUSCIDAE: PHAONIINAE *Ophyra aenescens* (Wiedemann) Beyer, 1925 MUSCIDAE: MUSCINAE *Musca domestica* Linnaeus Graham-Smith, 1909 Scott, 1917a MUSCIDAE: STOMOXYINAE *Stomoxys calcitrans* (Linnaeus) Graham-Smith, 1909 CALLIPHORIDAE: CALLIPHORINAE *Lucilia caesar* (Linnaeus) Beyer, 1925 Weil et al., 1933 *Phaenicia sericata* (Meigen) Weil et al., 1933 *Calliphora vicina* Robineau-Desvoidy Graham-Smith, 1909 *Calliphora vomitoria* (Linnaeus) Graham-Smith, 1909 SARCOPHAGIDAE: SARCOPHAGINAE *Sarcophaga haemorrhoidalis* (Fallén) Zmeev, 1943 *Sarcophaga hirtipes* Wiedemann Zmeev, 1943
Escherichia coli var. *communis* (Escherich) Breed (Synonyms: *Bacillus coli communis*, MacConkey's bacillus No. 34)	MUSCIDAE: MUSCINAE *Musca domestica* Linnaeus Nicoll, 1911a Cox et al., 1912 Buchanan, 1913

Organism	Fly and reference
Escherichia coli var. *communis* (cont.)	Duncan, 1926
	Tarasov et al., 1941
	House fly (presumably *Musca domestica* Linnaeus)
	Manning, 1902
	MUSCIDAE: STOMOXYINAE
	Stomoxys calcitrans (Linnaeus)
	Duncan, 1926
Escherichia coli var. *neapolitana* (Topley and Wilson) Yale (Synonyms: MacConkey's bacillus No. 72, *Bacillus neapolitanus*)	MUSCIDAE: MYDAEINAE
	Helina Robineau-Desvoidy
	Graham-Smith, 1909
	MUSCIDAE: FANNIINAE
	Fannia canicularis (Linnaeus)
	Graham-Smith, 1909
	MUSCIDAE: MUSCINAE
	Musca domestica Linnaeus
	Graham-Smith, 1909
	Nicoll, 1911a
	Cox et al., 1912
	Buchanan, 1913
	MUSCIDAE: STOMOXYINAE
	Stomoxys calcitrans (Linnaeus)
	Graham-Smith, 1909
	CALLIPHORIDAE: CALLIPHORINAE
	Calliphora vicina Robineau-Desvoidy
	Graham-Smith, 1909
	Calliphora vomitoria (Linnaeus)
	Graham-Smith, 1909
Escherichia freundii (Braak) Yale	ANTHOMYIIDAE: FUCELLINAE
	Fucellia rufitibia Stein
	O'Keefe et al., 1954
	ANTHOMYIIDAE: ANTHOMYIINAE
	Hylemya antiqua (Meigen)
	O'Keefe et al., 1954
	MUSCIDAE: PHAONIINAE
	Muscina stabulans (Fallén)
	O'Keefe et al., 1954
	MUSCIDAE: STOMOXYINAE
	Haematobia irritans (Linnaeus)
	Stirrat et al., 1955
	Stomoxys calcitrans (Linnaeus)
	Love et al., 1965
	CALLIPHORIDAE: CHRYSOMYINAE
	Phormia regina (Meigen)
	O'Keefe et al., 1954
Escherichia intermedia (Werkman and Gillen) Vaughn and Levine	MUSCIDAE: FANNIINAE
	Fannia canicularis (Linnaeus)
	Zmeev, 1943
	O'Keefe et al., 1954

Organism	Fly
(Synonyms: *Citrobacter intermedium, Escherichia intermedium*)	*Fannia leucosticta* (Meigen) Zmeev, 1943 *Fannia scalaris* (Fabricius) Zmeev, 1943 O'Keefe et al., 1954 MUSCIDAE: STOMOXYINAE *Haematobia irritans* (Linnaeus) Stirrat et al., 1955 *Stomoxys calcitrans* (Linnaeus) Love et al., 1965 SARCOPHAGIDAE: SARCOPHAGINAE *Sarcophaga haemorrhoidalis* (Fallén) Zmeev, 1943
Aerobacter Beijerinck (Synonyms: MacConkey's bacillus No. 67, MacConkey's bacillus No. 75, MacConkey's bacillus No. 102, MacConkey's bacillus No. 108a)	MUSCIDAE: FANNIINAE *Fannia leucosticta* (Meigen) Zmeev, 1943 *Fannia scalaris* (Fabricius) Zmeev, 1943 MUSCIDAE: PHAONIINAE *Muscina stabulans* (Fallén) Zmeev, 1943 MUSCIDAE: MUSCINAE *Musca* Linnaeus de la Paz, 1938 *Musca domestica* Linnaeus Nicoll, 1911a CALLIPHORIDAE: CHRYSOMYINAE *Chrysomya* Robineau-Desvoidy de la Paz, 1938 *Phormia regina* (Meigen) Knuckles, 1959 CALLIPHORIDAE: CALLIPHORINAE *Phaenicia sericata* (Meigen) Zmeev, 1943 SARCOPHAGIDAE: SARCOPHAGINAE *Sarcophaga* Meigen de la Paz, 1938 *Sarcophaga hirtipes* Wiedemann Zmeev, 1943
Aerobacter aerogenes (Kruse) Beijerinck (Synonym: *Bacillus lactis aerogenes*)	PIOPHILIDAE *Piophila casei* (Linnaeus) Gregor et al., 1960 MUSCIDAE: MYDAEINAE *Helina* Robineau-Desvoidy Graham-Smith, 1909 MUSCIDAE: FANNIINAE *Fannia canicularis* (Linnaeus) Zmeev, 1943 O'Keefe et al., 1954

Aerobacter aerogenes (cont.)

Gregor et al., 1960
Povolný et al., 1961
Fannia leucosticta (Meigen)
Zmeev, 1943
Fannia scalaris (Fabricius)
Zmeev, 1943

MUSCIDAE: PHAONIINAE
Hydrotaea dentipes (Fabricius)
Gregor et al., 1960
Hydrotaea irritans (Fallén)
Gregor et al., 1960
Ophyra leucostoma (Wiedemann)
Gregor et al., 1960
Muscina stabulans (Fallén)
Gregor et al., 1960
Havlik, 1964

MUSCIDAE: MUSCINAE
Musca domestica Linnaeus
Graham-Smith, 1909
Cox et al., 1912
Buchanan, 1913
Ostrolenk et al., 1942a
Hawley et al., 1951
Greenberg, 1959b
Povolný et al., 1961
Greenberg et al., 1963a
Havlik, 1964

MUSCIDAE: STOMOXYINAE
Haematobia irritans (Linnaeus)
Stirrat et al., 1955
Stomoxys calcitrans (Linnaeus)
Love et al., 1965

CALLIPHORIDAE: CHRYSOMYINAE
Protophormia terraenovae (Robineau-Desvoidy)
Gregor et al., 1960
Havlik, 1964

CALLIPHORIDAE: CALLIPHORINAE
Lucilia caesar (Linnaeus)
Gregor et al., 1960
Phaenicia sericata (Meigen)
Gregor et al., 1960
Havlik, 1964
Calliphora vicina Robineau-Desvoidy
Graham-Smith, 1909
Gregor et al., 1960
Greenberg, 1962
Havlik, 1964
Calliphora vomitoria (Linnaeus)
Graham-Smith, 1909

Organism	Fly associations
Aerobacter aerogenes (cont.)	Gregor et al., 1960 SARCOPHAGIDAE: SARCOPHAGINAE *Sarcophaga haemorrhoidalis* (Fallén) Zmeev, 1943
Aerobacter cloacae (Jordan) Bergey et al. (Synonyms: MacConkey's bacillus No. 108, *Bacillus cloacae*)	MUSCIDAE: MUSCINAE *Musca domestica* Linnaeus Nicoll, 1911a Povolný et al., 1961 MUSCIDAE: STOMOXYINAE *Haematobia irritans* (Linnaeus) Stirrat et al., 1955 *Stomoxys calcitrans* (Linnaeus) Love et al., 1965 CALLIPHORIDAE: CALLIPHORINAE *Phaenicia sericata* (Meigen) Michelbacher et al., 1932 *Calliphora vicina* Robineau-Desvoidy Povolný et al., 1961
Aerobacter oxytocum (Trevisan) Bergey et al. (Synonyms: MacConkey's bacillus No. 65, *Bacillus oxytocus perniciosus*)	MUSCIDAE: MUSCINAE *Musca domestica* Linnaeus Nicoll, 1911a
Klebsiella Trevisan	PIOPHILIDAE *Piophila casei* (Linnaeus) Lysenko et al., 1961 MUSCIDAE: MUSCINAE *Musca domestica* Linnaeus Prado et al., 1955 Lysenko et al., 1961 CALLIPHORIDAE: CHRYSOMYINAE *Protophormia terraenovae* (Robineau-Desvoidy) Lysenko et al., 1961 CALLIPHORIDAE: CALLIPHORINAE *Phaenicia sericata* (Meigen) Lysenko et al., 1961
Klebsiella cloacae (Jordan) Brisou	PIOPHILIDAE *Piophila casei* (Linnaeus) Gregor et al., 1960 SPHAEROCERIDAE *Leptocera ferruginata* (Stenhammar) Lysenko, 1958 MUSCIDAE: FANNIINAE *Fannia canicularis* (Linnaeus) Gregor et al., 1960 MUSCIDAE: PHAONIINAE *Hydrotaea dentipes* (Fabricius) Gregor et al., 1960

Organism	Fly and reference
Klebsiella cloacae (cont.)	*Hydrotaea irritans* (Fallén) Gregor et al., 1960 *Ophyra leucostoma* (Wiedemann) Gregor et al., 1960 *Muscina stabulans* (Fallén) Gregor et al., 1960 MUSCIDAE: MUSCINAE *Musca domestica* Linnaeus Lysenko, 1958 CALLIPHORIDAE: CHRYSOMYINAE *Protophormia terraenovae* (Robineau-Desvoidy) Gregor et al., 1960 CALLIPHORIDAE: CALLIPHORINAE *Lucilia caesar* (Linnaeus) Gregor et al., 1960 *Phaenicia sericata* (Meigen) Gregor et al., 1960 *Calliphora vicina* Robineau-Desvoidy Gregor et al., 1960 *Calliphora vomitoria* (Linnaeus) Gregor et al., 1960
Klebsiella pneumoniae (Schroeter) Trevisan (Synonyms: *Bacillus pneumoniae*, Friedlander's bacillus)	MUSCIDAE: MUSCINAE *Musca domestica* Linnaeus Buchanan, 1913 Shope, 1927
Klebsiella ?pneumoniae (Schroeter) Trevisan (Synonym: *Bacillus mucosus capsulatus*)	MUSCIDAE: MUSCINAE *Musca domestica* Linnaeus Testi, 1909
Cloaca Castellani and Chalmers	PIOPHILIDAE *Piophila casei* (Linnaeus) Lysenko et al., 1961 MUSCIDAE: FANNIINAE *Fannia canicularis* (Linnaeus) Zmeev, 1943 *Fannia leucosticta* (Meigen) Zmeev, 1943 *Fannia scalaris* (Fabricius) Zmeev, 1943 MUSCIDAE: PHAONIINAE *Phaonia querceti* (Bouché) Zmeev, 1943 *Muscina stabulans* (Fallén) Zmeev, 1943 MUSCIDAE: MUSCINAE *Musca domestica* Linnaeus Lysenko et al., 1961

Organism	Fly association
Cloaca (cont.)	CALLIPHORIDAE: CALLIPHORINAE *Phaenicia sericata* (Meigen) Zmeev, 1943 Lysenko et al., 1961 SARCOPHAGIDAE: SARCOPHAGINAE *Sarcophaga haemorrhoidalis* (Fallén) Zmeev, 1943 *Sarcophaga hirtipes* Wiedemann Zmeev, 1943
Paracolobactrum Borman, Stuart, and Wheeler	MUSCIDAE: MUSCINAE *Musca domestica* Linnaeus Prado et al., 1955 *Musca domestica vicina* Macquart Shterngol'd, 1949
Paracolobactrum aerogenoides Borman, Stuart, and Wheeler (Synonym: *Paraaerobacter*)	ANTHOMYIIDAE: FUCELLINAE *Fucellia rufitibia* Stein O'Keefe et al., 1954 ANTHOMYIIDAE: ANTHOMYIINAE *Hylemya antiqua* (Meigen) O'Keefe et al., 1954 MUSCIDAE: FANNIINAE *Fannia canicularis* (Linnaeus) Zmeev, 1943 *Fannia leucosticta* (Meigen) Zmeev, 1943 MUSCIDAE: PHAONIINAE *Muscina stabulans* (Fallén) O'Keefe et al., 1954 CALLIPHORIDAE: CALLIPHORINAE *Phaenicia sericata* (Meigen) Zmeev, 1943 *Calliphora vomitoria* (Linnaeus) O'Keefe et al., 1954
Paracolobactrum coliforme Borman, Stuart, and Wheeler (Synonym: *Bacterium paracoli*)	MUSCIDAE: MUSCINAE *Musca domestica* Linnaeus Jones, 1941 Radvan, 1960a, 1960b CALLIPHORIDAE: CHRYSOMYINAE *Protophormia terraenovae* (Robineau-Desvoidy) Radvan, 1960a, 1960b
Paracolobactrum intermedium Borman, Stuart, and Wheeler (Synonym: Coliform intermediates)	ANTHOMYIIDAE: FUCELLINAE *Fucellia rufitibia* Stein O'Keefe et al., 1954 MUSCIDAE: MUSCINAE *Musca domestica* Linnaeus Prado et al., 1955
Erwinia amylovora (Burrill) Winslow et al.	DROSOPHILIDAE: DROSOPHILINAE *Drosophila melanogaster* Meigen Ark et al., 1936

Erwinia amylovora (cont.)

MUSCIDAE: MUSCINAE
Musca domestica Linnaeus
  Ark et al., 1936
CALLIPHORIDAE: CALLIPHORINAE
Phaenicia sericata (Meigen)
  Ark et al., 1936

Serratia Bizio ....

OTITIDAE: ULIDIINAE
Physiphora demandata (Fabricius)
  Povolný et al., 1961
SEPSIDAE
Sepsis punctum (Fabricius)
  Povolný et al., 1961
PIOPHILIDAE
Piophila casei (Linnaeus)
  Povolný et al., 1961
UNDETERMINED SPHAEROCERIDAE
  Povolný et al., 1961
UNDETERMINED DROSOPHILIDAE
  Povolný et al., 1961
ANTHOMYIIDAE: SCATOPHAGINAE
Scatophaga stercoraria (Linnaeus)
  Povolný et al., 1961
ANTHOMYIIDAE: ANTHOMYIINAE
Hylemya cinerella (Fallén)
  Povolný et al., 1961
Anthomyia pluvialis (Linnaeus)
  Povolný et al., 1961
MUSCIDAE: MYDAEINAE
Mydaea urbana (Meigen)
  Povolný et al., 1961
MUSCIDAE: FANNIINAE
Fannia scalaris (Fabricius)
  Povolný et al., 1961
MUSCIDAE: PHAONIINAE
Hydrotaea dentipes (Fabricius)
  Povolný et al., 1961
Hydrotaea occulta (Meigen)
  Povolný et al., 1961
Lasiops simplex (Wiedemann)
  Povolný et al., 1961
Muscina stabulans (Fallén)
  Povolný et al., 1961
MUSCIDAE: MUSCINAE
Musca domestica Linnaeus
  Prado et al., 1955
  Povolný et al., 1961
MUSCIDAE: STOMOXYINAE
Stomoxys calcitrans (Linnaeus)
  Povolný et al., 1961

Serratia (cont.)

CALLIPHORIDAE: CALLIPHORINAE
Phaenicia sericata (Meigen)
Povolný et al., 1961
SARCOPHAGIDAE: SARCOPHAGINAE
Sarcophaga carnaria (Linnaeus)
Povolný et al., 1961
Sarcophaga haemorrhoidalis (Fallén)
Povolný et al., 1961
Sarcophaga melanura Meigen
Povolný et al., 1961

Serratia marcescens Bizio (Synonyms: *Bacterium prodigiosum*, *Micrococcus prodigiosus*, *Bacillus prodigiosus*)

SEPSIDAE
Sepsis Fallén
Nicholls, 1912
PIOPHILIDAE
Piophila casei (Linnaeus)
Gregor et al., 1960
SPHAEROCERIDAE
Limosina punctipennis (Wiedemann)
Nicholls, 1912
DROSOPHILIDAE: DROSOPHILINAE
Drosophila fenestrarum Fallén
Henneberg, 1902
Drosophila funebris (Fabricius)
Henneberg, 1902
Drosophila melanogaster Meigen
Nicholls, 1912
MUSCIDAE: FANNIINAE
Fannia canicularis (Linnaeus)
Gregor et al., 1960
MUSCIDAE: PHAONIINAE
Hydrotaea dentipes (Fabricius)
Gregor et al., 1960
Hydrotaea irritans (Fallén)
Gregor et al., 1960
Ophyra leucostoma (Wiedemann)
Gregor et al., 1960
Muscina stabulans (Fallén)
Gregor et al., 1960
Havlik, 1964
MUSCIDAE: MUSCINAE
Musca domestica Linnaeus
Graham-Smith, 1910, 1911a, 1911b
Orton et al., 1910
Herms, 1911
Ledingham, 1911
Duncan, 1926
Bychkov, 1932
Peppler, 1944
Vanni, 1946
Gerberich, 1951

Serratia marcescens (cont.)

Radvan, 1956, 1960a, 1960c
Stephens, 1963
Havlik, 1964

House fly (presumably *Musca domestica* Linnaeus)
Marpmann, 1884
Manning, 1902
Terry, 1912

MUSCIDAE: STOMOXYINAE

Stomoxys calcitrans (Linnaeus)
Duncan, 1926

CALLIPHORIDAE: CHRYSOMYINAE

Protophormia terraenovae (Robineau-Desvoidy)
Gregor et al., 1960
Radvan, 1960c
Havlik, 1964

CALLIPHORIDAE: CALLIPHORINAE

Lucilia caesar (Linnaeus)
Gregor et al., 1960

Phaenicia sericata (Meigen)
Michelbacher et al., 1932
Gregor et al., 1960
Havlik, 1964

Calliphora vicina Robineau-Desvoidy
Graham-Smith, 1911a, 1911b
Balzam, 1937
Gregor et al., 1960
Havlik, 1964

Calliphora vomitoria (Linnaeus)
Gregor et al., 1960

SARCOPHAGIDAE: SARCOPHAGINAE

Sarcophaga Meigen
Nicholls, 1912

Sarcophaga aurifinis Walker
Nicholls, 1912

Tricharaea Thomson
Nicholls, 1912

Proteus Hauser .... MUSCIDAE: MUSCINAE

Musca Linnaeus
de la Paz, 1938

Musca domestica Linnaeus
Hamilton, 1903
Reinstorf, 1923
Prado et al., 1955
Coutinho et al., 1957

Musca domestica vicina Macquart
Shterngol'd, 1949
Floyd et al., 1953

Proteus (cont.)

MUSCIDAE: STOMOXYINAE
Stomoxys calcitrans (Linnaeus)
Testi, 1909
CALLIPHORIDAE: CHRYSOMYINAE
Chrysomya Robineau-Desvoidy
de la Paz, 1938
SARCOPHAGIDAE: SARCOPHAGINAE
Sarcophaga Meigen
de la Paz, 1938
Sarcophaga sarraceniae Riley
Beyer, 1925

Proteus inconstans (Ornstein) Shaw and Clarke ....

PIOPHILIDAE
Piophila casei (Linnaeus)
Gregor et al., 1960
SPHAEROCERIDAE
Leptocera ferruginata (Stenhammar)
Lysenko, 1958
MUSCIDAE: FANNIINAE
Fannia canicularis (Linnaeus)
Gregor et al., 1960
Fannia incisurata (Zetterstedt)
Lysenko, 1958
MUSCIDAE: PHAONIINAE
Hydrotaea dentipes (Fabricius)
Lysenko, 1958
Gregor et al., 1960
Hydrotaea irritans (Fallén)
Gregor et al., 1960
Ophyra leucostoma (Wiedemann)
Lysenko, 1958
Gregor et al., 1960
Muscina stabulans (Fallén)
Gregor et al., 1960
MUSCIDAE: MUSCINAE
Musca domestica Linnaeus
Lysenko, 1958
CALLIPHORIDAE: CHRYSOMYINAE
Protophormia terraenovae (Robineau-Desvoidy)
Lysenko, 1958
Gregor et al., 1960
CALLIPHORIDAE: CALLIPHORINAE
Lucilia caesar (Linnaeus)
Gregor et al., 1960
Phaenicia sericata (Meigen)
Gregor et al., 1960
Calliphora vicina Robineau-Desvoidy
Gregor et al., 1960
Calliphora vomitoria (Linnaeus)
Gregor et al., 1960

Proteus mirabilis Hauser .... PIOPHILIDAE

Piophila casei (Linnaeus)
  Gregor et al., 1960
  Lysenko et al., 1961
Piophila pectiniventris Duda
  Lysenko, 1958

SPHAEROCERIDAE

Leptocera ferruginata (Stenhammar)
  Lysenko, 1958

MUSCIDAE: FANNIINAE

Fannia canicularis (Linnaeus)
  O'Keefe et al., 1954
  Gregor et al., 1960

MUSCIDAE: PHAONIINAE

Hydrotaea dentipes (Fabricius)
  Lysenko, 1958
  Gregor et al., 1960
Hydrotaea irritans (Fallén)
  Gregor et al., 1960
Ophyra leucostoma (Wiedemann)
  Gregor et al., 1960
Muscina stabulans (Fallén)
  Gregor et al., 1960

MUSCIDAE: MUSCINAE

Musca domestica Linnaeus
  Cao, 1906
  Greenberg, 1959b
  Lysenko et al., 1961

CALLIPHORIDAE: CHRYSOMYINAE

Chrysomya megacephala (Fabricius)
  Shimizu et al., 1965
Chrysomya pinguis (Walker)
  Shimizu et al., 1965
Protophormia terraenovae (Robineau-Desvoidy)
  Lysenko, 1958
  Gregor et al., 1960
  Lysenko et al., 1961
  Shimizu et al., 1965

CALLIPHORIDAE: CALLIPHORINAE

Lucilia caesar (Linnaeus)
  Cao, 1906
  Gregor et al., 1960
Phaenicia cuprina Wiedemann
  Shimizu et al., 1965
Phaenicia sericata (Meigen)
  Gregor et al., 1960
  Lysenko et al., 1961
Calliphora lata Coquillet
  Shimizu et al., 1965

Proteus mirabilis (cont.)

Calliphora vicina Robineau-Desvoidy
Gregor et al., 1960
Greenberg et al., 1963c
Greenberg, 1966
Calliphora vomitoria (Linnaeus)
Cao, 1906
Gregor et al., 1960
SARCOPHAGIDAE: SARCOPHAGINAE
Sarcophaga carnaria (Linnaeus)
Cao, 1906
UNDETERMINED SARCOPHAGIDAE
Shimizu et al., 1965

Proteus morganii ....
(Winslow, Kligler, and Rothberg) Yale
(Synonyms: *Morganella*, Morgan's bacillus, Morgan's bacillus No. 1, Morgan's bacillus No. 19, ?Pseudo No. 1 bacilli)

PIOPHILIDAE
Piophila casei (Linnaeus)
Gregor et al., 1960
Lysenko et al., 1961
MUSCIDAE: MYDAEINAE
Mydaea urbana (Meigen)
Povolný et al., 1961
MUSCIDAE: FANNIINAE
Fannia Robineau-Desvoidy
Shimizu et al., 1965
Fannia canicularis (Linnaeus)
O'Keefe et al., 1954
Gregor et al., 1960
Fannia scalaris (Fabricius)
O'Keefe et al., 1954
MUSCIDAE: PHAONIINAE
Hydrotaea dentipes (Fabricius)
Gregor et al., 1960
Hydrotaea irritans (Fallén)
Gregor et al., 1960
Ophyra leucostoma (Wiedemann)
Gregor et al., 1960
Povolný et al., 1961
Lasiops simplex (Wiedemann)
Povolný et al., 1961
Muscina stabulans (Fallén)
Gregor et al., 1960
Povolný et al., 1961
Havlik, 1964
MUSCIDAE: MUSCINAE
Orthellia caerulea Wiedemann
Shimizu et al., 1965
Musca Linnaeus
de la Paz, 1938
Shimizu et al., 1965
Musca domestica Linnaeus
Morgan et al., 1909
Nicoll, 1911a

Organism	Fly / Reference
Proteus morganii (cont.)	Cox et al., 1912
	Buchanan, 1913
	Tebbutt, 1913
	Jones, 1941
	Hawley et al., 1951
	Radvan, 1956, 1960a
	Lysenko, 1958
	Lysenko et al., 1961
	Povolný et al., 1961
	Havlik, 1964
	House fly (presumably *Musca domestica* Linnaeus)
	Nicoll, 1917b
	CALLIPHORIDAE: CHRYSOMYINAE
	Chrysomya Robineau-Desvoidy
	de la Paz, 1938
	Chrysomya megacephala (Fabricius)
	Shimizu et al., 1965
	Phormia regina (Meigen)
	Knuckles, 1959
	Protophormia terraenovae (Robineau-Desvoidy)
	Lysenko, 1958
	Gregor et al., 1960
	Lysenko et al., 1961
	Havlik, 1964
	Shimizu et al., 1965
	CALLIPHORIDAE: CALLIPHORINAE
	Lucilia caesar (Linnaeus)
	Gregor et al., 1960
	Phaenicia cuprina Wiedemann
	Shimizu et al., 1965
	Phaenicia sericata (Meigen)
	Gregor et al., 1960
	Lysenko et al., 1961
	Havlik, 1964
	Shimizu et al., 1965
	Aldrichina grahami (Aldrich)
	Shimizu et al., 1965
	Calliphora lata Coquillet
	Shimizu et al., 1965
	Calliphora vicina Robineau-Desvoidy
	Gregor et al., 1960
	Povolný et al., 1961
	Havlik, 1964
	Calliphora vomitoria (Linnaeus)
	Gregor et al., 1960
	UNDETERMINED LUCILIINI
	Shimizu et al., 1965

Proteus morganii (cont.)	UNDETERMINED CALLIPHORINI
	Shimizu et al., 1965
	SARCOPHAGIDAE: SARCOPHAGINAE
	Sarcophaga Meigen
	de la Paz, 1938
	Sarcophaga haemorrhoidalis (Fallén)
	Povolný et al., 1961
	Sarcophaga peregrina Robineau-Desvoidy
	Shimizu et al., 1965
	UNDETERMINED SARCOPHAGIDAE
	Shimizu et al., 1965
Proteus rettgeri (Hadley et al.) Rustigian and Stuart (Synonym: *Rettgerella*)	PIOPHILIDAE
	Piophila casei (Linnaeus)
	Gregor et al., 1960
	Lysenko et al., 1961
	Piophila pectiniventris Duda
	Lysenko, 1958
	MUSCIDAE: FANNIINAE
	Fannia canicularis (Linnaeus)
	Gregor et al., 1960
	Fannia incisurata (Zetterstedt)
	Lysenko, 1958
	MUSCIDAE: PHAONIINAE
	Hydrotaea dentipes (Fabricius)
	Lysenko, 1958
	Gregor et al., 1960
	Hydrotaea irritans (Fallén)
	Gregor et al., 1960
	Ophyra Robineau-Desvoidy
	Shimizu et al., 1965
	Ophyra leucostoma (Wiedemann)
	Lysenko, 1958
	Gregor et al., 1960
	Muscina stabulans (Fallén)
	Gregor et al., 1960
	MUSCIDAE: MUSCINAE
	Musca Linnaeus
	de la Paz, 1938
	Shimizu et al., 1965
	Musca domestica Linnaeus
	Lysenko et al., 1961
	MUSCIDAE: STOMOXYINAE
	Haematobia irritans (Linnaeus)
	Stirrat et al., 1955
	CALLIPHORIDAE: CHRYSOMYINAE
	Chrysomya megacephala (Fabricius)
	Shimizu et al., 1965
	Chrysomya pinguis (Walker)
	Shimizu et al., 1965

Organism	Fly / Reference
Proteus rettgeri (cont.)	*Protophormia terraenovae* (Robineau-Desvoidy)
	Lysenko, 1958
	Gregor et al., 1960
	Lysenko et al., 1960
	Shimizu et al., 1965
	CALLIPHORIDAE: CALLIPHORINAE
	Lucilia caesar (Linnaeus)
	Gregor et al., 1960
	Phaenicia cuprina Wiedemann
	Shimizu et al., 1965
	Phaenicia sericata (Meigen)
	Gregor et al., 1960
	Lysenko et al., 1961
	Shimizu et al., 1965
	Aldrichina grahami (Aldrich)
	Shimizu et al., 1965
	Calliphora lata Coquillet
	Shimizu et al., 1965
	Calliphora vicina Robineau-Desvoidy
	Gregor et al., 1960
	Calliphora vomitoria (Linnaeus)
	Gregor et al., 1960
	UNDETERMINED LUCILIINI
	Shimizu et al., 1965
	UNDETERMINED CALLIPHORINI
	Shimizu et al., 1965
	SARCOPHAGIDAE: SARCOPHAGINAE
	Sarcophaga haemorrhoidalis (Fallén)
	Povolný et al., 1961
	Sarcophaga peregrina Robineau-Desvoidy
	Shimizu et al., 1965
	UNDETERMINED SARCOPHAGIDAE
	Shimizu et al., 1965
Proteus vulgaris Hauser (Synonyms: *B. proteus vulgaris*, *Bacillus proteus*)	OTITIDAE: ULIDIINAE
	Physiphora demandata (Fabricius)
	Povolný et al., 1961
	SEPSIDAE
	Sepsis punctum (Fabricius)
	Povolný et al., 1961
	PIOPHILIDAE
	Piophila casei (Linnaeus)
	Gregor et al., 1960
	Lysenko et al., 1961
	Povolný et al., 1961
	Piophila pectiniventris Duda
	Lysenko, 1958
	SPHAEROCERIDAE
	Copromyza atra (Meigen)
	Povolný et al., 1961

Proteus vulgaris (cont.)

Leptocera ferruginata (Stenhammar)
Lysenko, 1958

UNDETERMINED SPHAEROCERIDAE
Povolný et al., 1961

UNDETERMINED DROSOPHILIDAE
Povolný et al., 1961

ANTHOMYIIDAE: FUCELLINAE

Fucellia rufitibia Stein
O'Keefe et al., 1954

ANTHOMYIIDAE: ANTHOMYIINAE

Hylemya antiqua (Meigen)
O'Keefe et al., 1954

Hylemya cinerella (Fallén)
Povolný et al., 1961

Anthomyia pluvialis (Linnaeus)
Povolný et al., 1961

MUSCIDAE: MYDAEINAE

Mydaea urbana (Meigen)
Povolný et al., 1961

MUSCIDAE: FANNIINAE

Fannia canicularis (Linnaeus)
O'Keefe et al., 1954
Gregor et al., 1960
Povolný et al., 1961

Fannia incisurata (Zetterstedt)
Lysenko, 1958

Fannia scalaris (Fabricius)
O'Keefe et al., 1954
Povolný et al., 1961

MUSCIDAE: PHAONIINAE

Hydrotaea dentipes (Fabricius)
Lysenko, 1958
Gregor et al., 1960
Povolný et al., 1961

Hydrotaea irritans (Fallén)
Gregor et al., 1960

Hydrotaea occulta (Meigen)
Povolný et al., 1961

Ophyra leucostoma (Wiedemann)
Gregor et al., 1960
Povolný et al., 1961

Muscina stabulans (Fallén)
O'Keefe et al., 1954
Gregor et al., 1960
Povolný et al., 1961
Havlik, 1964

MUSCIDAE: MUSCINAE

Orthellia caerulea Wiedemann
Shimizu et al., 1965

Proteus vulgaris (cont.)	*Musca domestica* Linnaeus
	Ficker, 1903
	Cao, 1906
	Testi, 1909
	Ledingham, 1911
	Buchanan, 1913
	Scott, 1917a
	Shope, 1927
	Gerberich, 1951
	Radvan, 1956, 1960a, 1960b, 1960c
	Greenberg, 1959b
	Lysenko et al., 1961
	Povolný et al., 1961
	Greenberg et al., 1963a
	Havlik, 1964
	Musca domestica vicina Macquart
	Shterngol'd, 1949
	Sukhova, 1954
	MUSCIDAE: STOMOXYINAE
	Stomoxys calcitrans (Linnaeus)
	Povolný et al., 1961
	CALLIPHORIDAE: CHRYSOMYINAE
	Chrysomya megacephala (Fabricius)
	Shimizu et al., 1965
	Chrysomya pinguis (Walker)
	Shimizu et al., 1965
	Phormia regina (Meigen)
	Maseritz, 1934
	O'Keefe et al., 1954
	Greenberg, 1962
	Protophormia terraenovae (Robineau-Desvoidy)
	Pavillard et al., 1957
	Lysenko, 1958
	Gregor et al., 1960
	Radvan, 1960a, 1960b, 1960c
	Lysenko et al., 1961
	Havlik, 1964
	Shimizu et al., 1965
	CALLIPHORIDAE: CALLIPHORINAE
	Lucilia caesar (Linnaeus)
	Cao, 1906
	Weil, 1933
	Gregor et al., 1960
	Phaenicia cuprina Wiedemann
	Shimizu et al., 1965
	Phaenicia sericata (Meigen)
	Hobson, 1932a, 1935
	Michelbacher et al., 1932
	Weil et al., 1933

Organism	Fly associations
Proteus vulgaris (cont.)	Simmons, 1935a, 1935b O'Keefe et al., 1954 Sukhova, 1954 Gregor et al., 1960 Lysenko et al., 1961 Povolný et al., 1961 Greenberg, 1962 Havlik, 1964 *Calliphora vicina* Robineau-Desvoidy Gregor et al., 1960 Povolný et al., 1961 Greenberg, 1962 Havlik, 1964 *Calliphora vomitoria* (Linnaeus) Cao, 1906 Wollman, 1911 Gregor et al., 1960 Povolný et al., 1961 *Cynomyopsis cadaverina* (Robineau-Desvoidy) O'Keefe et al., 1954 Greenberg, 1962 UNDETERMINED LUCILIINI Shimizu et al., 1965 SARCOPHAGIDAE: SARCOPHAGINAE *Sarcophaga carnaria* (Linnaeus) Cao, 1906 Testi, 1909 *Sarcophaga haemorrhoidalis* (Fallén) Sukhova, 1954 Povolný et al., 1961 UNDETERMINED SARCOPHAGIDAE Shimizu et al., 1965
Providencia Kauffmann (Synonym: *Providentia*)	OTITIDAE: ULIDIINAE *Physiphora demandata* (Fabricius) Povolný et al., 1961 PIOPHILIDAE *Piophila casei* (Linnaeus) Povolný et al., 1961 SEPSIDAE *Sepsis punctum* (Fabricius) Povolný et al., 1961 UNDETERMINED SPHAEROCERIDAE Povolný et al., 1961 UNDETERMINED DROSOPHILIDAE Povolný et al., 1961 ANTHOMYIIDAE: ANTHOMYIINAE *Hylemya cinerella* (Fallén) Povolný et al., 1961

Providencia (cont.)	*Anthomyia pluvialis* (Linnaeus) Povolný et al., 1961 MUSCIDAE: FANNIINAE *Fannia scalaris* (Fabricius) Povolný et al., 1961 MUSCIDAE: PHAONIINAE *Hydrotaea dentipes* (Fabricius) Povolný et al., 1961 *Hydrotaea occulta* (Meigen) Povolný et al., 1961 *Ophyra* Robineau-Desvoidy Shimizu et al., 1965 MUSCIDAE: MUSCINAE *Musca* Linnaeus Shimizu et al., 1965 *Musca domestica* Linnaeus Radvan, 1960b, 1960c Lysenko et al., 1961 Shimizu et al., 1965 *Musca domestica vicina* Macquart Floyd et al., 1953 CALLIPHORIDAE: CHRYSOMYINAE *Chrysomya megacephala* (Fabricius) Shimizu et al., 1965 *Chrysomya pinguis* (Walker) Shimizu et al., 1965 *Protophormia terraenovae* (Robineau-Desvoidy) Radvan, 1960a, 1960b, 1960c Lysenko et al., 1961 Shimizu et al., 1965 CALLIPHORIDAE: CALLIPHORINAE *Phaenicia cuprina* Wiedemann Shimizu et al., 1965 *Phaenicia sericata* (Meigen) Lysenko et al., 1961 Shimizu et al., 1965 UNDETERMINED LUCILIINI Shimizu et al., 1965 UNDETERMINED CALLIPHORINI Shimizu et al., 1965 UNDETERMINED SARCOPHAGIDAE Shimizu et al., 1965
Salmonella Lignières	SYRPHIDAE: MILESIINAE *Volucella obesa* (Fabricius) Alcivar et al., 1946 MUSCIDAE: PHAONIINAE *Ophyra aenescens* (Wiedemann) Greenberg et al., 1964

Organism	Fly associations
Salmonella (cont.)	MUSCIDAE: MUSCINAE *Musca domestica* Linnaeus Alcivar et al., 1946 Lindsay et al., 1953 Gabaldón et al., 1956 Bolaños, 1959 Greenberg et al., 1964 MUSCIDAE: STOMOXYINAE *Stomoxys calcitrans* (Linnaeus) Greenberg et al., 1964 CALLIPHORIDAE: CHRYSOMYINAE *Cochliomyia macellaria* (Fabricius) Alcivar et al., 1946 Greenberg et al., 1964 *Chrysomya megacephala* (Fabricius) Shimizu et al., 1965 CALLIPHORIDAE: CALLIPHORINAE *Phaenicia sericata* (Meigen) Alcivar et al., 1946 Greenberg et al., 1964 Shimizu et al., 1965 UNDETERMINED LUCILIINI Shimizu et al., 1965 SARCOPHAGIDAE: SARCOPHAGINAE *Sarcophaga* Meigen Alcivar et al., 1946
Salmonella alachua Kauffmann	CALLIPHORIDAE: CHRYSOMYINAE *Cochliomyia macellaria* (Fabricius) Greenberg et al., 1963b
Salmonella amersfoort Henning	SYRPHIDAE: MILESIINAE *Volucella obesa* (Fabricius) Alcivar et al., 1946 MUSCIDAE: MUSCINAE *Musca domestica* Linnaeus Alcivar et al., 1946 CALLIPHORIDAE: CHRYSOMYINAE *Cochliomyia macellaria* (Fabricius) Alcivar et al., 1946 CALLIPHORIDAE: CALLIPHORINAE *Phaenicia sericata* (Meigen) Alcivar et al., 1946 SARCOPHAGIDAE: SARCOPHAGINAE *Sarcophaga* Meigen Alcivar et al., 1946
Salmonella anatum (Rettger and Scoville) Bergey et al.	SYRPHIDAE: MILESIINAE *Volucella obesa* (Fabricius) Alcivar et al., 1946 SEPSIDAE *Paleosepsis* Duda Greenberg et al., 1963b

Organism	Fly

Salmonella anatum (cont.)	*Sepsis biflexuosa* Strobl
	  Greenberg et al., 1963b
	Sepsis punctum (Fabricius)
	  Greenberg et al., 1963b
	MUSCIDAE: PHAONIINAE
	Ophyra aenescens (Wiedemann)
	  Greenberg et al., 1963b
	MUSCIDAE: MUSCINAE
	Musca domestica Linnaeus
	  Alcivar et al., 1946
	  Bolaños, 1959
	  Greenberg et al., 1963b
	CALLIPHORIDAE: CHRYSOMYINAE
	Cochliomyia macellaria (Fabricius)
	  Alcivar et al., 1946
	  Greenberg et al., 1963b
	CALLIPHORIDAE: CALLIPHORINAE
	Phaenicia sericata (Meigen)
	  Alcivar et al., 1946
	SARCOPHAGIDAE: SARCOPHAGINAE
	Sarcophaga Meigen
	  Alcivar et al., 1946

Salmonella ballerup Kauffmann and Møller	SYRPHIDAE: MILESIINAE
	Volucella obesa (Fabricius)
	  Alcivar et al., 1946
	MUSCIDAE: MUSCINAE
	Musca domestica Linnaeus
	  Alcivar et al., 1946
	CALLIPHORIDAE: CHRYSOMYINAE
	Cochliomyia macellaria (Fabricius)
	  Alcivar et al., 1946
	CALLIPHORIDAE: CALLIPHORINAE
	Phaenicia sericata (Meigen)
	  Alcivar et al., 1946
	SARCOPHAGIDAE: SARCOPHAGINAE
	Sarcophaga Meigen
	  Alcivar et al., 1946

Salmonella blockley Kauffmann	MUSCIDAE: PHAONIINAE
	Ophyra aenescens (Wiedemann)
	  Greenberg et al., 1964
	MUSCIDAE: MUSCINAE
	Musca domestica Linnaeus
	  Greenberg et al., 1964
	MUSCIDAE: STOMOXYINAE
	Stomoxys calcitrans (Linnaeus)
	  Greenberg et al., 1964
	CALLIPHORIDAE: CHRYSOMYINAE
	Phormia regina (Meigen)
	  Greenberg et al., 1964

Organism	Fly / Reference
Salmonella blockley (cont.)	CALLIPHORIDAE: CALLIPHORINAE
	Phaenicia sericata (Meigen)
	Greenberg et al., 1964
Salmonella bovis-morbificans (Basenau) Schütze et al.	MUSCIDAE: PHAONIINAE
	Ophyra aenescens (Wiedemann)
	Greenberg et al., 1963b
Salmonella budapest Kauffmann	MUSCIDAE: MUSCINAE
	Musca domestica Linnaeus
	Prado et al., 1955
Salmonella chester Kauffmann	MUSCIDAE: PHAONIINAE
	Ophyra Robineau-Desvoidy
	Bulling et al., 1959
	UNDETERMINED MUSCIDAE
	Bulling et al., 1959
	CALLIPHORIDAE: CALLIPHORINAE
	Lucilia Robineau-Desvoidy
	Bulling et al., 1959
	Calliphora Robineau-Desvoidy
	Bulling et al., 1959
Salmonella cholerae-suis (Smith) Weldin (Synonym: bacillus of swine fever)	SYRPHIDAE: MILESIINAE
	Volucella obesa (Fabricius)
	Alcivar et al., 1946
	MUSCIDAE: MUSCINAE
	Musca Linnaeus
	de la Paz, 1938
	Musca domestica Linnaeus
	Scott, 1917a
	Alcivar et al., 1946
	CALLIPHORIDAE: CHRYSOMYINAE
	Cochliomyia macellaria (Fabricius)
	Alcivar et al., 1946
	Chrysomya Robineau-Desvoidy
	de la Paz, 1938
	Chrysomya putoria (Wiedemann)
	Brygoo et al., 1962
	CALLIPHORIDAE: CALLIPHORINAE
	Phaenicia sericata (Meigen)
	Alcivar et al., 1946
	Calliphora vomitoria (Linnaeus)
	Buchanan, 1907
	SARCOPHAGIDAE: SARCOPHAGINAE
	Sarcophaga Meigen
	Alcivar et al., 1946
Salmonella columbensis (Castellani) Castellani and Chalmers	MUSCIDAE: MUSCINAE
	Musca Linnaeus
	de la Paz, 1938
	CALLIPHORIDAE: CHRYSOMYINAE
	Chrysomya Robineau-Desvoidy
	de la Paz, 1938

Organism	Fly
Salmonella columbensis (cont.)	SARCOPHAGIDAE: SARCOPHAGINAE
	Sarcophaga Meigen
	de la Paz, 1938
Salmonella derby	SEPSIDAE
Warren and Scott	*Paleosepsis* Duda
	Greenberg et al., 1963b
	Sepsis biflexuosa Strobl
	Greenberg et al., 1963b
	Sepsis punctum (Fabricius)
	Greenberg et al., 1963b
	SPHAEROCERIDAE
	Leptocera ferruginata (Stenhammar)
	Greenberg et al., 1963b
	Leptocera vagans (Haliday)
	Greenberg et al., 1963b
	MUSCIDAE: PHAONIINAE
	Ophyra aenescens (Wiedemann)
	Greenberg et al., 1963b
	MUSCIDAE: MUSCINAE
	Musca domestica Linnaeus
	Prado et al., 1955
	Greenberg et al., 1963b
	Tacal et al., 1967
	CALLIPHORIDAE: CHRYSOMYINAE
	Cochliomyia macellaria (Fabricius)
	Greenberg et al., 1963b
	Phormia regina (Meigen)
	Greenberg et al., 1963b
	CALLIPHORIDAE: CALLIPHORINAE
	Calliphora Robineau-Desvoidy
	Tacal et al., 1967
Salmonella dublin	SYRPHIDAE: MILESIINAE
Kauffmann	*Volucella obesa* (Fabricius)
	Alcivar et al., 1946
	MUSCIDAE: MUSCINAE
	Musca domestica Linnaeus
	Alcivar et al., 1946
	CALLIPHORIDAE: CHRYSOMYINAE
	Cochliomyia macellaria (Fabricius)
	Alcivar et al., 1946
	CALLIPHORIDAE: CALLIPHORINAE
	Phaenicia sericata (Meigen)
	Alcivar et al., 1946
	SARCOPHAGIDAE: SARCOPHAGINAE
	Sarcophaga Meigen
	Alcivar et al., 1946

Organism	Fly associations
Salmonella enteritidis (Gaertner) Castellani and Chalmers (Synonyms: *Bacillus enteritidis* of Gaertner, *Bacillus enteritidis*)	SYRPHIDAE: MILESIINAE *Volucella obesa* (Fabricius) Alcivar et al., 1946 MUSCIDAE: MUSCINAE *Musca domestica* Linnaeus Graham-Smith, 1910 Cox et al., 1912 Buchanan, 1913 Ostrolenk et al., 1942a, 1942b Peppler, 1944 Alcivar et al., 1946 Gross et al., 1953 Prado et al., 1955 Greenberg, 1959a, 1959c CALLIPHORIDAE: CHRYSOMYINAE *Cochliomyia macellaria* (Fabricius) Alcivar et al., 1946 CALLIPHORIDAE: CALLIPHORINAE *Lucilia caesar* (Linnaeus) Graham-Smith, 1912 *Phaenicia sericata* (Meigen) Alcivar et al., 1946 Gross et al., 1953 *Calliphora vicina* Robineau-Desvoidy Graham-Smith, 1911a, 1911b, 1912 Gross et al., 1953 Greenberg, 1966 SARCOPHAGIDAE: SARCOPHAGINAE *Sarcophaga* Meigen Alcivar et al., 1946 *Sarcophaga haemorrhoidalis* (Fallén) Gross et al., 1953
Salmonella florida Cherry, Edwards, and Bruner	SYRPHIDAE: MILESIINAE *Volucella obesa* (Fabricius) Alcivar et al., 1946 MUSCIDAE: MUSCINAE *Musca domestica* Linnaeus Alcivar et al., 1946 CALLIPHORIDAE: CHRYSOMYINAE *Cochliomyia macellaria* (Fabricius) Alcivar et al., 1946 CALLIPHORIDAE: CALLIPHORINAE *Phaenicia sericata* (Meigen) Alcivar et al., 1946 SARCOPHAGIDAE: SARCOPHAGINAE *Sarcophaga* Meigen Alcivar et al., 1946
Salmonella gallinarum (Klein) Bergey	MUSCIDAE: MUSCINAE *Musca* Linnaeus de la Paz, 1938

Organism	Fly / Reference
Salmonella gallinarum (cont.)	*Musca domestica* Linnaeus
	Gerberich, 1951
Salmonella give Kauffmann	SYRPHIDAE: MILESIINAE
	Volucella obesa (Fabricius)
	Alcivar et al., 1946
	MUSCIDAE: MUSCINAE
	Musca domestica Linnaeus
	Alcivar et al., 1946
	Gross et al., 1953
	Bolaños, 1959
	CALLIPHORIDAE: CHRYSOMYINAE
	Cochliomyia macellaria (Fabricius)
	Alcivar et al., 1946
	Greenberg et al., 1963b
	Chrysomya putoria (Wiedemann)
	Brygoo et al., 1962
	Phormia regina (Meigen)
	Greenberg et al., 1963b
	CALLIPHORIDAE: CALLIPHORINAE
	Phaenicia sericata (Meigen)
	Alcivar et al., 1946
	Gross et al., 1953
	Calliphora vicina Robineau-Desvoidy
	Gross et al., 1953
	SARCOPHAGIDAE: SARCOPHAGINAE
	Sarcophaga Meigen
	Alcivar et al., 1946
	Sarcophaga haemorrhoidalis (Fallén)
	Gross et al., 1953
Salmonella hirschfeldii Weldin	MUSCIDAE: FANNIINAE
	Fannia canicularis (Linnaeus)
	Yao et al., 1929
	Fannia scalaris (Fabricius)
	Yao et al., 1929
	MUSCIDAE: MUSCINAE
	Musca domestica Linnaeus
	Yao et al., 1929
	CALLIPHORIDAE: CALLIPHORINAE
	Lucilia caesar (Linnaeus)
	Yao et al., 1929
	Calliphora vicina Robineau-Desvoidy
	Yao et al., 1929
	Calliphora vomitoria (Linnaeus)
	Yao et al., 1929
	SARCOPHAGIDAE: SARCOPHAGINAE
	Sarcophaga carnaria (Linnaeus)
	Yao et al., 1929
Salmonella kentucky Edwards	SYRPHIDAE: MILESIINAE
	Volucella obesa (Fabricius)
	Alcivar et al., 1946

Organism	Fly
Salmonella kentucky (cont.)	MUSCIDAE: MUSCINAE *Musca domestica* Linnaeus Alcivar et al., 1946 Greenberg et al., 1963b CALLIPHORIDAE: CHRYSOMYINAE *Cochliomyia macellaria* (Fabricius) Alcivar et al., 1946 CALLIPHORIDAE: CALLIPHORINAE *Phaenicia sericata* (Meigen) Alcivar et al., 1946 SARCOPHAGIDAE: SARCOPHAGINAE *Sarcophaga* Meigen Alcivar et al., 1946
Salmonella kottbus Schütze et al. (Synonym: *Salmonella cottbus*)	SYRPHIDAE: MILESIINAE *Volucella obesa* (Fabricius) Alcivar et al., 1946 MUSCIDAE: MUSCINAE *Musca domestica* Linnaeus Alcivar et al., 1946 UNDETERMINED MUSCIDAE Bulling et al., 1959 CALLIPHORIDAE: CHRYSOMYINAE *Cochliomyia macellaria* (Fabricius) Alcivar et al., 1946 CALLIPHORIDAE: CALLIPHORINAE *Lucilia* Robineau-Desvoidy Bulling et al., 1959 *Phaenicia sericata* (Meigen) Alcivar et al., 1946 *Calliphora* Robineau-Desvoidy Bulling et al., 1959 SARCOPHAGIDAE: SARCOPHAGINAE *Sarcophaga* Meigen Alcivar et al., 1946
Salmonella meleagridis Bruner and Edwards	SYRPHIDAE: MILESIINAE *Volucella obesa* (Fabricius) Alcivar et al., 1946 MUSCIDAE: MUSCINAE *Musca domestica* Linnaeus Alcivar et al., 1946 CALLIPHORIDAE: CHRYSOMYINAE *Cochliomyia macellaria* (Fabricius) Alcivar et al., 1946 Greenberg et al., 1963b CALLIPHORIDAE: CALLIPHORINAE *Phaenicia sericata* (Meigen) Alcivar et al., 1946 SARCOPHAGIDAE: SARCOPHAGINAE *Sarcophaga* Meigen Alcivar et al., 1946

Organism	Fly associations
Salmonella minnesota Edwards and Bruner	MUSCIDAE: MUSCINAE *Musca domestica* Linnaeus Prado et al., 1955
Salmonella muenchen Kauffmann	CALLIPHORIDAE: CHRYSOMYINAE *Phormia regina* (Meigen) Greenberg et al., 1963b
Salmonella narashino Kauffmann	SYRPHIDAE: MILESIINAE *Volucella obesa* (Fabricius) Alcivar et al., 1946 MUSCIDAE: MUSCINAE *Musca domestica* Linnaeus Alcivar et al., 1946 CALLIPHORIDAE: CHRYSOMYINAE *Cochliomyia macellaria* (Fabricius) Alcivar et al., 1946 CALLIPHORIDAE: CALLIPHORINAE *Phaenicia sericata* (Meigen) Alcivar et al., 1946 SARCOPHAGIDAE: SARCOPHAGINAE *Sarcophaga* Meigen Alcivar et al., 1946
Salmonella new brunswick Edwards	MUSCIDAE: PHAONIINAE *Ophyra aenescens* (Wiedemann) Greenberg et al., 1963b MUSCIDAE: MUSCINAE *Musca domestica* Linnaeus Greenberg et al., 1963b CALLIPHORIDAE: CHRYSOMYINAE *Cochliomyia macellaria* (Fabricius) Greenberg et al., 1963b *Phormia regina* (Meigen) Greenberg et al., 1963b CALLIPHORIDAE: CALLIPHORINAE *Phaenicia sericata* (Meigen) Greenberg et al., 1963b
Salmonella newport Schütze	SPHAEROCERIDAE *Leptocera ferruginata* (Stenhammar) Greenberg et al., 1963b *Leptocera vagans* (Haliday) Greenberg et al., 1963b MUSCIDAE: MUSCINAE *Musca domestica* Linnaeus Bolaños, 1959 CALLIPHORIDAE: CHRYSOMYINAE *Cochliomyia macellaria* (Fabricius) Greenberg et al., 1963b *Phormia regina* (Meigen) Greenberg et al., 1963b

Organism	Fly association
Salmonella onderstepoort Henning	SYRPHIDAE: MILESIINAE *Volucella obesa* (Fabricius) Alcivar et al., 1946 MUSCIDAE: MUSCINAE *Musca domestica* Linnaeus Alcivar et al., 1946 CALLIPHORIDAE: CHRYSOMYINAE *Cochliomyia macellaria* (Fabricius) Alcivar et al., 1946 CALLIPHORIDAE: CALLIPHORINAE *Phaenicia sericata* (Meigen) Alcivar et al., 1946 SARCOPHAGIDAE: SARCOPHAGINAE *Sarcophaga* Meigen Alcivar et al., 1946
Salmonella oranienburg Kauffmann	SYRPHIDAE: MILESIINAE *Volucella obesa* (Fabricius) Alcivar et al., 1946 MUSCIDAE: MUSCINAE *Musca domestica* Linnaeus Alcivar et al., 1946 CALLIPHORIDAE: CHRYSOMYINAE *Cochliomyia macellaria* (Fabricius) Alcivar et al., 1946 CALLIPHORIDAE: CALLIPHORINAE *Phaenicia sericata* (Meigen) Alcivar et al., 1946 SARCOPHAGIDAE: SARCOPHAGINAE *Sarcophaga* Meigen Alcivar et al., 1946
Salmonella oregon Edwards and Bruner	SYRPHIDAE: MILESIINAE *Volucella obesa* (Fabricius) Alcivar et al., 1946 MUSCIDAE: MUSCINAE *Musca domestica* Linnaeus Alcivar et al., 1946 CALLIPHORIDAE: CHRYSOMYINAE *Cochliomyia macellaria* (Fabricius) Alcivar et al., 1946 CALLIPHORIDAE: CALLIPHORINAE *Phaenicia sericata* (Meigen) Alcivar et al., 1946 SARCOPHAGIDAE: SARCOPHAGINAE *Sarcophaga* Meigen Alcivar et al., 1946
Salmonella panama Kauffmann	MUSCIDAE: MUSCINAE *Musca domestica* Linnaeus Bolaños, 1959

Salmonella panama (cont.)

CALLIPHORIDAE: CHRYSOMYINAE
Cochliomyia macellaria (Fabricius)
Greenberg et al., 1963b

Salmonella paratyphi A (Brion and Kayser) Castellani and Chalmers (Synonym: *Bacillus paratyphi* Type I) ....

MUSCIDAE: MUSCINAE
Musca domestica Linnaeus
Shope, 1927
Boikov, 1932
Gross et al., 1951, 1953
Prado et al., 1955

CALLIPHORIDAE: CALLIPHORINAE
Phaenicia sericata (Meigen)
Gross et al., 1953
Calliphora vicina Robineau-Desvoidy
Gross et al., 1953

SARCOPHAGIDAE: SARCOPHAGINAE
Sarcophaga haemorrhoidalis (Fallén)
Gross et al., 1953

Salmonella paratyphi B Castellani and Chalmers ....

SYRPHIDAE: MILESIINAE
Volucella obesa (Fabricius)
Alcivar et al., 1946

MUSCIDAE: MUSCINAE
Musca domestica Linnaeus
Nicoll, 1911a
Alcivar et al., 1946
Gross et al., 1951, 1953
Dishon, 1956
Radvan, 1956
Greenberg, 1959a

MUSCIDAE: STOMOXYINAE
Stomoxys calcitrans (Linnaeus)
Birk, 1932

CALLIPHORIDAE: CHRYSOMYINAE
Cochliomyia macellaria (Fabricius)
Alcivar et al., 1946

CALLIPHORIDAE: CALLIPHORINAE
Phaenicia sericata (Meigen)
Alcivar et al., 1946
Gross et al., 1953
Calliphora vicina Robineau-Desvoidy
Gross et al., 1953

SARCOPHAGIDAE: SARCOPHAGINAE
Sarcophaga Meigen
Alcivar et al., 1946
Pavan, 1949
Sarcophaga haemorrhoidalis (Fallén)
Gross et al., 1953

Salmonella paratyphi C Castellani and Chalmers ....

HELEOMYZIDAE: HELEOMYZINAE
Heleomyza modesta Meigen
Anonymous, 1952

Organism	Fly
Salmonella pullorum (Rettger) Bergey	MUSCIDAE: MUSCINAE *Musca domestica* Linnaeus Gwatkin et al., 1944 Gerberich, 1951, 1952 House fly (presumably *Musca domestica* Linnaeus) Gwatkin et al., 1952 UNDETERMINED CALLIPHORIDAE Blow flies Gwatkin et al., 1952
Salmonella reading Schütze	SYRPHIDAE: MILESIINAE *Volucella obesa* (Fabricius) Alcivar et al., 1946 MUSCIDAE: MUSCINAE *Musca domestica* Linnaeus Alcivar et al., 1946 Bolaños, 1959 CALLIPHORIDAE: CHRYSOMYINAE *Cochliomyia macellaria* (Fabricius) Alcivar et al., 1946 CALLIPHORIDAE: CALLIPHORINAE *Phaenicia sericata* (Meigen) Alcivar et al., 1946 SARCOPHAGIDAE: SARCOPHAGINAE *Sarcophaga* Meigen Alcivar et al., 1946
Salmonella saintpaul Edwards and Bruner	MUSCIDAE: MUSCINAE *Musca domestica* Linnaeus Bolaños, 1959
Salmonella sandiego Kauffmann	MUSCIDAE: MUSCINAE *Musca domestica* Linnaeus Bolaños, 1959
Salmonella schottmuelleri (Winslow, Kligler, and Rothberg) Bergey et al.	MUSCIDAE: MUSCINAE *Musca domestica* Linnaeus Hawley et al., 1951 Radvan, 1960a, 1960b, 1960c CALLIPHORIDAE: CHRYSOMYINAE *Phormia regina* (Meigen) Knuckles, 1959 *Protophormia terraenovae* (Robineau-Desvoidy) Radvan, 1960a, 1960b, 1960c CALLIPHORIDAE: CALLIPHORINAE *Calliphora vicina* Robineau-Desvoidy Greenberg, 1962
Salmonella typhi Warren and Scott (Synonyms: *Bacillus typhosus*, Typhoid, *Bacillus typhi abdominalis*, *Eberthella typhi*)	SEPSIDAE *Sepsis* Fallén Nicholls, 1912 SPHAEROCERIDAE *Leptocera* Macquart Boikov, 1932

Organism	Fly / Reference
Salmonella typhi (cont.)	*Limosina punctipennis* (Wiedemann)
	Nicholls, 1912
	DROSOPHILIDAE: DROSOPHILINAE
	Drosophila melanogaster Meigen
	Nicholls, 1912
	HELEOMYZIDAE: HELEOMYZINAE
	Heleomyza modesta Meigen
	Anonymous, 1952
	ANTHOMYIIDAE: ANTHOMYIINAE
	Hylemya Robineau-Desvoidy
	Anonymous, 1952
	MUSCIDAE: PHAONIINAE
	Muscina stabulans (Fallén)
	Anonymous, 1952
	MUSCIDAE: MUSCINAE
	Musca Linnaeus
	de la Paz, 1938
	Musca domestica Linnaeus
	Firth et al., 1902
	Ficker, 1903
	Hamilton, 1903
	Buchanan, 1907
	Graham-Smith, 1910
	Ledingham, 1911
	Cochrane, 1912
	Thomson, 1912
	Krontowski, 1913
	Tebbutt, 1913
	Teodoro, 1916
	Nicoll, 1917a
	Wollman, 1921
	Duncan, 1926
	Shope, 1927
	Ara et al., 1932
	Boikov, 1932
	Jones, 1941
	Curbelo et al., 1945
	Gross et al., 1951, 1953
	Prado et al., 1955
	Radvan, 1956, 1960a
	Webb et al., 1956
	Greenberg, 1959a, 1959c
	Greenberg et al., 1963a
	House fly (presumably *Musca domestica* Linnaeus)
	Celli et al., 1888
	Manning, 1902
	Bertarelli, 1909
	Ara, 1933
	Musca domestica vicina Macquart
	Floyd et al., 1953

Salmonella typhi (cont.)

MUSCIDAE: STOMOXYINAE

Stomoxys calcitrans (Linnaeus)
Duncan, 1926

CALLIPHORIDAE: CHRYSOMYINAE

Cochliomyia macellaria (Fabricius)
Beyer, 1925

Chrysomya Robineau-Desvoidy
de la Paz, 1938

Chrysomya megacephala (Fabricius)
Chow, 1940

Phormia regina (Meigen)
Maseritz, 1934
Greenberg, 1962

Protophormia terraenovae (Robineau-Desvoidy)
Webb et al., 1956

CALLIPHORIDAE: CALLIPHORINAE

Lucilia caesar (Linnaeus)
Krontowski, 1913
Teodoro, 1916
Wollman, 1921

Phaenicia sericata (Meigen)
Simmons, 1935a, 1935b
Anonymous, 1952
Gross et al., 1953
Sukhova, 1954
Greenberg, 1962

Eucalliphora lilaea (Walker)
Gwatkin et al., 1938

Calliphora terraenovae Macquart
Gwatkin et al., 1938

Calliphora vicina Robineau-Desvoidy
Graham-Smith, 1911a, 1911b
Teodoro, 1916
Gwatkin et al., 1938
Gross et al., 1953

Calliphora vomitoria (Linnaeus)
Wollman, 1921

Cynomyopsis cadaverina (Robineau-Desvoidy)
Gwatkin et al., 1938
Webb et al., 1956
Greenberg, 1962

Cynomya mortuorum (Linnaeus)
Krontowski, 1913

UNDETERMINED CALLIPHORIDAE

Blue bottle fly
Firth et al., 1902

Salmonella typhi (cont.)	SARCOPHAGIDAE: SARCOPHAGINAE *Sarcophaga* Meigen Nicholls, 1912 Pavan, 1949 *Sarcophaga aurifinis* Walker Nicholls, 1912 *Sarcophaga carnaria* (Linnaeus) Krontowski, 1913 Teodoro, 1916 Petragnani, 1925 *Sarcophaga haemorrhoidalis* (Fallén) Gross et al., 1953 *Tricharaea* Thomson Nicholls, 1912
Salmonella typhimurium (Loeffler) Castellani and Chalmers	SYRPHIDAE: MILESIINAE *Volucella obesa* (Fabricius) Alcivar et al., 1946 MUSCIDAE: PHAONIINAE *Ophyra* Robineau-Desvoidy Bulling et al., 1959 MUSCIDAE: MUSCINAE *Musca domestica* Linnaeus McNeil et al., 1944 Alcivar et al., 1946 Radvan, 1956, 1960a Greenberg, 1964 *Musca domestica vicina* Macquart Floyd et al., 1953 Brygoo et al., 1962 UNDETERMINED MUSCIDAE Bulling et al., 1959 CALLIPHORIDAE: CHRYSOMYINAE *Cochliomyia macellaria* (Fabricius) Alcivar et al., 1946 *Chrysomya putoria* (Wiedemann) Brygoo et al., 1962 *Phormia regina* (Meigen) Knuckles, 1959 *Protophormia terraenovae* (Robineau-Desvoidy) Pavillard et al., 1957 Ojala et al., 1966 CALLIPHORIDAE: CALLIPHORINAE *Lucilia* Robineau-Desvoidy Bulling et al., 1959 *Lucilia illustris* (Meigen) Nuorteva, 1965 Ojala et al., 1966

Organism	Fly / Reference
Salmonella typhimurium (cont.)	*Phaenicia sericata* (Meigen) Alcivar et al., 1946 Ojala et al., 1966 *Calliphora* Robineau-Desvoidy Bulling et al., 1959 *Calliphora vicina* Robineau-Desvoidy Greenberg et al., 1963c SARCOPHAGIDAE: SARCOPHAGINAE *Sarcophaga* Meigen Alcivar et al., 1946
Salmonella worthington Edwards and Bruner	MUSCIDAE: MUSCINAE *Musca domestica* Linnaeus Greenberg, 1963b
Salmonella zanzibar Kauffmann	SYRPHIDAE: MILESIINAE *Volucella obesa* (Fabricius) Alcivar et al., 1946 MUSCIDAE: MUSCINAE *Musca domestica* Linnaeus Alcivar et al., 1946 Bolaños, 1959 CALLIPHORIDAE: CHRYSOMYINAE *Cochliomyia macellaria* (Fabricius) Alcivar et al., 1946 CALLIPHORIDAE: CALLIPHORINAE *Phaenicia sericata* (Meigen) Alcivar et al., 1946 SARCOPHAGIDAE: SARCOPHAGINAE *Sarcophaga* Meigen Alcivar et al., 1946
Eberthella belfastiensis (Weldin and Levine) Bergey et al.	MUSCIDAE: MUSCINAE *Musca* Linnaeus de la Paz, 1938 CALLIPHORIDAE: CHRYSOMYINAE *Chrysomya* Robineau-Desvoidy de la Paz, 1938
Eberthella dubia (Chester) Bergey et al.	MUSCIDAE: PHAONIINAE *Musca* Linnaeus de la Paz, 1938
Eberthella kandiensis (Castellani) Bergey et al.	MUSCIDAE: MUSCINAE *Musca* Linnaeus de la Paz, 1938
Eberthella talavensis (Castellani) Bergey et al.	MUSCIDAE: MUSCINAE *Musca* Linnaeus de la Paz, 1938
Citrobacter Werkman and Gillen	MUSCIDAE: MUSCINAE *Musca domestica* Linnaeus Lysenko et al., 1961

Organism	Fly
Citrobacter freundii	PIOPHILIDAE
Werkman and Gillen	*Piophila casei* (Linnaeus)
(Synonyms: *Escherichia*	Gregor et al., 1960
anindolica, Escherichia freundii)	MUSCIDAE: FANNIINAE
	Fannia canicularis (Linnaeus)
	Zmeev, 1943
	Gregor et al., 1960
	Fannia incisurata (Zetterstedt)
	Lysenko, 1958
	Fannia leucosticta (Meigen)
	Zmeev, 1943
	Fannia scalaris (Fabricius)
	Zmeev, 1943
	MUSCIDAE: PHAONIINAE
	Hydrotaea dentipes (Fabricius)
	Gregor et al., 1960
	Hydrotaea irritans (Fallén)
	Gregor et al., 1960
	Ophyra leucostoma (Wiedemann)
	Lysenko, 1958
	Gregor et al., 1960
	Muscina stabulans (Fallén)
	Gregor et al., 1960
	MUSCIDAE: MUSCINAE
	Musca domestica Linnaeus
	Lysenko, 1958
	Greenberg, 1959b
	CALLIPHORIDAE: CHRYSOMYINAE
	Protophormia terraenovae (Robineau-Desvoidy)
	Gregor et al., 1960
	CALLIPHORIDAE: CALLIPHORINAE
	Lucilia caesar (Linnaeus)
	Gregor et al., 1960
	Phaenicia sericata (Meigen)
	Gregor et al., 1960
	Calliphora vicina Robineau-Desvoidy
	Gregor et al., 1960
	Calliphora vomitoria (Linnaeus)
	Gregor et al., 1960
	SARCOPHAGIDAE: SARCOPHAGINAE
	Sarcophaga haemorrhoidalis (Fallén)
	Zmeev, 1943
Bethesda	MUSCIDAE: MUSCINAE
Edwards, West, and Bruner	*Musca domestica vicina* Macquart
	Floyd et al., 1953
Shigella	SCIARIDAE
Castellani and Chalmers	*Sciara* Meigen
(Synonym: Dysentery bacillus)	Boikov, 1932

Shigella (cont.)

SPHAEROCERIDAE
Leptocera Macquart
  Boikov, 1932
DROSOPHILIDAE: DROSOPHILINAE
Drosophila funebris (Fabricius)
  Boikov, 1932
MUSCIDAE: FANNIINAE
Fannia scalaris (Fabricius)
  Zaidenov, 1961
MUSCIDAE: PHAONIINAE
Ophyra Robineau-Desvoidy
  Shimizu et al., 1965
Ophyra aenescens (Wiedemann)
  Greenberg et al., 1964
Muscina stabulans (Fallén)
  Zaidenov, 1961
MUSCIDAE: MUSCINAE
Musca domestica Linnaeus
  Boikov, 1932
  Lindsay et al., 1953
  Boyd, 1957
  Zaidenov, 1961
  Greenberg et al., 1964
MUSCIDAE: STOMOXYINAE
Stomoxys calcitrans (Linnaeus)
  Greenberg et al., 1964
CALLIPHORIDAE: CHRYSOMYINAE
Chrysomya megacephala (Fabricius)
  Shimizu et al., 1965
  How (original not seen)
Phormia regina (Meigen)
  Greenberg et al., 1964
Protophormia terraenovae (Robineau-Desvoidy)
  Zaidenov, 1961
CALLIPHORIDAE: CALLIPHORINAE
Calliphora vomitoria (Linnaeus)
  Zaidenov, 1961
UNDETERMINED LUCILIINI
  Shimizu et al., 1965

?*Shigella* Castellani and Chalmers ....

MUSCIDAE: PHAONIINAE
Muscina stabulans (Fallén)
  Sychevskaîa et al., 1959a
MUSCIDAE: MUSCINAE
Musca domestica vicina Macquart
  Sychevskaîa et al., 1959a
Musca larvipara Porchinskiĭ
  Sychevskaîa et al., 1959a
Musca osiris Wiedemann
  Sychevskaîa et al., 1959a

Organism	Fly / Reference
?Shigella (cont.)	*Musca sorbens* Wiedemann Sychevskai͡a et al., 1959a CALLIPHORIDAE: CALLIPHORINAE *Calliphora vicina* Robineau-Desvoidy Sychevskai͡a et al., 1959a SARCOPHAGIDAE: SARCOPHAGINAE *Sarcophaga haemorrhoidalis* (Fallén) Sychevskai͡a et al., 1959a
Shigella alkalescens (Andrewes) Weldin	SYRPHIDAE: MILESIINAE *Volucella obesa* (Fabricius) Alcivar et al., 1946 MUSCIDAE: MUSCINAE *Musca domestica* Linnaeus Alcivar et al., 1946 Gabaldón et al., 1956 CALLIPHORIDAE: CHRYSOMYINAE *Cochliomyia macellaria* (Fabricius) Alcivar et al., 1946 CALLIPHORIDAE: CALLIPHORINAE *Phaenicia sericata* (Meigen) Alcivar et al., 1946 SARCOPHAGIDAE: SARCOPHAGINAE *Sarcophaga* Meigen Alcivar et al., 1946
Shigella ambigua (Andrewes) Weldin	MUSCIDAE: MUSCINAE *Musca* Linnaeus de la Paz, 1938 CALLIPHORIDAE: CHRYSOMYINAE *Chrysomya* Robineau-Desvoidy de la Paz, 1938 SARCOPHAGIDAE: SARCOPHAGINAE *Sarcophaga* Meigen de la Paz, 1938
Shigella boydii Ewing	MUSCIDAE: MUSCINAE *Musca domestica* Linnaeus Bolaños, 1959
Shigella boydii Ewing 1, 2, 3, 4, 5, 6, 8, 9 (Synonym: *Shigella paradysenteriae*)	SYRPHIDAE: MILESIINAE *Volucella obesa* (Fabricius) Alcivar et al., 1946 MUSCIDAE: MUSCINAE *Musca domestica* Linnaeus Alcivar et al., 1946 CALLIPHORIDAE: CHRYSOMYINAE *Cochliomyia macellaria* (Fabricius) Alcivar et al., 1946 CALLIPHORIDAE: CALLIPHORINAE *Phaenicia sericata* (Meigen) Alcivar et al., 1946 SARCOPHAGIDAE: SARCOPHAGINAE *Sarcophaga* Meigen Alcivar et al., 1946

Shigella boydii Ewing 2 .... MUSCIDAE: PHAONIINAE
Ophyra Robineau-Desvoidy
Richards et al., 1961
Muscina Robineau-Desvoidy
Richards et al., 1961
MUSCIDAE: MUSCINAE
Musca domestica Linnaeus
Richards et al., 1961
CALLIPHORIDAE: CHRYSOMYINAE
Chrysomya Robineau-Desvoidy
Richards et al., 1961
CALLIPHORIDAE: CALLIPHORINAE
Phaenicia Robineau-Desvoidy
Richards et al., 1961
SARCOPHAGIDAE: SARCOPHAGINAE
Sarcophaga Meigen
Richards et al., 1961

Shigella boydii Ewing 5 .... MUSCIDAE: PHAONIINAE
Ophyra Robineau-Desvoidy
Richards et al., 1961
Muscina Robineau-Desvoidy
Richards et al., 1961
MUSCIDAE: MUSCINAE
Musca domestica Linnaeus
Richards et al., 1961
CALLIPHORIDAE: CHRYSOMYINAE
Chrysomya Robineau-Desvoidy
Richards et al., 1961
CALLIPHORIDAE: CALLIPHORINAE
Phaenicia Robineau-Desvoidy
Richards et al., 1961
SARCOPHAGIDAE: SARCOPHAGINAE
Sarcophaga Meigen
Richards et al., 1961

Shigella boydii Ewing 6 .... MUSCIDAE: PHAONIINAE
Ophyra Robineau-Desvoidy
Richards et al., 1961
Muscina Robineau-Desvoidy
Richards et al., 1961
MUSCIDAE: MUSCINAE
Musca domestica Linnaeus
Richards et al., 1961
CALLIPHORIDAE: CHRYSOMYINAE
Chrysomya Robineau-Desvoidy
Richards et al., 1961
CALLIPHORIDAE: CALLIPHORINAE
Phaenicia Robineau-Desvoidy
Richards et al., 1961

Organism	Fly associations
Shigella boydii Ewing 6 (cont.)	SARCOPHAGIDAE: SARCOPHAGINAE *Sarcophaga* Meigen Richards et al., 1961
Shigella ceylonensis (Weldin) Weldin	SYRPHIDAE: MILESIINAE *Volucella obesa* (Fabricius) Alcivar et al., 1946 MUSCIDAE: MUSCINAE *Musca domestica* Linnaeus Alcivar et al., 1946 CALLIPHORIDAE: CHRYSOMYINAE *Cochliomyia macellaria* (Fabricius) Alcivar et al., 1946 CALLIPHORIDAE: CALLIPHORINAE *Phaenicia sericata* (Meigen) Alcivar et al., 1946 SARCOPHAGIDAE: SARCOPHAGINAE *Sarcophaga* Meigen Alcivar et al., 1946
Shigella dispar (Andrewes) Bergey et al.	SYRPHIDAE: MILESIINAE *Volucella obesa* (Fabricius) Alcivar et al., 1946 MUSCIDAE: MUSCINAE *Musca domestica* Linnaeus Alcivar et al., 1946 Gabaldón et al. (in review), 1956 CALLIPHORIDAE: CHRYSOMYINAE *Cochliomyia macellaria* (Fabricius) Alcivar et al., 1946 CALLIPHORIDAE: CALLIPHORINAE *Phaenicia sericata* (Meigen) Alcivar et al., 1946 SARCOPHAGIDAE: SARCOPHAGINAE *Sarcophaga* Meigen Alcivar et al., 1946
Shigella dysenteriae (Shiga) Castellani and Chalmers (Synonyms: Shiga's bacillus, Bacillus dysenterique Shiga, Dysentery bacillus, Shiga-Kruse bacillus, *Shigella paradysenteriae* var. *Shiga-Kruse*)	DROSOPHILIDAE: DROSOPHILINAE *Drosophila ferruginea* Becker Beyer, 1925 ANTHOMYIIDAE: ANTHOMYIINAE *Hylemya* Robineau-Desvoidy Anonymous, 1952 MUSCIDAE: FANNIINAE *Fannia canicularis* (Linnaeus) Yao et al., 1929 *Fannia scalaris* (Fabricius) Yao et al., 1929 MUSCIDAE: PHAONIINAE *Ophyra aenescens* (Wiedemann) Beyer, 1925

Shigella dysenteriae (cont.)

MUSCIDAE: MUSCINAE

Synthesiomyia nudiseta (Wulp)
  Beyer, 1925

Musca Linnaeus
  de la Paz, 1938

Musca domestica Linnaeus
  Krontowski, 1913
  Manson-Bahr, 1919
  Paraf, 1920
  Wollman, 1921
  Reinstorf, 1923
  Duncan, 1926
  Yao et al., 1929
  R. M., 1937
  Emmel, 1949
  Hawley et al., 1951
  Boyd, 1957
  Stephens, 1963

House fly (presumably *Musca domestica* Linnaeus)
  Bahr, 1914

Musca domestica vicina Macquart
  Shterngol'd, 1949

MUSCIDAE: STOMOXYINAE

Stomoxys calcitrans (Linnaeus)
  Duncan, 1926

CALLIPHORIDAE: CHRYSOMYINAE

Chrysomya Robineau-Desvoidy
  de la Paz, 1938

Chrysomya megacephala (Fabricius)
  Chow, 1940

CALLIPHORIDAE: CALLIPHORINAE

Lucilia caesar (Linnaeus)
  Krontowski, 1913
  Wollman, 1921
  Beyer, 1925
  Yao et al., 1929

Calliphora vicina Robineau-Desvoidy
  Yao et al., 1929

Calliphora vomitoria (Linnaeus)
  Wollman, 1921
  Yao et al., 1929

Cynomya mortuorum (Linnaeus)
  Krontowski, 1913

SARCOPHAGIDAE: SARCOPHAGINAE

Sarcophaga Meigen
  de la Paz, 1938

Sarcophaga carnaria (Linnaeus)
  Krontowski, 1913
  Yao et al., 1929

Organism	Fly
Shigella dysenteriae (Shiga) Castellani and Chalmers 1 (Synonym: Grigor'ev-Shiga dysentery bacillus)	MUSCIDAE: MUSCINAE *Musca domestica vicina* Macquart Sychevskaia͡ et al., 1959a *Musca sorbens* Wiedemann Sychevskaia͡ et al., 1959a
Shigella dysenteriae (Shiga) Castellani and Chalmers 2 (Synonym: Schmitz-Shtutser dysentery bacillus)	MUSCIDAE: PHAONIINAE *Ophyra* Robineau-Desvoidy Richards et al., 1961 *Muscina* Robineau-Desvoidy Richards et al., 1961 MUSCIDAE: MUSCINAE *Musca domestica* Linnaeus Richards et al., 1961 *Musca domestica vicina* Macquart Floyd et al., 1953 Sychevskaia͡ et al., 1959a CALLIPHORIDAE: CHRYSOMYINAE *Chrysomya* Robineau-Desvoidy Richards et al., 1961 CALLIPHORIDAE: CALLIPHORINAE *Phaenicia* Robineau-Desvoidy Richards et al., 1961 *Phaenicia sericata* (Meigen) Sychevskaia͡ et al., 1959a *Calliphora vicina* Robineau-Desvoidy Sychevskaia͡ et al., 1959a SARCOPHAGIDAE: SARCOPHAGINAE *Sarcophaga* Meigen Richards et al., 1961 *Sarcophaga haemorrhoidalis* (Fallén) Sychevskaia͡ et al., 1959a
Shigella dysenteriae (Shiga) Castellani and Chalmers 3-7	MUSCIDAE: MUSCINAE *Musca domestica vicina* Macquart Floyd et al., 1953
Shigella flexneri Castellani and Chalmers (Synonyms: *Shigella paradysenteriae*, Flexner dysentery bacillus, *Bacillus dysenteriae* type "Y," *Pseudodysenteriae* A6154, *Pseudodysenteriae* D9112, *Shigella dysenteriae* Hiss-Flexner)	DROSOPHILIDAE: DROSOPHILINAE *Drosophila ananassae* Doleschall Gabaldón et al., 1956 MUSCIDAE: FANNIINAE *Fannia canicularis* (Linnaeus) Sychevskaia͡ et al., 1959a Arskiĭ et al., 1961 *Fannia scalaris* (Fabricius) Sychevskaia͡ et al., 1959a MUSCIDAE: PHAONIINAE *Muscina stabulans* (Fallén) Sychevskaia͡ et al., 1959a Arskiĭ et al., 1961

Organism	Fly / Reference
Shigella dysenteriae Castellani and Chalmers 3-7 (cont.)	MUSCIDAE: MUSCINAE
	Musca Linnaeus
	de la Paz, 1938
	Musca domestica Linnaeus
	Auché, 1906
	Tebbutt, 1913
	Emmel, 1949
	Gross et al., 1953
	Shura-Bura, 1955, 1957
	Gabaldón et al., 1956
	Greenberg, 1959c
	Reinstorf, 1923
	House fly (presumably *Musca domestica* Linnaeus)
	Auché, 1906
	Musca domestica vicina Macquart
	Shterngol'd, 1949
	Sukhova, 1954
	Shura-Bura, 1955
	Sychevskaia͡ et al., 1959a
	Arskiĭ et al., 1961
	Musca sorbens Wiedemann
	Sychevskaia͡ et al., 1959a
	CALLIPHORIDAE: CHRYSOMYINAE
	Chrysomya Robineau-Desvoidy
	de la Paz, 1938
	Chrysomya putoria (Wiedemann)
	Brygoo et al., 1962
	Phormia regina (Meigen)
	Greenberg, 1962
	CALLIPHORIDAE: CALLIPHORINAE
	Phaenicia sericata (Meigen)
	Gross et al., 1953
	Arskiĭ et al., 1961
	Greenberg, 1962
	Calliphora vicina Robineau-Desvoidy
	Gross et al., 1953
	Sychevskaia͡ et al., 1959a
	Greenberg, 1962
	Calliphora vomitoria (Linnaeus)
	Beyer, 1925
	Calliphora uralensis Villeneuve
	Sukhova, 1950
	Cynomyopsis cadaverina (Robineau-Desvoidy)
	Greenberg, 1962
	SARCOPHAGIDAE: SARCOPHAGINAE
	Sarcophaga Meigen
	de la Paz, 1938
	Pavan, 1949

Organism	Fly
Shigella dysenteriae Castellani and Chalmers 3-7 (cont.)	*Sarcophaga haemorrhoidalis* (Fallén) Gross et al., 1953 Arskiĭ et al., 1961 *Sarcophaga sarraceniae* Riley Beyer, 1925
Shigella ?flexneri Castellani and Chalmers (Synonym: *Shigella dysenteriae* Hiss-Flexner)	CALLIPHORIDAE: CALLIPHORINAE *Calliphora uralensis* Villeneuve Sukhova, 1950
Shigella flexneri Castellani and Chalmers 1a (Synonym: Flexner dysentery bacillus type "f")	MUSCIDAE: PHAONIINAE *Ophyra* Robineau-Desvoidy Richards et al., 1961 *Muscina* Robineau-Desvoidy Richards et al., 1961 *Muscina stabulans* (Fallén) Sychevskaîa et al., 1959b MUSCIDAE: MUSCINAE *Musca domestica* Linnaeus Sychevskaîa et al., 1959b Richards et al., 1961 *Musca domestica vicina* Macquart Sychevskaîa et al., 1959a, 1959b CALLIPHORIDAE: CHRYSOMYINAE *Chrysomya* Robineau-Desvoidy Richards et al., 1961 CALLIPHORIDAE: CALLIPHORINAE *Phaenicia* Robineau-Desvoidy Richards et al., 1961 *Phaenicia sericata* (Meigen) Sychevskaîa et al., 1959a *Calliphora vicina* Robineau-Desvoidy Sychevskaîa et al., 1959a SARCOPHAGIDAE: SARCOPHAGINAE *Ravinia striata* Fabricius Sychevskaîa et al., 1959b *Sarcophaga* Meigen Richards et al., 1961 *Sarcophaga haemorrhoidalis* (Fallén) Sychevskaîa et al., 1959a
Shigella flexneri Castellani and Chalmers 1a, 2a, Y (Synonym: *Shigella paradysenteriae* Flexner V, W, Y)	SYRPHIDAE: MILESIINAE *Volucella obesa* (Fabricius) Alcivar et al., 1946 MUSCIDAE: MUSCINAE *Musca domestica* Linnaeus Alcivar et al., 1946 CALLIPHORIDAE: CHRYSOMYINAE *Cochliomyia macellaria* (Fabricius) Alcivar et al., 1946

Organism	Fly / Reference
Shigella dysenteriae Castellani and Chalmers 1a, 2a, Y (cont.)	CALLIPHORIDAE: CALLIPHORINAE
	Phaenicia sericata (Meigen)
	Alcivar et al., 1946
	SARCOPHAGIDAE: SARCOPHAGINAE
	Sarcophaga Meigen
	Alcivar et al., 1946
Shigella flexneri Castellani and Chalmers 2a	MUSCIDAE: PHAONIINAE
	Ophyra Robineau-Desvoidy
	Richards et al., 1961
	Muscina Robineau-Desvoidy
	Richards et al., 1961
	MUSCIDAE: MUSCINAE
	Musca domestica Linnaeus
	Greenberg, 1959a
	Richards et al., 1961
	Evtodienko, 1968
	Musca domestica vicina Macquart
	Floyd et al., 1953
	CALLIPHORIDAE: CHRYSOMYINAE
	Chrysomya Robineau-Desvoidy
	Richards et al., 1961
	CALLIPHORIDAE: CALLIPHORINAE
	Phaenicia Robineau-Desvoidy
	Richards et al., 1961
	SARCOPHAGIDAE: SARCOPHAGINAE
	Sarcophaga Meigen
	Richards et al., 1961
Shigella flexneri Castellani and Chalmers 3	MUSCIDAE: PHAONIINAE
	Ophyra Robineau-Desvoidy
	Richards et al., 1961
	Muscina Robineau-Desvoidy
	Richards et al., 1961
	MUSCIDAE: MUSCINAE
	Musca domestica Linnaeus
	Richards et al., 1961
	CALLIPHORIDAE: CHRYSOMYINAE
	Chrysomya Robineau-Desvoidy
	Richards et al., 1961
	CALLIPHORIDAE: CALLIPHORINAE
	Phaenicia Robineau-Desvoidy
	Richards et al., 1961
	SARCOPHAGIDAE: SARCOPHAGINAE
	Sarcophaga Meigen
	Richards et al., 1961
Shigella flexneri Castellani and Chalmers 4a	MUSCIDAE: PHAONIINAE
	Ophyra Robineau-Desvoidy
	Richards et al., 1961
	Muscina Robineau-Desvoidy
	Richards et al., 1961

Organism	Fly association
Shigella flexneri (cont.)	MUSCIDAE: MUSCINAE *Musca domestica* Linnaeus Richards et al., 1961 CALLIPHORIDAE: CHRYSOMYINAE *Chrysomya* Robineau-Desvoidy Richards et al., 1961 CALLIPHORIDAE: CALLIPHORINAE *Phaenicia* Robineau-Desvoidy Richards et al., 1961 SARCOPHAGIDAE: SARCOPHAGINAE *Sarcophaga* Meigen Richards et al., 1961
Shigella flexneri .... Castellani and Chalmers 4b	MUSCIDAE: PHAONIINAE *Ophyra* Robineau-Desvoidy Richards et al., 1961 *Muscina* Robineau-Desvoidy Richards et al., 1961 MUSCIDAE: MUSCINAE *Musca domestica* Linnaeus Richards et al., 1961 CALLIPHORIDAE: CHRYSOMYINAE *Chrysomya* Robineau-Desvoidy Richards et al., 1961 CALLIPHORIDAE: CALLIPHORINAE *Phaenicia* Robineau-Desvoidy Richards et al., 1961 SARCOPHAGIDAE: SARCOPHAGINAE *Sarcophaga* Meigen Richards et al., 1961
Shigella flexneri .... Castellani and Chalmers 5	MUSCIDAE: MUSCINAE *Musca domestica vicina* Macquart Floyd et al., 1953
Shigella flexneri .... Castellani and Chalmers 6 (Synonyms: *Shigella paradysenteriae* Boyd 88, *Shigella newcastle*, Newcastle dysentery bacillus)	SYRPHIDAE: MILESIINAE *Volucella obesa* (Fabricius) Alcivar et al., 1946 MUSCIDAE: PHAONIINAE *Ophyra* Robineau-Desvoidy Richards et al., 1961 *Muscina* Robineau-Desvoidy Richards et al., 1961 MUSCIDAE: MUSCINAE *Musca domestica* Linnaeus Kuhns et al., 1944 Alcivar et al., 1946 Prado et al., 1955 Richards et al., 1961 *Musca domestica vicina* Macquart Sychevskaîa et al., 1959b

Organism	Fly
Shigella dysenteriae Castellani and Chalmers 6 (cont.)	CALLIPHORIDAE: CHRYSOMYINAE *Cochliomyia macellaria* (Fabricius) Alcivar et al., 1946 *Chrysomya* Robineau-Desvoidy Richards et al., 1961 CALLIPHORIDAE: CALLIPHORINAE *Phaenicia* Robineau-Desvoidy Richards et al., 1961 *Phaenicia sericata* (Meigen) Alcivar et al., 1946 SARCOPHAGIDAE: SARCOPHAGINAE *Sarcophaga* Meigen Alcivar et al., 1946 Richards et al., 1961 *Sarcophaga haemorrhoidalis* (Fallén) Sychevskaîa et al., 1959a
Shigella flexneri Castellani and Chalmers X (Synonym: Shigella variant "X" [flexner 7])	MUSCIDAE: MUSCINAE *Musca domestica vicina* Macquart Floyd et al., 1953
Shigella flexneri Castellani and Chalmers Y (Synonyms: Hiss and Russel Y bacillus, *Shigella paradysenteriae* var. Hiss-Russel)	MUSCIDAE: MUSCINAE House fly (presumably *Musca domestica* Linnaeus) Bahr, 1914 *Musca domestica vicina* Macquart Shterngol'd, 1949
Shigella giumai (Castellani) Hauduroy et al.	MUSCIDAE: MUSCINAE *Musca* Linnaeus de la Paz, 1938 CALLIPHORIDAE: CHRYSOMYINAE *Chrysomya* Robineau-Desvoidy de la Paz, 1938
Shigella madampensis (Castellani) Weldin	MUSCIDAE: MUSCINAE *Musca* Linnaeus de la Paz, 1938
Shigella pfaffi (Hadley et al.) Weldin	MUSCIDAE: MUSCINAE *Musca* Linnaeus de la Paz, 1938
Shigella sonnei (Levine) Weldin	MUSCIDAE: FANNIINAE *Fannia canicularis* (Linnaeus) Arskiĭ et al., 1961 MUSCIDAE: PHAONIINAE *Muscina stabulans* (Fallén) Arskiĭ et al., 1961 MUSCIDAE: MUSCINAE *Musca* Linnaeus de la Paz, 1938 *Musca domestica* Linnaeus Gross et al., 1953 Shura-Bura, 1955

Organism	Fly / References
Shigella sonnei (cont.)	Gabaldón et al., 1956
	Radvan, 1956, 1960a, 1960b, 1960c
	Evtodienko, 1968
	Musca domestica vicina Macquart
	Floyd et al., 1953
	Arskiĭ et al., 1961
	CALLIPHORIDAE: CHRYSOMYINAE
	Chrysomya Robineau-Desvoidy
	de la Paz, 1938
	Protophormia terraenovae (Robineau-Desvoidy)
	Radvan, 1960a, 1960b, 1960c
	CALLIPHORIDAE: CALLIPHORINAE
	Phaenicia sericata Meigen
	Gross et al., 1953
	Arskiĭ et al., 1961
	Calliphora vicina Robineau-Desvoidy
	Gross et al., 1953
	Greenberg, 1962
	SARCOPHAGIDAE: SARCOPHAGINAE
	Sarcophaga haemorrhoidalis (Fallén)
	Gross et al., 1953
	Arskiĭ et al., 1961
Shigella sonnei (Levine) Weldin 1	MUSCIDAE: PHAONIINAE
	Ophyra Robineau-Desvoidy
	Richards et al., 1961
	Muscina Robineau-Desvoidy
	Richards et al., 1961
	MUSCIDAE: MUSCINAE
	Musca domestica Linnaeus
	Richards et al., 1961
	CALLIPHORIDAE: CHRYSOMYINAE
	Chrysomya Robineau-Desvoidy
	Richards et al., 1961
	CALLIPHORIDAE: CALLIPHORINAE
	Phaenicia Robineau-Desvoidy
	Richards et al., 1961
	SARCOPHAGIDAE: SARCOPHAGINAE
	Sarcophaga Meigen
	Richards et al., 1961
UNDETERMINED ENTEROBACTERIACEAE	
Coliform (Synonym: *Coli-aerogenes*)	MUSCIDAE: MUSCINAE
	Musca domestica Linnaeus
	Ficker, 1903
	Peppler, 1944
	House fly (presumably *Musca domestica* Linnaeus)
	Geldreich et al., 1963

Organism	Fly
Coliform (cont.)	*Musca domestica vicina* Macquart McGuire et al., 1957 *Musca sorbens* Wiedemann McGuire et al., 1957 CALLIPHORIDAE: CALLIPHORINAE *Calliphora vomitoria* (Linnaeus) Testi, 1909
Klebsiella-Cloaca intermediate forms	PIOPHILIDAE *Piophila casei* (Linnaeus) Lysenko et al., 1961 MUSCIDAE: MUSCINAE *Musca domestica* Linnaeus Lysenko et al., 1961 CALLIPHORIDAE: CHRYSOMYINAE *Protophormia terraenovae* (Robineau-Desvoidy) Lysenko et al., 1961 CALLIPHORIDAE: CALLIPHORINAE *Phaenicia sericata* (Meigen) Lysenko et al., 1961
MacConkey's bacillus No. 7, unvalidated	MUSCIDAE: MUSCINAE *Musca domestica* Linnaeus Nicoll, 1911a
MacConkey's bacillus No. 36, unvalidated	MUSCIDAE: MUSCINAE *Musca domestica* Linnaeus Nicoll, 1911a
MacConkey's bacillus No. 36a, unvalidated	MUSCIDAE: MUSCINAE *Musca domestica* Linnaeus Nicoll, 1911a
MacConkey's bacillus No. 74, unvalidated	MUSCIDAE: MUSCINAE *Musca domestica* Linnaeus Nicoll, 1911a
MacConkey's bacillus No. 109, unvalidated	MUSCIDAE: MUSCINAE *Musca domestica* Linnaeus Nicoll, 1911a
Paraescherichia, unvalidated	MUSCIDAE: FANNIINAE *Fannia canicularis* (Linnaeus) Zmeev, 1943 MUSCIDAE: PHAONIINAE *Phaonia querceti* (Bouché) Zmeev, 1943 CALLIPHORIDAE: CALLIPHORINAE *Phaenicia sericata* (Meigen) Zmeev, 1943
BRUCELLACEAE	
Pasteurella boviseptica (Kruse) Holland	MUSCIDAE: STOMOXYINAE *Haematobia exigua* (de Meijere) Nieschulz et al., 1929

Pasteurella multocida (Lehmann and Neumann) Rosenbusch and Merchant (Synonyms: *Bacillus cuniculicida*, *Septicaemia haemorrhagica bubalorum*, bacillus of swine plague)	MUSCIDAE: MUSCINAE
	Musca domestica Linnaeus
	Scott, 1917a, 1917b
	Shope, 1927
	Skidmore, 1932
	Musca inferior Stein
	Nieschulz et al., 1929
	MUSCIDAE: STOMOXYINAE
	Stomoxys calcitrans (Linnaeus)
	Nieschulz et al., 1929
Pasteurella pestis (Lehmann and Neumann) Holland (Synonyms: plague bacillus, *Bacillus pestis*)	ANTHOMYIIDAE: ANTHOMYIINAE
	Hylemya Robineau-Desvoidy
	Anonymous, 1952
	MUSCIDAE: MUSCINAE
	Musca domestica Linnaeus
	Nuttall, 1897
	Hunter, 1906
	Gosio, 1925
	Duncan, 1926
	Wollman, 1927
	Russo, 1930
	MUSCIDAE: STOMOXYINAE
	Stomoxys calcitrans (Linnaeus)
	Duncan, 1926
	CALLIPHORIDAE: CHRYSOMYINAE
	Cochliomyia macellaria (Fabricius)
	Gosio, 1925
	CALLIPHORIDAE: CALLIPHORINAE
	Lucilia Robineau-Desvoidy
	Lang, 1940
	Lucilia caesar (Linnaeus)
	Russo, 1930
	Calliphora vomitoria (Linnaeus)
	Gosio, 1925
	Russo, 1930
	SARCOPHAGIDAE: SARCOPHAGINAE
	Sarcophaga carnaria (Linnaeus)
	Petragnani, 1925
	Russo, 1930
Pasteurella tularensis (McCoy and Chapin) Bergey et al. (Synonym: *B. tularense*)	MUSCIDAE: MUSCINAE
	Musca domestica Linnaeus
	Wayson, 1914
	MUSCIDAE: STOMOXYINAE
	Stomoxys calcitrans (Linnaeus)
	Wayson, 1914
	Somov et al., 1937
	Olsufiev, 1940
	Romanova, 1947
	Bozhenko et al., 1948

Organism	Fly
Brucella (Meyer and Shaw) Pribram	MUSCIDAE: MUSCINAE *Musca domestica* Linnaeus Antonov, 1945
Brucella abortus (Schmidt) Meyer and Shaw	MUSCIDAE: PHAONIINAE *Muscina stabulans* (Fallén) Ruhland et al., 1941 MUSCIDAE: MUSCINAE *Musca domestica* Linnaeus Duncan, 1926 Wollman, 1927 Ruhland et al., 1941 MUSCIDAE: STOMOXYINAE *Stomoxys calcitrans* (Linnaeus) Duncan, 1926 Ruhland et al., 1941 Wellmann, 1950/51, 1959 CALLIPHORIDAE: CALLIPHORINAE *Lucilia* Robineau-Desvoidy Ruhland et al., 1941 *Lucilia caesar* (Linnaeus) Wollman, 1927 *Eucalliphora lilaea* (Walker) Gwatkin et al., 1938 *Calliphora* Robineau-Desvoidy Ruhland et al., 1941 *Calliphora terraenovae* Macquart Gwatkin et al., 1938 *Calliphora vicina* Robineau-Desvoidy Gwatkin et al., 1938 *Cynomyopsis cadaverina* (Robineau-Desvoidy) Gwatkin et al., 1938
Brucella melitensis (Hughes) Meyer and Shaw (Synonym: *B. melitensis*)	MUSCIDAE: MUSCINAE *Musca domestica* Linnaeus Duncan, 1926 MUSCIDAE: STOMOXYINAE *Stomoxys calcitrans* (Linnaeus) Eyre et al., 1907 Duncan, 1926 Wellmann, 1950/51, 1959 SARCOPHAGIDAE: SARCOPHAGINAE *Sarcophaga* Meigen Pavan, 1949
Brucella suis Huddleson	MUSCIDAE: STOMOXYINAE *Stomoxys calcitrans* (Linnaeus) Wellmann, 1950/51, 1959
Haemophilus influenzae (Lehmann and Neumann) Winslow et al.	MUSCIDAE: MUSCINAE *Musca domestica* Linnaeus Nicolle et al., 1919 Wollman, 1927

Organism	Fly association
Haemophilus influenzae (cont.)	*Musca domestica vicina* Macquart McGuire et al., 1957 *Musca sorbens* Wiedemann McGuire et al., 1957 CALLIPHORIDAE: CALLIPHORINAE *Calliphora vicina* Robineau-Desvoidy Greenberg, 1962
Haemophilus ?influenzae (Lehmann and Neumann) Winslow et al.	MUSCIDAE: MUSCINAE *Musca sorbens* Wiedemann Sukhova, 1953
Moraxella bovis (Hauduroy et al.) Murray	MUSCIDAE: MUSCINAE *Musca autumnalis* De Geer Brown, 1965 Steve et al., 1965
Moraxella lacunata (Eyre) Lwoff	MUSCIDAE: MUSCINAE *Musca domestica* Linnaeus Wollman, 1927
Cillopasteurella delendae-muscae (Roubaud and Descazeaux) Prévot (Synonym: *Bacterium delendae-muscae*)	MUSCIDAE: MUSCINAE *Musca domestica* Linnaeus Roubaud et al., 1923 MUSCIDAE: STOMOXYINAE *Stomoxys calcitrans* (Linnaeus) Roubaud et al., 1923 CALLIPHORIDAE: CALLIPHORINAE ?*Calliphora vomitoria* (Linnaeus) Roubaud et al., 1923
BACTEROIDACEAE	
Bacteroides vulgatus Eggerth and Gagnon	MUSCIDAE: MUSCINAE *Musca domestica* Linnaeus Duncan, 1926 MUSCIDAE: STOMOXYINAE *Stomoxys calcitrans* (Linnaeus) Duncan, 1926
MICROCOCCACEAE	
Micrococcus Cohn	SYRPHIDAE: MILESIINAE *Eristalis tenax* (Linnaeus) Maddox, 1885b MUSCIDAE: MUSCINAE *Musca domestica* Linnaeus Smuidsinovicia, 1889 *Musca domestica* Linnaeus Prado et al., 1955 Coutinho et al., 1957 House fly (presumably *Musca domestica* Linnaeus) Ramirez, 1898 CALLIPHORIDAE: CALLIPHORINAE *Calliphora vomitoria* (Linnaeus) Maddox, 1885b Bogdanow, 1906, 1908

Organism	Fly association
Micrococcus candidus Cohn	MUSCIDAE: STOMOXYINAE *Haematobia irritans* (Linnaeus) Stirrat et al., 1955
Micrococcus caseolyticus Evans	MUSCIDAE: STOMOXYINAE *Haematobia irritans* (Linnaeus) Stirrat et al., 1955
Micrococcus rushmorei Brown	CALLIPHORIDAE: CALLIPHORINAE *Phaenicia sericata* (Meigen) Brown, 1927
Staphylococcus Rosenbach (beta-hemolytic)	MUSCIDAE: MUSCINAE *Musca domestica vicina* Macquart McGuire et al., 1957 *Musca sorbens* Wiedemann McGuire et al., 1957
Staphylococcus Rosenbach (non-hemolytic)	MUSCIDAE: MUSCINAE *Musca domestica vicina* Macquart McGuire et al., 1957 *Musca sorbens* Wiedemann McGuire et al., 1957
Staphylococcus Rosenbach (non-pathogenic)	MUSCIDAE: MUSCINAE *Musca domestica* Linnaeus Ewing, 1942
Staphylococcus Rosenbach (pathogenic)	MUSCIDAE: MUSCINAE *Musca domestica* Linnaeus Ewing, 1942 *Musca domestica vicina* Macquart Brygoo et al., 1962 MUSCIDAE: STOMOXYINAE *Stomoxys calcitrans* (Linnaeus) Brygoo et al., 1962 CALLIPHORIDAE: CHRYSOMYINAE *Chrysomya putoria* (Wiedemann) Brygoo et al., 1962
Staphylococcus Rosenbach	MUSCIDAE: MUSCINAE *Musca* Linnaeus de la Paz, 1938 *Musca domestica* Linnaeus Cox et al., 1912 Picado, 1935 CALLIPHORIDAE: CHRYSOMYINAE *Chrysomya* Robineau-Desvoidy de la Paz, 1938 *Phormia regina* (Meigen) Picado, 1935 CALLIPHORIDAE: CALLIPHORINAE *Phaenicia eximia* (Wiedemann) Picado, 1935

Staphylococcus (cont.)	SARCOPHAGIDAE: SARCOPHAGINAE *Sarcophaga* Meigen Picado, 1935 de la Paz, 1938
Staphylococcus afermentans Shaw, Stitt, and Cowan	PIOPHILIDAE *Piophila casei* (Linnaeus) Gregor et al., 1960 SPHAEROCERIDAE *Leptocera ferruginata* (Stenhammar) Lysenko, 1958 MUSCIDAE: FANNIINAE *Fannia canicularis* (Linnaeus) Gregor et al., 1960 MUSCIDAE: PHAONIINAE *Hydrotaea dentipes* (Fabricius) Lysenko, 1958 Gregor et al., 1960 *Hydrotaea irritans* (Fallén) Gregor et al., 1960 *Ophyra leucostoma* (Wiedemann) Gregor et al., 1960 *Muscina stabulans* (Fallén) Gregor et al., 1960 CALLIPHORIDAE: CHRYSOMYINAE *Protophormia terraenovae* (Robineau-Desvoidy) Gregor et al., 1960 CALLIPHORIDAE: CALLIPHORINAE *Lucilia caesar* (Linnaeus) Gregor et al., 1960 *Phaenicia sericata* (Meigen) Gregor et al., 1960 *Calliphora vicina* Robineau-Desvoidy Gregor et al., 1960 *Calliphora vomitoria* (Linnaeus) Gregor et al., 1960
Staphylococcus aureus Rosenbach (Synonyms: *Staphylococcus pyogenes, Micrococcus pyogenes* var. *aureus, Staphylococcus albus, Micrococcus aureus, Micrococcus citreus, Micrococcus pyogenes* var. *albus, Staphylococcus pyogenes* var. *aureus, Micrococcus albus*)	OTITIDAE: ULIDIINAE *Physiphora demandata* (Fabricius) Povolný et al., 1961 SEPSIDAE *Sepsis* Fallén Nicholls, 1912 *Sepsis punctum* (Fabricius) Povolný et al., 1961 PIOPHILIDAE *Piophila casei* (Linnaeus) Gregor et al., 1960 Povolný et al., 1961

Staphylococcus aureus (cont.)

SPHAEROCERIDAE
Limosina punctipennis (Wiedemann)
  Nicholls, 1912

UNDETERMINED SPHAEROCERIDAE
  Povolný et al., 1961

DROSOPHILIDAE: DROSOPHILINAE
Drosophila melanogaster Meigen
  Nicholls, 1912

UNDETERMINED DROSOPHILIDAE
  Povolný et al., 1961

CHLOROPIDAE: OSCINELLINAE
Hippelates flavipes Loew
  Nicholls, 1912
  Taplin et al., 1965
Siphunculina funicola de Miejere
  Syddiq, 1938

ANTHOMYIIDAE: ANTHOMYIINAE
Hylemya cinerella (Fallén)
  Povolný et al., 1961
Anthomyia pluvialis (Linnaeus)
  Povolný et al., 1961

MUSCIDAE: FANNIINAE
Fannia canicularis (Linnaeus)
  Gregor et al., 1960
  Povolný et al., 1961

MUSCIDAE: PHAONIINAE
Hydrotaea dentipes (Fabricius)
  Gregor et al., 1960
  Povolný et al., 1961
Hydrotaea irritans (Fallén)
  Gregor et al., 1960
Hydrotaea occulta (Meigen)
  Povolný et al., 1961
Ophyra leucostoma (Wiedemann)
  Gregor et al., 1960
Muscina stabulans (Fallén)
  Gregor et al., 1960

MUSCIDAE: MUSCINAE
Musca domestica Linnaeus
  Buchanan, 1907
  Herms, 1911
  Scott, 1917a
  Reinstorf, 1923
  Duncan, 1926
  Shope, 1927
  Hawley et al., 1951
  Gerberich, 1951
  Coutinho et al., 1957
  Radvan, 1960a, 1960b, 1960c

Staphylococcus aureus (cont.)

Povolný et al., 1961
Greenberg et al., 1963a
House fly (presumably *Musca domestica* Linnaeus)
Celli et al., 1888
Manning, 1902
MUSCIDAE: STOMOXYINAE
Stomoxys calcitrans (Linnaeus)
Duncan, 1926
CALLIPHORIDAE: CHRYSOMYINAE
Phormia regina (Meigen)
Maseritz, 1934
Greenberg, 1962
Protophormia terraenovae (Robineau-Desvoidy)
Pavillard et al., 1957
Gregor et al., 1960
Radvan, 1960a, 1960c
CALLIPHORIDAE: CALLIPHORINAE
Lucilia caesar (Linnaeus)
Beyer, 1925
Weil et al., 1933
Gregor et al., 1960
Phaenicia sericata (Meigen)
Michelbacher et al., 1932
Robinson et al., 1933
Weil et al., 1933
Robinson et al., 1934
Simmons, 1935a, 1935b
Gregor et al., 1960
Greenberg, 1962
Eucalliphora lilaea (Walker)
Gwatkin et al., 1938
Calliphora terraenovae Macquart
Gwatkin et al., 1938
Calliphora vicina Robineau-Desvoidy
Gwatkin et al., 1938
Gregor et al., 1960
Greenberg, 1962
Calliphora vomitoria (Linnaeus)
Wollman, 1911
Gregor et al., 1960
Cynomyopsis cadaverina (Robineau-Desvoidy)
Gwatkin et al., 1938
Greenberg, 1962
SARCOPHAGIDAE: SARCOPHAGINAE
Sarcophaga Meigen
Nicholls, 1912
Pavan, 1949

Organism	Fly
Staphylococcus aureus (cont.)	*Sarcophaga aurifinis* Walker Nicholls, 1912 *Sarcophaga carnaria* (Linnaeus) Testi, 1909 *Tricharaea* Thomson Nicholls, 1912
Staphylococcus epidermidis (Winslow and Winslow) Evans (Synonym: *Micrococcus epidermidis*)	MUSCIDAE: STOMOXYINAE *Haematobia irritans* (Linnaeus) Stirrat et al., 1955 CALLIPHORIDAE: CALLIPHORINAE *Lucilia caesar* (Linnaeus) Weil et al., 1933 *Phaenicia sericata* (Meigen) Weil et al., 1933
Staphylococcus flavus (Trevisan) Wood (Synonyms: *Micrococcus flavus*, *Micrococcus flavus liquefaciens*)	MUSCIDAE: STOMOXYINAE *Haematobia irritans* (Linnaeus) Stirrat et al., 1955 CALLIPHORIDAE: CHRYSOMYINAE *Phormia regina* (Meigen) Guyénot, 1907 Maseritz, 1934 CALLIPHORIDAE: CALLIPHORINAE *Lucilia caesar* (Linnaeus) Guyénot, 1907
Staphylococcus lactis Shaw, Stitt, and Cowan	PIOPHILIDAE *Piophila casei* (Linnaeus) Gregor et al., 1960 SPHAEROCERIDAE *Leptocera ferruginata* (Stenhammar) Lysenko, 1958 MUSCIDAE: FANNIINAE *Fannia canicularis* (Linnaeus) Gregor et al., 1960 MUSCIDAE: PHAONIINAE *Hydrotaea dentipes* (Fabricius) Gregor et al., 1960 *Hydrotaea irritans* (Fallén) Gregor et al., 1960 *Ophyra leucostoma* (Wiedemann) Gregor et al., 1960 *Muscina stabulans* (Fallén) Gregor et al., 1960 CALLIPHORIDAE: CHRYSOMYINAE *Protophormia terraenovae* (Robineau-Desvoidy) Lysenko, 1958 Gregor et al., 1960 Lysenko et al., 1961

Organism	Fly / Reference
Staphylococcus lactis (cont.)	CALLIPHORIDAE: CALLIPHORINAE *Lucilia caesar* (Linnaeus) Gregor et al., 1960 *Phaenicia sericata* (Meigen) Gregor et al., 1960 Lysenko et al., 1961 *Calliphora vicina* Robineau-Desvoidy Gregor et al., 1960 *Calliphora vomitoria* (Linnaeus) Gregor et al., 1960
Staphylococcus muscae Glaser	MUSCIDAE: MUSCINAE *Musca domestica* Linnaeus Glaser, 1924, 1938 Shope, 1927 House fly (presumably *Musca domestica* Linnaeus) Glaser, 1926
Staphylococcus saprophyticus ...... Fairbrother	PIOPHILIDAE *Piophila casei* (Linnaeus) Gregor et al., 1960 Lysenko et al., 1961 SPHAEROCERIDAE *Leptocera ferruginata* (Stenhammar) Lysenko, 1958 ANTHOMYIIDAE: ANTHOMYIINAE *Hylemya cinerella* (Fallén) Lysenko, 1958 MUSCIDAE: FANNIINAE *Fannia canicularis* (Linnaeus) Gregor et al., 1960 MUSCIDAE: PHAONIINAE *Hydrotaea dentipes* (Fabricius) Lysenko, 1958 Gregor et al., 1960 *Hydrotaea irritans* (Fallén) Gregor et al., 1960 *Ophyra leucostoma* (Wiedemann) Gregor et al., 1960 *Muscina stabulans* (Fallén) Gregor et al., 1960 MUSCIDAE: MUSCINAE *Musca domestica* Linnaeus Lysenko, 1958 Lysenko et al., 1961 CALLIPHORIDAE: CHRYSOMYINAE *Protophormia terraenovae* (Robineau-Desvoidy) Gregor et al., 1960 Lysenko et al., 1961

Organism	Fly association
Staphylococcus saprophyticus (cont.)	CALLIPHORIDAE: CALLIPHORINAE *Lucilia caesar* (Linnaeus) Gregor et al., 1960 *Phaenicia sericata* (Meigen) Gregor et al., 1960 Lysenko et al., 1961 *Calliphora vicina* Robineau-Desvoidy Gregor et al., 1960 *Calliphora vomitoria* (Linnaeus) Gregor et al., 1960
Gaffkya Trevisan	MUSCIDAE: MUSCINAE *Musca domestica vicina* Macquart McGuire et al., 1957 *Musca sorbens* Wiedemann McGuire et al., 1957
Gaffkya tetragena (Gaffky) Trevisan (Synonym: *Gaffkya tetragenus*)	MUSCIDAE: MUSCINAE *Musca domestica* Linnaeus Scott, 1917a CALLIPHORIDAE: CALLIPHORINAE *Lucilia caesar* (Linnaeus) Weil et al., 1933 *Phaenicia sericata* (Meigen) Weil et al., 1933 *Calliphora vomitoria* (Linnaeus) Testi, 1909
Sarcina Goodsir	MUSCIDAE: MUSCINAE *Musca domestica* Linnaeus Hamilton, 1903 Cox et al., 1912 Reinstorf, 1923 *Musca domestica vicina* Macquart Silverman et al., 1953 McGuire et al., 1957 *Musca sorbens* Wiedemann McGuire et al., 1957 CALLIPHORIDAE: CALLIPHORINAE *Phaenicia sericata* (Meigen) Michelbacher et al., 1932
Sarcina alba Zimmermann	MUSCIDAE: MUSCINAE House fly (presumably *Musca domestica* Linnaeus) Manning, 1902
Sarcina aurantiaca Flügge (Synonym: *Sarcina aurantia*)	MUSCIDAE: MUSCINAE House fly (presumably *Musca domestica* Linnaeus) Manning, 1902
Sarcina flava de Bary	MUSCIDAE: MUSCINAE *Musca domestica* Linnaeus Lysenko et al., 1961

Organism	Fly association
Sarcina flava (cont.)	CALLIPHORIDAE: CHRYSOMYINAE *Protophormia terraenovae* (Robineau-Desvoidy) Lysenko et al., 1961 CALLIPHORIDAE: CALLIPHORINAE *Phaenicia sericata* (Meigen) Lysenko et al., 1961
Sarcina loewenbergi Macé	MUSCIDAE: MUSCINAE *Musca domestica* Linnaeus Galli-Vallerio, 1908 CALLIPHORIDAE: CALLIPHORINAE *Calliphora vicina* Robineau-Desvoidy Galli-Vallerio, 1908
Sarcina lutea Schroeter	MUSCIDAE: MUSCINAE *Musca domestica* Linnaeus Hawley et al., 1951
Sarcina subflava Ravenel	CALLIPHORIDAE: CHRYSOMYINAE *Phormia regina* (Meigen) Beyer, 1925
NEISSERIACEAE	
Neiserria catarrhalis (Froschand and Kolle) Holland	MUSCIDAE: PHAONIINAE *Muscina stabulans* (Fallén) Beyer, 1925 MUSCIDAE: MUSCINAE *Musca domestica* Linnaeus Beyer, 1925 SARCOPHAGIDAE: SARCOPHAGINAE *Sarcophaga argyrostoma* (Robineau-Desvoidy) Beyer, 1925
Neisseria gonorrhoeae Trevisan	MUSCIDAE: MUSCINAE *Musca domestica vicina* Macquart McGuire et al., 1957 *Musca sorbens* Wiedemann McGuire et al., 1957
Neisseria luciliarum Brown	CALLIPHORIDAE: CALLIPHORINAE *Phaenicia sericata* (Meigen) Brown, 1927
BREVIBACTERIACEAE	
Brevibacterium ammoniagenes (Cooke and Keith) Breed (Synonym: *Bacterium ammoniagenes*)	MUSCIDAE: MUSCINAE *Musca domestica* Linnaeus Hawley et al., 1951
Kurthia zenkeri (Hauser) Bergey et al. (Synonym: *Proteus zenkeri*)	MUSCIDAE: MUSCINAE *Musca domestica* Linnaeus Cao, 1906 CALLIPHORIDAE: CALLIPHORINAE *Lucilia caesar* (Linnaeus) Cao, 1906

Kurthia zenkeri (cont.)	*Calliphora vomitoria* (Linnaeus) Cao, 1906 SARCOPHAGIDAE: SARCOPHAGINAE *Sarcophaga carnaria* (Linnaeus) Cao, 1906
Kurthia zopfii (Kurth) (Synonym: *zopfii*)	MUSCIDAE: MUSCINAE *Musca domestica* Linnaeus Hamilton, 1903
LACTOBACILLACEAE	
Diplococcus pneumoniae Weichselbaum (Synonym: *Pneumococcus*)	OTITIDAE: ULIDIINAE *Physiphora demandata* (Fabricius) Povolný et al., 1961 *Sepsis punctum* (Fabricius) Povolný et al., 1961 PIOPHILIDAE *Piophila casei* (Linnaeus) Povolný et al., 1961 UNDETERMINED SPHAEROCERIDAE Povolný et al., 1961 UNDETERMINED DROSOPHILIDAE Povolný et al., 1961 ANTHOMYIIDAE: ANTHOMYIINAE *Hylemya cinerella* (Fallén) Povolný et al., 1961 *Anthomyia pluvialis* (Linnaeus) Povolný et al., 1961 MUSCIDAE: FANNIINAE *Fannia scalaris* (Fabricius) Povolný et al., 1961 MUSCIDAE: PHAONIINAE *Hydrotaea dentipes* (Fabricius) Povolný et al., 1961 *Hydrotaea occulta* (Meigen) Povolný et al., 1961 MUSCIDAE: MUSCINAE *Musca domestica* Linnaeus Povolný et al., 1961 MUSCIDAE: STOMOXYINAE *Stomoxys calcitrans* (Linnaeus) Povolný et al., 1961 CALLIPHORIDAE: CHRYSOMYINAE *Phormia regina* (Meigen) Maseritz, 1934 CALLIPHORIDAE: CALLIPHORINAE *Phaenicia sericata* (Meigen) Gregor et al., 1960 Povolný et al., 1961 *Calliphora vicina* Robineau-Desvoidy Povolný et al., 1961

Organism	Fly association
Diplococcus pneumoniae (cont.)	SARCOPHAGIDAE: SARCOPHAGINAE *Sarcophaga carnaria* (Linnaeus) Povolný et al., 1961 *Sarcophaga haemorrhoidalis* (Fallén) Povolný et al., 1961
Diplococcus pneumoniae Weichselbaum I (Synonym: *Pneumococcus* I)	MUSCIDAE: MUSCINAE *Musca domestica* Linnaeus Shope, 1927 CALLIPHORIDAE: CHRYSOMYINAE *Protophormia terraenovae* (Robineau-Desvoidy) Pavillard et al., 1957
Diplococcus pneumoniae Weichselbaum II (Synonym: *Pneumococcus* II)	MUSCIDAE: MUSCINAE *Musca domestica* Linnaeus Shope, 1927
Streptococcus Rosenbach (alpha-hemolytic)	MUSCIDAE: MUSCINAE *Musca domestica vicina* Macquart McGuire et al., 1957 *Musca sorbens* Wiedemann McGuire et al., 1957 CALLIPHORIDAE: CALLIPHORINAE *Lucilia caesar* (Linnaeus) Weil et al., 1933 *Phaenicia sericata* (Meigen) Weil et al., 1932
Streptococcus Rosenbach (beta-hemolytic)	MUSCIDAE: MUSCINAE *Musca domestica vicina* Macquart McGuire et al., 1957 *Musca sorbens* Wiedemann McGuire et al., 1957 CALLIPHORIDAE: CALLIPHORINAE *Phaenicia sericata* (Meigen) Robinson et al., 1934
Streptococcus Rosenbach (non-hemolytic)	CHLOROPIDAE: OSCINELLINAE *Siphunculina funicola* de Meijere Syddiq, 1938 MUSCIDAE: MUSCINAE *Musca domestica vicina* Macquart McGuire et al., 1957 *Musca sorbens* Wiedemann McGuire et al., 1957
Streptococcus Rosenbach (Synonym: ?*Enterococcus*)	CHLOROPIDAE: OSCINELLINAE *Siphunculina funicola* de Meijere Syddiq, 1938 MUSCIDAE: PHAONIINAE *Ophyra leucostoma* (Wiedemann) Povolný et al., 1961 *Muscina stabulans* (Fallén) Povolný et al., 1961

Organism	Fly association
Streptococcus (cont.)	MUSCIDAE: MUSCINAE *Musca* Linnaeus de la Paz, 1938 *Musca domestica* Linnaeus Ledingham, 1911 Cox et al., 1912 Reinstorf, 1923 Ewing, 1942 Radvan, 1960c Povolný et al., 1961 House fly (presumably *Musca domestica* Linnaeus) Geldreich et al., 1963 MUSCIDAE: STOMOXYINAE *Haematobia irritans* (Linnaeus) Stirrat et al., 1955 *Stomoxys calcitrans* (Linnaeus) Schuberg et al., 1913, 1914 Love et al., 1965 CALLIPHORIDAE: CHRYSOMYINAE *Chrysomya* Robineau-Desvoidy de la Paz, 1938 *Protophormia terraenovae* (Robineau-Desvoidy) Radvan, 1960a, 1960c CALLIPHORIDAE: CALLIPHORINAE *Calliphora vicina* Robineau-Desvoidy Povolný et al., 1961 SARCOPHAGIDAE: SARCOPHAGINAE *Sarcophaga haemorrhoidalis* (Fallén) Povolný et al., 1961
Streptococcus Rosenbach C 54	MUSCIDAE: MUSCINAE *Musca domestica* Linnaeus Shope, 1927
Streptococcus Rosenbach C 55	MUSCIDAE: MUSCINAE *Musca domestica* Linnaeus Shope, 1927
Streptococcus Rosenbach 744	MUSCIDAE: MUSCINAE *Musca domestica* Linnaeus Shope, 1927
Streptococcus Rosenbach, Lancefield's Group A	CHLOROPIDAE: OSCINELLINAE *Hippelates currani* Aldrich Bassett, 1967 *Hippelates peruanus* Becker Bassett, 1967
Streptococcus agalactiae Lehmann and Neumann (Synonyms: bovine mastitis, *Streptococcus mastitidis*)	CHLOROPIDAE: OSCINELLINAE *Hippelates* Loew Sanders, 1940a

Organism	Fly associations
Streptococus agalactiae (cont.)	MUSCIDAE: MUSCINAE *Musca domestica* Linnaeus Sanders, 1940b Ewing, 1942 Schumann, 1961 CALLIPHORIDAE: CALLIPHORINAE *Eucalliphora lilaea* (Walker) Gwatkin et al., 1938 *Calliphora terraenovae* Macquart Gwatkin et al., 1938 *Calliphora vicina* Robineau-Desvoidy Gwatkin et al., 1938 *Cynomyopsis cadaverina* (Robineau-Desvoidy) Gwatkin et al., 1938
Streptococcus agalactiae Lehmann and Neumann Group B (Synonym: *Streptococcus haemolyticus*)	CALLIPHORIDAE: CHRYSOMYINAE *Protophormia terraenovae* (Robineau-Desvoidy) Pavillard et al., 1957
Streptococcus durans Sherman and Wing (Synonyms: *Streptococcus faecalis, Streptococcus faecalis* var. *durans*)	PIOPHILIDAE *Piophila casei* (Linnaeus) Gregor et al., 1960 Lysenko et al., 1961 ANTHOMYIIDAE: ANTHOMYIINAE *Hylemya cinerella* (Fallén) Lysenko, 1958 MUSCIDAE: FANNIINAE *Fannia canicularis* (Linnaeus) Gregor et al., 1960 MUSCIDAE: PHAONIINAE *Hydrotaea dentipes* (Fabricius) Lysenko, 1958 Gregor et al., 1960 *Hydrotaea irritans* (Fallén) Gregor et al., 1960 *Ophyra leucostoma* (Wiedemann) Gregor et al., 1960 *Muscina stabulans* (Fallén) Gregor et al., 1960 MUSCIDAE: STOMOXYINAE *Stomoxys calcitrans* (Linnaeus) Duncan, 1926 CALLIPHORIDAE: CHRYSOMYINAE *Phormia regina* (Meigen) Greenberg, 1962 *Protophormia terraenovae* (Robineau-Desvoidy) Gregor et al., 1960 Lysenko et al., 1961

Organism	Fly associations
Streptococcus durans (cont.)	CALLIPHORIDAE: CALLIPHORINAE *Lucilia caesar* (Linnaeus) Gregor et al., 1960 *Phaenicia sericata* (Meigen) Simmons, 1935a, 1935b Gregor et al., 1960 Lysenko et al., 1961 Povolný et al., 1961 Greenberg, 1962 *Cynomyopsis cadaverina* (Robineau-Desvoidy) Greenberg, 1962
Streptococcus faecalis Andrewes and Horder (Synonym: *Streptococcus faecalis* var. *faecalis*)	MUSCIDAE: PHAONIINAE *Hydrotaea dentipes* (Fabricius) Lysenko, 1958 MUSCIDAE: MUSCINAE *Musca domestica* Linnaeus Scott, 1917a Duncan, 1926 Gerberich, 1951 Lysenko et al., 1961 House fly (presumably *Musca domestica* Linnaeus) Geldreich et al., 1963 CALLIPHORIDAE: CALLIPHORINAE *Calliphora vicina* Robineau-Desvoidy Gregor et al., 1960 *Calliphora vomitoria* (Linnaeus) Gregor et al., 1960
Streptococcus faecalis Andrewes and Horder var. *faecalis*	ANTHOMYIIDAE: ANTHOMYIINAE *Hylemya cinerella* (Fallén) Lysenko, 1958
Streptococcus faecalis var. *liquefaciens* (Sternberg) Mattick	PIOPHILIDAE *Piophila casei* (Linnaeus) Lysenko et al., 1961 MUSCIDAE: MUSCINAE *Musca domestica* Linnaeus Lysenko et al., 1961 CALLIPHORIDAE: CHRYSOMYINAE *Protophormia terraenovae* (Robineau-Desvoidy) Lysenko et al., 1961 CALLIPHORIDAE: CALLIPHORINAE *Phaenicia sericata* (Meigen) Lysenko et al., 1961
Streptococcus faecium Orla-Jensen (Synonym: *Streptococcus faecalis faecium*)	MUSCIDAE: STOMOXYINAE *Stomoxys calcitrans* (Linnaeus) Love et al., 1965

Organism	Fly
Streptococcus lactis (Lister) Löhnis (Synonym: *Streptococcus* of the group *Faecalis lactis*)	MUSCIDAE: MUSCINAE *Musca domestica* Linnaeus Greenberg et al., 1963a CALLIPHORIDAE: CALLIPHORINAE *Calliphora vicina* Robineau-Desvoidy Greenberg, 1962 SARCOPHAGIDAE: SARCOPHAGINAE *Sarcophaga haemorrhoidalis* (Fallén) Zmeev, 1943
Streptococcus mitis Andrewes and Horder (Synonyms: *Streptococcus viridans, Streptococcus mitior*)	MUSCIDAE: MUSCINAE *Musca domestica* Linnaeus Radvan, 1960c CALLIPHORIDAE: CHRYSOMYINAE *Protophormia terraenovae* (Robineau-Desvoidy) Pavillard et al., 1957 Radvan, 1960c CALLIPHORIDAE: CALLIPHORINAE *Phaenicia sericata* (Meigen) Simmons, 1935a, 1935b
Streptococcus pyogenes Rosenbach (Synonyms: *hemolytic Streptococci* group A, *Streptococcus haemolyticus, Streptococcus pyogenes haemolyticus*)	CHLOROPIDAE: OSCINELLINAE *Hippelates currani* Aldrich Bassett, 1970 *Hippelates flavipes* Loew Bassett, 1970 *Hippelates peruanus* Becker Bassett, 1970 MUSCIDAE: MUSCINAE *Musca domestica* Linnaeus Scott, 1917a Wellmann, 1955 CALLIPHORIDAE: CHRYSOMYINAE *Protophormia terraenovae* (Robineau-Desvoidy) Pavillard et al., 1957 CALLIPHORIDAE: CALLIPHORINAE *Lucilia caesar* (Linnaeus) Weil et al., 1933 *Phaenicia sericata* (Meigen) Robinson et al., 1933 Weil et al., 1933 Simmons, 1935a, 1935b
Lactobacillus Beijerinck (Synonym: ?*Thermobakterien*)	DROSOPHILIDAE: DROSOPHILINAE *Drosophila fenestrarum* Fallén Henneberg, 1902 *Drosophila funebris* (Fabricius) Henneberg, 1902 MUSCIDAE: MUSCINAE *Musca domestica vicina* Macquart Silverman et al., 1953

Organism	Fly
CORYNEBACTERIACEAE	
Corynebacterium Lehmann and Neumann	MUSCIDAE: MUSCINAE *Musca domestica vicina* Macquart McGuire et al., 1957 *Musca sorbens* Wiedemann McGuire et al., 1957
?*Corynebacterium* Lehmann and Neumann (Synonym: Diphtheroids)	CHLOROPIDAE: OSCINELLINAE *Siphunculina funicola* de Meijere Syddiq, 1938 CALLIPHORIDAE: CALLIPHORINAE *Lucilia caesar* (Linnaeus) Weil et al., 1933 *Phaenicia sericata* (Meigen) Weil et al., 1933
Corynebacterium diphtheriae (Kruse) Lehmann and Neumann (Synonym: Diphtheria bacillus)	MUSCIDAE: MUSCINAE *Musca domestica* Linnaeus Graham-Smith, 1910 House fly (presumably *Musca domestica* Linnaeus) Smith, 1898
Corynebacterium pyogenes (Glage) Eberson	MUSCIDAE: PHAONIINAE *Hydrotaea irritans* (Fallén) Bahr, 1952, 1953
Listeria monocytogenes (Murray et al.) Pirie	MUSCIDAE: MUSCINAE *Musca domestica* Linnaeus Radvan, 1956, 1960a CALLIPHORIDAE: CHRYSOMYINAE *Protophormia terraenovae* (Robineau-Desvoidy) Radvan, 1960a
Erysipelothrix insidiosa (Trevisan) Langford and Hansen (Synonyms: ?Erysipelis, Rotlauf der Schweine)	MUSCIDAE: STOMOXYINAE *Stomoxys* Geoffroy Megnin, 1874, 1875 *Stomoxys calcitrans* (Linnaeus) Wellmann, 1948-49a, 1948-49b, 1949, 1950, 1959 Tolstīak, 1956
BACILLACEAE	
Bacillus Cohn (Synonyms: ?spore-forming microbes, Bacilli A, B1, 2 and 3, C, D and E, *Bacillus* A, Ledingham, non-lactose-fermenting bacillus)	OTITIDAE: ULIDIINAE *Physiphora demandata* (Fabricius) Povolný et al., 1961 SEPSIDAE *Sepsis punctum* (Fabricius) Povolný et al., 1961 PIOPHILIDAE *Piophila casei* (Linnaeus) Povolný et al., 1961 SPHAEROCERIDAE *Copromyza atra* (Meigen) Povolný et al., 1961

Bacillus (cont.)

UNDETERMINED SPHAEROCERIDAE
Povolný et al., 1961

DROSOPHILIDAE: DROSOPHILINAE
Drosophila melanogaster Meigen
Nicholls, 1912
Tatum, 1939

UNDETERMINED DROSOPHILIDAE
Povolný et al., 1961

ANTHOMYIIDAE: SCATOPHAGINAE
Scatophaga stercoraria (Linnaeus)
Povolný et al., 1961

ANTHOMYIIDAE: ANTHOMYIINAE
Hylemya cinerella (Fallén)
Povolný et al., 1961
Anthomyia pluvialis (Linnaeus)
Povolný et al., 1961
Calythea albicincta (Fallén)
Zmeev, 1943

MUSCIDAE: FANNIINAE
Fannia canicularis (Linnaeus)
Povolný et al., 1961
Fannia scalaris (Fabricius)
Povolný et al., 1961

MUSCIDAE: PHAONIINAE
Hydrotaea dentipes (Fabricius)
Povolný et al., 1961
Hydrotaea occulta (Meigen)
Povolný et al., 1961
Ophyra leucostoma (Wiedemann)
Povolný et al., 1961
Muscina stabulans (Fallén)
Povolný et al., 1961

MUSCIDAE: MUSCINAE
Musca domestica Linnaeus
Ledingham, 1911
Tebbutt, 1913
Povolný et al., 1961
House fly (presumably *Musca domestica* Linnaeus)
Ramirez, 1898
Graham-Smith, 1913
Musca domestica vicina Macquart
McGuire et al., 1957
Musca sorbens Wiedemann
McGuire et al., 1957

MUSCIDAE: STOMOXYINAE
Haematobia irritans (Linnaeus)
Stirrat et al., 1955

Bacillus (cont.)

CALLIPHORIDAE: CHRYSOMYINAE
Chrysomya rufifacies (Macquart)
Zmeev, 1943

CALLIPHORIDAE: CALLIPHORINAE
Phaenicia sericata (Meigen)
Povolný et al., 1961
Calliphora Robineau-Desvoidy
Zmeev, 1943
Calliphora vomitoria (Linnaeus)
Maddox, 1885a

SARCOPHAGIDAE: SARCOPHAGINAE
Sarcophaga Meigen
Nicholls, 1912
Sarcophaga aurifinis Walker
Nicholls, 1912
Sarcophaga haemorrhoidalis (Fallén)
Povolný et al., 1961
Sarcophaga melanura Meigen
Povolný et al., 1961
Tricharaea Thomson
Nicholls, 1912

Bacillus anthracis Cohn ....
(Synonyms: Anthrax or Anthrax bacillus, Bacillus anthrax, anthracis)

PIOPHILIDAE
Piophila casei (Linnaeus)
Legroux et al., 1945

ANTHOMYIIDAE: ANTHOMYIINAE
Hylemya Robineau-Desvoidy
Anonymous, 1952

MUSCIDAE: MUSCINAE
Musca domestica Linnaeus
Raimbert, 1869, 1870
Buchanan, 1907
Graham-Smith, 1910, 1911a, 1911b, 1912
Wollman, 1921
Duncan, 1926
Sen et al., 1944
Radvan, 1956, 1960a, 1960b, 1960c
House fly (presumably *Musca domestica* Linnaeus)
Dalrymple, 1912, 1914
Morris, 1920
Rinonpoli, 1930
Musca domestica vicina Macquart
Anonymous, 1952
Musca inferior Stein
Nieschulz, 1928a
Musca sorbens sorbens Wiedemann
Cleland, 1912, 1913

Organism	Fly and reference
Bacillus anthracis (cont.)	MUSCIDAE: STOMOXYINAE
	Haematobia exigua (de Meijere)
	Nieschulz, 1928a
	Haematobia irritans (Linnaeus)
	Morris, 1918, 1920
	Stomoxys Geoffroy
	Mégnin, 1874
	Terni, 1908a
	Stomoxys calcitrans (Linnaeus)
	Schuberg et al., 1912, 1913
	Mitzmain, 1914a, 1914c
	Schuberg et al., 1914
	Duncan, 1926
	Nieschulz, 1928a, 1928b
	Sen et al., 1944
	CALLIPHORIDAE: CHRYSOMYINAE
	Protophormia terraenovae (Robineau-Desvoidy)
	Radvan, 1960a, 1960b, 1960c
	CALLIPHORIDAE: CALLIPHORINAE
	Lucilia caesar (Linnaeus)
	Graham-Smith, 1912
	Calliphora vicina Robineau-Desvoidy
	Graham-Smith, 1911a, 1911b, 1912
	Sen et al., 1944
	Calliphora vomitoria (Linnaeus)
	Raimbert, 1870
	Calliphora vomitoria (Linnaeus)
	Buchanan, 1907
	UNDETERMINED CALLIPHORIDAE
	Blue bottle fly
	Morris, 1920
	SARCOPHAGIDAE: SARCOPHAGINAE
	Sarcophaga Meigen
	Sen et al., 1944
	Sarcophaga carnaria (Linnaeus)
	Petragnani, 1925
Bacillus ?anthracis Cohn	CALLIPHORIDAE: CALLIPHORINAE
	Calliphora vomitoria (Linnaeus)
	Maddox, 1885a
Bacillus bifermentans Weinberg and Séguin	SARCOPHAGIDAE: SARCOPHAGINAE
	Sarcophaga Meigen
	Picado, 1935
Bacillus cereus Frankland and Frankland	MUSCIDAE: MUSCINAE
	Musca domestica Linnaeus
	Hawley et al., 1951
	Briggs, 1960
	Harvey et al., 1960
	Radvan, 1960a, 1960b, 1960c
	Stephens, 1963

Organism	Fly association
Bacillus cereus (cont.)	CALLIPHORIDAE: CHRYSOMYINAE *Protophormia terraenovae* (Robineau-Desvoidy) Radvan, 1960b, 1960c
Bacillus cereus var. *mycoides* (Flügge) Smith et al. (Synonym: *Bacillus mycoides*)	PIOPHILIDAE *Piophila casei* (Linnaeus) Gregor et al., 1960 MUSCIDAE: FANNIINAE *Fannia canicularis* (Linnaeus) Gregor et al., 1960 MUSCIDAE: PHAONIINAE *Hydrotaea dentipes* (Fabricius) Gregor et al., 1960 *Hydrotaea irritans* (Fallén) Gregor et al., 1960 *Ophyra leucostoma* (Wiedemann) Gregor et al., 1960 *Muscina stabulans* (Fallén) Gregor et al., 1960 MUSCIDAE: MUSCINAE *Musca domestica* Linnaeus Hamilton, 1903 Duncan, 1926 MUSCIDAE: STOMOXYINAE *Haematobia irritans* (Linnaeus) Stirrat et al., 1955 *Stomoxys calcitrans* (Linnaeus) Duncan, 1926 CALLIPHORIDAE: CHRYSOMYINAE *Protophormia terraenovae* (Robineau-Desvoidy) Gregor et al., 1960 CALLIPHORIDAE: CALLIPHORINAE *Lucilia caesar* (Linnaeus) Gregor et al., 1960 *Phaenicia sericata* (Meigen) Gregor et al., 1960 *Calliphora vicina* Robineau-Desvoidy Gregor et al., 1960 *Calliphora vomitoria* (Linnaeus) Gregor et al., 1960
Bacillus coli mutabilis Neisser (Synonym: MacConkey's bacillus No. 8)	MUSCIDAE: MUSCINAE *Musca domestica* Linnaeus Nicoll, 1911a
Bacillus gasoformans Eisenberg (Synonym: MacConkey's bacillus No. 104 or *B. gasoformans non-liquefaciens*)	MUSCIDAE: MUSCINAE *Musca domestica* Linnaeus Nicoll, 1911a

Organism	Fly / Reference
Bacillus gruenthali Morgan (Synonym: MacConkey's bacillus No. 4, *Bacillus grünthal*)	MUSCIDAE: MUSCINAE *Musca domestica* Linnaeus Nicoll, 1911a
Bacillus lutzae Brown	CALLIPHORIDAE: CALLIPHORINAE *Phaenicia sericata* (Meigen) Brown, 1927
Bacillus megaterium de Bary (Synonym: *Bacillus megatherium*)	MUSCIDAE: MUSCINAE *Musca domestica* Linnaeus Hamilton, 1903 Hawley et al., 1951 CALLIPHORIDAE: CALLIPHORINAE *Lucilia caesar* (Linnaeus) Muzzarelli, 1925 *Calliphora vomitoria* (Linnaeus) Muzzarelli, 1925 SARCOPHAGIDAE: SARCOPHAGINAE *Sarcophaga carnaria* (Linnaeus) Muzzarelli, 1925 *Sarcophaga haemorrhoidalis* (Fallén) Zmeev, 1943
Bacillus mesentericus Trevisan	MUSCIDAE: MUSCINAE *Musca domestica* Linnaeus Hamilton, 1903 Duncan, 1926 MUSCIDAE: STOMOXYINAE *Stomoxys calcitrans* (Linnaeus) Duncan, 1926
Bacillus radiciformis Eberbach (Synonym: *B. radiciformis*)	CALLIPHORIDAE: CALLIPHORINAE *Lucilia caesar* (Linnaeus) Cao, 1906 *Calliphora vomitoria* (Linnaeus) Cao, 1906 SARCOPHAGIDAE: SARCOPHAGINAE *Sarcophaga carnaria* (Linnaeus) Cao, 1906
Bacillus ?radiciformis Eberbach (Synonym: *B. radiciformis*)	MUSCIDAE: MUSCINAE *Musca domestica* Linnaeus Cao, 1906
Bacillus subtilis Cohn (Synonyms: *B. subtilis*, ?hay bacillus group)	PIOPHILIDAE *Piophila casei* (Linnaeus) Lysenko et al., 1961 SPHAEROCERIDAE *Leptocera ferruginata* (Stenhammar) Lysenko, 1958 ANTHOMYIIDAE: ANTHOMYIINAE *Hylemya cinerella* (Fallén) Lysenko, 1958 MUSCIDAE: MUSCINAE *Musca domestica* Linnaeus Hamilton, 1903

Organism	Fly associations
Bacillus radiciformis (cont.)	Cao, 1906 Reinstorf, 1923 Duncan, 1926 Radvan, 1956, 1960a, 1960c Lysenko, 1958 Lysenko et al., 1961 Greenberg et al., 1963a *Musca domestica vicina* Macquart Silverman et al., 1953 MUSCIDAE: STOMOXYINAE *Haematobia irritans* (Linnaeus) Stirrat et al., 1955 *Stomoxys calcitrans* (Linnaeus) Duncan, 1926 CALLIPHORIDAE: CHRYSOMYINAE *Phormia regina* (Meigen) Greenberg, 1962 *Protophormia terraenovae* (Robineau-Desvoidy) Pavillard et al., 1957 Lysenko, 1958 Radvan, 1960a, 1960c Lysenko et al., 1961 CALLIPHORIDAE: CALLIPHORINAE *Lucilia caesar* (Linnaeus) Cao, 1906 *Phaenicia sericata* (Meigen) Michelbacher et al., 1932 Lysenko et al., 1961 Greenberg, 1962 *Calliphora vicina* Robineau-Desvoidy Balzam, 1937 Greenberg, 1962 *Calliphora vomitoria* (Linnaeus) Cao, 1906 *Cynomyopsis cadaverina* (Robineau-Desvoidy) Greenberg, 1962 SARCOPHAGIDAE: SARCOPHAGINAE *Sarcophaga carnaria* (Linnaeus) Cao, 1906
Bacillus ?subtilis Cohn (Synonym: *B. subtilis*)	CALLIPHORIDAE: CALLIPHORINAE *Calliphora vomitoria* (Linnaeus) Maddox, 1885a
Bacillus subtilis var. *niger* Smith et al.	MUSCIDAE: STOMOXYINAE *Haematobia irritans* (Linnaeus) Stirrat et al., 1955
Bacillus thuringiensis Berliner (pathogen)	MUSCIDAE: FANNIINAE *Fannia canicularis* (Linnaeus) Eversole et al., 1965

Bacillus thuringiensis (cont.)	MUSCIDAE: MUSCINAE *Musca autumnalis* De Geer Yendol et al., 1967 *Musca domestica* Linnaeus Hall et al., 1959 Figueiredo et al., 1960 Burns et al., 1961 Borgatti et al., 1963 Feigin, 1963 Greenwood, 1964 Harvey, 1964 Tonkonozhenko, 1967 CALLIPHORIDAE: CALLIPHORINAE *Phaenicia sericata* (Meigen) Greenwood, 1964 SARCOPHAGIDAE: SARCOPHAGINAE *Sarcophaga bullata* Parker McConnell et al., 1959
Bacillus thuringiensis var. *sotto* Heimpel and Angus (pathogen)	MUSCIDAE: MUSCINAE *Musca domestica* Linnaeus Briggs, 1960
Bacillus thuringiensis var. *thuringiensis* Heimpel and Angus (pathogen)	MUSCIDAE: MUSCINAE *Musca domestica* Linnaeus Briggs, 1960 Dunn, 1960 Burgerjohn et al., 1965 Gingrich, 1965 Harvey et al., 1965 MUSCIDAE: STOMOXYINAE *Haematobia irritans* (Linnaeus) Gingrich, 1965 *Stomoxys calcitrans* (Linnaeus) Gingrich, 1965
Bacillus vesiculosus Matzuschita (Synonyms: ?MacConkey's bacillus No. 5, *B. vesiculosus*)	MUSCIDAE: MUSCINAE *Musca domestica* Linnaeus Nicoll, 1911a
Clostridium Prazmowski	MUSCIDAE: MUSCINAE *Musca domestica vicina* Macquart McGuire et al., 1957 *Musca sorbens* Wiedemann McGuire et al., 1957
Clostridium bifermentans (Weinberg and Séguin) Bergey et al. (Synonym: *Bacillus foetidus*)	MUSCIDAE: MUSCINAE House fly (presumably *Musca domestica* Linnaeus) Marpmann, 1884
Clostridium botulinum (van Ermengem) Bergey et al. (Synonyms: *Bac. botulinus*, ?Botulism, ?limberneck)	SYRPHIDAE: MILESIINAE *Eristalis* Latreille Bail, 1900

Organism	Fly association
Clostridium bifermentans (cont.)	PIOPHILIDAE *Piophila casei* (Linnaeus) Legroux et al., 1945 MUSCIDAE: MUSCINAE *Musca domestica* Linnaeus Bail, 1900 MUSCIDAE: STOMOXYINAE *Stomoxys calcitrans* (Linnaeus) Bail, 1900 CALLIPHORIDAE: CALLIPHORINAE *Lucilia* Robineau-Desvoidy Bail, 1900 *Lucilia caesar* (Linnaeus) Saunders, 1921 SARCOPHAGIDAE: SARCOPHAGINAE *Sarcophaga* Meigen Bail, 1900
Clostridium botulinum Type C Spray	CALLIPHORIDAE: CALLIPHORINAE *Lucilia caesar* (Linnaeus) Saunders, 1921 Bengtson, 1922 *Phaenicia sericata* (Meigen) Bengtson, 1922
Clostridium chauvoei (Arloing, Cornevin, and Thomas) Scott	MUSCIDAE: MUSCINAE *Musca domestica* Linnaeus Sauer, 1908a UNDETERMINED MUSCIDAE Sauer, 1908a
Clostridium parabotulinum bovis Robinson	MUSCIDAE: MUSCINAE *Musca domestica* Linnaeus Theiler, 1927 CALLIPHORIDAE: CHRYSOMYINAE *Chrysomya albiceps* (Wiedemann) Theiler, 1927 *Chrysomya chloropyga* (Wiedemann) Theiler, 1927 *Chrysomya marginalis* (Wiedemann) Theiler, 1927
Clostridium perfringens (Veillon and Zuber) Holland (Synonyms: *Clostridium welchii,* Gas bacillus, *Bacillus perfringens*)	CALLIPHORIDAE: CHRYSOMYINAE *Phormia* Robineau-Desvoidy Green, 1953 *Protophormia terraenovae* (Robineau-Desvoidy) Pavillard et al., 1957 CALLIPHORIDAE: CALLIPHORINAE *Lucilia* Robineau-Desvoidy Green, 1953

Organism	Fly association
Clostridium perfringens (cont.)	*Phaenicia sericata* (Meigen) Michelbacher et al., 1932 Simmons, 1935a, 1935b *Calliphora* Robineau-Desvoidy Green, 1953 UNDETERMINED CALLIPHORIDAE Blow fly maggots Baer, 1929 Blow flies West, 1953 SARCOPHAGIDAE: SARCOPHAGINAE *Sarcophaga* Meigen Picado, 1935
Clostridium putrefaciens (McBryde) Sturges and Drake (Synonym: *B. putrefaciens*)	MUSCIDAE: MUSCINAE House fly (presumably *Musca domestica* Linnaeus) Purdy, 1906
Clostridium tetani (Flügge) Bergey et al.	MUSCIDAE: MUSCINAE *Musca domestica* Linnaeus Bail, 1900
MYCOBACTERIACEAE	
Mycobacterium leprae (Hansen) Lehmann and Neumann (Synonyms: Leprosy bacilli, bacillus of Hansen, *Bacillus leprae*)	MUSCIDAE: FANNIINAE *Fannia canicularis* (Linnaeus) Asami, 1934 MUSCIDAE: MUSCINAE *Musca domestica* Linnaeus Currie, 1910, 1911 Leboeuf, 1912, 1914 Minett, 1912 Asami, 1934 House fly (presumably *Musca domestica* Linnaeus) Römer, 1906 *Musca sorbens* Wiedemann Lamborn, 1935, 1937 UNDETERMINED MUSCIDAE Lamborn, 1936a CALLIPHORIDAE: CALLIPHORINAE *Lucilia* Robineau-Desvoidy Currie, 1910 Leboeuf, 1912a *Lucilia argyrocephala* Macquart Asami, 1934 *Calliphora lata* Coquillet Asami, 1934 SARCOPHAGIDAE: SARCOPHAGINAE *Sarcophaga argyrostoma* (Robineau-Desvoidy) Currie, 1910

Mycobacterium leprae (cont.)

UNDETERMINED TACHINIDAE
de Souza-Araujo, 1944

Mycobacterium ?leprae (Hansen) Lehmann and Neumann
(Synonyms: acid-fast bacillus, acid-fast leprae-like bacillus, Leprosy bacillus)

MUSCIDAE: PHAONIINAE
Muscina stabulans (Fallén)
Honeij et al., 1914
MUSCIDAE: MUSCINAE
Musca Linnaeus
Arizumi, 1934
Musca bezzii Patton and Cragg
de Mello et al., 1926
Musca domestica Linnaeus
Wherry, 1908a, 1908b
Sandes, 1911, 1912
Honeij et al., 1914
Barros et al., 1947
Musca sorbens Wiedemann
Lamborn, 1938
MUSCIDAE: STOMOXYINAE
Stomoxys calcitrans (Linnaeus)
Honeij et al., 1914
CALLIPHORIDAE: CALLIPHORINAE
Lucilia Robineau-Desvoidy
Honeij et al., 1914
Arizumi, 1934

Mycobacterium lepraemurium Marchoux and Sorel
(Synonym: Rat leprosy bacilli)

MUSCIDAE: MUSCINAE
Musca domestica Linnaeus
Wherry, 1908a
Marchoux, 1916
CALLIPHORIDAE: CALLIPHORINAE
Lucilia caesar (Linnaeus)
Wherry, 1908a
Calliphora vomitoria (Linnaeus)
Wherry, 1908a

Mycobacterium ?lepraemurium Marchoux and Sorel
(Synonym: ?Rat leprosy bacillus)

MUSCIDAE: MUSCINAE
Musca domestica Linnaeus
Wherry, 1908b
CALLIPHORIDAE: CALLIPHORINAE
Lucilia caesar (Linnaeus)
Wherry, 1908b
Calliphora vomitoria (Linnaeus)
Wherry, 1908b

Mycobacterium paratuberculosis Bergey et al.

CALLIPHORIDAE: CALLIPHORINAE
Lucilia caesar (Linnaeus)
Muzzarelli, 1925
Calliphora vomitoria (Linnaeus)
Muzzarelli, 1925
SARCOPHAGIDAE: SARCOPHAGINAE
Sarcophaga carnaria (Linnaeus)
Muzzarelli, 1925

Organism	Fly
Mycobacterium phlei Lehmann and Neumann (Synonym: acid-fast grass bacillus of Moeller)	SYRPHIDAE: MILESIINAE *Volucella obesa* (Fabricius) Currie, 1910 MUSCIDAE: MUSCINAE *Musca domestica* Linnaeus Currie, 1910 SARCOPHAGIDAE: SARCOPHAGINAE *Ravinia lherminieri* (Robineau-Desvoidy) Currie, 1910 *Sarcophaga* Meigen Pavan, 1949 *Sarcophaga argyrostoma* (Robineau-Desvoidy) Currie, 1910
Mycobacterium tuberculosis (Zopf) Lehmann and Neumann	MUSCIDAE: MUSCINAE *Musca domestica* Linnaeus Hayward, 1904 André, 1906 Buchanan, 1907 Graham-Smith, 1910 Wollman, 1921 Morellini et al., 1947, 1953 Tison, 1950 Morellini, 1952, 1956 House fly (presumably *Musca domestica* Linnaeus) Celli et al., 1888 Hofmann, 1888 Lord, 1904, 1905 André, 1908 *Musca sorbens* Wiedemann Lamborn, 1938, 1939 CALLIPHORIDAE: CHRYSOMYINAE *Chrysomya rufifacies* (Macquart) Morellini et al., 1947 CALLIPHORIDAE: CALLIPHORINAE *Lucilia caesar* (Linnaeus) Hayward, 1904 *Phaenicia sericata* (Meigen) Morellini et al., 1947 *Calliphora vicina* Robineau-Desvoidy Morellini et al., 1947 *Calliphora vomitoria* (Linnaeus) Wollman, 1921a SARCOPHAGIDAE: SARCOPHAGINAE *Sarcophaga argyrostoma* (Robineau-Desvoidy) Morellini et al., 1947, 1949 Levinsky et al., 1948

Organism	Fly
Mycobacterium tuberculosis (cont.)	*Sarcophaga carnaria* (Linnaeus) Petragnani, 1925
DERMATOPHILACEAE	
Dermatophilus congolensis van Saceghem	MUSCIDAE: MUSCINAE *Musca domestica* Linnaeus Richard et al., 1966 MUSCIDAE: STOMOXYINAE *Stomoxys calcitrans* (Linnaeus) Richard et al., 1966
SPIROCHAETACEAE	
Spirochaeta Ehrenberg (Synonym: Spirochete of shrew)	MUSCIDAE: MUSCINAE *Musca domestica* Linnaeus Wollman, 1927
Spirochaeta acuminata Castellani (Synonym: *Treponema acuminata*)	MUSCIDAE: MUSCINAE *Musca domestica* Linnaeus Castellani, 1907
Spirochaeta obtusa Castellani (Synonym: *Treponema obtusa*)	MUSCIDAE: MUSCINAE *Musca domestica* Linnaeus Castellani, 1907
Spirochaeta stomoxyae Jegen	MUSCIDAE: STOMOXYINAE *Stomoxys calcitrans* (Linnaeus) Jegen, 1924
TREPONEMATACEAE	
Borrelia anserina (Sakharoff) Bergey et al. (Synonym: *Spirochaeta gallinarum*)	MUSCIDAE: STOMOXYINAE *Stomoxys calcitrans* (Linnaeus) Schuberg et al., 1909
Borrelia berbera (Sergent and Foley) Bergey et al. (Synonym: spirochetes of relapsing fever)	MUSCIDAE: MUSCINAE *Musca domestica* Linnaeus Sergent et al., 1910 Wollman et al., 1928 MUSCIDAE: STOMOXYINAE *Stomoxys calcitrans* (Linnaeus) Wollman et al., 1928 CALLIPHORIDAE: CALLIPHORINAE *Lucilia caesar* (Linnaeus) Wollman et al., 1928
Borrelia recurrentis (Lebert) Bergey et al. (Synonyms: *Spirochaeta obermeieri*, relapsing fever)	MUSCIDAE: STOMOXYINAE *Stomoxys calcitrans* (Linnaeus) Schuberg et al., 1909, 1912 Fränkel, 1912
Borrelia refringens (Schaudinn and Hoffmann) Bergey et al. (Synonym: *Spirochaeta refringens*)	CHLOROPIDAE: OSCINELLINAE *Hippelates flavipes* Loew Kumm, 1935 Kumm et al., 1935
Treponema Schaudinn	DROSOPHILIDAE: DROSOPHILINAE *Drosophila fasciata* Dufour Ikeda, 1965

Organism	Fly
Treponema (cont.)	*Drosophila melanogaster* Meigen Ikeda, 1965 *Drosophila nebulosa* Sturtevant Poulson et al., 1961b *Drosophila willistoni* Sturtevant Poulson et al., 1961b Ikeda, 1965
Treponema carateum Brumpt	CHLOROPIDAE: OSCINELLINAE *Hippelates* Loew Leon Blanco et al., 1941 Soberón y Parra et al., 1944
Treponema pertenue Castellani	CHLOROPIDAE: OSCINELLINAE *Hippelates currani* Aldrich Kumm et al., 1936 *Hippelates illicis* Curran Kumm et al., 1936 *Hippelates flavipes* Loew Kumm, 1935 Kumm et al., 1935, 1936 *Hippelates peruanus* Becker Kumm et al., 1936 *Hippelates pusio* Loew Kumm et al., 1936 *Conioscinella mars* (Curran) Kumm et al., 1936 MUSCIDAE: COENOSIINAE *Atherigona orientalis* Schiner Satchell et al., 1953 MUSCIDAE: MUSCINAE *Musca domestica* Linnaeus Castellani, 1907, 1908 Oho, 1921 House fly (presumably *Musca domestica* Linnaeus) Yasuyama, 1928 *Musca domestica vicina* Macquart Satchell et al., 1953 *Musca sorbens* Wiedemann Thomson et al., 1934 Lamborn, 1936a, 1936b Satchell et al., 1953 UNDETERMINED CALLIPHORIDAE Blue bottle fly Yasuyama, 1928
Leptospira canicola Okell et al.	MUSCIDAE: MUSCINAE *Musca domestica* Linnaeus Kunert et al., 1952 Schmidtke, 1959

Leptospira canicola (cont.)	MUSCIDAE: STOMOXYINAE *Stomoxys calcitrans* (Linnaeus) Schmidtke, 1959 CALLIPHORIDAE: CALLIPHORINAE *Phaenicia sericata* (Meigen) Kunert et al., 1952 Schmidtke, 1959 *Calliphora vicina* Robineau-Desvoidy Kunert et al., 1952 Schmidtke, 1959
Leptospira grippotyphosa Topley and Wilson	MUSCIDAE: MUSCINAE *Musca domestica* Linnaeus Kunert et al., 1952 Schmidtke, 1959 MUSCIDAE: STOMOXYINAE *Stomoxys calcitrans* (Linnaeus) Schmidtke, 1959 CALLIPHORIDAE: CALLIPHORINAE *Phaenicia sericata* (Meigen) Kunert et al., 1952 Schmidtke, 1959 *Calliphora vicina* Robineau-Desvoidy Kunert et al., 1952 Schmidtke, 1959
Leptospira icterohaemorrhagiae (Inada and Ido) Noguchi (Synonym: Weil's disease)	MUSCIDAE: MUSCINAE *Musca domestica* Linnaeus Kunert et al., 1952 Schmidtke, 1959 MUSCIDAE: STOMOXYINAE *Stomoxys calcitrans* (Linnaeus) Reiter, 1917 Uhlenhuth et al., 1917 Schmidtke, 1959 CALLIPHORIDAE: CALLIPHORINAE *Phaenicia sericata* (Meigen) Kunert et al., 1952 Schmidtke, 1959 *Calliphora vicina* Robineau-Desvoidy Kunert et al., 1952 Schmidtke, 1959
BACTERIA OF UNCERTAIN AFFINITY	
Bacterium mathisi Roubaud and Treillard (unvalidated)	MUSCIDAE: MUSCINAE *Musca domestica* Linnaeus Roubaud et al., 1935 CALLIPHORIDAE: CALLIPHORINAE *Phaenicia sericata* (Meigen) Roubaud et al., 1935

Organism	Fly
Bacterium mathisi (cont.)	SARCOPHAGIDAE: SARCOPHAGINAE *Sarcophaga carnaria* (Linnaeus) Roubaud et al., 1935
RICKETTSIACEAE	
Rickettsia da Rocha-Lima	MUSCIDAE: MUSCINAE *Musca domestica* Linnaeus Hertig et al., 1924 MUSCIDAE: STOMOXYINAE *Stomoxys calcitrans* (Linnaeus) Hertig et al., 1924
Coxiella burneti (Derrick) Philip	MUSCIDAE: MUSCINAE *Musca domestica* Linnaeus Philip, 1948
CHLAMYDIACEAE	
Colesiota conjunctivae (Coles) Rake (Synonym: *Rickettsia conjunctivae*)	MUSCIDAE: MUSCINAE *Musca* Linnaeus Mitscherlich, 1941 *Musca domestica* Linnaeus Mitscherlich, 1943 MUSCIDAE: STOMOXYINAE *Stomoxys calcitrans* (Linnaeus) Mitscherlich, 1943
ANAPLASMATACEAE	
Anaplasma marginale Theiler	MUSCIDAE: STOMOXYINAE *Stomoxys calcitrans* (Linnaeus) Taylor, 1935 Anonymous, 1936
MUCORACEAE	
Mucor Micheli	SCIARIDAE *Sciara* Meigen Baumberger, 1919 MILICHIIDAE: MADIZINAE *Desmometopa m-nigrum* (Zetterstedt) Baumberger, 1919 DROSOPHILIDAE: DROSOPHILINAE *Drosophila melanogaster* Meigen Baumberger, 1919 Griffith, 1952
Mucor ambiguus Vuillemin	CALLIPHORIDAE: POLLENIINAE *Ophyra calcogaster* Wiedemann Usui, 1960
Mucor mucedo (Linnaeus) Brefeld	CHLOROPIDAE: OSCINELLINAE *Siphunculina funicola* de Meijere Syddiq, 1938 MUSCIDAE: MUSCINAE *Musca domestica* Linnaeus Reinstorf, 1923
Rhizopus Ehrenberg	DROSOPHILIDAE: DROSOPHILINAE *Drosophila melanogaster* Meigen Griffith, 1952

Rhizopus nigricans Ehrenberg .... DROSOPHILIDAE: DROSOPHILINAE
Drosophila melanogaster Meigen
Baumberger, 1919
MUSCIDAE: MUSCINAE
Musca domestica Linnaeus
Baumberger, 1919
CALLIPHORIDAE: CALLIPHORINAE
Lucilia caesar (Linnaeus)
Baumberger, 1919

ENTOMOPHTHORACEAE

Entomophthora Fresenius (Synonym: *Empusa*) (pathogen) .... ANTHOMYIIDAE: SCATOPHAGINAE
Scatophaga stercoraria (Linnaeus)
Bail, 1960
ANTHOMYIIDAE: ANTHOMYIINAE
Hylemya aestiva (Meigen)
Hammer, 1942
MUSCIDAE: FANNIINAE
Fannia canicularis (Linnaeus)
Graham-Smith, 1916
MUSCIDAE: PHAONIINAE
Hydrotaea dentipes (Fabricius)
Graham-Smith, 1916
MUSCIDAE: MUSCINAE
Musca domestica Linnaeus
Graham-Smith, 1916
CALLIPHORIDAE: CALLIPHORINAE
Calliphora vicina Robineau-Desvoidy
Graham-Smith, 1916

?*Entomophthora* Fresenius (Synonym: *Empusa*) (pathogen) .... ANTHOMYIIDAE: ANTHOMYIINAE
Anthomyia Meigen
Roubaud, 1911a
MUSCIDAE: MUSCINAE
Musca corvina Fabricius
Roubaud, 1911a
Musca domestica Linnaeus
Roubaud, 1911a
MUSCIDAE: STOMOXYINAE
Stomoxys brunnipes Grünberg
Roubaud, 1911a
Stomoxys calcitrans (Linnaeus)
Roubaud, 1911a
Stomoxys nigra Macquart
Roubaud, 1911a
CALLIPHORIDAE: CHRYSOMYINAE
Chrysomya putoria (Wiedemann)
Roubaud, 1911a
SARCOPHAGIDAE: SARCOPHAGINAE
Sarcophaga Meigen
Roubaud, 1911a

Organism	Fly host and reference
Entomophthora americana Thaxter (Synonym: *Empusa americana*) (pathogen)	MUSCIDAE: MUSCINAE *Musca domestica* Linnaeus Thaxter, 1888 CALLIPHORIDAE: CALLIPHORINAE *Phaenicia mexicana* (Macquart) Dresner, 1949 *Calliphora vomitoria* (Linnaeus) Rozsypal, 1957
Entomophthora ?americana Thaxter (pathogen)	CALLIPHORIDAE: POLLENIINAE *Pollenia rudis* (Fabricius) Judd, 1955
Entomophthora calliphorae Giard (pathogen)	CALLIPHORIDAE: CALLIPHORINAE *Calliphora* Robineau-Desvoidy Thaxter, 1888 *Calliphora vomitoria* (Linnaeus) Giard, 1888 *Calliphora vomitoria* var. *dunensis* Giard Giard, 1879
Entomophthora muscae Cohn (Synonym: *Empusa muscae*) (pathogen)	SYRPHIDAE: SYRPHINAE *Syrphus* Fabricius Thaxter, 1888 *Melanostoma mellinum* (Linnaeus) Rostrup, 1916 *Melanostoma scalare* Fabricius Rostrup, 1916 UNDETERMINED SYRPHIDAE Thaxter, 1888 DROSOPHILIDAE: DROSOPHILINAE *Drosophila melanogaster* Meigen Goldstein, 1927 *Drosophila repleta* Wollaston Goldstein, 1927 ANTHOMYIIDAE: SCATOPHAGINAE *Scatophaga furcata* (Say) Rostrup, 1916 *Scatophaga stercoraria* (Linnaeus) Rostrup, 1916 Graham-Smith, 1919 Hammer, 1942 ANTHOMYIIDAE: ANTHOMYIINAE *Hylemya cardui* (Meigen) Rostrup, 1916 *Hylemya coarctata* (Fallén) Rostrup, 1916 *Hylemya radicum* (Linnaeus) Graham-Smith, 1919 *Anthomyia* Meigen Thaxter, 1888

Entomophthora muscae (cont.)

MUSCIDAE: FANNIINAE

Fannia canicularis (Linnaeus)
Buchanan, 1913
Graham-Smith, 1919
Wilhelmi, 1919
Tischler, 1950
Steve, 1959

MUSCIDAE: PHAONIINAE

Hydrotaea dentipes (Fabricius)
Graham-Smith, 1919

Lasiops variabilis (Robineau-Desvoidy)
Rostrup, 1916

Muscina stabulans (Fallén)
Judd, 1955

MUSCIDAE: MUSCINAE

Musca corvina Fabricius
Graham-Smith, 1919

Musca domestica Linnaeus
Cohn, 1855
White, 1880
Thaxter, 1888
Cavara, 1899
Brefeld, 1908
Buchanan, 1913
Güssow, 1913
Picard, 1914
Bishopp et al., 1915a
Dove, 1916
Lodge, 1916
Graham-Smith, 1919
Lakon, 1919
Friederichs, 1920
Roubaud, 1922
Roubaud, et al., 1922b
Schweizer, 1936, 1947
de Salles et al., 1944
Judd, 1955
Baird, 1957
Krenner, 1961

House fly (presumably *Musca domestica* Linnaeus)
Cohn, 1857
Solms-Laubach, 1870
White et al., 1877
Goldstein, 1927
Yeager, 1939

MUSCIDAE: STOMOXYINAE

Stomoxys calcitrans (Linnaeus)
Surcouf, 1923

Organism	Fly associations
Entomophthora muscae (cont.)	CALLIPHORIDAE: CALLIPHORINAE *Lucilia caesar* (Linnaeus) Thaxter, 1888 Graham-Smith, 1919 *Phaenicia sericata* (Meigen) Dove, 1916 *Calliphora vomitoria* (Linnaeus) Thaxter, 1888 UNDETERMINED CALLIPHORIDAE Blow flies White et al., 1877 SARCOPHAGIDAE: MILTOGRAMMINAE *Pseudosarcophaga affinis* Fallén Baird, 1957 SARCOPHAGIDAE: SARCOPHAGINAE *Sarcophaga carnaria* (Linnaeus) Graham-Smith, 1919 UNDETERMINED TACHINIDAE de Souza-Araujo, 1944
*Entomophthora muscivora** Schroeter (pathogen)	RHAGIONIDAE: RHAGIONINAE *Rhagio lineola* Fabricius Rostrup, 1916 SCIOMYZIDAE: SCIOMYZINAE *Sciomyza* Fallén Rostrup, 1916 LAUXANIIDAE *Lauxania aenes* Fallén Rostrup, 1916 *Sapromyza obesa* Zetterstedt Rostrup, 1916
Entomophthora ovispora Nowakowski (pathogen)	UNDETERMINED SYRPHIDAE Thaxter, 1888 LAUXANIIDAE *Sapromyza* Fallén Thaxter, 1888 LONCHAEIDAE *Lonchaea chorea* (Fabricius) Thaxter, 1888
Entomophthora richteri (Bresadola and Staritz) Bubák (Synonym: *Tarichium richteri*) (pathogen)	LAUXANIIDAE *Lauxania aenea* Fallén Bubák, 1903 Lakon, 1915
Entomophthora sphaerosperma Fresenius (Synonym: *Empusa sphaerosperma*) (pathogen)	MUSCIDAE: MUSCINAE *Musca* Linnaeus Thaxter, 1888

* Treated as a synonym of *E. calliphorae* in D. M. MacLeod and E. Müller-Kögler, Mycologia, 62:33-66, 1970.

Organism	Fly host and reference
Zoophthora vomitoriae (Rozsypal)	CALLIPHORIDAE: CALLIPHORINAE *Calliphora vomitoria* (Linnaeus) Rozsypal, 1966
LABOULBENIACEAE	
Stygmatomyces baeri (Knoch) Peyritsch (pathogen)	MUSCIDAE: MUSCINAE *Musca planiceps* Wiedemann Senior-White et al., 1945
ENDOMYCETACEAE	
Coccidiascus legeri Chatton	DROSOPHILIDAE: DROSOPHILINAE *Drosophila funebris* (Fabricius) Chatton, 1913
Saccharomyces Hansen	DROSOPHILIDAE: DROSOPHILINAE *Drosophila melanogaster* Meigen Baumberger, 1919
Saccharomyces cerevisiae Hansen (Synonym: *Saccharomyces ellipsoideus*)	DROSOPHILIDAE: DROSOPHILINAE *Drosophila cellaris* Linnaeus Berlese, 1896 *Drosophila melanogaster* Meigen Baumberger, 1919 MUSCIDAE: MUSCINAE *Musca domestica* Linnaeus Peppler, 1944 CALLIPHORIDAE: CALLIPHORINAE *Calliphora vicina* Robineau-Desvoidy Berlese, 1896 SARCOPHAGIDAE: SARCOPHAGINAE *Sarcophaga carnaria* (Linnaeus) Berlese, 1896
Saccharomyces ?melli (Fabian and Quinet) (Synonym: *Saccharomyces mali*)	DROSOPHILIDAE: DROSOPHILINAE *Drosophila melanogaster* Meigen Delcourt et al., 1910
Saccharomyces pastorianus Hansen	DROSOPHILIDAE: DROSOPHILINAE *Drosophila cellaris* Linnaeus Berlese, 1896 CALLIPHORIDAE: CALLIPHORINAE *Calliphora vicina* Robineau-Desvoidy Berlese, 1896
Saccharomyces theobromae	DROSOPHILIDAE: DROSOPHILINAE *Drosophila melanogaster* Meigen Nicholls, 1912
HYPOCREACEAE	
Cordyceps dipterigena Berkeley and Broome	MUSCIDAE: MYDAEINAE *Mydaea* Robineau-Desvoidy Petch, 1923
AGARICACEAE	
Agaricus campestris Linnaeus (Synonym: *Agaricus citrinus*)	DROSOPHILIDAE: DROSOPHILINAE *Drosophila melanogaster* Meigen Baumberger, 1919 MUSCIDAE: PHAONIINAE *Muscina pascuorum* (Meigen) von Bremi, 1846

Organism	Fly
UNDETERMINED BASIDIOMYCETES	DROSOPHILIDAE: DROSOPHILINAE *Drosophila funebris* (Fabricius) Austin, 1933
MONILIACEAE	
Aspergillus (Micheli) Corda	DROSOPHILIDAE: DROSOPHILINAE *Drosophila* Fallén Blochwitz, 1929 *Drosophila melanogaster* Meigen Baumberger, 1919 MUSCIDAE: MUSCINAE *Musca domestica* Linnaeus Baumberger, 1919 MUSCIDAE: STOMOXYINAE *Haematobia irritans* (Linnaeus) Stirrat et al., 1955
Aspergillus clavatus Desmazières	DROSOPHILIDAE: DROSOPHILINAE *Drosophila melanogaster* Meigen Griffith, 1952
Aspergillus flavus Link	MUSCIDAE: MUSCINAE *Musca domestica* Linnaeus Amonkar et al., 1965 Beard et al., 1965 *Musca domestica vicina* Macquart Usui, 1960
Aspergillus niger van Tieghem	DROSOPHILIDAE: DROSOPHILINAE *Drosophila melanogaster* Meigen Griffith, 1952
Beauveria bassiana (Balsamo) Vuillemin (Synonym: *Botrytis bassiana*)	MUSCIDAE: MUSCINAE *Musca domestica* Linnaeus Dresner, 1949, 1950 Steinhaus et al., 1953 Dunn et al., 1963 CALLIPHORIDAE: CALLIPHORINAE *Calliphora vicina* Robineau-Desvoidy Rostrup, 1916
Blastomyces Costantin and Rolland	CALLIPHORIDAE: CALLIPHORINAE *Lucilia caesar* (Linnaeus) Muzzarelli, 1925 *Calliphora vomitoria* (Linnaeus) Muzzarelli, 1925 SARCOPHAGIDAE: SARCOPHAGINAE *Sarcophaga carnaria* (Linnaeus) Muzzarelli, 1925
?*Botrytis* Micheli	MUSCIDAE: MUSCINAE *Musca domestica* Linnaeus Gómez et al., 1888-9
Geotrichum Link	MUSCIDAE: STOMOXYINAE *Haematobia irritans* (Linnaeus) Stirrat et al., 1955

Organism	Fly
Geotrichum candidum Link (Synonym: *Oidium lactis*)	MUSCIDAE: MUSCINAE *Musca domestica* Linnaeus Ficker, 1903
Gliocladium Corda (Synonym: *Gleocladium*)	SCIARIDAE *Sciara* Meigen Baumberger, 1919
Metarrhizium anisopliae (Metschnikoff) Sorokin	MUSCIDAE: MUSCINAE *Musca domestica* Linnaeus Friederichs, 1919, 1920 Notini et al., 1944 House fly (presumably *Musca domestica* Linnaeus) Schaerffenberg, 1959
Monilia (Persoon) Saccardo	DROSOPHILIDAE: DROSOPHILINAE *Drosophila melanogaster* Meigen Griffith, 1952
Oidium Link	MUSCIDAE: STOMOXYINAE *Haematobia irritans* (Linnaeus) Stirrat et al., 1955
Penicillium Link	MUSCIDAE: MUSCINAE *Musca domestica* Linnaeus Baumberger, 1919 House fly (presumably *Musca domestica* Linnaeus) Cohn, 1857
	MUSCIDAE: STOMOXYINAE *Haematobia irritans* (Linnaeus) Stirrat et al., 1955
?*Penicillium*	CALLIPHORIDAE: CALLIPHORINAE *Calliphora vomitoria* (Linnaeus) Maddox, 1885b
Penicillium camemberti Thom	DROSOPHILIDAE: DROSOPHILINAE *Drosophila melanogaster* Meigen Griffith, 1952
Penicillium glaucum Link	DROSOPHILIDAE: DROSOPHILINAE *Drosophila melanogaster* Meigen Baumberger, 1919 Griffith, 1952
	MUSCIDAE: MUSCINAE *Musca domestica* Linnaeus Baumberger, 1919 Reinstorf, 1923
	CALLIPHORIDAE: CALLIPHORINAE *Lucilia caesar* (Linnaeus) Baumberger, 1919
Trichothecium Link	MUSCIDAE: STOMOXYINAE *Haematobia irritans* (Linnaeus) Stirrat et al., 1955

DEMATIACEAE

Alternaria Nees .... DROSOPHILIDAE: DROSOPHILINAE
Drosophila melanogaster Meigen
Griffith, 1952

Cladosporium Link .... MUSCIDAE: STOMOXYINAE
Haematobia irritans (Linnaeus)
Stirrat et al., 1955

Curvularia Boedijn .... DROSOPHILIDAE: DROSOPHILINAE
Drosophila melanogaster Meigen
Griffith, 1952

Hormodendrum Bonorden .... DROSOPHILIDAE: DROSOPHILINAE
Drosophila melanogaster Meigen
Griffith, 1952

Macrosporium Fries .... DROSOPHILIDAE: DROSOPHILINAE
Drosophila melanogaster Meigen
Griffith, 1952

Pullalaria Berkhout .... MUSCIDAE: STOMOXYINAE
Haematobia irritans (Linnaeus)
Stirrat et al., 1955

Torula Persoon .... DROSOPHILIDAE: DROSOPHILINAE
Drosophila cellaris Linnaeus
Berlese et al., 1897
MUSCIDAE: MUSCINAE
House fly (presumably *Musca domestica* Linnaeus)
Berlese, 1896

STILBACEAE

Isaria Persoon ex Fries .... MUSCIDAE: MUSCINAE
Musca domestica Linnaeus
Bail, 1860

TUBERCULARIACEAE

Fusarium Link .... SCIARIDAE
Sciara Meigen
Baumberger, 1919
DROSOPHILIDAE: DROSOPHILINAE
Drosophila melanogaster Meigen
Griffith, 1952

BLASTOCYSTIDACEAE
(uncertain position)

Blastocystis Alexeieff .... MUSCIDAE: MUSCINAE
Musca domestica Linnaeus
Buxton, 1920

CRYPTOCOCCACEAE

Candida Berkhout .... DROSOPHILIDAE: DROSOPHILINAE
(Synonym: *Dematium*)
Drosophila cellaris Linnaeus
Berlese, 1896
Berlese et al., 1897
MUSCIDAE: MUSCINAE
House fly (presumably *Musca domestica* Linnaeus)
Berlese, 1896

Organism	Fly association
Candida (cont.)	CALLIPHORIDAE: CALLIPHORINAE *Calliphora vicina* Robineau-Desvoidy Berlese, 1896 SARCOPHAGIDAE: SARCOPHAGINAE *Sarcophaga carnaria* (Linnaeus) Berlese, 1896
Candida albicans (Robin) Berkhout	CALLIPHORIDAE: CHRYSOMYINAE *Protophormia terraenovae* (Robineau-Desvoidy) Pavillard et al., 1957
Candida mycoderma (Reess) (Synonym: *Saccharomyces mycoderma*)	CALLIPHORIDAE: CALLIPHORINAE *Calliphora vicina* Robineau-Desvoidy Berlese, 1896
Kloeckera apiculata (Reess emend Klöcker) Janke (Synonym: *Saccharomyces apiculatus*)	DROSOPHILIDAE: DROSOPHILINAE *Drosophila cellaris* Linnaeus Berlese, 1896 CALLIPHORIDAE: CALLIPHORINAE *Calliphora vicina* Robineau-Desvoidy Berlese, 1896 SARCOPHAGIDAE: SARCOPHAGINAE *Sarcophaga carnaria* (Linnaeus) Berlese, 1896
FUNGI OF UNCERTAIN AFFINITY	
Streptotricea, unvalidated	CALLIPHORIDAE: CALLIPHORINAE *Lucilia caesar* (Linnaeus) Muzzarelli, 1925 *Calliphora vomitoria* (Linnaeus) Muzzarelli, 1925 SARCOPHAGIDAE: SARCOPHAGINAE *Sarcophaga carnaria* (Linnaeus) Muzzarelli, 1925
TRYPANOSOMATIDAE	
Crithidia Léger (Synonym: *Strigomonas*)	DROSOPHILIDAE: DROSOPHILINAE *Drosophila virilis* Sturtevant McGhee et al., 1965 MUSCIDAE: STOMOXYINAE *Stomoxys calcitrans* (Linnaeus) Taylor, 1930 CALLIPHORIDAE: CALLIPHORINAE *Lucilia* Robineau-Desvoidy Roubaud, 1908a *Phaenicia sericata* (Meigen) Thomson, 1933
Crithidia acanthocephali Hanson and McGhee	DROSOPHILIDAE: DROSOPHILINAE *Drosophila virilis* Sturtevant McGhee et al., 1965
Crithidia fasciculata Léger	DROSOPHILIDAE: DROSOPHILINAE *Drosophila virilis* Sturtevant McGhee et al., 1965

Organism	Fly / Reference
Crithidia haematopotae Jegen	MUSCIDAE: STOMOXYINAE *Stomoxys calcitrans* (Linnaeus) Jegen, 1924
Crithidia lesnei (Léger) (Synonym: *Herpetomonas lesnei*)	MUSCIDAE: MUSCINAE *Dasyphora pratorum* (Meigen) Léger, 1903
Crithidia luciliae (Strickland) Wallace and Clark	DROSOPHILIDAE: DROSOPHILINAE *Drosophila virilis* Sturtevant McGhee et al., 1965 MUSCIDAE: MUSCINAE *Musca domestica* Linnaeus Wallace et al., 1959 CALLIPHORIDAE: CALLIPHORINAE *Phaenicia sericata* (Meigen) Wallace et al., 1959
Crithidia oncopelti (Noguchi and Tilden) sensu M. and A. Lwoff	DROSOPHILIDAE: DROSOPHILINAE *Drosophila virilis* Sturtevant McGhee et al., 1965
Herpetomonas Kent	CHLOROPIDAE: OSCINELLINAE *Siphunculina funicola* de Meijere Patton, 1921c MUSCIDAE: FANNIINAE *Fannia canicularis* (Linnaeus) Dunkerly, 1911 MUSCIDAE: MUSCINAE *Musca domestica* Linnaeus Wenyon, 1911a Darling, 1912b *Musca domestica vicina* Macquart Chang, 1940 *Musca sorbens* Weidemann Chang, 1940 MUSCIDAE: STOMOXYINAE *Stomoxys* Geoffroy Wenyon, 1911a *Stomoxys calcitrans* (Linnaeus) Patton, 1912b *Stomoxys ?calcitrans* (Linnaeus) Gray, 1906 CALLIPHORIDAE: CHRYSOMYINAE *Chrysomya megacephala* (Fabricius) Chang, 1940 *Chrysomya putoria* (Wiedemann) Roubaud, 1912a CALLIPHORIDAE: CALLIPHORINAE *Lucilia* Robineau-Desvoidy Roubaud, 1908a *Lucilia caesar* (Linnaeus) Chang, 1940

Organism	Fly / Reference
Herpetomonas (cont.)	*Phaenicia sericata* (Meigen) Chang, 1940
	SARCOPHAGIDAE: SARCOPHAGINAE
	Sarcophaga Meigen Chang, 1940
	Sarcophaga carnaria (Linnaeus) Franchini, 1922
	Sarcophaga melanura Meigen Franchini, 1922
	Sarcophaga nurus Rondani Franchini, 1922
Herpetomonas caulleryi .... (Roubaud) Wenyon (Synonym: *Cercoplasma caulleryi*)	CALLIPHORIDAE: CALLIPHORINAE *Auchmeromyia luteola* (Fabricius) Roubaud, 1911c
Herpetomonas ?muscae .... (Carter)	OTITIDAE: OTITINAE *Melieria crassipennis* Fabricius Laveran et al., 1920
Herpetomonas muscarum .... (Leidy) Kent (Synonyms: *Herpetomonas muscae domesticae, Crithidia calliphorae, Leptomonas muscae domesticae, Herpetomonas homalomyiae, Crithidia muscae domesticae, Trypanosoma muscae domesticae, Herpetomonas luciliae, Herpetomonas sarcophagae, Herpetomonas calliphorae*)	SPHAEROCERIDAE
	Borborus Meigen Patton, 1921a
	Leptocera Macquart Becker, 1922-23
	EPHYDRIDAE: EPHYDRINAE
	Teichomyza fusca Macquart Léger, 1903
	DROSOPHILIDAE: DROSOPHILINAE
	Drosophila Fallén Patton, 1921a
	MUSCIDAE: FANNIINAE
	Fannia canicularis (Linnaeus) Dunkerly, 1911 Patton, 1921a
	Fannia scalaris (Fabricius) Léger, 1903 Brug, 1915
	MUSCIDAE: PHAONIINAE
	Muscina stabulans (Fallén) Becker, 1922-23
	MUSCIDAE: MUSCINAE
	Morellia micans (Macquart) Becker, 1924
	Musca corvina Fabricius Roubaud, 1909
	Musca domestica Linnaeus Prowazek, 1904 Carter, 1909 Roubaud, 1909 Rosenbusch, 1910 Flu, 1911

Herpetomonas muscarum (cont.)

Wenyon, 1911a
Cardamatis, 1912a
Chatton et al., 1913
Franchini et al., 1915
Root, 1921
Becker, 1922-23, 1924
Ross et al., 1924
Fantham et al., 1927
Vanni, 1946
Coutinho et al., 1957
Wallace et al., 1959
Kramer, 1961a

House fly (presumably *Musca domestica* Linnaeus)
Werner, 1909a
Wenyon, 1911b, 1913
Glaser, 1922

Musca domestica vicina Macquart
Pletneva, 1937

Musca nebulo Fabricius
Patton, 1910a, 1912a, 1921a

Musca sorbens Wiedemann
Patton, 1921a

MUSCIDAE: STOMOXYINAE

Stomoxys Geoffroy
Patton et al., 1909
Wenyon, 1911b

CALLIPHORIDAE: CHRYSOMYINAE

Cochliomyia macellaria (Fabricius)
Becker, 1922-23, 1924

Chrysomya marginalis (Wiedemann)
Roubaud, 1908b

Chrysomya putoria (Wiedemann)
Roubaud, 1908b, 1909

Phormia regina (Meigen)
Becker, 1922-23, 1924

CALLIPHORIDAE: CALLIPHORINAE

Lucilia Robineau-Desvoidy
Drbohlav, 1925

Lucilia argyricephala Macquart
Patton, 1912a, 1921a

Lucilia ?jeddensis Bigot
Yamasaki, 1927

Lucilia porphyrina (Walker)
Patton, 1921a

Phaenicia sericata (Meigen)
Becker, 1922-23, 1924
Drbohlav, 1926
Fantham et al., 1927
Thomson, 1933

Herpetomonas muscarum (cont.)	Bellosillo, 1937 Wallace et al., 1959 *Calliphora coloradensis* Hough Swingle, 1911 *Calliphora vicina* Robineau-Desvoidy Alexeieff, 1911 Laveran et al., 1920 Root, 1921 Becker, 1922-23, 1924 UNDETERMINED CALLIPHORINAE Patton, 1921a CALLIPHORIDAE: POLLENIINAE *Pollenia rudis* (Fabricius) Léger, 1903 SARCOPHAGIDAE: SARCOPHAGINAE *Sarcophaga bullata* Parker Becker, 1922-23, 1924 *Sarcophaga haemorrhoidalis* (Fallén) Laveran et al., 1920 *Sarcophaga nurus* Rondani Roubaud, 1909 *Sarcophaga securifera* Villeneuve Becker, 1922-23, 1924
Herpetomonas ?muscarum (Leidy) Kent (Synonyms: *Herpetomonas muscae domesticae*, *Herpetomonas luciliae*)	DRYOMYZIDAE *Dryomyza anilis* Fallén Mackinnon, 1910 ANTHOMYIIDAE: SCATOPHAGINAE *Scatophaga lutaria* (Fabricius) Mackinnon, 1910 MUSCIDAE: FANNIINAE *Fannia ?corvina* (Verrall) Mackinnon, 1910 CALLIPHORIDAE: CALLIPHORINAE *Lucilia* Robineau-Desvoidy Strickland, 1911 *Cynomyopsis cadaverina* (Robineau-Desvoidy) Wallace et al., 1964
Herpetomonas roubaudi (Chatton) Drbohlav (Synonym: *Leptomonas roubaudi*)	DROSOPHILIDAE: DROSOPHILINAE *Drosophila confusa* Staeger Chatton, 1912b
Herpetomonas rubrio-striatae Chatton and Leger (Synonym: *Leptomonas rubrio-striatae*)	DROSOPHILIDAE: DROSOPHILINAE *Drosophila rubrio-striata* Becker Chatton et al., 1911c, 1912
Leishmania Ross	SARCOPHAGIDAE: SARCOPHAGINAE *Sarcophaga carnaria* (Linnaeus) Franchini, 1922

Organism	Fly association

Leishmania (cont.)	*Sarcophaga melanura* Meigen
	  Franchini, 1922
	Sarcophaga nurus Rondani
	  Franchini, 1922

Leishmania donovani Laveran and Mesnil (Synonym: *Leishmania infantum*)	CHLOROPIDAE: OSCINELLINAE
	Siphunculina funicola de Meijere
	  Patton, 1921-22
	MUSCIDAE: MUSCINAE
	Musca nebulo Fabricius
	  Patton, 1921-22
	Musca sorbens Wiedemann
	  Patton, 1921-22
	  Thomson et al., 1934
	  Lamborn, 1935, 1955

Leishmania mexicana Biagi	MUSCIDAE: STOMOXYINAE *Stomoxys calcitrans* (Linnaeus) Lainson et al., 1965

Leishmania tropica (Wright) (Synonym: *Crithidia cunninghami*)	MUSCIDAE: MUSCINAE
	Musca domestica Linnaeus
	  Carter, 1909
	  Row, 1911
	  Laveran, 1915
	  Wollman, 1927
	  Wollman et al., 1928
	House fly (presumably *Musca domestica* Linnaeus)
	  Blanc et al., 1921
	Musca sorbens Wiedemann
	  Thomson et al., 1934
	  Lamborn, 1935
	MUSCIDAE: STOMOXYINAE
	Stomoxys calcitrans (Linnaeus)
	  Berberian, 1938, 1939
	CALLIPHORIDAE: CALLIPHORINAE
	Lucilia caesar (Linnaeus)
	  Wollman et al., 1928

Leishmania ?tropica (Wright)	MUSCIDAE: MUSCINAE
	Musca domestica Linnaeus
	  Cardamitis et al., 1911
	  Wenyon, 1911b

Leptomonas Kent	SPHAEROCERIDAE
	Leptocera hirtula var. *thalhammeri* Strobl
	  Chatton, 1912a
	DROSOPHILIDAE: DROSOPHILINAE
	Drosophila confusa Staeger
	  Chatton et al., 1912
	Drosophila rubrio-striata Becker
	  Chatton et al., 1912

Organism	Fly association
Leptomonas (cont.)	CHLOROPIDAE: OSCINELLINAE *Hippelates flavipes* Loew Kumm et al., 1935 MUSCIDAE: FANNIINAE *Fannia canicularis* (Linnaeus) Dunkerly, 1911 MUSCIDAE: MUSCINAE *Pyrellia* Robineau-Desvoidy Roubaud, 1912a CALLIPHORIDAE: CHRYSOMYINAE *Chrysomya putoria* (Wiedemann) Roubaud, 1909
Leptomonas ampelophilae Chatton and Léger	DROSOPHILIDAE: DROSOPHILINAE *Drosophila melanogaster* Meigen Chatton et al., 1911c, 1912
Leptomonas anthomyia (Franchini) Wenyon (Synonym: *Herpetomonas anthomyiae*)	ANTHOMYIIDAE: ANTHOMYIINAE *Hydrophoria ruralis* Meigen Franchini, 1922
Leptomonas craggi (Patton) Wenyon (Synonym: *Herpetomonas craggi*)	MUSCIDAE: MUSCINAE *Musca bezzii* Patton and Cragg Patton, 1921a
Leptomonas drosophilae Chatton and Alilaire (Synonym: *Trypanosoma drosophilae*)	DROSOPHILIDAE: DROSOPHILINAE *Drosophila confusa* Staeger Chatton et al., 1908, 1911a, 1911b, 1911c, 1912
Leptomonas gracilis (Léger) Wenyon (Synonym: *Herpetomonas gracilis*)	CALLIPHORIDAE: CALLIPHORINAE *Calliphora* Robineau-Desvoidy Alexeieff, 1911
Leptomonas legerorum Chatton	SPHAEROCERIDAE *Sphaerocera curvipes* Latreille Chatton, 1912a
Leptomonas lineata (Swingle) (Synonym: *Herpetomonas lineata*)	SARCOPHAGIDAE: SARCOPHAGINAE *Sarcophaga sarraceniae* Riley Swingle, 1911
Leptomonas mesnili Roubaud (Synonym: *Cercoplasma mesnili*)	CALLIPHORIDAE: CALLIPHORINAE *Lucilia* Robineau-Desvoidy Roubaud, 1908a *Phaenicia cluvia* Walker Roubaud, 1909 *Phaenicia sericata* (Meigen) Roubaud, 1908a, 1909, 1911c
Leptomonas mirabilis Roubaud (Synonyms: *Herpetomonas graphomyae*, *Herpetomonas mirabilis*, *Cercoplasma mirabilis*)	MUSCIDAE: MUSCINAE *Graphomya maculata* (Scopoli) Franchini, 1922 *Musca nebulo* Fabricius Patton, 1921a

Leptomonas mirabilis (cont.)	CALLIPHORIDAE: CHRYSOMYINAE *Chrysomya marginalis* (Wiedemann) Roubaud, 1908b, 1911c *Chrysomya megacephala* (Fabricius) Patton, 1921a *Chrysomya putoria* (Wiedemann) Roubaud, 1908b, 1909, 1912a *Chrysomya rufifacies* (Macquart) Patton, 1921a CALLIPHORIDAE: CALLIPHORINAE *Lucilia argyricephala* Macquart Patton, 1921a *Lucilia porphyrina* (Walker) Patton, 1921a *Cynomyopsis cadaverina* (Robineau-Desvoidy) Wallace et al., 1964 SARCOPHAGIDAE: SARCOPHAGINAE *Sarcophaga melanura* Meigen Franchini, 1922
Leptomonas pycnosomae Roubaud	CALLIPHORIDAE: CHRYSOMYINAE *Chrysomya putoria* (Wiedemann) Roubaud, 1909
Leptomonas pyrrhocoris (Zotta) Zotta	CALLIPHORIDAE: CALLIPHORINAE *Calliphora* Robineau-Desvoidy Zotta, 1921
Leptomonas soudanensis Roubaud	CALLIPHORIDAE: CHRYSOMYINAE *Chrysomya* Robineau-Desvoidy Roubaud, 1911d *Chrysomya putoria* (Wiedemann) Roubaud, 1912a
Leptomonas stomoxyae Jegen (Synonym: *Herpetomonas stomoxyae*)	MUSCIDAE: STOMOXYINAE *Stomoxys calcitrans* (Linnaeus) Jegen, 1924
Trypanosoma Gruby	MUSCIDAE: MUSCINAE *Musca domestica* Linnaeus Wenyon, 1911a MUSCIDAE: STOMOXYINAE *Haematobia* Lepeletier and Serville Krishna Iyer et al., 1935 *Stomoxys* Geoffroy Nabarro et al., 1905 Sergent et al., 1905, 1922a Wenyon, 1908 *Stomoxys calcitrans* (Linnaeus) Martini, 1903 Nieschulz, 1928c *Stomoxys nigra* Macquart Edington et al., 1907

Organism	Fly
Trypanosoma brucei Plimmer and Bradford (Synonym: *Trypanosoma pecaudi*)	MUSCIDAE: MUSCINAE
	Musca domestica Linnaeus
	Vanni, 1946
	Musca sorbens Wiedemann
	Lamborn, 1934
	Thomson et al., 1934
	Musca tempestiva Fallén
	Lamborn, 1934
	MUSCIDAE: STOMOXYINAE
	Haematobia angustifrons Malloch
	Lamborn, 1934
	Stomoxys boueti Roubaud
	Bouet et al., 1912b
	Stomoxys calcitrans (Linnaeus)
	Martin et al., 1908
	Roubaud, 1909
	Schuberg et al., 1909
	Taylor, 1930
	Stomoxys ?calcitrans (Linnaeus)
	Hall, 1927
	Stomoxys calcitrans (Linnaeus) var. *soudanense*
	Bouet et al., 1912
	Stomoxys nigra Macquart
	Martin et al., 1908
	Roubaud, 1909
	Bouet et al., 1912
Trypanosoma congolense Broden	MUSCIDAE: MUSCINAE
	Musca sorbens Wiedemann
	Lamborn, 1934
	Musca tempestiva Fallén
	Lamborn, 1934
	MUSCIDAE: STOMOXYINAE
	Haematobia angustifrons Malloch
	Lamborn, 1934
	Haematobia irritans (Linnaeus)
	Jack, 1917
	Stomoxys Geoffroy
	van Saceghem, 1922b
	Stomoxys calcitrans (Linnaeus)
	Roubaud, 1909
	Carmichael, 1934
	Stomoxys ?calcitrans (Linnaeus)
	Hall, 1927
	Stomoxys nigra Macquart
	Roubaud, 1909
Trypanosoma ?congolense Broden	MUSCIDAE: STOMOXYINAE
	Stomoxys Geoffroy
	Jowett, 1911

Organism	Fly / Reference
Trypanosoma dimorphon Laveran and Mesnil	MUSCIDAE: STOMOXYINAE
	Stomoxys Geoffroy
	Nabarro et al., 1905
Trypanosoma equinum Vages	MUSCIDAE: MUSCINAE
	Musca domestica Linnaeus
	Sivori et al., 1902
	MUSCIDAE: STOMOXYINAE
	Stomoxys calcitrans (Linnaeus)
	Sivori et al., 1902
	Lignières, 1903
Trypanosoma equiperdum Doflein	MUSCIDAE: STOMOXYINAE
	Stomoxys calcitrans (Linnaeus)
	Sieber et al., 1908
	Schuberg et al., 1909a
Trypanosoma evansi (Steel) (Synonyms: Surra, *Trypanosoma berberium, Trypanosoma annamense, Trypanosoma soudanense*)	MUSCIDAE: MUSCINAE
	Musca crassirostris Stein
	Nieschulz, 1927
	Musca domestica Linnaeus
	Musgrave et al., 1903
	Mitzmain, 1912b
	Musca inferior Stein
	Nieschulz, 1927
	Musca insignis (Austen)
	Fletcher, 1916
	MUSCIDAE: STOMOXYINAE
	Haematobia Lepeletier and Serville
	Mitzmain, 1912a
	Haematobia exigua (de Meijere)
	Schat, 1909
	Nieschulz, 1927
	Haematobia ?exigua (de Meijere)
	Schat, 1903
	Stomoxys Geoffroy
	Penning, 1904
	Nabarro et al., 1905
	Fraser, 1909
	Donatien et al., 1922, 1923
	Broudin, 1926
	Stomoxys bilineata Grünberg
	Bouet et al., 1912
	Stomoxys brunnipes Grünberg
	Nieschulz, 1930
	Stomoxys calcitrans (Linnaeus)
	Schat, 1903
	Leese, 1909
	Baldrey, 1911
	Bouet et al., 1912
	Mitzmain, 1912b
	Sergent et al., 1922b

Organism	Fly
Trypanosoma evansi (cont.)	Nieschulz, 1927, 1928b, 1929, 1930 Dieben, 1928 *Stomoxys ?calcitrans* (Linnaeus) Musgrave et al., 1903 *Stomoxys nigra* Macquart Bouet et al., 1912 Moutia, 1928
Trypanosoma ?evansi (Steel)	MUSCIDAE: STOMOXYINAE *Stomoxys calcitrans* (Linnaeus) Curry, 1902a
Trypanosoma gambiense Dutton	MUSCIDAE: MUSCINAE *Musca domestica* Linnaeus Beck, 1910 MUSCIDAE: STOMOXYINAE *Stomoxys* Geoffroy Greig et al., 1905 Gray, 1907 Dutton et al., 1908 Minchin, 1908 Beck, 1910 *Stomoxys bilineata* Grünberg Roubaud, 1909 *Stomoxys calcitrans* (Linnaeus) Roubaud, 1909 Schuberg et al., 1909 Duke, 1913 *Stomoxys ?calcitrans* (Linnaeus) Gray, 1906 *Stomoxys nigra* Macquart Roubaud, 1909 Duke, 1913
Trypanosoma hippicum Darling	MUSCIDAE: MUSCINAE *Musca domestica* Linnaeus Darling, 1912b, 1912c
Trypanosoma pecorum Bruce, Hamerton, Bateman, and Mackie	MUSCIDAE: STOMOXYINAE *Stomoxys* Geoffroy van Saceghem, 1921, 1922b
Trypanosoma rhodesiense Stephans and Fantham	MUSCIDAE: MUSCINAE *Musca sorbens* Wiedemann Lamborn, 1936a Lamborn et al., 1936 MUSCIDAE: STOMOXYINAE *Stomoxys* Geoffroy Duke, 1934 *Stomoxys calcitrans* (Linnaeus) Lamborn, 1933 *Stomoxys nigra* Macquart Lamborn, 1933

Organism	Fly
Trypanosoma suis Ochmann (Synonym: *Trypanosoma simiae*)	MUSCIDAE: MUSCINAE *Musca sorbens* Wiedemann Lamborn, 1934 *Musca tempestiva* Fallén Lamborn, 1934
Trypanosoma vivax Ziemann (Synonym: *Trypanosoma cazalboui*)	MUSCIDAE: STOMOXYINAE *Stomoxys* Geoffroy Bouffard, 1907a, 1907b *Stomoxys boueti* Roubaud Bouet et al., 1912 *Stomoxys calcitrans* (Linnaeus) Bouet et al., 1912 *Stomoxys ?calcitrans* (Linnaeus) Hall, 1927 *Stomoxys nigra* Macquart Bouet et al., 1912
BODONIDAE	
Bodo Ehrenberg (Synonym: *Prowazekia*)	MUSCIDAE: FANNIINAE *Fannia canicularis* (Linnaeus) Dunkerly, 1912 MUSCIDAE: MUSCINAE House fly (presumably *Musca domestica* Linnaeus) Flu, 1915 Attimonelli, ?1940b MUSCIDAE: STOMOXYINAE *Stomoxys calcitrans* (Linnaeus) Duke, 1913 *Stomoxys nigra* Macquart Duke, 1913
Rhynchoidomonas Patton	MUSCIDAE: PHAONIINAE *Muscina stabulans* (Fallén) Franchini, 1922 SARCOPHAGIDAE: SARCOPHAGINAE *Sarcophaga* Meigen Franchini, 1922
Rhynchoidomonas intestinalis (Roubaud) Alexeieff (Synonym: *Cystotrypanosoma intestinalis*)	CALLIPHORIDAE: CALLIPHORINAE *Phaenicia sericata* (Meigen) Roubaud, 1911b
Rhynchoidomonas luciliae (Patton) Patton	MUSCIDAE: MUSCINAE *Musca nebulo* Fabricius Patton, 1910c CALLIPHORIDAE: CALLIPHORINAE *Lucilia* Robineau-Desvoidy Alexeieff, 1911 *Lucilia argyricephala* Macquart Patton, 1910b *Calliphora vicina* Robineau-Desvoidy Alexeieff, 1911

Rhynchoidomonas roubaudi Roubaud (Synonym: *Cercoplasma drosophila*) .... DROSOPHILIDAE: DROSOPHILINAE
Drosophila Fallén
Roubaud, 1912b

Rhynchoidomonas siphunculinae Patton .... CHLOROPIDAE: OSCINELLINAE
Siphunculina funicola de Meijere
Patton, 1921c

TETRAMITIDAE

Tetramitus Perty .... MUSCIDAE: MUSCINAE
House fly (presumably *Musca domestica* Linnaeus)
Attimonelli, ?1940b

CHILOMASTIGIDAE

Chilomastix mesnili (Wenyon) .... OTITIDAE: ULIDIINAE
Physiphora Fallén
Harris et al., 1946

PLATYSTOMATIDAE: PLATYSTOMATINAE
Scholastes Loew
Harris et al., 1946

MUSCIDAE: COENOSIINAE
Atherigona Rondani
Harris et al., 1946

MUSCIDAE: FANNIINAE
Fannia canicularis (Linnaeus)
Root, 1921

MUSCIDAE: MUSCINAE
Musca Linnaeus
Harris et al., 1946
Musca domestica Linnaeus
Root, 1921
Musca domestica vicina Macquart
Chang, 1940
Musca sorbens Wiedemann
Chang, 1940

CALLIPHORIDAE: CHRYSOMYINAE
Cochliomyia macellaria (Fabricius)
Root, 1921
Chrysomya megacephala (Fabricius)
Chang, 1940
Harris et al., 1946

CALLIPHORIDAE: CALLIPHORINAE
Lucilia Robineau-Desvoidy
Harris et al., 1946
Lucilia caesar (Linnaeus)
Root, 1921
Chang, 1940
Phaenicia sericata (Meigen)
Chang, 1940

Chilomastix mesnili (cont.)	*Calliphora vicina* Robineau-Desvoidy Root, 1921 SARCOPHAGIDAE: SARCOPHAGINAE *Sarcophaga* Meigen Chang, 1940 Harris et al., 1946
HEXAMITIDAE	
Giardia Kunstler (Synonym: *Lamblia*)	MUSCIDAE: MUSCINAE House fly (presumably *Musca domestica* Linnaeus) Frye et al., 1932 Attimonelli, ?1940b
Giardia intestinalis (Lambl) (Synonym: *Lamblia intestinalis*)	MUSCIDAE: FANNIINAE *Fannia* Robineau-Desvoidy Wenyon et al., 1917a *Fannia canicularis* (Linnaeus) Root, 1921 MUSCIDAE: MUSCINAE *Musca* Linnaeus Wenyon et al., 1917a *Musca domestica* Linnaeus Roubaud, 1918 Buxton, 1920 Root, 1921 Jausion et al., 1923 Aleksander et al., 1935 Sieyro, 1942 Vanni, 1946 House fly (presumably *Musca domestica* Linnaeus) Wenyon et al., 1917a, 1917b *Musca domestica vicina* Macquart Pletneva, 1937 CALLIPHORIDAE: CHRYSOMYINAE *Cochliomyia macellaria* (Fabricius) Root, 1921 CALLIPHORIDAE: CALLIPHORINAE *Lucilia* Robineau-Desvoidy Wenyon et al., 1917a *Lucilia caesar* (Linnaeus) Root, 1921 *Calliphora vicina* Robineau-Desvoidy Root, 1921 SARCOPHAGIDAE: SARCOPHAGINAE *Sarcophaga* Meigen Wenyon et al., 1917b *Calliphora* Robineau-Desvoidy Wenyon et al., 1917a, 1917b

Organism	Fly
Giardia lamblia Stiles	OTITIDAE: ULIDIINAE
	Physiphora Fallén
	Harris et al., 1946
	PLATYSTOMATIDAE: PLATYSTOMATINAE
	Scholastes Loew
	Harris et al., 1946
	MUSCIDAE: COENOSIINAE
	Atherigona Rondani
	Harris et al., 1946
	MUSCIDAE: MUSCINAE
	Musca Linnaeus
	Harris et al., 1946
	Musca domestica Linnaeus
	Rendtorff et al., 1954
	Musca domestica vicina Macquart
	Chang, 1940
	Musca sorbens Wiedemann
	Chang, 1940
	CALLIPHORIDAE: CHRYSOMYINAE
	Chrysomya megacephala (Fabricius)
	Chang, 1940
	Harris et al., 1946
	CALLIPHORIDAE: CALLIPHORINAE
	Lucilia Robineau-Desvoidy
	Harris et al., 1946
	Lucilia caesar (Linnaeus)
	Chang, 1940
	Phaenicia sericata (Meigen)
	Chang, 1940
	SARCOPHAGIDAE: SARCOPHAGINAE
	Sarcophaga Meigen
	Chang, 1940
	Harris et al., 1946
TRICHOMONADIDAE	
Trichomonas Donné	MUSCIDAE: MUSCINAE
	Musca domestica Linnaeus
	Sieyro, 1942
	House fly (presumably *Musca domestica* Linnaeus)
	Wenyon et al., 1917b
	Attimonelli, ?1940b
	Akatov, 1955
Trichomonas hominis (Davaine)	OTITIDAE: ULIDIINAE
	Physiphora Fallén
	Harris et al., 1946
	PLATYSTOMATIDAE: PLATYSTOMATINAE
	Scholastes Loew
	Harris et al., 1946

Organism	Fly associations
Trichomonas hominis (cont.)	MUSCIDAE: COENOSIINAE *Atherigona* Rondani Harris et al., 1946 MUSCIDAE: MUSCINAE *Musca* Linnaeus Harris et al., 1946 *Musca domestica* Linnaeus Hegner, 1928 *Musca domestica vicina* Macquart Chang, 1940 *Musca sorbens* Wiedemann Chang, 1940 CALLIPHORIDAE: CHRYSOMYINAE *Chrysomya megacephala* (Fabricius) Chang, 1940 Harris et al., 1946 CALLIPHORIDAE: CALLIPHORINAE *Lucilia* Robineau-Desvoidy Harris et al., 1946 *Lucilia caesar* (Linnaeus) Chang, 1940 *Phaenicia sericata* (Meigen) Hegner, 1928 Chang, 1940 *Cynomyopsis cadaverina* (Robineau-Desvoidy) Hegner, 1928 SARCOPHAGIDAE: SARCOPHAGINAE *Sarcophaga* Meigen Chang, 1940 Harris et al., 1946
Trichomonas ?hominis (Davaine)	MUSCIDAE: MUSCINAE *Musca domestica* Linnaeus Simitch et al., 1937
Tritrichomonas foetus (Riedmüller) (Synonym: *Trichomonas foetus*)	MUSCIDAE: MUSCINAE *Musca domestica* Linnaeus Morgan, 1942 Holz, 1953
AMOEBIDAE	
Hartmannella hyalina Dangeard	MUSCIDAE: MUSCINAE *Musca domestica* Linnaeus Coutinho et al., 1957
ENDAMOEBIDAE	
Endolimax nana (Wenyon and O'Connor) (Synonym: *Amoeba limax*)	OTITIDAE: ULIDIINAE *Physiphora* Fallén Harris et al., 1946 PLATYSTOMATIDAE: PLATYSTOMATINAE *Scholastes* Loew Harris et al., 1946

Endolimax nana (cont.)	MUSCIDAE: COENOSIINAE *Atherigona* Rondani Harris et al., 1946 MUSCIDAE: FANNIINAE *Fannia canicularis* (Linnaeus) Root, 1921 Yao et al., 1929 *Fannia scalaris* (Fabricius) Yao et al., 1929 MUSCIDAE: MUSCINAE *Musca* Linnaeus Harris et al., 1946 *Musca domestica* Linnaeus Root, 1921 Yao et al., 1929 Coutinho et al., 1957 House fly (presumably *Musca domestica* Linnaeus) Werner, 1909b Frye et al., 1932 *Musca domestica vicina* Macquart Chang, 1940 *Musca sorbens* Wiedemann Chang, 1940 CALLIPHORIDAE: CHRYSOMYINAE *Cochliomyia macellaria* (Fabricius) Root, 1921 *Chrysomya megacephala* (Fabricius) Chang, 1940 Harris et al., 1946 CALLIPHORIDAE: CALLIPHORINAE *Lucilia* Robineau-Desvoidy Harris et al., 1946 *Lucilia caesar* (Linnaeus) Root, 1921 Yao et al., 1929 Chang, 1940 *Phaenicia sericata* (Meigen) Chang, 1940 *Calliphora vicina* Robineau-Desvoidy Root, 1921 Yao et al., 1929 *Calliphora vomitoria* (Linnaeus) Yao et al., 1929 SARCOPHAGIDAE: SARCOPHAGINAE *Sarcophaga* Meigen Chang, 1940 Harris et al., 1946 *Sarcophaga carnaria* (Linnaeus) Yao et al., 1929

Entamoeba coli (Grassi) .... OTITIDAE: ULIDIINAE

Physiphora Fallén
Harris et al., 1946

PLATYSTOMATIDAE:
PLATYSTOMATINAE

Scholastes Loew
Harris et al., 1946

MUSCIDAE: COENOSIINAE

Atherigona Rondani
Harris et al., 1946

MUSCIDAE: FANNIINAE

Fannia Robineau-Desvoidy
Wenyon et al., 1917a

Fannia canicularis (Linnaeus)
Root, 1921
Yao et al., 1929

Fannia scalaris (Fabricius)
Yao et al., 1929

MUSCIDAE: MUSCINAE

Musca Linnaeus
Wenyon et al., 1917a
Harris et al., 1946

Musca domestica Linnaeus
Roubaud, 1918
Buxton, 1920
Root, 1921
Yao et al., 1929
Aleksander et al., 1935
Vanni, 1946
Roberts, 1947
Rendtorff et al., 1954
Coutinho et al., 1957

House fly (presumably *Musca domestica* Linnaeus)
Wenyon et al., 1917a, 1917b
Frye et al., 1932

Musca domestica vicina Macquart
Pletneva, 1937
Chang, 1940

Musca sorbens Wiedemann
Chang, 1940

CALLIPHORIDAE: CHRYSOMYINAE

Cochliomyia macellaria (Fabricius)
Root, 1921

Chrysomya megacephala (Fabricius)
Chang, 1940
Harris et al., 1946

Entamoeba coli (cont.)

CALLIPHORIDAE: CALLIPHORINAE
Lucilia Robineau-Desvoidy
  Wenyon et al., 1917a
  Harris et al., 1946
Lucilia caesar (Linnaeus)
  Root, 1921
  Yao et al., 1929
  Chang, 1940
Phaenicia sericata (Meigen)
  Chang, 1940
Calliphora Robineau-Desvoidy
  Wenyon et al., 1917a, 1917b
Calliphora vicina Robineau-Desvoidy
  Root, 1921
  Yao et al., 1929
Calliphora vomitoria (Linnaeus)
  Yao et al., 1929
SARCOPHAGIDAE: SARCOPHAGINAE
Sarcophaga Meigen
  Wenyon et al., 1917b
  Chang, 1940
  Harris et al., 1946
Sarcophaga carnaria (Linnaeus)
  Yao et al., 1929

Entamoeba histolytica Schaudinn (Synonym: *Entamoeba tetragena*) ....

OTITIDAE: ULIDIINAE
Physiphora Fallén
  Harris et al., 1946
PLATYSTOMATIDAE: PLATYSTOMATINAE
Scholastes Loew
  Harris et al., 1946
MUSCIDAE: COENOSIINAE
Atherigona Rondani
  Harris et al., 1946
MUSCIDAE: FANNIINAE
Fannia Robineau-Desvoidy
  Wenyon et al., 1917a
Fannia canicularis (Linnaeus)
  Root, 1921
  Yao et al., 1929
Fannia scalaris (Fabricius)
  Yao et al., 1929
MUSCIDAE: MUSCINAE
Musca Linnaeus
  Wenyon et al., 1917a
  Harris et al., 1946
Musca domestica Linnaeus
  Kuenen et al., 1913
  Roubaud, 1918
  Buxton, 1920

Entamoeba histolytica (cont.)	Root, 1921
	Jausion et al., 1923
	Yao et al., 1929
	Aleksander et al., 1935
	Pipkin, 1942, 1949
	Sieyro, 1942
	Vanni, 1946
	Roberts, 1947
	House fly (presumably *Musca domestica* Linnaeus)
	Werner, 1909b
	Wenyon et al., 1917a, 1917b
	Rogers, 1929
	Frye et al., 1932
	Musca domestica vicina Macquart
	Pletneva, 1937
	Chang, 1940
	Musca sorbens Wiedemann
	Chang, 1940
	MUSCIDAE: STOMOXYINAE
	Stomoxys calcitrans (Linnaeus)
	Jausion et al.,1923
	CALLIPHORIDAE: CHRYSOMYINAE
	Cochliomyia macellaria (Fabricius)
	Root, 1921
	Pipkin, 1942, 1949
	Chrysomya Robineau-Desvoidy
	Anonymous Korean authors
	Chrysomya megacephala (Fabricius)
	Chang, 1940, 1945
	Harris et al., 1946
	Phormia regina (Meigen)
	Pipkin, 1942, 1949
	CALLIPHORIDAE: CALLIPHORINAE
	Lucilia Robineau-Desvoidy
	Wenyon et al., 1917a
	Harris et al., 1946
	Anonymous Korean authors
	Lucilia caesar (Linnaeus)
	Root, 1921
	Yao et al., 1929
	Chang, 1940
	Phaenicia pallescens (Shannon)
	Pipkin, 1942, 1949
	Phaenicia sericata (Meigen)
	Chang, 1940, 1945
	Calliphora Robineau-Desvoidy
	Wenyon et al., 1917a, 1917b

Fannia histolytica (cont.)

Calliphora vicina Robineau-Desvoidy
Root, 1921
Yao et al., 1929
Calliphora vomitoria (Linnaeus)
Jausion et al., 1923
Yao et al., 1929
SARCOPHAGIDAE: SARCOPHAGINAE
Sarcophaga Meigen
Wenyon et al., 1917b
Anonymous Korean authors
Chang, 1940, 1945
Sieyro, 1942
Harris et al., 1946
Sarcophaga carnaria (Linnaeus)
Yao et al., 1929
Sarcophaga misera Walker
Pipkin, 1942, 1949
Sarcophaga sarracenioides Aldrich
Pipkin, 1949

Iodamoeba bütschlii (Prowazek) ....

OTITIDAE: ULIDIINAE
Physiphora Fallén
Harris et al., 1946
PLATYSTOMATIDAE: PLATYSTOMATINAE
Scholastes Loew
Harris et al., 1946
MUSCIDAE: COENOSIINAE
Atherigona Rondani
Harris et al., 1946
MUSCIDAE: MUSCINAE
Musca Linnaeus
Harris et al., 1946
Musca domestica Linnaeus
Aleksander et al., 1935
Coutinho et al., 1957
Musca domestica vicina Macquart
Chang, 1940
Musca sorbens Wiedemann
Chang, 1940
CALLIPHORIDAE: CHRYSOMYINAE
Chrysomya megacephala (Fabricius)
Chang, 1940
Harris et al., 1946
CALLIPHORIDAE: CALLIPHORINAE
Lucilia Robineau-Desvoidy
Harris et al., 1946
Lucilia caesar (Linnaeus)
Chang, 1940
Phaenicia sericata (Meigen)
Chang, 1940

Organism	Fly association
Iodamoeba bütschlii (cont.)	SARCOPHAGIDAE: SARCOPHAGINAE *Sarcophaga* Meigen Chang, 1940 Harris et al., 1946
EIMERIIDAE	
Eimeria Schneider	MUSCIDAE: MUSCINAE *Musca domestica* Linnaeus Coutinho et al., 1957 House fly (presumably *Musca domestica* Linnaeus) Wenyon et al., 1917a
Eimeria acervalina (Tyzzer)	MUSCIDAE: MUSCINAE *Musca domestica* Linnaeus Roberts, 1947
Eimeria irresidua Kessel and Janciewicz	MUSCIDAE: MUSCINAE *Musca domestica* Linnaeus Metelkin, 1935 MUSCIDAE: STOMOXYINAE *Stomoxys calcitrans* (Linnaeus) Metelkin, 1935 CALLIPHORIDAE: CHRYSOMYINAE *Protophormia terraenovae* (Robineau-Desvoidy) Metelkin, 1935 CALLIPHORIDAE: CALLIPHORINAE *Lucilia caesar* (Linnaeus) Metelkin, 1935 *Calliphora vicina* Robineau-Desvoidy Metelkin, 1935 *Cynomya mortuorum* (Linnaeus) Metelkin, 1935
Eimeria perforans (Leuckart) Sluiter and Swellengrebel (Synonym: *Eimeria exigua*)	MUSCIDAE: MUSCINAE *Musca domestica* Linnaeus Metelkin, 1935 MUSCIDAE: STOMOXYINAE *Stomoxys calcitrans* (Linnaeus) Metelkin, 1935 CALLIPHORIDAE: CHRYSOMYINAE *Protophormia terraenovae* (Robineau-Desvoidy) Metelkin, 1935 CALLIPHORIDAE: CALLIPHORINAE *Lucilia caesar* (Linnaeus) Metelkin, 1935 *Calliphora vicina* Robineau-Desvoidy Metelkin, 1935 *Cynomya mortuorum* (Linnaeus) Metelkin, 1935

Organism	Fly host and references
Toxoplasma Nicolle and Manceaux	MUSCIDAE: PHAONIINAE *Musca stabulans* (Fallén) Kåss, 1954 MUSCIDAE: MUSCINAE *Musca domestica* Linnaeus Kåss, 1954 Schmidtke, 1959 MUSCIDAE: STOMOXYINAE *Stomoxys calcitrans* (Linnaeus) van Thiel et al., 1953 Schmidtke, 1959 CALLIPHORIDAE: CALLIPHORINAE *Phaenicia sericata* (Meigen) Schmidtke, 1959 *Calliphora vicina* Robineau-Desvoidy van Thiel, 1949 Schmidtke, 1959
Toxoplasma gondii Nicolle and Manceaux	MUSCIDAE: MUSCINAE *Musca domestica* Linnaeus Kunert et al., 1953 Varela et al., 1961 MUSCIDAE: STOMOXYINAE *Stomoxys calcitrans* (Linnaeus) Blanc et al., 1950 Kunert et al., 1953 Laarman, 1956, 1957 CALLIPHORIDAE: CALLIPHORINAE *Phaenicia sericata* (Meigen) Kunert et al., 1953 *Calliphora vicina* Robineau-Desvoidy Kunert et al., 1953 Schmidtke, 1955
NOSEMATIDAE	
Nosema apis Zander	MUSCIDAE: MUSCINAE *Musca domestica* Linnaeus Kramer, 1962 CALLIPHORIDAE: CHRYSOMYINAE *Phormia regina* (Meigen) Kramer, 1962 CALLIPHORIDAE: CALLIPHORINAE *Calliphora vicina* Robineau-Desvoidy Fantham et al., 1913 Kramer, 1962
Nosema bombycis Nägeli (Synonym: Pébrine)	MUSCIDAE: MUSCINAE *Musca domestica* Linnaeus Smuidsinovicia, 1889 Teodoro, 1926
Nosema kingi Kramer	DROSOPHILIDAE: DROSOPHILINAE *Drosophila willistoni* Sturtevant Kramer, 1964b

Organism	Fly and reference
Nosema kingi (cont.)	MUSCIDAE: MUSCINAE
	Musca domestica Linnaeus
	Kramer, 1964b
	CALLIPHORIDAE: CHRYSOMYINAE
	Phormia regina (Meigen)
	Kramer, 1964b
	CALLIPHORIDAE: CALLIPHORINAE
	Phaenicia cuprina Wiedemann
	Kramer, 1964b
Thelohania ovata Dunkerly	MUSCIDAE: FANNIINAE
	Fannia scalaris (Fabricius)
	Dunkerly, 1912
Thelohania thomsoni Kramer	MUSCIDAE: PHAONIINAE
	Muscina assimilis (Fallén)
	Kramer, 1961b
MRAZEKIIDAE	
Octosporea monospora Chatton and Krempf	DROSOPHILIDAE: DROSOPHILINAE
	Drosophila confusa Staeger
	Chatton et al., 1911d
	Drosophila plurilineata Villeneuve
	Chatton et al., 1911d
	MUSCIDAE: FANNIINAE
	Fannia scalaris (Fabricius)
	Brug, 1914
Octosporea muscae-domesticae Flu	DROSOPHILIDAE: DROSOPHILINAE
	Drosophila Fallén
	Kramer, 1965b
	Drosophila confusa Staeger
	Chatton et al., 1911d
	Drosophila funebris (Fabricius)
	Chatton et al., 1911d
	Drosophila melanogaster Meigen
	Chatton et al., 1911d
	Drosophila phalerata Meigen
	Chatton et al., 1911d
	Drosophila plurilineata Villeneuve
	Chatton et al., 1911d
	MUSCIDAE: MUSCINAE
	Musca domestica Linnaeus
	Flu, 1911
	Porter, 1953
	Fantham et al., 1958
	Kramer, 1964a, 1965b, 1966
	Musca sorbens Wiedemann
	Kramer, 1965b
	CALLIPHORIDAE: CHRYSOMYINAE
	Cochliomyia macellaria (Fabricius)
	Kramer, 1964a, 1965b

Organism	Fly / Reference
Octosporea muscae-domesticae (cont.)	*Phormia regina* (Meigen)
	Kramer, 1964a, 1965a, 1965b, 1965c, 1966
	CALLIPHORIDAE: CALLIPHORINAE
	Phaenicia sericata (Meigen)
	Kramer, 1964a, 1965b
	Calliphora vicina Robineau-Desvoidy
	Fantham et al., 1958
	Calliphora vomitoria (Linnaeus)
	Fantham et al., 1958
	Kramer, 1965b
	CALLIPHORIDAE: POLLENIINAE
	Pollenia rudis (Fabricius)
	Kramer, 1964a, 1965b
Octosporea ?muscae-domesticae Flu	MUSCIDAE: MUSCINAE
	Musca domestica Linnaeus
	Cardamatis, 1912b
Spiroglugea Léger and Hesse	MUSCIDAE: MUSCINAE
	Musca domestica Linnaeus
	Fantham et al., 1958
	CALLIPHORIDAE: CALLIPHORINAE
	Calliphora vicina Robineau-Desvoidy
	Porter, 1953
	Fantham et al., 1958
	Calliphora vomitoria (Linnaeus)
	Porter, 1953
	Fantham et al., 1958
Toxoglugea Léger and Hesse	CALLIPHORIDAE: CALLIPHORINAE
	Calliphora vicina Robineau-Desvoidy
	Porter, 1953
	Fantham et al., 1958
	Calliphora vomitoria (Linnaeus)
	Porter, 1953
	Fantham et al., 1958
TELOMYXIDAE	
Telomyxa Léger and Hesse	MUSCIDAE: MUSCINAE
	Musca domestica Linnaeus
	Fantham et al., 1958
	CALLIPHORIDAE: CALLIPHORINAE
	Calliphora vicina Robineau-Desvoidy
	Porter, 1953
	Fantham et al., 1958
	Calliphora vomitoria (Linnaeus)
	Porter, 1953
	Fantham et al., 1958
UNDETERMINED MICROSPORIDA	DROSOPHILIDAE: DROSOPHILINAE
	Drosophila Fallén
	Bell, 1952
	Stalker et al., 1963

Organism	Fly / Reference
UNDETERMINED MICROSPORIDA (cont.)	*Drosophila melanogaster* Meigen Wolfson et al., 1957 Stalker et al., 1963 *Drosophila paramelanica* Patterson Wolfson et al., 1957 *Drosophila parthenogenetica* Stalker Wolfson et al., 1957 *Drosophila robusta* Sturtevant Stalker et al., 1963 *Drosophila willistoni* Sturtevant Burnett et al., 1962 *Scaptomyza* Hardy Stalker et al., 1963 MUSCIDAE: MUSCINAE *Musca sorbens* Wiedemann Lamborn, 1935
HOLOPHRYIDAE	
Holophrya (Synonym: *Holophria*)	MUSCIDAE: MUSCINAE House fly (presumably *Musca domestica* Linnaeus) Attimonelli, ?1940b
HETEROPHYIDAE	
Heterophyes heterophyes (V. Siebold) Stile and Hassall	MUSCIDAE: MUSCINAE House fly (presumably *Musca domestica* Linnaeus) Wenyon et al., 1917a
SCHISTOSOMATIDAE	
Schistosoma Weinland (Synonym: *Bilharzia*)	MUSCIDAE: MUSCINAE House fly (presumably *Musca domestica* Linnaeus) Wenyon et al., 1917a
UNDETERMINED TREMATODA	MUSCIDAE: MUSCINAE House fly (presumably *Musca domestica* Linnaeus) Wenyon et al., 1917b
DIPHYLLOBOTHRIIDAE	
Bothriocephalus marginatus Krefft (Synonym: *Taenia marginata*)	MUSCIDAE: FANNIINAE *Fannia canicularis* (Linnaeus) Nicoll, 1911b MUSCIDAE: MUSCINAE *Musca domestica* Linnaeus Nicoll, 1911b CALLIPHORIDAE: CALLIPHORINAE *Calliphora vicina* Robineau-Desvoidy Nicoll, 1911b
Diphyllobothrium Cobbold	MUSCIDAE: MUSCINAE *Musca domestica* Linnaeus Pokrovskiĭ et al., 1938

Organism	Fly association
Diphyllobothrium (cont.)	CALLIPHORIDAE: CALLIPHORINAE *Lucilia* Robineau-Desvoidy Pokrovskiĭ et al., 1938
Diphyllobothrium latum (Linnaeus) Lühe (Synonym: *Dibothriocephalus latus*)	MUSCIDAE: MUSCINAE *Musca domestica* Linnaeus Pod"ıa͡pol'skaıa͡ et al., 1934 Aleksander et al., 1935 CALLIPHORIDAE: CALLIPHORINAE *Calliphora vicina* Robineau-Desvoidy Pod"ıa͡pol'skaıa͡ et al., 1934
ANOPLOCEPHALIDAE	
Moniezia expansa (Rudolphi)	MUSCIDAE: MUSCINAE *Musca domestica* Linnaeus Nicoll, 1911b
DAVAINEIDAE	
Davainea nana (Fuhrmann) (Synonym: *Taenia nana*)	MUSCIDAE: MUSCINAE House fly (presumably *Musca domestica* Linnaeus) Calandruccio, 1906
Raillietina cesticillus (Molin) (Synonyms: *Taenia infundibuliformis, Davainea cesticillus*)	MUSCIDAE: MUSCINAE *Musca domestica* Linnaeus Grassi et al., 1889 Gutberlet, 1916 Ackert, 1919 Neveu-Lemaire, 1936
Raillietina tetragona (Molin) (Synonym: *Davainea tetragona*)	MUSCIDAE: MUSCINAE *Musca domestica* Linnaeus Gutberlet, 1916 Ackert, 1920 Neveu-Lemaire, 1936
DILEPIDIDAE	
Choanotaenia infundibulum (Bloch) (Synonym: *Choanotaenia infundibuliformis*)	MUSCIDAE: MUSCINAE *Musca domestica* Linnaeus Gutberlet, 1916 Neveu-Lemaire, 1936 Wetzel, 1936 Reid et al., 1937 MUSCIDAE: STOMOXYINAE *Stomoxys calcitrans* (Linnaeus) Gutberlet, 1916
Dipylidium caninum (Linnaeus) Railliet	MUSCIDAE: MUSCINAE *Musca domestica* Linnaeus Nicoll, 1911b
HYMENOLEPIDIDAE	
Echinolepis carioca (Magalhães) Spassky and Spasskaja (Synonym: *Hymenolepis carioca*)	MUSCIDAE: STOMOXYINAE *Stomoxys calcitrans* (Linnaeus) Gutberlet, 1920 Neveu-Lemaire, 1936

Hymenolepis Weinland .... MUSCIDAE: MUSCINAE
Musca domestica Linnaeus
Pokrovskiĭ et al., 1938
Coutinho et al., 1957

Hymenolepis diminuta .... OTITIDAE: ULIDIINAE
(Rudolphi) Blanchard
Physiphora Fallén
Harris et al., 1946
PLATYSTOMATIDAE: PLATYSTOMATINAE
Scholastes Loew
Harris et al., 1946
MUSCIDAE: COENOSIINAE
Atherigona Rondani
Harris et al., 1946
MUSCIDAE: MUSCINAE
Musca Linnaeus
Harris et al., 1946
Musca domestica Linnaeus
Nicoll, 1911b
CALLIPHORIDAE: CHRYSOMYINAE
Chrysomya megacephala (Fabricius)
Harris et al., 1946
CALLIPHORIDAE: CALLIPHORINAE
Lucilia Robineau-Desvoidy
Harris et al., 1946
SARCOPHAGIDAE: SARCOPHAGINAE
Sarcophaga Meigen
Harris et al., 1946

Hymenolepis nana .... MUSCIDAE: PHAONIINAE
(V. Siebold) Blanchard
Muscina stabulans (Fallén)
Bogoi͡avlenskiĭ et al., 1928
Sychevskai͡a et al., 1959b
MUSCIDAE: MUSCINAE
Musca domestica Linnaeus
Buxton, 1920
Bogoi͡avlenskiĭ et al., 1928
Aleksander et al., 1935
Musca domestica vicina Macquart
Pletneva, 1937
Sychevskai͡a et al., 1959b

Vampirolepis fraterna .... MUSCIDAE: MUSCINAE
(Stiles) Spassky
(Synonym: *Hymenolepis fraterna*)
Musca domestica Linnaeus
Joyeux, 1920

TAENIIDAE

Echinococcus granulosus .... CALLIPHORIDAE: CALLIPHORINAE
(Batsch) Rudolphi
Lucilia Robineau-Desvoidy
Perez-Fontana et al., 1961
Calliphora Robineau-Desvoidy
Perez-Fontana et al., 1961

Organism	Fly association
Echinococcus granulosus (cont.)	SARCOPHAGIDAE: SARCOPHAGINAE *Sarcophaga tibialis* Macquart Heinz et al., 1955
Taenia pisiformis (Bloch) .... (Synonym: *Taenia serrata*)	MUSCIDAE: MUSCINAE *Musca domestica* Linnaeus Nicoll, 1911b
Taeniarhynchus saginatum .... (Goeze) (Synonym: *Taenia saginata*)	SYRPHIDAE: MILESIINAE *Eristalis aeneus* (Scopoli) Sychevskai͡a et al., 1958 MUSCIDAE: PHAONIINAE *Muscina stabulans* (Fallén) Bogoi͡avlenskiĭ et al., 1928 Sychevskai͡a et al., 1958 MUSCIDAE: MUSCINAE *Dasyphora asiatica* Zimin Sychevskai͡a et al., 1958 *Musca domestica* Linnaeus Buxton, 1920 Bogoi͡avlenskiĭ et al., 1928 Aleksander et al., 1935 House fly (presumably *Musca domestica* Linnaeus) Wenyon et al., 1917a *Musca domestica vicina* Macquart Zmeev, 1936 Sychevskai͡a et al., 1958 *Musca sorbens* Wiedemann Sychevskai͡a et al., 1958 CALLIPHORIDAE: CHRYSOMYINAE *Chrysomya albiceps* (Wiedemann) Round, 1961 *Chrysomya chloropyga* Wiedemann Round, 1961 *Chrysomya rufifacies* (Macquart) Sychevskai͡a et al., 1958 CALLIPHORIDAE: CALLIPHORINAE *Calliphora vicina* Robineau-Desvoidy Sychevskai͡a et al., 1958 CALLIPHORIDAE: POLLENIINAE *Pollenia rudis* (Fabricius) Sychevskai͡a et al., 1958 SARCOPHAGIDAE: SARCOPHAGINAE *Sarcophaga* Meigen Round, 1961 *Sarcophaga haemorrhoidalis* (Fallén) Sychevskai͡a et al., 1958
UNDETERMINED TAENIIDAE	MUSCIDAE: MUSCINAE *Musca domestica vicina* Macquart Sychevskai͡a et al., 1959b

UNDETERMINED CESTOIDEA	MUSCIDAE: MUSCINAE House fly (presumably *Musca domestica* Linnaeus) Wenyon et al., 1917b
Bassus laetatorius, unvalidated	MUSCIDAE: MUSCINAE *Musca domestica vicina* Macquart Sychevskaia͡, 1964b
TETRADONEMATIDAE	
Mermithonema entomophilum Goodey	SEPSIDAE *Sepsis cynipsea* (Linnaeus) Goodey, 1941
RHABDITIDAE	
Rhabditis axei (Cobbold)	MUSCIDAE: STOMOXYINAE *Stomoxys calcitrans* (Linnaeus) Hague, 1963
Rhabditis pellio (Schneider)	DROSOPHILIDAE: DROSOPHILINAE *Drosophila* Fallén Aubertot, 1923 MUSCIDAE: MUSCINAE *Musca domestica* Linnaeus Menzel, 1924
PANAGROLAIMIDAE	
Panagrellus zymosiphilus (Brunold) (Synonym: *Anguillula zymosiphila*)	DROSOPHILIDAE: DROSOPHILINAE *Drosophila* Fallén Brunold, 1950
STRONGYLOIDIDAE	
Strongyloides stercoralis (Babay)	MUSCIDAE: MUSCINAE *Musca domestica* Linaneus Buxton, 1920
TRICHURIDAE	
Trichuris Roederer (Synonym: *Trichocephalus*)	MUSCIDAE: MUSCINAE *Musca* Linnaeus Tao, 1936 CALLIPHORIDAE: CHRYSOMYINAE *Cochliomyia* Townsend Tao, 1936 *Chrysomya megacephala* (Fabricius) Chang, 1945 CALLIPHORIDAE: CALLIPHORINAE *Lucilia* Robineau-Desvoidy Tao, 1936 Perez-Fontana et al., 1961 *Phaenicia sericata* (Meigen) Chang, 1945 *Calliphora* Robineau-Desvoidy Perez-Fontana et al., 1961

Trichuris (cont.)

SARCOPHAGIDAE: SARCOPHAGINAE
Sarcophaga Meigen
Chang, 1945

Trichuris trichiura **(Linnaeus)** ......
(Synonyms: *Trichocephalus dispar, Trichocephalus trichiuris*)

OTITIDAE: ULIDIINAE
Physiphora Fallén
Harris et al., 1946

PLATYSTOMATIDAE: PLATYSTOMATINAE
Scholastes Loew
Harris et al., 1946

SPHAEROCERIDAE
Limosinae punctipennis (Wiedemann)
Nicholls, 1912

MUSCIDAE: COENOSIINAE
Atherigona Rondani
Harris et al., 1946

MUSCIDAE: PHAONIINAE
Muscina stabulans (Fallén)
Bogoi͡avlenskiĭ et al., 1928

MUSCIDAE: MUSCINAE
Musca Linnaeus
Harris et al., 1946
Musca domestica Linnaeus
Nicoll, 1911b
Buxton, 1920
Bogoi͡avlenskiĭ et al., 1928
Pod"i͡apol'skai͡a et al., 1934
Aleksander et al., 1935
House fly (presumably *Musca domestica* Linnaeus)
Wenyon et al., 1917a
Musca domestica vicina Macquart
Chang, 1940

CALLIPHORIDAE: CHRYSOMYINAE
Chrysomya megacephala (Fabricius)
Chang, 1940
Harris et al., 1946

CALLIPHORIDAE: CALLIPHORINAE
Lucilia Robineau-Desvoidy
Harris et al., 1946
Lucilia caesar (Linnaeus)
Chang, 1940
Phaenicia sericata (Meigen)
Chang, 1940
Calliphora vicina Robineau-Desvoidy
Pod"i͡apol'skai͡a et al., 1934

SARCOPHAGIDAE: SARCOPHAGINAE
Sarcophaga Meigen
Chang, 1940
Harris et al., 1946

ANCYLOSTOMATIDAE

Organism	Fly associations
Ancylostoma (Dubini) (Synonym: Hookworm)	OTITIDAE: ULIDIINAE *Physiphora* Fallén Harris et al., 1946 PLATYSTOMATIDAE: PLATYSTOMATINAE *Scholastes* Loew Harris et al., 1946 MUSCIDAE: COENOSIINAE *Atherigona* Rondani Harris et al., 1946 MUSCIDAE: MUSCINAE *Musca* Linnaeus Tao, 1936 Harris et al., 1946 *Musca domestica* Linnaeus Buxton, 1920 *Musca domestica vicina* Macquart Chang, 1940 *Musca sorbens* Wiedemann Chang, 1940 CALLIPHORIDAE: CHRYSOMYINAE *Cochliomyia* Townsend Tao, 1936 *Chrysomya megacephala* (Fabricius) Chang, 1940, 1945 Harris et al., 1946 CALLIPHORIDAE: CALLIPHORINAE *Lucilia* Robineau-Desvoidy Tao, 1936 Harris et al., 1946 *Lucilia caesar* (Linnaeus) Chang, 1940 *Phaenicia sericata* (Meigen) Chang, 1940, 1945 SARCOPHAGIDAE: SARCOPHAGINAE *Sarcophaga* Meigen Chang, 1940, 1945 Harris et al., 1946
Ancylostoma caninum (Ercolani)	MUSCIDAE: MUSCINAE *Musca domestica* Linnaeus Nicoll, 1911b Harada, 1953, 1954 CALLIPHORIDAE: CALLIPHORINAE *Calliphora* Robineau-Desvoidy Harada, 1953, 1954
Ancylostoma duodenale (Dubini) Creplin (Synonym: *Ankylostoma duodenale*)	MUSCIDAE: PHAONIINAE *Muscina stabulans* (Fallén) Bogoi͡avlenskiĭ et al., 1928

Ancylostoma duodenale (cont.)

MUSCIDAE: MUSCINAE
Musca domestica Linnaeus
Buxton, 1920
Bogoi͡avlenskiĭ et al., 1928
House fly (presumably *Musca domestica* Linnaeus)
Wenyon et al., 1917a

Necator americanus (Stiles) ....

SPHAEROCERIDAE
Limosina punctipennis (Wiedemann)
Nicholls, 1912
MUSCIDAE: MUSCINAE
Musca domestica Linnaeus
Buxton, 1920

STRONGYLIDAE

Stronglus equinus Mueller ....
(Synonym: *Sclerostomum equinum*)

MUSCIDAE: MUSCINAE
Musca domestica Linnaeus
Nicoll, 1911b

SYNGAMIDAE

Syngamus trachea (Montagu) ......

MUSCIDAE: MUSCINAE
Musca domestica Linnaeus
Clapham, 1939
CALLIPHORIDAE: CALLIPHORINAE
Phaenicia sericata (Meigen)
Clapham, 1939

TRICHOSTRONGYLIDAE

Trichostrongylus colubriformis .... (Giles)
(Synonym: *Trichostrongylus instabilis*)

MUSCIDAE: PHAONIINAE
Muscina stabulans (Fallén)
Bogoi͡avlenskiĭ et al., 1928
MUSCIDAE: MUSCINAE
Musca domestica Linnaeus
Bogoi͡avlenskiĭ et al., 1928

OXYURIDAE

Enterobius vermicularis .... (Linnaeus) Leach

MUSCIDAE: PHAONIINAE
Muscina stabulans (Fallén)
Bogoi͡avlenskiĭ et al., 1928
MUSCIDAE: MUSCINAE
Musca domestica Linnaeus
Bogoi͡avlenskiĭ et al., 1928
Pod"i͡apol'skai͡a et al., 1934
Aleksander et al., 1935
Musca domestica vicina Macquart
Pletneva, 1937
CALLIPHORIDAE: CALLIPHORINAE
Lucilia Robineau-Desvoidy
Pokrovskiĭ et al., 1938
Calliphora vicina Robineau-Desvoidy
Pod"i͡apol'skai͡a et al., 1934
SARCOPHAGIDAE: SARCOPHAGINAE
Sarcophaga Meigen
Pokrovskiĭ et al., 1938

ASCARIDIDAE

Organism	Fly associations
Ascaris Linnaeus	MUSCIDAE: MUSCINAE *Musca* Linnaeus Tao, 1936 *Musca domestica* Linnaeus Pod"îapol'skaîa et al., 1934 Pokrovskiĭ et al., 1938 CALLIPHORIDAE: CHRYSOMYINAE *Cochliomyia* Townsend Tao, 1936 *Chrysomya megacephala* (Fabricius) Chang, 1945 CALLIPHORIDAE: CALLIPHORINAE *Lucilia* Robineau-Desvoidy Tao, 1936 *Phaenicia sericata* (Meigen) Chang, 1945 *Calliphora vicina* Robineau-Desvoidy Pod"îapol'skaîa et al., 1934 SARCOPHAGIDAE: SARCOPHAGINAE *Sarcophaga* Meigen Chang, 1945
Ascaris lumbricoides Linnaeus (Synonym: *Ascaris suum*)	SYRPHIDAE: MILESIINAE *Eristalis* Latreille Sukhacheva, 1963 OTITIDAE: ULIDIINAE *Physiphora* Fallén Harris et al., 1946 PLATYSTOMATIDAE: PLATYSTOMATINAE *Scholastes* Loew Harris et al., 1946 SPHAEROCERIDAE *Limosina punctipennis* (Wiedemann) Nicholls, 1912 MUSCIDAE: COENOSIINAE *Atherigona* Rondani Harris et al., 1946 MUSCIDAE: FANNIINAE *Fannia* Robineau-Desvoidy Sukhacheva, 1963 MUSCIDAE: PHAONIINAE *Ophyra* Robineau-Desvoidy Sukhacheva, 1963 *Muscina stabulans* (Fallén) Bogoîavlenskiĭ et al., 1928 MUSCIDAE: MUSCINAE *Musca* Linnaeus Harris et al., 1946

Ascaris lumbricoides (cont.)	*Musca domestica* Linnaeus Stiles, 1889 (see Nuttall et al., 1909) Bogoi͡avlenskiĭ et al., 1928 Roberts, 1934 Aleksander et al., 1935 Sukhacheva, 1963 *Musca domestica vicina* Macquart Chang, 1940 Sychevskai͡a et al., 1958 *Musca sorbens* Wiedemann Chang, 1940 CALLIPHORIDAE: CHRYSOMYINAE *Chrysomya megacephala* (Fabricius) Chang, 1940 Harris et al., 1946 *Protophormia terraenovae* (Robineau-Desvoidy) Sukhacheva, 1963 CALLIPHORIDAE: CALLIPHORINAE *Lucilia* Robineau-Desvoidy Harris et al., 1946 *Lucilia caesar* (Linnaeus) Chang, 1940 *Phaenicia cuprina* Wiedemann Roberts, 1934 *Phaenicia sericata* (Meigen) Chang, 1940 *Calliphora* Robineau-Desvoidy Perez-Fontana et al., 1961 *Calliphora vicina* Robineau-Desvoidy Sukhacheva, 1963 SARCOPHAGIDAE: SARCOPHAGINAE *Sarcophaga* Meigen Chang, 1940 Harris et al., 1946 *Sarcophaga haemorrhoidalis* (Fallén) Sychevskai͡a et al., 1958
Ascaris ?lumbricoides Linnaeus	CALLIPHORIDAE: CALLIPHORINAE *Lucilia* Robineau-Desvoidy Perez-Fontana et al., 1961
Parascaris equorum (Goeze) .... (Synonym: *Ascaris megalocephala*)	MUSCIDAE: MUSCINAE *Musca domestica* Linnaeus Nicoll, 1911b
Toxascaris leonina (Linstow) ...... (Synonym: *Toxascaris limbata*)	MUSCIDAE: MUSCINAE *Musca domestica* Linnaeus Nicoll, 1911b

Organism	Fly
ACUARIIDAE	
Dispharynx ?nasuta (Rudolphi) (Synonym: *Dispharagus nasutus*)	MUSCIDAE: MUSCINAE *Musca domestica* Linnaeus Piana, 1896
Parabronema skrjabini Rassowska	MUSCIDAE: STOMOXYINAE *Haematobia titillans* (Bezzi) Ivashkin, 1959
SPIRURIDAE	
Habronema Diesing	MUSCIDAE: MUSCINAE *Musca domestica* Linnaeus Generali, 1886? Mello et al., 1934b Coutinho et al., 1957 *Musca sorbens sorbens* Wiedemann Johnston, 1912 Tryon, 1914 CALLIPHORIDAE: CALLIPHORINAE *Anastellorhina augur* (Fabricius) Johnston, 1920 Johnston et al., 1920a
Habronema megastoma (Rudolphi) Seurat	MUSCIDAE: FANNIINAE *Fannia canicularis* (Linnaeus) Roubaud et al., 1922b MUSCIDAE: PHAONIINAE *Muscina stabulans* (Fallén) Roubaud et al., 1922b MUSCIDAE: MUSCINAE *Orthellia* Robineau-Desvoidy Johnston, 1920 Johnston et al., 1920a Neveu-Lemaire, 1936 *Musca domestica* Linnaeus Bull, 1919 Hill, 1919 Johnston, 1920 Johnston et al., 1920a Roubaud et al., 1921, 1922a, 1922b Descazeaux et al., 1933 Neveu-Lemaire, 1936 *Musca fergusoni* Johnston and Bancroft Johnston, 1920 Johnston et al., 1920a Neveu-Lemaire, 1936 *Musca hilli* Johnston and Bancroft Johnston, 1920 Johnston et al., 1920a *Musca lusoria* Wiedemann Neveu-Lemaire, 1936

Organism	Fly
Habronema megastoma (cont.)	*Musca sorbens* Wiedemann
	Neveu-Lemaire, 1936
	Musca sorbens sorbens Wiedemann
	Johnston, 1920
	Johnston et al., 1920a
	Neveu-Lemaire, 1936
	Musca terraereginae Johnston and Bancroft
	Johnston, 1920
	Johnston et al., 1920a
	Neveu-Lemaire, 1936
	Musca ventrosa Wiedemann
	Neveu-Lemaire, 1936
	MUSCIDAE: STOMOXYINAE
	Stomoxys calcitrans (Linnaeus)
	Bull, 1919
	Hill, 1919
	SARCOPHAGIDAE: SARCOPHAGINAE
	Sarcophaga misera Walker
	Johnston, 1920
	Johnston et al., 1920a
Habronema ?megastoma (Rudolphi) Seurat	UNDETERMINED DROSOPHILIDAE
	Crawford, 1926
	MUSCIDAE: MUSCINAE
	Musca domestica Linnaeus
	Bull, 1918
Habronema microstoma (Schneider)	MUSCIDAE: MUSCINAE
	Orthellia Robineau-Desvoidy
	Johnston, 1920
	Musca domestica Linnaeus
	Bull, 1919
	Hill, 1919
	Johnston, 1920
	Roubaud, 1922b
	Neveu-Lemaire, 1936
	Musca fergusoni Johnston and Bancroft
	Johnston, 1920
	Musca hilli Johnston and Bancroft
	Johnston, 1920
	Musca sorbens sorbens Wiedemann
	Johnston, 1920
	Musca terraereginae Johnston and Bancroft
	Johnston, 1920
	MUSCIDAE: STOMOXYINAE
	Haematobia exigua (de Meijere)
	Neveu-Lemaire, 1936
	Stomoxys calcitrans (Linnaeus)
	Bull, 1919

Organism	Fly / Reference
Habronema microstoma (cont.)	Hill, 1919
	Johnston, 1920
	Johnston et al., 1920a
	Roubaud et al., 1922a, 1922b
	Neveu-Lemaire, 1936
	SARCOPHAGIDAE: SARCOPHAGINAE
	Sarcophaga melanura Meigen
	Neveu-Lemaire, 1936
	Sarcophaga misera Walker
	Johnston, 1920
Habronema muscae (Carter)	MUSCIDAE: FANNIINAE
(Synonym: *Filaria muscae*)	*Fannia canicularis* (Linnaeus)
	Roubaud et al., 1922b
	MUSCIDAE: PHAONIINAE
	Muscina stabulans (Fallén)
	Roubaud et al., 1922b
	MUSCIDAE: MUSCINAE
	Orthellia Robineau-Desvoidy
	Johnston, 1920
	Johnston et al., 1920a
	Neveu-Lemaire, 1936
	Musca domestica Linnaeus
	Carter, 1861
	Johnston, 1913, 1920
	Ransom, 1913
	van Saceghem, 1917, 1918a
	Bull, 1919
	Hill, 1919
	Johnston et al., 1920a
	Roubaud et al., 1922a, 1922b
	de Magarinos Torres et al., 1923
	Descazeaux et al., 1933
	Neveu-Lemaire, 1936
	Mello et al., 1943a
	House fly (presumably *Musca domestica* Linnaeus)
	Leidy, 1874
	Ransom, 1911
	Musca fergusoni Johnston and Bancroft
	Johnston, 1920
	Johnston et al., 1920a
	Neveu-Lemaire, 1936
	Musca hilli Johnston and Bancroft
	Johnston, 1920
	Johnston et al., 1920a
	Musca lusoria Wiedemann
	Neveu-Lemaire, 1936
	Musca sorbens Wiedemann
	Neveu-Lemaire, 1936

Habronema muscae (cont.)

Musca sorbens sorbens Weidemann
Johnston, 1920
Johnston et al., 1920a
Neveu-Lemaire, 1936

Musca terraereginae Johnston and Bancroft
Johnston, 1920
Johnston et al., 1920a
Neveu-Lemaire, 1936

Musca ventrosa Wiedemann
Neveu-Lemaire, 1936

MUSCIDAE: STOMOXYINAE
Stomoxys calcitrans (Linnaeus)
Johnston, 1913
Bull, 1919
Hill, 1919

SARCOPHAGIDAE: SARCOPHAGINAE
Sarcophaga misera Walker
Johnston, 1920
Johnston et al., 1920a
Neveu-Lemaire, 1936

Habronema ?muscae (Carter) ....

MUSCIDAE: MUSCINAE
Musca domestica Linnaeus
Piana, 1896
Musca sorbens sorbens Wiedemann
Johnston, 1913

THELAZIIDAE

Thelazia californiensis Price ....

MUSCIDAE: MUSCINAE
Musca autumnalis de Geer
Sabrosky, 1959

Thelazia gulosa .... Railliet and Henry

MUSCIDAE: MUSCINAE
Musca amica Zimin
Krastin, 1950, 1952
Musca autumnalis de Geer
Világiová, 1962
Musca larvipara Porchinskiĭ
Klesov, 1951
Világiová, 1962

Thelazia rhodesii (Desmarest) ....

MUSCIDAE: MUSCINAE
Musca autumnalis de Geer
Klesov, 1949a
Sabrosky, 1959
Világiová, 1962
Musca convexifrons Thomson
Krastin, 1949a, 1949b
Musca larvipara Porchinskiĭ
Klesov, 1949a, 1951
Világiová, 1962
Tukhmanyants et al., 1963

Organism	Fly host and reference
Thelazia skrjabini Erschov	MUSCIDAE: MUSCINAE *Musca amica* Zimin Krastin, 1950, 1952
DIPETALONEMATIDAE	
Oncocerca gibsoni (Cleland and Johnson) (Synonym: *Onchocerca gibsoni*)	MUSCIDAE: PHAONIINAE *Hydrotea australis* Malloch Henry, 1927 MUSCIDAE: MUSCINAE *Musca domestica* Linnaeus Henry, 1927 *Musca sorbens sorbens* Wiedemann Henry, 1927
FILARIIDAE	
?*Filaria Mueller* (Synonyms: ?*Dermatitis verminosa, vulnerosa bovis*)	MUSCIDAE: MUSCINAE *Musca domestica* Linnaeus Place, 1915 Iwanoff, 1934 *Musca sorbens sorbens* Wiedemann Place, 1915 MUSCIDAE: STOMOXYINAE *Stomoxys* Geoffroy Wenyon, 1911a *Stomoxys calcitrans* (Linnaeus) Place, 1915 Iwanoff, 1934
Setaria cervi (Rudolphi) (Synonyms: ?*Filaria labiato-papillosa, Setaria labiato-papillosa,* ?*Filaria stomoxeos*)	MUSCIDAE: STOMOXYINAE *Stomoxys* Geoffroy von Linstow, 1875 Noè, 1903 Neveu-Lemaire, 1936
STEPHANOFILARIIDAE	
Stephanofilaria assamensis Pande	MUSCIDAE: MUSCINAE *Musca conducens* Walker Srivastava et al., 1963 Patnaik, 1965
ALLANTONEMATIDAE	
Allantonema muscae (Roy and Mukherjee)	MUSCIDAE: MUSCINAE *Musca domestica vicina* Macquart Roy et al., 1937a
Allantonema stricklandi (Roy and Mukherjee)	MUSCIDAE: MUSCINAE *Musca domestica vicina* Macquart Roy et al., 1937b
Howardula aoronymphium Welch	DROSOPHILIDAE: DROSOPHILINAE *Drosophila kuntzei* Duda Welch, 1959 *Drosophila phalerata* Meigen Welch, 1959
Parasitylenchus diplogenus Welch	DROSOPHILIDAE: DROSOPHILINAE *Drosophila melanogaster* Meigen Welch, 1959

Organism	Fly host and reference
Parasitylenchus diplogenus (cont.)	*Drosophila obscura* Fallén Welch, 1959 *Drosophila silvestris* Basden Welch, 1959 *Drosophila subobscura* Collin Welch, 1959
UNDETERMINED ALLANTONEMATIDAE	DROSOPHILIDAE: DROSOPHILINAE *Drosophila confusa* Staeger Welch, 1959
CONTORTYLENCHIDAE	
Heterotylenchus Stoffolano and Nickle	MUSCIDAE: MUSCINAE *Morellia hortorum* (Fallén) Stoffolano, 1969 *Musca autumnalis* De Geer Stoffolano, Jr. et al, 1966 *Musca sorbens* Wiedemann Hughes et al., 1969
Heterotylenchus aberrans Bovien	ANTHOMYIIDAE: ANTHOMYIINAE *Hylemya antigua* (Meigen) Stoffolano, 1969
Heterotylenchus autumnalis Nickle	MUSCIDAE: PHAONIINAE *Hydrotaea meteorica* (Linnaeus) Világiová, 1968 MUSCIDAE: MUSCINAE *Morellia simplex* (Loew) Világiová, 1968 *Orthellia caesarion* (Meigen) Stoffolano, 1970 *Musca autumnalis* De Geer Jones et al., 1967 Nickle, 1967 Stoffolano, 1967, 1969, 1970 Treece et al., 1968 Világiová, 1968 *Musca domestica* Linnaeus Stoffolano, 1970 *Musca larvipara* Porchinskiĭ Világiová, 1968 *Musca tempestiva* Fallén Világiová, 1968 SARCOPHAGIDAE: SARCOPHAGINAE *Ravinia lherminieri* (Robineau-Desvoidy) Stoffolano, 1970
UNDETERMINED NEMATODES	CHLOROPIDAE: OSCINELLINAE *Hippelates flavipes* Loew Kumm et al., 1935

Organism	Fly
Heterotylenchus autumnalis (cont.)	MUSCIDAE: MUSCINAE *Musca domestica* Linnaeus Coutinho et al., 1957
ACHATINIDAE	
Burtoa nilotica (Pfeiffer)	MUSCIDAE: MYDAEINAE *Mydaea* Robineau-Desvoidy Rodhain et al., 1916
HELICIDAE	
Cryptomphalus aspersa (Müller)	MUSCIDAE: MUSCINAE *Musca domestica* Linnaeus Séguy, 1934
LUMBRICIDAE	
Allolobophora caliginosa (Savigny) subsp. *trapezoides*	CALLIPHORIDAE: CALLIPHORINAE *Calliphora clausa* Macquart Fuller, 1933
Helodrilus caliginosus (Savigny)	CALLIPHORIDAE: POLLENIINAE *Pollenia rudis* (Fabricius) Pimentel et al., 1960
Helodrilus chloroticus (Savigny) (Synonym: *Allolobophora chlorotica*)	CALLIPHORIDAE: POLLENIINAE *Pollenia rudis* (Fabricius) Keilin, 1909, 1911 Webb et al., 1916 Garrison, 1924 De Coursey, 1927
Helodrilus foetidus (Savigny)	CALLIPHORIDAE: POLLENIINAE *Pollenia rudis* (Fabricius) Pimentel et al., 1960
Helodrilus roseus (Savigny) (Synonym: *Eisenia rosea*)	CALLIPHORIDAE: POLLENIINAE *Pollenia rudis* (Fabricius) De Coursey, 1927 Pimentel et al., 1960
Lumbricus herculeus Savigny	CALLIPHORIDAE: POLLENIINAE *Pollenia rudis* (Fabricius) Barnes, 1924
Lumbricus rubellus Hoffmeister	CALLIPHORIDAE: POLLENIINAE *Pollenia rudis* (Fabricius) Pimentel et al., 1960
Lumbricus terrestris Linnaeus	CALLIPHORIDAE: POLLENIINAE *Pollenia rudis* (Fabricius) Pimentel et al., 1960
Octolasium lacteum (Orley)	CALLIPHORIDAE: POLLENIINAE *Pollenia rudis* (Fabricius) Pimentel et al., 1960
MEGASCOLECIDAE	
Diplocardia communis Garman	CALLIPHORIDAE: POLLENIINAE *Pollenia rudis* (Fabricius) Pimentel et al., 1960
Microscolex dubius (Fletcher)	CALLIPHORIDAE: CALLIPHORINAE *Calliphora clausa* Macquart Fuller, 1933

Organism	Fly association
UNDETERMINED OLIGOCHAETA	MUSCIDAE: COENOSIINAE *Allognota agromyzina* (Fallén) Keilin, 1917
UNDETERMINED ANNELIDA	CALLIPHORIDAE: POLLENIINAE *Pollenia rudis* (Fabricius) Blakitnaîâ, 1962
CHERNETIDAE	
Lamprochernes nodosus (Schrank) (Synonym: *Chelifer nodosus*)	MUSCIDAE: PHAONIINAE *Ophyra leucostoma* (Wiedemann) Graham-Smith, 1916 MUSCIDAE: MUSCINAE *Musca corvina* Fabricius Graham-Smith, 1916 *Musca domestica* Linnaeus Graham-Smith, 1916 MUSCIDAE: STOMOXYINAE *Stomoxys calcitrans* (Linnaeus) Graham-Smith, 1916
Pselaphochernes scorpioides (Hermann) (Synonym: *Chelifer scorpioides*)	MUSCIDAE: STOMOXYINAE *Stomoxys* Geoffroy Graham-Smith, 1916
Toxochernes panzeri (Koch) (Synonym: *Chelifer panzeri*)	MUSCIDAE: MUSCINAE *Musca domestica* Linnaeus Preudhomme de Borre, 1873
CHELIFERIDAE	
Chelifer Geoffroy	MUSCIDAE: MUSCINAE *Musca domestica* Linnaeus Stainton, 1864 Stevens, 1866 Hagen, 1867 Knab, 1897 CALLIPHORIDAE: CALLIPHORINAE *Calliphora vomitoria* (Linnaeus) Donovan, 1797
UNDETERMINED PSEUDOSCORPIONIDEA (Synonym: red false scorpion)	MUSCIDAE: MUSCINAE House fly (presumably *Musca domestica* Linnaeus) Manning, 1902
DICTYNIDAE	
Coenothele gregalis Simon	MUSCIDAE: MUSCINAE House fly (presumably *Musca domestica* Linnaeus) Diguet, 1909
THERIDIIDAE	
Latrodectus mactans (Fabricius) (predator)	DROSOPHILIDAE: DROSOPHILINAE *Drosophila* Fallén Herms et al., 1935

Organism	Fly and reference
	MUSCIDAE: MUSCINAE *Musca domestica* Linnaeus Herms et al., 1935
Latrodectus mactans (Fabricius) subsp. *tredecimguttatus*	MUSCIDAE: MUSCINAE *Musca domestica* Linnaeus Bettini, 1965
Steatoda Sundevall (predator)	MUSCIDAE: MUSCINAE *Musca domestica* Linnaeus Pavlovskiĭ, 1921
Teutana Simon (predator)	MUSCIDAE: MUSCINAE *Musca domestica* Linnaeus Pavlovskiĭ, 1921
Theridium Walckenaer (Synonym: *Teridium*) (predator)	MUSCIDAE: MUSCINAE *Musca domestica* Linnaeus Pavlovskiĭ, 1921
ERESIDAE	
Stegodyphus mimosarum Pavesi (predator)	MUSCIDAE: MUSCINAE *Musca domestica* Linnaeus Steyn, 1959
ARGIOPIDAE	
Epeira diadema Walker (predator)	MUSCIDAE: MUSCINAE *Musca domestica* Linnaeus Pavlovskiĭ, 1921
SALTICIDAE	
Salticus scenicus (Clerck) (predator)	MUSCIDAE: FANNIINAE *Fannia canicularis* (Linnaeus) Steve, 1959
ASCAIDAE	
Gamasellus Berlese	MUSCIDAE: PHAONIINAE *Hydrotaea armipes* (Fallén) Sychevskaı͡a, 1964a *Ophyra leucostoma* (Wiedemann) Sychevskaı͡a, 1964a
UNDETERMINED ASCAIDAE	MUSCIDAE: STOMOXYINAE *Haematobia titillans* (Bezzi) Pridantseva, 1959
PARASITIDAE	
Eulaelaps stabularis Koch	MUSCIDAE: STOMOXYINAE *Stomoxys calcitrans* (Linnaeus) Bouvier et al., 1944
Parasitus Latreille (Synonym: ?*Gamasus*)	CALLIPHORIDAE: CALLIPHORINAE *Calliphora vomitoria* (Linnaeus) Graham-Smith, 1916
Parasitus coleoptratorum (Linnaeus) (Synonym: *Gamasus coleoptratorum*)	MUSCIDAE: STOMOXYINAE *Stomoxys calcitrans* (Linnaeus) Wilhelmi, 1917b
Parasitus lunaris Berlese	SPHAEROCERIDAE *Copromyza* Fallén Sychevskaı͡a, 1964a

Parasitus lunaris (cont.) — MUSCIDAE: PHAONIINAE
Ophyra anthrax (Meigen)
Sychevskaia͡, 1964a

UNDETERMINED PARASITIDAE .... SCATOPSIDAE: SCATOPSINAE
Scatopse fuscipes Meigen
Sychevskaia͡, 1964a
ANTHOMYIIDAE: ANTHOMYIINAE
Calythea albicincta (Fallén)
Sychevskaia͡, 1964a
MUSCIDAE: MUSCINAE
Dasyphora gussakovskii Zimin
Sychevskaia͡, 1964a
Musca domestica vicina Macquart
Sychevskaia͡, 1964a
Musca larvipara Porchinskiĭ
Sychevskaia͡, 1964a
MUSCIDAE: STOMOXYINAE
Haematobia titillans (Bezzi)
Pridantseva, 1959

MACROCHELIDAE

Glyptholaspis confusa Foa (predator) .... MUSCIDAE: FANNIINAE
Fannia canicularis (Linnaeus)
Axtell, 1961a
MUSCIDAE: MUSCINAE
Musca domestica Linnaeus
Axtell, 1961a, 1961b, 1963a, 1963b
Rodriguez et al., 1962a

Macrocheles Latreille (predator) .... SEPSIDAE
Sepsis violacea Meigen
Sychevskaia͡, 1964a
OTITIDAE: OTITINAE
Ceroxys robusta Loew
Sychevskaia͡, 1964a
MUSCIDAE: FANNIINAE
Fannia scalaris (Fabricius)
Sychevskaia͡, 1964a
MUSCIDAE: PHAONIINAE
Hydrotaea irritans (Fallén)
Sychevskaia͡, 1964a
MUSCIDAE: MUSCINAE
Musca domestica Linnaeus
Axtell, 1961b
Musca domestica vicina Macquart
DeCoursey et al., 1956
Musca sorbens Wiedemann
DeCoursey et al., 1956

Macrocheles glaber (Müller) (predator) .... MUSCIDAE: FANNIINAE
Fannia canicularis (Linnaeus)
Filipponi, 1960

Organism	Fly associations
Macrocheles glaber (cont.)	MUSCIDAE: PHAONIINAE *Muscina stabulans* (Fallén) Sychevskaîa, 1964a MUSCIDAE: MUSCINAE *Musca domestica* Linnaeus Filipponi, 1960 Filipponi et al., 1963a *Musca domestica vicina* Macquart Sychevskaîa, 1964a MUSCIDAE: STOMOXYINAE *Stomoxys calcitrans* (Linnaeus) Filipponi, 1960 CALLIPHORIDAE: CALLIPHORINAE *Phaenicia sericata* (Meigen) Sychevskaîa, 1964a
Macrocheles merdarius (Berlese) (predator)	ANTHOMYIIDAE: ANTHOMYIINAE *Hylemya brassicae* Bouché Chant, 1960 MUSCIDAE: MUSCINAE *Musca domestica* Linnaeus Axtell, 1963b Filipponi et al., 1963b, 1964
Macrocheles muscaedomesticae (Scopoli) (Synonyms: ?*Gamasus musci*, *Macrocheles muscae*, *Macrocheles insectum*) (predator)	SCATOPSIDAE: SCATOPSINAE *Scatopse fuscipes* Meigen Sychevskaîa, 1964a SYRPHIDAE: MILESIINAE *Xylota pipiens* (Linnaeus) Sychevskaîa, 1964a PIOPHILIDAE *Piophila nigrimana* Meigen Sychevskaîa, 1964a SPHAEROCERIDAE *Leptocera fontinalis* (Fallén) Sychevskaîa, 1964a MILICHIIDAE: MADIZINAE *Desmometopa tarsalis* Loew Sychevskaîa, 1964a ANTHOMYIIDAE: ANTHOMYIINAE *Hylemya cinerella* (Fallén) Sychevskaîa, 1964a MUSCIDAE: FANNIINAE *Fannia canicularis* (Linnaeus) Buchanan, 1916 Steve, 1959 Filipponi, 1960 Axtell, 1961a Anderson, 1964 Singh et al., 1966 O'Donnell, 1967

Macrocheles muscaedomesticae (cont.)	*Fannia leucosticta* (Meigen) Sychevskaia͡, 1964a *Fannia scalaris* (Fabricius) Sychevskaia͡, 1964a MUSCIDAE: PHAONIINAE *Ophyra anthrax* (Meigen) Sychevskaia͡, 1964a *Ophyra leucostoma* (Wiedemann) Sychevskaia͡, 1964a *Dendrophaonia querceti* (Bouché) Sychevskaia͡, 1964a *Muscina stabulans* (Fallén) Anderson, 1964 Sychevskaia͡, 1964a MUSCIDAE: MUSCINAE *Musca autumnalis* De Geer Singh et al., 1966 *Musca domestica* Linnaeus Dugès, 1834 Berlese, 1882a, 1882b Leonardi, 1900 Oudemans, 1900, 1904 Ewing, 1913 Pereira et al., 1945, 1947 Filipponi, 1955, 1960, 1964, 1965 Rodriguez et al., 1960, 1961, 1962a Axtell, 1961a, 1961b, 1963a, 1963b Wade et al., 1961 Filipponi et al., 1963a, 1963b, 1964a Wallwork et al., 1963 Anderson, 1964 O'Donnell et al., 1965 Farish et al., 1966 Kinn, 1966 Singh et al., 1966 Willis et al., 1968 House fly (presumably *Musca domestica* Linnaeus) Schröck, 1686 Axtell, 1964 *Musca domestica vicina* Macquart Sychevskaia͡, 1964a MUSCIDAE: STOMOXYINAE *Stomoxys calcitrans* (Linnaeus) Filipponi, 1960 Axtell, 1964 Kinn, 1966 CALLIPHORIDAE: CHRYSOMYINAE *Phormia regina* (Meigen) Kinn, 1966

Macrocheles muscaedomesticae (cont.)	CALLIPHORIDAE: CALLIPHORINAE *Phaenicia sericata* (Meigen) Sychevskaîa, 1964a *Calliphora vomitoria* (Linnaeus) Trojan, 1908 SARCOPHAGIDAE: MILTOGRAMMINAE *Agria latifrons* Fallén Sychevskaîa, 1964a MUSCIDAE: SARCOPHAGINAE *Ravinia striata* Fabricius Sychevskaîa, 1964a
Macrocheles peniculatus Berlese (predator)	MUSCIDAE: MUSCINAE *Musca domestica* Linnaeus Filipponi et al., 1963a, 1964a, 1964b Filipponi, 1964
Macrocheles perglaber Filipponi and Pegazzano (predator)	MUSCIDAE: FANNIINAE *Fannia canicularis* (Linnaeus) Filipponi, 1960 MUSCIDAE: MUSCINAE *Musca domestica* Linnaeus Filipponi, 1960 Filipponi et al., 1963a, 1964 MUSCIDAE: STOMOXYINAE *Stomoxys calcitrans* (Linnaeus) Filipponi, 1960
Macrocheles plumiventris Hull (predator)	MUSCIDAE: FANNIINAE *Fannia leucosticta* (Meigen) Sychevskaîa, 1964a MUSCIDAE: MUSCINAE *Musca domestica* Linnaeus Rodriguez et al., 1960
Macrocheles robustulus Berlese (predator)	MUSCIDAE: MUSCINAE *Musca domestica* Linnaeus Axtell, 1961b, 1963b Filipponi et al., 1963a
Macrocheles scutatus (Berlese) (predator)	MUSCIDAE: MUSCINAE *Musca domestica* Linnaeus Filipponi et al., 1963a
Macrocheles subbadius Berlese (predator)	MUSCIDAE: FANNIINAE *Fannia canicularis* (Linnaeus) Filipponi, 1960 MUSCIDAE: MUSCINAE *Musca domestica* Linnaeus Filipponi, 1960 Axtell, 1961b, 1963b Filipponi et al., 1964 House fly (presumably *Musca domestica* Linnaeus) Axtell, 1964

Macrocheles subbadius (cont.)	MUSCIDAE: STOMOXYINAE *Stomoxys calcitrans* (Linnaeus) Filipponi, 1960 Axtell, 1964
GAMASOLAELAPTIDAE	
Digamasellus Berlese (predator)	MUSCIDAE: MUSCINAE *Musca domestica vicina* Macquart Sychevskaı͡a, 1964a CALLIPHORIDAE: CALLIPHORINAE *Phaenicia sericata* (Meigen) Sychevskaı͡a, 1964a
Saintdidieria sexclavatus (Oudemans) (Synonym: ?*Saintididicria sexclavatus*) (predator)	MUSCIDAE: PHAONIINAE *Muscina stabulans* (Fallén) Sychevskaı͡a, 1964a MUSCIDAE: MUSCINAE *Musca larvipara* Porchinskiĭ Sychevskaı͡a, 1964a
LAELAPTIDAE	
Alliphis halleri (Canestrini)	MUSCIDAE: MUSCINAE House fly (presumably *Musca domestica* Linnaeus) Heister, 1727
Holostaspis Kolenati (Synonym: *Holotaspis*)	MUSCIDAE: PHAONIINAE *Hydrotaea dentipes* (Fabricius) Graham-Smith, 1916 *Ophyra leucostoma* (Wiedemann) Graham-Smith, 1916 *Muscina stabulans* (Fallén) Graham-Smith, 1916 MUSCIDAE: MUSCINAE *Musca domestica* Linnaeus Graham-Smith, 1916
Holostaspis badius (Koch) Berlese (Synonym: *Macrocheles badius*)	ANTHOMYIIDAE: SCATOPHAGINAE *Scatophaga stercoraria* (Linnaeus) Weiss, 1915 MUSCIDAE: MUSCINAE *Musca domestica* Linnaeus Zanini, 1930 House fly (presumably *Musca domestica* Linnaeus) Scopoli, 1772
Holostaspis ?badius (Koch) Berlese	MUSCIDAE: MUSCINAE *Musca domestica* Linnaeus Scopoli, 1772 Goeze, 1776
Holostaspis marginatus (Hermann)	MUSCIDAE: STOMOXYINAE *Stomoxys calcitrans* (Linnaeus) Wilhelmi, 1917b

Organism	Fly / Reference
UROPODIDAE	
Fuscuropoda Vitzthum	MUSCIDAE: FANNIINAE *Fannia canicularis* (Linnaeus) Anderson, 1964
	MUSCIDAE: PHAONIINAE *Muscina stabulans* (Fallén) Anderson, 1964
	MUSCIDAE: MUSCINAE *Musca domestica* Linnaeus Anderson, 1964
Fuscuropoda vegetans (De Geer)	MUSCIDAE: FANNIINAE *Fannia canicularis* Linnaeus O'Donnell, 1967
	MUSCIDAE: MUSCINAE *Musca domestica* Linnaeus O'Donnell et al., 1965 Willis et al., 1968
PYEMOTIDAE	
Pygmephorus Kramer	ANTHOMYIIDAE: ANTHOMYIINAE *Hylemya cinerella* (Fallén) Sychevskaīā, 1964a
UNDETERMINED TARSONEMIDAE	MUSCIDAE: STOMOXYINAE *Haematobia titillans* (Bezzi) Pridantseva, 1959
TROMBIDIIDAE	
Allothrombium fuliginosum (Hermann)	SEPSIDAE *Sepsis cynipsea* (Linnaeus) Scopoli, 1763
Atomus inexpectatus (Oudemans) (Synonym: ?*Scarabaspis inexpectatus*)	MUSCIDAE: MUSCINAE *Musca larvipara* Porchinskiĭ Sychevskaīā, 1964a
Atomus parasiticus (De Geer) (Synonyms: ?*Acarus parasiticus*, *Trombidium parasiticum*)	MUSCIDAE: MUSCINAE House fly (presumably *Musca domestica* Linnaeus) Goeze, 1774 De Geer, 1778 Murray, 1877
Microtrombidium ?muscarum (Linnaeus) (Synonym: *Trombidium muscarum*)	MUSCIDAE: MUSCINAE *Musca domestica* Linnaeus Riley, 1877 Griffith, 1907 Ewing et al., 1918
Microtrombidium striaticeps (Oudemans) (Synonym: *Trombidium striaticeps*)	MUSCIDAE: STOMOXYINAE *Stomoxys calcitrans* (Linnaeus) Ewing, 1919

TROMBICULIDAE

Organism	Fly association
Ascoschoengastia indica (Hirst) (Synonym: *Neoschœngastia indica*)	DROSOPHILIDAE: DROSOPHILINAE *Drosophila* Fallén Wharton et al., 1946 *Drosophila melanogaster* Meigen Wharton, 1946
Euschoengastia peromysci (Ewing)	DROSOPHILIDAE: DROSOPHILINAE *Drosophila* Fallén Lipovsky, 1954 MUSCIDAE: MUSCINAE *Musca domestica* Linnaeus Lipovsky, 1954
Eutrombicula alfreddugesi (Oudemans) (Synonym: *Trombicula alfredugesi*)	DROSOPHILIDAE: DROSOPHILINAE *Drosophila* Fallén Lipovsky, 1954 MUSCIDAE: MUSCINAE *Musca domestica* Linnaeus Lipovsky, 1954
Eutrombicula splendens (Ewing) (Synonym: *Trombicula splendens*)	DROSOPHILIDAE: DROSOPHILINAE *Drosophila* Fallén Lipovsky, 1954 MUSCIDAE: MUSCINAE *Musca domestica* Linnaeus Lipovsky, 1954
Fonsecia gurneyi (Ewing) (Synonym: *Trombicula gurneyi*)	DROSOPHILIDAE: DROSOPHILINAE *Drosophila* Fallén Lipovsky, 1954 MUSCIDAE: MUSCINAE *Musca domestica* Linnaeus Lipovsky, 1954
Hannemania Oudemans	DROSOPHILIDAE: DROSOPHILINAE *Drosophila* Fallén Lipovsky, 1954 MUSCIDAE: MUSCINAE *Musca domestica* Linnaeus Lipovsky, 1954
Kayella lacerta (Brennan) (Synonym: *Euschöngastia lacerta*)	DROSOPHILIDAE: DROSOPHILINAE *Drosophila* Fallén Lipovsky, 1954 MUSCIDAE: MUSCINAE *Musca domestica* Linnaeus Lipovsky, 1954
Neoschoengastia americana (Hirst)	DROSOPHILIDAE: DROSOPHILINAE *Drosophila* Fallén Lipovsky, 1954 MUSCIDAE: MUSCINAE *Musca domestica* Linnaeus Lipovsky, 1954

Organism	Fly
Neotrombicula lipovskyi (Brennan and Wharton) (Synonym: *Trombicula lipovskyi*)	DROSOPHILIDAE: DROSOPHILINAE *Drosophila* Fallén Lipovsky, 1954 MUSCIDAE: MUSCINAE *Musca domestica* Linnaeus Lipovsky, 1954
Neotrombiculoides montanensis (Brennan) (Synonym: *Trombicula montanensis*)	DROSOPHILIDAE: DROSOPHILINAE *Drosophila* Fallén Lipovsky, 1954 MUSCIDAE: MUSCINAE *Musca domestica* Linnaeus Lipovsky, 1954
Pseudoschoengastia farneri Lipovsky	DROSOPHILIDAE: DROSOPHILINAE *Drosophila* Fallén Lipovsky, 1954 MUSCIDAE: MUSCINAE *Musca domestica* Linnaeus Lipovsky, 1954
Pseudoschoengastia hungerfordi Lipovsky	DROSOPHILIDAE: DROSOPHILINAE *Drosophila* Fallén Lipovsky, 1954 MUSCIDAE: MUSCINAE *Musca domestica* Linnaeus Lipovsky, 1954
UNDETERMINED TROMBICULIDAE	MUSCIDAE: STOMOXYINAE *Haematobia titillans* (Bezzi) Pridantseva, 1959
ACARIDAE	
Acarus Linnaeus	MUSCIDAE: STOMOXYINAE *Haematobia irritans* (Linnaeus) Wilhelmi, 1917a
Acarus siro Linnaeus (Synonym: *Tyroglyphus siro*)	MUSCIDAE: MUSCINAE *Musca domestica* Linnaeus Graham-Smith, 1916
ANOETIDAE	
Histiostoma laboratorium Hughes (phoresy)	DROSOPHILIDAE: DROSOPHILINAE *Drosophila* Fallén Hughes, 1950 Perron, 1954a *Drosophila melanogaster* Meigen Stolpe, 1938 Brown, 1965
Myianoetus Oudemans (phoresy)	SEPSIDAE *Sepsis violacea* Meigen Sychevskaia, 1964a MILICHIIDAE: MADIZINAE *Desmometopa tarsalis* Loew Sychevskaia, 1964a

Myianoetus (cont.)	ANTHOMYIIDAE: ANTHOMYIINAE *Hylemya cinerella* (Fallén) Sychevskaia͡, 1964a MUSCIDAE: FANNIINAE *Fannia scalaris* (Fabricius) Sychevskaia͡, 1964a MUSCIDAE: PHAONIINAE *Hydrotaea armipes* (Fallén) Sychevskaia͡, 1964a *Hydrotaea irritans* (Fallén) Sychevskaia͡, 1964a *Ophyra anthrax* (Meigen) Sychevskaia͡, 1964a *Ophyra leucostoma* (Wiedemann) Sychevskaia͡, 1964a *Muscina stabulans* (Fallén) Sychevskaia͡, 1964a MUSCIDAE: MUSCINAE *Musca autumnalis* De Geer Sychevskaia͡, 1964a
Myianoetus muscarum (Linnaeus) (Synonym: ?*Acarus muscarum*) (phoresy)	DROSOPHILIDAE: DROSOPHILINAE *Drosophila ?fenestrarum* Fallén Goeze, 1776 *Drosophila melanogaster* Meigen Greenberg et al., 1960 MUSCIDAE: FANNIINAE *Fannia canicularis* (Linnaeus) Greenberg et al., 1960 MUSCIDAE: PHAONIINAE *Muscina stabulans* (Fallén) Berlese, 1881 Greenberg et al., 1960 Greenberg, 1961 MUSCIDAE: MUSCINAE *Musca domestica* Linnaeus Menzel et al., 1683 De Geer, 1778 Greenberg et al., 1960 House fly (presumably *Musca domestica* Linnaeus) Gahrliep, 1696 Winterschmidt, 1765 De Geer, 1771 Schrank, 1776 Fabricius, 1780 Mohr, 1786 Walckenaer, 1802 MUSCIDAE: STOMOXYINAE *Stomoxys calcitrans* (Linnaeus) Greenberg et al., 1960

Organism	Fly association
Myianoetus muscarum (cont.)	CALLIPHORIDAE: CALLIPHORINAE *Phaenicia sericata* (Meigen) Greenberg et al., 1960
SARCOPTIDAE	
Sarcoptes Latreille	MUSCIDAE: MUSCINAE *Musca domestica* Linnaeus Newsad, 1930
SCUTIGERIDAE	
Scutigera coleoptrata (Linnaeus) (predator)	MUSCIDAE: MUSCINAE *Musca domestica* Linnaeus Pavlovskiĭ, 1921
SMINTHURIDAE	
Smithurus viridis (Linnaeus)	ANTHOMYIIDAE: SCATOPHAGINAE *Scatophaga stercoraria* (Linnaeus) Walters, 1966
LOCUSTIDAE	
Chortophaga viridifasciata (De Geer)	MUSCIDAE: PHAONIINAE *Muscina stabulans* (Fallén) Knutson, 1941
Locusta migratoria Linnaeus	SARCOPHAGIDAE: SARCOPHAGINAE *Sarcophaga destructor* Malloch Wood, 1933
Nomadacris septemfasciata Serville	MUSCIDAE: FANNIINAE *Fannia canicularis* (Linnaeus) Jack, 1935
Dociostaurus maroccanus (Thunberg)	SARCOPHAGIDAE: SARCOPHAGINAE *Blaesoxipha lineata* Fallén Baranov, 1924
Encoptolophus sordidus (Burmeister) subsp. *costalis*	MUSCIDAE: PHAONIINAE *Muscina stabulans* (Fallén) Knutson, 1941
	MUSCIDAE: PHAONIINAE *Muscina stabulans* (Fallén) Jack, 1935
Schistocerca gregaria Fórskal (Synonym: *Schistocerca peregrina*)	PHORIDAE: METOPININAE *Megaselia scalaris* (Loew) Nocedo, 1921
	MUSCIDAE: FANNIINAE *Fannia canicularis* (Linnaeus) Nocedo, 1921 Régnier, 1931
	MUSCIDAE: PHAONIINAE *Muscina stabulans* (Fallén) Nocedo, 1921 Régnier, 1931
	SARCOPHAGIDAE: SARCOPHAGINAE *Sarcophaga haemorrhoidalis* (Fallén) Régnier, 1931

Schistocerca paranensis Burmeister .... MUSCIDAE: STOMOXYINAE
Stomoxys calcitrans (Linnaeus)
Lahille, 1907

Xanthippus corallipes (Haldeman) subsp. *pantherinus* .... MUSCIDAE: PHAONIINAE
Muscina stabulans (Fallén)
Knutson, 1941

GRYLLIDAE

Nemobius fasciatus (De Geer) .... MUSCIDAE: STOMOXYINAE
Haematobia irritans (Linnaeus)
Bourne et al., 1967

LABIDURIDAE

Euborellia annulipes (Lucas) (predator) ...... OTITIDAE: ULIDIINAE
Euxesta annonae (Fabricius)
Illingworth, 1923a

MILICHIIDAE: MILICHIINAE
Milichiella lacteipennis (Loew)
Illingworth, 1923a

MUSCIDAE: FANNIINAE
Fannia pusio (Wiedemann)
Illingworth, 1923a

MUSCIDAE: PHAONIINAE
Ophyra nigra Wiedemann
Illingworth, 1923a

MUSCIDAE: MUSCINAE
Musca domestica Linnaeus
Illingworth, 1923a

MUSCIDAE: STOMOXYINAE
Stomoxys calcitrans (Linnaeus)
Illingworth, 1923a

CALLIPHORIDAE: CALLIPHORINAE
Phaenicia sericata (Meigen)
Illingworth, 1923a

SARCOPHAGIDAE: SARCOPHAGINAE
Sarcophaga fuscicauda Böttcher
Illingworth, 1923a

LABIIDAE

Labia minor (Linnaeus) .... MUSCIDAE: MUSCINAE
Musca domestica Linnaeus
Mourier et al., 1968

Labia pilicornis (Motschulsky) (predator) .... OTITIDAE: ULIDIINAE
Euxesta annonae (Fabricius)
Illingworth, 1923a

MILICHIIDAE: MILICHIINAE
Milichiella lacteipennis (Loew)
Illingworth, 1923a

MUSCIDAE: FANNIINAE
Fannia pusio (Wiedemann)
Illingworth, 1923a

Labia pilicornis (cont.)

MUSCIDAE: PHAONIINAE
Ophyra nigra Wiedemann
Illingworth, 1923a
MUSCIDAE: MUSCINAE
Musca domestica Linnaeus
Illingworth, 1923a
MUSCIDAE: STOMOXYINAE
Stomoxys calcitrans (Linnaeus)
Illingworth, 1923a
CALLIPHORIDAE: CALLIPHORINAE
Phaenicia sericata (Meigen)
Illingworth, 1923a
SARCOPHAGIDAE: SARCOPHAGINAE
Sarcophaga fuscicauda Böttcher
Illingworth, 1923a

Spingolabis hawaiiensis (Bormans) (predator) ....

OTITIDAE: ULIDIINAE
Euxesta annonae (Fabricius)
Illingworth, 1923a
MILICHIIDAE: MILICHIINAE
Milichiella lacteipennis (Loew)
Illingworth, 1923a
MUSCIDAE: FANNIINAE
Fannia pusio (Wiedemann)
Illingworth, 1923a
MUSCIDAE: PHAONIINAE
Ophyra nigra Wiedemann
Illingworth, 1923a
MUSCIDAE: MUSCINAE
Musca domestica Linnaeus
Illingworth, 1923a
MUSCIDAE: STOMOXYINAE
Stomoxys calcitrans (Linnaeus)
Illingworth, 1923a
CALLIPHORIDAE: CALLIPHORINAE
Phaenicia sericata (Meigen)
Illingworth, 1923a
SARCOPHAGIDAE: SARCOPHAGINAE
Sarcophaga fuscicauda Böttcher
Illingworth, 1923a

FORFICULIDAE

Forficula auricularia Linnaeus (predator) ....

MUSCIDAE: FANNIINAE
Fannia canicularis (Linnaeus)
Anderson, 1964
MUSCIDAE: PHAONIINAE
Muscina stabulans (Fallén)
Anderson, 1964
MUSCIDAE: MUSCINAE
Musca domestica Linnaeus
Anderson, 1964

Organism	Fly association
HAEMATOPINIDAE	
Haematopinus tuberculatus (Burmeister) (Synonym: *Haematopinus bituberculatus*)	MUSCIDAE: STOMOXYINAE *Haematobia* Lepeletier and Serville Mitzmain, 1912a
COENAGRIONIDAE	
Agria fumipennis (Burmeister) (predator)	MUSCIDAE: STOMOXYINAE *Stomoxys calcitrans* (Linnaeus) Wright, 1945
Enallagma durum Hagen (predator)	MUSCIDAE: STOMOXYINAE *Stomoxys calcitrans* (Linnaeus) Wright, 1945
Ischnura ramburii Selys (predator)	MUSCIDAE: STOMOXYINAE *Stomoxys calcitrans* (Linnaeus) Wright, 1945
LIBELLULIDAE	
Celithemis amanda Hagen (predator)	MUSCIDAE: STOMOXYINAE *Stomoxys calcitrans* (Linnaeus) Wright, 1945
Erythemis simplicicollis Say (predator)	MUSCIDAE: STOMOXYINAE *Stomoxys calcitrans* (Linnaeus) Wright, 1945
Libellula auripennis Burmeister (predator)	MUSCIDAE: STOMOXYINAE *Stomoxys calcitrans* (Linnaeus) Wright, 1945
Libellula pulchella Drury (predator)	MUSCIDAE: STOMOXYINAE *Stomoxys calcitrans* (Linnaeus) Wright, 1945
Libellula vibrans Fabricius (predator)	MUSCIDAE: STOMOXYINAE *Stomoxys calcitrans* (Linnaeus) Wright, 1945
Pachydiplax longipennis Burmeister (predator)	MUSCIDAE: STOMOXYINAE *Stomoxys calcitrans* (Linnaeus) Wright, 1945
Pantala flavescens Fabricius (predator)	MUSCIDAE: STOMOXYINAE *Stomoxys calcitrans* (Linnaeus) Wright, 1945
Tramea carolina Linnaeus (predator)	MUSCIDAE: STOMOXYINAE *Stomoxys calcitrans* (Linnaeus) Wright, 1945
AESCHNIDAE	
Aeschna cyanea (Müller) (predator)	CALLIPHORIDAE: CALLIPHORINAE *Calliphora* Robineau-Desvoidy Freeman, 1934
Anax junius Drury (predator)	MUSCIDAE: STOMOXYINAE *Stomoxys calcitrans* (Linnaeus) Wright, 1945

Coryphaeschna ingens Rambur (predator) .... MUSCIDAE: STOMOXYINAE
Stomoxys calcitrans (Linnaeus)
Wright, 1945

UNDETERMINED ODONATA (predators) .... MUSCIDAE: STOMOXYINAE
Stomoxys calcitrans (Linnaeus)
Brues, 1946

PHYMATIDAE

Phymata pennsylvanica americana Melin (predator) .... DROSOPHILIDAE: DROSOPHILINAE
Drosophila Fallén
Balduf, 1947
Drosophila melanogaster Meigen
Balduf, 1941, 1948
MUSCIDAE: MUSCINAE
Musca Linnaeus
Balduf, 1947
Musca domestica Linnaeus
Balduf, 1941, 1948

REDUVIIDAE

Apiomerus pilipes (Fabricius) (predator) .... MUSCIDAE: MUSCINAE
Musca domestica Linnaeus
Uribe, 1926

Pristhesancus papuensis Stål (predator) .... DROSOPHILIDAE: DROSOPHILINAE
Drosophila Fallén
Noble, 1936
UNDETERMINED CALLIPHORIDAE
Blow flies
Noble, 1936

Sinea diadema (Fabricius) (predator) .... DROSOPHILIDAE: DROSOPHILINAE
Drosophila Fallén
Balduf, 1947
MUSCIDAE: MUSCINAE
Musca Linnaeus
Balduf, 1947

APHIDAE

Myzus persicae (Sulzer) .... CALLIPHORIDAE: CALLIPHORINAE
Calliphora clausa Macquart
Fuller, 1933

HYPONOMEUTIDAE

Hyponomeuta padella Linnaeus (Synonym: *Hyponomeuta malinellus*) .... MUSCIDAE: MUSCINAE
Musca domestica Linnaeus
Séguy, 1934

OLETHREUTIDAE

Epiblema otiosana Clemens .... MUSCIDAE: PHAONIINAE
Muscina stabulans (Fallén)
Decker, 1932

Suleima helianthana Riley .... MUSCIDAE: PHAONIINAE
Muscina stabulans (Fallén)
Satterthwait, 1943

Organism	Fly
PYRALIDIDAE	
Epischnia incanella Holst (Synonym: *Epischnia canella*)	MUSCIDAE: FANNIINAE *Fannia canicularis* (Linnaeus) Heeger, 1848
NOCTUIDAE	
Agrotis castanea var. *neglecta* Hubner (Synonym: *Agrotis neglecta*)	MUSCIDAE: PHAONIINAE *Muscina stabulans* (Fallén) von Gercke, 1882
Agrotis segetum (Denis and Schiffermüller)	MUSCIDAE: PHAONIINAE *Muscina stabulans* (Fallén) Herold, 1923
Archanara oblonga Grote (Synonym: *Archanara subcarnea*)	MUSCIDAE: PHAONIINAE *Muscina stabulans* (Fallén) Cole, 1930
Heliothis obsoleta Fabricius	MUSCIDAE: PHAONIINAE *Muscina stabulans* (Fallén) Rodionov, 1927
Laphygma frugiperda (Smith, J. B.)	MUSCIDAE: PHAONIINAE *Muscina stabulans* (Fallén) Smith, 1921
Leucania unipuncta Haworth (Synonym: *Cirphis unipuncta*)	MUSCIDAE: PHAONIINAE *Muscina stabulans* (Fallén) Satterthwait, 1943
Macronoctua onusta Grote	MUSCIDAE: MYDAEINAE *Myospila meditabunda* (Fabricius) Breakey, 1929
	MUSCIDAE: PHAONIINAE *Muscina assimilis* (Fallén) Breakey, 1929
	Muscina stabulans (Fallén) Breakey, 1929
Papaipema nebris Guenée	MUSCIDAE: PHAONIINAE *Muscina stabulans* (Fallén) Decker, 1931
Peridroma margitosa form *saucia* Hubner (Synonym: *Peridromia saucia*)	MUSCIDAE: PHAONIINAE *Muscina stabulans* (Fallén) Fletcher, 1900
LIPARIDAE	
Liparis dispar Linnaeus (Synonym: *Porthetria dispar*)	MUSCIDAE: PHAONIINAE *Muscina stabulans* (Fallén) Levitt, 1935
	SARCOPHAGIDAE: MILTOGRAMMINAE *Pseudosarcophaga affinis* Fallén Sitowski, 1928
LYMANTRIIDAE	
Leucoma salicis Linnaeus (Synonym: *Stilpnotia salicis*)	SARCOPHAGIDAE: MILTOGRAMMINAE *Pseudosarcophaga affinis* Fallén Sitowski, 1928
Lymantria monacha Linnaeus (Synonym: *Bombyx monacha*)	MUSCIDAE: PHAONIINAE *Muscina pabulorum* (Fallén) Ratzeburg, 1844

Organism	Fly association
Lymantria monacha (cont.)	SARCOPHAGIDAE: MILTOGRAMMINAE *Pseudosarcophaga affinis* Fallén Sitowski, 1928
SPHINGIDAE	
Acherontia atropos Linnaeus	MUSCIDAE: FANNIINAE *Fannia canicularis* (Linnaeus) Keilin, 1917 MUSCIDAE: PHAONIINAE *Muscina stabulans* (Fallén) Keilin, 1917 MUSCIDAE: MUSCINAE *Musca domestica* Linnaeus Keilin, 1917 CALLIPHORIDAE: CALLIPHORINAE *Calliphora vicina* Robineau-Desvoidy Keilin, 1917
LASIOCAMPIDAE	
Dendrolimus pini Linnaeus (Synonym: *Bombyx pini*)	MUSCIDAE: PHAONIINAE *Muscina pabulorum* (Fallén) Ratzeburg, 1844 Sitowski, 1928 *Muscina stabulans* (Fallén) Keilin, 1917 Sitowski, 1928 MUSCIDAE: STOMOXYINAE *Stomoxys calcitrans* (Linnaeus) Sitowski, 1928 SARCOPHAGIDAE: MILTOGRAMMINAE *Pseudosarcophaga affinis* Fallén Sitowski, 1928
Malacosoma americana Fabricius (Synonym: *Malacosoma americanum*)	MUSCIDAE: PHAONIINAE *Muscina stabulans* (Fallén) Curran, 1942
CARABIDAE	
Carabus italicus Dejean (predator)	MUSCIDAE: PHAONIINAE *Phaonia signata* (Meigen) Giglio-Tos, 1892
UNDETERMINED CARABIDAE (predators)	MUSCIDAE: STOMOXYINAE *Haematobia titillans* (Bezzi) Pridantseva, 1959
HYDROPHILIDAE	
Cryptopleurum minutum Fabricius	OTITIDAE: ULIDIINAE *Euxesta annonae* (Fabricius) Illingworth, 1923a MILICHIIDAE: MILICHIINAE *Milichiella lacteipennis* (Loew) Illingworth, 1923a

Cryptopleurum minutum (cont.)

MUSCIDAE: FANNIINAE
Fannia pusio (Wiedemann)
Illingworth, 1923a
MUSCIDAE: PHAONIINAE
Ophyra nigra Wiedemann
Illingworth, 1923a
MUSCIDAE: MUSCINAE
Musca domestica Linnaeus
Illingworth, 1923a
MUSCIDAE: STOMOXYINAE
Stomoxys calcitrans (Linnaeus)
Illingworth, 1923a
CALLIPHORIDAE: CALLIPHORINAE
Phaenicia sericata (Meigen)
Illingworth, 1923a
SARCOPHAGIDAE: SARCOPHAGINAE
Sarcophaga fuscicauda Böttcher
Illingworth, 1923a

Dactylosternum abdominale Fabricius (Synonym: *Dactylospermum abdominale*) ....

OTITIDAE: ULIDIINAE
Euxesta annonae (Fabricius)
Illingworth, 1923a
MILICHIIDAE: MILICHIINAE
Milichiella lacteipennis (Loew)
Illingworth, 1923a
MUSCIDAE: FANNIINAE
Fannia pusio (Wiedemann)
Illingworth, 1923a
MUSCIDAE: PHAONIINAE
Ophyra nigra Wiedemann
Illingworth, 1923a
MUSCIDAE: MUSCINAE
Musca domestica Linnaeus
Illingworth, 1923a
MUSCIDAE: STOMOXYINAE
Stomoxys calcitrans (Linnaeus)
Illingworth, 1923a
CALLIPHORIDAE: CALLIPHORINAE
Phaenicia sericata (Meigen)
Illingworth, 1923a
SARCOPHAGIDAE: SARCOPHAGINAE
Sarcophaga fuscicauda Böttcher
Illingworth, 1923a

Sphaeridium scarabaeoides (Linnaeus) ....

MUSCIDAE: MUSCINAE
Orthellia caesarion (Meigen)
Hammer, 1942

SILPHIDAE

Nicrophorus nigritus Mannerheim (Synonym: *Necrophorus nigritis*) ....

CALLIPHORIDAE: CALLIPHORINAE
Phaenicia sericata (Meigen)
Illingworth, 1927

Organism	Fly associations
Nicrophorus nigritus (cont.)	SARCOPHAGIDAE: SARCOPHAGINAE *Blaesoxipha plinthopyga* (Wiedemann) Illingworth, 1927
UNDETERMINED SILPHIDAE	CALLIPHORIDAE: CALLIPHORINAE *Phaenicia sericata* (Meigen) Illingworth, 1927 SARCOPHAGIDAE: SARCOPHAGINAE *Blaesoxipha plinthopyga* (Wiedemann) Illingworth, 1927
STAPHYLINIDAE	
Aleochara Gravenhorst (predator)	MUSCIDAE: STOMOXYINAE *Haematobia irritans* (Linnaeus) Wolcott, 1922
Aleochara algarum Fauvel (predator)	COELOPIDAE *Coelopa frigida* (Fabricius) Lesne et al., 1922 *Coelopa pilipes* Haliday Lesne et al., 1922
Aleochara bimaculata Gravenhorst	MUSCIDAE: MUSCINAE *Musca domestica* Linnaeus Thomas et al., 1968
Aleochara curtula Goeze (predator)	SYRPHIDAE: MILESIINAE *Eristalis tenax* (Linnaeus) Balduf, 1935 ANTHOMYIIDAE: ANTHOMYIINAE *Pegomya hyoscyami* (Panzer) Kemner, 1926 CALLIPHORIDAE: CALLIPHORINAE *Lucilia caesar* (Linnaeus) Kemner, 1926 *Calliphora vicina* Robineau-Desvoidy Fuldner, 1964
Aleochara handschini Scheerpeltz (predator)	MUSCIDAE: STOMOXYINAE *Haematobia* Lepeletier and Serville Scheerpeltz, 1934 *Haematobia exigua* (de Meijere) Handschin, 1934a UNDETERMINED MUSCIDAE Handschin, 1934a
Aleochara laevigata Gyllenhall (Synonym: *Polychara laevigata*) (predator)	ANTHOMYIIDAE: ANTHOMYIINAE *Pegomya hyoscyami* (Panzer) Kemner, 1926
Aleochara taeniata Erichson (predator)	MUSCIDAE: MUSCINAE *Musca domestica* Linnaeus Legner et al., 1966a White et al., 1966

Organism	Fly association
Aleochara tristis Gravenhorst	MUSCIDAE: MUSCINAE *Musca autumnalis* De Geer Drea, 1966 Jones, 1969
Aleochara trivialis Kraatz (predator)	MUSCIDAE: MUSCINAE *Musca planiceps* Wiedemann Senior-White et al., 1945
Aleochara windredi Scheerpeltz (predator)	MUSCIDAE: STOMOXYINAE *Haematobia* Lepeletier and Serville Scheerpeltz, 1934 *Haematobia exigua* (de Meijere) Handschin, 1934a
Baryodma bimaculata Gravenhorst (predator)	MUSCIDAE: MUSCINAE *Orthellia* Robineau-Desvoidy Lindquist, 1936 SARCOPHAGIDAE: SARCOPHAGINAE *Sarcophaga* Meigen Lindquist, 1936
Baryodma ontarionis Casey (predator)	DROSOPHILIDAE: DROSOPHILINAE *Drosophila melanogaster* Meigen Colhoun, 1953 ANTHOMYIIDAE: ANTHOMYIINAE *Hylemya brassicae* Bouché Colhoun, 1953 MUSCIDAE: MUSCINAE *Musca domestica* Linnaeus Colhoun, 1953 SARCOPHAGIDAE: MILTOGRAMMINAE *Pseudosarcophaga affinis* Fallén Colhoun, 1953
Creophilus erythrocephalus Fabricius (predator)	CALLIPHORIDAE: CALLIPHORINAE *Calliphora vicina* Robineau-Desvoidy Froggatt, 1914a
Creophilus maxillosus Linnaeus (predator)	CALLIPHORIDAE: CALLIPHORINAE *Lucilia caesar* (Linnaeus) Porchinskiĭ, 1885
Creophilus maxillosus var. *villosus* Gravenhorst (Synonym: *Creophilus villosus*) (predator)	MUSCIDAE: MUSCINAE *Musca domestica* Linnaeus Mourier et al., 1968 CALLIPHORIDAE: CALLIPHORINAE *Phaenicia sericata* (Meigen) Illingworth, 1927 SARCOPHAGIDAE: SARCOPHAGINAE *Blaesoxipha plinthopyga* (Wiedemann) Illingworth, 1927
Emus hirtus Linnaeus (predator)	CALLIPHORIDAE: CALLIPHORINAE *Lucilia caesar* (Linnaeus) Porchinskiĭ, 1885

Organism	Fly
Ontholestes murinus Linnaeus (Synonym: *Leistotrophus murinus*) (predator)	CALLIPHORIDAE: CALLIPHORINAE *Lucilia caesar* (Linnaeus) Porchinskiĭ, 1885
Ontholestes tessellatus Fourcroy (Synonym: *Ontholestes nebulosus*) (predator)	ANTHOMYIIDAE: SCATOPHAGINAE *Scatophaga stercoraria* (Linnaeus) Hammer, 1942 MUSCIDAE: MUSCINAE *Orthellia caesarion* (Meigen) Hammer, 1942 *Musca domestica* Linnaeus Mourier et al., 1968
Oxytelus Gravenhorst (predator)	OTITIDAE: ULIDIINAE *Euxesta annonae* (Fabricius) Illingworth, 1923a MILICHIIDAE: MILICHIINAE *Milichiella lacteipennis* (Loew) Illingworth, 1923a MUSCIDAE: FANNIINAE *Fannia pusio* (Wiedemann) Illingworth, 1923a MUSCIDAE: PHAONIINAE *Ophyra nigra* Wiedemann Illingworth, 1923a MUSCIDAE: MUSCINAE *Musca domestica* Linnaeus Illingworth, 1923a MUSCIDAE: STOMOXYINAE *Stomoxys calcitrans* (Linnaeus) Illingworth, 1923a CALLIPHORIDAE: CALLIPHORINAE *Phaenicia sericata* (Meigen) Illingworth, 1923a SARCOPHAGIDAE: SARCOPHAGINAE *Sarcophaga fuscicauda* Böttcher Illingworth, 1923a
Oxytelus ocularis Fauvel (predator)	MUSCIDAE: STOMOXYINAE *Haematobia exigua* (de Meijere) Handschin, 1932
Oxytelus sculptus Gravenhorst	MUSCIDAE: MUSCINAE *Musca domestica* Linnaeus Mourier et al., 1968
Philonthus Curtis (predator)	ANTHOMYIIDAE: SCATOPHAGINAE *Scatophaga stercoraria* (Linnaeus) Hammer, 1942 MUSCIDAE: MUSCINAE *Orthellia caesarion* (Meigen) Hammer, 1942

Philonthus discoideus Gravenhorst (predator)	OTITIDAE: ULIDIINAE *Euxesta annonae* (Fabricius) Illingworth, 1923a MILICHIIDAE: MILICHIINAE *Milichiella lacteipennis* (Loew) Illingworth, 1923a MUSCIDAE: FANNIINAE *Fannia pusio* (Wiedemann) Illingworth, 1923a MUSCIDAE: PHAONIINAE *Ophyra nigra* Wiedemann Illingworth, 1923a MUSCIDAE: MUSCINAE *Musca domestica* Linnaeus Illingworth, 1923a MUSCIDAE: STOMOXYINAE *Stomoxys calcitrans* (Linnaeus) Illingworth, 1923a CALLIPHORIDAE: CALLIPHORINAE *Phaenicia sericata* (Meigen) Illingworth, 1923a SARCOPHAGIDAE: SARCOPHAGINAE *Sarcophaga fuscicauda* Böttcher Illingworth, 1923a
Philonthus fimetarius Gravenhorst (predator)	SYRPHIDAE: MILESIINAE *Eristalis tenax* (Linnaeus) Balduf, 1935
Philonthus longicornis Stephens (predator)	OTITIDAE: ULIDIINAE *Euxesta annonae* (Fabricius) Illingworth, 1923a MILICHIIDAE: MILICHIINAE *Milichiella lacteipennis* (Loew) Illingworth, 1923a MUSCIDAE: FANNIINAE *Fannia pusio* (Wiedemann) Illingworth, 1923a MUSCIDAE: PHAONIINAE *Ophyra nigra* Wiedemann Illingworth, 1923a MUSCIDAE: MUSCINAE *Musca domestica* Linnaeus Illingworth, 1923a MUSCIDAE: STOMOXYINAE *Stomoxys calcitrans* (Linnaeus) Illingworth, 1923a CALLIPHORIDAE: CALLIPHORINAE *Phaenicia sericata* (Meigen) Illingworth, 1923a

Organism	Fly associations
	SARCOPHAGIDAE: SARCOPHAGINAE *Sarcophaga fuscicauda* Böttcher Illingworth, 1923a
Philonthus politus Linnaeus (Synonyms: *Philonthus aeneus*, *Philonthus erythrocephala*) (predator)	SYRPHIDAE: MILESIINAE *Eristalis tenax* (Linnaeus) Balduf, 1935 MUSCIDAE: STOMOXYINAE ?*Haematobia* Lepeletier and Serville Fullaway, 1926 CALLIPHORIDAE: CALLIPHORINAE *Lucilia caesar* (Linnaeus) Porchinskiĭ, 1885
Tachyporus Gravenhorst (predator)	OTITIDAE: ULIDIINAE *Euxesta annonae* (Fabricius) Illingworth, 1923a MILICHIIDAE: MILICHIINAE *Milichiella lacteipennis* (Loew) Illingworth, 1923a MUSCIDAE: FANNIINAE *Fannia pusio* (Wiedemann) Illingworth, 1923a MUSCIDAE: PHAONIINAE *Ophyra nigra* Wiedemann Illingworth, 1923a MUSCIDAE: MUSCINAE *Musca domestica* Linnaeus Illingworth, 1923a MUSCIDAE: STOMOXYINAE *Stomoxys calcitrans* (Linnaeus) Illingworth, 1923a CALLIPHORIDAE: CALLIPHORINAE *Phaenicia sericata* (Meigen) Illingworth, 1923a SARCOPHAGIDAE: SARCOPHAGINAE *Sarcophaga fuscicauda* Böttcher Illingworth, 1923a
UNDETERMINED STAPHYLINIDAE (predators)	MUSCIDAE: FANNIINAE *Fannia canicularis* (Linnaeus) Anderson, 1964 MUSCIDAE: PHAONIINAE *Muscina stabulans* (Fallén) Anderson, 1964 MUSCIDAE: MUSCINAE *Musca domestica* Linnaeus Anderson, 1964 MUSCIDAE: STOMOXYINAE *Haematobia titillans* (Bezzi) Pridantseva, 1959

HISTERIDAE

Organism	Fly association
Pachylister chinensis Quensel (Synonym: *Platylister chinensis*) (predator)	MUSCIDAE: MUSCINAE *Musca domestica* Linnaeus Simmonds, 1958 MUSCIDAE: STOMOXYINAE *Haematobia exigua* (de Meijere) Bornemissza, 1968
Saprinus cyaneus Fabricius (Synonym: *Saprinus laetus*) (predator)	UNDETERMINED CALLIPHORIDAE Blow flies Gurney et al., 1926b
Saprinus detersus (Illiger) (predator)	SARCOPHAGIDAE: SARCOPHAGINAE *Sarcophaga carnaria* (Linnaeus) Pavlovskiĭ, 1921
Saprinus lugens Erichson (predator)	MUSCIDAE: MUSCINAE *Musca domestica* Linnaeus Vogt, 1948 CALLIPHORIDAE: CALLIPHORINAE *Phaenicia pallescens* (Shannon) Vogt, 1948 *Phaenicia sericata* (Meigen) Illingworth, 1927 SARCOPHAGIDAE: SARCOPHAGINAE *Blaesoxipha plinthopyga* (Wiedemann) Illingworth, 1927
Saprinus politus (Brahm) or *Saprinus aenus* (Fabricius) (Synonym: *Saprinus speculifer*) (predator)	SARCOPHAGIDAE: SARCOPHAGINAE *Sarcophaga carnaria* (Linnaeus) Pavlovskiĭ, 1921
Xerosaprinus fimbriatus (LeConte) (Synonym: *Saprinus fimbriatus*) (predator)	CALLIPHORIDAE: CALLIPHORINAE *Phaenicia sericata* (Meigen) Illingworth, 1927 SARCOPHAGIDAE: SARCOPHAGINAE *Blaesoxipha plinthopyga* (Wiedemann) Illingworth, 1927
Xerosaprinus lubricus (LeConte) (predator)	CALLIPHORIDAE: CALLIPHORINAE *Phaenicia sericata* (Meigen) Illingworth, 1927 SARCOPHAGIDAE: SARCOPHAGINAE *Blaesoxipha plinthopyga* (Wiedemann) Illingworth, 1927
UNDETERMINED HISTERIDAE	MUSCIDAE: STOMOXYINAE *Haematobia titillans* (Bezzi) Pridantseva, 1959 *Stomoxys calcitrans* (Linnaeus) Bishopp, 1913a

Organism	Fly
UNDETERMINED HISTERIDAE (cont.)	CALLIPHORIDAE: CALLIPHORINAE *Phaenicia sericata* (Meigen) Holdaway, 1930
CLERIDAE	
Necrobia rufipes De Geer	PIOPHILIDAE *Piophila casei* (Linnaeus) Simmons, 1922
DERMESTIDAE	
Dermestes ater De Geer	MUSCIDAE: MUSCINAE *Musca domestica* Linnaeus Vogt, 1948 CALLIPHORIDAE: CALLIPHORINAE *Phaenicia pallescens* (Shannon) Vogt, 1948
Dermestes caninus Germar	MUSCIDAE: MUSCINAE *Musca domestica* Linnaeus Vogt, 1948 CALLIPHORIDAE: CALLIPHORINAE *Phaenicia pallescens* (Shannon) Vogt, 1948
UNDETERMINED DERMESTIDAE	MUSCIDAE: MUSCINAE *Musca domestica* Linnaeus Packard, 1873 CALLIPHORIDAE: CALLIPHORINAE *Phaenicia sericata* (Meigen) Holdaway, 1930
OSTOMATIDAE	
Temnochila virescens (Fabricius) (predator)	CALLIPHORIDAE: CALLIPHORINAE *Phaenicia sericata* (Meigen) Struble, 1942
COCCINELLIDAE	
Adonia variegata Goeze	PHORIDAE: METOPININAE *Megaselia fasciata* (Fallén) Martelli, 1913
Coccinella septempunctata Linnaeus	PHORIDAE: METOPININAE *Megaselia fasciata* (Fallén) Rondani, du Buysson, 1917
Thea vigintiduopunctata Linnaeus	PHORIDAE: METOPININAE *Megaselia fasciata* (Fallén) Martelli, 1913 Lichtenstein, 1920
Vibidia duodecimguttata Poda	PHORIDAE: METOPININAE *Megaselia fasciata* (Fallén) Lichtenstein, 1920
SCARABAEIDAE	
Copris incertus Say	MUSCIDAE: MUSCINAE *Musca domestica* Linnaeus Simmonds, 1958

Organism	Fly association
Lachnosterna Hope	MUSCIDAE: FANNIINAE *Fannia canicularis* (Linnaeus) Davis, 1919
Melolontha melolontha Linnaeus	MUSCIDAE: PHAONIINAE *Muscina stabulans* (Fallén) Kamner, 1928
CERAMBYCIDAE	
Dorysthenes forficatus Fabricius	MUSCIDAE: PHAONIINAE *Muscina stabulans* (Fallén) Bouhélier et al., 1936
CHRYSOMELIDAE	
Gallerucella luteola Müller	MUSCIDAE: PHAONIINAE *Muscina stabulans* (Fallén) Curran, 1942
CURCULIONIDAE	
Conotrachelus retentus Say	MUSCIDAE: FANNIINAE *Fannia canicularis* (Linnaeus) Brooks, 1922
Pissodes strobi Peck	MUSCIDAE: PHAONIINAE *Muscina stabulans* (Fallén) MacAloney, 1930
Rhodobaenus tredecimpunctatus (Illiger)	MUSCIDAE: PHAONIINAE *Muscina stabulans* (Fallén) Satterthwait, 1943
TENTHREDINIDAE	
Tenthredo pectoralis Norton (Synonym: *Tenthredo variegatus*)	MUSCIDAE: MUSCINAE House fly (presumably *Musca domestica* Linnaeus) Venables, 1914
DIPRIONIDAE	
Diprion pallidus Klug (Synonym: *Lophyrus pallidus*)	MUSCIDAE: PHAONIINAE *Muscina stabulans* (Fallén) van der Wulp, 1869
Diprion pini (Linnaeus) (Synonym: *Tenthredo pini*)	MUSCIDAE: PHAONIINAE *Muscina stabulans* (Fallén) Keilin, 1917
ICHNEUMONIDAE	
Atractodes bicolor Gravenhorst (parasite)	CALLIPHORIDAE: CALLIPHORINAE *Calliphora vicina* Robineau-Desvoidy Graham-Smith, 1919
Atractodes gravidus Gravenhorst (parasite)	MUSCIDAE: PHAONIINAE *Hydrotaea dentipes* (Fabricius) Myers, 1929
	CALLIPHORIDAE: CALLIPHORINAE *Lucilia* Robineau-Desvoidy Myers, 1929
Atractodes ?gravidus Gravenhorst (parasite)	MUSCIDAE: PHAONIINAE *Ophyra leucostoma* (Wiedemann) Sychevskaia͡, 1964b

Organism	Fly associations
Atractodes mallyi Bridwell (parasite)	SARCOPHAGIDAE: SARCOPHAGINAE *Sarcophaga* Meigen Bridwell, 1919
Atractodes muiri Bridwell (parasite)	SARCOPHAGIDAE: SARCOPHAGINAE *Sarcophaga* Meigen Bridwell, 1919
Mesoleptus laevigatus (Gravenhorst) (Synonym: *Exolytus laevigatus*) (parasite)	SARCOPHAGIDAE: SARCOPHAGINAE *Sarcophaga peregrina* Robineau-Desvoidy Suenaga et al., 1963
Nemeritis canescens (Gravenhorst) (parasite)	CALLIPHORIDAE: CALLIPHORINAE *Calliphora vicina* Robineau-Desvoidy Salt, 1957
Opius nitidulator Nees (parasite)	MUSCIDAE: MUSCINAE *Musca domestica* Linnaeus Rambousek, 1929
Phygadeuon Gravenhorst	MUSCIDAE: MUSCINAE *Musca domestica* Linnaeus Legner et al., 1966a Mourier et al., 1968
Phygadeuon speculator Thomson (Synonym: *Phygadenon speculator*) (parasite)	CALLIPHORIDAE: CALLIPHORINAE *Calliphora vicina* Robineau-Desvoidy Graham-Smith, 1919
Stilpnus Gravenhorst (parasite)	MUSCIDAE: FANNIINAE *Fannia incisurata* (Zetterstedt) Sychevskaia͡, 1964b *Fannia scalaris* (Fabricius) Sychevskaia͡, 1964b MUSCIDAE: PHAONIINAE *Dendrophaonia querceti* (Bouché) Sychevskaia͡, 1964b
Stilpnus anthomyidiperda Viereck (parasite of maggots)	MUSCIDAE: FANNIINAE *Fannia canicularis* (Linnaeus) Legner et al. (pre-publication), 1965a Legner (pre-publication), 1966 Legner et al., 1966b
Theronia atalantae (Poda) (parasite)	CALLIPHORIDAE: CALLIPHORINAE *Calliphora vomitoria* (Linnaeus) Fahringer, 1922
UNDETERMINED ICHNEUMONIDAE (parasites)	SEPSIDAE *Saltella sphondylii* (Schrank) Hammer, 1942 *Sepsis pilipes* Wulp Hammer, 1942

Organism	Fly association
UNDETERMINED ICHNEUMONIDAE (cont.)	SPHAEROCERIDAE *Copromyza atra* (Meigen) Hammer, 1942 *Leptocera moesta* (Villeneuve) Hammer, 1942 ANTHOMYIIDAE: SCATOPHAGINAE *Scatophaga stercoraria* (Linnaeus) Hammer, 1942 MUSCIDAE: PHAONIINAE *Hydrotaea albipuncta* (Zetterstedt) Hammer, 1942 MUSCIDAE: MUSCINAE *Mesembrina meridiana* (Linnaeus) Hammer, 1942 *Morellia hortorum* (Fallén) Hammer, 1942 *Orthellia caesarion* (Meigen) Hammer, 1942 MUSCIDAE: STOMOXYINAE *Haematobia irritans* (Linnaeus) Hammer, 1942
BRACONIDAE *Alysia fossulata* Provancher (parasite)	CALLIPHORIDAE: CALLIPHORINAE *Lucilia* Robineau-Desvoidy Roberts, 1935 SARCOPHAGIDAE: SARCOPHAGINAE *Blaesoxipha plinthopyga* (Wiedemann) Roberts, 1935
Alysia manducator Panzer (parasite)	MUSCIDAE: FANNIINAE *Fannia* Robineau-Desvoidy Myers, 1927 MUSCIDAE: MUSCINAE *Musca domestica* Linnaeus Froggatt, 1922 Gurney et al., 1926a Legner et al., 1966a CALLIPHORIDAE: CHRYSOMYINAE *Chrysomya albiceps* (Wiedemann) Smit, 1929 *Chrysomya rufifacies* (Macquart) Miller, 1927 Newman, 1928 Morgan, 1929 *Protophormia terraenovae* (Robineau-Desvoidy) Altson, 1920 CALLIPHORIDAE: CALLIPHORINAE *Lucilia* Robineau-Desvoidy Myers, 1927

Alysia manducator (cont.)	*Lucilia caesar* (Linnaeus)
	Altson, 1920
	Phaenicia sericata (Meigen)
	Altson, 1920
	Gurney et al., 1926a
	Miller, 1927
	Newman, 1928
	Morgan, 1929
	Smit, 1929
	Holdaway et al., 1930, 1932
	Tillyard, 1930
	Salt, 1932
	Evans, 1933
	Calliphora Robineau-Desvoidy
	Myers, 1927
	Morgan, 1929
	Laing, 1937
	Calliphora icela (Walker)
	Miller, 1927
	Calliphora vicina Robineau-Desvoidy
	MacDougall, 1909
	Graham-Smith, 1916, 1919
	Altson, 1920
	Gurney et al., 1926a
	Miller, 1927
	Morgan, 1929
	Holdaway et al., 1932
	Evans, 1933
	Caudri, 1941
	Calliphora vomitoria (Linnaeus)
	Altson, 1920
	Holdaway et al., 1932
	Anastellorhina augur (Fabricius)
	Gurney et al., 1926a
	UNDETERMINED CALLIPHORIDAE
	Blow fly pupae
	Graham-Smith, 1916
	SARCOPHAGIDAE: SARCOPHAGINAE
	Sarcophaga Meigen
	Holdaway et al., 1932
Alysia ridibunda Say	MUSCIDAE: PHAONIINAE
(parasite)	*Ophyra leucostoma* (Wiedemann)
	Roberts, 1935
	MUSCIDAE: MUSCINAE
	Synthesiomyia nudiseta (Wulp)
	Roberts, 1935
	CALLIPHORIDAE: CHRYSOMYINAE
	Cochliomyia macellaria (Fabricius)
	Roberts, 1935

Alysia ridibunda (cont.)	CALLIPHORIDAE: CALLIPHORINAE *Phaenicia mexicana* (Macquart) Roberts, 1935 *Phaenicia sericata* (Meigen) Roberts, 1935 *Calliphora coloradensis* Hough Roberts, 1935 SARCOPHAGIDAE: SARCOPHAGINAE *Blaesoxipha plinthopyga* (Wiedemann) Roberts, 1935 *Sarcophaga sarracenioides* Aldrich Roberts, 1935
Aphaereta auripes (Provancher) (Synonym: *Aphaereta muscae*) (parasite)	UNDETERMINED CALLIPHORIDAE Blow flies Roberts, 1935
Aphaereta cephalotes Haliday (parasite)	CALLIPHORIDAE: CALLIPHORINAE *Calliphora vicina* Robineau-Desvoidy Graham-Smith, 1916 SARCOPHAGIDAE: SARCOPHAGINAE *Sarcophaga melanura* Meigen Graham-Smith, 1916
Aphaereta minuta Nees (parasite)	ANTHOMYIIDAE: SCATOPHAGINAE *Scatophaga stercoraria* (Linnaeus) Sychevskai͡a, 1964b CALLIPHORIDAE: CALLIPHORINAE *Phaenicia sericata* (Meigen) Holdaway et al., 1930 *Phaenicia sericata* (Meigen) Evans, 1933 *Calliphora vicina* Robineau-Desvoidy Evans, 1933 *Calliphora vomitoria* (Linnaeus) Evans, 1933 SARCOPHAGIDAE: SARCOPHAGINAE *Ravinia striata* Fabricius Sychevskai͡a, 1964b *Sarcophaga haemorrhoidalis* (Fallén) Sychevskai͡a, 1964b *Sarcophaga melanura* Meigen Sychevskai͡a, 1964b
Aphaereta pallipes Say (Synonym: *Aphaerta pallipes*) (parasite)	PIOPHILIDAE *Piophila casei* (Linnaeus) Beard, 1964 MUSCIDAE: MUSCINAE *Orthellia caesarion* (Meigen) Blickle, 1961 Benson et al., 1963

Organism	Fly association
Aphaereta pallipes (cont.)	Sanders et al., 1966
	Houser et al., 1967
	Musca autumnalis De Geer
	Blickle, 1961
	Benson et al., 1963
	Beard, 1964
	Sanders et al., 1966
	Houser et al., 1967
	Thomas et al., 1968
	Turner et al., 1968
	Musca domestica Linnaeus
	Blickle, 1961
	Beard, 1964
	CALLIPHORIDAE: CHRYSOMYINAE
	Phormia regina (Meigen)
	Beard, 1964
	CALLIPHORIDAE: CALLIPHORINAE
	Phaenicia sericata (Meigen)
	Beard, 1964
	SARCOPHAGIDAE: MILTOGRAMMINAE
	Pseudosarcophaga affinis Fallén
	Barlow, 1964
	Oxysarcodexia ventricosa (Wulp)
	Sanders et al., 1966
	Ravinia derelicta (Walker)
	Benson et al., 1963
	Ravinia latisetosa Parker
	Blickle, 1961
	Ravinia lherminieri (Robineau-Desvoidy)
	Blickle, 1961
	Ravinia querula (Walker)
	Blickle, 1961
	Sanders et al., 1966
	Houser et al., 1967
	Sarcophaga bullata Parker
	Beard, 1964
Aphaereta sarcophaga Gahan (parasite)	SARCOPHAGIDAE: SARCOPHAGINAE
	Sarcophaga Meigen
	Bridwell, 1919
Asobara tabida (Nees) (Synonyms: *Phaenocarpa tabida*, *Phaenocarpa tapida*) (parasite)	DROSOPHILIDAE: DROSOPHILINAE
	Drosophila melanogaster Meigen
	Jenni, 1951
	Walker, 1963
Aspilota concolor Nees (parasite)	DROSOPHILIDAE: DROSOPHILINAE
	Drosophila funebris (Fabricius)
	Austin, 1933
Aspilota nervosa Haliday (parasite)	PHORIDAE: METOPININAE
	Megaselia Brues
	Evans, 1933

Organism	Fly association
Aspilota nervosa (cont.)	CALLIPHORIDAE: CALLIPHORINAE *Phaenicia sericata* (Meigen) Evans, 1933
Idiasta lusoriae (Bridwell) (Synonym: *Alysia lusoriae*) (parasite)	MUSCIDAE: MUSCINAE *Musca lusoria* Wiedemann Bridwell, 1919
Opius nitidulator Nees (parasite)	CALLIPHORIDAE: CALLIPHORINAE *Lucilia caesar* (Linnaeus) Rambousek, 1929 *Calliphora vomitoria* (Linnaeus) Rambousek, 1929
CYNIPIDAE	
Bothrochacis stercoraria Bridwell (parasite)	MUSCIDAE: MUSCINAE *Orthellia cyanea* (Fabricius) Bridwell, 1919 *Musca lusoria* Wiedemann Bridwell, 1919
Cothonaspis nigricornis (Kieffer) (parasite)	SEPSIDAE *Sepsis* Fallén Sychevskaîa, 1964b MUSCIDAE: PHAONIINAE *Ophyra anthrax* (Meigen) Sychevskaîa, 1964b MUSCIDAE: MUSCINAE *Orthellia caesarion* (Meigen) Sychevskaîa, 1964b *Musca domestica ?vicina* Macquart Sychevskaîa, 1964b
Eucoila Westwood (Synonym: *Psilodora*) (parasite)	DROSOPHILIDAE: DROSOPHILINAE *Drosophila melanogaster* Meigen Jenni, 1947 MUSCIDAE: PHAONIINAE *Ophyra leucostoma* (Wiedemann) Roberts, 1935 MUSCIDAE: MUSCINAE *Synthesiomyia nudiseta* (Wulp) Roberts, 1935 *Orthellia caesarion* (Meigen) Blickle, 1961 Sanders et al., 1966 *Musca autumnalis* De Geer Blickle, 1961 *Musca domestica* Linnaeus Simmonds, 1958 Blickle, 1961 House fly (presumably *Musca domestica* Linnaeus) Simmonds, 1940

Organism	Fly association
Eucoila (cont.)	CALLIPHORIDAE: CHRYSOMYINAE *Cochliomyia macellaria* (Fabricius) Roberts, 1935 CALLIPHORIDAE: CALLIPHORINAE *Phaenicia sericata* (Meigen) Roberts, 1935 SARCOPHAGIDAE: SARCOPHAGINAE *Blaesoxipha plinthopyga* (Wiedemann) Roberts, 1935 *Ravinia latisetosa* Parker Blickle, 1961 *Ravinia lherminieri* (Robineau-Desvoidy) Blickle, 1961 *Ravinia querula* (Walker) Blickle, 1961 Sanders et al., 1966 *Sarcodexia pedunculata* (Hall) Roberts, 1935 *Sarcophaga* Meigen Sanders et al., 1966 *Sarcophaga bullata* Parker Roberts, 1935 *Sarcophaga sarracenioides* Aldrich Roberts, 1935
Eucoila drosophilae Kieffer (parasite)	DROSOPHILIDAE: DROSOPHILINAE *Drosophila* Fallén Bouché, 1936 *Drosophila melanogaster* Meigen Boche, 1939
Eucoila impatiens (Say) (parasite)	MUSCIDAE: MUSCINAE *Musca autumnalis* De Geer Thomas et al., 1968 Turner et al., 1968 SARCOPHAGIDAE: SARCOPHAGINAE *Sarcophaga fuscicauda* Böttcher Illingworth, 1923a SARCOPHAGIDAE: MILTOGRAMMINAE *Ravinia querula* (Walker) Turner et al., 1968 UNDETERMINED SARCOPHAGIDAE Illingworth, 1923
Hexacola Foerster (parasite)	CHLOROPIDAE: OSCINELLINAE *Hippelates* Loew Mulla, 1962 *Oscinella frit* (Linnaeus) Bay et al., 1963

Organism	Fly associations
Kleidotoma Westwood (parasite)	MUSCIDAE: PHAONIINAE *Hydrotaea dentipes* (Fabricius) James, 1928 MUSCIDAE: MUSCINAE *Musca domestica* Linnaeus James, 1928 CALLIPHORIDAE: CALLIPHORINAE *Lucilia caesar* (Linnaeus) James, 1928 *Phaenicia sericata* (Meigen) James, 1928 *Calliphora vicina* Robineau-Desvoidy James, 1928 SARCOPHAGIDAE: SARCOPHAGINAE *Sarcophaga carnaria* (Linnaeus) James, 1928
Kleidotoma certa Belizin (parasite)	SEPSIDAE *Sepsis* Fallén Sychevskaîâ, 1964b
Kleidotoma marshalli Marshall (parasite)	MUSCIDAE: PHAONIINAE *Hydrotaea dentipes* (Fabricius) James, 1928 MUSCIDAE: MUSCINAE *Musca domestica* Linnaeus James, 1928 CALLIPHORIDAE: CALLIPHORINAE *Lucilia caesar* (Linnaeus) James, 1928 *Phaenicia sericata* (Meigen) James, 1928 *Calliphora vicina* Robineau-Desvoidy James, 1928 SARCOPHAGIDAE: SARCOPHAGINAE *Sarcophaga carnaria* (Linnaeus) James, 1928
Pseudeucoila bochei Weld (parasite)	DROSOPHILIDAE: DROSOPHILINAE *Drosophila busckii* Coquillett Jenni, 1951 *Drosophila funebris* (Fabricius) Jenni, 1951 *Drosophila hydei* Sturtevant Jenni, 1951 *Drosophila littoralis* Meigen Jenni, 1951 *Drosophila melanogaster* Meigen Jenni, 1951 Schlegel-Oprecht, 1953 Nøstvik, 1954 Walker, 1959, 1961, 1963

Organism	Fly associations
Pseudeucoila bochei (cont.)	*Drosophila phalerata* Meigen Jenni, 1951 *Drosophila subobscura* Collin Jenni, 1951
Pseudeucoila trichopsila (Hartig) (Synonym: *Eucoila trichopsila*) (parasite)	ANTHOMYIIDAE: SCATOPHAGINAE *Scatophaga stercoraria* (Linnaeus) Sychevskaîa, 1964b CALLIPHORIDAE: CALLIPHORINAE *Phaenicia sericata* (Meigen) Sychevskaîa, 1964b SARCOPHAGIDAE: SARCOPHAGINAE *Ravinia striata* Fabricius Sychevskaîa, 1964b, 1966 *Sarcophaga argyrostoma* (Robineau-Desvoidy) Sychevskaîa, 1964b *Sarcophaga haemorrhoidalis* (Fallén) Sychevskaîa, 1964b, 1966 *Sarcophaga melanura* Meigen Sychevskaîa, 1964b *Sarcophaga parkeri* (Rohdendorf) Sychevskaîa, 1964b
Trybliographa Foerster (parasite)	CHLOROPIDAE: OSCINELLINAE *Hippelates* Loew Legner et al., 1964 *Hippelates collusor* (Townsend) Legner et al., 1969b *Hippelates pusio* Loew Legner et al., 1965d
FIGITIDAE	
Figites Latreille (parasite)	MUSCIDAE: MUSCINAE *Musca domestica* Linnaeus Legner (pre-publication), 1966
Figites anthomyiarum (Bouché) (parasite)	MUSCIDAE: PHAONIINAE *Hydrotaea dentipes* (Fabricius) James, 1928 MUSCIDAE: MUSCINAE *Musca domestica* Linnaeus James, 1928 CALLIPHORIDAE: CALLIPHORINAE *Lucilia caesar* (Linnaeus) James, 1928 *Phaenicia sericata* (Meigen) James, 1928 *Calliphora vicina* Robineau-Desvoidy James, 1928 SARCOPHAGIDAE: SARCOPHAGINAE *Sarcophaga carnaria* (Linnaeus) James, 1928

Figites discordis Belizin (parasite)	MUSCIDAE: PHAONIINAE *Ophyra anthrax* (Meigen) Sychevskai͡a, 1964b SARCOPHAGIDAE: SARCOPHAGINAE *Ravinia striata* Fabricius Sychevskai͡a, 1964b *Sarcophaga haemorrhoidalis* (Fallén) Sychevskai͡a, 1964b
Figites sarcophagarum Belizin (parasite)	SARCOPHAGIDAE: SARCOPHAGINAE *Sarcophaga parkeri* (Rohdendorf) Sychevskai͡a, 1964b
Figites scutellaris (Rossi) (parasite)	SARCOPHAGIDAE: SARCOPHAGINAE *Ravinia striata* Fabricius Sychevskai͡a, 1964b *Sarcophaga melanura* Meigen Sychevskai͡a, 1964b
Figites striolatus Hartig (parasite)	MUSCIDAE: PHAONIINAE *Hydrotaea dentipes* (Fabricius) Myers, 1929 SARCOPHAGIDAE: SARCOPHAGINAE *Ravinia striata* Fabricius Sychevskai͡a, 1964b
Neralsia Cameron (Synonym: *Xyalosema*) (parasite)	CALLIPHORIDAE: CHRYSOMYINAE *Cochliomyia macellaria* (Fabricius) Roberts, 1935 SARCOPHAGIDAE: SARCOPHAGINAE *Blaesoxipha plinthopyga* (Wiedemann) Roberts, 1935 *Sarcophaga* Meigen Roberts, 1935
Neralsia armata (Say) (Synonym: *Xyalosema armata*) (parasite)	SARCOPHAGIDAE: SARCOPHAGINAE *Blaesoxipha plinthopyga* (Wiedemann) Roberts, 1935 *Sarcophaga bullata* Parker Roberts, 1935
Neralsia bifoveolata (Cresson) (Synonym: *Xyalosema bifoveolata*) (parasite)	MUSCIDAE: STOMOXYINAE *Haematobia irritans* (Linnaeus) Wolcott, 1922
Xyalophora quinquelineata (Say) (parasite)	MUSCIDAE: MUSCINAE *Orthellia caesarion* (Meigen) Blickle, 1961 *Musca autumnalis* De Geer Blickle, 1961 *Musca domestica* Linnaeus Blickle, 1961

Xyalophora quinquelineata (cont.)	SARCOPHAGIDAE: SARCOPHAGINAE *Ravinia latisetosa* Parker Blickle, 1961 *Ravinia lherminieri* (Robineau-Desvoidy) Blickle, 1961 *Ravinia querula* (Walker) Blickle, 1961
CHALCIDIDAE	
Brachymeria argentifrons (Ashmead) (parasite)	SARCOPHAGIDAE: SARCOPHAGINAE *Sarcophaga dux* Thomson Roy et al., 1939 *Sarcophaga dux* var. *tuberosa* Thomson Roy et al., 1939
Brachymeria calliphorae (Froggatt) (Synonym: *Chalcis calliphorae*) (parasite)	CALLIPHORIDAE: CALLIPHORINAE *Anastellorhina augur* (Fabricius) Froggatt, 1916 Johnston et al., 1921
Brachymeria fonscolombei (Dufour) (Synonym: *Chalcis fonscolombei*) (parasite)	MUSCIDAE: MUSCINAE *Synthesiomyia nudiseta* (Wulp) Roberts, 1933a, 1933b, 1935 *Musca* Linnaeus Masi, 1916 CALLIPHORIDAE: CHRYSOMYINAE *Cochliomyia macellaria* (Fabricius) Bishopp, 1929-30 Marlatt, 1931 Roberts, 1933a, 1933b, 1935 *Phormia regina* (Meigen) Marlatt, 1931 Roberts, 1933b, 1935 CALLIPHORIDAE: CALLIPHORINAE *Lucilia* Robineau-Desvoidy Masi, 1916 Parker, 1923 *Phaenicia eximia* (Wiedemann) Roberts, 1933a *Phaenicia mexicana* (Macquart) Roberts, 1933a, 1933b, 1935 *Phaenicia sericata* (Meigen) Roberts, 1933b, 1935 *Calliphora coloradensis* Hough Roberts, 1933b, 1935 SARCOPHAGIDAE: SARCOPHAGINAE *Blaesoxipha impar* (Aldrich) Roberts, 1933b, 1935

Brachymeria fonscolombei (cont.)	*Blaesoxipha plinthopyga* (Wiedemann) Roberts, 1933b, 1935 *Sarcophaga* Meigen Masi, 1916 Parker, 1923 Roberts, 1933a *Sarcophaga carnaria* (Linnaeus) Stefani, 1889 Roberts, 1935 Nikol'skaı͡a, 1960b *Sarcophaga haemorrhoidalis* (Fallén) von Dalla Torre, 1898 Roberts, 1935 *Sarcophaga peregrina* Robineau-Desvoidy Suenaga et al., 1963 *Sarcophaga ruficornis* Fabricius Roy et al., 1939 TACHINIDAE: TACHININAE *Linnaemya haemorrhoidalis* Fallén Nikol'skaı͡a, 1960b
Brachymeria fulvitarsis (Cameron) (parasite)	CALLIPHORIDAE: CHRYSOMYINAE *Chrysomya megacephala* (Fabricius) Roy et al., 1939
Brachymeria minuta (Fabricius) (parasite)	MUSCIDAE: PHAONIINAE *Dendrophaonia querceti* (Bouché) Sychevskaı͡a, 1966 *Muscina stabulans* (Fallén) Sychevskaı͡a, 1966 MUSCIDAE: MUSCINAE *Musca domestica vicina* Macquart Sychevskaı͡a, 1966 CALLIPHORIDAE: CALLIPHORINAE *Phaenicia sericata* (Meigen) Sychevskaı͡a, 1966 *Calliphora vicina* Robineau-Desvoidy Sychevskaı͡a, 1964b, 1966 SARCOPHAGIDAE: MILTOGRAMMINAE *Pseudosarcophaga affinis* Fallén Nikol'skaı͡a, 1960a SARCOPHAGIDAE: SARCOPHAGINAE *Ravinia striata* Fabricius Sychevskaı͡a, 1964b, 1966 *Sarcophaga argyrostoma* (Robineau-Desvoidy) Sychevskaı͡a, 1964b *Sarcophaga haemorrhoidalis* (Fallén) Nikol'skaı͡a, 1960a Sychevskaı͡a, 1964b, 1966

Organism	Fly associations
Brachymeria minuta (cont.)	*Sarcophaga hirtipes* Wiedemann Nikol'skaîa, 1960 Sychevskaîa, 1964b *Sarcophaga melanura* Meigen Sychevskaîa, 1964b *Sarcophaga parkeri* (Rohdendorf) Nikol'skaîa, 1960a
Brachymeria obtusata Foerster (parasite)	CALLIPHORIDAE: CALLIPHORINAE *Lucilia caesar* (Linnaeus) Nikol'skaîa, 1960a
Brachymeria vicina Walker (parasite)	SYRPHIDAE: MILESIINAE *Eristalis tenax* (Linnaeus) Nikol'skaîa, 1960a
Dirhinoides vlasovi Nikol'skaîa (parasite)	SARCOPHAGIDAE: SARCOPHAGINAE *Ravinia striata* Fabricius Sychevskaîa, 1964b UNDETERMINED SARCOPHAGIDAE Nikol'skaîa, 1960a
Dirhinoides wohlfahrtiae Ferrière (parasite)	SARCOPHAGIDAE *Wohlfahrtia nuba* Wiedemann Nikol'skaîa, 1960b
Dirhinus Dalman (parasite of fly pupae)	MUSCIDAE: MUSCINAE *Musca domestica* Linnaeus Simmonds, 1940
Dirhinus luzonensis Rohwer (parasite)	MUSCIDAE: MUSCINAE House fly (presumably *Musca domestica* Linnaeus) Dresner, 1954
Dirhinus pachycerus Masi (parasite)	DROSOPHILIDAE: DROSOPHILINAE *Drosophila melanogaster* Meigen Roy et al., 1940 MUSCIDAE: MUSCINAE *Musca inferior* Stein Roy et al., 1939 *Musca domestica vicina* Macquart Roy et al., 1940 *Musca nebulo* Fabricius Roy et al., 1940 MUSCIDAE: STOMOXYINAE *Stomoxys calcitrans* (Linnaeus) Roy et al., 1940 CALLIPHORIDAE: CHRYSOMYINAE *Chrysomya megacephala* (Fabricius) Roy et al., 1939 SARCOPHAGIDAE: SARCOPHAGINAE *Sarcophaga dux* Thompson var. *tuberosa* Roy et al., 1939 *Sarcophaga ruficornis* Fabricius Roy et al., 1939

Dirhinus sarcophagae Froggatt (parasite)	CALLIPHORIDAE: CHRYSOMYINAE *Chrysomya rufifacies* (Macquart) Johnston et al., 1921 *Chrysomya varipes* (Macquart) Johnston et al., 1921 CALLIPHORIDAE: CALLIPHORINAE *Lucilia* Robineau-Desvoidy Johnston et al., 1921 *Phaenicia sericata* (Meigen) Johnston et al., 1921 *Calliphora stygia* (Fabricius) Froggatt, 1921, 1922 SARCOPHAGIDAE: SARCOPHAGINAE *Sarcophaga aurifrons* Macquart Froggatt, 1919, 1922 *Sarcophaga impatiens* (Walker) Johnston et al., 1921
Dirhinus texanus (Ashmead) (Synonym: *Eniaca texana*) (parasite)	SARCOPHAGIDAE: SARCOPHAGINAE *Blaesoxipha plinthopyga* (Wiedemann) Roberts, 1935
Euchalcidia blanda Nikol'skaîa (parasite)	SARCOPHAGIDAE: SARCOPHAGINAE *Ravinia striata* Fabricius Nikol'skaîa, 1960b *Sarcophaga haemorrhoidalis* (Fallén) Nikol'skaîa, 1960a Sychevskaîa, 1964b
Stenomalus muscarum Linnaeus (parasite)	MUSCIDAE: MUSCINAE *Musca* Linnaeus Waterston, 1916
UNDETERMINED CHALCIDIDAE (parasites)	MUSCIDAE: STOMOXYINAE *Stomoxys calcitrans* (Linnaeus) Surcouf, 1923 CALLIPHORIDAE: CALLIPHORINAE *Calliphora stygia* (Fabricius) Froggatt, 1914 *Calliphora vicina* Robineau-Desvoidy Froggatt, 1914a, 1914b
PTEROMALIDAE	
Dibrachys affinis Masi (parasite)	CALLIPHORIDAE: CALLIPHORINAE *Calliphora vicina* Robineau-Desvoidy Voukassovitch, 1924a *Calliphora vomitoria* (Linnaeus) Voukassovitch, 1924a
Dibrachys cavus (Walker) (parasite)	MUSCIDAE: PHAONIINAE *Muscina stabulans* (Fallén) Graham-Smith, 1919

Organism	Fly association
Dibrachys cavus (cont.)	CALLIPHORIDAE: CHRYSOMYINAE *Protophormia terraenovae* (Robineau-Desvoidy) Graham-Smith, 1919
Eupteromales hemipterus (Walker) (parasite)	CHLOROPIDAE: OSCINELLINAE *Hippelates collusor* (Townsend) Legner et al., 1965, 1966
Eupteromales nidulans (Thomson)	CHLOROPIDAE: OSCINELLINAE *Hippelates collusor* (Townsend) Bay (unpublished)
Habrocytus obscurus Dalman (parasite)	CALLIPHORIDAE: CALLIPHORINAE *Lucilia caesar* (Linnaeus) Voukassovitch, 1924b *Calliphora vomitoria* (Linnaeus) Voukassovitch, 1924b
Muscidifurax raptor Girault and Sanders (pupal parasite)	MUSCIDAE: FANNIINAE *Fannia canicularis* (Linnaeus) Legner (pre-publication), 1966 *Fannia femoralis* (Stein) Legner et al. (pre-publication), 1965a Legner (pre-publication), 1966 Legner et al., 1966b MUSCIDAE: PHAONIINAE *Ophyra leucostoma* (Wiedemann) Legner et al. (pre-publication), 1965a Legner (pre-publication), 1966 Legner et al., 1966b MUSCIDAE: MUSCINAE *Musca autumnalis* De Geer Thomas et al., 1968 Turner et al., 1968 *Musca domestica* Linnaeus Girault et al., 1910a, 1910b Roberts, 1935 Puerto Rico, USDA Experim. Sta., 1938 Bartlett, 1939 Legner et al. (pre-publication), 1965a Legner et al., 1965e, 1966 Legner (pre-publication), 1966 Mourier et al., 1968 Legner, 1969 MUSCIDAE: STOMOXYINAE *Haematobia* Lepeletier and Serville Fullaway, 1917b

Muscidifurax raptor (cont.)	*Stomoxys calcitrans* (Linnaeus) Bartlett, 1939 Legner (pre-publication), 1966 CALLIPHORIDAE: CHRYSOMYINAE *Cochliomyia macellaria* (Fabricius) Girault et al., 1910b *Phormia regina* (Meigen) Girault et al., 1910b Legner (pre-publication), 1966 SARCOPHAGIDAE: MILTOGRAMMINAE *Ravinia querula* (Walker) Turner et al., 1968
Nasonia Ashmead (Synonym: *Mormoniella*) (parasite)	CALLIPHORIDAE: CALLIPHORINAE *Lucilia* Robineau-Desvoidy Hardy, 1924 SARCOPHAGIDAE: SARCOPHAGINAE *Sarcophaga* Meigen Feng, 1933
?*Nasonia* Ashmead (parasite)	CALLIPHORIDAE: CALLIPHORINAE *Phaenicia sericata* (Meigen) Feng, 1933
Nasonia vitripennis (Walker) (Synonyms: *Mormoniella vitripennis, Nasonia brevicornis, Mormoniella brevicornis*)	PIOPHILIDAE *Piophila casei* (Linnaeus) Roberts, 1935 MUSCIDAE: FANNIINAE *Fannia femoralis* (Stein) Roberts, 1933a Legner et al. (pre-publication), 1965a Legner (pre-publication), 1966 Legner et al., 1966b MUSCIDAE: PHAONIINAE *Ophyra aenescens* (Wiedemann) Roberts, 1933a *Ophyra leucostoma* (Wiedemann) Roberts, 1933a Legner (pre-publication), 1966 *Ophyra nigra* Wiedemann Froggatt, 1919 Johnston et al., 1921 *Muscina stabulans* (Fallén) Séguy, 1929 MUSCIDAE: MUSCINAE *Synthesiomyia nudiseta* (Wulp) Roberts, 1933a *Musca* Linnaeus Johnston et al., 1921 Parker et al., 1928

Nasonia vitripennis (cont.)

Musca autumnalis De Geer
Beard, 1964
Hair et al., 1965

Musca domestica Linnaeus
Girault et al., 1910a
Roubaud, 1917
Altson, 1920
Johnston et al., 1920c, 1921
Froggatt, 1921, 1922
Séguy, 1929
Fujita, 1932
Smirnov et al., 1933, 1934
Smirnov, 1934
Vladimirova et al., 1934
DeBach et al., 1947
Edwards, 1954
Wylie, 1962, 1965a, 1965b, 1965c
Nagel et al., 1963
Pimentel, 1963
Pimentel et al., 1963, 1965a
Beard, 1964
Madden, 1964
Hair et al., 1965
Madden et al., 1965
Legner (pre-publication), 1966
Legner et al., 1966a

Musca hilli Johnston and Bancroft
Johnston et al., 1920b

Musca sorbens sorbens Wiedemann
Johnston et al., 1920b

Musca terraereginae Johnston and Bancroft
Johnston et al., 1920b

MUSCIDAE: STOMOXYINAE

Stomoxys calcitrans (Linnaeus)
Séguy, 1929

CALLIPHORIDAE: CHRYSOMYINAE

Cochliomyia macellaria (Fabricius)
Girault et al., 1910a
Marlatt, 1931
Roberts, 1933a, 1935

Chrysomya Robineau-Desvoidy
Roubaud, 1917

Chrysomya albiceps (Wiedemann)
Anonymous, 1925
Smit, 1929
Ullyett, 1950

Chrysomya chloropyga (Wiedemann)
Ullyett, 1950

Nasonia vitripennis (cont.)

Chrysomya rufifacies (Macquart)
Froggatt et al., 1917
Froggatt, 1919, 1921, 1922
Johnston et al., 1921
Gurney et al., 1926a
Chrysomya varipes (Macquart)
Froggatt, 1917, 1919
Johnston et al., 1921
Phormia regina (Meigen)
Girault et al., 1910a
Marlatt, 1931
Roberts, 1935
Legner (pre-publication), 1966
Phormia sordida Zetterstedt
Roubaud, 1917
Protophormia terraenovae (Robineau-Desvoidy)
Altson, 1920
Smirnov, 1934
Smirnov et al., 1934
Vladimirova et al., 1934
CALLIPHORIDAE: CALLIPHORINAE
Lucilia Robineau-Desvoidy
Roubaud, 1917
Johnston et al., 1921
Hardy, 1925
Parker et al., 1928
Myers, 1929
Cousin, 1933
Roberts, 1935
Lucilia caesar (Linnaeus)
Altson, 1920
Smirnov, 1934
Smirnov et al., 1934
Vladimirova et al., 1934
Phaenicia eximia (Wiedemann)
Roberts, 1933a
Phaenicia mexicana (Macquart)
Roberts, 1933a
Phaenicia sericata (Meigen)
Froggatt, 1919
Altson, 1920
Johnston et al., 1921
Anonymous, 1925
Smit, 1929
Cousin, 1933
Evans, 1933
Ullyett, 1950
Pimentel et al., 1963
Calliphora Robineau-Desvoidy
Parker et al., 1928

Nasonia vitripennis (cont.)

Roberts, 1935
Edwards, 1954

Calliphora stygia (Fabricius)
Froggatt et al., 1914
Froggatt, 1919
Johnston et al., 1921

Calliphora vicina Robineau-Desvoidy
Froggatt et al., 1914
Graham-Smith, 1916
Roubaud, 1917
Altson, 1920
Johnston et al., 1921
Séguy, 1929
Cousin, 1933
Smirnov et al., 1933, 1934
Smirnov, 1934
Vladimirova et al., 1934
Velthuis et al., 1965

Calliphora vicina Robineau-Desvoidy
= [*Calliphora eyrthrocephala*]
Johnston et al., 1920b

Calliphora vicina Robineau-Desvoidy
= [*Calliphora rufifacies*]
Johnston et al., 1920b

Calliphora vomitoria (Linnaeus)
Altson, 1920

Neopollenia stygia (Fabricius)
Johnston et al., 1920b

Anastellorhina augur (Fabricius)
Froggatt et al., 1914
Froggatt, 1919
Johnston et al., 1920b, 1921

Cynomyopsis cadaverina (Robineau-Desvoidy)
Girault et al., 1910a

UNDETERMINED CALLIPHORIDAE

Blow flies
Gurney et al., 1926b

SARCOPHAGIDAE: SARCOPHAGINAE

Blaesoxipha plinthopyga (Wiedemann)
Roberts, 1935

Ravinia striata Fabricius
Sychevskaia͡, 1964b

Sarcophaga Meigen
Girault et al., 1910a
Roubaud, 1917
Parker et al., 1928
Fujita, 1932
Roberts, 1933a, 1935

Organism	Fly / Reference
Nasonia vitripennis (cont.)	*Sarcophaga aurifrons* Macquart Froggatt, 1919 *Sarcophaga carnaria* (Linnaeus) Cousin, 1933 *Sarcophaga haemorrhoidalis* (Fallén) Anonymous, 1925 Smit, 1929 *Sarcophaga misera* Walker Johnston et al., 1921 *Sarcophaga parkeri* (Rohdendorf) Sychevskaīā, 1964b *Sarcophaga peregrina* Johnston et al., 1921
Nasonia ?vitripennis (Walker) (Synonym: *Nasonia brevicornis*) (pupal parasite)	CALLIPHORIDAE: CHRYSOMYINAE *Phormia regina* (Meigen) Bishopp et al., 1915b CALLIPHORIDAE: CALLIPHORINAE *Lucilia caesar* (Linnaeus) Bishopp et al., 1915b *Phaenicia sericata* (Meigen) Bishopp et al., 1915b SARCOPHAGIDAE: SARCOPHAGINAE *Sarcophaga* Meigen Bishopp et al., 1915b
Pachycrepoideus vindemmiae Rondani (Synonym: *Pachycrepoideus dubius*) (pupal parasite)	PIOPHILIDAE *Piophila casei* (Linnaeus) Simmons, 1922 Crandell, 1939 MUSCIDAE: FANNIINAE *Fannia canicularis* (Linnaeus) Steve, 1959 *Fannia scalaris* (Fabricius) Steve, 1959 MUSCIDAE: MUSCINAE *Musca domestica* Linnaeus Bartlett, 1939 McCoy, 1963 Legner et al., 1965e, 1966a MUSCIDAE: STOMOXYINAE *Haematobia* Lepeletier and Serville Fullaway, 1917b *Haematobia exigua* (de Meijere) Handschin, 1934a *Stomoxys calcitrans* (Linnaeus) Bartlett, 1939 UNDETERMINED MUSCIDAE Handschin, 1934a
Pachyneuron vindemmiae (Synonym: *Pachineuron vindemmiae*) (parasite)	DROSOPHILIDAE: DROSOPHILINAE *Drosophila funebris* (Fabricius) Milani, 1947

Pachyneuron vindemmiae (cont.)	*Drosophila immigrans* Sturtevant
	Milani, 1947
	Drosophila melanogaster Meigen
	Milani, 1947
	Drosophila similis Williston
	Milani, 1947
Prospalangia platensis Brèthes	MUSCIDAE: MUSCINAE
(parasite)	*Musca domestica* Linnaeus
	Brèthes, 1915
	MUSCIDAE: STOMOXYINAE
	Stomoxys calcitrans (Linnaeus)
	Brèthes, 1915
Spalangia Latreille	CHLOROPIDAE: OSCINELLINAE
(pupal parasite)	*Hippelates pusio* Loew
	Legner et al., 1965d
	MUSCIDAE: FANNIINAE
	Fannia canicularis (Linnaeus)
	Sychevskai͡a, 1964b
	Fannia leucosticta (Meigen)
	Sychevskai͡a, 1964b
	Fannia scalaris (Fabricius)
	Sychevskai͡a, 1964b
	MUSCIDAE: PHAONIINAE
	Ophyra anthrax (Meigen)
	Sychevskai͡a, 1964b
	MUSCIDAE: MUSCINAE
	Musca domestica Linnaeus
	Girault et al., 1910a
	Simmonds, 1922
	Vandenberg, 1930
	Legner et al., 1966a
	Musca domestica vicina Macquart
	Roy et al., 1939
	Sychevskai͡a, 1964b
	MUSCIDAE: STOMOXYINAE
	Haematobia irritans (Linnaeus)
	Wolcott, 1922
	Stomoxys calcitrans (Linnaeus)
	Roy et al., 1939
	CALLIPHORIDAE: CHRYSOMYINAE
	Chrysomya megacephala (Fabricius)
	Roy et al., 1939
	SARCOPHAGIDAE: SARCOPHAGINAE
	Sarcophaga Meigen
	Roy et al., 1939
	Sarcophaga haemorrhoidalis (Fallén)
	Sychevskai͡a, 1964b
Spalangia cameroni Perkins	MUSCIDAE: FANNIINAE
(pupal parasite)	*Fannia canicularis* (Linnaeus)
	Legner (pre-publication), 1966

Spalangia cameroni (cont.)

Fannia femoralis (Stein)
Legner et al. (pre-publication), 1965a
Legner (pre-publication), 1966
Legner et al., 1966b

Fannia leucosticta (Meigen)
Bouček, 1963

Fannia scalaris (Fabricius)
Bouček, 1963

MUSCIDAE: PHAONIINAE

Hydrotaea dentipes (Fabricius)
Bouček, 1963

Ophyra anthrax (Meigen)
Bouček, 1963

Ophyra leucostoma (Wiedemann)
Legner (pre-publication), 1966

Phaonia querceti (Bouché)
Bouček, 1963

MUSCIDAE: MUSCINAE

Musca Linnaeus
Bouček, 1963

Musca domestica Linnaeus
Legner (pre-publication), 1966
Legner et al., 1966a, 1969a
Gerling et al., 1968
Mourier et al., 1968
Legner, 1969

House fly (presumably *Musca domestica* Linnaeus)
Simmonds, 1929

Musca domestica vicina Macquart
Bouček, 1963

MUSCIDAE: STOMOXYINAE

Haematobia Lepeletier and Serville
Fullaway, 1917b

Haematobia exigua (de Meijere)
Bouček, 1963

Haematobia irritans (Linnaeus)
Bouček, 1963

Stomoxys calcitrans (Linnaeus)
Legner (pre-publication), 1966
Legner et al., 1969a

SARCOPHAGIDAE: SARCOPHAGINAE

Ravinia striata Fabricius
Bouček, 1963

Sarcophaga haemorrhoidalis (Fallén)
Bouček, 1963

UNDETERMINED SARCOPHAGIDAE
Simmonds, 1929

Spalangia chontalensis (Cameron) (parasite)	MUSCIDAE: MUSCINAE
	Musca domestica Linnaeus
	Legner et al., 1966a
Spalangia drosophilae Ashmead (pupal parasite)	DROSOPHILIDAE: DROSOPHILINAE
	Drosophila Fallén
	Ashmead, 1887, 1900
	Riley et al., 1891
	Richardson, 1913a
	Viereck, 1916
	Drosophila melanogaster Meigen
	Simmonds, 1944
	Bouček, 1963
	CHLOROPIDAE: OSCINELLINAE
	Hippelates Loew
	Legner et al., 1964
	Hippelates collusor (Townsend)
	Bay et al., 1963
	Bay, unpublished
	Legner, 1967, 1969
	Hippelates pusio Loew
	Legner et al., 1965d
	Oscinella frit (Linnaeus)
	Bay et al., 1963
	MUSCIDAE: MUSCINAE
	Orthellia Robineau-Desvoidy
	Lindquist, 1936
	Musca domestica Linnaeus
	Lindquist, 1936
	MUSCIDAE: STOMOXYINAE
	Haematobia irritans (Linnaeus)
	Lindquist, 1936
	Bartlett, 1939
	Bouček, 1963
	SARCOPHAGIDAE: SARCOPHAGINAE
	Sarcophaga Meigen
	Lindquist, 1936
Spalangia endius Walker (pupal parasite)	MUSCIDAE: FANNIINAE
	Fannia canicularis (Linnaeus)
	Legner (pre-publication), 1966
	Fannia femoralis (Stein)
	Legner et al. (pre-publication), 1966
	Legner et al., 1966b
	Fannia leucosticta (Meigen)
	Bouček, 1963
	Fannia scalaris (Fabricius)
	Bouček, 1963
	MUSCIDAE: PHAONIINAE
	Hydrotaea australis Malloch
	Bouček, 1963

Organism	Fly / Reference
Spalangia endius (cont.)	*Ophyra leucostoma* (Wiedemann) Legner et al. (pre-publication), 1965a Legner (pre-publication), 1966 Legner et al., 1966b *Muscina stabulans* (Fallén) Legner (pre-publication), 1966 MUSCIDAE: MUSCINAE *Orthellia* Robineau-Desvoidy Bouček, 1963 *Musca domestica* Linnaeus Legner et al. (pre-publication), 1965a Legner et al., 1965c Legner (pre-publication), 1966 Legner et al., 1966a *Musca domestica vicina* Macquart Bouček, 1963 *Musca sorbens* Wiedemann Bouček, 1963 MUSCIDAE: STOMOXYINAE *Haematobia exigua* (de Meijere) Bouček, 1963 *Haematobia irritans* (Linnaeus) Bouček, 1963 *Stomoxys calcitrans* (Linnaeus) Legner (pre-publication), 1966 CALLIPHORIDAE: CHRYSOMYINAE *Chrysomya albiceps* (Wiedemann) Bouček, 1963 *Chrysomya megacephala* (Fabricius) Bouček, 1963 SARCOPHAGIDAE: SARCOPHAGINAE *Blaesoxipha impar* (Aldrich) Bouček, 1963 *Ravinia effrenata* (Walker) Bouček, 1963 *Ravinia striata* Fabricius Bouček, 1963 *Ravinia sueta* (Wulp) Bouček, 1963 *Sarcophaga haemorrhoidalis* (Fallén) Bouček, 1963
Spalangia ?endius Walker (pupal parasite)	MUSCIDAE: PHAONIINAE *Phaonia querceti* (Bouché) Bouček, 1963
Spalangia erythromera brachyceps Bouček (pupal parasite)	ANTHOMYIIDAE: ANTHOMYIINAE *Hylemya cinerea* Fallén Bouček, 1963

Spalangia erythromera (cont.)	*Hylemya platura* (Meigen) Bouček, 1963 *Pegomya* Robineau-Desvoidy Bouček, 1963
Spalangia haematobiae Ashmead (parasite)	MUSCIDAE: STOMOXYINAE *Haematobia irritans* (Linnaeus) Riley et al., 1891 Ashmead, 1894 von Dalla Torre, 1898 Ashmead, 1900 Schmiedeknecht, 1909 Viereck, 1909, 1916 Richardson, 1913b Bartlett, 1939 Bouček, 1963
Spalangia hirta Haliday (parasite)	MUSCIDAE: MUSCINAE *Musca domestica* Linnaeus Richardson, 1913a
Spalangia longepetiolata Bouček	MUSCIDAE: MUSCINAE *Musca domestica* Linnaeus Legner et al., 1969a MUSCIDAE: STOMOXYINAE *Stomoxys calcitrans* (Linnaeus) Legner et al., 1969a
Spalangia muscae Howard (pupal parasite)	MUSCIDAE: MUSCINAE *Musca domestica* Linnaeus Howard, 1911 Girault, 1916 Peck, 1951 MUSCIDAE: STOMOXYINAE *Stomoxys calcitrans* (Linnaeus) Bishopp, 1913a Girault, 1921
Spalangia muscidarum Richardson (parasite)	MUSCIDAE: MUSCINAE *Orthellia caesarion* (Meigen) Pinkus, 1913 *Musca* Linnaeus Johnston et al., 1921 *Musca domestica* Linnaeus Pinkus, 1913 Richardson, 1913a, 1913b Johnston et al., 1920b, 1921 Froggatt, 1922 Bartlett, 1939 Simmonds, 1958 House fly (presumably *Musca domestica* Linnaeus) Simmonds, 1940 *Musca hilli* Johnston and Bancroft Johnston et al., 1920b

Spalangia muscidarum (cont.)

Musca sorbens sorbens Wiedemann
Johnston et al., 1920b
Musca terraereginae Johnston and Bancroft
Johnston et al., 1920b
MUSCIDAE: STOMOXYINAE
Haematobia irritans (Linnaeus)
Pinkus, 1913
Stomoxys calcitrans (Linnaeus)
Pinkus, 1913
Richardson, 1913b
Johnston et al., 1920b, 1921
Bartlett, 1939
CALLIPHORIDAE: CHRYSOMYINAE
Chrysomya megacephala (Fabricius)
Johnston et al., 1920b, 1921
Chrysomya rufifacies (Macquart)
Johnston et al., 1920b, 1921
Chrysomya varipes (Macquart)
Johnston et al., 1920b, 1921
CALLIPHORIDAE: CALLIPHORINAE
Lucilia Robineau-Desvoidy
Johnston et al., 1921
Calliphora stygia (Fabricius)
Johnston et al., 1921
Neopollenia stygia (Fabricius)
Johnston et al., 1920b
Anastellorhina augur (Fabricius)
Johnston et al., 1920b, 1921
SARCOPHAGIDAE: SARCOPHAGINAE
Helicobia quadrisetosa Coquillett
Pinkus, 1913
Sarcophaga Meigen
Johnston et al., 1920b
Sarcophaga aurifrons Macquart
Johnston et al., 1920
Sarcophaga ?frontalis Doleschall
Johnston et al., 1920b
Sarcophaga misera Walker
Johnston et al., 1920b, 1921

Spalangia muscophaga Girault .... (pupal parasite)

MUSCIDAE: MUSCINAE
Musca gibsoni Patton and Cragg
Bouček, 1963

Spalangia nigra Latreille .... (parasite)

PHORIDAE: METOPININAE
Megaselia ?iroquoiana (Malloch)
Roberts, 1935
MUSCIDAE: MUSCINAE
Musca autumnalis De Geer
Turner et al., 1968

Organism	Fly host / references
Spalangia nigra (cont.)	*Musca domestica* Linnaeus Richardson, 1913b Parker, 1924 Parker et al., 1928 Sanders, 1942 Legner, 1969 MUSCIDAE: STOMOXYINAE *Haematobia irritans* (Linnaeus) Bouček, 1963 *Stomoxys calcitrans* (Linnaeus) Bishopp, 1913b, 1939 Bouček, 1963 CALLIPHORIDAE: CALLIPHORINAE *Phaenicia sericata* (Meigen) Roberts, 1935 SARCOPHAGIDAE: MILTOGRAMMINAE *Ravinia querula* (Walker) Turner et al., 1968 SARCOPHAGIDAE: SARCOPHAGINAE *Sarcophaga* Meigen Parker et al., 1928
Spalangia nigripes Curtis (pupal parasite)	MUSCIDAE: FANNIINAE *Fannia leucosticta* (Meigen) Bouček, 1963 *Fannia scalaris* (Fabricius) Bouček, 1963 MUSCIDAE: MUSCINAE *Musca domestica* Linnaeus Bouček, 1963 Legner et al., 1969a *Musca domestica vicina* Macquart Bouček, 1963 MUSCIDAE: STOMOXYINAE *Stomoxys calcitrans* (Linnaeus) Legner et al., 1969a CALLIPHORIDAE: CALLIPHORINAE *Lucilia* Robineau-Desvoidy Bouček, 1963
Spalangia nigroaenea Curtis (pupal parasite)	PHORIDAE: METOPININAE *Megaselia iroquoiana* (Malloch) Bouček, 1963 MUSCIDAE: FANNIINAE *Fannia canicularis* (Linnaeus) Legner (pre-publication), 1966 *Fannia femoralis* (Stein) Legner et al. (pre-publication), 1965a Legner (pre-publication), 1966 Legner et al., 1966b

Spalangia nigroaenea (cont.)	MUSCIDAE: PHAONIINAE *Ophyra leucostoma* (Wiedemann) Legner et al. (pre-publication), 1965a Legner (pre-publication), 1966 Legner et al., 1966b *Phaonia corbetti* Malloch Bouček, 1963 *Phaonia querceti* (Bouché) Bouček, 1963 MUSCIDAE: MUSCINAE *Musca domestica* Linnaeus Legner et al., 1965c, 1966a, 1969a Legner (pre-publication), 1966 *Musca domestica vicina* Macquart Sychevskaia͡, 1964b *Musca fergusoni* Johnston and Bancroft Bouček, 1963 *Musca hilli* Johnston and Bancroft Bouček, 1963 *Musca sorbens* Wiedemann Bouček, 1963 *Musca sorbens sorbens* Wiedemann Bouček, 1963 *Musca terraereginae* Johnston and Bancroft Bouček, 1963 MUSCIDAE: STOMOXYINAE *Haematobia exigua* (de Meijere) Bouček, 1963 *Haematobia irritans* (Linnaeus) Bouček, 1963 *Stomoxys calcitrans* (Linnaeus) Legner (pre-publication), 1966 Legner et al., 1969a CALLIPHORIDAE: CHRYSOMYINAE *Chrysomya megacephala* (Fabricius) Bouček, 1963 *Chrysomya rufifacies* (Macquart) Bouček, 1963 *Chrysomya varipes* (Macquart) Bouček, 1963 CALLIPHORIDAE: CALLIPHORINAE *Lucilia* Robineau-Desvoidy Bouček, 1963 *Phaenicia sericata* (Meigen) Bouček, 1963 *Neopollenia stygia* (Fabricius) Bouček, 1963

Spalangia nigroaenea (cont.)	*Anastellorhina augur* (Fabricius) Bouček, 1963 SARCOPHAGIDAE: SARCOPHAGINAE *Ravinia striata* Fabricius Bouček, 1963 *Sarcophaga dux* Thompson Bouček, 1963 *Sarcophaga frontalis* Doleschall Bouček, 1963 *Sarcophaga impatiens* (Walker) Bouček, 1963
Spalangia orientalis Graham (pupal parasite)	MUSCIDAE: PHAONIINAE *Hydrotaea australis* Malloch Mackerras, 1932 MUSCIDAE: STOMOXYINAE *Haematobia exigua* (de Meijere) Handschin, 1932, 1934a, 1934b UNDETERMINED MUSCIDAE Handschin, 1932, 1934b
Spalangia philippinensis Fullaway (pupal parasite)	MUSCIDAE: MUSCINAE *Musca domestica* Linnaeus Bartlett, 1939 House fly (presumably *Musca domestica* Linnaeus) Fullaway, 1917a MUSCIDAE: STOMOXYINAE *Haematobia* Lepeletier and Serville Fullaway, 1917b *Haematobia irritans* (Linnaeus) Bartlett, 1939 *Stomoxys calcitrans* (Linnaeus) Bartlett, 1939 SARCOPHAGIDAE: SARCOPHAGINAE *Tricharaea* Thomson Bartlett, 1939
Spalangia platensis (Brèthes) (pupal parasite)	MUSCIDAE: MUSCINAE *Musca domestica* Linnaeus Bouček, 1963 MUSCIDAE: STOMOXYINAE *Stomoxys calcitrans* (Linnaeus) Bouček, 1963
Spalangia rugosicollis Ashmead (pupal parasite)	ANTHOMYIIDAE: ANTHOMYIINAE *Hylemya antiqua* (Meigen) Perron, 1954b MUSCIDAE: STOMOXYINAE *Stomoxys calcitrans* (Linnaeus) Girault, 1920 Peck, 1951

Spalangia rugulosa Foerster .... (pupal parasite)
- MUSCIDAE: PHAONIINAE
- *Muscina stabulans* (Fallén)
  - Bouček, 1963

Spalangia simplex Perkins .... (pupal parasite)
- MUSCIDAE: FANNIINAE
- *Fannia femoralis* (Stein)
  - Legner et al. (pre-publication), 1965a
- MUSCIDAE: MUSCINAE
- *Musca domestica* Linnaeus
  - Legner et al., 1966a

Spalangia stomoxysiae Girault .... (Synonym: *Spalangia muscidarum stomoxysiae*) (pupal parasite)
- MUSCIDAE: PHAONIINAE
- *Ophyra anthrax* (Meigen)
  - Bouček, 1963
- MUSCIDAE: MUSCINAE
- *Orthellia* Robineau-Desvoidy
  - Lindquist, 1936
  - Peck, 1951
- *Musca domestica* Linnaeus
  - Lindquist, 1936
  - Legner et al., 1965c
- *Musca domestica vicina* Macquart
  - Sychevskaîâ, 1964b
- MUSCIDAE: STOMOXYINAE
- *Haematobia irritans* (Linnaeus)
  - Lindquist, 1936
  - Peck, 1951
- *Stomoxys calcitrans* (Linnaeus)
  - Girault, 1916, 1920
  - Peck, 1951
- SARCOPHAGIDAE: SARCOPHAGINAE
- *Blaesoxipha impar* (Aldrich)
  - Lindquist, 1936
  - Peck, 1951
- *Ravinia effrenata* (Walker)
  - Lindquist, 1936
  - Peck, 1951
- *Ravinia sueta* (Wulp)
  - Lindquist, 1936
  - Peck, 1951
- *Sarcophaga* Meigen
  - Lindquist, 1936

Spalangia sundaica Graham .... (pupal parasite)
- MUSCIDAE: STOMOXYINAE
- *Haematobia exigua* (de Meijere)
  - Handschin, 1932, 1934a, 1934b
  - Lever, 1936
- *Stomoxys* Geoffroy
  - Handschin, 1932, 1934a
- UNDETERMINED MUSCIDAE
  - Handschin, 1932, 1934a, 1934b

Organism	Fly associations
Sphegigaster Spinola	MUSCIDAE: MUSCINAE *Musca domestica* Linnaeus Legner et al., 1969a MUSCIDAE: STOMOXYINAE *Stomoxys calcitrans* (Linnaeus) Legner et al., 1969a
Stenomolina muscarum (Linnaeus) (Synonym: *Pteromalus muscarum*) (parasite)	MUSCIDAE: PHAONIINAE *Muscina stabulans* (Fallén) Ratzeburg, 1844
UNDETERMINED PTEROMALIDAE (parasites)	CHLOROPIDAE: OSCINELLINAE *Hippelates pusio* Loew Legner et al., 1965d MUSCIDAE: MUSCINAE *Musca domestica* Linnaeus Bishopp, 1913a MUSCIDAE: STOMOXYINAE *Stomoxys calcitrans* (Linnaeus) Pinkus, 1913 Bishopp, 1913a SARCOPHAGIDAE: SARCOPHAGINAE *Sarcophaga* Meigen Sanders et al., 1966
ENCYRTIDAE	
Australencyrtus giraulti Johnston and Tiegs (parasite)	CALLIPHORIDAE: CHRYSOMYINAE *Chrysomya megacephala* (Fabricius) Johnston et al., 1921 *Chrysomya rufifacies* (Macquart) Johnston et al., 1921 *Chrysomya varipes* (Macquart) Johnston et al., 1921 CALLIPHORIDAE: CALLIPHORINAE *Lucilia* Robineau-Desvoidy Johnston et al., 1921 *Phaenicia sericata* (Meigen) Johnston et al., 1921 *Anastellorhina augur* (Fabricius) Johnston et al., 1921 SARCOPHAGIDAE: SARCOPHAGINAE *Sarcophaga* Meigen Johnston et al., 1921
Cerchysius lyperosae Ferrière (parasite)	MUSCIDAE: STOMOXYINAE *Haematobia exigua* (de Meijere) Handschin, 1934a UNDETERMINED MUSCIDAE Handschin, 1934a
Homalotylus eitelweinii Ratzeberg	PHORIDAE: METOPININAE *Megaselia fasciata* (Fallén) du Buysson, 1921

Organism	Fly association
Ooencyrtus submetallicus (Howard) (parasite)	CHLOROPIDAE: OSCINELLINAE *Hippelates pusio* Loew Legner et al., 1965b, 1965d
Stenoterys fulvoventralis Dodd (parasite)	CALLIPHORIDAE: CHRYSOMYINAE *Chrysomya rufifacies* (Macquart) Newman et al., 1930
Tachinaephagus Ashmead (parasite)	CALLIPHORIDAE: CALLIPHORINAE *Lucilia* Robineau-Desvoidy Hardy, 1924
Tachinaephalus giraulti Johnston and Tiegs (parasite)	MUSCIDAE: MUSCINAE *Musca domestica* Linnaeus Legner et al., 1966a MUSCIDAE: STOMOXYINAE *Haematobia exigua* (de Meijere) Handschin, 1934a UNDETERMINED MUSCIDAE Handschin, 1934a
EUPELMIDAE	
Mesocomys pulchriceps Cameron (parasite)	CALLIPHORIDAE: CHRYSOMYINAE *Chrysomya chloropyga* (Wiedemann) Anonymous, 1925
EULOPHIDAE	
Synthomosphyrum albiclavus Kerrich (parasite)	MUSCIDAE: MUSCINAE *Musca domestica* Linnaeus Saunders, 1964 CALLIPHORIDAE: CALLIPHORINAE *Lucilia* Robineau-Desvoidy Saunders, 1964 *Calliphora* Robineau-Desvoidy Saunders, 1964 SARCOPHAGIDAE: SARCOPHAGINAE *Sarcophaga* Meigen Saunders, 1964
Syntomosphyrum glossinae Waterston (parasite)	MUSCIDAE: MUSCINAE *Musca domestica* Linnaeus Saunders, 1964 *Musca nebulo* Fabricius Lamborn, 1925 CALLIPHORIDAE: CHRYSOMYINAE *Chrysomya albiceps* (Wiedemann) Nash, 1933 *Chrysomya chloropyga* (Wiedemann) Nash, 1933 *Chrysomya marginalis* (Wiedemann) Nash, 1933 *Chrysomya putoria* (Wiedemann) Lamborn, 1925 Nash, 1933 CALLIPHORIDAE: CALLIPHORINAE *Lucilia* Robineau-Desvoidy Saunders, 1964

Syntomosphyrum glossinae (cont.)	*Calliphora* Robineau-Desvoidy Saunders, 1964 SARCOPHAGIDAE: SARCOPHAGINAE *Sarcophaga* Meigen Lamborn, 1925 Saunders, 1964 *Sarcophaga haemorrhoidalis* (Fallén) Nash, 1933
Trichospilus pupivora Ferrière (parasite)	MUSCIDAE: STOMOXYINAE *Haematobia exigua* (de Meijere) Handschin, 1934a UNDETERMINED MUSCIDAE Handschin, 1934a
DIAPRIIDAE	
Diapria conica (Fabricius) (parasite)	SYRPHIDAE: MILESIINAE *Eristalis tenax* (Linnaeus) Sychevskai͡a, 1964b SARCOPHAGIDAE: SARCOPHAGINAE *Ravinia striata* Fabricius Sychevskai͡a, 1964b
Hemilexomyia abrupta Dodd (parasite)	MUSCIDAE: PHAONIINAE *Ophyra nigra* Wiedemann Dodd, 1920 Johnston et al., 1921 MUSCIDAE: MUSCINAE *Musca domestica* Linnaeus Johnston et al., 1921 CALLIPHORIDAE: CALLIPHORINAE *Phaenicia cuprina* Wiedemann Dodd, 1920 *Calliphora stygia* (Fabricius) Dodd, 1920 Froggatt, 1921, 1922 Johnston et al., 1921
Paraspilomicrus Johnston and Tiegs (parasite)	CALLIPHORIDAE: CALLIPHORINAE *Lucilia* Robineau-Desvoidy Hardy, 1924
Paraspilomicrus froggatti Johnston and Tiegs (parasite)	CALLIPHORIDAE: CALLIPHORINAE *Lucilia* Robineau-Desvoidy Johnston et al., 1921
Phaenopria Ashmead (parasite)	MUSCIDAE: PHAONIINAE *Hydrotaea australis* Malloch Mackerras, 1932 MUSCIDAE: STOMOXYINAE *Haematobia exigua* (de Meijere) Handschin, 1932 UNDETERMINED MUSCIDAE Handschin, 1934a
Phaenopria fimicola Ferrière (parasite)	MUSCIDAE: STOMOXYINAE *Haematobia exigua* (de Meijere) Handschin, 1934a

Organism	Fly / Reference
Phaenopria occidentalis Fouts (parasite)	CHLOROPIDAE: OSCINELLINAE *Hippelates collusor* (Townsend) Bay et al., 1963 Bay, unpublished Legner et al., 1966 Legner et al., 1969a
Trichopria Ashmead (Synonym: *Ashmeadopria*) (parasite)	MUSCIDAE: FANNIINAE *Fannia canicularis* (Linnaeus) Legner (pre-publication), 1966 *Fannia femoralis* (Stein) Legner (pre-publication), 1966 MUSCIDAE: MUSCINAE *Musca domestica* Linnaeus Bartlett, 1939 Legner et al., 1965c, 1966, 1969a Legner (pre-publication), 1966 MUSCIDAE: STOMOXYINAE *Haematobia exigua* (de Meijere) Handschin, 1932 *Stomoxys calcitrans* (Linnaeus) Bartlett, 1939 Legner et al., 1969a
Trichopria commoda Muesebeck (parasite)	MUSCIDAE: MUSCINAE *Musca domestica* Linnaeus Muesebeck, 1961
Trichopria hirticollis (Ashmead) (parasite)	UNDETERMINED CALLIPHORIDAE Blow flies Roberts, 1935a SARCOPHAGIDAE: SARCOPHAGINAE *Blaesoxipha plinthopyga* (Wiedemann) Roberts, 1935a
FORMICIDAE	
Camponotus pennsylvanicus De Geer (predator)	MUSCIDAE: FANNIINAE *Fannia canicularis* (Linnaeus) Steve, 1959
Crematogaster lineolata (Say) (predator)	CALLIPHORIDAE: CHRYSOMYINAE *Cochliomyia hominivorax* (Coquerel) Lindquist, 1942
Eciton Latreille (predator)	CALLIPHORIDAE: CHRYSOMYINAE *Cochliomyia hominivorax* (Coquerel) Strong, 1938
Forelius foetidus (Buckley) (Synonym: *Forelius maccooki*) (predator)	CALLIPHORIDAE: CHRYSOMYINAE *Cochliomyia hominivorax* (Coquerel) Lindquist, 1942
Iridomyrmex humilis (Mayr)	DROSOPHILIDAE: DROSOPHILINAE *Drosophila funebris* (Fabricius) Pavan, 1952 *Drosophila virilis* Sturtevant Pavan, 1952

Organism	Fly association
	MUSCIDAE: MUSCINAE *Musca domestica* Linnaeus Pavan, 1952
Labidus coecus (Latreille) (Synonym: *Eciton coecum*) (predator)	CALLIPHORIDAE: CHRYSOMYINAE *Cochliomyia hominivorax* (Coquerel) Lindquist, 1942
	UNDETERMINED CALLIPHORIDAE Blow flies Lindquist, 1942
Monomorium pharaonis (Linnaeus) (Synonym: *Tetramorium pharaonis*) (predator)	CHLOROPIDAE: OSCINELLINAE *Hippelates pusio* Loew Legner et al., 1965d
	UNDETERMINED MUSCIDAE Handschin, 1934a
Paratrechina longicornis (Latreille) (predator)	MUSCIDAE: MUSCINAE *Musca domestica* Linnaeus Pimentel, 1955
	CALLIPHORIDAE: CHRYSOMYINAE *Cochliomyia macellaria* (Fabricius) Pimentel, 1955
	CALLIPHORIDAE: CALLIPHORINAE *Phaenicia* Robineau-Desvoidy Pimentel, 1955
	SARCOPHAGIDAE: SARCOPHAGINAE *Sarcophaga* Meigen Pimentel, 1955
Pheidole (predator)	CALLIPHORIDAE: CHRYSOMYINAE *Cochliomyia hominivorax* (Coquerel) Strong, 1938 Lindquist, 1942
Pheidole megacephala (Fabricius) (predator)	OTITIDAE: ULIDIINAE *Euxesta annonae* (Fabricius) Illingworth, 1923a
	MILICHIIDAE: MILICHIINAE *Milichiella lacteipennis* (Loew) Illingworth, 1923a
	MUSCIDAE: FANNIINAE *Fannia pusio* (Wiedemann) Illingworth, 1923a
	MUSCIDAE: PHAONIINAE *Ophyra nigra* Wiedemann Illingworth, 1923a
	MUSCIDAE: MUSCINAE *Musca domestica* Linnaeus Bridwell, 1918 Illingworth, 1923a Phillips, 1934 Simmonds, 1958

Pheidole megacephala (cont.)

House fly (presumably *Musca domestica* Linnaeus)
Illingworth, 1915

MUSCIDAE: STOMOXYINAE
Stomoxys calcitrans (Linnaeus)
Illingworth, 1923a

CALLIPHORIDAE: CALLIPHORINAE
Phaenicia sericata (Meigen)
Illingworth, 1923a

UNDETERMINED CALLIPHORIDAE
Blow flies
Phillips, 1934

SARCOPHAGIDAE: SARCOPHAGINAE
Sarcophaga fuscicauda Böttcher
Illingworth, 1923a

Pheidole morrisii var. *impexa* Wheeler (predator) ....

CALLIPHORIDAE: CHRYSOMYINAE
Cochliomyia hominivorax (Coquerel)
Lindquist, 1942

Pheidole subarmata borinquenensis Wheeler (predator) ....

MUSCIDAE: MUSCINAE
Musca domestica Linnaeus
Pimentel, 1955

Pheidole oceanica Mayr (predator) ....

MUSCIDAE: STOMOXYINAE
Haematobia exigua (de Meijere)
Lever, 1936

Pheidologeton affinis (Jerd.) ....

MUSCIDAE: MUSCINAE
Musca domestica Linnaeus
Pimentel et al., 1969

Pogonomyrmex barbatus var. *fuscatus* Emery (predator) ....

CALLIPHORIDAE: CHRYSOMYINAE
Cochliomyia hominivorax (Coquerel)
Lindquist, 1942

Pogonomyrmex barbatus var. *molefaciens* (Buckley) (predator) ....

CALLIPHORIDAE: CHRYSOMYINAE
Cochliomyia hominivorax (Coquerel)
Lindquist, 1942

Ponera perkinsi Forel (predator) ....

OTITIDAE: ULIDIINAE
Euxesta annonae (Fabricius)
Illingworth, 1923a

MILICHIIDAE: MILICHIINAE
Milichiella lacteipennis (Loew)
Illingworth, 1923a

MUSCIDAE: FANNIINAE
Fannia pusio (Wiedemann)
Illingworth, 1923a

MUSCIDAE: PHAONIINAE
Ophyra nigra Wiedemann
Illingworth, 1923a

MUSCIDAE: MUSCINAE
Musca domestica Linnaeus
Illingworth, 1923a

Ponera perkinsi (cont.)	MUSCIDAE: STOMOXYINAE *Stomoxys calcitrans* (Linnaeus) Illingworth, 1923a CALLIPHORIDAE: CALLIPHORINAE *Phaenicia sericata* (Meigen) Illingworth, 1923a SARCOPHAGIDAE: SARCOPHAGINAE *Sarcophaga fuscicauda* Böttcher Illingworth, 1923a
Solenopsis corticalis Forel (predator)	MUSCIDAE: MUSCINAE *Musca domestica* Linnaeus Pimentel, 1955 CALLIPHORIDAE: CALLIPHORINAE *Phaenicia* Robineau-Desvoidy Pimentel, 1955
Solenopsis geminata (Fabricius) (predator)	CHLOROPIDAE: OSCINELLINAE *Hippelates pusio* Loew Legner et al., 1965d MUSCIDAE: MUSCINAE *Musca domestica* Linnaeus Pimentel, 1955 CALLIPHORIDAE: CHRYSOMYINAE *Cochliomyia hominivorax* (Coquerel) Lindquist, 1942 *Cochliomyia macellaria* (Fabricius) Pimentel, 1955 CALLIPHORIDAE: CALLIPHORINAE *Phaenicia* Robineau-Desvoidy Pimentel, 1955 SARCOPHAGIDAE: SARCOPHAGINAE *Sarcophaga* Meigen Pimentel, 1955
Tapinoma melanocephalum (Fabricius) (predator)	CHLOROPIDAE: OSCINELLINAE *Hippelates pusio* Loew Legner et al., 1965d MUSCIDAE: MUSCINAE *Musca domestica* Linnaeus Pimentel, 1955 CALLIPHORIDAE: CHRYSOMYINAE *Cochliomyia* Townsend Pimentel, 1955
Tetramorium guineense (Fabricius) (predator)	CHLOROPIDAE: OSCINELLINAE *Hippelates pusio* Loew Legner et al., 1965d
Wasmamia auropunctata (Roger) (predator)	CHLOROPIDAE: OSCINELLINAE *Hippelates pusio* Loew Legner et al., 1965d

Organism	Fly association
UNDETERMINED FORMICIDAE (predators)	CALLIPHORIDAE: CALLIPHORINAE *Lucilia* Robineau-Desvoidy Pavlovskiĭ, 1921
SPHECIDAE	
Bembix Fabricius (Synonym: *Bembex*) (predator)	MUSCIDAE: MUSCINAE *Musca* Linnaeus Peckham et al., 1905
Bembix lunata (Fabricius) (Synonym: *Bembex lunata*) (predator)	MUSCIDAE: MUSCINAE *Musca* Linnaeus Ramakrishna, Ayyar, 1920 MUSCIDAE: STOMOXYINAE *Haematobia* Lepeletier and Serville Ramakrishna, Ayyar, 1920 *Stomoxys* Geoffroy Ramakrishna, Ayyar, 1920
Bembix oculata Latreille (Synonym: *Bembex oculata*) (predator)	MUSCIDAE: STOMOXYINAE *Stomoxys calcitrans* (Linnaeus) Farre, 1914
Bembex olivata Dahlbom (Synonym: *Bembex olivata*) (predator)	CALLIPHORIDAE: CALLIPHORINAE *Phaenicia sericata* (Meigen) Ullyett et al., 1941 CALLIPHORIDAE: CALLIPHORINAE *Auchmeromyia luteola* (Fabricius) Roubaud, 1913
Bembix spinolae (Lepeletier) (predator)	MUSCIDAE: MUSCINAE *Orthellia caesarion* (Meigen) Parker, 1917 *Musca domestica* Linnaeus Parker, 1917 SARCOPHAGIDAE: SARCOPHAGINAE *Sarcophaga* Meigen Parker, 1917
Crabro advenus Smith	ANTHOMYIIDAE: ANTHOMYIINAE *Pegomya finitima* Stein Kurczewski et al., 1968 MUSCIDAE: COENOSIINAE *Coenosia tigrina* (Fabricius) Kurczewski et al., 1968 MUSCIDAE: FANNIINAE *Fannia scalaris* (Fabricius) Kurczewski et al., 1968 MUSCIDAE: PHAONIINAE *Muscina assimilis** (Fallén) Kurczewski et al., 1968 MUSCIDAE: MUSCINAE *Musca autumnalis* (De Geer) Kurczewski et al., 1968

* Authors: *Musca* [*sic*] *assimilis* (Fallén).

Organism	Fly associations
	CALLIPHORIDAE: CALLIPHORINAE *Phaenicia sericata* (Meigen) Kurczewski et al., 1968 CALLIPHORIDAE: POLLENIINAE *Pollenia rudis* (Fabricius) Kurczewski et al., 1968 SARCOPHAGIDAE: SARCOPHAGINAE *Sarcophaga scoparia* Pandellé Kurczewski et al., 1968
Crabro alpinus Imhof (Synonym: *Thyreopus alpinus*) (predator)	MUSCIDAE: STOMOXYINAE *Lyperosiops stimulans* (Meigen) Vergne, 1931
Crabro cavifrons Thomson (Synonym: *Clytochrysus cavifrons*) (predator)	SYRPHIDAE: SYRPHINAE *Syrphus albimanus* Fabricius Hamm et al., 1926 *Syrphus peltatus* (Meigen) Hamm et al., 1926 CALLIPHORIDAE: CALLIPHORINAE *Lucilia caesar* (Linnaeus) Hamm et al., 1926 *Calliphora vicina* Robineau-Desvoidy Hamm et al., 1926 SARCOPHAGIDAE: SARCOPHAGINAE *Sarcophaga* Meigen Hamm et al., 1926
Crabro clypeatus (Schreber) (Synonym: *Thyreus clypeatus*) (predator)	ANTHOMYIIDAE: ANTHOMYIINAE *Anthomyia* Meigen Hamm et al., 1926 MUSCIDAE: PHAONIINAE *Muscina* Robineau-Desvoidy Hamm et al., 1926 MUSCIDAE: STOMOXYINAE *Stomoxys* Geoffroy Hamm et al., 1926
Crabro cribraria (Linnaeus) (Synonym: *Thyreopus cribrarius*) (predator)	ANTHOMYIIDAE: ANTHOMYIINAE *Hylemya strigosa* (Fabricius) Hamm et al., 1926 MUSCIDAE: MUSCINAE *Morellia hortorum* (Fallén) Hamm et al., 1926 MUSCIDAE: STOMOXYINAE *Stomoxys calcitrans* (Linnaeus) Hamm et al., 1926 CALLIPHORIDAE: CALLIPHORINAE *Calliphora vicina* Robineau-Desvoidy Hamm et al., 1926 *Calliphora vomitoria* (Linnaeus) Hamm et al., 1926 *Onesis cognata* Meigen Hamm et al., 1926

Crabro cribraria (cont.)	CALLIPHORIDAE: POLLENIINAE *Pollenia rudis* (Fabricius) Hamm et al., 1926
Crabro elongatus (Provancher) .... (Synonym: *Crossocerus elongatulus*) (predator)	PHORIDAE: METOPININAE *Megaselia pulicaria* (Fallén) Hamm et al., 1926 DROSOPHILIDAE: DROSOPHILINAE *Scaptomyza graminum* (Fallén) Hamm et al., 1926
Crabro leucostoma (Linnaeus) .... (Synonym: *Blepharipus leucostomus*) (predator)	EMPIDIDAE: EMPIDINAE *Hilara litorea* Fallén Hamm et al., 1926 SEPSIDAE *Sepsis cynipsea* (Linnaeus) Hamm et al., 1926 *Sepsis nigripes* Meigen Hamm et al., 1926 ANTHOMYIIDAE: ANTHOMYIINAE *Hylemya* Robineau-Desvoidy Hamm et al., 1926 *Hylemya fugax* (Meigen) Hamm et al., 1926 *Hylemya radicum* (Linnaeus) Hamm et al., 1926 *Anthomyia pluvialis* (Linnaeus) Hamm et al., 1926 MUSCIDAE: COENOSIINAE *Coenosia lineatipes* (Zetterstedt) Hamm et al., 1926 *Coenosia sexnotata* Meigen Hamm et al., 1926 MUSCIDAE: MYDAEINAE *Helina duplicata* (Meigen) Hamm et al., 1926 MUSCIDAE: FANNIINAE *Fannia armata* (Meigen) Hamm et al., 1926 *Fannia kowarzi* (Verrall) Hamm et al., 1926 *Fannia manicata* (Meigen) Hamm et al., 1926 *Fannia polychaeta* Stein Hamm et al., 1926 *Fannia similis* Stein Hamm et al., 1926 MUSCIDAE: PHAONIINAE *Hydrotaea albipuncta* (Zetterstedt) Hamm et al., 1926 *Hydrotaea armipes* (Fallén) Hamm et al., 1926

Organism	Fly associates
Crabro leucostoma (cont.)	*Hydrotaea irritans* (Fallén) Hamm et al., 1926
	Hydrotaea meteorica (Linnaeus) Hamm et al., 1926
	Ophyra leucostoma (Wiedemann) Hamm et al., 1926
	Phaonia Robineau-Desvoidy Hamm et al., 1926
	MUSCIDAE: MUSCINAE
	Musca corvina Fabricius Hamm et al., 1926
	CALLIPHORIDAE: POLLENIINAE
	Pollenia rudis (Fabricius) Hamm et al., 1926
Crabro palmarius (Schreber) (Synonym: *Crossocerus palmarius*) (predator)	DROSOPHILIDAE: DROSOPHILINAE
	Scaptomyza graminum (Fallén) Hamm et al., 1926
	MUSCIDAE: COENOSIINAE
	Coenosia tricolor (Zetterstedt) Hamm et al., 1926
	MUSCIDAE: MYDAEINAE
	Helina duplicata (Meigen) Hamm et al., 1926
	Helina impuncta (Fallén) Hamm et al., 1926
Crabro peltarius (Schreber) (Synonym: *Thyreopus peltarius*) (predator)	ANTHOMYIIDAE: ANTHOMYIINAE
	Hylemya nigrimana (Meigen) Hamm et al., 1926
	Hylemya radicum (Linnaeus) Hamm et al., 1926
	Hylemya variata (Fallén) Hamm et al., 1926
	Anthomyia Meigen Hamm et al., 1926
	MUSCIDAE: COENOSIINAE
	Coenosia tigrina (Fabricius) var. *leonina* Hamm et al., 1926
	MUSCIDAE: MYDAEINAE
	Helina depuncta (Fallén) Hamm et al., 1926
	Helina duplicata (Meigen) Hamm et al., 1926
	Helina impuncta (Fallén) Hamm et al., 1926
	Helina lucorum (Fallén) Hamm et al., 1926
	Helina quadrum (Fabricius) Hamm et al., 1926

Organism	Fly associations
Crabro peltarius (cont.)	MUSCIDAE: FANNIINAE *Fannia armata* (Meigen) Hamm et al., 1926 MUSCIDAE: PHAONIINAE *Hydrotaea irritans* (Fallén) Hamm et al., 1926 *Ophyra leucostoma* (Wiedemann) Hamm et al., 1926 MUSCIDAE: MUSCINAE *Musca corvina* Fabricius Hamm et al., 1926 CALLIPHORIDAE: CALLIPHORINAE *Calliphora vicina* Robineau-Desvoidy Hamm et al., 1926 *Onesia sepulchralis* (Meigen) Hamm et al., 1926 CALLIPHORIDAE: POLLENIINAE *Pollenia rudis* (Fabricius) Hamm et al., 1926
Crabro pubescens Shuckard .... (Synonym: *Blepharipus pubescens*) (predator)	MUSCIDAE: FANNIINAE *Fannia polychaeta* Stein Hamm et al., 1926
Crabro serripes Panzer .... (Synonym: *Cuphopterus serripes*) (predator)	CALLIPHORIDAE: POLLENIINAE *Pollenia rudis* (Fabricius) Hamm et al., 1926
Crabro signatus Panzer .... (Synonym: *Blepharipus signatus*) (predator)	MUSCIDAE: FANNIINAE *Fannia scalaris* (Fabricius) Chevalier, 1923b MUSCIDAE: PHAONIINAE *Phaonia querceti* (Bouché) Chevalier, 1923b, 1926
Crabro vagus (Linnaeus) .... (Synonym: *Solenius vagus*) (predator)	MUSCIDAE: MUSCINAE *Musca* Linnaeus Hamm et al., 1926 CALLIPHORIDAE: CALLIPHORINAE *Lucilia caesar* (Linnaeus) Hamm et al., 1926 *Calliphora* Robineau-Desvoidy Hamm et al., 1926 *Calliphora vomitoria* (Linnaeus) Hamm et al., 1926 CALLIPHORIDAE: POLLENIINAE *Pollenia* Robineau-Desvoidy Hamm et al., 1926 SARCOPHAGIDAE: SARCOPHAGINAE *Sarcophaga* Meigen Hamm et al., 1926

Crabro venator (Rohwer)	ANTHOMYIIDAE: ANTHOMYIINAE *Hylemya cinerella* Fallén Kurczewski et al., 1968 MUSCIDAE: MYDAEINAE *Myospila meditabunda* (Fabricius) Kurczewski et al., 1968 MUSCIDAE: MUSCINAE *Orthellia caesarion* (Meigen) Kurczewski et al., 1968
Crossocerus lentus (Fox) (Synonym: *Crabro lentus*) (predator)	MUSCIDAE: MUSCINAE *Musca* Linnaeus Peckham et al., 1905
Crossocerus quadrimaculatus (Fabricius) (Synonym: *Hoplocrabro quadrimaculatus*) (predator)	EMPIDIDAE: EMPIDINAE *Hilara flavipes* Meigen Hamm et al., 1926 *Hilara litorea* Fallén Hamm et al., 1926 MUSCIDAE: MYDAEINAE *Helina clara* (Meigen) Hamm et al., 1926 *Helina quadrum* Fabricius Hamm et al., 1926 MUSCIDAE: FANNIINAE *Fannia canicularis* (Linnaeus) Hamm et al., 1926 *Fannia incisurata* (Zetterstedt) Hamm et al., 1926 *Fannia scalaris* (Fabricius) Hamm et al., 1926
Dinetus pictus (Fabricius) (parasitoid)	ANTHOMYIIDAE: ANTHOMYIINAE *Leucophora albiseta* (von Roser) Chevalier, 1923c
Diodontus minutus (Fabricius) (Synonym: ?*Dinetus minutus*) (parasitoid)	ANTHOMYIIDAE: ANTHOMYIINAE *Leucophora albiseta* (von Roser) Chevalier, 1923c
Ectemnius chrysostomus Lepeletier and Brulle (Synonym: *Clytochrysus chrysotomus*) (predator)	MUSCIDAE: PHAONIINAE *Phaonia vagans* (Fallén) Hamm et al., 1926
Ectemnius quadricinctus (Fabricius) (Synonym: *Metacrabro quadricinctus*) (predator)	SYRPHIDAE: SYRPHINAE *Platycheirus radicum* Fabricius Hamm et al., 1926 MUSCIDAE: MYDAEINAE *Helina anceps* (Zetterstedt) Hamm et al., 1926 *Helina impuncta* (Fallén) Hamm et al., 1926

Ectemnius quadricinctus (cont.)	*Mydaea detrita* Zetterstedt Hamm et al., 1926 *Mydaea urbana* (Meigen) Hamm et al., 1926 MUSCIDAE: PHAONIINAE *Hydrotaea irritans* (Fallén) Hamm et al., 1926 *Phaonia basalis* (Zetterstedt) Hamm et al., 1926 *Phaonia errans* (Meigen) Hamm et al., 1926 *Phaonia erratica* (Fallén) Hamm et al., 1926 *Phaonia querceti* (Bouché) Hamm et al., 1926 *Phaonia scutellaris* (Fallén) Hamm et al., 1926 *Phaonia signata* (Meigen) Hamm et al., 1926 *Phaonia variegata* (Meigen) Hamm et al., 1926 *Muscina pabulorum* (Fallén) Hamm et al., 1926 *Muscina stabulans* (Fallén) Hamm et al., 1926 MUSCIDAE: MUSCINAE *Polietes lardaria* Fabricius Hamm et al., 1926 *Mesembrina meridiana* (Linnaeus) Hamm et al., 1926 *Morellia* Robineau-Desvoidy Hamm et al., 1926 *Musca autumnalis autumnalis* De Geer Hamm et al., 1926 CALLIPHORIDAE: CHRYSOMYINAE *Protocalliphora azurea* (Fallén) Hamm et al., 1926 CALLIPHORIDAE: CALLIPHORINAE *Lucilia caesar* (Linnaeus) Hamm et al., 1926 *Phaenicia sericata* (Meigen) Hamm et al., 1926 *Calliphora vicina* Robineau-Desvoidy Hamm et al., 1926 *Calliphora vomitoria* (Linnaeus) Hamm et al., 1926 *Onesia sepulchralis* (Meigen) Hamm et al., 1926

Organism	Fly
Ectemnius quadricinctus (cont.)	*Onesia townsendi* Hall Hamm et al., 1926 CALLIPHORIDAE: POLLENIINAE *Pollenia rudis* (Fabricius) Hamm et al., 1926 SARCOPHAGIDAE: SARCOPHAGINAE *Sarcophaga* Meigen Hamm et al., 1926 *Sarcophaga carnaria* (Linnaeus) Hamm et al., 1926
Gorytes Latreille (predator)	CALLIPHORIDAE: CALLIPHORINAE *Calliphora vicina* Robineau-Desvoidy Froggatt, 1914a
Mellinus Fabricius (predator of adult flies)	MUSCIDAE: MUSCINAE *Orthellia caesarion* (Meigen) Hammer, 1942 *Musca autumnalis* De Geer Hammer, 1942 MUSCIDAE: STOMOXYINAE *Lyperosiops stimulans* (Meigen) Hammer, 1942
Mellinus arvensis (Linnaeus) (predator)	MUSCIDAE: MUSCINAE *Musca domestica* Linnaeus Herold, 1922 CALLIPHORIDAE: POLLENIINAE *Pollenia rudis* (Fabricius) Chevalier, 1923
Nysson Latreille (predator)	MUSCIDAE: MUSCINAE *Musca corvina* Fabricius Froggatt, 1917 CALLIPHORIDAE: CHRYSOMYINAE *Chrysomya varipes* (Macquart) Froggatt, 1917
Oxybelus Latreille (predator of adult flies)	MUSCIDAE: MUSCINAE *Musca amica* Zimin Pridantseva, 1959 *Musca tempestiva* Fallén Pridantseva, 1959 MUSCIDAE: STOMOXYINAE *Haematobia titillans* (Bezzi) Pridantseva, 1959 *Stomoxys* Geoffroy Roubaud, 1911a Barotte, 1925
Oxybelus bipunctatus Olivier (predator of adult flies)	STRATIOMYIDAE: BERIDINAE *Chorisops tibialis* Meigen Chevalier, 1926 OTITIDAE: ULIDIINAE *Physiphora demandata* (Fabricius) Chevalier, 1926

Organism	Fly and reference
Oxybelus bipunctatus (cont.)	ANTHOMYIIDAE: ANTHOMYIINAE *Hylemya radicum* (Meigen) Chevalier, 1926 MUSCIDAE: FANNIINAE *Fannia polychaeta* Stein Chevalier, 1926 MUSCIDAE: PHAONIINAE *Hydrotaea meteorica* (Linnaeus) Chevalier, 1926
Oxybelus quattuordecimnotatus .... Jurine (predator of adult flies)	ANTHOMYIIDAE: ANTHOMYIINAE *Leucophora albiseta* (von Roser) Chevalier, 1926 MUSCIDAE: PHAONIINAE *Hydrotaea dentipes* (Fabricius) Chevalier, 1926 *Ophyra leucostoma* (Wiedemann) Chevalier, 1926 CALLIPHORIDAE: CALLIPHORINAE *Onesia sepulchralis* (Meigen) Chevalier, 1926
Oxybelus uniglumis (Linnaeus) .... (predator of adult flies)	OTITIDAE: ULIDIINAE *Physiphora demandata* (Fabricius) Chevalier, 1926 MUSCIDAE: FANNIINAE *Fannia canicularis* (Linnaeus) Chevalier, 1926 *Fannia scalaris* (Fabricius) Chevalier, 1926 MUSCIDAE: PHAONIINAE *Phaonia querceti* (Bouché) Chevalier, 1926 MUSCIDAE: MUSCINAE *Musca domestica* Linnaeus Chevalier, 1926 CALLIPHORIDAE: POLLENIINAE *Pollenia rudis* (Fabricius) Chevalier, 1926
Sericophorus sydneyi Rayment .... (predator)	ANTHOMYIIDAE: ANTHOMYIINAE *Hylemya urbana* Malloch Rayment, 1954 MUSCIDAE: MUSCINAE *Musca corvina* Fabricius Rayment, 1954 *Musca sorbens sorbens* Wiedemann Rayment, 1954
Sericophorus teliferopodus .... Rayment (predator)	MUSCIDAE: MUSCINAE *Musca convexifrons* Thomson Rayment, 1954

Organism	Fly association
Sericophorus teliferopodus (cont.)	CALLIPHORIDAE: CHRYSOMYINAE *Paralucilia fulvipes* (Macquart) Rayment, 1954 CALLIPHORIDAE: CALLIPHORINAE *Lucilia argyricephala* Macquart Rayment, 1954 *Neopollenia stygia* (Fabricius) Rayment, 1954
Sericophorus viridis roddi Rayment (predator)	MUSCIDAE: MUSCINAE *Musca convexifrons* Thomson Rayment, 1954 *Musca sorbens sorbens* Wiedemann Rayment, 1954 CALLIPHORIDAE: CHRYSOMYINAE *Paralucilia fulvipes* (Macquart) Rayment, 1954 CALLIPHORIDAE: CALLIPHORINAE *Lucilia argyricephala* Macquart Rayment, 1954
Stictia denticornis (Handlirsch) (Synonym: *Monedula denticornis*) (predator)	STRATIOMYIDAE: HERMETINAE *Hermetia illucens* Bodkin, 1917 MUSCIDAE: STOMOXYINAE *Stomoxys calcitrans* (Linnaeus) Bodkin, 1917
Stizus Latreille (predator)	MUSCIDAE: MUSCINAE *Musca corvina* Fabricius Froggatt, 1917 CALLIPHORIDAE: CHRYSOMYINAE *Chrysomya varipes* (Macquart) Froggatt, 1917
Stizus turneri Froggatt (predator)	MUSCIDAE: MUSCINAE *Musca corvina* Fabricius Froggatt, 1917 Rayment, 1954 CALLIPHORIDAE: CHRYSOMYINAE *Chrysomya varipes* (Macquart) Froggatt, 1917
MUTILLIDAE	
Mutilla glossinae Turner (parasite)	SARCOPHAGIDAE: SARCOPHAGINAE *Sarcophaga* Meigen Lamborn, 1925
VESPIDAE	
Polistes hebraeus (Fabricius) (predator)	MUSCIDAE: MUSCINAE *Musca domestica* Linnaeus Jepson, 1915
Vespa Linnaeus (predator)	MUSCIDAE: MUSCINAE *Orthellia caesarion* (Meigen) Hammer, 1942

Organism	Fly
Vespa (cont.)	*Musca autumnalis* De Geer Hammer, 1942 MUSCIDAE: STOMOXYINAE *Lyperosiops stimulans* (Meigen) Hammer, 1942
Vespa germanica Fabricius (predator)	MUSCIDAE: FANNIINAE *Fannia canicularis* (Linnaeus) Kühlhorn, 1961 MUSCIDAE: MUSCINAE *Musca domestica* Linnaeus Herold, 1922 Kühlhorn, 1961 MUSCIDAE: STOMOXYINAE *Stomoxys calcitrans* (Linnaeus) Kühlhorn, 1961
Vespa vulgaris Linnaeus (predator)	MUSCIDAE: FANNIINAE *Fannia canicularis* (Linnaeus) Kühlhorn, 1961 MUSCIDAE: MUSCINAE *Musca domestica* Linnaeus Kühlhorn, 1961 MUSCIDAE: STOMOXYINAE *Stomoxys calcitrans* (Linnaeus) Kühlhorn, 1961
APIDAE	
Bombus agrorum Fabricius	MUSCIDAE: PHAONIINAE *Muscina pabulorum* (Fallén) Verhoeff, 1891
Bombus terrestris Linnaeus	MUSCIDAE: FANNIINAE *Fannia canicularis* (Linnaeus) Hewitt, 1912
TRICHOCERIDAE	
Trichocera Meigen	MUSCIDAE: PHAONIINAE *Phaonia variegata* (Meigen) Thomson, 1937
TIPULIDAE	
Ula macroptera (Macquart)	MUSCIDAE: COENOSIINAE *Allognota agromyzina* (Fallén) Keilin, 1917
UNDETERMINED TIPULIDAE	MUSCIDAE: MUSCINAE *Graphomya maculata* (Scopoli) Keilin, 1917
PTYCHOPTERIDAE	
Ptychoptera contaminata (Linnaeus)	MUSCIDAE: MUSCINAE *Graphomya maculata* (Scopoli) Keilin, 1917
CULICIDAE	
Aedes geniculatus Olivier	MUSCIDAE: PHAONIINAE *Phaonia exoleta* Meigen Tate, 1935

Culex pipiens Linnaeus	MUSCIDAE: PHAONIINAE *Phaonia exoleta* Meigen Tate, 1935
UNDETERMINED CULICINAE	CALLIPHORIDAE: CHRYSOMYINAE *Cochliomyia macellaria* (Fabricius) Zepeda, 1913
UNDETERMINED CULICIDAE	MUSCIDAE: STOMOXYINAE *Stomoxys calcitrans* (Linnaeus) Surcouf, 1923
UNDETERMINED CERATOPOGONIDAE	MUSCIDAE: PHAONIINAE *Phaonia cincta* (Zetterstedt) Keilin, 1917
UNDETERMINED CHIRONOMIDAE	ANTHOMYIIDAE: ANTHOMYIINAE *Paraprosalpia billbergi* (Zetterstedt) Hobby, 1934a MUSCIDAE: STOMOXYINAE *Stomoxys calcitrans* (Linnaeus) Surcouf, 1923
SIMULIIDAE	
Simulium damnosum Theobald	MUSCIDAE: LIMNOPHORINAE *Xenomyia oxycera* Emden Crosskey et al., 1962
ANISOPODIDAE	
Mycetobia pallipes Meigen	MUSCIDAE: PHAONIINAE *Phaonia cincta* (Zetterstedt) Keilin, 1917
MYCETOPHILIDAE	
Mycetophila lineola Meigen	MUSCIDAE: MYDAEINAE *Mydaea tincta* (Zetterstedt) Keilin, 1917
Mycetophila ornata Stephens	MUSCIDAE: PHAONIINAE *Phaonia variegata* (Meigen) Thomson, 1937
SCIARIDAE	
Sciara militaris Nowicki	MUSCIDAE: PHAONIINAE *Muscina pabulorum* (Fallén) Beling, 1868 Porchinskiĭ, 1913
STRATIOMYIDAE	
Hermetia illucens Linnaeus	MUSCIDAE: MUSCINAE *Musca domestica* Linnaeus Furman et al., 1959 Vasquez-Gonzales et al., 1962-63
TABANIDAE	
Tabanus striatus Fabricius (predator of larvae of other species)	MUSCIDAE: STOMOXYINAE *Stomoxys* Geoffroy Mitzmain, 1913

Organism	Fly
UNDETERMINED TABANIDAE	MUSCIDAE: MUSCINAE *Graphomya maculata* (Scopoli) Keilin, 1917
ASILIDAE	
Isopogon brevirostris (Meigen) (predator of adults of other species)	MUSCIDAE: FANNIINAE *Fannia aërea* (Zetterstedt) Hobby, 1934b
Lasiopogon cinctus (Fabricius) (predator of adults of other species)	MUSCIDAE: MUSCINAE *Musca autumnalis* De Geer Hobby, 1934b
Laphria flava (Linnaeus) (predator of adults of other species)	MUSCIDAE: PHAONIINAE *Hydrotaea irritans* (Fallén) Hobby, 1934b
Efferia aestuans (Linnaeus) (Synonym: *Erax aestuans*) (predator of adults of other species)	MUSCIDAE: MUSCINAE House fly (presumably *Musca domestica* Linnaeus) Bromley, 1946
Efferia pogonias (Wiedemann) (Synonym: *Erax rufibarbis*) (predator of adults of other species)	MUSCIDAE: MUSCINAE House fly (presumably *Musca domestica* Linnaeus) Bromley, 1946
Neoitamus flavofemoratus (Hine) (Synonym: *Asilus flavofemoratus*) (predator of adults of other species)	MUSCIDAE: MUSCINAE House fly (presumably *Musca domestica* Linnaeus) Bromley, 1946
Proctacanthus philadelphicus Macquart (predator of adults of other species)	MUSCIDAE: MUSCINAE House fly (presumably *Musca domestica* Linnaeus) Bromley, 1946
Regasilus notatus (Wiedemann) (Synonym: *Asilus notatus*) (predator of adults of other species)	MUSCIDAE: MUSCINAE House fly (presumably *Musca domestica* Linnaeus) Bromley, 1946
Regasilus sadyates (Walker) (Synonym: *Asilus sadyates*) (predator of adults of other species)	MUSCIDAE: MUSCINAE House fly (presumably *Musca domestica* Linnaeus) Bromley, 1946
UNDETERMINED ASILIDAE (predators of adults of other species)	MUSCIDAE: STOMOXYINAE *Stomoxys calcitrans* (Linnaeus) Bishopp, 1913a
BOMBYLIIDAE	
Thyridanthrax abruptus Loew (pupal parasite)	CALLIPHORIDAE: RHININAE *Rhyncomyia pictifacies* Bigot McDonald, 1958

Organism	Fly
EMPIDIDAE	
Empis grisea Fallén	ANTHOMYIIDAE: ANTHOMYIINAE *Paraprosalpia silvestris* (Fallén) Hobby, 1934a
SYRPHIDAE	
Eristalis Latreille	MUSCIDAE: MUSCINAE *Graphomya maculata* (Scopoli) Keilin, 1917
UNDETERMINED ERISTALINI	MUSCIDAE: PHAONIINAE *Phaonia cincta* (Zetterstedt) Keilin, 1917
UNDETERMINED MILESIINAE (Synonym: ERISTALINAE)	MUSCIDAE: PHAONIINAE *Phaonia variegata* (Meigen) Keilin, 1917
SPHAEROCERIDAE	
Borborus Meigen	MUSCIDAE: MUSCINAE *Polietes albolineata* (Fallén) Porchinskiĭ, 1910
Copromyza equina (Fallén) (Synonym: *Borborus equinus*)	ANTHOMYIIDAE: SCATOPHAGINAE *Scatophaga stercoraria* (Linnaeus) Cotterell, 1920
DROSOPHILIDAE	
Drosophila melanogaster Meigen (interspecific larval competition)	DROSOPHILIDAE: DROSOPHILINAE *Drosophila simulans* Sturtevant Moore, 1952
CHLOROPIDAE	
Hippelates collusor (Townsend)	*Hippelates collusor* (Townsend) Legner, 1966
CLUSIIDAE	
Clusiodes albimanus (Meigen) (Synonym: *Heteroneura albimana*)	MUSCIDAE: PHAONIINAE *Phaonia goberti* (Mik) Keilin, 1917
AULACIGASTRIDAE	
Aulacigaster leucopeza (Meigen) (Synonym: *Aulacigaster rufitarsis*)	MUSCIDAE: PHAONIINAE *Phaonia cincta* (Zetterstedt) Keilin, 1917
ANTHOMYIIDAE	
Scatophaga merdaria (Fabricius) (predator of adults of other species)	SCATOPSIDAE: SCATOPSINAE *Scatopse notata* Linnaeus Hewitt, 1914
Scatophaga stercoraria (Linnaeus) (predator of adults of other species)	SEPSIDAE *Sepsis cynipsea* (Linnaeus) Hobby, 1934b ANTHOMYIIDAE: ANTHOMYIINAE *Hylemya aestiva* (Meigen) Hobby, 1934b

Scatophaga stercoraria (cont.)	MUSCIDAE: FANNIINAE *Fannia canicularis* (Linnaeus) Hewitt, 1914 Steve, 1959 *Fannia ?serena* (Fallén) Hobby, 1934b MUSCIDAE: PHAONIINAE *Hydrotaea dentipes* (Fabricius) Hobby, 1934b *Hydrotaea ?irritans* (Fallén) Hobby, 1934b MUSCIDAE: MUSCINAE *Orthellia caesarion* (Meigen) Hewitt, 1914 *Musca domestica* Linnaeus Hewitt, 1914 MUSCIDAE: STOMOXYINAE *Stomoxys calcitrans* (Linnaeus) Hewitt, 1914 CALLIPHORIDAE: CALLIPHORINAE *Calliphora vicina* Robineau-Desvoidy Hewitt, 1914 MUSCIDAE: POLLENIINAE *Pollenia rudis* (Fabricius) Hewitt, 1914
UNDETERMINED SCATOPHAGINAE .... (Synonym: SCATOPHAGIDAE)	MUSCIDAE: MUSCINAE *Graphomya maculata* (Scopoli) Keilin, 1917
Hylemya brassicae (Bouché) ...... (Synonym: *Chortophila brassicae*)	MUSCIDAE: PHAONIINAE *Phaonia trimaculata* (Bouché) Bouché, 1834
MUSCIDAE	
Coenosia tigrina .... (Zetterstedt) (predator of adults of other species)	SPHAEROCERIDAE *Leptocera humida* Haliday Hobby, 1934b MUSCIDAE: PHAONIINAE *Hydrotaea parva* Meade Hobby, 1934b
Hebecnema umbratica .... (Meigen) (predator of larvae of other species)	MUSCIDAE: MUSCINAE *Orthellia caesarion* (Meigen) Thomson, 1937
Myospila meditabunda .... (Fabricius) (Synonym: *Myiospila meditabunda*) (predator of larvae of other species)	ANTHOMYIIDAE: ANTHOMYIINAE *Hylemya strigosa* (Fabricius) Porchinskiĭ, 1910 MUSCIDAE: MUSCINAE *Dasyphora cyanella* (Meigen) Thomson, 1937

Organism	Associated flies
Myospila meditabunda (cont.)	*Orthellia caesarion* (Meigen) Porchinskiĭ, 1910 Thomson, 1937 *Musca autumnalis autumnalis* De Geer Porchinskiĭ, 1910 MUSCIDAE: STOMOXYINAE *Lyperosiops stimulans* (Meigen) Thomson, 1937 CALLIPHORIDAE: CHRYSOMYINAE *Boreellus atriceps* (Zetterstedt) Porchinskiĭ, 1910
Mydaea urbana (Meigen) (predator of larvae of other species)	MUSCIDAE: MUSCINAE *Dasyphora cyanella* (Meigen) Thomson, 1937 *Orthellia caesarion* (Meigen) Thomson, 1937 MUSCIDAE: STOMOXYINAE *Lyperosiops stimulans* (Meigen) Thomson, 1937
Mydaea pagana (Fabricius) (cannibalism among larvae)	MUSCIDAE: MYDAEINAE *Mydaea pagana* (Fabricius) Thomson, 1937
Hydrotaea dentipes (Fabricius)	MUSCIDAE: PHAONIINAE *Muscina stabulans* (Fallén) Porchinskiĭ, 1913 MUSCIDAE: MUSCINAE *Musca domestica* Linnaeus Porchinskiĭ, 1911 Derbeneva-Ukhova, 1935, 1961 MUSCIDAE: STOMOXYINAE *Haematobia irritans* (Linnaeus) Wilhelmi, 1917a *Stomoxys calcitrans* (Linnaeus) Derbeneva-Ukhova, 1961
Ophyra leucostoma (Wiedemann) (predator of larvae of other species)	MUSCIDAE: FANNIINAE *Fannia canicularis* (Linnaeus) Anderson, 1964 *Fannia femoralis* (Stein) Legner et al. (pre-publication), 1965a MUSCIDAE: PHAONIINAE *Muscina stabulans* (Fallén) Anderson, 1964 MUSCIDAE: MUSCINAE *Musca domestica* Linnaeus Anderson, 1964 Anderson et al., 1964

Phaonia variegata (Meigen) (predator of larvae of other species) ....
ANTHOMYIIDAE: ANTHOMYIINAE
Pegomya haemorrhoa (Zetterstedt)
Keilin, 1917
Pegomya winthemi (Meigen)
Keilin, 1917

Muscina assimilis (Fallén) (predator of larvae of other species) ....
PHORIDAE: PHORINAE
Phora Latreille
Keilin, 1917
PHORIDAE: METOPININAE
Megaselia rufipes (Meigen)
Keilin, 1917
DROSOPHILIDAE: DROSOPHILINAE
Drosophila confusa Staegel
Keilin, 1917
MUSCIDAE: FANNIINAE
Fannia canicularis (Linnaeus)
Keilin, 1917

Muscina pabulorum (Fallén) (predator of larvae of other species) ......
CALLIPHORIDAE: CALLIPHORINAE
Lucilia Robineau-Desvoidy
Thomson, 1937
Calliphora Robineau-Desvoidy
Thomson, 1937

Muscina stabulans (Fallén) (predator of larvae of other species) ....
HELEOMYZIDAE: SUILLIINAE
Suillia lineata Robineau-Desvoidy
Porchinskiĭ, 1913
Suillia ustulata Meigen
Porchinskiĭ, 1913
HELEOMYZIDAE: HELEOMYZINAE
Heleomyza Fallén
Porchinskiĭ, 1913
Heleomyza gigantea Meigen
Porchinskiĭ, 1913
MUSCIDAE: PHAONIINAE
Hydrotaea dentipes (Fabricius)
Porchinskiĭ, 1913
MUSCIDAE: MUSCINAE
Musca domestica Linnaeus
Porchinskiĭ, 1913

Polietes albolineata (Fallén) (predator of larvae of other species) ....
MUSCIDAE: PHAONIINAE
Hydrotaea dentipes (Fabricius)
Porchinskiĭ, 1910
Muscina stabulans (Fallén)
Porchinskiĭ, 1913
MUSCIDAE: MUSCINAE
Orthellia caesarion (Meigen)
Porchinskiĭ, 1910
Musca corvina Fabricius
Porchinskiĭ, 1910
MUSCIDAE: STOMOXYINAE
Lyperosiops stimulans (Meigen)
Porchinskiĭ, 1910

Polietes hirticrura Meade (predator of larvae of other species)	MUSCIDAE: MUSCINAE *Orthellia caesarion* (Meigen) Thomson, 1937
Mesembrina meridiana Linnaeus (predator of larvae of other species)	MUSCIDAE: MUSCINAE *Dasyphora cyanella* (Meigen) Thomson, 1937 *Orthellia caesarion* (Meigen) Thomson, 1937 MUSCIDAE: STOMOXYINAE *Lyperosiops stimulans* (Meigen) Thomson, 1937
Musca domestica Linnaeus (interspecific larval competition)	CALLIPHORIDAE: CALLIPHORINAE *Phaenicia sericata* (Meigen) Pimentel et al., 1965b
CALLIPHORIDAE	
Chrysomya Robineau-Desvoidy (Synonym: *Chrysomyia*) (interspecific larval competition)	CALLIPHORIDAE: CALLIPHORINAE *Phaenicia sericata* (Meigen) Salt, 1932
Chrysomya rufifacies (Macquart) (Synonyms: *Chrysomyia rufifacies, Chrysomyia albiceps*) (predator of larvae of other species)	CALLIPHORIDAE: CALLIPHORINAE *Phaenicia sericata* (Meigen) Holdaway, 1930 Mackerras, 1930 *Calliphora stygia* (Fabricius) Mackerras, 1930 *Anastellorhina augur* Mackerras, 1930 SARCOPHAGIDAE: SARCOPHAGINAE *Sarcophaga argyrostoma* (Robineau-Desvoidy) Illingworth, 1923b
Chrysomya villeneuvei Patton (Synonym: *Chrysomyia villeneuvii*) (predator of larvae of other species)	CALLIPHORIDAE: CHRYSOMYINAE *Chrysomya rufifacies* (Macquart) Patton, 1922 SARCOPHAGIDAE: SARCOPHAGINAE *Sarcophaga* Meigen Patton, 1922
Protophormia terraenovae (Robineau-Desvoidy) (Synonym: *Phormia groenlandica*) (interspecific larval competition)	MUSCIDAE: MUSCINAE *Musca domestica* Linnaeus Vladimirova et al., 1938
Phaenicia cuprina Wiedemann (Synonym: *Lucilia cuprina*) (hybridization)	CALLIPHORIDAE: CALLIPHORINAE *Phaenicia sericata* (Meigen) Mackerras, 1933
Calliphora Robineau-Desvoidy (interspecific larval competition)	CALLIPHORIDAE: CALLIPHORINAE *Phaenicia sericata* (Meigen) Salt, 1932
SARCOPHAGIDAE	
Ravinia striata Fabricius (interspecific larval competition)	MUSCIDAE: FANNIINAE *Fannia canicularis* (Linnaeus) Sychevskaîa, 1954

Organism	Associated flies
Ravinia striata (cont.)	*Fannia leucosticta* (Meigen) Sychevskaia͡, 1954 *Fannia scalaris* (Fabricius) Sychevskaia͡, 1954 *Fannia subscalaris* Zimin Sychevskaia͡, 1954
Sarcophaga Meigen (interspecific larval competition)	CALLIPHORIDAE: CALLIPHORINAE *Phaenicia sericata* (Meigen) Salt, 1932
Sarcophaga haemorrhoidalis (Fallén) (interspecific larval competition)	MUSCIDAE: FANNIINAE *Fannia canicularis* (Linnaeus) Sychevskaia͡, 1954 *Fannia leucosticta* (Meigen) Sychevskaia͡, 1954 *Fannia scalaris* (Fabricius) Sychevskaia͡, 1954 *Fannia subscalaris* Zimin Sychevskaia͡, 1954
UNDETERMINED SARCOPHAGIDAE (interspecific larval competition)	CALLIPHORIDAE: CALLIPHORINAE *Phaenicia sericata* (Meigen) Holdaway, 1930
CUTEREBRIDAE	
Dermatobia hominis (Linnaeus) (phoresy: *D. hominis* eggs carried by other species of flies)	ANTHOMYIIDAE: ANTHOMYIINAE *Anthomyia lindigii* Schiner Lutz, 1917 MUSCIDAE: LIMNOPHORINAE *Limnophora* Robineau-Desvoidy Dunn, 1930 Neel et al., 1955 MUSCIDAE: MUSCINAE *Synthesiomyia nudiseta* (Wulp) Lutz, 1917 *Sarcopromusca arcuata* Townsend Lins de Almeida, 1933 *Musca* Linnaeus Townsend, 1922 *Musca domestica* Linnaeus Neiva et al., 1917 Neel et al., 1955 Zeledón, 1957 MUSCIDAE: STOMOXYINAE *Stomoxys calcitrans* (Linnaeus) Neiva et al., 1917 Neel et al., 1955 Zeledón, 1957 UNDETERMINED MUSCIDAE Neel et al., 1955

Organism	Fly
Dermatobia hominis (cont.)	CALLIPHORIDAE: CHRYSOMYINAE *Cochliomyia macellaria* (Fabricius) Lins de Almeida, 1933 Neel et al., 1955
	CALLIPHORIDAE: CALLIPHORINAE *Phaenicia eximia* (Wiedemann) Blanchard, 1899
	SARCOPHAGIDAE: SARCOPHAGINAE *Blaesoxipha plinthopyga* (Wiedemann) Blanchard, 1899
	Sarcophaga wiedemanni Aldrich Blanchard, 1899
DISCOGLOSSIDAE	
Discoglossus pictus Otth (predator of adult flies)	MUSCIDAE: STOMOXYINAE *Stomoxys calcitrans* (Linnaeus) Surcouf, 1923
BUFONIDAE	
Bufo mauritanicus Schlegel (predator of adult flies)	MUSCIDAE: STOMOXYINAE *Stomoxys calcitrans* (Linnaeus) Surcouf, 1923
Bufo viridis Laurenti (predator of adult flies)	MUSCIDAE: STOMOXYINAE *Stomoxys calcitrans* (Linnaeus) Surcouf, 1923
HYLIDAE	
Hyla viridis Laurenti var. *grisea* (predator of adult flies)	MUSCIDAE: STOMOXYINAE *Stomoxys calcitrans* (Linnaeus) Surcouf, 1923
RANIDAE	
Rana ridibunda Pallas (Synonym: *Rana viridis* var. *rudibunda*) (predator of adult flies)	MUSCIDAE: STOMOXYINAE *Stomoxys calcitrans* (Linnaeus) Surcouf, 1923
AGAMIDAE	
Agama agilis Olivier (predator of adult flies)	MUSCIDAE: STOMOXYINAE *Stomoxys calcitrans* (Linnaeus) Surcouf, 1923
SCINCIDAE	
Chalcides tridactylus Gray (Synonym: *Sceps tridactylites*) (predator of adult flies)	MUSCIDAE: STOMOXYINAE *Stomoxys calcitrans* (Linnaeus) Surcouf, 1923
LACERTIDAE	
Acanthodactylus boskianus (Lichtenstein) (predator of adult flies)	MUSCIDAE: STOMOXYINAE *Stomoxys calcitrans* (Linnaeus) Surcouf, 1923
Lacerta agilis Linnaeus (predator of adult flies)	MUSCIDAE: MUSCINAE *Musca domestica* Linnaeus Antonov, 1945

PHASIANIDAE

Gallus gallus Linnaeus (predator) ....
- MUSCIDAE: FANNIINAE
- *Fannia* Robineau-Desvoidy
  - Rodriguez et al., 1962b
- MUSCIDAE: PHAONIINAE
- *Ophyra* Robineau-Desvoidy
  - Rodriguez et al., 1962b
- *Muscina* Robineau-Desvoidy
  - Rodriguez et al., 1962b
- MUSCIDAE: MUSCINAE
- *Musca domestica* Linnaeus
  - Bushnell et al., 1924
  - Rodriguez et al., 1962b
- MUSCIDAE: STOMOXYINAE
- *Stomoxys calcitrans* (Linnaeus)
  - Bushnell et al., 1924

HIRUNDINIDAE

Delichon urbica (Linnaeus) subsp. *urbica* (Synonym: *Hirundo urbica*) (maggots breeding in nest) ....
- SCENOPINIDAE
- *Scenopinus fenestralis* (Linnaeus)
  - Séguy, 1929
- MUSCIDAE: FANNIINAE
- *Fannia scalaris* (Fabricius)
  - Séguy, 1929
- MUSCIDAE: PHAONIINAE
- *Ophyra leucostoma* (Wiedemann)
  - Séguy, 1929
- *Muscina stabulans* (Fallén)
  - Séguy, 1929
- MUSCIDAE: MUSCINAE
- *Musca domestica* Linnaeus
  - Séguy, 1929
- CALLIPHORIDAE: CHRYSOMYINAE
- *Phormia regina* (Meigen)
  - Séguy, 1929
- CALLIPHORIDAE: CALLIPHORINAE
- *Calliphora vicina* Robineau-Desvoidy
  - Séguy, 1929

Hirundo rustica Linnaeus subsp. *rustica* (Synonym: *Chelidon rustica*) (maggots breeding in nest) ....
- SCENOPINIDAE
- *Scenopinus albicincta* (Rossi)
  - Séguy, 1929
- *Scenopinus fenestralis* (Linnaeus)
  - Séguy, 1929
- *Scenopinus senilis* (Fabricius)
  - Séguy, 1929
- MUSCIDAE: FANNIINAE
- *Fannia canicularis* (Linnaeus)
  - Séguy, 1929
- *Fannia scalaris* (Fabricius)
  - Séguy, 1929

Organism	Fly
Hirundo rustica rustica (cont.)	MUSCIDAE: PHAONIINAE *Ophyra leucostoma* (Wiedemann) Séguy, 1929 *Muscina stabulans* (Fallén) Séguy, 1929 MUSCIDAE: MUSCINAE *Musca domestica* Linnaeus Séguy, 1929 MUSCIDAE: STOMOXYINAE *Stomoxys calcitrans* (Linnaeus) Séguy, 1929 CALLIPHORIDAE: CHRYSOMYINAE *Phormia regina* (Meigen) Séguy, 1929 CALLIPHORIDAE: CALLIPHORINAE *Calliphora vicina* Robineau-Desvoidy Séguy, 1929 SARCOPHAGIDAE: SARCOPHAGINAE *Sarcophaga melanura* Meigen Séguy, 1929
PARIDAE	
Aegithalus caudatus (Linnaeus) (maggots breeding in nest)	ANTHOMYIIDAE: ANTHOMYIINAE *Anthomyia pluvialis* (Linnaeus) Séguy, 1929
Parus caeruleus Linnaeus subsp. *caeruleus* (Synonym: *Parus caeruleus*) (maggots breeding in nest)	PHORIDAE: METOPININAE *Megaselia rufipes* (Meigen) Séguy, 1929 ANTHOMYIIDAE: ANTHOMYIINAE *Anthomyia pluvialis* (Linnaeus) Séguy, 1929 MUSCIDAE: MYDAEINAE *Mydaea pertusa* Meigen Séguy, 1929
UNDETERMINED PARIDAE (Synonym: Titmouse) (maggots breeding in nest)	MUSCIDAE: FANNIINAE *Fannia incisurata* (Zetterstedt) Séguy, 1929 MUSCIDAE: PHAONIINAE *Phaonia querceti* (Bouché) Séguy, 1929
TURDIDAE	
Luscinia megarhynchos Brehm (Synonym: Nightingale) (maggots breeding in nest)	MUSCIDAE: FANNIINAE *Fannia canicularis* (Linnaeus) Séguy, 1929
MUSCICAPIDAE	
Leucocira leucophrys (Latham) (Synonym: *Rhipidura tricolor*) (predator)	UNDETERMINED CALLIPHORIDAE Blow flies Froggatt et al., 1917

Organism	Fly
STURNIDAE	
Sturnus vulgaris Linnaeus (maggots breeding in nest)	SCENOPINIDAE *Scenopinus glabrifrons* Meigen Séguy, 1929 ANTHOMYIIDAE: ANTHOMYIINAE *Anthomyia pluvialis* (Linnaeus) Séguy, 1929 MUSCIDAE: FANNIINAE *Fannia lineata* Stein Séguy, 1929 MUSCIDAE: PHAONIINAE *Muscina stabulans* (Fallén) Séguy, 1929
UNDETERMINED STURNIDAE (Synonym: Mynah birds) (maggots breeding in nest)	MUSCIDAE: STOMOXYINAE *Stomoxys* Geoffroy Fletcher, 1920
MELIPHAGIDAE	
Melliphaga penicillata Gould (Synonym: *Ptilotis penicillata*) (predator)	UNDETERMINED CALLIPHORIDAE Blow flies Froggatt et al., 1917
Myzantha melanocephala (Synonym: *Myzantha garrula*) (predator)	UNDETERMINED CALLIPHORIDAE Blow flies Froggatt et al., 1917
FRINGILLIDAE	
Fringilla coelebs Linnaeus (Synonym: *Fringilla caelebs*) (maggots breeding in nest)	ANTHOMYIIDAE: ANTHOMYIINAE *Anthomyia pluvialis* (Linnaeus) Séguy, 1929
PLOCEIDAE	
Passer domesticus (Linnaeus) (maggots breeding in nest)	SCENOPINIDAE *Scenopinus bouvieri* (Séguy) Séguy, 1929 *Scenopinus senilis* (Fabricius) Séguy, 1929 ANTHOMYIIDAE: ANTHOMYIINAE *Anthomyia pluvialis* (Linnaeus) Séguy, 1929 MUSCIDAE: FANNIINAE *Fannia scalaris* (Fabricius) Séguy, 1929
UNDETERMINED PLOCEIDAE (Synonym: Sparrow) (maggots breeding in nest)	CALLIPHORIDAE: CALLIPHORINAE *Lucilia illustris* (Meigen) Séguy, 1929
MURIDAE	
Apodemus sylvaticus Linnaeus (Synonym: *Mus silvaticus*) (maggots breeding in mouse nest)	SCENOPINIDAE *Scenopinus fenestralis* (Linnaeus) Séguy, 1929
UNDETERMINED GLIRIDAE (Synonym: Dormouse) (maggots in Dormouse cage)	MUSCIDAE: FANNIINAE *Fannia canicularis* (Linnaeus) Hewitt, 1912

Bibliography

Ackert, J. E.
1919. On the life cycle of the fowl cestode, *Davainea cesticillus* (Molin). Jour. Parasit., *5*:41-43.

Ackert, J. E.
1920. On the life-history of *Davainea tetragona* (Molin), a fowl tapeworm. Jour. Parasit., *6*:28-34.

Ackert, J. E.
1937. See Reid, W. M., and Ackert, J. E.

Akatov, V. A.
1955. O dlitel'nosti sokhraneniia trikhomonad u domashnei mukhi. [On the length of survival of trichomonads in house flies.] Veternariia, *32*(8):84.

Alcivar, Z. C., and Campos, R. F.
1946. Las moscas, como agentes vectores de enfermedades entericas en Guayaquil. Rev. Ecuatoriana Hig. Med. Trop., *3*:3-14.

Aleksander, L. A., and Dansker, V. N.
1935. Rol' mukh v rasprostranenii kishechnykh parazitov. [Role of flies in the dissemination of intestinal parasites.] Trudy Leningrad. Inst. Epidem. Bacter. imeni Paster, *2*:169-179. (Engl. summ. pp. 233-234.)

Alessandrini, G.
1912. Vitalità del vibrione colerigeno nelle mosche. Ann. Ig. (Sper.), new series, *22*:634-650.

Alessandrini, G., and Sampietro, G.
1912. Sulla vitalità del vibrione colerigeno nel latte e nelle mosche. Ann. Ig. (Sper.), new series, *22*:623-633.

Alexeieff, A.
1911. Sur les cercomonadines intestinales de *Calliphora erythrocephala* Mg. et de *Lucilia* sp. Compt. Rend. Soc. Biol., *71*:379-382.

Alessi, G.
1888. See Celli, A., and Alessi, G.

Al-Hafidh, R.
1965. See Pimentel, D., and Al-Hafidh, R.

Alilaire, E.
1908. See Chatton, E., and Alilaire, E.

Altson, A. M.
1920. The life history and habits of two parasites of blowflies. Proc. Zool. Soc. London, *15*:195-243.

Amonkar, S. V.
1967. Mechanism of pathogenicity of *Pseudomonas* in the house fly. Jour. Invert. Pathol., *9*:235-240.

Amonkar, S. V., and Nair, K. K.
1965. Pathogenicity of *Aspergillus flavus* Link to *Musca domestica nebulo* Fabricius. Jour. Invert. Pathol., *7*:513-514.

André, C.
1906. Dissémination du bacille tuberculeux par les mouches. Compt. Rend. Soc. Med. Hôp. Lyon, *5*:321-327.

André, C.
1908. Les mouches comme agents de dissémination du bacilli de Koch. VI Intl. Congr. Tuberculosis, pp. 162-166.

Anderson, C.
1928. See Wollman, E., et al.

Anderson, J. F., and Frost, W. H.
1912. Transmission of poliomyelitis by means of the stable fly (*Stomoxys calcitrans*). Publ. Health Rept., *27*:1733-1735.

Anderson, J. R.
1964. The behavior and ecology of various flies associated with poultry ranches in northern California. Proc. & Papers 32nd Ann. Conf. California Mosquito Control Assoc., Inc., pp. 30-34.

Anderson, J. R., and Poorbaugh, J. H.
1964. Biological control possibility for house flies. California Agric., *18*:2-4.

Anderson, T. G.
1944. See Kuhns, D. M., and Anderson, T. G.

Andrewartha, H. G.
1930. See Newman, L. J., and Andrewartha, H. G.

Anonymous, "Korean authors et al."
See Sukhova, M. N., Zool. Zhurnal, 1951, *30*:188-190.

Anonymous
1925. Parasites of sheep maggot-flies. Jour. Dept. Agric. Union So. Africa, *10*:380.

Anonymous
1936. Blood parasites of cattle. Australian Council Scient. Indust. Res. Rept., *9*:31.

Anonymous
1952. Report of the international commission for the investigation of the facts concerning bacterial warfare in Korea and China. Peking, 665 pp.

Antonov, V. K.
1945. Éksperimental'nyĭ brutsellëz reptiliĭ i amphibiĭ. Tezisy doklada na i͡ubileĭnoĭ konferentsii Almaatinskogo Zoovetinstituta i Nauchno-issledovatel'skogo Instituta Veterinarii Kazfiliala VASKhNIL, posvi͡ashennoĭ XXV-leti͡iu Kaz S.S.R. i XX-leti͡iu Nauchno-issled. Vetinstituta, 25-28, Okti͡abria. [Experimental brucellosis of reptiles and amphibians. "Thesis of a report" to the Jubilee Conference of the Alma-Ata Zoo-Vet-Institute. 25-28 Oct. 1945.]

Ara, F.
1933. Ulteriori ricerche sull' importanza della mosca nella diffusione della febbre tifoide. Ig. Moderna, *26*:327-335.

Ara, F., and Marengo, U.
1932. The importance of the housefly in the spread of typhoid fever. Sull' importanza della mosca nella diffusione della febbre tifoide (Sper.), *7*:150-154.

Arakawa, K. Y.
1959. See Hall, I. M., and Arakawa, K. Y.

Arango, M. C.
1945. See Curbelo, A., and Arango, M. C.

Arizumi, S.
1934. On the potential transmission of *Bacillus leprae* by certain insects. Taiwan Igakkai Zasshi (Jour. Med. Assoc. Formosa), *33*:634-661 [Japanese].

Ark, P. A., and Thomas, H. E.
1936. Persistence of *Erwinia amylovora* in certain insects. Phytopathology, *26*:375-381.

Arskiĭ, V. G., Gadzhei, E. F., Zatsepin, N. I., and Yasinskiĭ, A. V.
1961. The role of flies in the seasonal character of dysentery. Jour. Microb. Epidem. Immunob., *32*:1013-1020. (Same as Zh. Mikrob., *32*:27-32.)

Asahina, S., Ogata, K., Noguchi, Y., Uchida, S., and Murata, M.
1963. Detection of polioviruses from flies and cockroaches captured during 1961 epidemics in Kumamoto prefecture. Jap. Jour. Sanit. Zool., *14*:28-31 [Eisei dobutsu].

Asami, S.
1934. Concerning the carrying of the leprosy bacillus by insects. Intl. Jour. Leprosy, *2*:465-469. (Also: Leprosy [Osaka], *3*:17 [5], 1932.)

Ashmead, W. H.
1887. Trans. Amer. Entom. Soc., *14*:199.

Ashmead, W. H.
1894. Proc. Ent. Soc. Washington, *3*:36-37.

Ashmead, W. H.
1900. Ann. Rept. New Jersey St. Bd. Agric., p. 559.

Attimonelli, R.
1940a. Reperti parassitologici nell'acqua di lavaggio esterno delle mosche. Pathologica, *32*:111-112.

Attimonelli, R.
?1940b. Di un metodo semplice per la colorazione degli organi di motilità dei protozoi. Pathologica, *32*:44.

Aubertot, M.
1923. Sur la dissémination et la transport de nématodes du genre *Rhabditis* par les diptères. Compt. Rend. Acad. Sci., *176*:1257-1260.

Auché, A.
1906. Transport des bacilles dysentériques par les mouches. Compt. Rend. Soc. Biol., *61*:450-452.

Austin, M. D.
1933. The insect and allied fauna of cultivated mushrooms. Entom. Monthly Mag., *69*:16-19.

Axtell, R. C.
1961a. Mites–enemies of house flies. Farm Res., *27*:4-5.

Axtell, R. C.
1961b. New records of North American Macrochelidae (Acarina: Mesostigmata) and their predation rates on the house fly. Ann. Entom. Soc. Amer., *54*:748.

Axtell, R. C.
1963a. Effect of Macrochelidae (Acarina: Mesostigmata) on house fly production from dairy cattle manure. Jour. Econ. Entom., *56*:317-321.

Axtell, R. C.
1963b. Manure-inhabiting Macrochelidae (Acarina: Mesostigmata) predaceous on the house fly. In: Naegele, J. A. (ed.), Advances in Acarology, Vol. 1, xii + 480 pp., 64 figs., refs. Ithaca, New York, Cornell Univ. Press, pp. 55-59.

Axtell, R. C.
1964. Phoretic relationship of some common manure-inhabiting Macrochelidae (Acarina: Mesostigmata) to the house fly. Ann. Entom. Soc. Amer., *57*:584-587.

Axtell, R. C.
1965. See O'Donnell, A. E., and Axtell, R. C.

Axtell, R. C.
1966. See Farish, D. J., and Axtell, R. C.

Bacot, A. W.
1911a. On the persistence of bacilli in the gut of an insect during metamorphosis. Trans. Entom. Soc. London, Part II, pp. 497-500.

Bacot, A. W.
1911b. The persistence of *Bacillus pyocyaneus* in pupae and imagines of *Musca domestica* raised from larvae experimentally infected with the bacillus. Parasitology, *4*:68-74.

Baer, W. S.
1929. Sacro-Iliac Joint—Arthritis Deformans viable antiseptic in chronic osteomyelitis. Proc. Inter-State Postgrad. Med. Assoc. N. America, *5*:365-372.

Bahr, L.
1952. Nogle undersogelser vedrorende "sommermastitis." Dansk Maanedsskr. Dyrlaeger, *62*:367-394.

Bahr, L.
1953. "Summer-Mastitis" (Preliminary Rept.; summary). Proc. XV Intl. Veter. Cong. Stockholm, Vol. 2, Part 1, pp. 849-852; Part 2, p. 348.

Bahr, P. H.
1914. A study of epidemic dysentery in the Fiji Islands. British Med. Jour. London, *1*:294-296.

Bail, O.
1900. Versuche über eine Möglichkeit der Entstehung von Fleischvergiftungen. Hyg. Rundsch., *10*:1017-1020.

Bail, T.
1860. Über Krankheiten der Insecten durch Pilze. Verhandlungen der 35te Naturforschenden Versammlung in Königsberg, Botan., 1860.

Bailey, S. F.
1935. See Herms, W. B., et al.

Baird, R. B.
1957. Notes on a laboratory infection of Diptera caused by the fungus *Empusa muscae* Cohn. Canadian Entom., *89*:432-435.

Bakri, G.
1959. See Bulling, E., et al.

Baldrey, F.S.H.
1911. The evolution of *Trypanosoma evansi* through the fly: *Tabanus* and *Stomoxys*. Jour. Trop. Veter. Science, *6*:271-282.

Balduf, W. V.
1935. The bionomics of entomophagous Coleoptera. St. Louis: John S. Swift Co., Inc., 220 pp.

Balduf, W. V.
1941. Quantitative dietary studies on *Phymata*. Jour. Econ. Entom., *34*:614-620.

Balduf, W. V.
1947. The weights of *Phymata pennsylvanica americana* Melin. Ann. Entom. Soc. Amer., *40*:576-587.

Balduf, W. V.
1948. A summary of studies on the ambush bug, *Phymata pennsylvanica americana* Melin (Phymatidae, Hemiptera). Trans. Illinois St. Acad. Science, *41*:101-106.

Balzam, N.
1937. Destin de la flore bactérienne pendant la métamorphose de la mouche à viande (*Calliphora erythrocephala*). Ann. Inst. Pasteur, *58*:181-212.

Bancroft, M. J.
1920a. See Johnston, T. H., and Bancroft, M. J.

Bancroft, M. J.
1920b. See Johnston, T. H., and Bancroft, M. J.

Bang, F. B., and Glaser, R. W.
1943. The persistence of poliomyelitis virus in flies. Amer. Jour. Hyg., *37*: 320-324.

Baranov, N.
1924. Glasnik [of] Yugoslav. Ministry of Agric., *2*:40-52 (not seen).

Barlow, J. S.
1964. Personal communication to R. L. Beard: "Barlow . . . has stated that the host (*Agria affinis*) he uses for rearing *Aphaereta* [*pallipes*] must be reared on fresh pork liver for most effective parasitization. Frozen liver, or beef liver, cannot be substituted."

Barnes, H. F.
1924. Some facts about *Pollenia rudis* Fabr. Vasculum, *10*:34-38.

Barotte, J.
1925. Les trypanosomiases de l'Afrique du nord. Mém. Soc. Sciences Nat. Maroc., *11*:117-122.

Barros, R. Donoso, and Urzúa, L. Ferrada
1947. Discovery of flies naturally infected by acid-alcohol resistant bacilli (probably Hansen group) in leprosarium of Pascua Island. Hallazgo de moscas naturalmente infectadas por bacilos acido-alcohol-resistentes (probablemente grupo Hansen) en leproserías de Isla de Pascua, *7*: 133-134.

Bartlett, K. A.
1939. Introduction into Puerto Rico of beneficial insects for control of the horn fly. Puerto Rico Agric. Exp. St., USDA, Agric. Notes, No. 88, 6 pp.

Bassett, D.C.J.
1967. (Personal communication), Univ. of West Indies, Trinidad Regional Virus Lab.

Bassett, D.C.J.
1970. *Hippelates* flies and streptococcal skin infection in Trinidad. Trans. Roy. Soc. Trop. Med. Hyg., *64*:138-147.

Baumberger, J. P.
1919. A nutritional study of insects, with special reference to microorganisms and their substrata. Jour. Exp. Zool., *28*:1-81.

Bay, E. C.
1964. See Legner, E. F., and Bay, E. C.

Bay, E. C.
1965b. See Legner, E. F., and Bay, E. C.

Bay, E. C.
1965c. See Legner, E. F., et al.

Bay, E. C.
1965d. See Legner, E. F., and Bay, E. C.

Bay, E. C., and Legner, E. F.
1963. The prospect for the biological control of *Hippelates collusor* (Townsend) in southern California. Proc. & Papers 31st Conf. California Mosquito Contr. Assoc., Inc., pp. 76-79.

Bay, E. C.
Unpublished data. Bay found that excessive numbers of males in cultures did not diminish the oviposition of females of *Spalangia drosophilae*, *Phaenopria occidentalis* or *Eupteromales nidulans* when these were being reared on pupae of *Hippelates collusor*.

Beard, R. L.
1964. Parasites of muscoid flies. Bull. Wld. Health Org., *31*:491-493.

Beard, R. L., and Walton, G. S.
1965. An *Aspergillus* toxin lethal to larvae of the house fly. Jour. Invert. Pathol., *7*:522-523.

Beck, M.
1910. Experimentelle Beiträge zur Infektion mit *Trypanosoma gambiense* und zur Heilung der menschlichen Trypanosomiasis. Arb. Kaisl. Gesundheit., *34*:318-376.

Becker, E. R.
1922-3. Observations on the morphology and life history of *Herpetomonas muscae-domesticae* in North American muscoid flies. Jour. Parasit., *9*: 199-213.

Becker, E. R.
1924. Transmission experiments on the specificity of *Herpetomonas muscae-domesticae* in muscoid flies. Jour. Parasit., *10*:25-34.

Beesley, W. N.
1968. Observations on the biology of the ox warble-fly (*Hypoderma*: Diptera, Oestridae). Ann. Trop. Med. Parasit., *62*:8-12.

Beling
1868. Der Heerwurm. Der zoologische Garten, 9. (Quoted from Beling by Thomson, R.C.M., 1937.)

Bell, C. R.
1953. See Steinhaus, E. A., and Bell, C. R.

Bell, R. L.
1952. An Indiana microsporidian parasitic on *Drosophila*. Proc. Indiana Acad. Science, *62*:297.

Bellosillo, G. C.
1937. *Herpetomonas muscularum* (Leidy) in *Lucilia sericata* Meigen. Philippine Jour. Science, *63*:285-305.

Bengtson, I. A.
1922. Preliminary note on a toxin-producing anaerobe isolated from the larvae of *Lucilia caesar*. Publ. Health Rept., *37*:164-170.

Benson, O. L., and Wingo, C. W.
1963. Investigations of the face fly in Missouri. Jour. Econ. Entom., *56*:251-258.

Bequaert, J.
1916. See Rodhain, J., and Bequaert, J.

Berberian, D. A.
1938. Successful transmission of cutaneous leishmaniasis by the bites of *Stomoxys calcitrans*. Proc. Soc. Exp. Biol. Med., *38*:254-256.

Berberian, D. A.
1939. A second note on successful transmission of oriental sore by the bites of *Stomoxys calcitrans*. Ann. Trop. Med. Parasit., *33*:95-96.

Berkaloff, A., Bregliano, J., and Ohanessian, A.
1965. Mise en évidence de virions dans des drosophiles infectées par le virus héréditaire. Compt. Rend. Séances Acad. Sciences Paris, *260*:5956-5959.

Berlese, Am.
1896. Rapporti fra la vite ed i saccaromicete. Ricerche sui mezzi di trasporto dei fermenti alcoolici. Riv. Patolog. Vegetale Zimologia, *3*:295-342; *4*:162-180; *5*:184-204.

Berlese, Am.
1897. See Berlese, An., and Berlese, Am.

Berlese, An.
1881. Indagini sulle metamorfosi di alcuni acari insetticoli. Atti Ist. Veneto Sci. Lettere Arti, *8*:43-46.

Berlese, An.
1882a. Il polimorfismo e la partenogenese di alcuni acari (Gamasidi). Bull. Soc. Entom. Italiana, *14*:88-141.

Berlese, An.
1882b. Polymorphisme et parthénogénèse de quelques acariens (Gamasides). Arch. Italiennes Biol., *2*:108-129.

Berlese, An., and Berlese, Am.
1897. See Gigliogi, I. Insects and yeasts. Nature, *56*:575-577.

Bernard, J.
1964. Propagation du virus héréditaire de la Drosophile, σ, au niveau de disques imaginaux transplantés. Compt. Rend. Acad. Sci., *259*:4879-4881.

Bernard, J.
1968. Contribution à l'étude du virus héréditaire de la Drosophile, sigma. Exptl. Cell Res., *50*:117-126.

Bertarelli, E.
1909. Diffusione del tifo, colle mosche e mosch. portatrici di bacilli specifici nella case dei tifosi. Boll. Soc. Med. Parma, *2*:262-272.

Berti, A. L.
1956. See Gabaldón, A., et al.

Bettini, S.
1965. Acquired immune response of the house fly, *Musca domestica* (Linnaeus), to injected venom of the spider *Latrodectus mactans tredecimguttatus* (Rossi). Jour. Invert. Pathol., *7*:378-383.

Beyer, G. E.
1925. The bacteriology of market flies of New Orleans. Quart. Louisiana St. Bd. Health, *16*:110-116.

Bhalla, S. C., and Sokal, R. R.
1964. Competition among genotypes in the house fly at varying densities and proportions (The *green* strain). Evolution, *18*:312-330.

Birk, W.
1932. Die Uebertragung des Paratyphus Breslau durch die *Stomoxys calcitrans*. Zentralbl. Bakt. Parasit. Infekt. Hyg., I *Abt.*, *124*:280-300.

Bishop, M. B.
1943. See Power, M. E., et al.

Bishopp, F. C.
1913a. The stable-fly (*Stomoxys calcitrans* L.) an important live-stock pest. Jour. Econ. Entom., *6*:112-126.

Bishopp, F. C.
1913b. Farmers' Bull., U. S. Dept. Agric., No. 540, 23 pp.

Bishopp, F. C.
1929-1930 In Marlatt, C. L. Report [1929-30] of the Chief of the Bureau of Entomology. Washington, D.C.: U. S. Dept. Agric., 1930, 76 pp.

Bishopp, F. C., Dove, W. E., and Parman, D. C.
1915a. Notes on certain points of economic importance in the biology of the house fly. Jour. Econ. Entom., *8*:54-71.

Bishopp, F. C., and Laake, E. W.
1915b. A preliminary statement regarding wool maggots of sheep in the United States. Jour. Econ. Entom., *8*:466-474.

Bishopp, M. B.
1939. Farmers' Bull., U. S. Dept. Agric., No. 1097, 15 pp.

Blakitnaîa, L. P.
1962. On wintering and some other aspects of the biology and ecology of synanthropic flies in northern regions of the Kirghiz Republic. Med. Parazit. Moscow, *31*:424-429.

Blanc, G.
1919. See Nicolle, C., et al.

Blanc, G., and Caminopetros, J.
1921. Enquête sur le bouton d'orient en Crète. Réflexions qu'elle suggère sur l'étiologie et le mode de dispersion de cette maladie. Ann. Inst. Pasteur, *35*:151-166.

Blanc, G., Bruneau, J., and Chabaud, A.
1950. Quelques essais de transmission de la toxoplasmose par arthropodes piqueurs. Ann. Inst. Pasteur, *78*:277-280.

Blanchard, R.
1899. Bull. Soc. Entom. France, *65*:641 (not seen).

Blickle, R. L.
1961. Parasites of the face fly, *Musca autumnalis*, in New Hampshire. Jour. Econ. Entom., *54*:802.

Blochwitz, A.
1929. Schimmelpilze als Tierparasiten. Berichte Deutsch. Botan. Ges., *47*: 31-34.

Boche, R. D.
1939. Hymenopteran parasitism of *Drosophila*. Genetics, *24*:95.

Bodkin, G. E.
1917. Cowfly tigers. An account of the hymenopterous family Bembecidae in British Guiana. Jour. Bd. Agric. British Guiana, *10*:119-125.

Bogdanow, E. A.
1906. Über das Züchten der Larven der gewöhnlichen Fleischfliege (*Calliphora vomitoria*) in sterilisierten Nahrmitteln. Arch. Gesammte: Physiol., *113*: 97-105.

Bogdanow, E. A.
1908. Über die Abhängigkeit des Wachstums der Fliegenlarven von Bakterien und Fermenten und über Variabilität und Vererbung bei den Fleischfliegen. Archiv. Physiol., Abt. Arch. Anat. Physiol. (Suppl.), pp. 173-200.

Bogoi͡avlenskiĭ, N. A., and Demidova, A. I͡a.
1928. Rol' mukh v perenesenii i͡aits paraziticheskikh cherveĭ. [Role of flies in dissemination of parasitic worm eggs.] Vrachebnai͡a Gazeta, No. 16, pp. 1101-1103.

Boikov, B. V.
1932. Rol' mukh v rasprostranenii briushnogo tifa i drugikh zheludochnokishechnykh zabolevaniĭ. [Role of flies in dissemination of typhoid and other stomach and intestinal diseases.] Zhurnal Mikrob. Epidem. Immunol., No. 7-8, pp. 26-39.

Böing, W.
1913. See Schuberg, A., and Böing, W.

Böing, W.
1914. See Schuberg, A., and Böing, W.

Bolaños, R.
1959. Frecuencia de *Salmonella* y *Shigella* en moscas domésticas colectadas en la ciudad de San José. Rev. Biol. Trop., *7*:207-210.

Borel, F.
1905. See Chantemesse, A., and Borel, F.

Borgatti, A. L., and Guyer, G. E.
1963. The effectiveness of commercial formulations of *Bacillus thuringiensis* Berliner on house-fly larvae. Jour. Insect Path., *5*:377-384.

Bornemissza, G. F.
1968. Studies on the histerid beetle *Pachylister chinensis* in Fiji, and its possible value in the control of buffalo-fly in Australia. Aust. J. Zool., *16*: 673-688.

Bornstein, A.
1963b. See Greenberg, B., et al.

Bornstein, A. A.
1964. See Greenberg, B., and Bornstein, A. A.

Bos, A.
1932. Overbrengingsproeven van hoenderpokken door *Anopheles maculipennis* Mg., *Theobaldia annulata* Schr. en *Stomoxys calcitrans*. Tijdschr. Diergeneesk., *59*:191-194.

Bossink, G.A.H.
1965. See Velthuis, H.H.W., et al.

Bouček, Z.
1963. A taxonomic study in *Spalangia* Latr. (Hymenoptera, Chalcidoidea). Acta Entom. Musei Nationalis Prag, *35*:492-512.

Bouché, P. F.
1834. Naturgeschichte der Insekten, besonders ihrer ersten Zustände als Larven und Puppen. Berlin (book; not seen).

Boudreau, F. G., Brain, C. K., and McCampbell, E. F.
1914. Acute poliomyelitis—with special reference to the disease in Ohio and certain transmission experiments. Monthly Bull., Ohio St. Bd. Health, *4*:23-62, 175-203, 335-357.

Bouet, G., and Roubaud, E.
1912. Expériences de tran[s]mission des trypanosomiases animales de l'Afrique occidental française, par les stomoxes. Bull. Soc. Path. Exot. Paris, *5*:544-550.

Bouffard, G.
1907a. Sur l'étiologie de la souma, trypanosomiase du Soudan français. Compt. Rend. Soc. Biol., *62*:71-73.

Bouffard, G.
1907b. La souma, trypanosomiase du Soudan français. Note préliminaire. Ann. Inst. Pasteur, *21*:587-592.

Bouhélier, R., and Hudault, E.
1936. Un dangereux parasite de la vigne au Maroc *Dorysthenes forficatus* F. (Col., Céramb.). Rev. Zool. Agric., *35*:145-153, 173-176.

Bouillant, A., Lee, V. H., and Hanson, R. P.
1965. Epizootiology on mink enteritis: II, *Musca domestica* L. as a possible vector of virus. Canadian Jour. Comp. Med. Veter. Science, *29*:148-152.

Bourke, A.T.C.
1964. An evaluation of the role of *Phormia regina* in transmission of eastern encephalitis. Publ. Health Rept., *79*:522-524.

Bourne, J. R., and Nielsson, R. J.
1967. *Nemobius fasciatus*—a predator on horn fly pupae. Jour. Econ. Entom., *60*:272-274.

Bouvier, G., and Gaschen, H.
1944. Sur quelques parasites de diptères piqueurs. Mitt. Schweiz. nt. Ges., *19*:191-197.

Boyd, J.S.K.
1957. Dysentery: some personal experiences and observations. Trans. Roy. Soc. Trop. Med. Hyg., *51*:471-487.

Bozhenko, V. P., and Kni͡azevskiĭ, A. N.
1948. Osenni͡ai͡a mukha-zhigalka *Stomoxys calcitrans* L. kak perenoschik tularemii. [The autumn biting fly *Stomoxys calcitrans* L. as a carrier of tularemia.] Izvest. Akad. Nauk Kazakh. SSR, Serii͡a Parasit., *6*:62-66.

Brain, C. K.
1914. See Boudreau, F. G., et al.

Brauns, W.
1955. See Heinz, H. J., and Brauns, W.

Breakey, E. P.
1929. Notes on the natural enemies of the iris borer, *Macronoctua onusta* Grote (Lepidoptera). Ann. Entom. Soc. Amer., *22*:459-464.

Brefeld, O. Von
1908. Untersuchungen aus dem Gesamtgebiete der Mykologie. XIV Band. Die Kultur der Pilze. Münster: see pp. 91, 92, 119-122.

Bregliano, J. C.
1965. Étude de la quantité de virus sigma infectieux contenue dans les cystes ovariens de drosophiles stabilisées et inoculées. Ann. Inst. Pasteur, *109*:638-651.

Bregliano, J. C.
1965. See Berkaloff, A., et al.
von Bremi
1846. Beitrag zur Kunde der Dipteren. Isis, *3*:164-175.
Brèthes, J.
1915. Sur la *Prospalangia platensis* (n. gen., n. sp.) (Hymén.) et sa biologie. Ann. Soc. Cient. Argentina, *79*:314-320.
Brethour, J. R.
1960. See Harvey, T. L., and Brethour, J. R.
Bridwell, J. C.
1918. Certain aspects of medical and sanitary entomology in Hawaii. Trans. 25th and 26th Ann. Mtgs. Med. Soc. Hawaii (1916 and 1917), pp. 27-32.
Bridwell, J. C.
1919. Descriptions of new species of hymenopterous parasites of muscoid Diptera with notes on their habits. Proc. Hawaiian Entom. Soc., *4*: 166-179.
Briggs, J. D.
1960. Reduction of adult house-fly emergence by the effects of *Bacillus* spp. on the development of immature forms. Jour. Insect Path., *2*:418-432.
Brooks, F. E.
1922. Curculios that attack the young fruits and shoots of walnut and hickory. U. S. Dept. Agric., Bull. 1066, Washington, D.C., 16 pp.
Broudin, Peytavin, Nguyên-Van-Dên, Kiêu-Thièn-Thè, and Nguyên-Trung-Truyên.
1926. Contribution à l'étude du surra d'Indochine. Bull. Soc. Path. Exot. Paris, *19*:746-761.
Brown, F. M.
1927. Descriptions of new bacteria found in insects. Amer. Mus. Novit., No. 251, pp. 1-11.
Brown, G. C.
1948. See Francis, T., Jr., et al.
Brown, J. F.
1965. Preliminary investigations on transmission of *Moraxella bovis* by the face fly, *Musca autumnalis*. Thesis, Clemson Univ. Dept. Entom. Zool., 44 pp.
Brown, R. V.
1965. Control of *Histiostoma laboratorium* in *Drosophila* cultures. Jour. Econ. Entom., *58*:156-157.
Brues, C. T.
1912a. See Rosenau, M. J., and Brues, C. T.
Brues, C. T.
1912b. See Rosenau, M. J., and Brues, C. T.
Brues, C. T.
1946. Dragonflies as predatory enemies of the stable fly (*Stomoxys calcitrans*). Psyche, *53*:50-51.
Brug, S. L.
1914. *Octosporea monospora* (Chatton u. Krempf). Arch. Protist., *35*:127-138.
Brug, S. L.
1915. *Herpetomonas homalomyiae* n. sp. Arch. Protist., *35*:119-126.
Brun, G., L'Héritier, P., and Plus, N.
1955b. Le virus héréditaire de la drosophile. Atti VI Cong. Intl. Microb., Roma, *5*:517-519.
Brun, G., and Sigot, A.
1955a. Étude de la sensibilité héréditaire au gaz carbonique chez la drosophile. II. Installation du virus σ dans la lignée germinale à la suite d'une inoculation. Ann. Inst. Pasteur, *88*:488-513.

Bruneau, J.
1950. See Blanc, G., et al.

Brunold, E.
1950. Über eine neue nematodenart der gattung *Anguillula* aus *Drosophila*—Nährböden. Vierteljahresschr. Naturforsch. Ges. Zürich, *95*:148-150.

Brydon, H. W.
1965a. See Legner, E. F., and Brydon, H. W.

Brydon, H. W.
1966. See Legner, E. F., and Brydon, H. W.

Brygoo, E. R., Sureau, P., and Le Noc, P.
1962. Virus et germes fécaux des mouches de l'agglomération urbaine de Tananarive. Bull. Soc. Path. Exot., *55*:866-881.

Bubák, F.
1903. Beitrag zur kenntnis einiger phycomyceten. Hedwigia, *42*:100.

Buchanan, R. M.
1907. The carriage of infection by flies. Lancet, *2*:216-218.

Buchanan, R. M.
1913. *Empusa muscae* as a carrier of bacterial infection from the house-fly. British Med. Jour., *2*:1369.

Buchanan, R. M.
1916. Insects in relation to disease. Glasgow Med. Jour., *85*:1-24.

Bull, L. B.
1918. See Hill, G. F., 1919.

Bull, L. B.
1919. A contribution to the study of habronemiasis: A clinical, pathological, and experimental investigation of a granulomatous condition of the horse—habronemic granuloma. Trans. Roy. Soc. South Australia, *43*: 85-141.

Bulling, E., Bakri, G., and Kirchberg, E.
1959. Necrophage Fliegen als Salmonellenverbreiter. Intern. Sympos. über Schädliche Fliegen. Zeitsch. Angew. Zool., *46*:331-332.

Burdette, W. J., and Jong Sik Yoon
1967. Mutations, chromosomal aberrations, and tumors in insects treated with oncogenic virus. Science, *154*:340-341.

Burgerjon, A., and Galichet, P. F.
1965. The effectiveness of the heat-stable toxin of *Bacillus thuringiensis* var. *thuringiensis* Berliner on larvae of *Musca domestica* Linnaeus. Jour. Invert. Pathol., *7*:263-264.

Burkman, A.
1963a. See Greenberg, B., and Burkman, A.

Burnett, R. G., and King, R. C.
1962. Observations on a microsporidian parasite of *Drosophila willistoni* Sturtevant. Jour. Insect Path., *4*:104-112.

Burns, E. C., Wilson, B. H., and Tower, B. A.
1961. Effect of feeding *Bacillus thuringiensis* to caged layers for fly control. Jour. Econ. Entom., *54*:913-915.

Bushnell, L. D., and Hinshaw, W. R.
1924. Prevention and control of poultry diseases. Kansas Agric. Exp. Stat., Circ. 106, Manhattan, Kansas, 78 pp.

Bussereau, F.
1964. Essai d'extraction de virus infectieux à partir des spermatozoïdes de drosophiles mâles stabilisées. Compt. Rend. Acad. Sci., *259*:3888-3891.

Buxton, P. A.
1920. The importance of the house-fly as a carrier of *E. histolytica*. British Med. Jour., *1*:142-144.

Bychkov, V. A.
1932. O dlitel'nosti khraneniı͡a mukhami *Bacterium prodigiosum.* [The duration of the persistence of *Bacterium prodigiosum* in flies.] Parasit. Sborn. Zool. Inst. Akad. Nauk SSSR, *3*:149-159.

Cabral, J.
1926. See de Mello, F., and Cabral, J.

Calandruccio, S.
1906. Ulteriori ricerche sulla *Taenia nana.* Bull. Accad. Gioenia Catania, *89*: 15-19. (Also in: Bull. Soc. Zool. Ital. con sede in Roma, 1906, *7*:65-69.)

Caminopetros, J.
1921. See Blanc, G., and Caminopetros, J.

Campos, R. F.
1946. See Alcivar, Z. C., and Campos, R. F.

Canon, G. B.
1966. Intraspecies competition, viability, and longevity in experimental populations. Evolution, *20*:116-130.

Cao, G.
1906. Sul passaggio dei germi a traverso le larve di alcuni insetti. Ann. Ig. Sper., *16*:645-664.

Cardamatis, J. P., and Mélissidis, A.
1911. Du rôle probable de la mouche domestique dans la transmission des "*Leishmania.*" Bull. Soc. Path. Exot., *4*:459-461.

Cardamatis, J. P.
1912a. Des flagellaires dans la mouche domestique. Centralbl. Bakt., Parasit. Infekt. I. Abt., Orig., *65*:66-77.

Cardamatis, J. P.
1912b. De quelques microsporidies chez la mouche domestique. Centralbl. Bakt., Parasit. Infekt., I. Abt., Orig., *65*:77-79.

Carmichael, J.
1934. Annual Report, 1933, of the Veterinary Pathol., Entebbe. Research. (3) Trypanosomiasis. Uganda Prot., Ann. Rept. Veter. Dept. Year ended 1933, pp. 29-45.

Carpenter, P. D.
1960. See Greenberg, B., and Carpenter, P. D.

Carson, H. L.
1957. See Wolfson, M., et al.

Carson, H. L.
1963. See Stalker, H. D., and Carson, H. L.

Carter, H. J.
1861. On a bisexual nematoid worm which infests the common house-fly (*Musca domestica*) in Bombay. Ann. Nat. Hist., Ser. 3, *7*:29-33.

Carter, R. M.
1909. Oriental sore of northern India a protozoal infection: A preliminary communication on the etiology of the disease and the extra-corporeal cycle of the parasite. British Med. Jour., *2*:647-650.

Carvalho, G. G.
1960. See Malogolowkin, C., et al.

Carver, R. K.
1946. See Wharton, G. W., and Carver, R. K.

Casida, J.
1954. See Jenkins, D. W., et al.

Castellani, A.
1907. Experimental investigations on *Framboesia tropica* (Yaws). Jour. Hyg., *7*:558-569.

Castellani, A.
1908. *Framboesia tropica* (Yaws, Pian, Bouba). Jour. Cutan. Dis., *26*:211-224.

Castro, L. E.
1957. See Cavalcanti, A.G.L., et al.
Castro, L. E.
1958. See Cavalcanti, A.G.L., et al.
Cattaneo, C.
1947. See Morellini, M., and Cattaneo, C.
Cattani, G.
1886. Studj [Studi] sul colera. Gazz. Ospedali Milano, *7*:611-612.
Catts, E. P.
1959. See Furman, D. P., et al.
Caudri, L.W.D.
1941. The braconid, *Alysia manducator* Panzer, in its relation to the blow-fly, *Calliphora erythrocephala* Meig. Dissert., Univ. of Leiden.
Cavalcanti, A.G.L., Falcão, D. N., and Castro, L. E.
1957. Sex-ratio in *Drosophila prosaltans*, a character due to interaction between nuclear genes and a cytoplasmic factor. Amer. Naturalist, *91*:321-325.
Cavalcanti, A.G.L., Falcão, D. N., and Castro, L. E.
1958. The interaction of nuclear and cytoplasmic factors in the inheritance of the "sex-ratio" character in *Drosophila prosaltans*. Univ. Brasil, Publ. fac. nacl. filosof. Ser. Cient, *1*:25-54.
Cavara, F.
1947. Osservazioni citologiche sulle "Entomophthoreae." Nuovo Giorn. Botan. Ital. Nuova serie, *6*:411-466.
Celli, A., and Alessi, G.
1888. Trasmissibilitá dei germi patogeni mediante le dejezioni delle mosche. Bull. Soc. Lancisiana Ospedali Roma, *1*:5-8.
Chabaud, A.
1950. See Blanc, G., et al.
Chang, K.
1940. Domestic flies as mechanical carriers of certain human intestinal parasites in Chengtu. West China Border Res. Soc. Jour., Ser. B, *12*:92-98.
Chang, K.
1945. Domestic flies as mechanical carriers of certain human intestinal parasites in Chengtu. Trop. Dis. Bull., *42*:759-760. (Review. Original not seen, possibly same as Chang, K., 1940, above.)
Chant, D. A.
1960. An unusual instance of phoresy in Acarina. Entom. News, *71*:270-271.
Chantemesse, A., and Borel, F.
1905. Mouches et choléra. Bull. Acad. Med. Paris, *54*:252-259.
Chatton, E.
1912a. *Leptomonas* de deux Borborinae (Muscides). Evolution de *Leptomonas legerorum* n. sp. Compt. Rend. Soc. Biol., *73*:286-289.
Chatton, E.
1912b. *Leptomonas roubaudi* n. sp., parasites des tubes de Malpighi de *Drosophila confusa* Staeger. Compt. Rend. Soc. Biol., *73*:289.
Chatton, E.
1913. *Coccidiascus legeri*, n.g., n. sp., levure ascosporée parasite des cellules intestinales de *Drosophila funebris* Fabr. Compt. Rend. Soc. Biol., *75*: 117-120.
Chatton, E., and Alilaire, E.
1908. Coexistence d'un *Leptomonas* (*Herpetomonas*) et d'un *Trypanosoma* chez un muscide non vulnérant, *Drosophila confusa* Staeger. Compt. Rend. Soc. Biol., *64*:1004-1006.
Chatton, E., and Krempf, A.
1911d. Sur le cycle évolutif et la position systématique des protistes du genre *Octosporea* Fl., parasites des muscides. Bull. Soc. Zool. France, *36*:172-179.

Chatton, E., and Leger, A.
1911a. Eutrypanosomes, *Leptomonas* et Leptotrypanosomes chez *Drosophila confusa* Staeger (Muscide). Compt. Rend. Soc. Biol., *70*:34-36.

Chatton, E., and Leger, A.
1911b. Sur l'autonomie spécifique du *Trypanosoma drosophilae* Chatton et Alilaire, et sur les eutrypanosomes des muscides non sanguivores. Compt. Rend. Soc. Biol., *71*:573-575.

Chatton, E., and Leger, A.
1911c. Documents en faveur de la pluralité des espèces chez les *Leptomonas* des Drosophiles. Remarques sur leur morphologie. Compt. Rend. Soc. Biol., *71*:663-666.

Chatton, E., and Leger, M.
1912. Trypanosomides et membrane péritrophique chez les Drosophiles. Culture et évolution. Compt. Rend. Soc. Biol., *72*:453-456.

Chatton, E., and Leger, M.
1913. L'autonomie des trypanosomes propres aux muscides démontrée par les élevages purs indéfinis. Compt. Rend. Soc. Biol., *74-75*:549-551.

Chaïkin, V. N.
1941. See Tarasov, V. A., and Chaïkin, V. N.

Chevalier, L.
1923a. Bull. Soc. Sciences Seine-et-Oise, Série II, *3* (not seen).

Chevalier, L.
1923b. Études sur *Blepharipus signatus*, Hyménoptère mangeur de mouches. Bull. Soc. Sciences Seine-et-Oise, Série II, *3*:38-39.

Chevalier, L.
1923c. Observations sur *Dinetus pictus*, Hyménoptère mangeur de punaises et sur son parasite naturel *Anthomya albescens*. Bull. Soc. Sciences Seine-et-Oise, Série II, *1*:12-14.

Chevalier, L.
1926. Note sur la biologie et la manière de vivre de trois espèces d'Oxybèles. Bull. Soc. Sciences Seine-et-Oise, *2*:1-13.

Chow, C. Y.
1940. The common blue-bottle fly, *Chrysomyia megacephala*, as a carrier of pathogenic bacteria in Peiping, China. Chinese Med. Jour., *57*:145-153.

Clapham, P. A.
1939. On flies as intermediate hosts of *Syngamus trachea*. Jour. Helminth., *17*: 61-64.

Clark, P. F.
1911. See Flexner, S., and Clark, P. F.

Clark, T. B.
1959. See Wallace, F. G., and Clark, T. B.

Clegg, M. T.
1903. See Musgrave, W. E., and Clegg, M. T.

Cleland, J. B.
1912. The relationship of insects to disease in man in Australia. Second Rept. Govt. Bur. Microb. for 1910-11, pp. 141-158.

Cleland, J. B.
1913. Insects and their relationship to disease in man in Australia. Trans. Australasian Med. Cong. (Sydney, 1911), *1*:548-570.

Cochrane, E.W.W.
1912. A small epidemic of typhoid fever in connection with specifically infected flies. Jour. Roy. Army Med. Corps., *18*:271-276.

Coffey, J. H.
1954. See Melnick, J. L., et al.

Cohn, F.
1855. *Empusa muscae* und die Krankheit der Stubenfliege. Nova Acta Acad. Caes. Leopold. Carol. Nat. Cur., xxv, Part 1, 300-360.

Cohn, F.
1857. *Empusa muscae*, and the disease of the common house-fly. A contribution towards the knowledge of epidemics characterised by the presence of parasitic fungi. Quart. Jour. Microscop. Science, *5*:154-160.

Colas-Belcour, J.
1928. See Wollman, E., et al.

Cole, A. C.
1930. *Muscina stabulans* Fall. (Dipt., Muscidae) parasitic on *Arachnara subcarnea* Kell. (Lepidopt., Noctuidae). Entom. News, *41*:112.

Colhoun, E. H.
1953. Notes on the stages and the biology of *Baryodma ontarionis* Casey (Coleoptera: Staphylinidae), a parasite of the cabbage maggot, *Hylemya brassicae* Bouché (Diptera: Anthomyiidae). Canadian Entom., *85*:1-8.

Cook, B. H.
1953. See Floyd, T. M., and Cook, B. H.

Cotterell, G. S.
1920. The life-history and habits of the yellow dung fly (*Scatophaga stercoraria*); a possible blow-fly check. Proc. Zool. Soc. London, *4*:629-647.

Counce, S. J.
1959. See Poulson, D. F., and Counce, S. J.

Counce, S. J., and Poulson, D. F.
1961. Developmental effects of hereditary infections in *Drosophila.* Amer. Zool., *1*:443.

Cousin, G.
1933. Étude biologique d'un chalcidien: *Mormoniella vitripennis* Walk. Bull. Biol. France et Belgique, *67*:371-400.

Coutinho, J. O., Taunay, A. de E., and Penna, L.
1957. Importancia de *Musca domestica* como vector de agentes pathogênicos para o homen. Rev. Inst. Adolpho Lutz. São Paulo, *17*:5-23.

Coutinho, J. M.
1960. See Figueiredo, M. B., and Coutinho, J. M.

Coutts, J. M.
1907. See Edington, A., and Coutts, J. M.

Cova García, P.
1956. Las moscas problema de salud publica y organizacion del servicio de aseo urbano y domicilianio en la ciudad de Valencia. Div. Malar. Direccion Salud Publica, Ministerio Sanidad y Asistencia Social, pp. 1-43.

Cox, G. L., Lewis, F. C., and Glynn, E. F.
1912. The number and varieties of bacteria carried by the common house fly in sanitary and unsanitary city areas. Jour. Hyg., *12*:290-319.

Craig, T. C.
1894. The transmission of the cholera spirillum by the alimentary contents and intestinal dejecta of the common house-fly. Med. Rec. New York, *46*: 38-39.

Crandell, H. A.
1939. The biology of *Pachycrepoideus dubius* Ashmead (Hymenoptera), a pteromalid parasite of *Piophila casei* Linne. Ann. Entom. Soc. Amer., *32*:632-654.

Crawford, M.
1926. Development of *Habronema* larvae in drosophilid flies. Jour. Compt. Path. Therap., *39*:321-323.

Crosskey, R. W., and Davies, J. B.
1962. *Xenomyia oxycera* Emden, a muscid predator on *Simulium damnosum Theobald* in northern Nigeria. Proc. Roy. Entom. Soc. London (A), *37*:22-26.
Cuenod, A.
1919. See Nicolle, C., et al.
Cuocolo, R.
1943a. See Mello, M. J., and Cuocolo, R.
Cuocolo, R.
1943b. See Mello, M. J., and Cuocolo, R.
Curbelo, A., and Arango, M. C.
1945. *Eberthella typhi* en la *Musca domestica.* Inst. Finlay (Instituto Nacional de Higiene), Havana, Seccion de Publ. Cientificas Biblioteca y Museo, pp. 1-43.
Curran, C. H.
1942. The parasitic habits of *Muscina stabulans* Fabricius. Jour. New York Entom. Soc., *50*:355-356.
Currie, D. H.
1910. Flies in relation to the transmission of leprosy. Publ. Health Bull., U. S. Publ. Health Serv., Treas. Dept., *39*:21-42.
Currie, D. H.
1911. Mosquitoes and flies in relation to the transmission of leprosy. Jour. Trop. Med. Hyg., *14*:138-142.
Curry, J. J.
1902. "Surra" or nagana. A report of an acute, fatal epidemic disease affecting horses and other animals; with studies on the mode of transmission, etc. Amer. Med., *4*:95-99.
Cuthbertson, E.
1962. See Gear, J., et al.
da Fonseca, O.
1923. See de Magarinos Torres, C. B., et al.
von Dalla Torre, K. W.
1898. Catalogus hymenopterorum (Cat. hym. hucusque descriptorum systematicus et synonymicus) Lipsiae [Leipzig], *2*:Sect. 4.
Dalrymple, W. H.
1912. Anthrax and tick fever. Amer. Veter. Rev., *40*:601-610.
Dalrymple, W. H.
1914. Some of the more important insects affecting our farm animals. Amer. Veter. Rev., *42*:419-427.
Dansker, V. N.
1935. See Aleksander, L. A., and Dansker, V. N.
da Paz, M. C.
1960. See Malogolowkin, C., et al.
Darling, S. T.
1912a. Experimental infection of the mule with *Trypanosoma hippicum* by means of *Musca domestica.* Jour. Exp. Med., *15*:365-369.
Darling, S. T.
1912b. The part played by flies and other insects in the spread of infectious diseases in the tropics with special reference to ants and to the transmission of *Tr. hippicum* by *Musca domestica.* Trans. 15th Intl. Cong. Hyg. Demogr., Washington, D.C., pp. 182-185.
Davé, K. H., and Wallis, R. C.
1965. Survival of Type 1 and Type 3 polio vaccine virus in blowflies (*Phaenicia sericata*) at 40°C. Proc. Soc. Exp. Biol. Med., *119*:121-124.
Davies, J. B.
1962. See Crosskey, R. W., and Davies, J. B.

Davis, J. J.
1919. Contributions to a knowledge of the natural enemies of *Phyllophaga*. State Illinois Nat. Hist. Survey Bull. Urbana, *13* (Art. 5): 55-133.

de Alba, J.
1955. See Neel, W. W., et al.

de Arêa Leao, A. E.
1923. See de Magarinos Torres, C. B., et al.

De Bach, P., and Smith, H. S.
1947. Effects of parasite population density on rate of change of host and parasite populations. Ecology, *28*:290-298.

de Balsac, H. H.
1961. See Dhennin, L., et al.

DeCapito, T. M.
1961. See Richards, C. S., et al.

De Castro, M. P.
1945. See Pereira, C., and De Castro, M. P.

De Castro, M. P.
1947. See Pereira, C., and De Castro, M. P.

Decker, G. C.
1931. The biology of the stalk borer, *Papaipema nebris* (Gn.). Res. Bull. Iowa Agric. Exp. Stat., Ames, Iowa, No. 143, pp. 289-351.

Decker, G. C.
1932. Biology of the Bidens borer, *Epiblema otiosana* (Clemens) (Lepidoptera, Olethreutidae). Jour. New York Entom. Soc., *40*:503-509.

De Coursey, J. D., and Otto, J. S.
1956. Flies on the faces of Egyptian children. Jour. New York Entom. Soc., *64*:129-135.

De Coursey, R. M.
1927. A bionomical study of the cluster fly, *Pollenia rudis* (Fabr.). Ann. Entom. Soc. Amer., *20*:368-384.

De Geer, C.
1771. Mémoires pour servir à l'histoire des insectes, *2* (Part 1): 85. Stockholm.

De Geer, C.
1778. Mémoires pour servir à l'histoire des insectes, *7*:115. Stockholm.

De Geer, C.
1782. Abhandlungen zur geschichte der insekten, *4*:38. Tr. into Germ. by Götze, German edition, Nürnberg. (Also excerpted in: Oken, Allgemeine Naturgeschichte, *5* [Part 1]: 788.) "The fly disease was first mentioned by de Geer. . . . He described only the outward symptoms . . . as to the cause, merely supposed it was a poison ingested with the flies' food. In the body he found only an oily fluid."

Dekester, M.
1923. See Jausion, H., and Dekester, M.

de la Paz, G. C.
1938. The bacterial flora of flies caught in foodstores in the city of Manila. Monthly Bull. Bureau Health, *18* (No. 8): 1-20.

Delcourt, A., and Guyénot, E.
1910. De la possibilité d'étudier certains diptères en milieu défini. Compt. Rend. Acad. Sci., *151*:255-257.

de Lestrange, M.-T.
1954. Action de la température sur le virus responsable de la sensibilité à l'anhydride carbonique chez la drosophile. Compt. Rend. Acad. Sci., *239*:1159-1162.

de Magarinos Torres, C. B., da Fonseca, O., and de Arêa Leao, A. E.
1923. Sur la "Esponja" (Habronémose cutanée des équidés). Du parasitisme des mouches par l'*Habronema muscae* Carter. Compt. Rend. Soc. Biol., *89*:767-768.

de Mello, F., and Cabral, J.
1926. Les insectes sont-ils susceptibles de transmettre la lèpre? Bull. Soc. Path. Exot., *19*:774-778.

Demidova, A. Ia.
1928. See Bogoīavlenskiĭ, N. A., and Demidova, A. Īa.

Derbeneva-Ukhova, V. P.
1935. On the number of generations of *Musca domestica*. Med. Parazit. Moscow, *4*:404-407.

Derbeneva-Ukhova, V. P.
1961. K sravnitel'noĭ èkologii sinantropnykh vidov semeĭstv (Muscidae i Calliphoridae [Diptera]). [On the comparative ecology of synanthropic species belonging to the families Muscidae and Calliphoridae (Diptera).] Med. Parazit. Moscow, *30*:27-37.

de Salles, J. F., and Hathaway, C. R.
1944. Nota sôbre a infestação de *Musca domestica*, Linneu, 1758 por um ficomiceto do genêro *Empusa* (*). Mem. Inst. Oswaldo Cruz, *41*:95-99.

Descazeaux, J.
1921. See Roubaud, E., and Descazeaux, J.

Descazeaux, J.
1922a. See Roubaud, E., and Descazeaux, J.

Descazeaux, J.
1922b. See Roubaud, E., and Descazeaux, J.

Descazeaux, J.
1923. See Roubaud, E., and Descazeaux, J.

Descazeaux, J., and Morel, R.
1933. Diagnostic biologique (xénodiagnostic) des habronémoses gastriques du cheval. Bull. Soc. Path. Exot., *26*:1010-1014.

de Souza-Araujo, H. C.
1944. Verificação da infecção de moscas da familia Tachinidae pela *Empusa* Cohn 1855. Essas moscas, sugando ulceras lepróticas, se infestaram com o bacilo de Hansen. Mem. Inst. Oswaldo Cruz, *41*:201-203.

de Vries, A. H.
1941. See Ullyett, G. C., and de Vries, A. H.

Dhennin, L.
1961. See Dhennin, L., et al.

Dhennin, L., de Balsac, H. H., Verge, J. and Dhennin, L.
1961. Du rôle des parasites dans la transmission naturelle et expérimentale du virus de la fièvre aphteuse. Rec. Méd. Véter., *137*:95-104.

Dick, R. J.
1955. See Ferris, D., et al.

di Delupis, G. D.
1963a. See Filipponi, A., and di Delupis, G. D.

di Delupis, G. D.
1964b. See Filipponi, A., and di Delupis, G. D.

Dieben, C.P.A.
1928. Enkele surraoverbrengingsproeven met *Stomoxys calcitrans* en *Ctenocephalus canis*. [Some surra-transmission experiments with *S. calc.* and *Cten. canis.*] Nederl.-Indie. Bl. Diergeneesk., *40*:57-83.

Diguet, L.
1909. Le mosquero. Bull. Soc. Nationale d'Acclimatation France, October, pp. 368-375. (Also: Compt. Rend. Acad. Sci. Paris, 1909, pp. 735-736.)

Dishon, T.
1956. Ph.D. Thesis, Hebrew University of Jerusalem (not seen). Quoted from Theodor, O., Israel Jour. Exp. Med., *11* (2), 1963 (no pages given).

Dobson, R. C.
1966. See Sanders, D. P., and Dobson, R. C.

Dodd, A. P.
1920. Two new Hymenoptera of the superfamily Proctotrypidae from Australia. Proc. Linn. Soc. New South Wales, Sydney, *45*:443-446.

Dodd, W. L.
1910. See Orton, S. T., and Dodd, W. L.

Donatien, A.
1922a. See Sergent, E., and Donatien, A.

Donatien, A.
1922b. See Sergent, E., and Donatien, A.

Donatien, A., and Lestoquard, F.
1923. Le *debab* naturel du chien. Transmission par les stomoxes. Bull. Soc. Path. Exot., *16*:168-170.

Donatien, A., and Parrot, L.
1922. Trypanosomiase naturelle du chien au Sahara. Bull. Soc. Path. Exot., *15*:549-551.

Donovan, E.
1797. The natural history of British insects. Vol. 6, see p. 84.

Dove, W. E.
1915a. See Bishopp, F. C., et al.

Dove, W. E.
1916. Some notes concerning overwintering of the house-fly, *Musca domestica*, at Dallas, Texas. Jour. Econ. Entom., *9*:528-538.

Dow, R. P.
1953. See Melnick, J. L., and Dow, R. P.

Dowden, P. B.
1935. *Brachymeria intermedia* (Nees), a primary parasite, and *B. compsilurae* (Cwfd.) a secondary parasite of the Gypsy moth. Jour. Agr. Res., *50*: 495-523.

Down, H. A.
1946. See Harris, A. H., and Down, H. A.

Downey, T. W.
1963. Polioviruses and flies: studies on the epidemiology of enteroviruses in an urban area. Yale Jour. Biol. Med., *35*:341-352.

Drbohlav, J. J.
1925. Studies on the relation of insect herpetomonad and crithidial flagellates to leishmaniasis. Amer. Jour. Hyg., *5*:580-621.

Drbohlav, J. J.
1926. The cultivation of *Herpetomonas muscarum* (Leidy, 1866) Kent, 1881 from *Lucilia sericata.* Jour. Parasit., *12*:183-190.

Drea, J. J.
1966. Studies of *Aleochara tristis* (Coleoptera: Staphylinidae), a natural enemy of the face fly. Jour. Econ. Entom., *59*:1368-1373.

Dresner, E.
1949. Culture and use of entomogenous fungi for the control of insect pests. Contr. Boyce Thompson Inst., *15*:319-335.

Dresner, E.
1950. The toxic effect of *Beauveria bassiana* (Bals.) Vuill., on insects. Jour. New York Entom. Soc., *58*:269-278.

Dresner, E.
1954. Observations on the biology and habits of pupal parasites of the Oriental fruit fly. Proc. Hawaiian Entom. Soc., *15*:299-309.

du Buysson, H.
1917. Observations sur des nymphes de *Coccinella septempunctata* L. (Col.) parasitées par le *Phora fasciata* Fallén. (Dipt.). Bull. Soc. Entom. France, *15*:249-250.

du Buysson, H.
1921. *Phora fasciata* Fall. (Dipt.) parasité par *Homalotylus eitelweinii* Ratzb. (Hymenoptera). Miscellanea Entom. (Uzès), *25*:66-67.

Duca, E.
1958. See Duca, M., et al.

Duca, M., Duca, E., Tomescu, E., and Oana, C.
1958. Cercetări asupra rolului mustei de casă in transmiterea infectiei cu virus Coxsackie. Studii Cercetări Inframicrob., *9*:31-39.

Dugès, A.
1834. Remarques sur la famille des Gamases (3e. famille de l'ordre). Troisième mémoire, article premier. Ann. Sciences Naturelles, Séconde série, Zool., *2*:18-36.

Duhamel, C., and Plus, N.
1956. Phénomène d'interférence entre deux variants du virus de la Drosophile. Compt. Rend. Acad. Sci., *242*:1540-1543.

Duke, H. L.
1913. Some attempts to transmit *Trypanosoma gambiense* by wild *Stomoxys*; with a note on the intestinal fauna of these flies. Rept. Sleeping Sickness Comm. Roy. Soc. London, *13*:89-93.

Duke, H. L.
1934. Annual report of the human trypanosomiases research institute (Uganda) for the year ended 31 Dec. 1933 (see p. 3).

Duncan, J. T.
1926. On a bactericidal principle present in the alimentary canal of insects and arachnids. Parasitology, *18*:238-252.

Dunkerly, J. S.
1911. On some stages in the life history of *Leptomonas muscae domesticae*, with some remarks on the relationships of the flagellate parasites of insects. Quart. Jour. Microscop. Science, *56*:645-655.

Dunkerly, J. S.
1912. On the occurrence of *Thelohania* and *Prowazekia* in anthomyid flies. Centralbl. Bakt., Parasit. Infekt., I. Abt., Orig., *62*:136-137.

Dunn, L. H.
1930. Rearing the larvae of *Dermatobia hominis* Linn. in man. Psyche, *37*: 327-342.

Dunn, P. H.
1960. Control of house flies in bovine feces by a feed additive containing *Bacillus thuringiensis* var. *thuringiensis* Berliner. Jour. Insect Path., *2*:13-16.

Dunn, P. H., and Mechalas, B. J.
1963. The potential of *Beauveria bassiana* (Balsamo) Vuillemin as a microbial insecticide. Abstr. Symp. Pap., 10th Pacific Science Congress, Pac. Science Assoc., Honolulu, 1961, pp. 190-191.

Durán Borda, G.
1888. See Gómez, P., and Durán Borda, G.

Durant, R. C.
1957. See McGuire, C. D., and Durant, R. C.

Dutt, S. C.
1963. See Srivastava, H. D., and Dutt, S. C.

Dutton, J. E., Todd, J. L., and Hanington, J.W.R.
1908. Trypanosome transmission experiments. Ann. Trop. Med. Parasit., *1*: 201-229.

Dzenis, L.
1952. See Gwatkin, R., and Dzenis, L.
Edin, I. M.
1909. See Patton, W. S., and Edin, I. M.
Edington, A., and Coutts, J. M.
1907. A note on a recent epidemic of trypanosomiasis at Mauritius. Lancet, Oct. 5, pp. 952-955.
Edwards, R. L.
1954. The host-finding and oviposition behavior of *Mormoniella vitripennis* (Walker) (Hym., Pteromalidae), a parasite of muscoid flies. Behavior, *17*:88-112.
Emmel, L.
1949. Die Rolle der Fliegen als Krankheits-Überträger, Untersuchungen zur Frage der Bedeutung der *Musca domestica* bei der Übertragung der Bakterienruhr. Zeitsch. Hyg., *129*:288-302.
Emmons, J.
1954. See Melnick, J. L., et al.
Enan, O. H.
1965. Laboratory studies on the effect of crowding on percent emergence, size of adult, and period of development of the house fly *Musca domestica*. Jour. Egyptian Publ. Health Assoc., *40*:177-183.
Epstein, B.
1960. See Pimentel, D., and Epstein, B.
Eskina, G. V.
1963. See Tukhmanyants, A. A., et al.
Evans, A. C.
1930. See Holdaway, F. G., and Evans, A. C.
Evans, A. C.
1933. Comparative observations on the morphology and biology of some hymenopterous parasites of carrion-infesting Diptera. Bull. Entom. Res., *24*:385-405.
Eversole, J. W., Lilly, J. H., and Shaw, F. R.
1965. Comparative effectiveness and persistence of certain insecticides in poultry droppings against larvae of the little house fly. Jour. Econ. Entom., *58*:704-709.
Evtodienko, V. G.
1968. K voprosu o vyzhivaemosti dizenteriinykh mikrobov na poverkhnosti tela i v kishechnike mukh. Kishechnye Infek. Resp. Mezhvedomsb., *2*: 80-81.
Ewing, H. E.
1913. A new parasite of the house fly (Acarina, Gamasoidea). Entom. News, *24*:452-456.
Ewing, H. E.
1919. Stable-flies and chiggers. Jour. Econ. Entom., *12*:466.
Ewing, H. E.
1942. The relation of flies (*Musca domestica* Linn.) to the transmission of bovine mastitis. Amer. Jour. Veter. Res., *3*:295-299.
Ewing, H. E., and Hartzell, A.
1918. The chigger-mites affecting man and domestic animals. Jour. Econ. Entom., *11*:256-264.
Eyre, J.W.H., McNaught, J. C., Kennedy, J. C., and Zammit, T.
1907. Reports of the commission for the investigation of Mediterranean fever (Part vi). I. Report upon the bacteriological and experimental investigations during the summer of 1906, pp. 96-99.
Fabre, J.H.C.
1914. Souvenirs Entomologiques, Série 1, Édition Définitive, see pp. 257, 273.

Fabricius, O.
1780. Fauna groenlandica . . . etc. Hafniae, Lipsiae [Leipzig]; p. 223 (not seen).
Fahringer, J.
1922. Beiträge zur Kenntnis der Lebensweise einiger Schmarotzerwespen unter besonderer Berücksichtigung ihrer Bedeutung für Biolog. Bekämpfung von Schädlingingen. Zeitsch. Angew. Entom., *8*:325-388.
Falcão, D. N.
1957. See Cavalcanti, A.G.L., et al.
Falcão, D. N.
1958. See Cavalcanti, A.G.L., et al.
Fallis, A. M.
1938. See Gwatkin, R., and Fallis, A. M.
Fantham, H. B., and Porter, A.
1913. The pathogenicity of *Nosema apis* to insects other than hive bees. Ann. Trop. Med. Hyg., *7*:569-579.
Fantham, H. B., and Porter, A.
1958. Some pathogenic bacteriform Microsporidia from Crustacea and Insecta. Proc. Zool. Soc. London, *130*:153-168.
Fantham, H. B., and Robertson, K. G.
1927. Some parasitic protozoa found in South Africa. So. African Jour. Science, *24*:441-449.
Farish, D. J., and Axtell, R. C.
1966. Sensory functions of the palps and first tarsi of *Macrocheles muscaedomesticae* (Acarina: Macrochelidae) a predator of the house fly. Ann. Entom. Soc. Amer., *59*:165-170.
Feigin, J. M.
1963. Exposure of the house fly to selection by *Bacillus thuringiensis*. Ann. Entom. Soc. Amer., *56*:878-879.
Feinberg, E. H.
1965. See Pimentel D., et al.
Feng, Lan-chou
1933. Some parasites of mosquitoes and flies found in China. Lingnan Science Jour., *12*:23-31.
Fernier, L.
1934. See Parisot, J., and Fernier, L.
Ferris, D., Hanson, R. P., Dick, R. J., and Roberts, R. H.
1955. Experimental transmission of vesicular stomatitis virus by Diptera. Jour. Infect. Dis., *96*:184-192.
Ficker, M.
1903. Typhus und Fliegen. Vorläufige Mitteilung. Archiv. Hyg., *46*:274-283.
Figueiredo, M. B., Coutinho, J. M., and Orlando, A.
1960. Novas perspectivas para o contrôle biológico de algumas pragas com *Bacillus thuringiensis*. Arq. Inst. Biol. São Paulo, *27*:77-85.
Filipponi, A.
1955. Sulla natura dell'associazione tra *Macrocheles muscaedomesticae* e *Musca domestica*. Riv. Parassit., *16*:83-102.
Filipponi, A.
1960. Macrochelidi (Acarina, Mesostigmata) foretici di mosche. Risultati parziali di una indagine ecoligica in corso nell'agro pontino. Parassitologia, *2*:167-172.
Filipponi, A.
1964. The intrinsic rate of increase of a Macrochelid mite (Acari: Mesostigmata), compared with that of the housefly one of the species on which it preys. First. Intl. Cong. Parasit. Rome (no pagination).

Filipponi, A.
1965. Facultative viviparity in *Macrochelidae* (Acari: Mesostigmata). Proc. XII Intl. Cong. Entom. London, 1964, pp. 309-310.

Filipponi, A., and di Delupis, G. D.
1963a. Sul regime dietetico di alcuni macrochelidi (Acari: Mesostigmata), associati in natura a muscidi di interesse sanitario. Riv. Parassit., *24*: 277-288.

Filipponi, A., and di Delupis, G. D.
1964b. Sulla biologia e capacitá riproduttiva di *Macrocheles peniculatus* Berlese (Acari: Mesostigmata) in condizioni sperimentali di laboratorio. Riv. Parassit., *25*:93-111.

Filipponi, A., and Francaviglia, G.
1964a. Larviparità facoltativa in alcuni macrochelidi (Acari: Mesostigmata) associati a muscidi di interesse sanitario. Parassitologia, *6*:99-113.

Filipponi, A., and Pegazzano, F.
1963b. Specie Italiane del grupo-*subbadius* (Acarina, Mesostigmata, Macrochelidae). Redia, *48*:69-91.

Firth, R. H., and Horrocks, W. H.
1902. An inquiry into the influence of soil, fabrics, and flies in the dissemination of enteric infection. British Med. Jour., *2*:936-943.

Fletcher, J.
1900. Rept. of the Entomologist and Botanist. Exp. Farms Rept., Dept. Agric. of Canada, Ottawa, pp. 225-226.

Fletcher, T. B.
1916. Report of the Imperial Pathological Entomologist. Rept. Agric. Res. Inst., Pusa, 1915-16, pp. 78-84.

Fletcher, T. B.
1920. Sci. Repts., Agric. Res. Inst., Pusa, 1919-20, Calcutta, pp. 95-108 (not seen).

Flexner, S., and Clark, P. F.
1911. Contamination of the fly with poliomyelitis virus. Jour. Amer. Med. Assoc., *56*:1717-1718.

Flocken, C. F., and Howard, L. O.
? Unpublished reports U. S. Bur. Animal Indust. (Seen in Stein, C. D. [Jour. Amer. Veter. Med. Assoc., *87*:312-324, 1935] who states that they "were able to transmit the disease with the stable fly [*Stomoxys calcitrans*] in one test but failed in the second.")

Florencio Gomes, J.
1917. See Neiva, A., and Florencio Gomes, J.

Floyd, T. M., and Cook, B. H.
1953. The housefly as a carrier of pathogenic human enteric bacteria in Cairo. Jour. Egyptian Publ. Health Assoc., *28*:75-85.

Flu, P. C.
1911. Studien über die im Darm der Stubenfliege, *Musca domestica* vorkommenden protozoären Gebilde. Centralbl. Bakt., Parasit. Infekt., I. Abt., Orig., *57*:522-535.

Flu, P. C.
1915. Epidemiologische studiën over de cholera te Batavia, 1909-1915. Geneeskundig Tijdschrift voor Nederlansch-Indië, *55*:863-925.

Foley, H.
1910. See Sergent, E., and Foley, H.

Francaviglia, G.
1964a. See Filipponi, A., and Francaviglia, G.

Franchini, G.
1922. Protozoaires de muscides divers capturés sur des euphorbes. Bull. Soc. Path. Exot., *15*:970-978.

Franchini, G., and Mantovani, M.
1915. Infection expérimentale du rat et de la souris par *Herpetomonas muscae domesticae*. Bull. Soc. Path. Exot., *8*:109-111.

Francis, E.
1914. An attempt to transmit poliomyelitis by the bite of *Lyperosia irritans*. Jour. Infect. Dis., *15*:1-5.

Francis, T., Jr.
1943. See Rendtorff, R. C., and Francis, T., Jr.

Francis, T., Jr., Brown, G. C., and Penner, L. R.
1948. Search for extra-human sources of poliomyelitis virus. Jour. Amer. Med. Assoc., *136*:1088-1093.

Fränkel, L.
1912. Zur Biologie der Rekurrensfäden. Virchows. Arch., *209*:97-125.

Fraser, H.
1909. Surra in the federated Malay States. Jour. Trop. Veter. Science, *4*:345-389.

Freeman, J. R.
1934. Prey of *Aeshna cyanea* Müll. (Odon.). Jour. Soc. British Entom., *1*:35.

Friederichs, K.
1919. Studien über Nashornkäfer als Schädlinge der Kokospalme. Monogr. Angew. Entom., No. 4, 116 pp.

Friederichs, K.
1920. Über die Pleophagie des Insektenpilzes *Metarrhizium anisopliae* (Metsch.) Sor. Centralbl. Bakt., Parasit., II, Abt., *50*:335-356.

Froggatt, J. L.
1917. See Froggatt, W. W., and Froggatt, J. L.

Froggatt, J. L.
1919. An economic study of *Nasonia brevicornis*, a hymenopterous parasite of muscid Diptera. Bull. Entom. Res., *9*:257-262.

Froggatt, W. W.
1914a. Sheep-maggot flies in Australia. Bull. Entom. Res., *5*:37-39.

Froggatt, W. W.
1914b. The sheep-maggot fly (*Calliphora rufifacies*) and its parasite. Agric. Gaz. New South Wales, Sydney, Misc. Publ., No. 1716, pp. 107-111.

Froggatt, W. W.
1916. A new parasite on sheep-maggot flies. Notes and description of a chalcid parasite (*Chalcis calliphorae*). Agric. Gaz. New South Wales, Sydney, *27*:505-507.

Froggatt, W. W.
1917. "Policemen flies." Fossorial wasps that catch flies. Agric. Gaz. New South Wales, Sydney, *28*:667-669.

Froggatt, W. W.
1919. The digger chalcid parasite (*Dirhinus sarcophagae*, sp. n. on *Sarcophaga aurifrons*). Agric. Gaz. New South Wales, Sydney, *30*:853-855.

Froggatt, W. W.
1921. Sheep-maggot flies and their parasites. Agric. Gaz. New South Wales, Sydney, *32*:725-731, 807-813.

Froggatt, W. W.
1922. Sheep-maggot flies: No. 5. Methods of control recommended by the Department. Dept. Agric. New South Wales, Farmers' Bull. No. 144, pp. 15-32.

Froggatt, W. W., and Froggatt, J. L.
1917. Sheep-maggot flies: No. 3. Dept. Agric. New South Wales, Farmers' Bull. No. 113, 37 pp.

Froggatt, W. W., and McCarthy, T.
1914. The parasite of the sheep-maggot fly (*Nasonia brevicornis*). Notes and observations in the field and laboratory. Agric. Gaz. New South Wales, Sydney, *25*:759-764.

Frost, W. H.
1912. See Anderson, J. F., and Frost, W. H.

Frye, W. W., and Meleney, H. E.
1932. Investigations of *Endamoeba histolytica* and other intestinal Protozoa in Tennessee: IV. A study of flies, rats, mice and some domestic animals as possible carriers of the intestinal Protozoa of man in a rural community. Amer. Jour. Hyg., *16*:729-749.

Fujita, S.
1932. On the parasitic wasp of the pupa of *Tricholyga bombycis* Bech. I. [In Japanese.] Rep. Dept. Seric. Korea Agric. Expt. Stat., *3*:23-38.

Fukuda, M.
1963. See Suenaga, O., and Fukuda, M.

Fuldner, D.
1964. Geruchsorientierung einer parasitischen Käferlarve (*Aleochara curtula* Goeze) am Puparium des Wirtes (*Calliphora*). Die Naturwissenschaften, *51*:345-346.

Fullaway, D. T.
1917a. Description of a new species of *Spalangia*. Proc. Hawaiian Entom. Soc., *3*:292-294.

Fullaway, D. T.
1917b. Report on beneficial insects. Rept. Div. Entom. for the Bien. Period ending Dec. 31, 1916, Territory of Hawaii Bd. Agric. Forestry, Honolulu, pp. 105-109.

Fullaway, D. T.
1926. Ann. Rept. [of the Entomologist] 1925. Hawaiian Forester and Agric., *23*:47-48.

Fuller, M. E.
1933. The life history of *Onesia accepta* Malloch (Diptera, Calliphoridae). Parasitology, *25*:342-352.

Furman, D. P., Young, R. D., and Catts, E. P.
1959. *Hermetia illucens* (Linnaeus) as a factor in the natural control of *Musca domestica* Linnaeus. Jour. Econ. Entom., *52*:917-921.

Gabaldón, A.
1955. Enseñanzas para la accion sanitaria en la America Latina derivadas de la lucha antimalarica en Venezeula. Bol. Ofic. Sanit. Panamer., *38*:259-265.

Gabaldón, A., Berti, A. L., and Jove, J. A.
1956. El saneamiento en la lucha contra la gastroenteritis y colitis. I Congreso Venezolano de Salud Publica y III Conferencia de Unidades Sanitarias, 55 pp.

Gadzhei, E. F.
1961. See Arskiĭ, V. G., et al.

Gahrliep
1696. De musca, innumerorum minorum reptilium nutritia. Misc. Cur. sive Ephem. Med.-Phys. Germ. Acad. Caes. Leopold Nat. Cur., Dec. 3. Ann. 3, p. 299.

Galichet, P. F.
1965. See Burgerjon, A., and Galichet, P. F.

Galli-Valerio, B.
1908. Recherches expérimentales sur une sarcine pathogène. Centralbl. Bakt., Parasit. Infekt., I. Abt., Orig., *47*:177-186.

Ganon, J.
1908. Cholera en vliegen. Geneeskundig Tijdschrift voor Nederlandsch-Indië, *48*:227-233.

Garrison, G. L.
1924. Rearing records of *Pollenia rudis* Fab. (Dipt., Muscidae). Entom. News, *35*:135-138.

Gaschen, H.
1944. See Bouvier, G., and Gaschen, H.

Gear, J., Cuthbertson, E., and Ryan, J.
1962. A study of South African strains of trachoma virus in experimental animals. Ann. New York Acad. Sciences, *98* (Art. 1): 197-200.

Geldreich, E. E., Kenner, B. A., and Kabler, P. W.
1963. Occurrence of coliforms, fecal coliforms, and streptococci on vegetation and insects. Appl. Microb., *12*:63-69.

Generali, G.
1886? Una larva di nematode della mosca comune. Atti. Soc. Nat. Modena, Ser. 3, *2*:88-89.

Gerberich, J. B.
1951. Transmission of bacteria through the metamorphosis of the housefly and the longevity of such an association. Ph.D. Dissertation. Ohio State Univ., Columbus, Ohio, 82 pp.

Gerberich, J. B.
1952. The housefly (*Musca domestica* Linn.), as a vector of *Salmonella pullorum* (Retteger) Bergy, the agent of white diarrhea of chickens. Ohio Jour. Science, *52*:287-290.

Gercke, G. von
1882. Über die Metamorphose einiger Dipteren. Verhdlg. Vereins Naturwiss. Unterhaltung Hamburg, *5*:68-80.

Gerling, D., and Legner, E. F.
1968. Developmental history and reproduction of *Spalangia cameroni*, parasite of synanthropic flies. Ann. Entom. Soc. Amer., *61*:1436-1443.

Giard, A.
1879. Deux espèces d'*Entomophthora* nouvelles pour la flore française, et présence de la forme *Tarichium* sur une muscide. Bull. Scientifique Dept. du Nord. Sér. II (No. 11):353-363.

Giard, A.
1888. Note sur deux types remarquables d'entomophthorées, *Empusa fresenii* Now. et *Basidiobolus ranarum* Eid., suivie de la description de quelques espèces nouvelles. Compt. Rend. Soc. Biol., Nov. 24, pp. 783-787.

Giglio-Tos
1892. Parasitismo di una larva *Aricia* in un Crabro. Ann. R. Acad. Agric. Torina, *34* (not seen).

Gill, G. D.
1965a. See Love, J. A., and Gill, G. D.

Gill, G. D.
1965b. See Love, J. A., and Gill, G. D.

Gingrich, R. E.
1965. *Bacillus thuringiensis* as a feed additive to control dipterous pests of cattle. Jour. Econ. Entom., *58*:363-364.

Gip, L., and Svensson, S. A.
1968. Can flies cause the spread of dermatophytosis? Acta Derm.-venereol., *48*:26-29.

Girault, A. A.
1916. Soc. Entom., *31*:57-60 (not seen).

Girault, A. A.
1920. Proc. U. S. Nat. Mus., *58*:213 (not seen).

Girault, A. A.
1921. New serphidoid, cynipoid, and chalcidoid Hymenoptera. Proc. U. S. Nat. Mus., *58*:177-216.

Girault, A. A., and Sanders, G. E.
1910a. The chalcidoid parasites of the common house or typhoid fly (*Musca domestica* Linn.) and its allies. Psyche, *16*:9-28.

Girault, A. A., and Sanders, G. E.
1910b. The chalcidoid parasites of the common house or typhoid fly (*Musca domestica* Linn.) and its allies. III. Description of a new North American genus and species of the family Pteromalidae from Illinois, parasitic on *M. domestica* Linn., with biological notes. Psyche, *16*:145-160.

Glaser, R. W.
1922. *Herpetomonas muscae-domesticae*, its behavior and effect in laboratory animals. Jour. Parasit., *8*:99-108.

Glaser, R. W.
1924. A bacterial disease of adult house flies. Amer. Jour. Hyg., *4*:411-415.

Glaser, R. W.
1926. Further experiments on a bacterial disease of adult flies with a revision of the etiological agent. Ann. Entom. Soc. Amer., *19*:193-198.

Glaser, R. W.
1938. Test of a theory on the origin of bacteriophage. Amer. Jour. Hyg., *27*:311-315.

Glaser, R. W.
1943. See Bang, F. B., and Glaser, R. W.

Glynn, E. F.
1912. See Cox, G. L., et al.

Gnedina, M. P.
1934. See Pod"i͡apol'skai͡a, V. P., and Gnedina, M. P.

Goeze, J.A.E.
1774. Des Herrn von Geer I. Discurs von den Insekten . . . etc., mit Fussnoten. Der Naturforscher *3*.

Goeze, J.A.E.
1776. Insecten an Thieren und Selbst an Insecten. Beschäftigungen Berliner Ges. Naturforsch. Freunde, *2*:253-285.

Goldstein, B.
1927. An *Empusa* disease of *Drosophila*. Mycologia, *19*:97-109.

Gómez, P., and Durán Borda, G.
1888-9. Sobre la causa de la muerte de las moscas en Bogotá. Rev. Med., *12*:65-74.

Gonder, R.
1908. See Sieber, H., and Gonder, R.

Goodey, T.
1941. On the morphology of *Mermithonema entomophilum* n.g., n. sp., a nematode parasite of the fly, *Sepsis cynipsea* L. Jour. Helminth., *19*: 105-114.

Gordon, F. B.
1943. Studies on the survival of poliomyelitis virus in insects. Jour. Bact., *45*:77.

Gosio, B.
1925. Über die Verbreitung der Bubonenpesterreger durch Insektenlarven. Archiv. Schiffs. Trop Hyg., *29* (Beiheft 1): 134-139.

Graham, H. M.
1956. See Webb, J. E., Jr., and Graham, H. M.

Graham-Smith, G. S.
1909. Preliminary note on examinations of flies for the presence of colon bacilli. Repts. Loc. Govt. Bd. Publ. Health Med. Subjs. London, n.s., *16*:9-13.

Graham-Smith, G. S.
1910. Observations on the ways in which artificially infected flies (*Musca domestica*) carry and distribute pathogenic and other bacteria. Repts. Loc. Govt. Bd. Publ. Health Med. Subjs. London, n.s., *40* (3):1-40.

Graham-Smith, G. S.
1911a. Further observations of the ways in which artificially infected flies (*Musca domestica* and *Calliphora erythrocephala*) carry and distribute pathogenic and other bacteria. Repts. Loc. Govt. Bd. Publ. Health Med. Subjs. London, n.s., *53*:31-48.

Graham-Smith, G. S.
1911b. Further reports on flies as carriers of infection. 4. Further observations on the ways in which artificially infected flies (*Musca domestica* and *Calliphora erythrocephala*) carry and distribute pathogenic and other bacteria. Repts. Loc. Govt. Bd. Publ. Health Med. Subjs. London, n.s., *16*:31-48.

Graham-Smith, G. S.
1912. An investigation into the possibility of pathogenic microorganisms being taken up by the larva and subsequently distributed by the fly. 41st Ann. Rept. Loc. Govt. Bd., Supp. Rept. Med. Officer, 1911-1912, pp. 330-335.

Graham-Smith, G. S.
1913. Further observation on non-lactose fermenting bacilli in flies, and the sources from which they are derived with special reference to Morgan's bacillus. Repts. Loc. Govt. Bd. Publ. Health Med. Subjs. London, n.s., *85*; further rept. (No. 6) on flies as carriers of infection, pp. 43-46.

Graham-Smith, G. S.
1916. Observations on the habits and parasites of common flies. Parasitology, *8*:440-544.

Graham-Smith, G. S.
1919. Further observations on the habits and parasites of common flies. Parasitology, *11*:347-384.

Grassi, B., and Rovelli, G.
1889. Embryologische forschungen an cestoden. Centralbl. Bakt., Parasit. Infekt. I. Abt., Orig., *5*:370-377.

Gray, A.C.H.
1905. See Greig, E.D.W., and Gray, A.C.H.

Gray, A.C.H.
1906. Some notes on a *Herpetomonas* found in the alimentary tract of *Stomoxys* (*calcitrans*?) in Uganda. Proc. Roy. Soc., *78* (Ser. B):254-257. (See also: Jour. Roy. Army Med. Corps, *7*:581-583.)

Gray, A.C.H.
1907. Some notes on a *Herpetomonas* found in the alimentary tract of *Stomoxys* (*calcitrans*?) in Uganda. Repts. S. S. Comm. Roy. Soc. No. 8 pp. 133-135.

Green, A. A.
1953. Blowflies: A community problem. The Sanitarian, June, 7pp.

Greenberg, B.
1959a. Persistence of bacteria in the developmental stages of the housefly. 1. Survival of enteric pathogens in the normal and aseptically reared host. Amer. Jour. Trop. Med. Hyg., *8*: 405-411.

Greenberg, B.
1959b. Persistence of bacteria in the developmental stages of the housefly. 2. Quantitative study of the host-contaminant relationship in flies breeding under natural conditions. Amer. Jour. Trop. Med. Hyg., *8*:412-416.

Greenberg, B.
1959c. Persistence of bacteria in the developmental stages of the housefly. 4. Infectivity of the newly emerged adult. Amer. Jour. Trop. Med. Hyg., *8*:618-622.

Greenberg, B.
1961. Mite orientation and survival on flies. Nature, *190*:107-108.

Greenberg, B.
1962. Host-contaminant biology of muscoid flies. 3. Effect of hibernation, diapause, and larval bactericides on normal flora of blow-fly prepupae. Jour. Insect Path., *4*:415-428.

Greenberg, B.
1964. Experimental transmission of *Salmonella typhimurium* by houseflies to man. Amer. Jour. Hyg., *80*:149-156.

Greenberg, B.
1966. Bacterial interactions in gnotobiotic flies. Symposium on Gnotobiology. IX Intl. Congr. Microb. Moscow, 1966, pp. 371-380.

Greenberg, B.
1968. Model for destruction of bacteria in the midgut of blow fly maggots. Jour. Med. Entom., *5*:31-38.

Greenberg, B.
1969. *Salmonella* suppression by known populations of bacteria in flies. Jour. Bact., *99*:629-635.

Greenberg, B., and Bornstein, A. A.
1964. Fly dispersion from a rural Mexican slaughterhouse. Amer. Jour. Trop. Med. Hyg., *13*:881-886.

Greenberg, B., and Burkman, A.
1963a. Effect of B-vitamins and a mixed flora on the longevity of germ-free adult houseflies, *Musca domestica* L. Jour. Cell. Comp. Physiol., *62*: 17-22.

Greenberg, B., and Carpenter, P. D.
1960. Factors in phoretic association of a mite and fly. Science, *132*:738-739.

Greenberg, B., and Miggiano, V.
1963c. Host-contaminant biology of muscoid flies. 4. Microbial competition in a blowfly. Jour. Infect. Dis., *112*:37-46.

Greenberg, B., Varela, G., Bornstein, A., and Hernandez, H.
1963b. Salmonellae from flies in a Mexican slaughterhouse. Amer. Jour. Hyg., *77*:177-183.

Greenwood, E. S.
1964. *Bacillus thuringiensis* in the control of *Lucilia sericata* and *Musca domestica.* New Zealand Jour. Science, *7*:221-226.

Gregor, F., and Povolný, D.
1960. Zur Chorologie und hygienisch-epidemiologischen Rolle synanthroper Fliegen in Mitteleuropa. Proc. 11th Intl. Congr. Entom., *2*:419-422.

Greig, E.D.W.
1905. See Nabarro, O., and Greig, E.D.W.

Greig, E.D.W., and Gray, A.C.H.
1932. Reports on the sleeping sickness. Comm. Roy. Soc. London, No. 6. Continuation Rept. on sleeping sickness in Uganda, pp. 203-209.

Griffith, B. T.
1952. A study of antibiosis between wind-borne molds and insect larvae from wind-borne eggs. Jour. Allergy, *23*:375-382.

Griffith, F.
1907. Description of a housefly parasite. Med. Brief, St. Louis, *35*:59-63.

Gross, H., and Preuss, U.
1951. Infektionsversuche an Fliegen mit darmpathogenen Keimen. I. Mitteilung. Versuche mit Typhus und Paratyphusbakterien. Zentralbl. Bakt., I. Abt., Orig., *156*:371-377.

Gross, H., and Preuss, U.
1953. Infektionsversuche an Fliegen mit Darmpathogenen Keimen. II. Mitteilung. Zentralbl. Bakt., Parasit. Infekt. Hyg., I. Abt., Orig., *160*:526-529.

Grudtzina, M. V.
1959a. See Sychevskaia͡, V. I., et al.

Gudnadóttir, M. G.
1960. Studies of the fate of type 1 polioviruses in flies. Jour. Exp. Med., *113*:159-176.

Gudnadóttir, M., and Paul, J. R.
1960. Studies on the fate of type 1 polioviruses in flies. Bact. Proc. 60th Ann. Mtg., p. 104.

Gurney, W. B., and Woodhill, A. R.
1926a. Investigations on sheep blowflies. Part 1. Range of flight and longevity. Agric. Dept. New South Wales, Science Bull., No. 27, pp. 4-28.

Gurney, W. B., and Woodhill, A. R.
1926b. Reports of the external parasites in sheep committee of the departmental research council. No. 3. Biological notes on sheep blowflies in the Moree district. Agric. Gaz. New South Wales, Sydney, *37*:135-144.

Güssow, H. T.
1913. *Empusa muscae* and the extermination of the housefly. Rept. Loc. Govt. Bd. Publ. Health and Med. Subj., n.s., No. 85, pp. 10-14.

Gutberlet, J. E.
1916. Studies on the transmission and prevention of cestode infection in chickens. Jour. Amer. Veter. Med. Assoc., *49*:218-237.

Gutberlet, J. E.
1920. On the life history of the chicken cestode, *Hymenolepis carioca* (Magalhaes). Jour. Parasit., *6*:35-38.

Guyénot, E.
1907. L'appareil digestif et la digestion de quelques larves de mouches. Bull. Sci. France et Belgique, *41* (6 Série): 353-370.

Guyénot, E.
1910. See Delcourt, A., and Guyénot, E.

Guyer, G. E.
1963. See Borgatti, A. L., and Guyer, G. E.

Gwatkin, R., and Dzenis, L.
1952. Studies in pullorum disease. XXIX. Bacteriological examination of blowflies (*Lucilia* sp.) and houseflies (*Musca domestica*) which during their larval stage had fed on chicks infected with *Salmonella pullorum*. Canadian Jour. Comp. Med. Veter. Science, *16*:148-150.

Gwatkin, R., and Fallis, A. M.
1938. Bactericidal and antigenic qualities of the washings of blowfly maggots. Canadian Jour. Res., *16*:343-352.

Gwatkin, R., and Mitchell, C. A.
1944. Transmission of *Salmonella pullorum* by flies. Canadian Jour. Publ. Health, *35*:281-285.

Ghosal, S. C.
1939. See Lal, R., et al.

Hagen, H.
1867. Proc. Boston Soc. Nat. Hist., *11*:323.

Hague, N.G.M.
1963. The influence of *Rhabditis (Rhabditella) axei* (Rhabditinae) on the development of *Stomoxys calcitrans.* Nematologica, *9*:181-184.

Hair, J. A., and Turner, E. C., Jr.
1965. Attempted propagation of *Nasonia vitripennis* on the face fly. Jour. Econ. Entom., *58*:159-160.

Hall, G. N.
1927. Res. Div. Ann. Rept. Rept. Veter. Dept. Uganda year ended 1926, pp. 12-16 (Transmission experiments).

Hall, I. M., and Arakawa, K. Y.
1959. The susceptibility of the house fly, *Musca domestica* Linnaeus, to *Bacillus thuringiensis* var. *thuringiensis* Berliner. Jour. Insect Path., *1*:351-355.

Hamilton, A.
1903. The fly as a carrier of typhoid. An inquiry into the part played by the common house fly in the recent epidemic of typhoid fever in Chicago. Jour. Amer. Med. Assoc., *40*:576-583.

Hamm, A. H., and Richards, O. W.
1926. The biology of the British Crabronidae. Trans. Entom. Soc. London, *74*:297-331.

Hammer, O.
1942. Biological and ecological investigations on flies associated with pasturing cattle and their excrement. Videnskabelige Meddelelser fra Dansk naturhistorisk forening i København, *105*:141-393 [appeared as separate on 15 Nov. 1941].

Handschin, E.
1932. A preliminary report on investigations on the buffalo fly (*Lyperosia exigua* de Meij.) and its parasites in Java and Northern Australia. Australian Council Science Indust. Res. Pamphlet, *31*:24 pp.

Handschin, E.
1934a. Studien an *Lyperosia exigua* Meijere und ihren parasiten. II. Die natürlichen feinden von *Lyperosia.* Rev. Suisse Zool., *41*:1-71.

Handschin, E.
1934b. Studien an *Lyperosia exigua* Meijere und ihren Parasiten. III: Die Anziehung von *Spalangia* zu ihrem Wirte. Rev. Suisse Zool., *41*:267-297.

Hanington, J.W.B.
1908. See Dutton, J. E., et al.

Hanson, R. P.
1955. See Ferris, D., et al.

Hanson, R. P.
1965. See Bouillant, A., et al.

Hanson, W. L.
1965. See McGhee, R. B., et al.

Harada, F.
1953. On the fly as a carrier of hookworm larvae. Med. Biol., *29*:28-30.

Harada, F.
1954. Investigations of hookworm larvae. IV. On the fly as a carrier of infective larvae. Yokohama Med. Bull., *5*:282-286.

Hardy, G. H.
1924. A blowfly and some parasites. Queensland Agric. Jour., *22*:349-350.

Hardy, G. H.
1925. The fecundity of *Mormoniella* and some problems in parthenogenesis. Queensland Agric. Jour., *24*:347-348.

Harris, A. H., and Down, H. A.
1946. Studies of the dissemination of cysts and ova of human intestinal parasites by flies in various localities on Guam. Amer. Jour. Trop. Med., *26*:789-800.

Harrison, R. A.
1953. See Satchell, G. H., and Harrison, R. A.

Hartzell, A.
1918. See Ewing, H. E., and Hartzell, A.

Harvey, T. L.
1964. House fly resistance to *Bacillus thuringiensis* Berliner, a microbial insecticide. Dissert. Abst., *25* (4): 2463.

Harvey, T. L., and Brethour, J. R.
1960. Feed additives for control of house fly larvae in livestock feces. Jour. Econ. Entom., *53*:774-776.

Harvey, T. L., and Howell, D. E.
1965. Resistance of the house fly to *Bacillus thuringiensis* Berliner. Jour. Invert. Pathol., *7*:92-100.

Hashimoto, M.
1965. See Shimizu, F., et al.

Hathaway, C. R.
1944. See de Salles, J. F., and Hathaway, C. R.

Havlik, B.
1964. Sanitarny problem synantropijnych much (Diptera) Wielkiej Pragi. [Sanitary problem of synanthropic flies of Greater Prague.] Wiad. Parazit., *10*:588-589.

Hawley, J. E., Penner, L. R., Wedberg, S. E., and Kulp, W. L.
1951. The role of the house fly, *Musca domestica*, in the multiplication of certain enteric bacteria. Amer. Jour. Trop. Med., *31*:572-582.

Hayes, J. T.
1965. See Pimentel, D., et al.

Hayward, E. H.
1904. The fly as a carrier of tuberculous infection. New York and Philadelphia Med. Jour., *80*:643-644.

Heeger
1848. See Hewitt, C. G., 1912.

Hegner, R.
1928. Experimental studies on the viability and transmission of *Trichomonas hominis*. Amer. Jour. Hyg., *8*:16-34.

Heinz, H. J., and Brauns, W.
1955. The ability of flies to transmit ova of *Echinococcus granulosus* to human foods. So. African Jour. Med. Science, *20*:131-132.

Heister
1727. De pediculis sive pulicibus muscarum. Acta Phys. Med. Acad. Caes. Nat. Cur., *1*:409.

Henneberg, W.
1902. Essigfliegen (*Drosophila fenestrarum* Fall. und *funebris* Fabr.). Deutsche Essigindustrie Berlin, *6*:333-336.

Henry, M.
1927. Investigations into onchocerciasis in New South Wales. II. Final report of the special committee. Rept. Director-General Publ. Health New South Wales for the year 1925, pp. 195-207.

Herms, W. B.
1911. The house fly in its relation to public health. Univ. California, Agric. Exp. Stat. Bull., No. 215, pp. 513-548.

Herms, W. B.
1932. See Michelbacher, A. E., et al.

Herms, W. B. Bailey, S. F., and McIvor, B.
1935. The black widow spider. Univ. California, Agric. Exp. Stat. Bull., No. 591, pp. 1-30.

Hernandez, H.
1936b. See Greenberg, B., et al.

Herold, W.
1922. Beobachtungen an zwei Feinden der Stubenfliege: *Mellinus arvensis* L. und *Vespa germanica* Fabr. Zeitsch. Angew. Entom., *8*:459.

Herold, W.
1923. Zur Kenntnis von *Agrotis segetum* Schiff (Saateule). III. Feinde u. Krankheiten. Zeitsch. Angew. Entom., *9*:306-332.

Herreng, F.
1967. Étude de la multiplication de l'arbovirus "Sindbus" chez la drosophile. Compt. Rend. Acad. Sci., *264*:2854-2857.

Hertig, M., and Wolbach, S. B.
1924. Studies on rickettsia-like micro-organisms in insects. Jour. Med. Res., *44*:329 and on (see p. 359).

Hewitt, C. G.
1912. An account of the bionomics and the larvae of the flies *Fannia (Homalomyia) canicularis* L. and *F. scalaris* Fab., and their relation to myiasis of the intestinal and urinary tract. Repts. Loc. Govt. Bd. Publ. Health Med. Subjs. London, n. s., *66*:15-21.

Hewitt, C. G.
1914. On the predaceous habits of *Scatophaga stercoraria*: a new enemy of *Musca domestica.* Canadian Entom., *46*:2-3.

Hill, G. F.
1919. Relationship of insects to parasitic diseases in stock. 1. The life history of *Habronema muscae, Habronema microstoma,* and *Habronema megastoma.* Proc. Roy. Soc. Victoria, n.s., *31*:11-107.

Hills, G. J.
1961. See Smith, K. M., et al.

Hinshaw, W. R.
1924. See Bushnell, L. D., and Hinshaw, W. R.

Hinshaw, W. R.
1944. See McNeil, E., and Hinshaw, W. R.

Hobby, B. M.
1934a. Notes on predaceous Anthomyiidae and Cordyluridae. Entom. Monthly Mag., *70*:185-190.

Hobby, B. M.
1934b. Predaceous Diptera and their prey. Jour. Soc. British Entom., *1*:35-39.

Hobson, R. P.
1932a. Studies on the nutrition of blow-fly larvae. 2. Rôle of the intestinal flora in digestion. Jour. Exp. Biol., *9*:128-138.

Hobson, R. P.
1932b. Studies on the nutrition of blow-fly larvae. 4. The normal rôle of micro-organisms in larval growth. Jour. Exp. Biol., *9*:366-377.

Hobson, R. P.
1935. Growth of blow-fly larvae on blood and serum. II. Growth in association with bacteria. Biochem. Jour., *29*:1286-1291.

Hoffmann, S.
1950. Die hygienische Bedeutung der Fliegen. Travaux Chimie alimentaire Hyg. Berne, *41*:189-222.

Hofmann, E.
1888. Ueber die Gefahr der Verbreitung der Tuberculose durch unsere Stubenfliege. Correspondenzbl. d. ärztl. Kreis- und Bezirksvereine im Königr. Sachsen, *44*:130-133.

Holdaway, F. G.
1930. Field populations and natural control of *Lucilia sericata.* Nature, *126*: 648-649.

Holdaway, F. G.
1932. Fly strike of sheep: A natural phenomenon. Jour. Coun. Sci. Ind. Res. Australia, *5*:205-211.

Holdaway, F. G., and Evans, A. C.
1930. Parasitism a stimulus to pupation: *Alysia manducator* in relation to the host *Lucilia sericata.* Nature, *125*:598-599.

Holdaway, F. G., and Smith, H. F.
1932. A relation between size of host puparia and sex ratio of *Alysia manducator* Pantzer. Australia Jour. Exp. Biol. Med. Science, *9*:247-259.

Holt, C. J.
1954. See Rendtorff, R. C., and Holt, C. J.

Holz, J.
1953. Die Bedeutung von *Musca domestica* als Überträgen von *Trichomonas foetus.* Tierärztl. Umschau, *8*:396-397.

Honeij, J. A., and Parker, R. R.
1914. Leprosy: Flies in relation to the transmission of the disease. Jour. Med. Res., *30*:127-130.

Horrocks, W. H.
1902. See Firth, R. H., and Horrocks, W. H.

Horstmann, D. M.
1962. See Paul, J. R., et al.

Hoskins, M.
1933. An attempt to transmit yellow fever virus by dog fleas (*Ctenocephalides canis* Curt.) and flies (*Stomoxys calcitrans* Linn.). Jour. Parasit., *19*: 299-303.

Hoskins, W. M.
1932. See Michelbacher, A. E., et al.

Houser, E. C., and Wingo, C. W.
1967. *Aphaereta pallipes* as a parasite of the face fly in Missouri, with notes on laboratory culture and biology. Jour. Econ. Entom., *60*:731-733.

How
? Cited by Sukhova, M. N., 1951, Zool. Zhurnal, *60*:180-190.

Howard, C. W.
1917. Insect transmission of infectious anemia of horses. Jour. Parasit., *4*:70-79.

Howard, L. O.
1891. See Riley, F. V., and Howard, L. O.

Howard, L. O.
? See Flocken, C. F., and Howard, L. O.

Howard, L. O.
1911. The house fly, disease carrier; an account of its dangerous activities and means of destroying it. New York, Fred. A. Stokes Co., 2nd edn. 312 pp.

Howat, C. H.
1936. See Lamborn, W. A., and Howat, C. H.

Howell, D. E.
1965. See Harvey, T. L., and Howell, D. E.

Hudault, E.
1936. See Bouhélier, R., and Hudault, D.

Huddleson, J. F.
1941. See Ruhland, H. H., and Huddleson, J. F.

Hughes, R. D.
1950. The genetics laboratory mite *Histiostoma laboratorium* n. sp. (Anoetidae). Jour. Washington Acad. Sciences, *40*:177-183.

Hughes, R. D., and Nicholas, W. L.
1969. *Heterotylenchus* spp. parasitising the Australian bush fly; additional information on the origin of the parasite of the face fly. Jour. Econ. Entom., *62*:520-521.

Huie, D.
1929. See Yao, H. Y., et al.

Hunter, W.
1906. The spread of plague infection by insects. Centralbl. Bakt. Orig., *40*:43-55

Hurlbut, H. S.
1950. The recovery of poliomyelitis virus after parenteral introduction into cockroaches and houseflies. Jour. Infect. Dis., *86*:103-104.

Hussain, M.
1924. See Ross, W. C., and Hussain, M.

Hutchison, H. H.
1916. See Webb, J. L., and Hutchison, H. H.

Ihle, J. A.
1683. See Menzel, C., and Ihle, J. A.

Ikeda, H.
1965. Interspecific transfer of the "sex-ratio" agent of *Drosophila willistoni* in *Drosophila bifasciata* and *Drosophila melanogaster*. Science, *147*: 1147-1148.

Illingworth, J. F.
1915. Hen fleas, *Xestopsylla gallinacea*, Westw. Hawaiian Forester and Agric., *12*:130-132.

Illingworth, J. F.
1923a. Insect fauna of hen manure. Proc. Hawaiian Entom. Soc., *5*:270-273.

Illingworth, J. F.
1923b. Insects attracted to carrion in Hawaii. Proc. Hawaiian Entom. Soc., *5*:280-281.

Illingworth, J. F.
1927. Insects attracted to carrion in Southern California. Proc. Hawaiian Entom. Soc., *6*:397-401.

Ingram, R. L.
1954a. See Jenkins, D. W., et al.

Ingram, R. L., Larsen, J. R., Jr., and Pippen, W. F.
1956. Biological and bacteriological studies on *Escherichia coli* in the housefly, *Musca domestica*. Amer. Jour. Trop. Med. Hyg., *5*:820-830.

Ingrao, F.
1949. See Morellini, M., et al.

Ivashkin, V. M.
1959. Epizootologiya parabronematoza zhvachnykh. [The epizootiology of *Parabronema* of ruminants.] Tr. Gel'mintol. Lab. Akad. Nauk SSSR, *9*:97-105.

Iwanoff, X.
1934. Über Sommerwunden beim Rinde. Arch. Tierheilk, *67*:261-270.

Jack, R. W.
1917. Natural transmission of trypanosomiasis (*T. pecorum* group) in the absence of tsetse-fly. Bull. Entom. Res., *8*:35-41.

Jack, R. W.
1935. The report of the Chief Entomologist for the year ending 31 December 1934. Agricultural. Rhodesia Agric. Jour., *32*:558-566; also in Bull. Minist. Agric. [So. Rhodesia, No. 962, pp. 1-9.]

Jackson, W. B.
1961. See Richards, C. S., et al.

James, H. C.
1928. On the life histories and economic status of certain cynipid parasites of dipterous larvae, with descriptions of some new larval forms. Ann. Appl. Biol., *15*:287-316.

Jausion, H., and Dekester, M.
1923. Sur la transmission comparée des kystes d' "*Entamoeba dysenteriae*" et de "*Giardia intestinalis*" par les mouches. Arch. Inst. Past. Algérie, *1*: 154-155; also Arch. Inst. Past. l'Afrique du Nord, *3*:154-155.

Jegen, G.
1924. Die protozoäre Parasitenfauna der Stechfliege *Stomoxys calcitrans.* Zool. Jahrb, Abt. Anat., *46*:389-472.

Jenkins, D. W., Casida, J., Ingram, R. L., and Larsen, J. R.
1954. Unpublished data discussed in Exp. Parasit., *3*:474-490.

Jenni, W.
1947. Beziehung zwischen Geschlechtsverhältnis und Parasitierungsgrad einer in Drosophilalarven schmarotzenden Gallwespe (*Eucoila* sp.). Rev. Suisse Zool., *54*:252-258.

Jenni, W.
1951. Beitrag zur Morphologie und Biologie der Cynipide *Pseudeucoila bochei* Weld, eines Larvenparasiten von *Drosophila melanogaster* Meig. Acta Zool., *32*:177-254.

Jepson, F. P.
1915. Report of the Entomologist. Ann. Rept. Dept. Agric. Fiji for 1914; 3. Division of Entomology, pp. 17-27.

Jimenez, L. A.
1955. See Prado, E., and Jimenez, L. A.

Johnston, T. H.
1913. Notes on some Entozoa. Proc. Roy. Soc. Queensland, *24*:63-91.

Johnston, T. H.
1920. Flies as transmitters of certain worm parasites of horses. Science and Industry, Melbourne, *2*:369-372.

Johnston, T. H., and Bancroft, M. J.
1920a. The life history of *Habronema* in relation to *Musca domestica* and native flies in Queensland. Proc. Roy. Soc. Queensland, *32*:61-88.

Johnston, T. H., and Bancroft, M. J.
1920b. Notes on the chalcid parasites of muscoid flies in Australia. Proc. Roy. Soc. Queensland, *32*:19-30.

Johnston, T. H. and Tiegs, O. W.
1921. On the biology and economic significance of the chalcid parasites of Australian sheep-maggot flies. Proc. Roy. Soc. Queensland, *33*:99-128.

Jones, C. M.
1967. *Aleochara tristis,* a natural enemy of face fly. Jour. Econ. Entom., *60*: 816-817.

Jones, C. M., and Perdue, J. M.
1967. *Heterotylenchus autumnalis,* a parasite of the face fly. Jour. Econ. Entom., *60*:1393-1395.

Jones, E. B.
1941. A fly-borne epidemic of enteric fever. Med. Officer, *65*:65-67.

Jong Sik Yoon
1967. See Burdette, W. J., and Jong Sik Yoon.

Jove, J. A.
1956. See Gabaldón, A., et al.

Jowett, W.
1911. Further note on a cattle trypanosomiasis of Portuguese East Africa. Jour. Comp. Path. Therap., *24*:21-40.

Joyeux, C.
1920. Cycle évolutif de quelques cestodes. Suppl. Bull. Biol. France et Belgique, pp. 146-147.

Judd, W. W.
1955. Mites (Anoetidae), fungi (*Empusa* sp.) and pollinia of milkweed (*Asclepias syriaca*) transported by calyptrate flies. Canadian Entom., *87*:366-369.

Kabler, P. W.
1963. See Geldreich, E. E., et al.

Kakizawa, H.
1965. See Shimizu, F., et al.

Kamner, A.
1928. Die Maikäferschwärmjahre in Siebenbürgen. [The swarming years of May beetles in Transylvania.] Verhdl. Siebenburg, Verein Naturwiss., *78*: (Wiss. Teil I): 11-28, Hermannstadt.

Kamyszek, F.
1965. Wplyw niektórych czynników na możliwość biernego przenoszenia patogennych grzybów przez muchy domowe (*Musca domestica*). Wiadomości Parazyt., *11*:567-572.

Kaneko, K.
1965. See Shimizu, F., et al.

Kaneko, K., and Kano, R.
1960. Experimental transmission of *Salmonella pullorum* with larvae of *Sarcophaga peregrina*. Jap. Jour. Sanit. Zool., *11*:66-71.

Kano, R.
1960. See Kaneko, K. and Kano, R.

Kano, R.
1965. See Shimizu, F., et al.

Kåss, E.
1954. Undersøkelser over *Toxoplasma* og toxoplasmose. 101 pp. (see pp. 59-77). Bakter. Inst. Rikshospitalet, Oslo.

Keilin, D.
1909. Sur le parasitisme de la larve de *Pollenia rudis* Fab. dans *Allolobophora chlorotica* Savigny. Compt. Rend. Soc. Biol., *67*:201-203.

Keilin, D.
1911. On the parasitism of the larvae of *Pollenia rudis* (Fabr.) in *Allolobophora chlorotica* Savigny. Proc. Entom. Soc. Washington, D.C., *13*:182-184.

Keilin, D.
1917. Recherches sur les anthomyides à larves carnivores. Parasitology, *9*:325-445.

Kemner, N. A.
1926. Zur Kenntnis der Staphyliniden-Larven. II. Die Lebensweise und die parasitische Entwicklung der echten Aleochariden. Entom. Tidskr., *47*: 133-170.

Kennedy, J. C.
1907. See Eyre, J. W., et al.

Kenner, B. A.
1963. See Geldreich, E. E., et al.

Kiêu-Thièn-Thè
1926. See Broudin et al.

King, R. C.
1962. See Burnett, R. G., and King, R. C.

King, W. E.
1966. See Singh, P., et al.

Kinn, D. N.
1966. Predation by the mite, *Macrocheles muscaedomesticae* (Acarina: Macrochelidae), on three species of flies. Jour. Med. Entom., 3:155-158.

Kirchberg, E.
1959. See Bulling, E., et al.

Klesov, M. D.
1949. Izuchenie biologii nematody *Thelazia rhodesi* Desm. [Biological studies on the nematode *Thelazia rhodesi* Desm.] Zool. Zhurnal, *28*:515-522.

Klesov, M. D.
1951. On the question of the biology of nematodes of the genus *Thelazia* Bosc., 1819. Veterinariia͡, *28*:22-25. [In Russian.]

Knab, F.
1897. Entom. News, *8*:13.

Kniâzevskiĭ, A. N.
1948. See Bozhenko, V. P., and Kniâzevskiĭ, A. N.

Knuckles, J. L.
1959. Studies on the role of *Phormia regina* (Meigen) as a vector of certain enteric bacteria. Dissert. Abst., *20*(4).

Knutson, H.
1941. The occurrence of larvae of the stable fly, *Muscina stabulans* (Zett.) in living nymphs of the grasshopper *Xanthippus corallipes pantherinus* Sc. Jour. Parasit., *27*:90-91.

Kostitch, D.
1937. See Simitch, T., and Kostitch, D.

Kramer, J. P.
1961a. *Herpetomonas muscarum* (Leidy) in the haemocoele of larval *Musca domestica* L. Entom. News, *72*:165-166.

Kramer, J. P.
1961b. *Thelohania thomsoni* n. sp., a microsporidian parasite of *Muscina assimilis* (Fallén) (Diptera, Muscidae). Jour. Insect Path., *3*:259-265.

Kramer, J. P.
1962. The fate of spores of *Nosema apis* Zander ingested by muscoid flies. Colloque International sur la Pathologie des Insectes et la Lutte Microbiologique, Paris.

Kramer, J. P.
1964a. The microsporidian *Octosporea muscaedomesticae* Flu, a parasite of calypterate muscoid flies in Illinois. Jour. Insect Path., *6*:331-342.

Kramer, J. P.
1964b. *Nosema kingi*, sp. n., a microsporidian from *Drosophila willistoni* Sturtevant, and its infectivity for other muscoids. Jour. Insect Path., *6*:491-499.

Kramer, J. P.
1965a. Generation time of the microsporidian *Octosporea muscaedomesticae* Flu in adult *Phormia regina* (Meigen) (Diptera, Calliphoridae). Zeitsch. Parasit., Bd. *25* (Heft 4): 309-313.

Kramer, J. P.
1965b. The microsporidian *Octosporea muscaedomesticae* Flu, a little-known pathogen of muscoid flies. WHO/EBL/53.65, pp. 1-8.

Kramer, J. P.
1965c. Effects of an octosporeosis on the locomotor activity of adult *Phormia regina* (Meigen) (Dipt. Calliphoridae). Entomophaga (Paris), *10*:339-342.

Kramer, J. P.
1966. On the octosporeosis of muscoid flies caused by *Octosporea muscaedomesticae* Flu (Microsporidia). Amer. Midland Naturalist, *75*:214-220.

Kraneveld, F. C.
1929. See Nieschultz, O., and Kraneveld, F. C.

Krastin, N. I.
1949a. Epizootologia tel͡iazioza krupnogo rogatogo skota i biologi͡ia *Thelazia rhodesi* (Desmarest, 1827). Veterinari͡ia, *26*:6-8.

Krastin, N. I.
1949b. The decipherment of the cycle of development of the nematode *Thelazia rhodesi* (Desmarest, 1827), parasitizing the eyes of cattle. Dokl. Akad. Nauk SSSR, *64*:885-887.

Krastin, N. I.
1950. The decipherment of the cycle of development of the nematode *Thelazia gulosa* (Railliet et Henry, 1910), a parasite of the eyes of cattle. Dokl. Akad. Nauk SSSR, *70*:549-551.

Krastin, N. I.
1952. The decipherment of the cycle of development of the nematode, *Thelazia skrjabini* Erschow, 1928, a parasite of the eyes of cattle. Dokl. Akad. Nauk SSSR, n.s., *82*:829-831.

Krempf, A.
1911d. See Chatton, E., and Krempf, A.

Krenner, J. A.
1961. Studies in the field of the microscopic fungi: On *Entomophthora aphidis* H. Hofm., with special regard to the family of the *Entomophthoraceae* in general. Acta Bot. Acad. Sci. Hungaricae Budapest, *7*:345-376.

Krishna Iyer, P. R., and Sarwar, S. M.
1935. Bovine surra in India, with a description of a recent outbreak. Indian Jour. Veter. Science, *5*:158-170.

Krontowski, A.
1913. K voprosu o rasprostranenie tifa i dizenterii mukhami. Kiev: Tipografi͡ia Korchak-Novitskogo: 1913 (i.e. as separate pamphlet). Referat: Vrachebna͡ia Gazeta, 1913, No. 11, p. 417. [Zur Frage über die Typhus und Dysenterieverbreitung durch Fliegen.] Centralbl. Bakt., Parasit. Infekt., *68*:586-590.

Kudo, R.
1917. See Noguchi, A., and Kudo, R.

Kuenen, W. A., and Swellengrebel, W. H.
1913. Die entamöben des menschen und ihre practische bedeutung. Zentrabl. Bakt., *71*:378 and on (see pp. 401-403).

Kuhn, P.
1909. See Schuberg, A., and Kuhn, P.

Kuhn, P.
1912. See Schuberg, A., and Kuhn, P.

Kuhn, P.
1917. See Uhlenhuth, P., and Kuhn, P.

Kühlhorn, F.
1961. Über das Verhalten sozialer Faltenwespen (Hymenoptera: Vespidae) beim Stalleinflug, innerhalb von Viehställen und beim Fliegenfang. Untersuchungen überdie Insektenfauna von Raümen: 5. Zeitsch. Angew. Zool., *48*:405-422.

Kuhns, D. M., and Anderson, T. G.
1944. A fly-borne bacillary dysentery epidemic in a large military organization. Amer. Jour. Publ. Health, *34*:750-755.

Kulp, W. L.
1951. See Hawley, J. E., et al.

Kumm, H. W.
1935. The natural infection of *Hippelates pallipes* Loew, with the spirochaetes of yaws. Trans. Roy. Soc. Trop. Med. Hyg., *29*:265-272.

Kumm, H. W., Turner, T. B., and Peat, A. A.
1935. The duration of motility of the spirochaetes of yaws in a small West Indian fly—*Hippelates pallipes* Loew. Amer. Jour. Trop. Med., *15*:209-223.

Kumm, H. W., and Turner, T. B.
1936. The transmission of yaws from man to rabbits by an insect vector, *Hippelates pallipes* Loew. Amer. Jour. Trop. Med., *16*:245-262.

Kunert, H., and Schmidtke, L.
1952. Die Bedeutung der nichtstechenden Fliegen fur die Verschleppung von Leptospiren. Zeitsch. Tropenmed., *3*:475-486.

Kunert, H., and Schmidtke, L.
1953. Zur Übertragung von *Toxoplasma gondii* durch Fliegen. Zeitsch. Hyg., *136*:163-173.

Kunike, G.
1927a. Untersuchungen über die Rolle der Fliegen als Übertrager der Maul-und Klauenseuche auf Meerschweinchen. Zeitsch. Desinfekt. Gsndhtw., *19*: 115-117.

Kunike, G.
1927b. Experimentelle Untersuchungen über die Möglichkeit der Uebertragung der Maul- und Klauenseuche durch Fliegen. Centralbl. Bakt., Orig., *102*: 68-81. (See also: Berlin Tierarztl. Wschr., *43*:123-127.)

Kurczewski, F. E., and Acciavatti, R. E.
1968. A review of the nesting behaviors of the Nearctic species of *Crabro*, including observations on *C. advenus* and *C. latipes* (Hymenoptera: Sphecidae). Jour. New York Entom. Soc., *76*:196-212.

Kuzina, O.
1933. See Smirnov, E. S., and Kuzina, O.

Laake, E. W.
1915b. See Bishopp, F. C., and Laake, E. W.

Laarman, J. J.
1953. See van Thiel, P. H., and Laarman, J. J.

Laarman, J. J.
1956. Transmission of experimental toxoplasmosis by *Stomoxys calcitrans*. Documenta Medicina Geographica Tropica, *8*:293-298.

Laarman, J. J.
1957. Transmission of experimental toxoplasmosis by *Stomoxys calcitrans*. Acta Leidensia, *26* (or *27*):116-122.

Lahille, F.
1907. Anales del Minist. de Agric., *3* (4):115 (not seen).

Laing, J.
1937. Host-finding by insect parasites. 1. Observations on the finding of hosts by *Alysia manducator, Mormoniella vitripennis* and *Trichogramma evanescens*. Jour. Animal Ecol., *6*:298-317.

Lainson, R., and Southgate, B. A.
1965. Mechanical transmission of *Leishmania mexicana* by *Stomoxys calcitrans*. Roy. Soc. Trop. Med. Hyg., *59* (6):716.

Lakon, G.
1915. Zur systematik der entomophthoreengattung *Tarichium*. Zeitsch. Pflanzenkrankh., *25*:257-272.

Lakon, G.
1919. Notiz über die Körperstellung der an *Empusa muscae* verendeten gemeinen Stubenfliegen. Zeitsch. Angew. Entom., *5*:126-127.

Lal, R. B., Ghosal, S. C., and Mukherji, B.
1939. Investigations on the variation of vibrios in the house fly. Indian Jour. Med. Res., *26*:597-609.

Lamborn, W. A.
1925. An attempt to control *Glossina morsitans* by means of *Syntomosphyrum glossinae*, Waterston. Bull. Entom. Res., *15*:303-309.

Lamborn, W. A.
1933. Annual report of the medical entomologist for 1932. Ann. Med. Rept. Nyasaland, p. 59.

Lamborn, W. A.
1934. Annual report of the medical entomologist for 1933. Ann. Med. Rept. Nyasaland, 1933, pp. 61-63.

Lamborn, W. A.
1934. See Thomson, J. G., and Lamborn, W. A.

Lamborn, W. A.
1935. Annual report of the medical entomologist for 1934. Ann. Med. Rept. Nyasaland, 1935, pp. 65-69.

Lamborn, W. A.
1936a. The experimental transmission to man of *Treponema pertenue* by the fly *Musca sorbens* Wd. Jour. Trop. Med. Hyg., *39*:235-239.

Lamborn, W. A.
1936b. Annual report of the medical entomologist for 1936. Ann. Med. Sanit. Rept. Nyasaland, 1936, pp. 43-44.

Lamborn, W. A.
1937. The haematophagous fly *Musca sorbens*, Weid., in relation to the transmission of leprosy. Jour. Trop. Med. Hyg., *40*:37-42.

Lamborn, W. A.
1938. Annual report of the medical entomologist for 1938. Ann. Med. Sanit. Rept. Nyasaland, 1938, pp. 40-48.

Lamborn, W. A.
1955. The hematophagous fly as a possible vector of *Leishmania*. Bull. Endemic Dis. Bagdad, *1*:239-249.

Lamborn, W. A., and Howat, C. H.
1936. A possible reservoir host to *Trypanosoma rhodesiense*. British Med. Jour., *1*:1153-1155.

Lang, N. N.
1940. K voprosu o sokhranenii chumnykh mikrobov v razvivai͡ushchikhsi͡a lichinkakh mukh. [On question of the preservation of *B. pestis* in developing fly larvae.] Vestnik Mikr. Epid. Parazit., *19*:96-97.

Larsen, J. R.
1954a. See Jenkins, D. W., et al.

Larsen, J. R.
1956. See Ingram, R. L., et al.

Laveran, A.
1915. Comment le Bouton d'Orient se propage-t-il? Ann. Inst. Pasteur, *29*: 415-439.

Laveran, A., and Franchini, G.
1920. Contribution à l'étude des flagellés des culicides, des muscides, des phlébotomes et de la blatte orientale. Bull. Soc. Path. Exot., *13*:138-143.

LeBailly, C.
1924. Les mouches ne jouent pas de rôle dans la dissémination de la fièvre aphteuse. Compt. Rend. Acad. Sci., *179*:1225-1227.

Leboeuf, A.
1908. See Martin, G., et al.

Leboeuf, A.
1912. Dissémination du bacille de Hansen par la mouche domestique. Bull. Soc. Path. Exot., *5*:860-868.

Leboeuf, A.
1914. La lèpre en Nouvelle-Caledonie et dépendances. Ann. Hyg. Med. Colon., *17*:177-197.

Lecler, E.
1902. See Sivori, F., and Lecler, E.

Ledingham, J.C.G.
1909. See Morgan, H. de R., and Ledingham, J.C.G.

Ledingham, J.C.G.
1911. On the survival of specific microorganisms in pupae and imagines of *Musca domestica* raised from experimentally infected larvae. Experiments with *B. typhosus*. Jour. Hyg., *11*:333-340.

Lee, V. H.
1965. See Bouillant, A., et al.

Leese, A. S.
1909. Experiments regarding the natural transmission of surra carried out in Mohand in 1908. Jour. Trop. Veter. Science, *4*:107-132.

Leger, A.
1911a. See Chatton, E., and Leger, A.

Leger, A.
1911b. See Chatton, E., and Leger, A.

Leger, A.
1911c. See Chatton, E., and Leger, A.

Léger, L.
1903. Sur quelques cercomonadines nouvelles ou peu connues parasites de l'intestin des insectes. Arch. Protistenk., *2*:180-189.

Leger, M.
1912. See Chatton, E., and Leger, M.

Leger, M.
1913. See Chatton, E., and Leger, M.

Legner, E. F.
1963a. See Bay, E. C., and Legner, E. F.

Legner, E. F.
1966 (pre-publication).
Parasites of *Musca domestica* Linnaeus and other nuisance filth-breeding Diptera in Southern California.

Legner, E. F.
1966. See White, E. B., and Legner, E. F.

Legner, E. F.
1966. Competition among larvae of *Hippelates collusor* (Diptera: Chloropidae) as a natural control factor. Jour. Econ. Entom., *59*:1315-1321.

Legner, E. F.
1967. Two exotic strains of *Spalangia drosophilae* merit consideration in biological control of *Hippelates collusor* (Diptera: Chloropidae). Ann. Entom. Soc. Amer., *60*:458-462.

Legner, E. F.
1969. Adult emergence interval and reproduction in parasitic Hymenoptera influenced by host size and density. Ann. Entom. Soc. Amer., *62*:220-226.

Legner, E. F., and Bay, E. C.
1964. Natural exposure of *Hippelates* eye gnats to field parasitization and the discovery of one pupal and two larval parasites. Ann. Entom. Soc. Amer., *57*:767-769.

Legner, E. F., and Bay, E. C.
1965b. *Ooencyrtus submetallicus* Howard in an extraordinary host relationship with *Hippelates pusio* Loew. Canadian Entom., *97*:556-557.

Legner, E. F., Bay, E. C., and McCoy, C. W.
1965c. Parasitic natural regulatory agents attacking *Musca domestica* L. in Puerto Rico. Jour. Agric. Univ. Puerto Rico, *49*:368-376.

Legner, E. F., and Bay, E. C.
1965d. Predatory and parasitic agents attacking the *Hippelates pusio* complex in Puerto Rico. Jour. Agric. Univ. Puerto Rico, *49*:377-385.

Legner, E. F., Bay, E. C., and Medved, R. A.
1966. Behavior of three native pupal parasites of *Hippelates collusor* in controlled systems. Ann. Entom. Soc. Amer., *59*:977-984.

Legner, E. F., and Brydon, H. W.
1965a (pre-publication).
Suppression of dung-inhabiting fly populations by pupal parasites. [The manuscript was sent to us for inclusion in this book and is the same as the next publication cited.]

Legner, E. F., and Brydon, H. W.
1966b. Suppression of dung-inhabiting fly populations by pupal parasites. Ann. Entom. Soc. Amer., *59*:638-651.

Legner, E. G., and Greathead, D. J.
1969a. Parasitism of pupae in East African populations of *Musca domestica* and *Stomoxys calcitrans.* Ann. Entom. Soc. Amer., *62*:128-133.

Legner, E. F., and McCoy, C. W.
1966a. The house fly, *Musca domestica* Linnaeus, as an exotic species in the western hemisphere incites biological control studies. Canadian Entom., *98*:243-248.

Legner, E. F., and Olton, G. S.
1968. Activity of parasites from Diptera: *Musca domestica, Stomoxys calcitrans,* and species of *Fannia, Muscina,* and *Ophyra.* II. At sites in the eastern hemisphere and Pacific area. Ann. Entom. Soc. Amer., *61*: 1306-1314.

Legner, E. F., and Olton, G. S.
1969b. Migrations of *Hippelates collusor* larvae from moisture and trophic stimuli and their encounter by *Trybliographa* parasites. Ann. Entom. Soc. Amer., *62*:136-141.

Legroux, R., and Second, L.
1945. La spore botulique dans la mouche *Piophila casei* L. Ann. Inst. Pasteur, *71*:464-466.

Leidy, J.
1874. On a parasitic worm of the house-fly. Proc. Acad. Nat. Science Philadelphia, *26*:139-140.

Le Noc, P.
1962. See Brygoo, E. R., et al.

León y Blanco, F.
1944. See Soberón y Parra, G., and León y Blanco, F.

León y Blanco, F., and Soberón y Parra, G.
1941. Nota sobre la trasmisión experimental del mal del pinto por medio de una mosca del género *Hippelates.* Gac. Med. Mexico, *71*:534-539.

Leonardi, G.
1900. Storia naturale degli acari insetticoli. Bull. Soc. Entom. Italiana, *32*:1-76.

Lesne, P., and Mercier, L.
1922. Un staphylinide des muscides fucicoles *Aleochara* [*Polystoma*] *algarum* Fauvel. Caractères adaptatifs de la larvae à la vie parasitaire. Ann. Entom. Soc. France, *91*:351-358.

Lestoquard, F.
1923. See Donatien, A., and Lestoquard, F.

Lever, R. A.
1936. Report of entomologist for year 1935-36. [Tulagi] Brit. Solomon Island Prot. Agric. Comm.

Levinsky, L., and Morellini, M.
1948. Reperti microscopici dei bacilli tubercolari nei casi di infezione artificiale delle larve di *Galleria mellonella* e di *Sarcophaga falculata.* Ann. Inst. "Carlo Forlanini," *9*:1-21.

Levinson, Z. H.
1960. Food of housefly larvae. Nature, *188*:427-428.

Levitt, M. M.
1935. Variability of pupae and of the fecundity of adults of the gypsy moth (*Porthetria dispar*). [In Ukrainian.] Rech. Ecol. Anim. Terr. No. 2, pp. 135-170. Kiev, Vidaun. Vseukr. Akad. Nauk. (With summaries in Russian pp. 165-167, and English pp. 168-170.)

Levkovich, E. N., and Sukhova, M. N.
1957. Duration of retention and excretion of poliomyelitis virus by synanthropic flies and its relation to dissemination and prevention of poliomyelitis. Med. Parazit., Moscow, *26*:343-347.

Lewis, F. C.
1912. See Cox, G. L., et al.

L'Héritier, P.
1955. See Brun, G., et al.

L'Héritier, P.
1957. Le virus héréditaire de la drosophile. Rev. Path. Gen. Compt., *57*: 1471-1486.

L'Héritier, P., and Teissier, G.
1937. Une anomalie physiologique héréditaire chez la drosophile. Compt. Rend. Acad. Sci., *205*:1099-1101.

L'Héritier, P., and Teissier, G.
1933. Étude d'une population de drosophiles en équilibre. Compt. Rend. Acad. Sci., *197*:1765-1767.

L'Héritier, P., and Teissier, G.
1934. Sur quelques facteurs de succès dans la concurrence larvaire chez *Drosophila melanogaster.* Compt. Rend. Soc. Biol., *2*:306-308.

Lichtenstein, J. L.
1920. Le parasitisme de *Aphiochaeta* (*Phora*) *fasciata* Fallén. Compt. Rend. Acad. Sci., *170*:531-534.

Lignières, J.
1903. De la trypanosomose des équidés Sud-Américains. Bull. Soc. Centr. Méd. Véter. (See chapter IV, pp. 177-181.)

Lilly, J. H.
1965. See Eversole, J., et al.

Lilly, J. H.
1965. See Steve, P. C., and Lilly, J. H.

Lindquist, A. W.
1936. Parasites of horn fly and other flies breeding in dung. Jour. Econ. Entom., *29*:1154-1158.

Lindquist, A. W.
1942. Ants as predators of *Cochliomyia americana.* Jour. Econ. Entom., *35*:850-852.

Lindsay, D. R., Stewart, W. H., and Watt, J.
1953. Effects of fly control on diarrheal diseases in an area of moderate morbidity. Publ. Health Rept., *68*:361-367.

Lins de Almeida, J.
1933. Nouveaux agents de transmission de la berne (*Dermatobia hominis* L.) au Brazil. Soc. Biol. Compt. Rend., *113*:1274-1275.

Linstow von
1875. See Hill, G. F., 1919.

Lipovsky, L. J.
1954. Studies of the food habits of postlarval stages of chiggers (Acarina, Trombiculidae). Univ. Kansas Science Bull., *36*:943-958.

Lodge, O. C.
1916. Fly investigation reports. IV. Some enquiry into the question of baits and poisons for flies, being a report on the experimental work carried out during 1915 for the Zoological Society of London. Proc. Zool. Soc. London, *2*:481-518.

Lord, F. T.
1904. Flies and tuberculosis. Boston Med. Surg. Jour., *151*:651-654.

Lord, F. T.
1905. Flies and tuberculosis. Publications Massachusetts General Hospital Boston, *1*:118-125.

Lotze, J. C.
1942. See Stein, C. D., et al.

Love, J. A., and Gill, G. D.
1965. Incidence of coliforms and enterococci in field populations of *Stomoxys calcitrans* (Linnaeus). Jour. Invert. Pathol., *7*:430:436.

Lührs
1919. Die ansteckende blutarmut der pferde. Zeitsch. Veterinärkunde, *31*:369-440, 450-464.

Lutz, A.
1917. Contribuições ao conhecimento dos oestrideos brazileiros. Mem. Inst. Oswaldo Cruz, *9*:94-113.

Lysenko, O.
1958. Mikroflora některých našich much. [Microflora of some flies of Czechoslovakia.] Československá Mikrob., *3*:51-53.

Lysenko, O., and Povolný, D.
1961. The microflora of synanthropic flies in Czechoslovakia. Folia Microb., *6*:27-32.

MacAloney, H. J.
1930. The white pine weevil (*Pissodes strobi* Peck), its biology and control. Bull. New York State Coll. Forestry, *3* (Tech. Publ. No. 28), pp. 1-87.

MacDougall, R. S.
1909. Sheep maggot and related flies: Their classification, life-history, and habits. Trans. Highl. Agric. Soc. Scotland, *21*:135-174.

Mackerras, I. M.
1930. Recent developments in blowfly research. Jour. Australian Council Science Indust. Res., *4*:212-219.

Mackerras, I. M.
1932. Buffalo fly investigations. A note on the occurrence of *Hydrotaea australis* Malloch in northern Australia. Jour. Australian Council Science Indust. Res., *5*:253-254.

Mackerras, M. J.
1933. Observations on the life-histories, nutritional requirements and fecundity of blowflies. Bull. Entom. Res., *24*:353-362.

Mackinnon, D. L.
1910. Herpetomonads from the alimentary tract of certain dung-flies. Parasitology, *3*:255-274.

Madden, J. L.
1964. Ecological studies of the parasite, *Nasonia vitripennis* (Walk.) and the housefly host, *Musca domestica* Linn. Dissert. Abst., *24*:3508-3509.

Madden, J. L.
1963. See Pimentel, D., et al.

Madden, J. L., and Pimentel, D.
1965. Density and spatial relationships between a wasp parasite and its housefly host. Canadian Entom., *97*:1031-1037.

Maddox, R. L.
1885a. Experiments on feeding some insects with the curved or "comma" bacillus, and also with another bacillus (*B. subtilis*?). Jour. Roy. Microscop. Soc., Series II, *5*:602-607.

Maddox, R. L.
1885b. Further experiments on feeding insects with the curved or "comma" bacillus. Jour. Roy. Microscop. Soc., Series II, *5*:941-952.

Magni, G. E.
1954. Thermic cure of cytoplasmic sex-ratio in *Drosophila bifasciata*. Carylogia (Suppl.), *6*:1213-1216.

Maier, P. P.
1961. See Richards, C. S., et al.

Malogolowkin, C.
1958. Maternally inherited "sex-ratio" conditions in *Drosophila willistoni* and *Drosophila paulistorum*. Genetics, *43*:274-286.

Malogolowkin, C., Carvalho, G. G., and da Paz, M. C.
1960. Interspecific transfer of the "sex-ratio" condition in *Drosophila*. Genetics, *45*:1553-1557.

Malogolowkin, C., and Poulson, D. F.
1957. Infective transfer of maternally inherited abnormal sex-ratio in *Drosophila willistoni*. Science, *126*:32.

Malogolowkin, C., Poulson, D. F., and Wright, E. V.
1959. Experimental transfer of maternally inherited abnormal sex-ratio in *Drosophila willistoni*. Genetics, *44*:59-74.

Manning, J.
1902. A preliminary report on the transmission of pathogenic germs by the housefly. Jour. Amer. Med. Assoc., *38*:1291-1294.

Manson-Bahr, P. H.
1919. Bacillary dysentery. Trans. Roy. Soc. Trop. Med. Hyg., *13*:64-72.

Mantovani, M.
1915. See Franchini, G., and Mantovani, M.

Marchoux, E.
1916. Transmission de la lèpre par les mouches (*Musca domestica*). Ann. Inst. Pasteur, *30*:61-68.

Marengo, U.
1932. See Ara, F., and Marengo, U.

Marlatt, C. L.
1931. Report [1930-31] of the Chief of the Bureau of Entomology. U. S. Dept. Agric., Washington, D. C., 87 pp.

Marpmann, G.
1884. Die Verbreitung von Spaltpilzen durch Fliegen. Archiv. Hyg., *2*:360-363.

Martelli, G.
1913. La *Thea 22-punctata* L. è solamente micofaga. Giorn. Agric. Merid. Messina, *6*:189-195.

Martin, G., Leboeuf, A., and Roubaud, E.
1908. Expériences de transmission du "nagana" par les stomoxes et par les moustiques du genre *Mansonia*. Bull. Soc. Path. Exot., *1*:355-358.

Martini, E.
1903. Ueber die Entwickelung der Tsetseparasiten in Säugethieren. Zeitsch. Hyg. Infect., *42*:341-349.

Maseritz, I. H.
1934. Digestion of bone by larvae of *Phormia regina*. Its relationship to bacteria. Arch. Surg., *28*:589-607.

Masi, L.
1916. Chalcididi del giglio, prima serie. Materiali per una fauna dell'Arcipelago Toscano, XI. Annali Museo Civico di Storia Naturale Genova, *7* (47), Serie 3, pp. 1-66.

Mathlein, R.
1944. See Notini, G., and Mathlein, R.

Matsuo, K.
1962. Effects of water contents of medium and population density on the larval development of *Musca domestica vicina* and *Sarcophaga peregrina*. Endemic Dis. Bull. Nagasaki Univ., *4*:74-81.

McCampbell, E. F.
1914. See Boudreau, F. G., et al.

McCarthy, T.
1914. See Froggatt, W. W., and McCarthy, T.

McConnell, E., and Richards, A. G.
1959. The production by *Bacillus thuringiensis* Berliner of a heat-stable substance toxic for insects. Canadian Jour. Microb., *5*:161-168.

McCoy, C. W.
1963. Unpublished MS Thesis, University of Nebraska, Lincoln, Nebraska.

McCoy, C. W.
1965c. See Legner, E. F., et al.

McCoy, C. W.
1966a. See Legner, E. F., and McCoy, C. W.

McDonald, W. A.
1958. A calliphorid host of *Thyridanthrax abruptus* (Lw.) in Nigeria (Diptera, Bombyliidae). Bull. Entom. Res., *48*:533.

McGhee, R. B., Schmittner, S. M., and Hanson, W. L.
1965. Speciation in the genus *Crithidia*. Org. Mond. de la Santé, WHO/EBL, 45.65, pp. 1-7.

McGuire, C. D., and Durant, R. C.
1957. The role of flies in the transmission of eye disease in Egypt. Amer. Jour. Trop. Med. Hyg., *6*:569-575.

McIvor, B.
1935. See Herms, W. B., et al.

McLintock, J.
1955. See Stirrat, J. H., and McLintock, J.

McNaught, J. C.
1907. See Eyre, J. W., et al.

McNeil, E., and Hinshaw, W. R.
1944. Snakes, cats, and flies as carriers of *Salmonella typhimurium*. Poultry Science, *23*:456-457.

Mechalas, B. J.
1963. See Dunn, P. H., and Mechalas, B. J.

Mégnin, J. P.
1874. Du transport et de l'inoculation du virus charbonneux et autres par les mouches. Compt. Rend. Acad. Sci., *79*:1338-1340.

Mégnin, J. P.
1875. Mémoire sur la question du transport et de l'inoculation des virus par les mouches. Jour. Anat. Physiol. Paris, *11*:121-133.

Melissidis, A.
1911. See Cardamatis, J. P., and Melissidis, A.

Mello, M. J., and Cuocolo, R.
1943a. Alguns aspetos das relações do *Habronema muscae* (Carter, 1861) com a mosca doméstica. Arch. Inst. Biol. São Paulo, *14*:227-234.

Mello, M. J., and Cuocolo, R.
1943b. Técnica para o xenodiagnóstico da habronemose gastrica dos equideos. Arquiv. Inst. Biol., *14*:217-226.

Melnick, J. L., Power, M. E., et al.
1943. (With the exception of flies belonging to the genera *Hylemya, Fannia, Helina, Scatophaga,* and *Sarcophaga,* flies in this study were identified to species.)

Melnick, J. L., Trask, J. D., et al.
1943. (With the exception of *Musca domestica* and *Stomoxys calcitrans,* flies in this study were assigned to genus but not to species.)

Melnick, J. L.
1949. Isolation of poliomyelitis virus from single species of flies collected during an urban epidemic. Amer. Jour. Hyg., *49*:8-16.

Melnick, J. L.
1950. Personal communication, in Curnen, E. C.

Melnick, J. L., and Dow, R. P.
1953. Poliomyelitis in Hidalgo County, Texas, 1948. Poliomyelitis and coxsackie viruses from flies. Amer. Jour. Hyg., *58*:288-309.

Melnick, J. L., Emmons, J., Coffey, J. H., and Schoof, H.
1954. Seasonal distribution of coxsackie viruses in urban sewage and flies. Amer. Jour. Hyg., *59*:164-184.

Melnick, J. L., and Penner, L. R.
1947. Experimental infection of flies with human poliomyelitis virus. Proc. Soc. Exp. Biol. Med., *65*:342-346.

Melnick, J. L., and Penner, L. R.
1952. The survival of poliomyelitis and coxsackie viruses following their ingestion by flies. Jour. Exp. Med., *96*:255-271.

Melnick, J. L., Shaw, E. W., and Curnen, E. C.
1949. A virus isolated from patients diagnosed as non-paralytic poliomyelitis or aseptic meningitis. Proc. Soc. Exp. Biol. Med., *71*:344-349.

Melnick, J. L., and Ward, R.
1945. Susceptibility of vervet monkeys to poliomyelitis virus in flies collected at epidemics. Jour. Infect. Dis., *77*:249-252.

Meleney, H. E.
1932. See Frye, W. W., and Meleney, H. E.

Menzel, C., and Ihle, J. A.
1683. De muscis quibusdam culiciformibus, pediculosis. . . . Misc. Cursive Ephem. Med.-Phys. Germ. Acad. Caes. Leopold Nat. Cur., Dec. 2, Ann. 1682, p. 71, obs. 30, Nürnberg.

Menzel, R.
1924. Über die verbreitung von rhabditis-larven durch dipteren. Zool., Anz., *58*:345-349.

Mercier, L.
1922. See Lesne, P., and Mercier, L.

Merk, A.
1910. Vaccine und Fliegen. Hyg. Rundsch., *20*:233-235.

Metelkin, A. I.
1935. Significance of flies in the spread of coccidioses among animals and man. Med. Parazit., *4*:75-82.

Michelbacher, A. E., Hoskins, W. M., and Herms, W. B.
1932. The nutrition of flesh fly larvae, *Lucilia sericata* (Meigen). Jour. Exp. Zool., *64*:109-128.

Miggiano, V.
1963c. See Greenberg, B., and Miggiano, V.

Milani, R.
1947. Wasp parasites of *Drosophila,* and the possibility of utilizing them for genetical research. Drosophila Inform. Serv., *21*:87.

Miller, D.
1927. Parasitic control of sheep maggot-flies. New Zealand Jour. Agric., *25*: 219-220.

Miller, R. S.
1964. Interspecies competition in laboratory populations of *Drosophila melanogaster* and *D. simulans.* Amer. Naturalist, *98*:221-238.

Minchin, E. A.
1908. Investigations on the development of trypanosomes in tsetse-flies and other Diptera. Quart. Jour. Microscop. Science, *52*:159-260 (see pp. 180, 184).

Minett, E. P.
1912. The question of flies as leprosy carriers. Jour. London School Trop. Med., *1*:31-35.

Minett, F. C.
1944. See Sen, S. K., and Minett, F. C.

Mitchell, C. A.
1944. See Gwatkin, R., and Mitchell, C. A.

Mitscherlich, E.
1941. Die aetiologische Bedeutung von *Rickettsia conjunctivae* (Coles 1931) für die spezifische Kerato-Konjunktivitis der Schafe in Deutsch-Südwestafrika. Zeitsch. Infekt. Parasit. Krankh. Hyg. Haustiere, *57*:271-287.

Mitscherlich, E.
1943. Die Übertragung der Kerato-Conjunctivitis Infectiosa des Rindes durch Fliegen und die Tenazität von *Rickettsia conjunctivae* in der Aussenwelt. Deutsch. Tropenmed. Zeitsch., *47*:57-64.

Mitzmain, M. B.
1912a. Collected notes on the insect transmission of surra in carabaos. Philippine Agric. Rev., *5*:670-681.

Mitzmain, M. B.
1912b. The role of *Stomoxys calcitrans* in the transmission of *Trypanosoma evansi.* Philippine Jour. Science, Sect. B, *7*:475-519.

Mitzmain, M. B.
1913. The biology of *Tabanus striatus* Fabr., the horsefly of the Philippines. Philippine Jour. Science, *8* (Ser. B., No. 3):197-218.

Mitzmain, M. B.
1914a. Experimental insect transmission of anthrax. Publ. Health Rept., *29*: 75-77.

Mitzmain, M. B.
1914b. An experiment with *Stomoxys calcitrans* in an attempt to transmit a filaria of horses in the Philippines. Amer. Jour. Trop. Dis. Prevent., Med., *2*:759-763.

Mitzmain, M. B.
1914c. Summary of experiments in the transmission of anthrax by biting flies. Treas. Dept., U. S. Publ. Health Serv., Hyg. Lab. Bull., *94*:41-48.

Miyamoto, K.
1965. See Shimizu, F., et al.

Mohler, J. R.
1920. Annual Reports of the Dept. of Agric. for the year 1919. Report of the Chief of the Bureau of Animal Industry. U. S. Dept. Agric. Washington, D. C. (see pp. 125-126).

Mohr, N.
1786. Forsög til en Islandsk naturhistorie. Kjøbenhavn, No. 228, p. 104.

Moore, J. A.
1952. Competition between *Drosophila melanogaster* and *Drosophila simulans.* II. The improvement of competitive ability through selection. Proc. U. S. Nat. Acad. Sciences, *38*:813-817.

Morel, R.
1933. See Descazeaux, J., and Morel, R.

Morellini, M.
1948. See Levinsky, L., and Morellini, M.

Morellini, M.
1952. Micobatteri tubercolari e paratubercolari nella mosca domestica adulta. Nuovi Ann. Ig. Microb., *3*:305-320.

Morellini, M.
1956. Rapporti fra mosca domestica e infezione tubercolare. Contributi di Pneumotisiologia, Federazione Italiana Contro La Tubercolosi, pp. 1-10.

Morellini, M., and Cattaneo, L.
1947. Il comportamento della *Galleria mellonella* e di alcune mosche di fronte al bacillo di Koch. Ann. Ist. "Carlo Forlanini," *10*:161-175.

Morellini, M., Ingrao, F., and Vella, L.
1949. Comportamento dell'infezione tubercolare sperimentale nella mosca (*Sarcophaga falculata*). Ann. Ist. "Carlo Forlanini," *11*:508-511.

Morellini, M., and Saccá, G.
1953. Alcuni aspetti del meccanismo di diffusione del b. tubercolare da parte di *M. domestica* (studio sperimentale). Rend. Ist. Superiore Sanitá, *16*:267-285.

Morgan, B. B.
1942. The viability of *Trichomonas foetus* (Protozoa) in the house fly (*Musca domestica*). Proc. Helminth. Soc. Washington, D.C., *9*:17-20.

Morgan, H. de R., and Ledingham, J.C.G.
1909. The bacteriology of summer diarrhoea. Proc. Roy. Soc. Med., *2*:133-158.

Morgan, W. L.
1929. *Alysia manducator* Pz. An introduced parasite of the sheep blowfly maggot. Agric. Gaz. New South Wales, Sydney, *40*:818-829.

Morris, H.
1918. Bloodsucking insects as transmitters of anthrax or charbon. Louisiana State Univ. Agric. Exp. Stat. Baton Rouge, Bull. No. 163, pp. 3-15.

Morris, H.
1920. Some carriers of anthrax infection. Jour. Amer. Veter. Med. Assoc., *56*:606-608.

Mourier, H., and ben Hannine, S.
1968. Survey of the importance of the natural enemies of houseflies in Denmark. Government Pest Infestation Laboratory Annual Report. Lyngby, Denmark.

Moutia, A.
1928. Surra in Mauritius and its principal vector *Stomoxys nigra.* Bull. Entom. Res., *19*:211-216.

Mott, L. O.
1942. See Stein, C. D., et al.

Muesebeck, C.F.W.
1961. A new Japanese *Trichopria* parasitic on the house fly (Hymenoptera: Diapriidae). Mushi, *35*:1-2.

Mukherjee, P. K.
1937a. See Roy, D. N., and Mukherjee, P. K.

Mukherjee, P. K.
1937b. See Roy, D. N., and Mukherjee, P. K.

Mukherjee, S. P.
1940. See Roy, D. N., et al.

Mukherji, B.
1939. See Lal, R. B., et al.

Mulla, M. S.
1962. Recovery of a cynipoid parasite from *Hippelates* pupae. Mosquito News, *22*:301-302.

Murata, M.
1963. See Asahina, S., et al.

Murray, A.
1877. Economic entomology. Aptera. London, Chapman and Hall, p. 129. (Also, New York: Scribner, Welford, and Armstrong.)

Musgrave, W. E., and Clegg, M. T.
1903. *Trypanosoma* and trypanosomiasis, with special reference to surra in the Philippine Islands. Dept. Interior, Bur. Govt. Labs., Biol. Lab. Manila, *5*:1-248.

Muzzarelli, E.
1925. Sul passaggio di alcuni microrganismi patogeni attraverso lo intestino delle mosche carnarie. Ann. Ig., *35*:219-228.

Myers, J. G.
1927. The habits of *Alysia manducator* (Hym., Braconidae). Bull. Entom. Res., *17*:219-229.

Myers, J. G.
1929. Further notes on *Alysia manducator* and other parasites (Hym.) of muscoid flies. Bull. Entom. Res., *19*:357-360.

Nabarro, D., and Greig, E.D.W.
1905. Further observations on the trypanosomiases (Human and Animal) in Uganda. Rept. Sleeping Sickness Comm. Roy. Soc. London, *5*:8-47.

Nagel, W. P.
1963. See Pimentel, D., et al.

Nagel, W. P., and Pimentel, D.
1963. Some ecological attributes of a pteromalid parasite and its housefly host. Canadian Entom., *95*:208-213.

Nair, K. K.
1965. See Amonkar, S. V., and Nair, K. K.

Nash, T.A.M.
1933. The ecology of *Glossina morsitans* and two possible methods for its destruction. Part 2. Bull. Entom. Res., *24*:163-195.

Neel, W. W., Urbina, D., Viale, E., and De Alba, J.
1955. Ciclo biológico del tórsalo (*Dermatobia hominis* L. Jr.) en Turrialba, Costa Rica. Turrialba, *5*:91-104.

Nelson, J. M.
1968. Parasites and symbionts of nests of *Polistes* wasps. Ann. Entom. Soc. Amer., *61*:1528-1538. (Noted by title only.)

Neiva, A., and Florencio Gomes, J.
1917. Biologia da mosca do berne (*Dermatobia hominis*) observada em todas as suas phases. Ann. Paulistas Med. Cirurgia, *8*:197-209.

Neveu-Lemaire, M.
1936. Traité d'helminthologie médicale et vétérinaire. Vigor frères, éditeurs, Paris, 1514 pp. (see pp. 1458-59).

Newman, L. J.
1928. Sheep maggot fly parasite (*Alysia manducator*). Jour. Dept. Agric. West. Australia, *5*:150-151.

Newman, L. J., and Andrewartha, H. G.
1930. Blowfly parasite. The red-legged chalcid. *Stenoterys fulvoventralis* (Dodd). Jour. Dept. Agric. West. Australia, *7*:89-95.

Newsad, A.
1930. Sarcoptesmilben auf den Stubenfliegen. [House flies as carriers of *Sarcoptes* mites.] Preliminary report. Arch. Schiffs. Tropen-Hyg. Path. Ther. Exot. Krank., *34*:399-400.

Nicholls, L.
1912. The transmission of pathogenic microorganisms by flies in Saint Lucia. Bull. Entom. Res., *3*:81-88.

Nicoll, W.
1911a. On the varieties of *Bacillus coli* associated with the house-fly (*Musca domestica*). Jour. Hyg., *11*:381-389.

Nicoll, W.
1911b. Further reports (No. 4) on flies as carriers of infection. 3. On the part played by flies in the dispersal of the eggs of parasitic worms. Rept. Loc. Govt. Bd. Publ. Health and Med. Subj., London, n.s., No. 16, pp. 13-30.

Nicoll, W.
1917a. Flies and typhoid. Jour. Hyg., *15*:505-526.

Nicoll, W.
1917b. Flies and bacillary enteritis. British Med. Jour., No. 2948, pp. 870-872.

Nicolle, C., Cuenod, A., and Blanc, G.
1919. Demonstration expérimentale du rôle des mouches dans la propagation du trachôme (Conjonctivite granuleuse). Compt. Rend. Acad. Sci., *169*:1124-1126.

Nickle, W. R.
1966. See Stoffolano, Jr., J. G., and Nickle, W. R.

Nickle, W. R.
1967. *Heterotylenchus autumnalis* sp. n. (Nematoda: Sphaerulariidae), a parasite of the face fly, *Musca autumnalis.* De Geer. Jour. Parasit., *53*:398-401.

Nieschulz, O.
1927. Zoologische bijdragen tot het surraprobleem. XIX. Overbrengingsproeven met *Stomoxys, Lyperosia, Musca* en *Stegomyia.* [Zoological contribution to the surra problem. XIX. Transmission experiments with *Stomoxys, Lyperosia, Musca* and *Stegomyia.*] Ned. Ind. Blad. Diergeneesk., *39*: 371-390.

Nieschulz, O.
1928a. Enkele miltvuuroverbrengingsproeven met tabaniden, musciden en muskieten. Ned. Ind. Blad. Diergeneesk., *40*:355-377.

Nieschulz, O.
1928b. Zoologische Beitrage zum Surraproblem. XXII. Uebertragungsversuche mit *Anopheles fuliginosus* Gil. Centralbl. Bakt., Parasit. Infekt., I. Abt., Orig., *109*:327-330.

Nieschulz, O.
1929. Zoologische Beiträge zum Surraproblem: XXV. Ueber den Einfluss verschiedener Versuchtiere auf das Ergebnis von Surraübertragungsversuchen mit *Stomoxys calcitrans.* Zentralbl. Bakt., Parasit. Infekt., *113*: 80-89.

Nieschulz, O.
1930. Surraübertragungsversuche auf Java und Sumatra. Veeartsenijk, Meded. Ned.-Ind., No. 75, pp. 1-295. (Summary, see Archiv. Schiffs. Tropen-Hyg., *33*:257-266, 1929.)

Nieschulz, O., and Kraneveld, F. C.
1929. Experimentelle Untersuchungen über die Uebertragung der Büffelseuche durch Insekten. Zentrabl. Bakt., Parasit. Infekt., I. Abt., Orig., *113*: 403-417.

Nikol'skaia͡, M. N.
1960a. Khal'tsidy semeĭstva Chalcididae i Leucospidae. Fauna SSSR, Pereponchatokrylye, Part 5, pp. 1-220. [Chalcids . . . , *in* Fauna of the USSR, Hymenoptera, Vol. VII, Part 5.]

Nikol'skaia͡, M. N.
1960b. Fauna and ecology of the insects of Turkmen SSR. Results of expeditions 1951-53, Part 1. Chalcididae i Leucospidae Srednei Azii (Hymenoptera, Chalcidoidea). Trudy Zool. Inst. AN SSSR, *27*:220-247.

Nguyèn-Trung-Truyèn
1926. See Broudin et al.

Nguyèn-Van-Dên
1926. See Broudin et al.

Noble, N. S.
1936. *Pristhesancus papuensis* Stål, an "assassin" bug. Jour. Australian Inst. Agric. Science, *2*:124-126.

Nocedo, C.
1921. Dipteros nuevos parásitos de la langosta (*Schistocerca peregrina*). [New dipterous parasites of the locust (*S. peregrina*).] Rev. Agric. Mexico, *2*: 132-135, 183-186.

Noè, G.
1903. Studî sul ciclo evolutivo della *Filaria labiato-papillosa* Alessandrini. Atti Reale Accad. Lincei, Ser. 5, *12*:387-393.

Noguchi, H., and Kudo, R.
1917. The relation of mosquitoes and flies to the epidemiology of acute poliomyelitis. Jour. Exp. Med., *26*:49-57.

Noguchi, Y.
1963. See Asahina, S., et al.

Norwood, V. H.
1933. See Robinson, W., and Norwood, V. H.

Norwood, V. H.
1934. See Robinson, W., and Norwood, V. H.

Nøstvik, E.
1954. A study of *Pseudeucoila bochei* Weld. and its relationship to *Drosophila melanogaster* Meig. Genetica Entom., *2* (not seen).

Notini, G., and Mathlein, R.
1944. Grönmykos förorsakad av *Metarrhizium anisopliae* (Metsch.) Sorok: I: Grönmykosen som biologiskt insektbekämpningsmedel. Meddelander Statens Växtskyddsanstalten Stockholm, No. 43, pp. 1-58 (with English summary).

Nuorteva, P.
1966. See Ojala, O., and Nuorteva, P.

Nuorteva, P.
1965. Personal communication.

Nuttall, G.H.F.
1897. Zur Aufklärung der Rolle, welche die Insekten bei der Verbreitung der Pest spielen. Ueber die Empfindlichkeit verschiedener Tiere für dieselbe. Centrabl. Bakt., *22*:87-97.

Oana, C.
1958. See Duca, M., et al.

O'Connor, F. W.
1917a. See Wenyon, C. M., and O'Connor, F. W.

O'Connor, F. W.
1917b. See Wenyon, C. M., and O'Connor, F. W.

O'Donnell, A. E., and Axtell, R. C.
1965. Predation by *Fuscuropoda vegetans* (Acarina: Uropodidae) on the house fly (*Musca domestica*). Ann. Entom. Soc. Amer., *58*:403-404.

O'Donnell, A. E., and Nelson, E. L.
1967. Predation by *Fuscuropoda vegetans* (Acarina: Uropodidae) and *Macrocheles muscaedomesticae* (Acarina: Macrochelidae) on the eggs of the little house fly, *Fannia canicularis*. Jour. Kansas Entom. Soc., *40*:441-443.

Ogata, K.
1963. See Asahina, S., et al.

Ohanessian, A.
1965. See Berkaloff, A., et al.

Ohanessian, A., and Echalier, G.
1967. Multiplication du virus *Sindbus* chez *Drosophila melanogaster* (Insecte diptère), en conditions expérimentales. Compt. Rend. Acad. Sci., *264*: 1356-1358.

Oho, O.
1921. Über die Framboesie in Formosa. Far Eastern Assoc. Trop. Med. Proc. Fourth Congr., *2*:138-148.

Ojala, O., and Nuorteva, P.
1966. Isolering av *Salmonella*-bakterier hos flugor i södra Finland. Proc. 10th Congr. (1966) Nordic Veterinarians, Stockholm. [Offprint paged 1-6.]

O'Keefe, W. B., and Schorsch, C. B.
1954. Studies on some enteric bacteria associated with flies; Diptera: Calliphoridae, Muscidae, Anthomyidae. Med. Technicians Bull. (Suppl. to U. S. Armed Forces Med. Jour.), *5*:15-17.

Olsufiev, N. G.
1940. Rôle of *Stomoxys calcitrans* L. in the transmission and preservation of tularemia. Arkhiv. Biolog. Nauk, *58*:25-31.

Oota, W.
1965. See Shimizu, F., et al.

Orlando, A.
1960. See Figueiredo, M. B., et al.

Orton, S. T., and Dodd, W. L.
1910. Experiments on transmission of bacteria by flies with special relation to an epidemic of bacillary dysentery at the Worcester State Hospital, Massachusetts. Boston Med. Surg. Jour., *163*:863-868.

Ostrolenk, M., and Welch, H.
1942a. The common house fly (*Musca domestica*) as a source of pollution in food establishments. Food Res., *7*:192-200.

Ostrolenk, M., and Welch, H.
1942b. The house fly as a vector of food poisoning organisms in food producing establishments. Amer. Jour. Publ. Health, *32*:487-494.

Otto, J. S.
1956. See De Coursey, J. D., and Otto, J. S.

Oudemans, A. C.
1900. Further notes on acari. Tijdschr. Entom., *43*:109-128.

Oudemans, A. C.
1904. Notes on acari. XI. Tijdschr. Entom., *46*:93-134, + esp. xi to xiii.

Oudemans, A. C.
1926. Kritisch historisch overzicht der acarologie. Eerste gedeelte, 850 v. Chr. tot 1758. Tijdschr. Entom. V., *69*: Suppl. viii + 500 pp.

Oudemans, A. C.
1929. Kritisch historisch overzicht der acarologie. Tweede gedeelte, 1759-1804. Tijdschr. Entom. V., *72*: Suppl. xvii + 1097 pp.

Packard, A. S.
1873. On the transformations of the common house fly, with notes on allied forms. Proc. Boston Soc. Nat. Hist., *160*:136-150.

Paraf, J.
1920. Étude expérimentale du rôle des mouches dans la propagation de la dysenterie bacillaire. Rev. Hyg. Police. Sanit., *42*:241-244.

Parisot, J., and Fernier, L.
1934. The best methods of treating manure-heaps to prevent the hatching of flies. League of Nations Quart. Bull. Health Org., *3*:1-31.

Parker, H. L.
1923. Contribution à la connaissance de *Chalcis fonscolombei* Dufour. Bull. Soc. Entom. France, Nov. 28, pp. 238-240.

Parker, H. L., and Thompson, W. R.
1928. Contribution à la biologie des chalcidiens entomophages. Ann. Soc. Entom. France, *97*:425-465.

Parker, J. B.
1917. A revision of the bembicine wasps of America north of Mexico. Proc. U. S. Nat. Mus., *52*:1-155.

Parker, R. R.
1914. See Honeij, J. A., and Parker, R. R.

Parker, R. R.
1924. Ann. Soc. Entom. France, *93*; see pp. 267, 291, 310, 311, 320, 323, 325, 328.

Parman, D. C.
1915. See Bishopp, F. C., et al.

Parrot, L.
1922. See Donatien, A., and Parrot, L.

Patnaik, B.
1965. Personal communication.

Patton, W. S.
1910a. Experimental infection of the Madras bazaar fly, *Musca nebulo* Fabricius with *Herpetomonas muscae domesticae* (Burnett). Bull. Soc. Path. Exot., *3*:264-274.

Patton, W. S.
1910b. *Rhynchomonas luciliae*, nov. gen.; nov. spec. A new flagellate parasitic in the malpighian tubes of *Lucilia serenissima.* Walk. Bull. Soc. Path. Exot., *3*:433.

Patton, W. S.
1910c. *Rhynchomonas luciliae*, nov. gen.; nov. spec. A new flagellate parasitic in the malpighian tubes of *Lucilia serenissima* Walk. Bull. Soc. Path. Exot., *3*:300-303.

Patton, W. S.
1912a. Studies on the flagellates of the genera *Herpetomonas, Crithidia* and *Rhynchoidomonas.* 1. The morphology and life history of *Herpetomonas culicis* Novy, MacNeal and Torrey. Sci. Mem. Off. Med. Sanit. Dept. Govt. India. (new series?), *57*:1-21.

Patton, W. S.
1912b. Preliminary report on an investigation into the etiology of oriental sore in Cambay. India. Med. Dept. Scientific Mem. Off. (new series?), No. 50, pp. 1-21.

Patton, W. S.
1921a. Studies on the flagellates of the genera *Herpetomonas, Crithidia* and *Rhynchoidomonas.* No. 7. Indian Jour. Med. Res., *9*:230-239.

Patton, W. S.
1921b. Notes on the myiasis-producing Diptera of man and animals. Bull. Entom. Res., *12*:239-261.

Patton, W. S.
1921c. Studies on the flagellates of the genera *Herpetomonas, Crithidia* and *Rhynchoidomonas.* No. 3. The morphology and life history of *Rhyn-*

choidomonas siphunculinae sp. nov., parasitic in the malpighian tubes of *Siphunculina funicola* de Meijere. Indian Jour. Med. Res., *8*:603-612.

Patton, W. S.
1921-2. Some reflections on the kala-azar and oriental sore problems. Indian Jour. Med. Res., *9*:496-532.

Patton, W. S.
1922. Some notes on Indian Calliphorinae. IV. *Chrysomyia albiceps* Wied. (*rufifacies*, Froggatt); one of the Australian sheep maggot flies and *Chrysomyia villeneuvii*, sp. nov. Indian Jour. Med. Res., *9*:561-574.

Patton, W. S., and Edin, I. M.
1909. The life cycle of a species of *Crithidia* parasitic in the intestinal tracts of *Tabanus hilarius* and *Tabanus* sp. Arch. Protist., *15*:333-362.

Paul, J. R.
1943. See Trask, J. D., et al.

Paul, J. R.
1960. See Gudnadóttir, M., and Paul, J. R.

Paul, J. R.
1961. See Riordan, J. T., et al.

Paul, J. R., Horstmann, D. M., Riordan, J. T., Opton, E. M., Niederman, J. C., Isacson, E. P., and Green, R. H.
1962. An oral poliovirus vaccine trial in Costa Rica. Bull. Wld. Health Org., *26*:311-329.

Paul, J. R., Trask, J. D., Bishop, M. B., Melnick, J. L., and Casey, A. E.
1941. The detection of poliomyelitis virus in flies. Science, *94*:395-396.

Pavan, M.
1949. Ricerche sugli antibiotici di origine animale. La Ricerca Scientifica, *19*: 1011-1017.

Pavan, M.
1952. "Iridomyrmecin" as insecticide. Trans. Ninth Intl. Congr. Entom., *1*: 321-327.

Pavillard, E. R., and Wright, E. A.
1957. An antibiotic from maggots. Nature, *180*:916-917.

Pavlovskiĭ, E. N.
1921. Flies, their structure and life, their spreading of contagious diseases as parasites of man, and the battle against them. Publ. by People's Commissariat of Public Health, USSR, 100 pp.

Peat, A. A.
1935. See Kumm, H. W., et al.

Peck, O.
1951. In: Muesebeck, C.F.W., et al., Agric. Monog. U. S. Dept. Agric., No. 2, p. 535.

Peckham, E. G.
1905. See Peckham, G. W., and Peckham, E. G.

Peckham, G. W., and Peckham, E. G.
1905. Wasps, social and solitary. Boston: Houghton, Mifflin and Co., 311 pp.

Pegazzano, F.
1963b. See Filipponi, A., and Pegazzano, F.

Penna, L.
1957. See Coutinho, J. O., et al.

Penner, L. R.
1947. See Melnick, J. L., and Penner, L. R.

Penner, L. R.
1948. See Francis, T., Jr., et al.

Penner, L. R.
1951. See Hawley, J. E., et al.

Penner, L. R.
1952. See Melnick, J. L., and Penner, L. R.
Penning, C. A.
1904. Les trypanosomoses aux Indes Néerlandaises. Janus, *9*:514-522, 557-564, 620-626.
Peppler, H. J.
1944. Usefulness of microorganisms in studying dispersion of flies. Bull. U. S. Army Med. Dept., *75*:121-122.
Pereira, C., and De Castro, M. P.
1945. Contribuição para o conhecimento da espécie tipo de "*Macrocheles* Latr." ("Acarina"): "*M. muscaedomesticae* (Scopoli, 1772)" emend. Arch. Inst. Biol., *16*:153-186.
Pereira, C., and De Castro, M. P.
1947. Forese e partenogênese arrenótoca em "*Macrocheles muscaedomesticae*" (Scopoli) ("Acarina: Macrochelidae") e sua significação ecológia. Arch. Inst. Biol., *18*:71-89.
Perez-Fontana, U., and Severino-Brea, R.
1961. The transmission of hydatidosis by flies. Arch. Intl. Hidatidosis, *20*:283-286.
Perron, R.
1954a. Untersuchungen über Bau, Entwicklung und Physiologie der Milbe *Histostoma laboratorium* Hughes. Acta Zool., Stockholm, *34*:71-176.
Perron, R.
1954b. Canadian Entom., *86*:222.
Petch, T.
1923. Parasites of scale-insect fungi. Trans. British Mycol. Soc., *8*:206-212.
Petragnani, G.
1925. La mosca è sempre batterifera sin dalla nascita. Ig. Moderna, *18*:33-41.
Petrova, T. A.
1958. See Sychevskaīā, V. I., and Petrova, T. A.
Petrova, Z. F.
1959. See Sychevskaīā, V. I., et al.
Peytavin
1926. See Broudin, et al.
Philip, C. B.
1948. Observations on experimental Q fever. Jour. Parasit., *34*:457-464.
Phillips, J. S.
1934. The biology and distribution of ants in Hawaiian pineapple fields. Hawaii Pineapple Prod. Stat. Bull., *15*:1-57.
Piana, G. P.
1896. Osservazioni sul *Dispharagus nasutus* Rud. dei polli e sulle larve nematoelmintiche delle mosche e dei porcellioni. Atti. Soc. Ital. Sci. Nat. Milano, *36*:239-262.
Picado, C.
1935. Sur le principe bactéricide des larves des mouches (Myiases des plaies et myiases des fruits). Bull. Biol. France et Belgique, *69*:409-438.
Picard, F.
1914. Les champignons parasites des insectes et leur utilisation agricole. Ann. École Nat. Agric. Montpelier, n.s., *13*:121-248.
Pier, A. C.
1966. See Richard, J. L., and Pier, A. C.
Pimentel, D.
1955. Relationship of ants to fly control in Puerto Rico. Jour. Econ. Entom., *48*:28-30.
Pimentel, D.
1963. See Nagel, W. P., and Pimentel, D.

Pimentel, D.
1963. Natural population regulation and interspecies evolution. Proc. Intl. Congr. Zool., *3*:329-336.

Pimentel, D.
1965. See Madden, J. L., and Pimentel, D.

Pimentel, D., and Al-Hafidh, R.
1965a. Ecological control of a parasite population by genetic evolution in the parasite-host system. Ann. Entom. Soc. Amer., *58*:1-6.

Pimentel, D., and Epstein, B.
1960. The cluster fly, *Pollenia rudis* (Diptera: Calliphoridae). Ann. Entom. Soc. Amer., *53*:553-554.

Pimentel, D., Feinberg, E. H., Wood, P. W., and Hayes, J. T.
1965b. Selection, spatial distribution, and the coexistence of competing fly species. Amer. Naturalist, *99*:97-109.

Pimentel, D., Nagel, W. P., and Madden, J. L.
1963. Space-time structure of the environment and the survival of parasite-host systems. Amer. Naturalist, *47*:141-167.

Pimentel, D., and Uhler, L.
1969. Ants and the control of house flies in the Philippines. Jour. Econ. Entom., *62*:248.

Pinkus, H.
1913. The life-history and habits of *Spalangia muscidarum* Richardson, a parasite of the stable fly. Psyche, *20*:148-158.

Pipkin, A. C.
1942. Filth flies as transmitters of *Entamoeba histolytica*. Proc. Soc. Exp. Biol. Med., *49*:46-48.

Pipkin, A. C.
1949. Experimental studies on the role of filth flies in the transmission of *Endamoeba histolytica*. Amer. Jour. Hyg., *49*:255-275.

Pippen, W. F.
1956. See Ingram, R. L., et al.

Pirone, P. P.
1947. See Toomey, J. A., et al.

Place, F. E.
1915. Flies: A factor in a phase of filariasis in the horse. Thesis, Univ. Melbourne, 7 pp. (Also in Veter. Record, *28*:120.)

Pletneva, N. A.
1937. Rol'mukh kak mekhanicheskikx perenoschikov tsist Protozoa i i͡aitz glist Ashkhabade. [The role of flies as the mechanical carriers of Protozoa and of the eggs of the tapeworm in Ashkhabad.] Problemy parazitologii i fauna Turkmenii, pp. 117-120.

Plus, N.
1950. L'invasion d'un organisme par un virus héréditaire. Recherches quantitatives sur la vitesse d'acquisition de la sensibilité au CO_2 par les drosophiles. Exp. Cell Res., *1*:217-230.

Plus, N.
1954. Étude de la multiplication du virus de la sensibilité au gaz carbonique chez la drosophile. Bull. Biol. France et Belgique, *88*:248-293.

Plus, N.
1955a. Étude de la sensibilité héréditaire au gaz carbonique chez la drosophile. I. Multiplication du virus σ et passage à la descendance après inoculation ou transmission héréditaire, Ann. Inst. Pasteur, *88*:347-364.

Plus, N.
1955b. See Brun, G., et al.

Plus, N.
1956. See Duhamel, C., and Plus, N.

Pod"i͡apol'skai͡a, V. P., and Gnedina, M. P.
1934. O roli mukhi v èpidemiologii glistnykh zabolevaniĭ. [On the role of flies in the epidemiology of tapeworm infestation.] Med. Parazit., *3*:179-185.

Pokrovskiĭ, S. N., and Zima, G. G.
1938. Mukhi kak perenoschiki i͡aits glist v estestvennykh uslovii͡akh. [Flies as carriers of tapeworm eggs under natural conditions.] Med. Parazit., *7*: 262-264. (Rev. Appl. Entom., *26*:244.)

Poorbaugh, J. H.
1964. See Anderson, J. R., and Poorbaugh, J. H.

Porchinskiĭ, I. A.
1885. O razlichnykh "formakh" razmnozhenii͡a i o sokrashchennom" sposobe razvitii͡a u nekotorykh" obyknovenneĭshikh" vidov mukh." [On the different forms of reproduction and abbreviated modes of development in some of the most common species of flies.] Trudy Russk, Entom. Obshchestva, *19*:210-244. Also has as title: Muscarum cadaverinarum stercorariarumque biologia comparata. (English résumé under Osten-Sacken, 1887, Berlin Entom. Zeitsch., *31*:17-28.)

Porchinskiĭ, I. A.
1910. Osenni͡ai͡a zhigalka (*Stomoxys calcitrans*), ei͡a biologii͡a v" svi͡azi s" drugimi mukhami i bor'ba s" nei͡u. [The "autumn gadfly," *Stomoxys calcitrans*, its biology in relation to that of other flies, and its control.] Trudy Bi͡uro po Èntomologii . . . Glavnago Upravlenii͡a Zemleustroistva Zemledelii͡a, St. Petersburg, *8*, 90 pp.

Porchinskiĭ, I. A.
1911. *Hydrotaea dentipes*, its biology and the destruction by its larvae of the larvae of *Musca domestica*. Mem. Bur. Entom. Sci. Comm. Centr. Bd. Land Adminstr. Agric., St. Petersburg, *9* (5):1-30.

Porchinskiĭ, I. A.
1913. Domovai͡a mukha (*Muscina stabulans* Fall.), ei͡a znachenie dli͡a cheloveka i ego khozi͡aĭstva i otnoshenie ei͡a k" komnatnoĭ mukhe. [The "house" fly (*Muscina stabulans* Fall.), its significance for man and his household, and its relation to the "room" fly.] Trudy Bi͡uro po Èntomologii . . . Glavnago Upravlenii͡a Zemleustroistva Zemledelii͡a, St. Petersburg, *10*, No. 1, 39 pp.

Porter, A.
1913. See Fantham, H. B., and Porter, A.

Porter, A.
1953. Report of the honorary parasitologist for the year 1952. Proc. Zool. Soc. London, *123*:253-257.

Porter, A.
1958. See Fantham, H. B., and Porter, A.

Poulson, D. F.
1957. See Malogolowkin, C., and Poulson, D. F.

Poulson, D. F.
1959. See Malogolowkin, C., et al.

Poulson, D. F.
1959. See Sakaguchi, B., and Poulson, D. F.

Poulson, D. F.
1960. See Sakaguchi, B., and Poulson, D. F.

Poulson, D. F.
1961. See Counce, S. J., and Poulson, D. F.

Poulson, D. F.
1961. See Sakaguchi, B., and Poulson, D. F.

Poulson, D. F., and Counce, S. J.
1959. Effects of "sex-ratio" particles in embryonic and adult males of *Drosophila willistoni*. Anat. Record, *134*:625-626.

Poulson, D. F., and Sakaguchi, B.
1961a. Hereditary infections in *Drosophila.* [Abstract.] Genetics, *46*:890-891.

Poulson, D. F., and Sakaguchi, B.
1961b. Nature of "sex-ratio" agent in *Drosophila.* Science, *133*:1489-1490.

Povolný, D.
1960. See Gregor, F., and Povolný, D.

Povolný, D.
1961. See Lysenko, O., and Povolný, D.

Povolný, D., and Přívora, M.
1961. Kritische Bewertung mikrobiologischer Befunde bei synanthropen Fliegen in Mitteleuropa. Angew. Parasit., *2*:66-74.

Power, M. E., Melnick, J. L., and Bishop, M. B.
1943. A study of the 1942 fly population of New Haven. Yale Jour. Biol. Med., *15*:693-705.

Prado, E., and Jimenez, L. A.
1955. Estudio bacteriológico de la flora enterica transportada por las moscas de la ciudad de Santiago. Biol. Inst. Bact. Chile, *8*:14-18.

Preudhomme de Borre, A.
1873. *Musca domestica* with *Chelifer panzeri* Koch attached parasitically. S. B. Zool.-Bot. Ges. Wien, *23*:26.

Preuss, U.
1951. See Gross, H., and Preuss, U.

Preuss, U.
1953. See Gross, H., and Preuss, U.

Pridantseva, E. A.
1959. The biology of *Lyperosia titillans* Bezzi (Diptera, Muscidae), an intermediate host of the nematode *Parabronema skrjabini.* Entom. Rev., *38*: 129-138.

Přívora, M.
1961. See Povolný, D., and Přívora, M.

Prowazek, S.
1904. Die Entwicklung von *Herpetomonas.* Arb. Kaisl. Gesundheit., *20*:440-452.

Puerto Rico. USDA Experiment Station
1938. Report, p. 107.

Purdy, J. S.
1909. Flies and fleas as factors in the dissemination of disease. The effects of petroleum as an insecticide. Jour. Roy. Sanit. Inst., *30*:496-503.

R. M.
1937. Sotsialist. Nauk i Tekhnika, *5*: No. 9. (Seen as: Mouches et infections intestinales. Bull. Inst. Pasteur, *37*:418, 1939.)

Radvan, R.
1956. Přežívání bakterií při metamorfose *Musca domestica* L. Sborník Vědeckých Prací Voj. Lék. Akad. Jevp., *4*:104-111.

Radvan, R.
1960a. Persistence of bacteria during development in flies. I. Basic possibilities of survival. Folia Microb., *5*:50-56.

Radvan, R.
1960b. Persistence of bacteria during development in flies. II. The number of surviving bacteria. Folia Microb., *5*:85-92.

Radvan, R.
1960c. Persistence of bacteria during development in flies. III. Localization of the bacteria and transmission after emergence of the fly. Folia Microb., *5*:149-156.

Railway, B. N.
1945. See Senior-White, R., and Railway, B. N.

Raimbert, A.
1869. Recherches expérimentales sur la transmission du charbon par les mouches. Compt. Rend. Acad. Sci., *69*:805-812.

Raimbert, A.
1870. Recherches expérimentales sur la transmission du charbon par les mouches. L'Union Médicale Paris, *9*:209-210, 350-352, 507-509, 709-710

Ramakrishna Ayyar, T. V.
1920. Note on a musciphagous wasp (*Bembex lunata*). Rept. Proc. 3rd Entom. Mtg. Calcutta, Supt. Govt. Printing, *3*:909-910.

Rambousek, F.
1929. Die Rübenschädlinge im Jahre 1927 und 1928. Z. Zuckerind, Čsl. Repub., *54*:105-114.

Ramirez-Genel, M.
1962-3. See Vasquez-Gonzalez, J., et al.

Ramirez, R.
1898. The Diptera from a hygienic point of view. Publ. Health Rept., *24*:260-262.

Ransom, B. H.
1911. The life-history of a parasitic nematode, *Habronema muscae*. Science, n.s., *34*:690-692.

Ransom, B. H.
1913. The life history of *Habronema muscae* (Carter), a parasite of the horse transmitted by the house fly. Bull. U. S. Bur. Anim. Indust., No. 163, pp. 1-36.

Ratzeburg, J.T.C.
1844. Die Forst-Insekten, dritte Teil (Zweiflügler, Fliegen und Mücken, p. 175). (Quoted from Keilin, 1917.)

Rayment, T.
1954. Taxonomy, morphology and biology of sericophorine wasps. Mem. Nat. Museum Victoria, *19*:11-105.

Reeves, W. W.
1877. See White, T. C., and Reeves, W. W.

Régnier, P. R.
1931. Les invasions d'acridiens au Maroc de 1927 à 1931. [Morocco]: Dir. gén. Agric. Comm. Colonis., Défense des Cultures, No. 3, v + 139 pp.

Reid, W. M., and Ackert, J. E.
1937. The cysticercoid of *Choantaenia infundibulum* and the housefly as its host. Trans. Amer. Microscop. Soc., *56*:99-104.

Reinstorf, A.P.N.
1923. Uebertragung der Ruhr durch Fliegen. Inaug. Dissert. Giessen. Berlin, Morenhoven, 21 pp.

Reiter, H.
1917. Zur Kenntnis der Weilschen Krankheit. Deutsche med. Wochenschrift, *43*:552-554.

Rendtorff, R. C., and Francis, T., Jr.
1943. Survival of the Lansing strain of poliomyelitis virus in the common house-fly, *Musca domestica* L. Jour. Infect. Dis., *73*:198-205.

Rendtorff, R. C., and Holt, C. J.
1954. The experimental transmission of human intestinal protozoan parasites. III. Attempts to transmit *Endamoeba coli* and *Giardia lamblia* cysts by flies. Amer. Jour. Hyg., *60*:320-326.

Richard, J. L., and Pier, A. C.
1966. Transmission of *Dermatophilus congolensis* by *Stomoxys calcitrans* and *Musca domestica*. Amer. Jour. Veter. Res., *27*:419-423.

Richards, A. G.
1959. See McConnell, E., and Richards, A. G.

Richards, O. W.
1926. See Hamm, A. H., and Richards, O. W.

Richards, C. S., Jackson, W. B., DeCapito, T. M., and Maier, P. P.
1961. Studies on rates of recovery of *Shigella* from domestic flies and from humans in southwestern United States. Amer. Jour. Trop. Med. Hyg., *10*:44-48.

Richardson, C. H.
1913a. Studies on the habits and development of a hymenopterous parasite, *Spalangia muscidarum* Richardson. Jour. Morph., *24*:513-549.

Richardson, C. H.
1913b. An undescribed hymenopterous parasite of the housefly. Psyche, *20*: 38-39.

Rico, M.
1964. Proporcion entre sexos (S. R.) en *Drosophila melanogaster*. 1. Una evidencia mas sobre la herencia citoplasmica. Bol. Inst. Nac. Invest. Agron., *24*:85-95.

Riehl, L. A.
1962b. See Rodriguez, J. L., and Riehl, L. A.

Riley, C. V.
1877. Rept. U. S. Entom. Comm. (not seen).

Riley, C. V., and Howard, L. O.
1891. Insect Life, No. 4, p. 123.

Rinonapoli, G.
1930. Contributo allo studio epidemiologico della infezione carbonchiosa. Med. Prat., *15*:281-287.

Riordan, J. T., Paul, J. R., Yoshioka, I., and Horstmann, D. M.
1961. The detection of poliovirus and other enteric viruses in flies. Results of tests carried out during an oral poliovirus vaccine trial. Amer. Jour. Hyg., *74*:123-136.

Riordan, J. T.
1962. See Paul, J. R., et al.

Rivers, C. F.
1961. See Smith, K. M., et al.

Roberg, D. N.
1915. I. The rôle played by the insects of the dipterous family Phoridae in relation to the spread of bacterial infections. II. Experiments on *Aphiochaeta ferruginea* Brunetti with the cholera vibrio. Philippine Jour. Science, Sec. B., Trop. Med., *10*:309-336.

Roberts, E. W.
1947. The part played by the faeces and vomit-drop in the transmission of *Entamoeba histolytica* by *Musca domestica*. Ann. Trop. Med. Parasit., *41*:129-142.

Roberts, F.H.S.
1934. The large roundworm of pigs, *Ascaris lumbricoides* L., 1758: Its economic importance in Queensland, life cycle, and control. Animal Health Stat., Yeerongpilly Bull. No. 1, pp. 1-81.

Roberts, R. A.
1933a. Activity of blowflies and associated insects at various heights above the ground. Ecology, *14*:306-314.

Roberts, R. A.
1933b. Biology of *Brachymeria fonscolombei* (Dufour), a hymenopterous parasite of blowfly larvae. U. S. Dept. Agric., Washington, D. C., Tech. Bull. No. 365, 21 pp.

Roberts, R. A.
1935. Some North American parasites of blowflies. Jour. Agric. Res., *50*:479-494.

Roberts, R. H.
1955. See Ferris, D., et al.

Robinson, W., and Norwood, V. H.
1933. The role of surgical maggots in the disinfection of osteomyelitis and other infected wounds. Jour. Bone Joint Surg., *15*:409-412.

Robinson, W., and Norwood, V. H.
1934. Destruction of pyogenic bacteria in the alimentary tract of surgical maggots implanted in infected wounds. Jour. Lab. Clin. Med., *19*:581-586.

Rodhain, J., and Bequaert, J.
1916. Matériaux pour une étude monographique des diptères parasites de l'Afrique. Première partie: Histoire de *Passeromyia heterochaeta* Villen. et de *Stasia* (*Cordylobia*) *rodhaini* Ged. Bull. Sci. France et Belge, 7e Série, *49*:236-289.

Rodionov, Z. S.
1927. Pests of cotton. Part 1. [In Russian.] Défense des plantes, *4*:28-59, Leningrad.

Rodriguez, J. G.
1961. See Wade, C. F., and Rodriguez, J. G.

Rodriguez, J. G.
1963. See Wallwork, J. H., and Rodriguez, J. G.

Rodriguez, J. G.
1966. See Singh, P., et al.

Rodriguez, J. G., and Wade, C. F.
1960. Preliminary studies on biological control of housefly eggs using macrochelid mites. Proc. XIth Intl. Congr. Entom. Vienna.

Rodriguez, J. G., and Wade, C. F.
1961. The nutrition of *Macrocheles muscaedomesticae* (Acarina: Macrochelidae) in relation to its predatory action on the house fly egg. Ann. Entom. Soc. Amer., *54*:782-788.

Rodriguez, J. G., Wade, C. F., and Wells, C. N.
1962a. Nematodes as a natural food for *Macrocheles muscaedomesticae* (Acarina: Macrochelidae), a predator of the housefly egg. Ann. Entom. Soc. Amer., *55*:507-511.

Rodriguez, J. L., and Riehl, L. A.
1962b. Control of flies in manure of chickens and rabbits by cockerels in south California. Jour. Econ. Entom., *55*:473-477.

Rogers, L.
1929. Recent advances in tropical medicine. Philadelphia: P. Blakiston's Son and Co., see pp. 243-245.

Römer, R.
1906. La lèpre. Bull. Acad. Roy. Méd. Belgique, 27 Jan., pp. 99-165 (and Lepra, *6*:258).

Romanova, V.
1937. See Somov, P., and Romanova, V.

Rondani
1913. See Martelli, G.

Root, F. M.
1921. Experiments on the carriage of intestinal protozoa by flies. Amer. Jour. Hyg., *1*:131-153.

Rosenau, M. J., and Brues, C. T.
1912a. Some experimental observations upon monkeys concerning the transmission of poliomyelitis through the agency of the *Stomoxys calcitrans*. 15th Intl. Congr. Hyg. Demogr., *1*:616-623.

Rosenau, M. J., and Brues, C. T.
1912b. Some experimental observations upon monkeys concerning the transmission of poliomyelitis through the agency of *Stomoxys calcitrans*. Monthly Bull. St. Bd. Health Massachusetts, *7*:314-317.

Rosenbusch, F.
1910. Ueber eine neue Encystierung bei *Crithidia muscae domesticae*. Centralbl. Bakt., Parasit. Infekt., I. Abt., Orig., *53*:387-393.

Ross, W. C., and Hussain, M.
1924. On the life history of *Herpetomonas muscae-domesticae*. A preliminary note. Indian Med. Gaz., *59*:614-615.

Rostrup, O.
1916. Bidrag til Danmarks svampeflora. I. Dansk Botanisk Arkiv. Bd. 2, No. 5, pp. 1-56.

Roubaud, E.
1908. See Martin G., et al.

Roubaud, E.
1908a. *Leptomonas mesnili* n. sp., nouveau flagellé à formes trypanosomes de l'intestin de muscides non piqueurs. Compt. Rend. Soc. Biol., *65*:39-41.

Roubaud, E.
1908b. Sur un nouveau flagellé, parasite de l'intestin des muscides, au Congo Français. Compt. Rend. Soc. Biol., *64*:1106-1108.

Roubaud, E.
1909. La maladie du sommeil au Congo Français. Non spécificité de la culture intestinale, etc. Rapport Mission d'Études, 1906-08, Paris, see pp. 542-625.

Roubaud, E.
1911a. Études sur les stomoxydes du Dahomey. Bull. Soc. Path. Exot., *4*:122-132.

Roubaud, E.
1911b. *Cystotrypanosoma intestinalis* n. sp.; trypanosome vrai à reproduction kystique, de l'intestin des mouches vertes (Lucilies) de l'Afrique tropicale. Comp. Rend. Soc. Biol., *71*:306-308.

Roubaud, E.
1911c. *Cercoplasma* (n. gen.) *caulleryi* (n. sp.) nouveau flagellé à formes typanosomiennes de l'intestin d'*Auchmeromyia luteola* Fabr. (Muscide). Compt. Rend. Soc. Biol., *71*:503-505.

Roubaud, E.
1911d. Sur un type nouveau de leptomonades intestinales des muscides, *Leptomonas soudanensis* n. sp., parasite des Pycnosomes Africains. Compt. Rend. Soc. Biol., *71*:570-573.

Roubaud, E.
1912. See Bouet, G., and Roubaud, E.

Roubaud, E.
1912a. Expériences de transmission de flagellés divers chez les muscides Africains du genre *Pycnosoma*. Compt. Rend. Soc. Biol., *72*:508-510.

Roubaud, E.
1912b. Sur un nouveau flagellé à forme trypanosome des drosophiles d'Afrique, *Cercoplasma drosophilae* n. sp. Compt. Rend. Soc. Biol., *72*:554-556.

Roubaud, E.
1913. Recherches sur les Auchmeromyies, Calliphorines à larves suceuses de sang de l'Afrique tropicale. Bull. Sci. France et Belgique, *47*:105-202.

Roubaud, E.
1917. Observations biologiques sur *Nasonia brevicornis* Ashm., chalcidide parasite des pupes de muscides. Bull. Sci. France et Belgique, *50*:425-439.

Roubaud, E.
1918. Le rôle des mouches dans la dispersion des amibes dysentériques et autres protozoaires intestinaux. Bull. Soc. Path. Exot., *11*:166-171.

Roubaud, E.
1922. Recherches sur la fécondité et la longévité de la mouche domestique. Ann. Inst. Pasteur, *36*:765-783.

Roubaud, E., and Descazeaux, J.
1921. Contribution à l'histoire de la mouche domestique comme agent vecteur des habronémoses d'équidés. Cycle évolutif et parasitisme de l'*Habronema megastoma* (Rudolphi, 1819) chez la mouche. Bull. Soc. Path. Exot., *14*:471-506.

Roubaud, E., and Descazeaux, J.
1922a. Évolution de l'*Habronema muscae* Carter chez la mouche domestique et de l'*H. microstomum* Schneider chez le stomoxe. (Note prélim.) Bull. Soc. Path. Exot., *15*:572-574.

Roubaud, E., and Descazeaux, J.
1922b. Deuxième contribution à l'étude des mouches, dans leurs rapports avec l'évolution des habronèmes d'équidés. Bull. Soc. Path. Exot., *15*:978-1001.

Roubaud, E., and Descazeaux, J.
1923. Sur un agent bactérien pathogène pour les mouches communes: *Bacterium delendae-muscae* n. sp. Compt. Rend. Acad. Sci., *177*:716-717.

Roubaud, E., and Treillard, M.
1935. Un coccobacille pathogène pour les mouches tsétsés. Compt. Rend. Acad. Sci., *201*:304-306.

Round, M. C.
1961. Observations on the possible role of filth flies in the epizootiology of bovine cysticercosis in Kenya. Jour. Hyg., *59*:505-513.

Rovelli, G.
1889. See Grassi, B., and Rovelli, G.

Row, R.
1911. *Leishmania tropica* and the oriental sore of Cambay. Proc. Bombay Branch British Med. Assoc. Jan. 29, pp. 828-831.

Roy, D. N., and Mukherjee, P. K.
1937a. *Allantonema muscae* sp. nov., a new parasitic nematode of the family Rhabditidae from the haemocoele of *Musca vicina.* Ann. Trop. Med. Parasit., *31*:449-451.

Roy, D. N., and Mukherjee, P. K.
1937b. *Allantonema stricklandi* sp. nov., a parasitic nematode of house-flies, *Musca vicina.* Ann. Trop. Med. Parasit., *31*:453-456.

Roy, D. N., and Siddons, L. B.
1939. A list of Hymenoptera of superfamily Chalcidoidea, parasites of calyptrate Muscoidea. Rec. Indian Mus., *41*:223-224.

Roy, D. N., Siddons, L. B., and Mukherjee, S. P.
1940. The bionomics of *Dirhinus pachycerus* Masi (Hymenoptera: Chalcidoidea), a pupal parasite of muscoid flies. Indian Jour. Entom., *2*:229-240.

Rozsypal, J.
1957. Houbovā nākaza masovē Līhnē bzučivky obecnē (*Calliphora vomitoria* L.). [Mycoses in *Calliphora vomitoria* L. flies in places of their mass multiplication.] Zool. Listy, Praha, *6*:12-16.

Rozsypal, J.
1966. A new fungal parasite of calyptrate flies from Europe *Zoophthora vomitoriae* sp. nov. (Entomophthoraceae). Acta Mycol., *2*:23-24.

Ruhland, H. H., and Huddleson, J. F.
1941. The rôle of one species of cockroach and several species of flies in the dissemination of *Brucella.* Amer. Jour. Veter. Res., *2*:371-372.

Russo, C.
1930. Recherches expérimentales sur l'épidémiogenése de la peste bubonique par les insectes. Bull. Men. Off. Intl. Hyg. Publ., *22*:2108-2120.

Ryan, J.
1962. See Gear, J., et al.

Sabin, A. B., and Ward, R.
1942. Insects and epidemiology of poliomyelitis. Science, *95*:300-301.

Sabrosky, C. W.
1959. *Musca autumnalis* in the Central States. Jour. Econ. Entom., *52*:1030-1031.

Saccá, G.
1953. See Morellini, M., and Saccá, G.

Saceghem, R. Van
1917. Contribution à l'étude de la dermite granuleuse des équidés. Bull. Soc. Path. Exot., *10*:726-729.

Saceghem, R. Van
1918. Cause étiologique et traitement de la dermite granuleuse. Bull. Soc. Path. Exot., *11*:575-578.

Saceghem, R. Van
1921. La trypanosomiase du Ruanda. Compt. Rend. Soc. Biol. Paris, *84*:283-286. (Practically identical with Ann. Méd. Vétér., *66*:305-311, 1921; and Bull. Agric. Congo Belge., *12*:296 and on, 1921.)

Saceghem, R. Van
1922a. Mécanisme de la propagation des trypanosomiases par les stomoxes. Ann. Soc. Belge Méd. Trop., *2*:161-164.

Saceghem, R. Van
1922b. Mécanisme de la propagation des trypanosomiases par les stomoxes. Bull. Agric. Congo Belge, *13*:606-619.

Sakaguchi, B.
1961a. See Poulson, D. F., and Sakaguchi, B.

Sakaguchi, B.
1961b. See Poulson, D. F., and Sakaguchi, B.

Sakaguchi, B., and Poulson, D. F.
1959. Distribution of the "sex-ratio" agents in the tissues of adults of *Drosophila willistoni.* Ann. Rept. Natl. Inst. Genet., Misima, Japan, *10*:27-28.

Sakaguchi, B., and Poulson, D. F.
1960. Transfer of the "sex-ratio" condition from *Drosophila willistoni* to *Drosophila melanogaster.* Anat. Rec., *138*:381 (abstracts).

Sakaguchi, B., and Poulson, D. F.
1961. Distribution of "sex-ratio" agent in tissues of *Drosophila willistoni.* Genetics, *46*:1665-1676.

Salt, G.
1930. The natural control of *Lucilia sericata.* Nature, *125*:203.

Salt, G.
1932. The natural control of the sheep-blowfly, *Lucilia sericata* Meigen. Bull. Entom. Res., *23*:235-245.

Salt, G.
1957. Experimental studies in insect parasitism. X. The reactions of some endopterygote insects to an alien parasite. Proc. Roy. Soc. B, *147*:167-184.

Sampietro, G.
1912. See Alessandrini, G., and Sampietro, G.

Sanders, D. A.
1940a. *Hippelates* flies as vectors of bovine mastitis (Preliminary Report). Jour. Amer. Veter. Med. Assoc., *97*:306-308.

Sanders, D. A.
1940b. *Musca domestica* a vector of bovine mastitis (Preliminary Report). Jour. Amer. Veter. Med. Assoc., *97*:120-123.

Sanders, D. A.
1942. Pests, *10* (3), pp. 23, 26.

Sanders, D. P., and Dobson, R. C.
1966. The insect complex associated with bovine manure in Indiana. Ann. Entom. Soc. Amer., *59*:955-959.

Sanders, G. E.
1910a. See Girault, A. A., and Sanders, G. E.

Sanders, G. E.
1910b. See Girault, A. A., and Sanders, G. E.

Sandes, T. L.
1911. The mode of transmission of leprosy. British Med. Jour., *2*:469-470.

Sandes, T. L.
1912. The mode of transmission of leprosy. Lepra. (Leipzig), *12*:65-69.

Sarwar, S. M.
1935. See Krishna Iyer, P. R., and Sarwar, S. M.

Satchell, G. H., and Harrison, R. A.
1953. II. Experimental observations on the possibility of transmission of yaws by wound-feeding Diptera, in western Samoa. Trans. Roy. Soc. Trop. Med. Hyg., *47*:148-153.

Satterthwait, A. F.
1943. Notes on the parasitic habits of *Muscina stabulans* (Fall.) (Diptera, Muscidae). Jour. New York Entom. Soc., *50*:233-234.

Sauer, E.
1908. Können ohne veterinär-polizeiliche Bedenken die Häute Rauschbrand-Kranker Tiere zu Gerbereizwecken verwendet werden? Zeitsch. Tiermed., *12*:34-71.

Saunders, D. S.
1964. Rearing tsetse-fly parasites in blowfly puparia. Bull. Wld. Health Org., *31*:509-510.

Saunders, E. W.
1921. The paralysis fly *Lucilia caesar*. Some of the difficulties encountered in obtaining its toxi virulent larvae for experimental purposes. Jour. Missouri St. Med. Assoc., *18*:160-161.

Savchenko, I. G.
1892a. Die Bezihung der Fliegen zur Verbreitung der Cholera. Centralbl. Bakt., Parasit., Orig., *12*:893-898.

Savchenko, I. G.
1892b. Materials on the etiology of cholera. The role of the fly in spreading cholera infection. Vrach, *45*:1131-1132.

Schaeffer, M.
1947. See Toomey, J. A., and Schaeffer, M.

Schaerffenberg, B.
1959. Zur Biologie und Okologie des insektentotenden Pilzes *Metarrihzium anisopliae* (Metsch.) Sorok. (Entwecklung. Kultur, Lebensanspruche, Infektionsverlauf, praktische Bedeutung.) Zeitsch. Angew. Entom., *44*: 264-271.

Schat, P. T.
1903. Verdere mededeelingen over "sura." Meded. Proefstat. Oost-Java., Ser. 3, No. 44, pp. 1-19.

Schat, P. T.
1909. Beiträge zu den Untersuchungen uber die *Trypanosoma evansi* und zur Bekampfung der Surra unter dem Hornvieh auf Java. Inaugural dissertation, Universität Bern. 102 pp.

Scheerpeltz, O.
1934. Zwei neue Arten der Gattung *Aleochara* Gravh. (Coleopt. Staphylinidae), die aus den Puppen von *Lyperosia* (Dipt.) als Parasiten gezogen wurden. Rev. Suisse Zool., *41*:131-147.

Schlegel-Oprecht, E.
1953. Versuche zur Auslösung von Mutationen bei der zoophagen Cynipide *Pseudeucoila bochei* Weld und Befunde über die stammspezifische Abwehrreaktion des wirtes *Drosophila melanogaster*. Z. Indukt. Abstamm. Vereblehre, *85*:245-281.

Schmidtke, L.
1952. See Kunert, H., and Schmidtke, L.

Schmidtke, L.
1953. See Kunert, H., and Schmidtke, L.

Schmidtke, L.
1955. Histologishe Untersuchungen an toxoplas-mainfizierten Insekten (*Calliphora erythrocephala, Periplaneta americana*). Zentralbl. Bakt., Parasit. Infekt. Hyg., *164*:508-513.

Schmidtke, L.
1959. Experimentelle Übertragung von Krankheitserregern durch Fliegen. Über negativ verlaufene Versuche. Zeitsch. Angew. Zool., *46*:332-338.

Schmiedeknecht, O.
1909. In: Genera Insectorum, No. 97, p. 387 (not seen).

Schmit-Jensen, H. O.
1927. Eksperimentelle undersøgelser over tovingede insekters (Diptera's) betydning som smittespredere ved mund- og klovesyge. I. Virus' skaebne i stikfluen, *Stomoxys calcitrans*' organisme, belyst ved forsøg paa marsvin. Maanedsskrift Dyrlaeger, *39*:1-39.

Schmittner, S. M.
1965. See McGhee, R. B., et al.

Schoof, H.
1954. See Melnick, J. L., et al.

Schorsch, C. B.
1954. See O'Keefe, W. B., and Schorsch, C. B.

Schrank, F.V.P.
1776. Beyträge zur Naturgeschichte. Augsburg.

Schröck
1686. Pediculi in muscis vulgaribus. Misc. Cur. sive Ephem. Med.-Phys. Germ. Acad. Caes. Leopold Nat. Cur., Dec. 2, Ann. 4, p. 41.

Schuberg, A., and Böing, W.
1913. Weitere Üntersuchungen über die Übertragung von Krankheitserregern durch einheimische Stechfliegen. Centralbl. Bakt., Parasit. Infekt., I. Abt., *57*:301-303.

Schuberg, A., and Böing, W.
1914. Über die Übertragung von Krankheiten durch einheimische stechende Insekten. III. Teil. Arb. Kaisl. Gesundheit., *47*:491-512.

Schuberg, A., and Kuhn, P.
1909. Über die Übertragung von Krankheiten durch einheimische stechende Insekten. I. Teil. Arb. Kaisl. Gesundheit., *31*:377-393.

Schuberg, A., and Kuhn, P.
1912. Über die Übertragung von Krankheiten durch einheimische stechende Insekten. II. Teil. Arb. Kaisl. Gesundheit., *40*:209-234.

Schumann, H.
1961. Die Bedeutung symboviner Fliegen als Verbreiter von Mastitis-Erregern. Mh. Vetmed., 16 Pt., pp. 624-626.

Schweizer, G.
1936. Der Pilz *Empusa muscae* und seine Bedeutung bei der Fliegenbekämpfung. Natur und Kultur, *33*:149-152.

Schweizer, G.
1947. Über die Kultur von *Empusa muscae* Cohn und anderen Entomophthoraceen auf kalt sterilisierten Nährböden. Planta, *35*:132-175.

Scopoli, J. A.
1763. Entomologia Carniolica . . . Wien. No. 1066, p. 389.

Scopoli, J. A.
1772. Annus V. Historico Naturalis, Leipzig, No. 157, p. 125.

Scott, J. R.
1917a. Studies upon the common house-fly (*Musca domestica* Linn.). I. A general study of the bacteriology of the house-fly in the District of Columbia. Jour. Med. Res., *37*:101-119.

Scott, J. R.
1917b. Studies upon the common house-fly (*Musca domestica* Linn.). II. Isolation of *B. cuniculicida*, a hitherto unreported isolation. Jour. Med. Res., *37*:121-124.

Scott, J. W.
1917. Some experiments on the transmission of swamp fever by insects. Anat. Rec., *11*:540-541.

Scott, J. W.
1920. Experimental transmission of swamp fever or infectious anemia by means of insects. Jour. Amer. Veter. Med. Assoc., n.s. *9*:448-454.

Scott, J. W.
1922. Insect transmission of swamp fever or infectious anemia of horses. Univ. Wyoming Agric. Exp. St. Bull. No. 133, pp. 57-137.

Second, L.
1945. See Legroux, R., and Second, L.

Seecof, R. L.
1965. Resistance to sigma virus infection in *Drosophila.* Nature, *207*:887-888.

Seecof, R. L.
1969. Sigma virus multiplication in whole-animal culture of *Drosophila.* Virology, *38*:134-139.

Séguy, E.
1929. Étude sur les Diptères à larves commensales ou parasites des oiseaux de l'Europe occidentale. Encycl. Entom., Sér. B, Diptera, *5*(fasc.2): 63-82.

Séguy, E.
1934. Contribution à l'étude des mouches phytophages de l'Europe Occidentale: II (1). Encycl. Entom., Sér. B, Diptera, *7*:167-264.

Sen, S. K.
1925-6. Experiments on the transmission of rinderpest by means of insects. Mem. Dept. Agric. India, Entom. Ser., *9*:59-63, 166-181, 184-185.

Sen, S. K., and Minett, F. C.
1944. Experiments on the transmission of anthrax through flies. Indian Jour. Veter. Science and Anim. Husb., *14*:149-158.

Senior-White, R., and Railway, B. N.
1945. Some notes on the life-history of *Musca planiceps* Weid. Indian Jour. Veter. Science, *14*:123-125.

Sergent, Ed., and Donatien, A.
1922a. Les stomoxes, propagateurs de la trypanosomiase des dromadaires. Compt. Rend. Acad. Sci., *174*:582-584.

Sergent, Ed., and Donatien, A.
1922b. Transmission naturelle et expérimentale de la trypanosomiase des dromadaires par les stomoxes. Arch. Inst. Pasteur Afrique du Nord, *2*: 291-315.

Sergent, Ed., and Foley, H.
1910. Recherches sur la fièvre récurrente et son mode de transmission, dans une épidémie algérienne. Ann. Inst. Pasteur, *24*:337-373.

Sergent, Ed., and Sergent, Et.
1905. El-debab, trypanosomiase des dromadaires de l'Afrique du Nord. Ann. Inst. Pasteur, *19*:17-48.

Sergent, Et.
1905. See Sergent, Ed., and Sergent, Et.

Severino-Brae, R.
1961. See Perez-Fontana, V., and Severino-Brae, R.

Shakhurina, E. A.
1963. See Tukhmanyants, A. A., et al.

Shaw, E. W.
1949. See Melnick, J. L., et al.

Shaw, F. R.
1965. See Eversole, J. W., et al.

Shimizu, F., Hashimoto, M., Taniguchi, H., Oota, W., Kakizawa, H., Takada, R., Kano, R., Tange, H., Kaneko, K., Shinonaga, S., and Miyamoto, K.
1965. Epidemiological studies on fly-borne epidemics. Report I. Significant role of flies in relation to intestinal disorders. Jap. Jour. Sanit. Zool., *16*: 201-211.

Shinonaga, S.
1965. See Shimizu, F., et al.

Shope, R. E.
1927. Bacteriophage isolated from the common house-fly (*Musca domestica*). Jour. Exp. Med., *45*:1037-1044.

Shterngol'd, E. I͡a.
1949. Komnatnye mukhi kak perenoschiki kishechnykh infektsii. [Roomflies as carriers of intestinal infections.] Trudy Uzbekskogo Inst. Epidem. Mikrob., *3*:172-180.

Shura-Bura, B. L.
1952. Zagri͡aznenie fruktov synantropnymi mukhami. [Soiling of fruit by flies.] Entom. Obozrenie Moscow, *32*:117-125.

Shura-Bura, B. L.
1955. On the occurrence of naturally infected flies in the environment of dysenteric patients. Report given at the fourth lecture in honor of N. A. Kholodkovskii, 3 March 1951. Moscow/Leningrad, pp. 29-52.

Shura-Bura, B. L.
1957. Emploi des isotopes radioactifs dans l'étude du rôle épidémiologique des mouches. Jour. Hyg. Epidem. Microb. Immun., *1*:249-255.

Siddons, L. B.
1939. See Roy, D. N., and Siddons, L. B.

Siddons, L. B.
1940. See Roy, D. N., and Siddons, L. B.

Sieber, H., and Gonder, R.
1908. Übertragung von *Trypanosoma equiperdum*. Archiv. Schiffs. Tropen-Hyg., *12*:646.

Sieyro, L.
1942. Die Hausfliege (*Musca domestica*) als Überträger von *Entamoeba histolytica* und anderen Darmprotozoen. Deutsche Tropenmed. Zeitsch., *46*: 361-372.

Sigot, A.
1955a. See Brun, G., and Sigot, A.
Silverman, L.
1953. See Silverman, P. H., and Silverman, L.
Silverman, P. H., and Silverman, L.
1953. Growth measurements on *Musca vicina* (Macq.) reared with a known bacterial flora. Riv. Parassit., *14*:89-95.
Simitch, T., and Kostitch, H. D.
1937. Rôle de la mouche domestique dans la propagation du "*Trichomonas intestinalis*" chez l'homme. Ann. Parasit., *15*:323-325.
Simmonds, F. G.
1950-65. See Thompson, W. R., and Simmonds, F. G.
Simmonds, H. W.
1922. Entomological notes. Agric. Circ. Fiji, *3*:24.
Simmonds, H. W.
1929. Introduction of *Spalangia cameroni*, parasite of the housefly, into Fiji. Agric. Jour. Fiji, *2*:35.
Simmonds, H. W.
1940. Investigations with a view to the biological control of houseflies in Fiji. Trop. Agric., *17*:197-199.
Simmonds, H. W.
1944. Bull. Entom. Res., *35*:222.
Simmonds, H. W.
1958. House-fly problem in Fiji and Samoa. So. Pacific Comm. Quart. Bull., *8*:29, 30, 47.
Simmons, P.
1922. Controlling the ham or cheese skipper. Separate from National Provisioner, Chicago, 2pp.
Simmons, S. W.
1935a. A bactericidal principle in excretions of surgical maggots which destroys important etiological agents of pyogenic infections. Jour. Bact., *30*: 253-267.
Simmons, S. W.
1935b. The bactericidal properties of excretions of the maggot of *Lucilia sericata*. Bull. Entom. Res., *26*:559-563.
Simon, R. J.
1933. See Weil, G. C., et al.
Singh, P., King, W. E., and Rodriguez, J. G.
1966. Biological control of muscids as influenced by host preference of *Macrocheles muscaedomesticae*. Jour. Med. Entom., *3*:78-81.
Sitowski, L.
1928. O pasorzytach barczatki (*Dendrolimus pini* L.) mniszki (*Lymantria monacha* L.). [Parasites of *D. pini* L. and *L. monacha* L.] Rocz. Nauk Rol. Lesn., *19*, reprint 12, Poznan.
Sivori, F., and Lecler, E.
1902. La surra americana ou mal de caderas. Buenos Aires, 79 pp., see pp. 42, 43, 58-61.
Skidmore, L. V.
1932. The transmission of fowl cholera to turkeys by the common housefly (*Musca domestica* Linn.), with brief notes on the viability of fowl cholera microorganisms. Cornell Veter., *22*:281-285.
Skopina, N. P.
1959b. See Sychevskaia͡, V. I., et al.
Smirnov, E. S.
1934. Issledovaniia͡ po èkologii *Mormoniella vitripennis*, parazita sinantropnykh mukh. Med. Parasit. Parasitarn. Bolezni, *3*:330-335.

Smirnov, E. S.
1934. See Vladimirova, M. S., and Smirnov, E. S.

Smirnov, E. S.
1938. See Vladimirova, M. S., and Smirnov, E. S.

Smirnov, E. S., and Kuzina, O.
1933. Èxperimental' no èkologicheskie issledovaniia nad parazitami mukh. Zool. Zhurnal, *12*:96-109.

Smirnov, E. S., and Vladimirova, M.
1934. Studien über die Vermehrungsfähigkeit der Pteromalide *Mormoniella vitripennis* Wlk. Zeitsch. Wiss. Zool., *145*:507-522.

Smit, B.
1929. The biological control of sheep blowflies in South Africa. So. African Jour. Science, *26*:441-448.

Smith, F.
1898. Flies as carriers of diphtheria bacilli. Publ. Health, London, *11*:222-223.

Smith, H. F.
1932. See Holdaway, F. G., and Smith, H. F.

Smith, H. S.
1947. See De Bach, P., and Smith, H. S.

Smith, R. C.
1921. Observations on the fall army worm (*Laphygma frugiperda* Smith and Abbott) and some control experiments. Jour. Econ. Entom., *14*:300-305.

Smith, K. M., Hills, G. J., and Rivers, C. F.
1961. Studies on the cross-inoculation of the *Tipula* iridescent virus. Virology, *13*:233-241.

Smuidsinovicia, V. I.
1889. *Musca domestica* as a disseminator of silkworm disease. Ann. Caucasian Silkworm Stat., *1*.

Soberón y Parra, G.
1941. See León y Blanco, F., and Soberón y Parra, G.

Soberón y Parra, G., and León y Blanco, F.
1944. Las moscas del genero *Hippelates* como posibles vectores del mal del pinto. Ciencia, *4*:299-300.

Sokal, R. R.
1964. See Bhalla, S. C., and Sokal, R. R.

Sokal, R. R., and Sullivan, R. L.
1963. Competition between mutant and wild-type house fly strains at varying densities. Ecology, *44*:314-322.

Solms-Laubach, Graf zu
1870. Über die herbstliche Pilzkrankheit der Stubenfliege. Abhdlg. der Naturforsch. Gesellchaft zu Halle, p. 37.

Somov, P., Romanova, V., and Romanova, V.
1937. Bull. Azov-Black Sea District Inst. Microb. Epid. at Rostov-on-Don., Issue 16, pp. 91-100 (original not seen).

Southgate, B. A.
1965. See Lainson, R., and Southgate, B. A.

Srivastava, H. D., and Dutt, S. C.
1963. Studies on the life history of *Stephanofilaria assamensis*, the causative parasite of "humpsore" of Indian cattle. Indian Jour. Veter. Science, *33*:173-177.

Stainton, H. T.
1864. Proc. Entom. Soc. (3), II, 112.

Stalker, H. D.
1957. See Wolfson, M., et al.

Stalker, H. D., and Carson, H. L.
1963. A very serious parasite of laboratory *Drosophila.* Second report, Drosophila Inform. Serv., *38*:96.

Stefani, T. de
1889. Una nota sulla *Chalcis dalmanni* Thms. Nat. Siciliano, *9*:11-12.

Stein, C. D., Lotze, J. C., and Mott, L. O.
1942. Transmission of equine infectious anemia or swamp fever by the stablefly, *Stomoxys calcitrans,* the horsefly, *Tabanus sulcifrons* (Macquart) and by injection of minute amounts of virus. Amer. Jour. Veter. Res., *3*:183-193.

Steinhaus, E. A., and Bell, C. R.
1953. The effect of certain microorganisms and antibiotics on stored-grain insects. Jour. Econ. Entom., *46*:582-598.

Stephens, J. M.
1963. Bactericidal activity of hemolymph of some normal insects. Jour. Insect Path., *5*:61-65.

Steve, P. C.
1959. Parasites and predators of *Fannia canicularis* (L.) and *Fannia scalaris* (F.). Jour. Econ. Entom., *52*:530-531.

Steve, P. C., and Lilly, J. H.
1965. Investigations on transmissability of *Moraxella bovis* by the face fly. Jour. Econ. Entom., *58*:444-446.

Stevens, S.
1866. Proc. Entom. Soc. London (3), v, p. 27.

Stewart, W. H.
1953. See Lindsay, D. R., et al.

Steyn, J. J.
1963. Use of social spiders against gastrointestinal infections spread by house flies. So. African Med. Jour., *33*:730-731.

Stiles
Personal communication to G.H.F. Nuttall, 1889; in Nuttall, G.H.F., and Jepson, F. P., Great Britain Rept. Publ. Health, *16*:13-41, see p. 28.

Stirrat, J. H., McLintock, J., Schwindt, G. W., and Depner, K. R.
1955. Bacteria associated with wild and laboratory-reared horn flies, *Siphona irritans* (L.) (Diptera: Muscidae). Jour. Parasit., *41*:398-406.

Stoffolano, Jr., J. G.
1967. The synchronization of the life cycle of diapausing face flies, *Musca autumnalis,* and of the nematode, *Heterotylenchus autumnalis.* Jour. Invert. Pathol., *9*:395-397.

Stoffolano, Jr., J. G.
1969. Nematode parasites of the face fly and the onion maggot in France and Denmark. Jour. Econ. Entom., *62*:792-795.

Stoffolano, Jr., J. G.
1970. Experimental parasitization of *Musca autumnalis* De Geer, *Musca domestica* L., *Orthellia caesarion* Meig., and *Ravinia lherminieri* R.-D. by the nematode *Heterotylenchus autumnalis* Nickle with special reference to host reactions, hemocytic involvement, anal organ, and the hemocytopoietic organ. Dissertation, Univ. Connecticut.

Stoffolano, Jr., J. G., and Nickle, W. R.
1966. Nematode parasite (*Heterotylenchus* sp.) of face fly in New York State. Jour. Econ. Entom., *59*:221-222.

Stolpe, S. G.
1938. The life cycle of tyroglyphid mites infesting cultures of *Drosophila melanogaster.* Anat. Rec., *72* (Suppl.):133-134.

Strickland, C.
1911. Description of a *Herpetomonas* parasitic in the alimentary tract of the common green bottle fly, *Lucilia* sp. Parasitology, *4*:222-236.

Strong, L. A.
1938. Report of the Chief of the Bureau of Entomology and Plant Quarantine. U. S. Dept. of Agric.

Struble, G. R.
1942. Laboratory propagation of two predators of the mountain pine beetle. Jour. Econ. Entom., *35*:841-844.

Suenaga, O., and Fukuda, M.
1963. Ecological studies of flies. 7. On the species and seasonal prevalence of flies breeding out from a privy and a urinary pit in a farm village. Endemic Dis. Bull. Nagasaki Univ., *5*:72-80.

Sukhacheva, E. I.
1963. Rol'nekotorykh vidov lichinok nasekomykh v èpidemiologii askaridoza. [The role of certain insect larvae in the epidemiology of ascariasis.] Med. Parazit. Parazitarn. Bolezni, *32*:600-604.

Sukhova, M. N.
1950. Novye dannye po ekologii i èpidemiologicheskomu znacheniiu sinikh miasnykh mukh *Calliphora uralensis* Vill. and *Calliphora erythrocephala* Meig. (Diptera, Calliphoridae). [New data on the ecology and epidemiological significance of the blue meat flies, *Calliphora uralensis* Vill. and *Calliphora erythrocephala* Meig.] Entom. Obozrenie, *31*:90-94.

Sukhova, M. N.
1953. O znachenie bazarnoĭ mukhi (*Musca sorbens* Wied.) v èpidemiologii ostrogo èpidemicheskogo kon'iunktivita v zapadnoi Turkmenii. [On the significance of the bazaar fly (*M. sorbens* Wied.) in the epidemiology of acute epidemic conjunctivitis in western Turkmenia.] Gigiena Sanitaria, *7*:40-42.

Sukhova, M. N.
1954. Voprosy kommunal' noĭ gigeny v usloviiakh zharkogo klomata sredneĭ azii. Medgiz: Moscow, pp. 126-141 (book). [Flies in population centers in west Turkmenia, their sanitary and epidemiological significance and their control. Anthology . . . in: Problems of communal hygiene in the hot-climate conditions of Central Asia.]

Sukhova, M. N.
1957. See Levkovich, E. N., and Sukhova, M. N.

Sullivan, R. L.
1963. See Sokal, R. R., and Sullivan, R. L.

Sullivan, R. L., and Sokal, R. R.
1963. The effects of larval density on several strains of the house fly. Ecology, *44*:120-130.

Sullivan, R. L., and Sokal, R. R.
1965. Further experiments on competition between strains of house flies. Ecology, *46*:172-182.

Surcouf, J.M.R.
1923. Deuxième note sur les conditions biologiques du *Stomoxys calcitrans* L. Bull. Mus. Hist. Nat., Paris, *29*:168-172.

Sureau, P.
1962. See Brygoo, E. R., et al.

Sweadner, W. R.
1933. See Weil, G. C., et al.

Swellengrebel, W. H.
1913. See Kuenen, W. A., and Swellengrebel, W. H.

Swingle, L. D.
1911. The transmission of *Trypanosoma lewisi* by rat fleas (*Ceratophyllus* sp. and *Pulex* sp.) with short descriptions of three new herpetomonads. Jour. Infect. Dis., *8*:123-146.

Sychevskaia͡, V. I.
1954. Materialy k biologii i èkologii sinantropnych mukh roda *Fannia* R. D. v Samarkande. [Material on the biology and ecology of synanthropic flies of the genus *Fannia* R. D. in Samarkand.] Med. Parazit. Parazitarn. Bolezni, *23*:45-54.

Sychevskaia͡, V. I.
1964a. Mites found on synanthropic flies in Uzbekistan. Med. Parazit. Parazitarn. Bolezni, *5*:557-560.

Sychevskaia͡, V. I.
1964b. Rev. Entom. URSS, *43*:391-404.

Sychevskaia͡, V. I.
1966. The biology of *Brachymeria minuta* (L.). (Hymenoptera, Chalcidoidea), a parasitoid of synanthropic flies of the family Sarcophagidae. Wld. Health Org. WHO/EBL/66.60, 3 pp.

Sychevskaia͡, V. I., Grudtzina, M. V., and Vyrvikhvost, L. A.
1959a. The epidemiological significance of synanthropic flies (Diptera) in Bukhara. Entom. Obozreni, *38*:568-578.

Sychevskaia͡, V. I., and Petrova, T. A.
1958. O roli mukh rasprostranenii yaits gel'mintov v Uzbekistane. [The role of flies in spreading the eggs of helminths in Uzbekistan.] Zool. Zhurnal, *37*:563-569.

Sychevskaia͡, V. I., Skopina, N. P., and Petrova, Z. F.
1959b. Contamination of synanthropic flies with dysentery bacilli and helminth eggs in Fergana. Trudy Uzbekistanskogo Instituta Malia͡rii i Meditsinskoĭ Parazitologii, *4*:225-235. In: Works of Uzbek Inst. Malaria and Parasit. Uzbek Dept. Publ. Health, Samarkand.

Syddiq, M. M.
1938. *Siphunculina funicola* (eye-fly). Indian Med. Gaz., *73*:17-19.

Tacal, Jr., J. V., and Menez, C. F.
1967. *Salmonella* studies in the Philippines: VII. The isolation of *Salmonella derby* from abattoir flies and chicken ascarids. Philippine Jour. Vet. Med., *6*:106-111.

Takacs, W. S.
1941. See Toomey, J. A., et al.

Takacs, W. S.
1947. See Toomey, J. A., et al.

Takada, R.
1965. See Shimizu, F., et al.

Tange, H.
1965. See Shimizu, F., et al.

Taniguchi, H.
1965. See Shimizu, F., et al.

Tao, C. S.
1936. Transmission of helminths ova by flies. Jour. Shanghai Science Inst., Sect. 4, *2*:109-116.

Taplin, D., Zaias, N., and Rebell, G.
1965. Environmental influences on the microbiology of the skin. Arch. Environ. Health, *11*:546-550.

Tarasov, V. A., and Chaĭkin, V. N.
1941. On the frequency of occurrence of enteric bacilli in the room fly. Zhurn. Mikrob. Epidem. Immun., *9*:23-27.

Tate, P.
1935. The larvae of *Phaonia mirabilis* Ringdahl, predatory on mosquito larvae (Diptera, Anthomyidae). Parasitology, *27*:556-560.

Tatum, E. L.
1939. Development of eye-colors in *Drosophila*: Bacterial synthesis of v^+ hormone. Proc. U. S. Nat. Acad. Sciences, *25*:486-490.

Taunay, A. de E.
1957. See Coutinho, J. O., et al.

Taylor, A. W.
1930. Experiments on the mechanical transmission of West African strains of *Trypanosoma brucei* and *T. gambiense* by *Glossina* and other biting flies. Trans. Roy. Soc. Trop. Med. Hyg., *24*:289-303.

Taylor, E. L.
1935. An attempt to transmit anaplasmosis by British biting flies. Veter. Jour., *91*:4-11.

Tebbutt, H.
1913. On the influence of the metamorphosis of *Musca domestica* upon bacteria administered in the larval stage. Jour. Hyg., *12*:516-526.

Teodoro, G.
1916. Persistenza e resistenza del bacilli dell'ileo tifo nell'apparato digerente delle mosche. Atti Reale Ist. Veneto, *75*:1559-1567.

Teodoro, G.
1926. Mosche e pebrina. Ann. Ig., *8*:585-587.

Terni, C.
1908a. Bericht XIV Intl. Kongr. Hyg. Demogr., *4*:132 (discussion).

Terni, C.
1908b. Mouches domestiques et *Stomoxys* dans l'étiologie de la variole et du vaccin animal. Bericht XIV Intl. Kongr. Hyg. Demogr., *4*:133-135.

Terni, C.
1909. Contribution à l'étude de la variole et du vaccin et des autres maladies similaires. Centralbl. Bakt., Parasit. Infekt., I. Abt., Orig., *50*:23 and on (see pp. 23, 32).

Terry, C. E.
1912. Extermination of the house fly in cities, its necessity and possibility. Amer. Jour. Publ. Health, *2*:14-22.

Testi, F.
1909. Ricerche batteriologiche sull'intestino degli insetti. Riv. Ig. Sanità Publ. *20*:491-498.

Thaxter, R.
1888. The Entomophthoreae of the United States. Mem. Boston Soc. Nat. Hist., *4*:133-201.

Theiler, A.
1927. Lamsiekte (Parabotulism) in cattle in South Africa. 11th & 12th Repts. Dir. Veter. Educ. and Res., Union So. Africa, Part 2, pp. 821-1361.

van Thiel, P. H.
1949. The transmission of toxoplasmosis and rôle of *Calliphora erythrocephala* Meig. Doc. Neerl. Indo. Morb. Trop., *1*:264-269.

van Thiel, P. H., and Laarman, J. J.
1953. Therapy and transmission of experimental toxoplasmosis. Atti. VI Congr. Intl. Microb. Roma, *5*:445-448.

Thomas, G. D., and Wingo, C. W.
1968. Parasites of the face fly and two other species of dung-inhabiting flies in Missouri. Jour. Econ. Entom., *61*:147-152.

Thomas, H. E.
1936. See Ark, P. A., and Thomas, H. E.

Thompson, W. R.
1928. See Parker, H. L., and Thompson, W. R.

Thompson, W. R., and Simmonds, F. G.
1950-65. A catalogue of the parasites and predators of insect pests. Commonwealth Inst. Biol. Control, Ottawa, Canada (various volumes were used).

Thomson, F. W.
1912. The house-fly as a carrier of typhoid infection. Jour. Trop. Med. Hyg., *15*:273-277.

Thomson, J. G.
1933. The natural occurrence of flagellates of the subgenus *Strigomonas* M. and A. Lwoff in the gut of *Tabanus africanus* from Nyassaland and *Lucilia sericata* in England. Jour. Trop. Med. Hyg., *36*:361-365.

Thomson, J. G., and Lamborn, W. A.
1934. Mechanical transmission of trypanosomiasis, leishmaniasis, and yaws through the agency of non-biting haematophagous flies. British Med. Jour., *2*:506-509.

Thomson, R.C.M.
1937. Observations on the biology and larvae of the Anthomyidae. Parasitology, *29*:273-358.

Tiegs, O. W.
1921. See Johnston, T. H., and Tiegs, O. W.

Tillyard, R. J.
1930. The work of the division of Economic Entomology for the year 1929/30. Canberra, Council Science Ind. Res., 34 pp.

Tischer, L. A.
1941. See Toomey, J. A., et al.

Tischler, W.
1950. Biozönotische Untersuchungen bei Hausfliegen. Zeitsch. Angew. Entom., *32*:195-207.

Tison, F.
1950. Transmission de la tuberculose par les mouches. Ann. Inst. Pasteur, *79*: 454.

Todd, J. L.
1908. See Dutton, J. E., et al.

Todd, S. R.
1964. See Wallace, F. G., and Todd, S. R.

Tolstiak, I. E.
1956. Peredacha rozhi svineĭ cherez uqus mukhi-zhigalki. Veterinariia (Moskva), *33*:73-75.

Tomescu, E.
1958. See Duca, M., et al.

Tonkonozhenko, A. P.
1967. Toksichnost' nekotorykh entomopatogennykh bakterri dlya komnatnoi mukhi i deistvie termostabil'nogo ekzotoksina *Bac. thuringiensis* na lichinki mukh. Tr. Vses. Nauch-Issled Inst. Vet. Sanit., *28*:332-337.

Toomey, J. A., Takacs, W. S., and Tischer, L. A.
1941. Poliomyelitis virus from flies. Proc. Soc. Exp. Biol. Med., *48*:637-639.

Toomey, J. A., Pirone, P. P., Takacs, W. S., and Schaeffer, M.
1947. Can *Drosophila* flies carry poliomyelitis virus? Jour. Infect. Dis., *81*: 135-138.

Tower, B. A.
1961. See Burns, E. C., et al.

Trask, J. D.
1941. See Paul, J. R., and Trask, J. D.

Trask, J. D., Paul, J. R., and Melnick, J. L.
1943. The detection of poliomyelitis virus in flies collected during epidemics of poliomyelitis. I. Methods, results, and types of flies involved. Jour. Exp. Med., *77*:531-544.

Treece, R. E., and Miller, T. A.
1968. Observations on *Heterotylenchus autumnalis* in relation to the face fly. Jour. Econ. Entom., *61*:454-456.

Treillard, M.
1935. See Roubaud, E., and Treillard, M.

Trojan, E.
1908. *Holostaspis sita*, eine neue Acarine. Arch. Naturgesch., *74*:1-12.

Tryon, H.
1914. Stock and animal ecto-parasites. Ann. Rept. Dept. Agric. Queensland, see p. 116.

Tukhmanyants, A. A., Shakhurina, E. A., and Eskina, G. V.
1963. K ekologii *Musca larvipara* (Portsch., 1910) promezhutochnogo khozyaina *Thelazia rhodesi* (Desmarest, 1827) krupnogo rogatogo skota. [Contribution to the ecology of *Musca larvipara* (Portsch., 1910), an intermediate host of *Thelazia rhodesi* (Desmarest, 1827) from horned cattle.] Uzbeksk. Biol. Zhur., *7*:57-62.

Turner, Jr., E. C.
1965. See Hair, J. A., and Turner, E. C., Jr.

Turner, Jr., E. C., Burton, R. P., and Gerhardt, R. R.
1968. Natural parasitism of dung-breeding Diptera: a comparison between native hosts and an introduced host, the face fly. Jour. Econ. Entom., *61*:1012-1015.

Turner, T. B.
1935. See Kumm, H. W., et al.

Turner, T. B.
1936. See Kumm, H. W., and Turner, T. B.

Uchida, S.
1963. See Asahina, S., et al.

Uhlenhuth, P., and Kuhn, P.
1917. Experimentelle Übertragung der Weilschen Krankheit durch die Stallfliege (*Stomoxys calcitrans*). Zeitsch. Hyg. Infekt. Krankh., *84*:517-540.

Ullyett, G. C.
1947. Competition for food and allied phenomena in sheep blowfly populations. Phil. Trans. Roy. Soc. London B, *234*:77-174.

Ullyett, G. C.
1950. Pupation habits of sheep blowflies in relation to parasitism by *Mormoniella vitripennis* Wlk. Bull. Entom. Res., *40*:533-537.

Ullyett, G. C., and de Vries, A. H.
1941. *Bembix* wasps as enemies of sheep blow-flies. Farming in So. Africa, *16*:19-20.

Urbina, D.
1955. See Neel, W. W., et al.

Uribe, C.
1926. A new invertebrate host of *Trypanosoma cruzi* Chagas. Jour. Parasit., *12*:213-215.

Urzúa, L. F.
1947. See Barros, R. D., and Urzúa, L. F.

Usui, Y.
1960. *Aspergillus* and *Mucor* found on house fly and *Ophyra calcogaster* Wi[e]demann. Jap. Jour. Sanit. Zool., *11*:54-55 [Eisei dōbutsu].

Vandenberg, S. R.
1930. Report of the entomologist. House-fly parasites. Guam Agric. Exp. Stat. (1928), pp. 23-31.

Vanni, V.
1946. Sul meccanismo infettante della mosca domestica. Ann. Ig., *56*:151-155.

Varela, G.
1963. See Greenberg, B., et al.

Varela, G., and Zavala, J.
1961. Ensayos de transmisión de la toxoplasmosis por insectos. Rev. Inst. Salubr. Enferm. Trop., México, *21*:141-148.

Vasquez-Gonzalez, J., Young, W. R., and Ramirez-Genel, M.
1962-3. Reducción de la poblacion de mosca domestica en gallinaza por la mosca soldado en el tropico. Agric. Tec. Mexico, *2*:53-57.

Vella, L.
1949. See Morellini, M., et al.

Velthuis, H.H.W., Velthuis-Kluppell, F. M., and Bossink, G.A.H.
1965. Some aspects of the biology and population dynamics of *Nasonia vitripennis* (Walker) (Hymenoptera: Pteromalidae). Entom. Exp. Appl., *8*:205-227.

Velthuis-Kluppell, F. M.
1965. See Velthuis, H.H.W., et al.

Venables, E. P.
1914. A note upon the food habits of adult Tenthredinidae. Canadian Entom., *46*:121.

Verge, J.
1961. See Dhennin, L., et al.

Vergne, M.
1931. Note sur *Thyreopus alpinus* Imhoff, sphégide prédateur de diptères. Bull. Soc. Entom. France, pp. 83-87.

Verhoeff, C.
1891. Biologische Aphorismen über einige Hymenopteren, Dipteren und Coleopteren. Verhdlg. Naturhist. Vereins Pr. Rheinl., Bonn, *48*:1-80.

Viale, E.
1955. See Neel, W. W., et al.

Viereck, H. L.
1909. Ann. Rept. New Jersey State Mus., p. 641.

Viereck, H. L.
1916. Bull. Connecticut State Geol. Nat. Hist. Surv., No. 22, p. 485.

Vigier, P.
1961. Le problème de l'accumulation du virus infectieux dans une *Drosophile* sensible au gaz carbonique, stabilisée. Compt. Rend. Acad. Sci., *252*: 2953-2955.

Világiová, I.
1962. Význam múch pre vývin očných parazitov—pôvodcov teláziózy hovädzieho dobytka. [The importance of flies for the development of eye parasites—the causal agents of thelaziosis of cattle.] Biológia, *17*:297-299.

Világiová, I.
1968. *Heterotylenchus autumnalis* Nickle (1967): A parasite of pasture flies. Biológia (Bratislava), *23*:397-400.

Vladimirova, M. S.
1934. See Smirnov, E. S., and Vladimirova, M. S.

Vladimirova, M. S., and Smirnov, E.
1934. Über das Verhalten der Schlupfwespe *Mormoniella vitripennis* Wlk. zu verschiedenen Fliegenarten. Zool. Anz., *107*:85-89.

Vladimirova, M. S., and Smirnov, E. S.
1938. Vnutrividovaiа̑ i mezhvidovaiа̑ konkurentsiiа̑ *Musca domestica* L. i *Phormia groenlandica* Zett. [Intra-species and inter-species competition of *Musca domestica* L. and *Phormia groenlandica* Zett.] Med. Parazit. Parazitarn. Bolezni, *7*:755-777.

Vogt, G. B.
1948. *Dermestes* and *Saprinus* as predators and pests in fleshfly rearing. Jour. Econ. Entom., *41*:826-827.

Voukassovitch, P.
1924a. Sur la biologie du *Dibrachys affinis* Masi, parasite de l'Eudémis. Rev. Zool. Agric. Appl., *23*:92-98, 119-131.

Voukassovitch, P.
1924b. Sur la biologie d'un chalcide, hyperparasite des tachinaires d'*Oenophthira pilleriana* Schiff. Bull. Soc. Hist. Nat. Toulouse, *53*:1-4.

Vyrvikhvost, L. A.
1959a. See Sychevskaiа̑, V. I., et al.

Wade, C. F.
1960. See Rodriguez, J. G., and Wade, C. F.

Wade, C. F.
1961. See Rodriguez, J. G., and Wade, C. F.

Wade, C. F.
1962. See Rodriguez, J. G., et al.

Wade, C. F., and Rodriguez, J. G.
1961. Life history of *Macrocheles muscaedomesticae* (Acarina: Macrochelidae), a predator of the house fly. Ann. Entom. Soc. Amer., *54*:776-781.

Walker, I.
1959. Die Abwehrreaktion des Wirtes *Drosophila melanogaster* gegen die zoophage Cynipide *Pseudeucoila bochei* Weld. Rev. Suisse Zool., *66*:569-632.

Walker, I.
1961. *Drosophila* und *Pseudeucoila* II. Schwierigkeiten beim Nachweis eines Selektionserfolges. Rev. Suisse Zool., *68*:252-263.

Walker, I.
1963. The relationship between *Drosophila melanogaster* and its parasite *Pseudeucoila bochei* Weld. Proc. Intl. Congr. Zool., *16*:213.

Wallace, F. G., and Clark, T. B.
1959. Flagellate parasites of the fly, *Phaenicia sericata* (Meigen). Jour. Protoz., *6*:58-61.

Wallace, F. G., and Todd, S. R.
1964. *Leptomonas mirabilis* Roubaud 1908 in a Central American blowfly. Jour. Protoz., *11*:502-505.

Wallis, R. C.
1965. See Davé, K. H., and Wallis, R. C.

Wallwork, J. H., and Rodriguez, J. G.
1963. Effects of ammonia on the predation rate of *Macrocheles muscaedomesticae* on house fly eggs. Adv. Acarol., *1*:60-69.

Walters, M. C.
1966. The yellow dung-fly *Scatophaga stercoraria* (L.) (Diptera: Cordyluridae), as a predator of the lucerne springtail, *Sminthurus viridis* (L.). South African Jour. Agr. Sci., *9*:739-740.

Walton, G. S.
1965. See Beard, R. L., and Walton, G. S.

Ward, R.
1942. See Sabin, A. B., and Ward, R.

Ward, R.
1945. See Melnick, J. L., and Ward, R.

Waterston, J.
1916. On the occurrence of *Stenomalus muscarum* (Linn.) in company with hibernating flies. Scottish Naturalist, Edinburgh, *54*:140-142.

Watt, J.
1953. See Lindsay, D. R., et al.

Wayson, N. E.
1914. Plague and plague-like diseases, a report on their transmission by *Stomoxys calcitrans* and *Musca domestica.* U. S. Publ. Health Rept., *29*: 3390-3393.

Webb, J. E., Jr., and Graham, H. M.
1956. Observations on some filth flies in the vicinity of Fort Churchill, Manitoba, Canada, 1953-54. Jour. Econ. Entom., *49*:595-600.

Webb, J. L., and Hutchinson, H. H.
1916. A preliminary note on the bionomics of *Pollenia rudis* Fabr. in America. Proc. Entom. Soc. Washington, D. C., *18*:197-199.

Wedberg, S. E.
1951. See Hawley, J. E., et al.

Weil, G. C., Simon, R. J., and Sweadner, W. R.
1933. A biological, bacteriological and clinical study of larval or maggot therapy in the treatment of acute and chronic pyogenic infections. Amer. Jour. Surg., *19*:36-48.

Weiss, H. B.
1915. Preliminary list of New Jersey Acarina. Entom. News, *26*:149-152.

Welch, H.
1942a. See Ostrolenk, M., and Welch, H.

Welch, H.
1942b. See Ostrolenk, M., and Welch, H.

Welch, H. E.
1959. Taxonomy, life cycle, development and habits of two new species of Allantonematidae (Nematoda) parasitic in drosophilid flies. Parasitology, *49*:83-103.

Wellmann, G.
1948-9a. Die gemeine Stechfliege (*Stomoxys calcitrans*) als Überträger des Rotlaufs. Zentralbl. Bakt., Parasit., I. Abt., Orig., *153*:185-199.

Wellmann, G.
1948-9b. Übertragung des Rotlaufs auf Schweine durch den gemeinen Wadenstecher (*Stomoxys calcitrans*). Zentralbl. Bakt., Parasit, I. Abt., Orig., *153*:200-203.

Wellmann, G.
1949. Die Übertragung des Schweinerotlaufs durch den Saugakt der gemeinen Stechfliege (*Stomoxys calcitrans*) und ihre epidemiologische Bedeutung. Berliner Münchener Tierärztl. Wschr., No. 4, 39-46.

Wellmann, G.
1950. Rotlaufübertragung durch verschiedene blutsaugende Insektenarten auf Tauben. Zentralbl. Bakt., Parasit., I. Abt., Orig., *155*:109-115.

Wellmann, G.
1950-1. Blutsaugende Insekten als mechanische Brucellenüberträger. Zentralbl. Bakt., Parasit. Infekt. Hyg., I. Abt., Orig., *156*:414-426.

Wellmann, G.
1955. Die Übertragung der Schweinerotlaufinfektion durch die Stubenfliege (*Musca domestica*). Zentralbl. Bakt., Parasit. Infekt. Hyg., I. Abt., Orig., *162*:261-264.

Wellmann, G.
1959. Stechfliegen als Überträger von Zoonosen. Intl. Symposium über schädliche Fliegen. Zeitsch. Angew. Zool., *46*:328-331.

Wells, C. N.
1962a. See Rodriguez, J. G., et al.

Wenyon, C. M.
1908. Report of travelling pathologist and protozoologist. Third report, Wellcome Res. Lab., Gordon Memorial Coll., Khartoum, p. 146.

Wenyon, C. M.
1911a. Oriental sore in Bagdad, together with observations on a gregarine in *Stegomyia fasciata*, the haemogregarine of dogs, and the flagellates of house flies. Parasitology, *4*:273-340.

Wenyon, C. M.
1911b. Report of six months' work of the expedition to Bagdad on the subject of oriental sore. Jour. Trop. Med. Hyg., *14*:103-109.

Wenyon, C. M.
1913. Observation on *Herpetomonas muscae-domesticae* and some allied flagellates. Arch. Protist., *31*:1-36.

Wenyon, C. M., and O'Connor, F. W.
1917a. The carriage of cysts of *Entamoeba histolytica* and other intestinal protozoa and eggs of parasitic worms by house-flies with some notes on the resistance of cysts to disinfectants and other agents. Jour. Army Med. Corps, *28*:522-527.

Wenyon, C. M., and O'Connor, F. W.
1917b. An inquiry into some problems affecting the spread and incidence o intestinal protozoal infections of British troops and natives in Egyp with special reference to the carrier question, diagnosis and treatment c amoebic dysentery, and an account of three new human intestinal Protozoa. Part IV. Jour. Roy. Army Med. Corps, *28*:686-698.

Werner, H.
1909a. Über eine eingeisselige Flagellatenform im Darm der Stubenfliege. Arch. Protist., *13*:19-22.

Werner, H.
1909b. Studies regarding pathogenic amoebae. Indian Med. Gaz., pp. 241-245.

West, L. S.
1953. Fly control in the Eastern Mediterranean and elsewhere. Report of a survey and study. WHO/Insecticides, *19*:139 pp, May 12.

Wetzel, R.
1936. Neuere Ergebnisse über die Entwicklung von Hühnerbandwürmern. Verhandl. Deutschen Zool. Ges., pp. 195-200.

Wharton, G. W.
1946. Observations on *Ascoschöngastia indica* (Hirst, 1915). Ecol. Monogr., *16*:151-184, see pp. 165, 167, 171, 179, 180.

Wharton, G. W., and Carver, R. K.
1946. Food of nymphs and adults of *Neoschöngastia indica* (Hirst, 1915). Science, *104*:76-77.

Wherry, W. B.
1908a. Notes on rat-leprosy and on the fate of human and rat lepra bacilli in flies. Publ. Health Rept. U. S. Marine Hosp. Serv., *23*:1481-1487.

Wherry, W. B.
1908b. Further notes on rat leprosy and on the fate of human and rat lepra bacilli in flies. Jour. Infect. Dis., *5*:507-514.

White, E. B., and Legner, E. F.
1966. Notes on the life history of *Aleochara taeniata*, a staphylinid parasite of the house fly, *Musca domestica*. Ann. Entom. Soc. Amer., *59*:573-577.

White, T. C.
1880. Contagion from flies. British Med. Jour., *2*:766.

White, T. C., and Reeves, W. W.
1877. *Empusa muscae*. Jour. Monthly Microsc. London, *17*:253-254.

Wilhelmi, J.
1917a. Zur Biologie der kleinen Stechfliege, *Lyperosia irritans* L. Sitz.-Ber. Ges. Naturf. Freunde Berlin, *3*:510-516.

Wilhelmi, J.
1917b. Über *Stomoxys calcitrans* L. Sitz.-Ber. Ges. Naturf. Freunde Berlin, *3*: 179-195.

Wilhelmi, J.
1919. Zur Biologie der kleinen Stubenfliege *Fannia canicularis*. Zeitsch. Angew. Entom., *5*:261-266.

Wilhelmi, J.
1927. Untersuchungen über die Übertragung der Maul- und Klauenseuche durch die Stechfliege *Stomoxys calcitrans* L. Zeitsch. Desinfekt. Gsndhtsw., *19*:104-111.

Willis, R. R., and Axtell, R. C.
1968. Mite predators of the house fly: A comparison of *Fuscuropoda vegetans* and *Macrocheles muscae-domesticae*. Jour. Econ. Entom., *61*:1669-1674.

Wilson, B. H.
1961. See Burns, E. C., et al.

Wingo, C. W.
1963. See Benson, O. L., and Wingo, C. W.

Winterschmidt, A. W.
1765. Abbildung und Beschreibung einer Stubenfliege mit vielen sehr kleinen Insekten geplagt. (As appendix to: Ledermüller, M. F., Nachlese seiner Mikrosk. Gemüths-u. Augen- Ergötz . . . etc. 1762-5.)

Wolbach, S. B.
1924. See Hertig, M., and Wolbach, S. B.

Wolcott, G. N.
1922. Insect parasite introduction in Porto Rico. Jour. Dept. Agric. Porto Rico, *6*:5-20.

Wolfson, M., Stalker, H. D., and Carson, H. L.
1957. A serious parasite of laboratory *Drosophila*. Drosophila Information Service, *31*:170.

Wollman, E.
1911. Sur l'élevage des mouches stériles. Contribution a la connaissance du rôle des microbes dans les voies digestives. Ann. Inst. Pasteur, *25*:79-88.

Wollman, E.
1921. Le rôle des mouches dans le transport des germes pathogènes étudié par la méthode des élevages aseptiques. Ann. Inst. Pasteur, *35*:431-449.

Wollman, E.
1927. Le rôle des mouches dans le transport de quelques germes importants pour la pathologie tunisienne. Arch. Inst. Pasteur Tunis, *16*:347-364.

Wollman, E., Anderson, C., and Colas-Belcour, J.
1928. Recherches sur la conservation des virus hémophiles chez les insectes. Arch. Inst. Pasteur Tunis, *17*:229-232.

Wood, A. H.
1933. Notes on some dipterous parasites of *Schistocerca* and *Locusta* in the Sudan. Bull. Entom. Res., *24*:521-530.

Wood, P. W.
1965. See Pimentel, D., et al.

Woodhill, A. R.
1926a. See Gurney, W. B., and Woodhill, A. R.

Woodhill, A. R.
1926b. See Gurney, W. B., and Woodhill, A. R.

Wright, E. A.
1957. See Pavillard, E. R., and Wright, E. A.

Wright, E. V.
1959. See Malogolowkin, C., et al.
Wright, M.
1945. Dragonflies predaceous on the stable fly *Stomoxys calcitrans* (L.). Florida Entom., *28*:31-32.
Wulp, van der
1869. Note sur le parasitisme de *Malonophora roralis* L. dans *Asopia farinalis* L. et de *Cyrotoneura stabulans* Fall. dans *Lophyrus pallidus* Klug. Tijdschr. Entom., 2e Serie, Part 4, pp. 184-185.
Wylie, H. G.
1962. An effect of host age on female longevity and fecundity in *Nasonia vitripennis* (Walk.) (Hymenoptera: Pteromalidae). Canadian Entom., *94*:990-993.
Wylie, H. G.
1965a. Effects of superparasitism on *Nasonia vitripennis* (Walk.) (Hymenoptera: Pteromalidae). Canadian Entom., *97*:326-331.
Wylie, H. G.
1965b. Some factors that reduce the reproductive rate of *Nasonia vitripennis* (Walk.) at high adult population densities. Canadian Entom., *97*:970-977.
Wylie, H. G.
1965c. Discrimination between parasitized and unparasitized house fly pupae by females of *Nasonia vitripennis* (Walk.) (Hymenoptera: Pteromalidae). Canadian Entom., *97*:279-286.
Yamasaki, S.
1927. Über eine zweigeisselige Flagellatenart im Darm von *Lucilia* sp. (*Jedensis*?). Abhandl. Gebiet Auslandskunde Reihe D., Med. Veterinar. Med. *26*:605-607.
Yao, H. Y., Yuan, I. C., and Huie, D.
1929. The relation of flies, beverages and well water to gastro-intestinal diseases in Peiping. Natl. Med. Jour. China, *15*:410-418.
Yasinskiĭ, A. V.
1961. See Arskiĭ, V. G., et al.
Yasuyama, K.
1928. Viability of *Treponema pertenue* outside of the body and its significance in the transmission of yaws. Philippine Jour. Science, *30*:333-349.
Yeager, C. C.
1939. *Empusa* infections of the housefly in relation to moisture conditions of northern Idaho. Mycologia, *31*:154-156.
Yendol, W. G., and Miller, E. M.
1967. Susceptibility of the face fly to commercial preparations of *Bacillus thuringiensis*. Jour. Econ. Entom., *60*:860-864.
Yoshioka, I.
1961. See Riordan, J. T., et al.
Young, R. D.
1959. See Furman, D. P., et al.
Young, W. R.
1962-3. See Vasquez-Gonzalez, J., et al.
Yuan, I. C.
1929. See Yao, H. Y., et al.
Zaidenov, A. M.
1961. Opyt izucheniia͡ epidemológicheskogo znacheniia͡ cinantropnykh mukh usloviia͡kh goroda. [The experience of the study of the epidemiological role of synanthropic flies under the conditions of a town.] Entom. Obozrenie, *40*:554-567.

Zammit, T.
1907. See Eyre, J.W.H., et al.

Zanini, E.
1930. È l'*Holostaspis badius* (Koch) parassita della mosca domestica? Contributo alla lotta contro la mosca domestica. Boll. Lab. Zool. Agric. Bachic., Milano (1928-9), *1*:59-73.

Zardi, O.
1964. Importanza di *Musca domestica* nella trasmissione dell'agente del tracoma. Nuovi Ann. Ig. Microb., *15*:587-590.

Zatsepin, N. I.
1961. See Arskiĭ, V. G., et al.

Zavala, J.
1961. See Varela, G., and Zavala, J.

Zeledón, R.
1957. Algunas observaciones sobre la biologia de la *Dermatobia hominis* (L. Jr.) y el problema del tórsalo en Costa Rica. Rev. Biol. Trop., *5*:63-74.

Zepeda, P.
1913. Nouvelle note concernant les moustiques qui propagent les larves de *Dermatobia cyaniventris* et de *Chrysomia macellaria* et peut-ètre celle de Lund, et de la *Cordilobia antropophaga*. Rev. Med. Hyg. Trop., *10*: 93-95.

Zmeev, G. I͡a.
1936. O znachenie nekotorykh sinantropnykh nasekomykh kak perenoschikov i promezhutochnykh khozi͡aev parasiticheskikh cherveĭ v Tadzhikistane. [On the importance of certain synanthropic insects as carriers and intermediate hosts of parasitic worms in Tadzhikistan.] In: Malaria and other parasitological problems of Southern Tadzhikistan, Trudy (Procs.) Tadzhikskoi Bazy Akad. Nauk SSSR, *6*:241-248.

Zmeev, G., I͡a.
1943. Opyt bakteriologicheskogo obsledovanii͡a razlichnykh vidov mukh s tsel'i͡u otsenki ikh epidemiológicheskogo znachenii͡a. [Experimental observation of various species of flies to evaluate their epidemiologic importance.] In book: Problemy kishechnykh infektsiĭ, Izd. An SSSR, Stalinabad (Dushanbe), pp. 118-122.

Zotta, G.
1921. Sur la transmission expérimentale du *Leptomonas pyrrhocoris* Z. chez des insectes divers. Compt. Rend. Soc. Biol., *85*:135-137.

Index

(The letter "g" after an entry indicates that the item is in a drawing or graph.)

Date Due